Soil & Environment

VOLUME 2

Contaminated Soil '93

*Fourth International KfK/TNO Conference on Contaminated Soil,
3–7 May 1993, Berlin, Germany*

edited by

F. ARENDT
Program Management
"Low Pollution and Low Waste Processes",
Karlsruhe Nuclear Research Center (KfK),
Karlsruhe, Germany

G.J. ANNOKKÉE
R. BOSMAN
W.J. VAN DEN BRINK

Netherlands Organization for
Applied Scientific Research TNO,
Delft/Apeldoorn, The Netherlands

Volume II

KLUWER ACADEMIC PUBLISHERS
DORDRECHT / BOSTON / LONDON

A C.I.P. Catalogue record for this book is available from the Library of Congress.

ISBN 0-7923-2326-2 (Vol. I)
ISBN 0-7923-2327-0 (Vol. II)
ISBN 0-7923-2328-9 (Set)

Published by Kluwer Academic Publishers,
P.O. Box 17, 3300 AA Dordrecht, The Netherlands.

Kluwer Academic Publishers incorporates
the publishing programmes of
D. Reidel, Martinus Nijhoff, Dr W. Junk and MTP Press.

Sold and distributed in the U.S.A. and Canada
by Kluwer Academic Publishers,
101 Philip Drive, Norwell, MA 02061, U.S.A.

In all other countries, sold and distributed
by Kluwer Academic Publishers Group,
P.O. Box 322, 3300 AH Dordrecht, The Netherlands.

Printed on acid-free paper

Printed in the Netherlands

Fourth International KfK/TNO Conference on Contaminated Soil

organized by
the Karlsruhe Nuclear Research Center (KfK)
the Netherlands Organization for Applied Scientific Research (TNO)

in cooperation with
the German Federal Environmental Agency (UBA)
the Association of technical Engineers for waste dump sites (ITVA)
the Technical University of Hamburg-Harburg
the Environmental Research Center Leipzig/Halle (UFZ)
the U.S. Environmental Protection Agentcy (EPA)
the German Society for Chemical Instrumentation, Chemical Technology and Biotechnology (DECHEMA)

supported by
the Commission of the European Communities
the German Federal Ministry of Research and Technology (BMFT)
the German Federal Ministry of the Environment, Nature Conservation and Nuclear Safety (BMU)
The Netherlands Ministry of Housing, Physical Planning and the Environment (VROM)
the U.S. Department of Energy (DOE)
the U.S. Environmental Protection Agency (EPA)
the City of Berlin

<u>CONTENTS - OVERVIEW</u>

VOLUME 1

VOLUME 2

Contents (Volume 1 & 2)

CONTENTS

VOLUME 1

1 STRATEGIES AND POLICIES; LEGAL, ECONOMIC AND SOCIAL ASPECTS

1.1 International, national and regional programmes

1.2 Legal, economic and social aspects

1.3 Short communications

2 RISKS

2.1 Sources and behaviour of contaminants

2.4 Safety of workers

2.5 Short communications

3 MANAGEMENT

3.1 Site investigation; surveying and samping strategies

3.5   Short communications

VOLUME 2

4   STATE OF THE ART OF APPLIED REMEDIATING TECHNOLOGIES (CASE STUDIES, EXPERIENCES AND EVALUATION)

<u>4.1   Emphasis on: In-situ technologies</u>

4.2 Emphasis on: Ex-situ technologies

4.3 Short communications

5 RESEARCH AND DEVELOPMENT, EMERGING TECHNOLOGIES

5.1 Emphasis on: Biological and Chemical Treatment

5.3 Short communications

6   PREVENTION OF SOIL CONTAMINATION

<u>6.1 Waste disposal and agricultural activities</u>

Further contributions

For technical reasons, the following contributions could be published in the German edition of these Proceedings (*Altlastensanierung '93*, ISBN 0-7923-2331-9) only.

2.5 Risks - Short communications

Leitfaden zur Erarbeitung von Konzepten zur Gefahrenabwehr
Beine, R.A.; Lassl, M.; Hoffmann, B.

3.3 Control and monitoring

Auffindung und Abgrenzung von Kontaminationen auf Altlasten, Deponien und Industriestandorten
Haupt, M.

3.4 Reuse of cleaned sites and/or soil

Vegetationsentwicklung auf dekontaminierten Böden - nach Pflanzensoziologischen Gesichtspunkten künstlich herbeigeführt.
Jochimsen, M.E.

4. STATE OF THE ART OF APPLIED REMEDIATING TECHNOLOGIES (CASE STUDIES, EXPERIENCES AND EVALUATION)

REMEDIATION OF CONTAMINATED SOIL: STATE OF THE ART AND DESIRABLE FUTURE DEVELOPMENTS

W.H. RULKENS*, J.T.C. GROTENHUIS*, E.R. SOCZÓ**

* Wageningen Agricultural University, Department of Environmental Technology, P.O. Box 8129, 6700 EV Wageningen, The Netherlands
** National Institute of Public Health and Environmental Protection, P.O. Box 1, 3720 BA Bilthoven, The Netherlands

ABSTRACT

For about ten years soil treatment techniques have been developed and used to clean-up contaminated sites in Europe as well as the USA and Canada. Techniques successfully applied in practice for the clean-up of excavated soil are thermal treatment, extraction/classification and bioremediation. During the past years the approach of soil remediation has been shifting more and more from clean-up after excavation to in-situ clean-up. Techniques which at present are successfully applied for in-situ treatment are biorestoration, extraction and soil venting. It is a necessity and also a technological challenge to improve techniques and to develop alternative techniques for the treatment of soils that cannot be treated with existing techniques. Another challenge is the development of treatment systems for the clean-up of soil in temporary disposal sites and for the clean-up of very large, diffuse polluted areas. These techniques may be partly based on existing treatment principles and partly on new principles such as the use of plants.

1. INTRODUCTION

During the last decade a tremendous effort in the highly industrialized Western countries has been made into the development of remedial action techniques for contaminated soil (1,2,3). For application in practice several types of remediation techniques are now available: clean-up of excavated soil, in-situ clean-up of soil and polluted groundwater, isolation or immobilization of contaminated sites, and excavation of polluted soil followed by disposal in a controlled disposal site. Regarding clean-up, most of the attention has been paid to the development of clean-up techniques for excavated soil. However, due to the strong increase in the number of polluted sites that has been identified during the past years, the approach of soil remediation by clean-up is shifting more and more from complete clean-up of excavated soil to complete or partial clean-up in-situ.

The aim of this paper is to discuss briefly several principles of clean-up, and to survey the most important techniques for clean-up of excavated soil and in-situ treatment. Practical experience with these techniques and the bottlenecks observed in practical application, will also be given. On the basis of this overview the possibilities for improvement of existing techniques and broadening their field of operation are considered. The need for new clean-up techniques and the possibilities for the development of new techniques are also discussed.

F. Arendt, G.J. Annokkée, R. Bosman and W.J. van den Brink (eds.), Contaminated Soil '93, 1007–1018.
© 1993 *Kluwer Academic Publishers. Printed in the Netherlands.*

2. CLEAN-UP PRINCIPLES AND RELEVANT FACTORS IN CLEAN-UP

Clean-up is a remediation method aimed at removal of the contaminants from the soil. Major factors governing the removal possibility of a contaminant from soil are:
- The type and concentration of the pollutant
- The physical state of the pollutant
- The chemical, physical and (micro)biological properties of the pollutant
- The type of soil (sand, clay, loam, peat)
- The size and history of the polluted site.

The type, physical state, and concentration of the pollutants can vary strongly. Due to the fact that soil pollution has been caused mainly by industrial activities, the types of pollutants observed in different countries look very similar. Main pollutants are:
- Volatile and non-volatile aliphatic and aromatic hydrocarbons such as petrol, benzene and polycyclic aromatic hydrocarbons
- Volatile and non-volatile halogenated organic compounds such as trichloroethylene, perchloroethylene, pesticides and PCBs
- Heavy metals and heavy metal compounds
- Free and complex cyanides.

Besides the type and concentration of the pollutants their physical state is a very important factor. The following physical states are observed:
- Particulate pollutants of a size equal to, smaller or larger than the non-polluted soil particles
- Pollutants present as a liquid film around soil particles
- Pollutants adsorbed at the surface of soil particles
- Pollutants absorbed in (organic) soil particles
- Pollutants present as a solid or liquid phase in the pores of the soil particles
- Pollutants dissolved in the aqueous phase in the pores of the soil particles.

Regarding the history of the polluted site, three factors have relevancy for soil clean-up, especially for in-situ clean-up of soil: the way the site has been polluted, the way the site has been used after it has been polluted and the time interval between the moment of pollution and clean-up.

As already mentioned clean-up of contaminated soil is aimed at removal of the pollutants from the soil. Basically, there are four different principles for their removal:
- Molecular separation
- Phase separation
- Chemical destruction
- Biodegradation.

A clean-up technique is not always based on one removal principle. Very often more than one principle is involved in a treatment process. All types of clean-up methods make use of the specific differences in properties between the pollutants (or polluted particles) and the soil particles. Properties of the pollutants, which may be the basis for a cleaning technique, are:
- Volatility of pollutants
- Solubility of pollutants in water or in an organic solvent
- Chemical/thermal instability of pollutants

- Biodegradability of pollutants
- Adsorption/absorption behaviour of pollutants
- Magnetic or electric properties of pollutants
- Size, shape or density of particulate pollutants.

3. CLEAN-UP TECHNIQUES FOR EXCAVATED SOIL

3.1 General

To clean-up excavated soil three major groups of techniques are available:
- Extraction/classification using aqueous extracting agents
- Thermal treatment by incineration or infrared heating
- Biological treatment by landfarming and bioreactor techniques.

Each group of techniques comprises several modifications, most of them used intensively on a practical scale.
Besides these three groups there is a group of alternative treatment processes which are either still in the development stage or which at the moment have to be considered as too expensive for large-scale application.

3.2 Extraction/classification using aqueous extracting agents

Extraction/classification is a treatment method in which the pollutants are removed from the soil by means of an extracting agent. For practical application at present aqueous extracting agents are almost always used. Basically, the extraction process consists of three major treatment steps:
- Intensive mixing of contaminated soil and extracting agent
- Separation of extracting agent and clean soil particles
- Treatment of extracting agent.

The intensive mixing step is aimed at promoting the transfer of contaminants from the soil into the aqueous extracting agent. Two removal mechanisms can be distinguished. The first is aimed at dissolving pollutants into the extracting agent. This process may be enhanced by the addition of chemicals such as inorganic acids, complexing agents or sodium hydroxide. The second mechanism is aimed at the dispersion of polluted particles into the extracting agent. Chemicals that can be used to promote this process are sodium hydroxide and several types of detergents and surfactants. High pressure soil washing systems and mechanical vibration scrubbing systems can improve the efficiency.
For the separation of the extracting agent, which contains the dissolved and suspended pollutants, from the clean soil particles, several techniques are available. These are based on differences in particle size, shape and density, as well as in settling velocity, and magnetic and surface properties. Separation techniques and systems often used in extracting installations are hydrocyclones, fluidized bed separation systems, flotation, Humphrey spiral separation techniques, shaking tables, jig techniques, tilting frame techniques and mechanical vibration.
Several physical and chemical wastewater treatment systems are available for cleaning the separated extracting agent. Treatment of the extracting agent results in a relatively large amount of sludge. This sludge consists mainly of soil substances. In spite of the fraction

of polluted compounds being relatively small, the sludge has to be considered as a hazardous waste due to the presence of these pollutants. Further treatment of the sludge using thermal methods or disposal of the sludge in a controlled disposal site is necessary.

Extraction/classification of excavated soil is a clean-up process primarily suited to treat sandy soils or sandy soils with a clay and humic content of less than 10 to 15%. The reason for this limitation is that the small clay particles and humic substances are concentrated to a large extent in the sludge.
With extraction/classification it is possible to remove both organic and heavy metal pollutants. Also the physical state of the contaminant (ad(b)sorbed to soil particles, or present as a separate particulate pollutant) presents no real limitation. However, it will be clear that the modification of the process and the process conditions which have to be applied depend heavily on the type of contaminated soil and the properties of the contaminants. In fact, for each type of contaminated soil a tailor made process has to be applied.

Extraction/classification of excavated contaminated soil is practised on a large scale in treating sandy soils (4,5). There is considerable experience in the removal of all types of contaminants using the various modifications. Major bottlenecks at present are the large amounts of sludge and the fact that a complete clean-up is not always achieved.

3.3 <u>Thermal treatment</u>

Thermal decontamination of soil consists, in general, of two treatment steps. The first step is aimed at evaporation of the contaminants from the soil. This evaporation occurs at temperatures between 150 and 700 °C. Depending on the temperature chemical destruction of pollutants into volatile pollutants may also occur. Organic soil substances may also be destructed and volatilized. The organic pollutants in the gas phase are completely oxidized in a separate incineration step.
Several modifications of the thermal treatment process are available. The most important one is treatment in a rotating kiln provided with a burner for directly heating the soil. Due to the rotation of the kiln soil particles come into intensive contact with the hot gases of the burner. The gases leaving the kiln are passed to an afterburner where a complete oxidation of all organic volatiles is achieved. The gases from the afterburner are cooled by means of a heat exchanger and are then passed through pollution control equipment consisting of a particles separation system and a gas scrubber.
There are three alternatives for a rotating kiln with a direct soil heating system: a rotating kiln provided with an internal or external heat exchanger, a fluidized bed thermal treatment system, and an infrared thermal treatment system.

Thermal treatment is suitable for the removal of all types of organic pollutants and can be applied to all types of soil. The temperature necessary for removal of the contaminants depends highly on the type of contaminants. Petrol and diesel oil can be removed at temperatures of approximately 300 °C; high molecular polycyclic aromatics need temperatures of about 450 to 500 °C; complex iron cyanides can be destructed at temperatures of about 450 °C. It is also possible to remove halogenated organic compounds from the soil. However, to prevent emission of dioxins, temperatures of more than 1000 to 1100 °C in the afterburner are necessary. Formation of dioxins can be suppressed by careful control of process parameters. Mercury and mercury compounds can also be removed by evaporation from the soil. However, removal of evaporated mercury and mercury compounds from the gas phase leaving the kiln requires a special pollution control system, different from the ones normally used in thermal treatment of soil contaminated with organic pollutants.

Thermal soil treatment in rotating kilns and in infrared-heated treatment installations is applied on a large scale (4,5). However, the practical experience is mainly limited to soils contaminated with non-halogenated organic pollutants.

3.4 <u>Biodegradation</u>

In biodegradation processes microorganisms are used for conversion of contaminants into non-polluting compounds such as water and carbon dioxide. Natural biodegradation of pollutants occurs in most soil environments, however, in general, the conditions for biodegradation are not favourable for achieving an effective clean-up. Bioremediation techniques are aimed at improving the conditions for microbiological degradation. Conditions which have to be optimized may be the temperature, the moisture content of the soil, pH, redox potential, concentration of oxygen, type of electron acceptor, presence of desired microorganisms, and the bioavailability of the pollutant to the microorganisms. Biodegradation can occur under aerobic or anaerobic conditions. In general, conditions are aerobic. However, in biodegradation of chlorinated hydrocarbons a combination of aerobic and anaerobic treatment steps is required for complete conversion of the pollutants.

With respect to biological treatment of soils contaminated with organic pollutants, there are, at present, two major types of biological techniques available: landfarming and bioreactors (10).
In landfarming the polluted soil is spread out in a layer of 0.5 to 1 m in thickness on top of a specially constructed impermeable sublayer provided with a drainage system. The aerobic microbiological degradation of pollutants in this layer is stimulated by regularly cultivating the layer to enhance oxygen transfer, and adding water, nutrients and trace elements if necessary. Polluted drainage water is collected and can be recirculated over the layer. A further acceleration and improvement of the process can be obtained by the use of greenhouse techniques, forced aeration and increasing the temperature of the layer. Basically, landfarming is a relatively simple method and makes treatment times of 1 to 3 years acceptable. Landfarming is practised for the treatment of soil contaminated with easily biodegradable pollutants.
Bioreactor techniques involve the treatment of a slurry of contaminated soil in a compact reactor system to which oxygen and nutrients are supplied. The slurry is intensively mixed to promote better contact between microorganisms, pollutants, oxygen and nutrients. Due to this intensive contact and the better possibilities for controlling the process, substantially higher biodegradation rates can be achieved than in landfarming. However, the process is more complicated than landfarming and requires a rather complex and expensive treatment installation. A special form of bioreactor is the composting reactor in which the soil is treated at relatively dry free-flowing conditions.

With respect to the practical application of bioremediation, a large number of soil pollutants are biodegradable. However, at present, the practical application of bioremediation is limited to the removal of easily biodegradable compounds such as petrol and low molecular aromatic hydrocarbons (2,6). The main reason is, in fact, the low bioavailability of the contaminants. This bioavailability is primarily determined by the desorption of the contaminants from the soil particles and not by microbial activity. A low bioavailability means that the biodegradation rate is too low for practical application and/or that the residual concentration of the contaminants in the treated soil usually remains above the standards for clean soil.

There are several prospects for improving bioremediation techniques:
- Inoculation with especially cultivated microorganisms mainly in the case of the degradation of chlorinated compounds
- Separation of the soil into an easily biodegradable fraction and a poorly biodegradable fraction
- Application of physical methods to reduce mass transfer limitations
- Use of chemical detoxification techniques to convert poorly biodegradable pollutants into easily biodegradable pollutants.

A special type of bioremediation technique is the bioleaching of heavy metals by *Thiobacilli*. In this process an aqueous solution of heavy metals is obtained which can be treated by physical/chemical methods to remove the heavy metals. The process is still in the development phase.

3.5 Alternatives

Besides the clean-up techniques already mentioned, which can be considered the most important, there are several alternative clean-up techniques:
- Extraction of pollutants with organic solvents
- Supercritical extraction with CO_2
- Oxidation of organic pollutants in a water phase at subcritical or supercritical conditions (wet air oxidation)
- Vitrification (melting of the soil by electrical heating)
- Removal of ions and small colloidal particles under the influence of an electrical current (electroreclamation)
- Dehalogenation of chlorinated organic compounds using an alkali polyethylene glycol
- Chemical reduction or oxidation of contaminants
- Steam or air stripping (volatilization of contaminants)
- Pyrolytic incineration
- Plasma torch techniques (use of a mixture of ions and electrons generated by an electrical arch)
- Microwave heating

The state of the art of these techniques varies (7,8). Some are very sophisticated and have been developed for the treatment of hazardous waste. Others are still in the development stage. Some have already been developed for large-scale soil clean-up. However, low cleaning efficiency or high treatment costs prevent large-scale applications.

4. IN-SITU CLEAN-UP TECHNIQUES FOR SOIL

4.1 General

Also three major groups of techniques can be distinguished for in-situ treatment of soil:
- In-situ extraction (soil washing)
- Soil venting (soil vapour extraction)
- Biorestoration.

These three groups of clean-up techniques are applied in practice. Besides these three groups there are some alternatives.

4.2 In-situ extraction

In-situ extraction consists basically of percolation of an aqueous extracting agent into the contaminated site. This percolation can be achieved by means of surface trenches, horizontal drains or vertical deep wells, or by combinations of these three methods. Soluble contaminants present in the soil will dissolve in the percolate. By means of a special withdrawal system the percolate is pumped up and treated. After reconditioning, the extracting agent is reused. The process is continued until the residual concentrations in the soil satisfy the standards for clean soil.

The following aspects need attention if in-situ extraction is to be considered for practical application: the geohydrological properties of the contaminated site, mechanical disturbance of the soil, origin of the pollution, solubility of the contaminants, treatment of the extracting agent withdrawn and the conditioning of the extracting agent.

With respect to the geohydrological properties of the contaminated site, a high permeability of the soil layer is required. In fact this means that only sandy soils are suitable for in-situ treatment. The presence of clay layers is not a real bottleneck but requires special measures to achieve an appropriate percolation of the extracting agent. It is important that the extracting agent can percolate into the complete area where the contaminants may be present. Mechanical disturbance of the surface layer of the contaminated site can cause a spread of the contaminants to areas in the site which cannot be percolated easily.

It will be clear that the origin of the pollution strongly governs the possibilities for in-situ clean-up. If the pollutants have been originally dissolved in a water phase, the chances for successful in-situ extraction are high. However, if the soil has been polluted by non-soluble or slightly soluble particulate pollutants, then the applicability of an in-situ extraction is doubtful.

In-situ extraction is only possible if the solubility of the contaminant in the extracting agent is high enough and the bond between soil particles and contaminants is not too strong. Increasing the solubility of the contaminants or the dispersion of polluted colloidal particles into the extracting agent is possible by the addition of such chemicals as inorganic acids, organic acids, sodium hydroxide, complexing agents, oxidizing or reducing agents, surfactants and detergents. It should be noted that some of the chemicals added to the water phase will be adsorbed to the soil particles.

In-situ extraction is already applied on a limited scale, however, with some exceptions, the residual concentrations of the contaminants are, in general, above the standards for clean soil. This is a real bottleneck. Another is the mass transfer limitation in the site resulting in a long treatment period (5).

4.3 In-situ soil vapour extraction

Soil vapour extraction is a treatment method that makes use of the volatility of a contaminant. The concentration of the pollutants in the gas phase between the soil particles in a contaminated site is more or less in equilibrium with the pollutants adsorbed at or absorbed into the soil particles. Flushing the soil with a gas phase makes complete removal of the volatile pollutants from the soil possible. For practical application of this clean-up principle the contaminated site is provided with a number of vertically placed soil vapour extraction wells. The vapour extraction wells are connected with a vapour treatment system in which the contaminants are removed from the vapour. This can be done, for example, by adsorption on activated carbon or by catalytic incineration. Air infiltration into the soil is caused by the negative pressure gradients in the unsaturated soil in the vicinity of the extraction wells. It is also possible to increase the stripping process using forced injection of air into the saturated zone. There are several modifications of this treatment system. The optimal modification highly depends on the type of soil, structure of the contaminated site and the type and concentration of the pollutants.

Vapour extraction is applicable for in-situ treatment of soil contaminated with volatile organic compounds, such as petrol, trichloroethylene and perchloroethylene (5,9). Only sandy soils are treatable. The permeability of claylike soils is, in general, too low for achieving an efficient stripping process of the contaminants.

4.4. Biorestoration

In-situ biological treatment is aimed at the improvement of process conditions for microbiological degradation in polluted sites by the addition of oxygen (or other electron acceptors), nutrients, and, if necessary, also microorganisms and surfactants. These additions are achieved by infiltration of a water phase via drains, trenches or wells. The infiltrated water is pumped up, treated to remove pollutants, conditioned and refed to the soil. Oxygen supply can also be achieved by soil venting. In-situ treatment of contaminated soil is only feasible if the permeability of the soil is high enough. At present, only sandy soils are suitable for in-situ treatment. As for biological treatment of excavated soil, the practical application of biorestoration techniques is limited to the removal of easily biodegradable compounds such as petrol and aromatic hydrocarbons (3).

In some cases in-situ biorestoration is combined with two other in-situ treatment systems: in-situ extraction with a water phase and soil vapour extraction. The injection of air results in an efficient oxygen transfer for the bioremediation process, enhancing the biodegradation rate. The use of an aqueous extracting agent also makes the removal of non-volatile, non-biodegradable compounds (e.g. intermediates) possible (5).

4.5 Alternatives

Besides the in-situ clean-up techniques already mentioned, which can be considered the

most important, there are some alternatives:
- Electroreclamation (for removal of heavy metals from clayey soils)
- Vitrification
- Microwave heating
- Steam stripping.

In spite of the cleaning results obtained with these techniques being sometimes very promising, large-scale application is still pending, mainly due to the high costs (7,8).

5. DISCUSSION

Several methods are available to clean-up a contaminated soil (5). The practical applicability of these methods strongly depends on the type and the concentration of the contaminants, physical state of the contaminants, soil structure, and size and history of the contaminated site.

Clean-up of excavated soil

For excavated soil the applicability, practical experience and need for further improvement of the existing techniques can be summarized as follows:
- Extraction/classification with aqueous extracting agents
  * applicable to sandy soils
  * applicable to all types of pollutants
  * large-scale practical experience
  * further improvement focused on:
    - less sludge production
    - treatment of the sludge
    - increase of clean-up efficiency (both for heavy metals and organic pollutants).
- Thermal treatment:
  * applicable to all types of soil
  * applicable to organic pollutants
  * large-scale practical experience
  * further improvement focused on:
    - cleaning of the gas phase for chlorinated organic compounds
    - decrease of energy consumption.
- Biological treatment (landfarming, bioreactors):
  * applicable to all types of soil
  * applicable to easily biodegradable compounds
  * large-scale practical experience with landfarming
  * further improvement focused on:
    - increase of biodegradation rate and clean-up efficiency (improvement of bioavailability)
    - improvement of biodegradability of not easily biodegradable compounds such as chlorinated organic pollutants
    - application of a combination of biological and physical/chemical treatment
    - application of fungi.

For the above-mentioned types of polluted soils alternative treatment techniques are either still in the development stage or have already been developed, but are, at present, too expensive for practical application.

Excavated soils that cannot be treated in practice with one of the above-mentioned techniques are:
- Sandy soils with relatively large amounts of clay, loam or humic substances, and polluted with heavy metals or a combination of heavy metals and organic pollutants
- Clay-like, loamy and peaty soils, polluted with heavy metals or a combination of heavy metals and organic pollutants.

Practical applicable techniques for these types of soil are still in the development stage.

In-situ clean-up of soil

For treatment in-situ the applicability, practical experience, and the need for further improvement of existing techniques can be summarized as follows:
- In-situ extraction:
  * applicable to sandy soils
  * applicable to soluble pollutants
  * practical experience on a limited scale
  * further improvement focused on:
    - increase of clean-up efficiency (increasing solubility of the pollutants, and improvement of infiltration techniques)
    - development of models for better process control.
- In-situ vapour extraction:
  * applicable to sandy soils
  * applicable to volatile pollutants
  * practical experience on a large scale
  * further improvement focused on:
    - improvement of clean-up efficiency
    - development of models for better process control.
- Biorestoration:
  * applicable to sandy soils
  * applicable to easily biodegradable pollutants
  * practical experience on a limited scale
  * further improvement focused on:
    - increase of bioavailability and clean-up efficiency using physical/chemical methods
    - development of models for better process control
    - removal of not easily biodegradable compounds such as chlorinated organic compounds.

Also for the above-mentioned types of polluted soils, alternative in-situ treatment techniques are possible. However, at the moment these techniques are either too expensive or still in the development stage.

Contaminated soils which cannot be treated sufficiently in-situ at present are:
- Clay-like and loamy soils

- Sandy soils containing non-soluble particulate pollutants
- Very heterogeneous soils.

Potential treatment methods for soil in temporary disposal sites

Large amounts of excavated contaminated soil are temporarily stored at present in controlled soil disposal sites. Temporary storage can be necessary if no practical clean-up method is available at the moment of excavation or if treatment capacity is not available. Another reason may be that treatment by means of existing techniques is too expensive. Temporary storage of the soil in a controlled disposal site for a long period of time offers several unique possibilities for complete or partial clean-up. Techniques that can be considered are based either on in-situ treatment techniques or on treatment techniques for excavated soil, for instance:
- Extraction of inorganic or organic pollutants using relatively mild extracting conditions
- Chemical degradation under relatively mild conditions.
- Soil venting (vapour extraction)
- Biological degradation by techniques similar to biorestoration, soil slurry reactor techniques or landfarming techniques
- Bioleaching techniques based on the use of *Thiobacilli* for removal of heavy metals.

As far as it is known the treatment of soil in a soil disposal site is still in the first stages of development. Further research is necessary to develop potential clean-up principles into practical techniques.

Clean-up of large diffuse polluted areas

The clean-up techniques mentioned in the foregoing have been developed especially for treating strongly polluted areas of limited size. In general these techniques are aimed at complete or nearly complete removal of the pollutants. The costs of these techniques are relatively high. For a large number of contaminated sites of limited size these treatment techniques can be used to clean-up the site within the constraints of acceptable treatment costs.
In several regions, both in highly industrialized Western countries and in Eastern and Central Europe, there are large areas at which the top layer of the soil is diffuse polluted with inorganic and/or organic pollutants. Excavation of the soil by means of the usual excavation techniques, followed by intensive treatment in a treatment installation or intensive in-situ treatment (using the usual expensive treatment techniques developed for small sites) is not possible. For these areas a completely different clean-up approach is necessary. The following constraints for such a clean-up techniques have to be considered:
- Relatively simple
- Treatment time of several years
- Extensive process with no negative environmental impacts
- Relatively cheap
- Not always necessary to achieve the standards for completely clean soil.

Basically there are several clean-up principles that might satisfy the above-mentioned constraints. Some are directly derived from the experience obtained from intensive treatment techniques for excavated soil or from in-situ treatment. The following potential clean-up methods for large areas may be considered:

- The use of green plants which can accumulate the pollutants. After harvesting, these plants have to be treated in order to concentrate or destruct the pollutants. It is known that some plants can accumulate heavy metals to a very high concentration.
- The use of plants with a root structure making controlled removal of the plant roots as plant sods possible. The layer removed, containing the plants and the contaminated soil, has to be treated in order to concentrate or destruct the pollutants.
- The use of small amounts of chemicals which may directly promote microbiological degradation of pollutants or isomerisation of non-biodegradable pollutants to biodegradable pollutants.
- Cultivating the top layer of the soil to promote natural photochemical conversion and/or biological degradation.

The principles of these methods are known: however, there is still a long way to go to develop these principles for practicable applicability.

6. LITERATURE

1. Arendt, F., Hinsenveld, M., Brink, W.J. van den (eds.). Contaminated Soil '90, Volume I and II. Kluwer Academic Publishers, Dordrecht, Boston, London, 1990.
2. Hinchee, R.E., Olfenbuttel, R.F. (eds.). On-site Bioreclamation, Processes for Xenobiotic and Hydrocarbon Treatment. Butterworth-Heinemann, Massachusetts, USA, 1991.
3. Hinchee, R.E., Olfenbuttel, R.F. (eds.). In-situ Bioreclamation, Applications and Investigations for Hydrocarbon and Contaminated Site Remediation. Butterworth-Heinemann, Massachusetts, USA, 1991.
4. Soczó, E.R., Versluijs, C.W. Review of soil treatment techniques in The Netherlands, in Proc. of the 8th Annual HazMat Conference Int. Atlantic City, New Jersey, June 5-7, 1990.
5. Olfenbuttel, R.F. (ed.). Demonstration of Remedial Action Technologies for Contaminated Land and Groundwater. Final report NATO/CCMS Pilot Study, Committee on the Challenges of Modern Society, Number 190, Brussels, Belgium, 1992.
6. Soczó, E.R., Visscher, K. Research and development programs for biological hazardous waste treatment in the Netherlands, in Sayler A.S., et al. (ed.): Environmental Biotechnology for Waste Treatment. New York, Plenum Press, 1991.
7. Leur, G.J. van de, et al. Study on alternative physico-chemical and thermal treatment of contaminated soil. RIVM report no. 736102003, Bilthoven, The Netherlands, 1990.
8. Innovative Treatment Technologies: Semi-Annual Status Report (Third Edition). EPA/540/2-91/001, U.S. EPA Office of Solid Waste and Emergency Response, Washington, DC 20460, U.S.A.
9. Soil vapor extraction technology: Reference Handbook. EPA/540/2-91/003, U.S. EPA Risk Reduction Engineering Laboratory, Cincinnati, Ohio, U.S.A., 1991.
10. Staps, J.J.M. International evaluation of in-situ biorestoration of contaminated soil and groundwater. RIVM report no. 738708006, Bilthoven, The Netherlands, 1990.

REMEDIAL OPERATION FOR SUBSURFACE POLLUTION DUE TO VOLATILE ORGANOCHLORINE USING SOIL VENTILATION AND GROUNDWATER EXTRACTION

Tatemasa HIRATA and Osami NAKASUGI

National Institute for Environmental Studies, Tsukuba, Ibaraki 305 Japan

1. ABSTRACT

The soil-gas survey was applied as a remote geochemical technique to delineate the pollution boundary and determine the sites for making boreholes. Considering the gaseous plume, 14 borings were completed in a groundwater region contaminated with trichloroethylene. It resulted in revealing the existing form of trichloroethylene standing over both sides of the groundwater table. In addition, the vacuum extraction method was employed to remove trichloroethylene from the vadose zone, and the groundwater extraction was determined for the groundwater cleaning. The soil ventilation gained the trichloroethylene recovery rate of 0.60 kg/hr, and the groundwater extraction did 0.15 kg/hr.

2. INTRODUCTION

The groundwater pollution due to halogenated hydrocarbons has been increasingly becoming a great environmental issue in Japan and developed nations. The volatile organochlorines like trichloroethylene and tetrachloroethylene are widely and effectively used in cleansing processes by many and various industries producing huge amounts of fine products. The nationwide groundwater survey in 1982 revealed that both chemicals were detected in one in three groundwater samples taken from the urban area (Hirata et al., 1991). In addition three percentage of groundwater samples overshot the provisional standard for the drinking water i.e. tri-chloroethylene < 0.03mg/l and tetrachloroethylene < 0.01mg/l.

The groundwater pollution by volatile organochlorines holds several insidious features. Many organochlorines are very persistent to biodeg-radation in subsurface environment. Therefore, such contaminants should be removed, otherwise the groundwater is threatened to lose the value as the natural water resource for many years. In addition the wide-range usage of the volatile organochlorines in many industries makes the groundwater pollution issue much complicated and difficult to identify the pollutant source.

The success of remedial operation totally depends on how much information concerning the contaminant existing form and location in subsurface environment and the areal extent of pollution can be obtained. This is because from the cost-beneficial point of view the remedial opera-tion must be taken place close to the pollutant source to reduce the volume of contaminated soil, water and gas to be treated. However, the all-core borings over the polluted area to investigate the contaminant concentration in soil column are so costly. In this context, surface soil-gas monitoring, utilizing volatilization of the contaminants, is able to be an option of remote geochemical reconnaissance to determine the boring sites and also to delineate the pollution boundary (Marrin and Thompson, 1987).

The paper describes first the interrelation between surface soil-gas

F. Arendt, G.J. Annokkée, R. Bosman and W.J. van den Brink (eds.), Contaminated Soil '93, 1019–1028.

concentration and soil content, based on the areal distribution of the volatile contaminant. Second, a remedial procedure from identifying the pollutant source to picking up suitable remedial technique is introduced, taking an example on the groundwater pollution site with trichloroethylene.

3. SURFACE SOIL-GAS EXPLORATION

The volatile organochlorines staying in the vadose zone and groundwater tend to become gaseous phase in the air(soil-gas) of the vadose zone and migrate upward to the groundsurface. The volatile organochlorines in the vadose zone is basically in a situation well partitioned among soil, soil-water and soil-gas. In order to utilize the soil-gas monitoring as a remote geochemical technique, the understandings concerning gaseous behavior of the contaminants, for example interrelation between surface soil-gas concentration and soil content, and the interrelation between soil-gas and groundwater concentrations, are required.

The soil-gas survey was designed to detect pollution boundary and high points of the gaseous plume in one of our study sites, the groundwater of which is contaminated with tetrachloroethylene. The study site is covered by the gravel until 5 m depth, from which the bedrock is underlain. The groundwater table appears at 4 m depth, so the aquifer depth of the groundwater specified is so shallow of 1 m.

At the first stage, a small hole of 2 cm in diameter and 2 m in depth was drilled to take contaminated soil from the hole bottom. Then, stainless steel tube of 3 mm in diameter was installed into the hole bottom, then burying back the hole space with the dug out soil. After ensuring that the soil-gas concentration reached steady state, approximately in 6 hours from achievement of the hole, soil-gas sample was collected through a diaphragm pump into a bag. The soil-gas collection was completed at 19 locations at the soil-gas flowrate of 500 ml per minute. The volatile contaminants in soil were extracted using n-Hexane, and the chemical determinations of soil content and soil-gas were made with the gaschromatograph-ECD.

Fig. 1 shows the relationship between tetrachloroethylene concentrations in soil and soil-gas. It results in making approximately a linear relationship between both concentrations on full-log scale. Such linear correlation is also obtained between soil-gas and groundwater concentrations of tetrachloroethylene (Yoshioka et al., 1992). In addition the surface soil-gas concentration distribution can be recognized in Fig. 2 to extend in a form of the concentric circles around a suspicious pollutant source, where a laundry firm is running.

On the basis of the areal distribution of the soil-gas concentration, the dissipation of the tetrachloroethylene concentration in gaseous plume to the distance from the suspicious pollutant source is illustrated in Fig. 3. The dissipation rate of the soil-gas concentration within 15 m from the suspicious pollutant source takes -0.7 on the full-log scale. As the exact solution on the gaseous diffusion emitted continuously from the point source gives -1.0 to the distance dissipation rate, the soil-gas concentration near the pollutant source is likely to be affected by the volatility and the gaseous diffusion from the high accumulations of the contaminant in the vadose zone. With being away from the pollutant source, the soil-gas concentration seems to be becoming influenced by the volatility from the groundwater table. As a result, the soil-gas survey is recognized to be an effective technology as a remote geochemical exploration for delineating the pollutant distribution and identifying the pollutant source in subsurface environment.

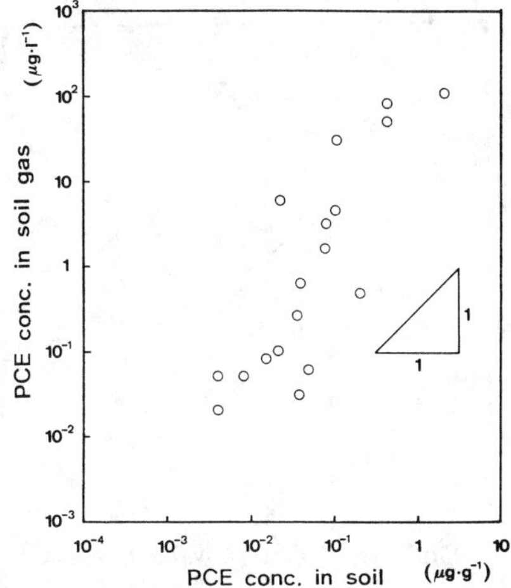

Fig. 1 Relationship between tetrachloroethylene(PCE) concentrations in soil and soil-gas.

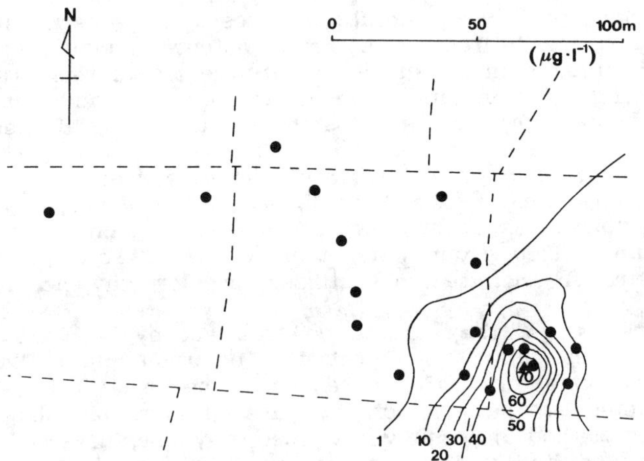

Fig. 2 Gaseous plume of tetrachloroethylene at 1 m depth in the shallow groundwater region. The closed circles are the locations for sampling.

4. A CASE STUDY OF REMEDIAL OPERATION

The feasible options at present for repairing the subsurface environment contaminated with volatile organochlorines are limited to polluted soil excavation, groundwater extraction and soil ventilation. Other technologies such as bioremediation and chemical decomposition procedures are under development, however, the degradation process in particular mass balance and intermediate products should be solved. This is because sometimes chemicals stronger in toxicity are produced from less harmful chemicals, for example 1,1-dichloroethylene production from 1,1,1-trichloroethane in the subsurface environment.

The single remedial operation picked up gives the limited effective-

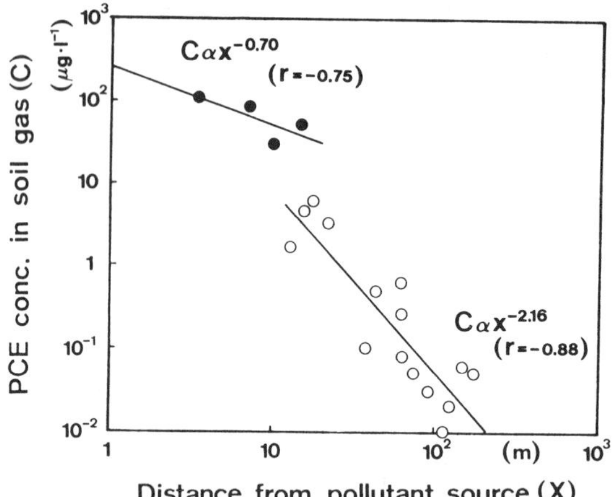

Fig. 3 Dissipation of tetrachloroethylene concentration in soil-gas with distance from suspicious pollutant source. The closed circles are the data taken inside the laundry firm.

ness in many cases of the restoration process. In general, suitable remedial techniques are desired to be combined, corresponding to the existing form of contaminants in subsurface environment. In this study, remedial operation using soil ventilation and groundwater extraction was practiced in unconfined groundwater contaminated with trichloroethylene.

4.1. *Study site description and surface soil-gas survey*

The study area is heavily polluted with trichloroethylene, and is covered by volcano ash over 60 m depth overlying on impermeable clay layer. The unconfined groundwater table appeared approximately at 42 m depth. A firm in the area had utilized trichloroethylene as solvent for many years.

At first stage, the surface soil-gas survey was conducted as a preliminary investigation to detect the pollution plume boundary and determine the sites for making boreholes. The result using the gas-tube method depicted two patterns of the gaseous plume as shown in Fig. 4. The gas-tube method is possibly the most easy one, however, the application is strictly limited to the gaseous concentration over 1 ppmv. Taking soil-gas through a gas-tube from a small hole, color of the granular chemicals packed in the gas-tube turns.

In order to ensure the gaseous plume in Fig. 4, another soil-gas survey of the Finger Print method developed by NERI-Petrex was also applied to the same area. In this method, activated carbon glued on a wire collects volatile chemicals. The test piece placed in a vial is buried at the depth of 30 cm under the groundsurface during a couple of weeks, and then the volatile chemicals coming up from the vadose zone are analyzed with mass-spectrometer. The result in Fig. 5 exhibits only one pattern of the gaseous plume contrary to Fig. 4. The soil-gas survey is a very convenient tool to detect the subsurface pollution with volatile organochlorine. It is real that the gaseous plume in the shallow groundwater region reflects the existing form of the contaminants to a considerable extent as shown in Fig. 2. However, when the depth of the vadose zone is large to be 40 m, detailed investigation is required to reveal what the volatile constituents in soil-gas reflect, the contaminant intruding path or the existing form in the subsurface environment.

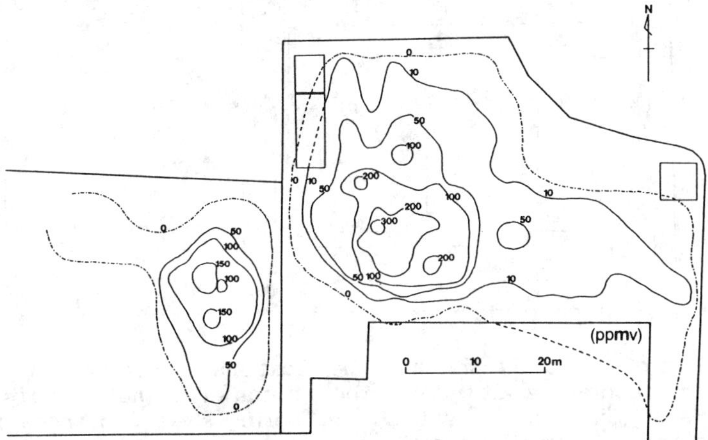

Fig. 4 Gaseous plume of trichloroethylene detected by the gas-tube method at 1 m depth. The concentration unit is in ppmv.

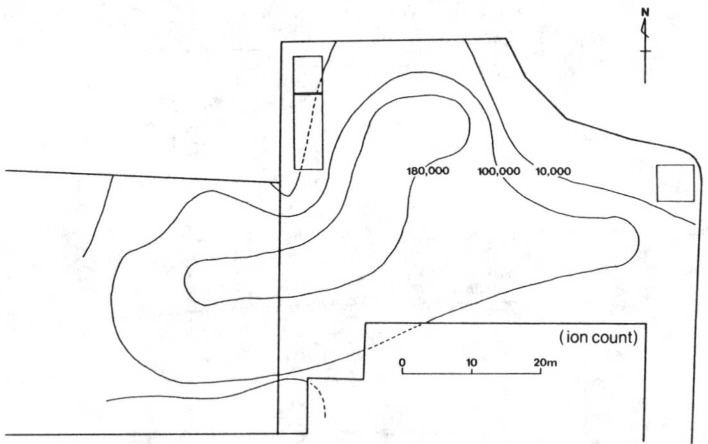

Fig. 5 Gaseous plume of trichloroethylene detected by the Finger Print method at 0.3 m depth. The concentration unit is in the ion count.

4.2. *Vertical boring and existing form of contaminant*

Taking into consideration of the gaseous plume, soil borings over 60 m depth were completed at 14 locations including the site of the highest soil-gas concentration. Fig. 6 shows the boring sites. The chemical determination of the undisturbed soil column taken from the boreholes revealed the existing form of trichloroethylene in the subsurface environment as illustrated in Fig. 7, which was drawn along the boring sites through B-7, B-5, B-6, B-13 to B-9. The maximum concentration in soil reaches 138 mg/kg at 46.1 m depth at B-6 site, and that in the groundwater does 131 mg/l at 50 m depth at B-6 site. Judging from the trichloroethylene existing form in the subsurface environment, the gaseous plume obtained by the Finger Print method seems to be acceptable.

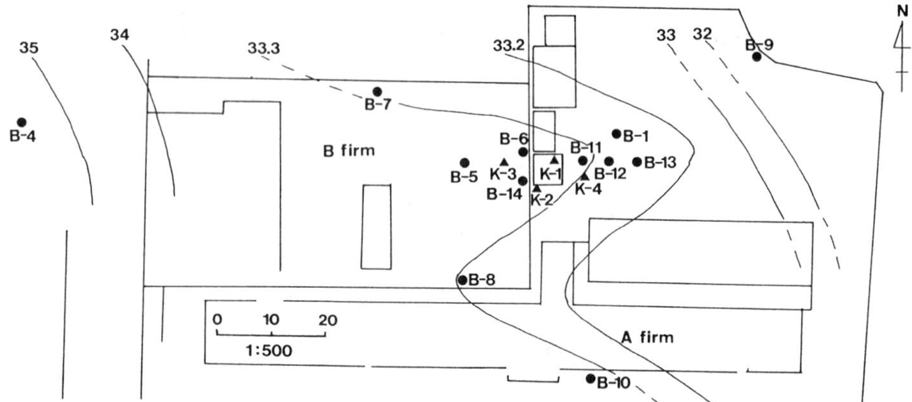

Fig. 6 Outline of the study site and the locations of boreholes.   The lines in the figure denote the altitude of the groundwater table from sea level. The symbols of closed circles indicate the boring sites, and those of closed triangle do the dual extraction wells

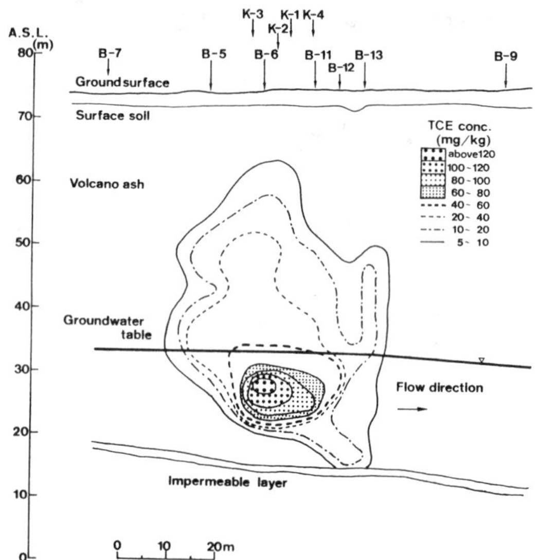

Fig. 7 The concentration contours of trichloroethylene in soil.

In addition, highly accumulated part of trichloroethylene is clearly recognized to be a lump like onion standing over both sides of the groundwater table.   Then, considering such existing form of trichloroethylene, soil venting method was employed to remove the contaminant from the vadose zone, and the groundwater extraction was also determined for the cleanup of the groundwater region.

4.3. *Construction of dual extraction wells*

In order to extract the soil-gas and groundwater from each well at the same time, the dual-extraction wells were constructed at 4 locations seen in Fig. 6. The first dual-extraction well of K-1 was set up in November 1991, and the second one was K-2 in December 1991. The diameter of K-1 is 10 cm and that of K-2 doubles the K-1 well. The depth of both wells are 60 m, and the screen to extract the soil-gas and groundwater was installed between GL-30 m and GL-60 m. Therefore, the soil-gas in the vadose zone is collected from the vertical span of 12 m. The other two wells of K-3 and K-4, the dimensions of which are exactly the same as the K-2 well, were provided in September 1992.

The soil-gas extracted by the vacuum unit with a blower inflows first into the gas-liquid separator, and then the trichloroethylene in the soil-gas is removed in two canisters packing the activated granular carbon. In addition, groundwater abstracted is treated with the air-stripping system, and also the trichloroethylene contained in the air is recovered with the activated carbon canister prior to the emission into the atmosphere.

4.4. *Physical response of subsurface environment to soil-gas extraction*

The soil-gas flowrate and its trichloroethylene concentration were checked in use of the K-1 and K-2 wells to determine the operation condition for the soil ventilation system. Fig. 8 demonstrates a part of the results concerning the response of the groundwater level to the extraction pressure and the changes of the trichloroethylene concentration in the soil-gas extracted. The soil-gas extraction was set in motion at 1400 on 11th December 1991, using the K-1 well. At the beginning of the test, the extraction pressure was adjusted to 0.51 atm. It resulted in raising the groundwater level inside the K-1 well by about 5 m to offset the depression of the air pressure. Then, the vacuum system was down at 1800 on

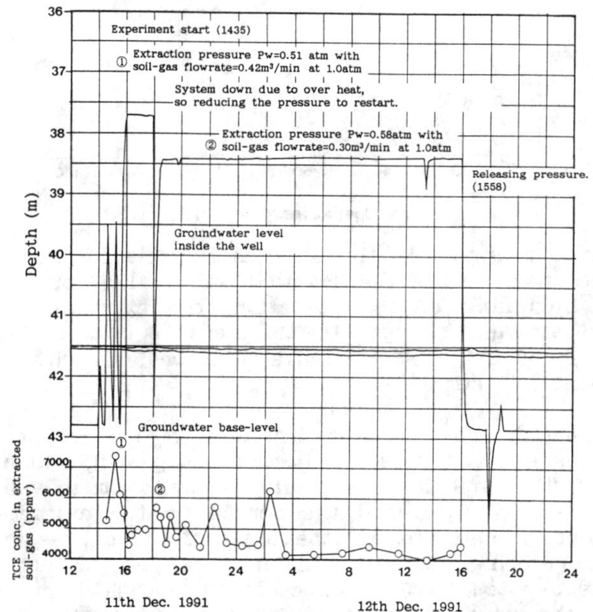

Fig. 8 The response of the groundwater level inside the K-1 well to the soil-gas extraction and the trichloroethylene concentration in soil-gas extracted. In this test, any groundwater was not abstracted.

11th due to the over heat, and immediately after that the groundwater level began coming down very quickly. In addition, the restart of the vacuum system at the extraction pressure of 0.58 atm lifted the groundwater level by about 4 m. During the soil-gas extraction test, the trichloroethylene concentration kept the considerably high value between 7,300 and 3,600 ppmv.

The soil-gas flowrate is basically proportional to the increase of the difference between the extraction and atmospheric pressures. In this test, the soil-gas flowrate was recognized to exponentially rise according to the pressure difference between them. Fig. 9 illustrates the time-varied changes of the soil-gas flowrate in the K-1, K-2 and K-4 wells. The difference between the soil-gas pressures inside and outside the well at the initial stage of the operation is the same as the extraction pressure exerted, and then the propagation of the extraction pressure into the vadose zone is going to ease the gradient of the pressure difference. As a result, the rapid decrease of the soil-gas flowrate took place in the early stage of the soil-gas extraction.

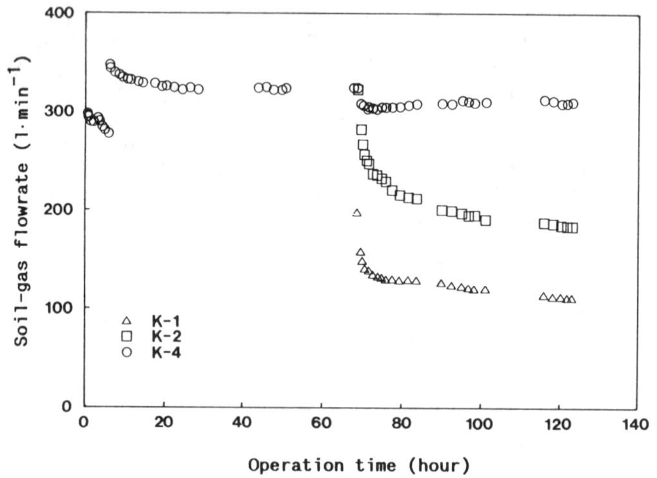

Fig. 9 Time-varied changes of the soil-gas flowrate in three dual extraction wells. The onset time of the K-4 well was at 1252 on 13th September 1992. At the beginning of the operation, the extraction pressure was adjusted to 0.79 atm, and changed to 0.61 atm in 6 hours. The soil ventilation in the K-1 and K-2 wells started at 0940 on 16th with the extraction pressure of about 0.61 atm.

With respect to the radius of influence of the extraction well, the depression of the soil-gas pressure in monitoring well, which is assigned around the extraction well, is able to offer an approximate value. Provided the parallel soil-gas flow toward the screen on the extraction well, the radius of influence is estimated on the basis of the depression of soil-gas pressure in a monitoring well and a distance between extraction and monitoring wells, as denoted by Johnson et al.(1990). The soil ventilation in the K-2 well at the extraction pressure of 0.61 atm gave rise to the pressure depression of 0.027 atm in the K-3 well and 0.095 atm in the B-14 well. The distances of the K-3 and B-14 wells from the K-2 well are 6.8 and 2.82 m. Such conditions made the radius of influence to be 10.1 and

10.9 m respectively.

4.5. *Recovery rate of trichloroethylene by soil ventilation*

The soil-gas flowrate and concentration are basically time-dependent and decreasing with passing the operation time, even when the extraction well is constructed at the central part of the high accumulation of contaminant in vadose zone.   In this context, the influence of the soil-gas flowrate upon the trichloroethylene concentration and recovery rate was examined at the early stage of the remedial operation during the period of December 1991 and January 1992, using the K-1 and K-2 wells.   As illustrated in Fig. 10, with respect to the result of the K-1 well the trichloroethylene concentration tends to be reduced with increasing the soil-gas flowrate.   Contrary to this, the recovery rate remains in a positive relation with the soil-gas flowrate.   This is because the increasing rate of the soil-gas flowrate is overcoming the decreasing rate of the concentration. The recovery rate of trichloroethylene by means of the soil ventilation is clearly recognized to be at the magnitude of about 0.60 kg/hr.

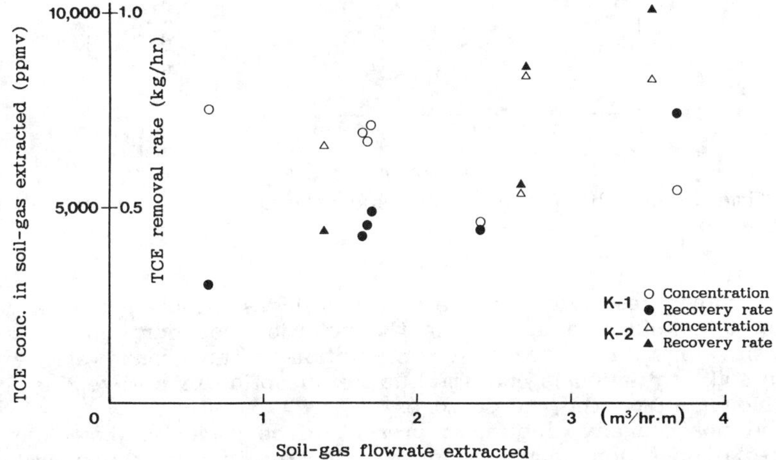

Fig. 10 Trichloroethylene concentration and removal rate vs soil-gas flowrate extracted.

Fig. 11 shows the changes of the recovery rate of the soil ventilation obtained in September 1992.   The trichloroethylene concentration in the soil-gas extracted has continued to decline.   Coupling with the decrease of the soil-gas flowrate illustrated in Fig. 9, the recovery rate of trichloroethylene is also reduced in three extraction wells, the total recovery rate of which declines from 0.99 kg/hr at the beginning to 0.47 kg/hr in 55 hours.

4.6. *Recovery rate of trichloroethylene by groundwater extraction*

The groundwater was extracted for two purposes to remove the contaminant and offset the groundwater level rise due to the soil-gas extraction.   The total volume of groundwater pumped up from the K-1 and K-2 wells counted for 1.5 m³ per hour during the remedial operation in January 1992.   The trichloroethylene concentration in the groundwater ranged between 100 and 150 mg/l, which indicated the recovery rate to be approximately 0.15 kg/hr.

The trichloroethylene concentration in the groundwater also declines with the operation time. From September 1992, the groundwater extraction at the volume of 3 m³ per hour has lasted using the K-4 well. The trichloroethylene concentration of 20 through 50 mg/l indicated the recovery rate is between 0.06 and 0.15 kg/hr.

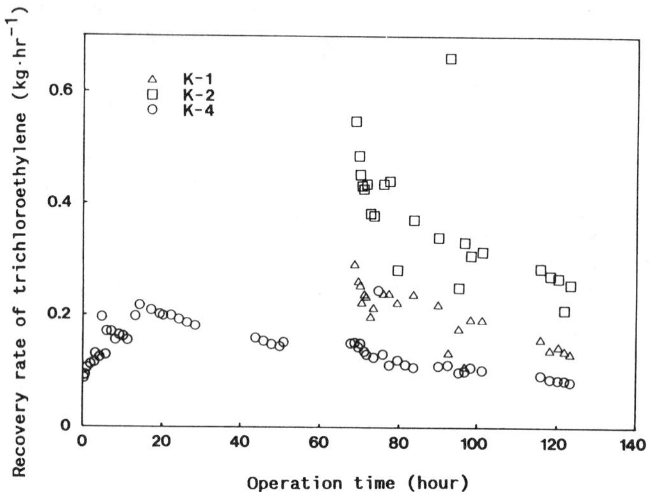

Fig. 11 Time-varied changes of the trichloroethylene recovery rate in three dual extraction wells.

5. SUMMARY

The paper describes the role of the soil-gas exploration as a remote geochemical technique in delineating the pollution boundary and determining the site for making boreholes. In addition to the groundwater extraction, the soil ventilation is confirmed to be an option to remove the volatile organochlorines from the vadose zone.

The most urgent problem at present to be solved is how long time the remedial operation should be done. The duration of the operation is going to gradually reduce the contaminant concentration both in soil-gas and groundwater extracted and also recovery rate. In this context, the suitable standard to finish the operation and the method to evaluate the effect of the remedial operation are desired.

6. REFERENCES

Hirata, T., Nakasugi, O., Yoshioka, M. and Sumi, K. (1991). Groundwater pollution by volatile organochlorines in Japan and related phenomena in subsurface environment. *1st IAWPRC Internat. Sympo. on Hazard Assess. and Cntl. of Environ. Contaminant in Water*, Ohtsu, Japan, 25-28.

Marrin, D.L. and Thompson, G.M. (1987). Gaseous behavior of TCE overlying a contaminated aquifer. *Ground Water 25(1)*, 21-27.

Yoshioka, M.,Yamasaki, T., Okuno, T., Hirata, T. and Nakasugi, O. (1992).Soil gas monitoring for survey on groundwater pollution due to volatile halogenated hydrocarbons. *J. Japan Society Water Environ. 15(10)*, 719-725 (in Japanese).

Johnson, P.C., Kemblowski, M.W. and Colthart, J.D. (1990). Quantitative analysis for the cleanup of hydrocarbon-contaminated soils by in-situ soil venting. *Ground Water 28(3)*, 413-429.

IN - SITU SOIL REMEDIATION PROJECT FOR THE REHABILITATION OF LOW-PERMEABILITY SOIL
CONTAMINATED BY CARBURETTOR GASOLINE: CONSIDERATION OF REACTION ENGINEERING
ASPECTS, BIODEGRADATION TESTS AND OPERATING RESULTS

DR. GREINER, DR. SCHWARZ, MERCEDES-BENZ AG, STUTTGART
DR. ECKARDT. UMWELT-MESSTECHNIK GMBH, STUTTGART

SUMMARY
During building work in keuper rock with an average k_f value of approximately 10^{-7} m/s,
contamination by carburettor gasoline of some 1 g/kg and PAH of around 3 mg/kg was discovered.
The possibilities to implement in-situ soil remediation were investigated by a combination of
theoretical and practical testing within the framework of industrial operating possibilities. In a 2 x 2 m
grid pattern, 99 boreholes were sunk. These permitted the soil to be rehabilitated to a large degree by
means of soil air extraction during a period of 20 months. By means of C14 measurements, laboratory
testing and soil analysis, it was possible to prove the feasibility of effecting biodegradation of the
contaminants in situ despite the limits imposed by conditions on site. However, it proved impossible
to find a satisfactory solution to the problem of supplying sufficient oxygen, biologically available
nitrogen and water during the time available for testing. Complete biodegradation of the residual
contamination left after completion of the soil air extraction process was thus not succesful in the
whole area.

1. CAUSE AND SCOPE OF THE DAMAGE

In the course of building work to extend a production hall, soil contaminated by carburettor gasoline
was discovered. The contamination, which had been caused by spillages in a refuelling station, was
limited to unsaturated soil zones. Fig. 1 offers a survey of the in-situ recovery area.
The maximum concentrations discovered were:

Total aromatic compounds	250	mg / kg
Total hydrocarbons	1030	mg / kg
PAH	3.5	mg / kg
Lead	40	mg / kg

Table 1: Soil contamination

The heating of the soil caused by the sampling method (augering) resulted in a considerable
expulsion of pollutants. For this reason, the actual maximum values have been most probably
considerably higher. The lead content is not above the average level for this type of soil.
Due primarily to the presence of polycyclic aromatic hydrocarbons (PAHs), which most probably
arose as a result of etching of the tank casing, the authorities ordered recovery of the entire soil
volume.
The time pressure arising due to the impending building work meant that no time could be lost in
holding lengthy negotiations about the actual necessity for recovery measures of what in fact turned
out to be only minor contamination damage.
The majority of the contaminated soil was therefore disposed of as hazardous waste. Only the
30 x 20 x 2 m volume depicted in Fig. 1 was treated in the course of the in-situ remediation project
described here. As Franki piles had already been driven into the area at the start of construction, soil
exchange was no longer possible in this area.

F. Arendt, G.J. Annokkée, R. Bosman and W.J. van den Brink (eds.), Contaminated Soil '93, 1029–1036.

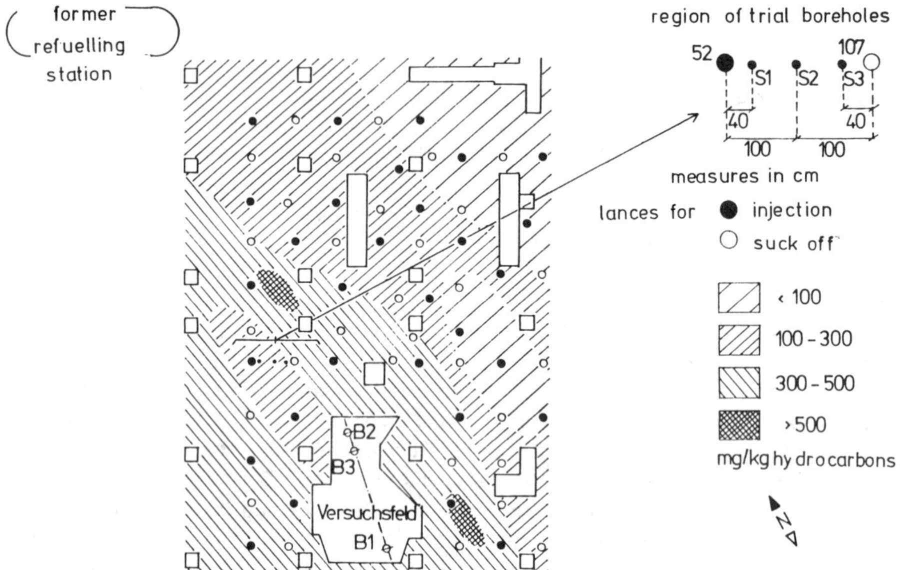

Figure 1: Remediation site and distribution of contaminants

The situation in the building excavation is indicated by Fig. 2. The head of one of the Franki piles and the fine horizontal stratification are clearly recognizable.

Figure 2 :Situation in the foundation - trench

2. GEOLOGY

The contamination is located in the so-called dark red marls, a strata group of the gypsum keuper which demonstrates a changing fine stratification with thicknesses in the mm to the cm range. The lower limit of this strata is formed by the Bochinger horizon, whose ground water content prevented the contamination from penetrating to a greater depth.

The grain size distribution and mineral composition were ascertained at the Institute of Applied Geology at the University of Karlsruhe. A particularly significant feature is the content of clay minerals with swelling capacity, which averages 12 % by weight. This effects a lift of some 1% with a maximum swelling pressure of 300 kN / m^2. Although this should not affect static stability, it can exercise a great influence on soil permeability.

Permeability was determined on undisturbed small samples approximately 96 mm in diameter and appr. 150 mm in length (Table 2).

As expected, a marked permeability anisotrophism took place. In addition, the k_f value is around 160 times higher when using nitrogen than when using water.

This means that permeability to water is lower by approximately the factor three than would normally be expected in view of the ratio between the viscosities (see below). This is most probably due to the influence of the swelling capability of the clay materials. The deterioration of permeability to water proved to be reversible. However, it took 48 hours at a pressure difference of 0.4 bar until the previous permeability properties had been restored after draining through the small sample with water.

Direction of flow	Medium	k_f [m/s]
Horizontal to stratification	Nitrogen	1.4×10^{-5} - $5.8 \cdot 10^{-5}$
Horizontal to stratification	Water	8.5×10^{-8} - 1.4×10^{-7}
Vertical to stratification	Nitrogen	3.9×10^{-6} - 6.1×10^{-6}
Vertical to stratification	Water	1.2×10^{-8}

Table 2: Permeability of the soil (Measurements by the Chair of Soil Mechanics and Foundation Engineering at the University of Karlsruhe)

The low k_f value for water of appr. 1×10^{-7} m/s excluded the possibility of hydraulic recovery right from the outset, as this method can only be applied up to values of 10^{-4} m/s.

3. REACTION ENGINEERING CONSIDERATIONS

For in-situ recovery projects, there are two basic important conditions: The elimination mechanism and the transportation of matter. In the case under discussion, where k_f values are low and the pollutants involved are highly volatile and easily biodegradable, the transport of matter represents the limiting factor. In the present case, the following aspects substantiate the superiority of recovery from the gas phase of the soil over a hydraulic technique (the following specifications refer to 15°C, about the temperature of the soil in the case under discussion).

1.   The viscosity of water is 1.14 cP, of air 0.0181 cP. One litre of water can dissolve 500 mg/l of toluene, while air saturated with toluene vapour contains some 90 mg/l. As the viscosity of air is around 60 times lower than that of water, the quantity which can be transported in the air is around 10 times higher than in water.

2.   In gases, the coefficients of diffusion lie at around 0.1 cm^2/s, while in water they are around 10^{-5} cm^2/s. The diffusive transportation of matter in the air is therefore greater than in water by a factor of around 10^4.

3.	Using pure oxygen, it is possible to dissolve appr. 48 mg per litre of water. In contrast, air contains appr. 300 mg/litre oxygen, i.e. around 6 times more. When using pure oxygen, this factor increases even more.

On the other hand, however, water is a necessary component of biological reaction systems, and if recovery is not possible without biological reactions, it must therefore be available in sufficient quantities.

4. TESTING PRELIMINARY TO BIOLOGICAL RECOVERY

It was not possible to take samples under sterile conditions. However, despite this the results summarized in Table 3 indicate that a sufficient quantity of hydrocarbon-degrading micro-organisms was available to permit the microbiological degradation of the pollutants.

Type	Number /g. soil
Total no. of colonies	appr. 10^6
Hydrocarbon decomposer	appr. 5×10^3
Fungi	appr. 5×10^2

Table 3: Microbiological population of the contaminated soil
(tests by the Engler-Bunte Institute, University of Karlsruhe).

In the course of preliminary testing, a test was developed which, although simple, approximated the actual soil conditions:

50 g of contaminated earth from the recovery site was weighed into sealable 250 ml Erlenmeyer flasks with glass stopper and glass stopcock and the following treatments performed single or in combination:

-	Water and nutrients were added directly through the aperture. The flask was then flushed with oxygen and sealed.

-	After flushing the flask with oxygen, ammonia was injected in through the stopcock using a gas-tight syringe.

-	In order to add NO, the flask was first flushed with nitrogen, after which the necessary quantity of NO was added through the stopcock using a gas-tight syringe. After leaving to stand for one day, the flask was flushed with oxygen.

The flasks treated in this way were stored in the dark at room temperature. The biodegradation of the highly volatile gasoline components was measured by taking air samples using a gas-tight syringe through the stopcock. The hydrocarbons were then determined using gas chromatography.

Table 4 provides an example of a successful biodegradation test

Concentration in the gas space (g/m^3)

Test duration [days]	Toluene	Ethyl benzene	Xylene	3-Ethyl toluene	1.2.4. Trimethyl benzene
0	0.237	0.278	1.114	0.413	0.508
3	<0.01	0.054	0.064	0.157	0.100
10	<0.01	<0.01	0.043	0.043	0.053
15	<0.01	<0.01	<0.01	<0.01	<0.01

Table 4: Biological degradation in a sample fertilized with NO.

These tests indicated the presence of phosphorus, sulphur, magnesium and trace elements in sufficient quantity, as well as the possibility of biological degradation under the following conditions:

1.	The added nitrogen must be in a biologically available form. Around 5 mg are required per kg soil. Suitable compounds are for example nitrates, urea, ammonia and nitrogen oxides.

2.	A hydrogen acceptor must be added. However, the substitution of elementary oxygen by nitrate poses something of a problem. Depending on whether a reduction to nitrite or nitrogen is effected, for a concentration of 1 g of hydrocarbon per kg, between 8 - 22 g of potassium nitrate are required per kg of soil. However, the high salt concentrations necessary bring the degradation process to a standstill.

3.	The water content of the soil must be over 16 - 18%. The level fell below this threshold only one year after closing the roof of the building. However, this problem situation only became evident during the course of the recovering project and was therefore not taken sufficiently into consideration. In a patch of the recovery site measuring 7 x 5 m (cf. Fig. 1), the untreated soil remained freely exposed for the duration of the recovery project. Here, several tests were carried out to check the flow-through capacity of the soil (which does not form part of this report), as were the in-situ experiments described below.

Initially, pure oxygen was fed into a borehole sunk through the entire stratum in this untreated soil patch, until a significant increase of the oxygen content of the soil air was registered. Following treatment with pure oxygen, a CO_2 increase from 3 to 9 % by volume, accompanied by O_2 depletion from 25 to 10 % by volume was measured within three weeks on one horizon only, at a depth of 1.0 to 1.5 m.

In the course of this test, the C^{14} content in the CO_2 was also determined. The recorded ΔC^{14} value of 64.1%, compared to the "pre-bomb" value, corresponds to an age of 8,230 years and can be considered proof that the majority of the CO_2 originates from petrified material, i.e. from a crude oil product. The δC^{13} - value of - 2.915 % showed, that the original material had been of biological origin. (Measurements by the Institute of Environmental Physics at the University of Heidelberg.) Corresponding to the biological degradation tests, preliminary experiments were carried out regarding nitrogen fertilization using ammonia or NO from the gas phase and using a nitrate solution (plant fertilizer). Only the latter produced a clear effect with an obvious reduction of the organic carbon content in the soil air from 25 to 2 g/m^3 within 40 days.

The significance of the water content in the soil only became evident at a later juncture when the moisture in the soil had diminished after closure of the building roof. It may be assumed that the failure of fertilization using nitrogen monoxide was due to an insufficient water content in the soil. The reason for the lack of success achieved when supplying nitrogen with ammonia was its good water solubility.

5. REMEDIAL ACTIONS

In a 2 x 2 m grid pattern, a total of 99 soil injection lances were installed in the recovery site (cf. Figs. 1 and 3). These permitted a total air stream of 1.5 m^3/h to be extracted with feed pressures of 0.2 bar and extraction pressures of 0.4 bar.

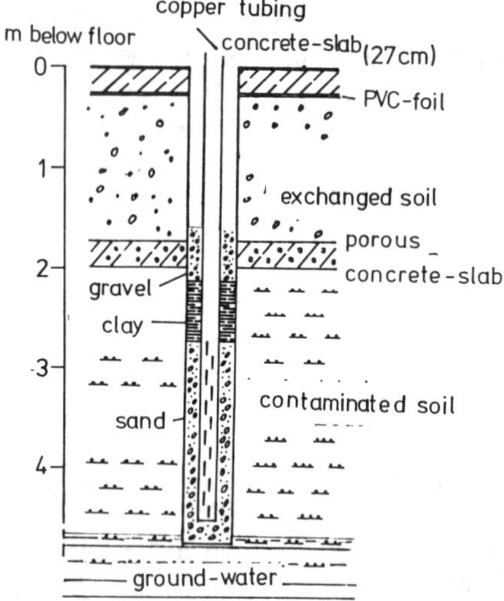

Figure 3: Lance

This level of permeability occurred only after some time. At the onset of the recovery project, the soil was moistened through to such an extent following the winter months in the open excavated site that it was almost impermeable even to air. In some areas, the soil remained impermeable throughout the entire recovery period. Fig. 4 indicates the progress of concentration in the extracted soil air applied to the extracted air volume. From the initiation of the recovery project on August 31, 1987 until completion on July 5, 1989, a total of almost 900 kg of organic carbon were extracted by stripping. The progress of time is indicated by specification of dates for important measures.

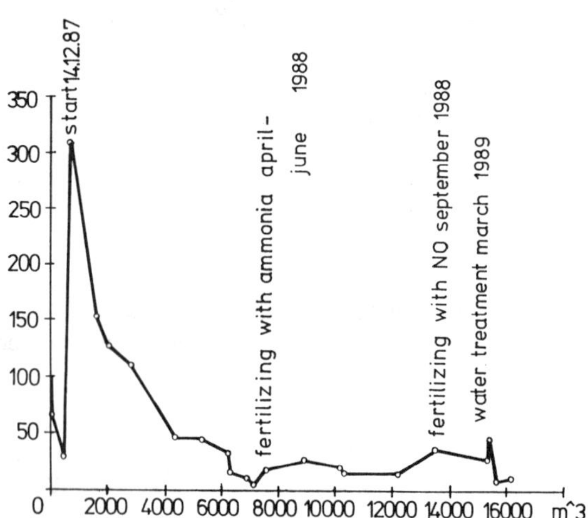

Figure 4: Recovery progress

Once it became evident that, alongside fertilization with nitrogen, also the addition of water was important, 106 m^3 of water and dissolved plant fertilizer with a total of 15 kg of nitrogen were added, not withstanding the possible blockage of the air passage that this might cause. The effect of this measure is indicated, despite the markedly reduced extraction, by the clearly reduced concentration of organically bound carbon in the exhaust air. However, the CO_2 content of 2 - 2.5% remained unaffected.

At this time, the recovery project had to be broken off prematurely to permit continuation of building work. It was therefore no longer possible to continue testing of the measures to support biological degradation.

Trial boreholes sunk after completion of the recovery project offered an indication of the recovery success achieved up until this point (cf. Fig. 5).

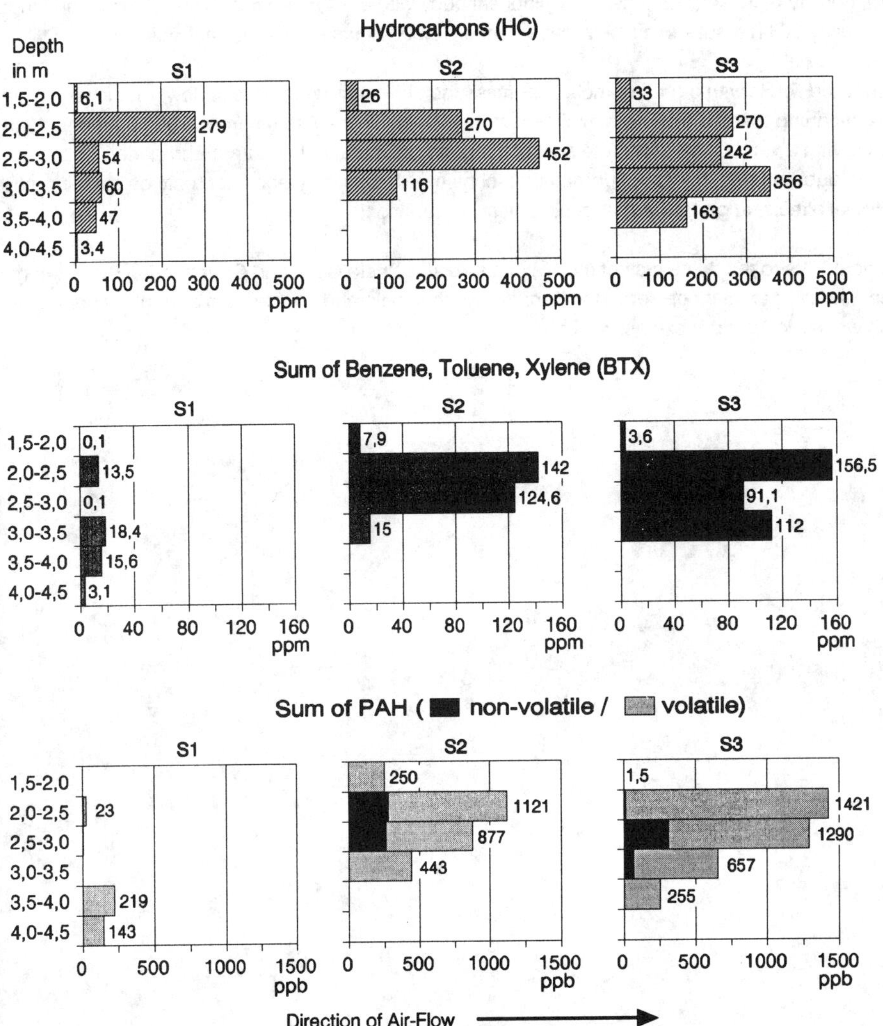

Figure 5: Recovery success

The continuous increase in the residual concentration from the injection well towards the extraction well would appear to indicate that the contamination was removed in the main by evaporation and convective discharge. However, the concentration development of the polycyclic aromatic compounds cannot be explained in this way:

The stated value is the sum of naphthaline*, acenaphthene*, fluorene, phenanthrene*, anthracene*, fluoranthene*, pyrene*, benzo(a)anthracene, chrysene*, benzo(a)fluoranthene, benzo(k)fluoranthene and benzo(a)pyrene. (For the substances marked with an asterisk, vapor pressure values are provided in Hoyer und Peperle Z.f. Elektrochemie 62(1958), page 61.)

An estimate carried out for phenanthrene indicates that this and all more volatile substances could have been removed within the available time simply by stripping to the observed extent. (Vapor pressure of phenanthrene 4×10^{-5} Torr at 15°C, reduction of the phenanthrene concentration by 200 μg/kg, density of the soil 1.8 tons/m^3, cylindrical body with a 0.4 metre radius and 2 metres in length, flow velocity 50 l/h results in a time requirement of 18,000 hours if the air is saturated with phenanthrene).

All identified PAHs with a greater molecular mass than phenanthrene have far lower vapor pressure levels, meaning that the marked diminishment, particularly in the stratum from 2.5 to 3 metres depth, of PAHs with a very low volatility (the sum of fluoranthene, pyrene, benzo(a)anthracene, benzo(a)fluoranthene, benzo(k)fluoranthene, chrysene, benzo(a)pyrene), can only be explained by additional effects, in particular that of biological degradation.

We should like to thank the staff of the Materials Testing, Installation and Construction Departments of the Sindelfingen plant of Mercedes-Benz AG for their help and cooperation during the period in which we carried out our extensive work.

GROUNDWATER AND SOIL REMEDIATION ON THE LOCATION OF THE CHEMICAL
PLANT LEUNA - CONCEPT AND FIRST RESULTS

Wolfgang Schneider[1], Ludwig Luckner[2], Claus Nitsche[2],
Jürgen Kayser[1], Dieter Eichhorn[3] und Jürgen Hensel[4]

[1]LEUNA-WERKE AG; [2]Dresdner Grundwasserforschungszentrum an der
TU Dresden e.V.; [3]Dresdner Grundwasser Consulting GmbH;
[4]Institut für Bioanalytik, Umwelttoxikologie und Biotechnologie
GmbH

1. Introduction

An intensive and environmental stressing chemical production has
been running on the location of the chemical production plant
Leuna for several decades. Important landmarks of the
technological development of the plant are:

1916 Ammonia producing plant Merseburg, using high pressure
     synthesis of Haber-Bosch, 1928 largest production capacity
     of the world
1919 Ammonium sulphate production
1923 World's first high pressure methanol synthesis according to
     M. Pier
1925 World's first production of synthetic gas, using F.
     Winkler's gasification technique of carbonised lignite
1926 World's first production of gasoline using Bergius
     hydration of coal with the large scale technology of M.
     Pier
1938 Large scale production of caprolactam acc. to Paul Schenk
1940 Large scale surfactant production as basis for production
     of detergents
1949 Urea synthesis as basis for the production of adhesives
1959 Start of building of plant section II, basis of oil
     chemistry in central Germany
1966 Replacement of coal hydration by oil processing
1986 Start of operation of low pressure methanol synthesis
     (Capacity 2,000 metric tones per day)

The chemical production plant Leuna is characterised by
interlacement of oil refinery and chemical production (Fig. 1).
It is the aim of the Treuhandanstalt (THA, an federal Holding)

F. Arendt, G.J. Annokkée, R. Bosman and W.J. van den Brink (eds.), Contaminated Soil '93, 1037–1046.
© 1993 *Kluwer Academic Publishers. Printed in the Netherlands.*

and the management of the enterprise, to keep these specific connections working during the process of privatisation of the Leuna Werke AG. The existing refinery Leuna (capacity 4.5 million metric tones per day) will be replaced by a new, larger refinery with a capacity of 10 million metric tones per day by 1996, according to an agreement with the french consortium Elf Aquitaine/Thyssen Handelsunion. This agreement is a breakthrough in the efforts to secure the production at this location, as well as the raw material supply. At present the privatisation of the chemical production is on the way - with existing prospects for almost all parts.

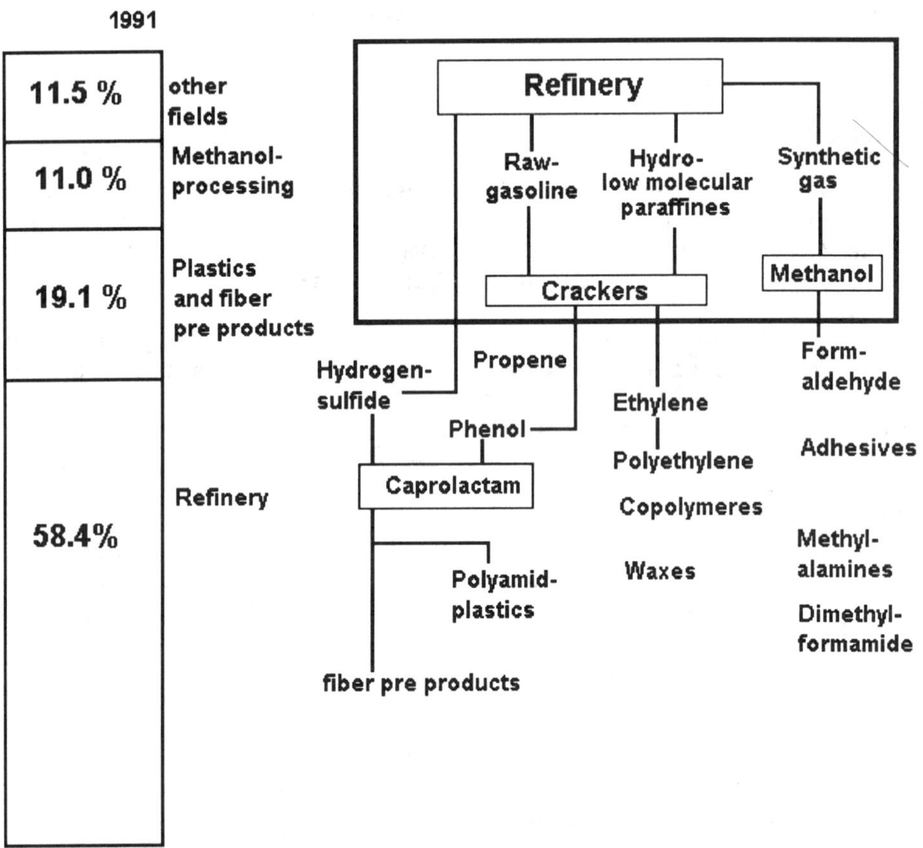

Fig.1.: main production fields of the enterprise and their interlacement

The area of the chemical production plant is located south of Merseburg, stretching some 5 km along railway tracks Halle-Erfurt and the federal highway Bundesstraße 91, with an total area of about 11 km2. On the territory of the plant are 135 km streets, 220 km railway tracks, 130 km sewage pipes, more than 500 reservoirs and tanks, and 2300 km main pipes.
The installation of the equipment was made using state of the art techniques at the given time, and the object of the internal regulations of the plant was to assure a high security and safety. Despite that, a possible leakage, caused by leaky equipment or carelessness in handling and service, resulting in a pollution of soil and ground water, can not be excluded. The processing of about 4.5 to 5 million metric tones oil per year, over a time range of 30 years gives a total of about 135 to 150 million metric tones raw product, which has been transported into the refinery via railway tracks and pipelines. The same amount of final product has been transported within the plant. Even if you calculate the leakage to be one ten thousandth of a percent (0.0001 %), the losses sum up to some 300 metric tones.
In addition, in 1944 and 1945, 22 air raids had been flown against Leuna by Anglo-American bomber formations. About 10 000 demolition bombs and mines and a large number of incendiary bombs had been dropped. A large area of the plant, with about 80 % of the production constructions, was destroyed. The effects have never been quantified, so there is no balance of the leakage caused by these destruction.
In the middle of the 1980th., a hydro-geological expertise, ordered by the Leuna Werke, has been made at the University of Technology in Dresden. Based on that expertise and additional investigations, made by competent mining institutions, in 1990/91 the number of ground water observation wells had been expanded from 42 to 96, the number of larger wells from 2 to 5. Furthermore, several ram core tests had been made. This, and the implementation of refined analytical methods and computer based data processing and evaluation, largely increased the knowledge of both, the underground of the production plant and the area in the down stream region of the ground water. At present some 18.000 data are saved. According to these data, there are five suspicious areas at the territory of the plant, where the soil and the ground water is contaminated with phenol and mineral oil (Fig. 2). These sites are especially in the area of the refinery, including the tank areas and regions with former and partially still working carbo-chemical production.
One of these potentially contaminated areas goes beyond the plant's border. Due to the direction of the ground water flow, from south south west to north north east, and the immediately

bordered town of Leuna a high potential of danger and a constraint for remediation exists. To do so, a protective well has been installed, where the contaminated ground water is pumped continuously and treated in a biological ground water treatment plant.

In September 1991 a small path of a ground water contamination with a water-soluble gasoline-additive MTBE (methyl tert butyl ether) has been discovered in the area of the community of Spergau, which is located next to the tank storage area. The cause of this contamination has been discovered and resolved.

Two more on-site ground water treatment plants for decontamination of the area are in the planning stage. But the effects of such treatment plants are only limited, and not suitable as single measure to resolve the remediation problems of a chemical production location.

A remediation of the location by soil exchange and on-site ground water treatment is largely restricted, due to economical reasons and the high construction density within the area of the Leuna Werke AG. So the emphasis of the sanitation work should be directed toward in-situ treatment of the soil and ground water to be cleaned.

After a successful application for financial support, in the beginning of 1992 the BMFT[1] supported research project "Biological in-situ remediation of contaminated ground water of the Leuna Werke AG" has been started. This is a joint research program with an extend of 10 million Deutschmark (about 6.5 million US$) in a co-operation between the Leuna Werke AG, DGFZ, DGC, and IFB. The objective is, to develop contaminant-specific in-situ ground water cleaning procedures for contaminated regions within the area of the Leuna Werke AG. These procedures have to guarantee that the sanitation goal will be attained, reliable and environmentally compatible. Starting with the selection of centres of contaminations, which can not be cleaned up satisfactory by use of traditional water treatment technologies, the hydro-geological and micro-biotical conditions of a specific location will be modelled in laboratory technology research. Based on that, new, especially physico-chemical and bio-chemical technologies of ground water treatment will be developed. The transition of the laboratory results will be done by in-situ demonstration tests.

---

[1]Bundesministerium für Forschung und Technologie
Federal department of research and technology

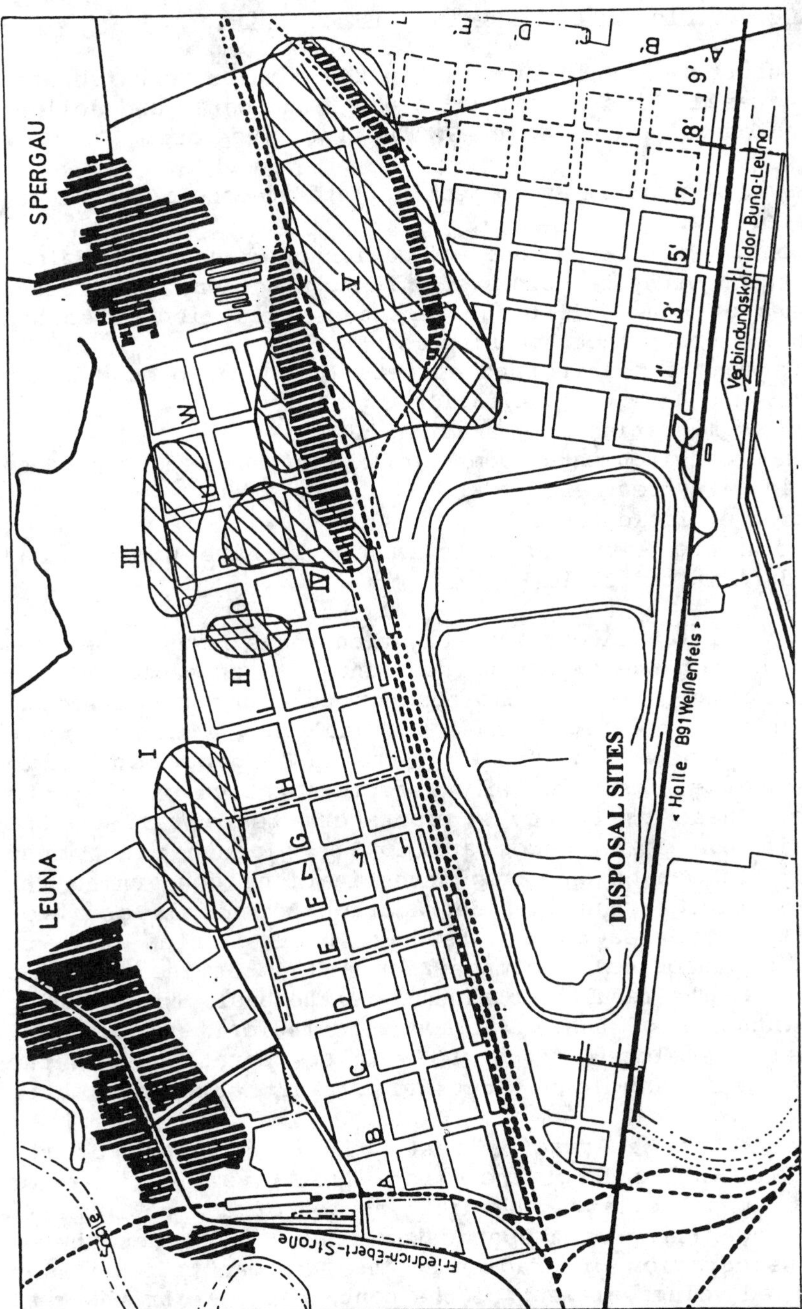

Fig.2.: suspicious areas at the territory of the plant

2. Selection and planning of the laboratory tests

Laboratory tests is a main part within the complex research work for in-situ remediation of contaminated ground water and soil at the location of the Leuna Werke AG. The objectives are:

- development of problem and object oriented technologies of in-situ restoration
- development, verification and validation of transport models
- determination of hysteretic system state functions and process parameters of non-equilibrium processes (reaction constants e.g., for 1. order exchange kinetic)
- demonstration of results and effects (e.g., of ground water treatment, ground water sanitation)
- prognosis of migration of contaminants
- valuation of environmental compatibility of the measures to be used in in-situ treatment
- preparation of field tests
- preparation and setup of a regime to measure process and control data in the field

Accordingly, all laboratory investigations have been chosen and planed, to reflect the important compounds and processes of the field test systems. In the bio-reactor test the bio-chemical processes only of the fluid phase are analysed, with no regard of the solid phase (soil), i.e. the REV (representative elementary volume) contains only a representative part of the pore space. These small scale laboratory tests reflect the processes in the pore spaces and in the planned ditch test system (Fig. 3). Based on basic investigations concerning the viability of micro-organisms, degradation potential, and bio-availability, bio-reactor tests have been carried out, to optimise the biological decomposition and determine limiting milieu factors. The results obtained, are the basis for planning and implementation of the REV fluid circulation test for soil core samples, developed by the DGFZ in co-operation with the University of Technology Stuttgart, Institute of Water construction.
Compared with the bio-reactor test, the solid phase of the underground is included for the first time. Investigations for improvement of solubility, the distribution of the pollutants, and the degradability of these contaminants can be done here, without consideration of gradients. The REV can be seen as a point oriented volume element of the container test or the real test systems (Fig. 4). In the third test and scaling level soil column filter tests will be carried out, based on results of the

REV fluid circulation test. The consideration of gradients corresponds to the one-dimensional processes in the container test and real test systems (Fig. 4). The inclusion of the formation of zones in the test volume achieved herewith, is a basis for development and laboratory testing of suitable procedures for in-situ sanitation of contaminated ground water and soil. Based on this, field test systems can be planned and operated.

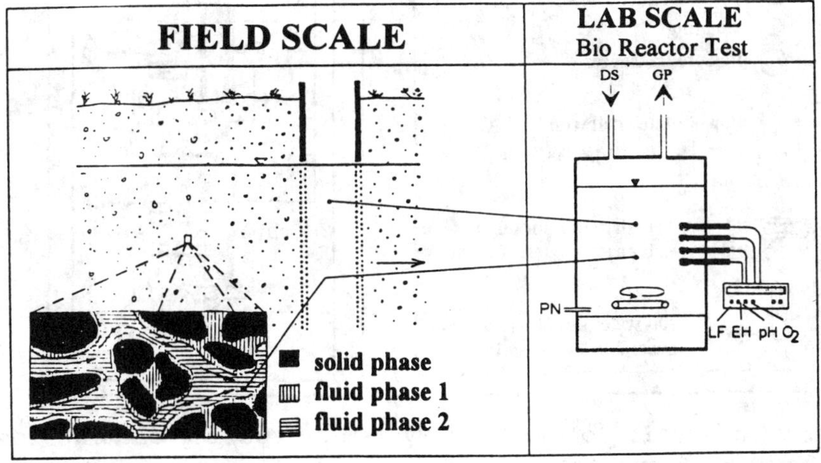

Fig. 3. Bio reactor test as scaling element for field test systems

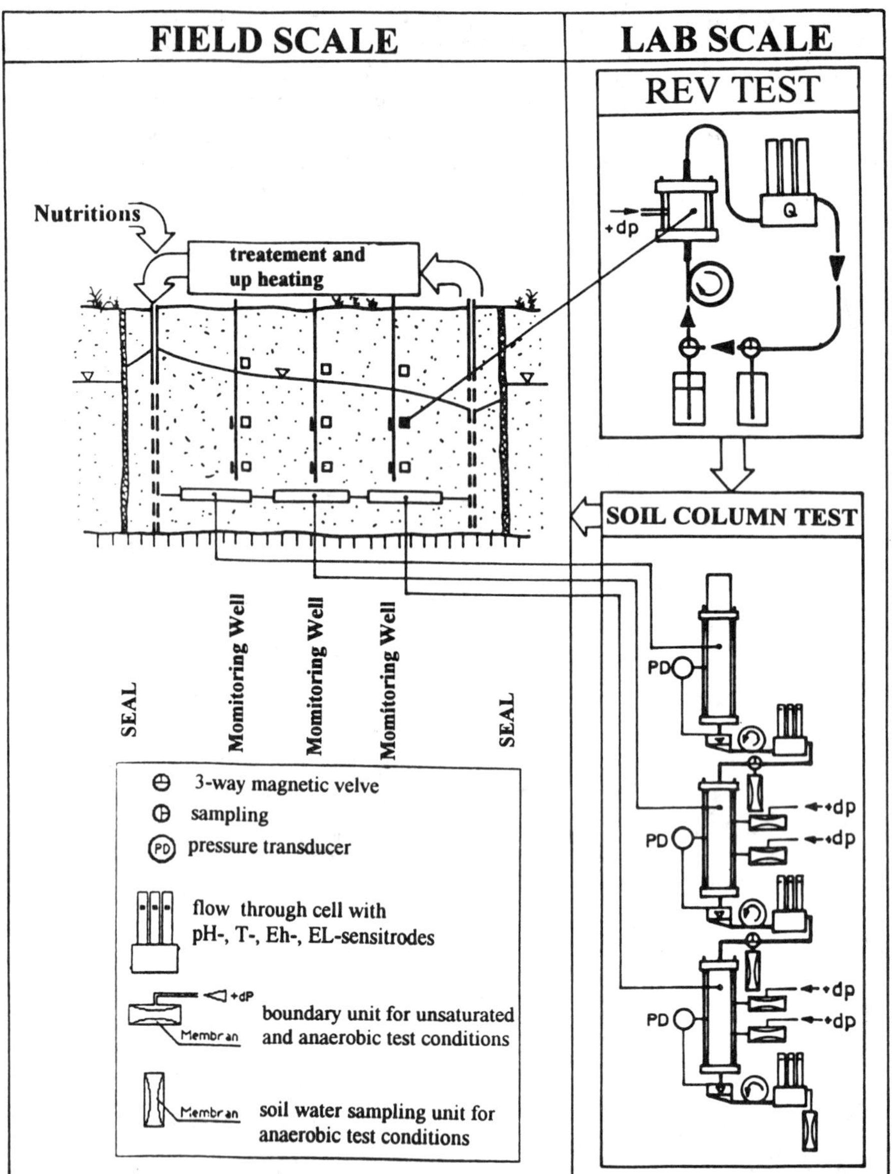

Fig.4.: REV- and Soil Column Test System as scaling element for field test systems

3. <u>First results</u>

To get used to the problems, first investigations of ground water, taken from wells within the test areas, has been carried out. The mixture of ground water samples had an average pH-value of 7, BOD_5 values of about 1,200 mg/l, and COD values of about 1,700 mg/l.

From the area around the wells 12 strains of bacteria and 2 strains of fungi have been isolated, which can be classified into different groups, according to morphology and staining (Gram). The exact classification of the strains of the micro-organisms is not finished yet.

At present, investigations are being made in a small laboratory fermentation reactor (reactor volume 500 ml, operation temperature 30 °C, pH=7, rotation speed 750 rpm, aeration rate 1l air per litre and minute), which is not optimised in the parameters yet. The micro-organisms (mixed population) totally degrade the phenols, which are in the mixed ground water samples from the wells, with a rate of dilution of 0.02 per hour (phenol intake about 300 mg/l, outflow <0.01 mg/l, COD intake 1,360 mg/l, outflow 66.6 mg/l, i.e. 95.1% reduction).

According to results obtained so far, the strains in this mixed population degrade the phenols contained in the ground water, using the ß-ketoadipin acid way, but meta-activities can not be excluded. This implicated consequences on the regulatory aspect of the substratum utilisation by the strains of micro-organisms isolated here. In general, the genetic stability of organisms using ortho-splitting enzyme systems to metabolise aromatic substrata, is higher, compared with the strains using 2,3-oxygeases. On the other hand, the substratum specification of the key enzymes of the first group of micro-organism is considerably higher, i.e. the adaptability toward mixtures of an other substrata combination is considerably lower. Additional tests to describe the stability of micro-organism population will be executed.

An acknowledged method to evaluate the toxicity of environmental contaminants is the bioluminescence test (DIN 38 412, T 34). This method has been used to assess the effectivity of the contaminant reduction by the mixed population in the laboratory fermentation chamber (bio-reactor test). While the G_L-value (Toxicity in the bioluminescence test = dilution, with an inhibition of <20 %) of the intake water of the chamber was 800, the G_L-value was reduced to 4 at the outflow.

From all investigations concerning the biological decomposability of the components of the ground water contamination done so far, it can be concluded; a biological

restoration and clean-up of appropriate areas is in principle
possible, under the use of qualified constructive devices which
can guarantee the conditions necessary for effective biological
processes.

4. References

Luckner, L: Dresdner Konzept der mathematischen Modellierung der
Mehrphasen-/ Mehrmigrantenprozesse im Untergrund. Karlsruhe, DFG
Kolloquium "Modellierung hydrochemischer Reaktions- und
Transportprozesse im Grundwasserbereich", Universität Karlsruhe,
Institut für Hydromechanik, 1990.

Luckner, L. & Eichhorn, D.: Arbeiten des Dresdner Grundwasser-
forschungszentrums zur Untergrundwasserbehandlung; Stuttgarter
Berichte zur Siedlungs-wasserwirtschaft Nr. 115, Oldenburg
Verlag, 1991.

Luckner,L. & Schestakow, W.: Migration Process in the Soil and
Groundwater Zone; Lewis Publishers, Inc., 1991, Catalog No.
L302LAEH.

Nitsche, C. u. Luckner, L.: Entwicklung und Musterbau einer
modernen Bodensäulentestanlage für dynamische Migrationsversuche
im Labor. Forschungsteilbericht; TU Dresden, Sektion Wasserwesen, Bereich Wassererschließung, 1988.

Nitsche, C. u.a.: Entwicklung und Musterbau einer modernen REV-
Fluidzirkulationsanlage zur laborativen Ermittlung von System-
zustandsfunktionen des Untergrundes. Forschungsteilbericht; TU
Dresden, Sektion Wasserwesen, Bereich Wassererschließung, 1988.

Nitsche, C.: Verfahren und Vorrichtungen zur Erfassung von
Systemzuständen im Boden- und Grundwasserbereich auf der Grundlage von Bodenwasserproben. 31. Darmstädter Wasserbaukolloquium
an der TU Darmstadt, 1991.

Rößner, U.: Bodensäulen- und Batchversuche zur laborativen
Simulation von biochemischen Umwandlungsprozessen im Boden- und
Grundwasserbereich. 31. Darmstädter Wasserbauikolloquium an der
TU Darmstadt, 1991.

Roy, W.R., Krapac, I.G., Chou, S.F.J., and R.A. Griffin: Batch-
Type Adsorption Procedures for Estimating Soil Attenuation of
Chemicals. Draft Technical Resource Document for Public Comment;
Illinois State Geological Survey; 1987

Autorenkollektiv: Zwischenbericht für das BMFT-Projekt "Biologische in-situ Reinigung verunreinigter Grundwässer und Böden
auf dem Gebiet der LEUNA-Werke AG"; Dresden, 1992

BIOLOGICAL IN SITU-REMEDIATION OF SANDY GRAVELLY GASWORKS SUBSOILS

o. Prof. Dr. Ing. G. Gudehus, Dr. Ing. J. Świniański, Dipl. Biol. H. Würdemann
Dept. of Soil Mechanics and Foundation Engineering, University of Karlsruhe (FRG)

Introduction

With its research and development, the Department of Soil Mechanics and Foundation Engineering of the University of Karlsruhe is contributing to safety and economy in geotechnical engineering. Concerning remediation of contaminated sites, we are concentrating our activities on developing biological in situ methods of practice and assessment giving consideration to the safety concepts of geotechnical engineering. A start was made with laboratory tests of undisturbed large samples spiked with artificial mineral oil contamination (*Lund 1990 and 1991*). The microbiological decontamination process was brought close to an optimum and a complete and controlled balance of contaminants was achieved.

For about two years the proposed biological in situ-remediation method has been tested at a former gasworks site in Karlsruhe on mainly gravelly and sandy ground. Partners in this large scale field experiment are the City of Karlsruhe and the special contractor Bauer GmbH, Schrobenhausen. We are conducting accompanying laboratory research, as also are the following institutes of Karlsruhe University: DVGW Research Department, Institute of Microbiology, Institute of Applied Geology.

Field experiment

The experimental site is located at the centre of contamination at a gasworks where the tar-ammonia separating sump used to be. It has been isolated from its surroundings by sealing walls which reach down to a depth of 17 m into an impervious layer of clay. About 1350 m^3 of contaminated soil are contained within the isolated area. The maximum contamination zone is located at a depth of between 5 and 7 m below the surface. Within the area of confinement the water table can be lowered to the limit of contamination.

Geological investigations have revealed the following general subsoil profile: directly below the surface there is a 2 - 3 m fill layer, below which is a natural, gravel and sand aquifer with layers of varying permeability (average value $5 \cdot 10^{-3}$ m/s) extending down to an impermeable layer of clay.

F. Arendt, G.J. Annokkée, R. Bosman and W.J. van den Brink (eds.), Contaminated Soil '93, 1047–1056.
© 1993 *Kluwer Academic Publishers. Printed in the Netherlands.*

The surface and subsurface installations employed in the remediation of the experimental site are illustrated in Fig. 1. In order to obtain a more or less horizontal airflow in the drained, contaminated soil, a network of airlances was sunk into the ground to serve the inflow of injected air and the outflow of spent air. The spent air is cleaned by passing it through an active carbon filter.

On the surface of the site a water sprinkling system was installed in order to produce a vertical flow of percolating process water. The process water, together with leakage water from the sealing walls, is pumped back up to the surface from two peripheral wells and treated in a water treatment plant. Before being resprinkled, inorganic nutrient salts are added. Corresponding to the flux of leakage water, 20 m^3/d are cleaned in an active carbon filter and discharged.

The biological remedation process consists of the drained contaminated subsoil being flushed with a horizontal stream of air and a vertical stream of water. The vertical flow of percolating water is used to transport ammonia and phosphate into the subsoil and to keep it moist. During the aerobic decomposition of hydrocarbons the oxygen of the soil water will be consumed. This consumption is replenished continiuously by diffusion from the airflow: the horizontal air-stream flows in the unsatured pores and transports O_2 into the soil. The oxygen diffuses from the air-stream into the soil water.

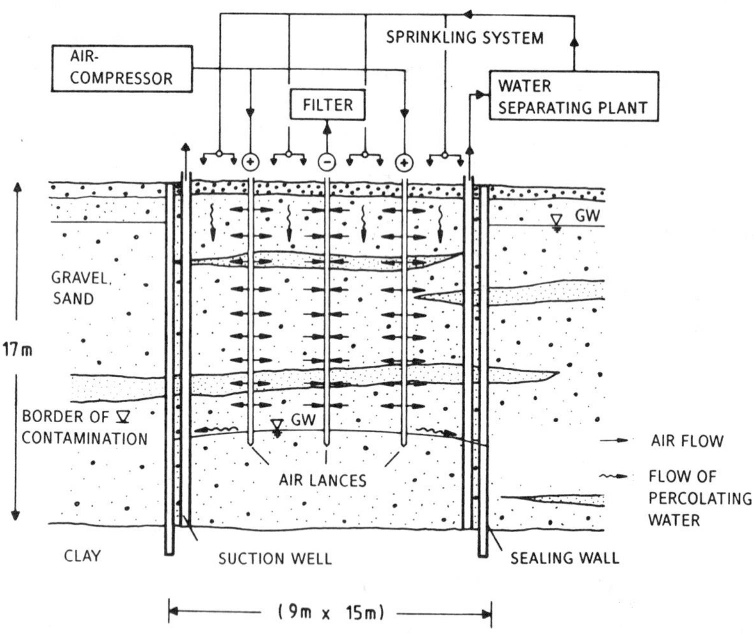

Fig. 1 Illustration of principle surface and subsurface installations at the experimental site (vertical cross-section)

At widely spaced intervals the continuous remediation process of concurrent circulation of air and process water was interrupted for one to two weeks, in order to conduct large-scale flushing of the body of soil with the aim of gaining information on the remaining concentration of elutable contaminants (large-scale elution test). Using one of the peripheral wells as an infiltration well and the other as a suction well, the pores of contaminated soil were repeatedly flushed (10 to 20 times) with nutrient-rich water. In order to raise soil temperature, the flushing water was heated in the first year using solar energy. In the second year an oil-fired heater was also used so as produce a greater increase in temperature.

Laboratory experiments

Laboratory experiments are conducted with undisturbed and disturbed samples of soil in order to gain data for optimising biodegradation in the field experiment. As a simulation of in situ conditions, large undisturbed soil samples ($\phi = 0.6$ m, h $= 1.0$ m) where extracted from the site using a ground freezing technique and then placed in a modified triaxial cell (bioreactor) for experimentation. Experiments with disturbed soil samples (20 kg) are conducted to obtain an idea of the maximum achievable degradation rate and the total degradation in the soil.

The determined initial concentration of contaminants in the sand and gravel soil was as high as 60 g lipophilic organics (CS_2 used as extractant), 21 g hydrocarbons determined by infrared spectroscopy (IR-HC) or 13.5 g of the 16 polycyclic aromatic hydrocarbons (PAH - contained in the EPA list) per kg soil (particle size < 2 mm). As in the field experiment, the samples are aerated and moistened and the O_2 and CO_2 concentrations in the spent air are continuously monitored. The samples are regularly flushed with water (so called "elution test") in order to effect a better distribution of nutrients.

In Fig. 2, as an example of the data obtained, biodegradation rate and soil temperature are plotted against time for the large undisturbed sample. An initially high degree of biological activity decreased considerably with the reduction of the bioavailable organic compounds. The results of elution tests confirmed the removal of contaminants available for microorganisms. After 100 days of bioremediation the concentration of IR-hydrocarbons (max. initial concentration 16 mg/l) and of PAH (EPA) (max. initial concentration 5 mg/l) in the water used for flushing had decreased to below the detection limit. The concentration of the 6 PAH of the TWVO (german regulations for drinking water) was reduced from 300 μg/l to 3 μg/l after 300 days of biological treatment. The collective parameters, DOC and COD, showed a reduction in water contamination of 70 %.

Fig. 2 also illustrates the strong influence of soil temperature on the process of biodegradation. Together with O_2-supply temperature proved to be the most important optimising factor.

Even in the laboratory experiments there are problems in estimating the efficiency of the remediation process. The uneven distribution of contamination in the undisturbed soil sample made it difficult to obtain representative analytical results, so that the initial amounts of contaminants in it can only be roughly estimated. As in the field experiment, continuous monitoring of O_2 and CO_2 in the spent air was very important for deducing the progress of biodegradation. According to data thus obtained, about 2 500 g of hydrocarbons were mineralized during the first 620 days of this experiment assuming that

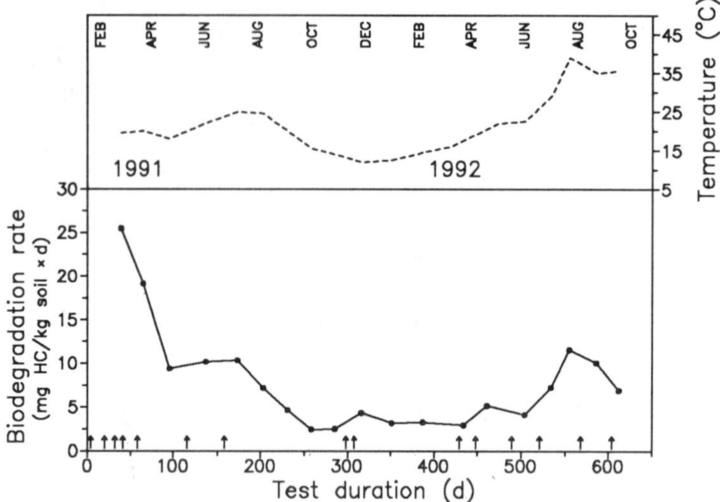

Fig. 2 Biodegradation rates and soil temperature during the large scale experiment (arrows: conducted elution tests)

3 kg of oxygen are necessary to mineralize 1 kg of hydrocarbons. In the course of the hydraulic treatments about 210 g of organic carbon (DOC) were washed out of the soil. The sum of the two amounts to about 35 % of assumed initial contamination.

In the experiment with the mixed soil sample - which is still running - about 550 g of CO_2 were produced in 530 days, corresponding to about 45 % elimination of the initial contamination.

A comparison of the mean biodegradation rates of specific gasworks hydrocarbons with some other common organic pollutants will serve to illustrate the low bioavailability of the former (Tab. 1). More soluble pollutants, such as mineral oil hydrocarbons, have even at lower level of contamination biodegradation rates which are up to 4 times higher than those associated with gasworks.

The low bioavailability of gasworks hydrocarbons considerably hampers their biological degradation. However, by raising the soil temperature from 15 to 38 °C it was possible to increase the biodegradation rates by a factor of 4 (due to increased solubility and metabolic activity).

The data obtained from water analyses shows a significant reduction in contaminant emissions via the water phase, despite relatively high concentrations of contaminants still contained in the soil. The toxicity of the effluent water also decreased from 95 to 20 %, as shown in the bacterial bioluminescence test.[1]

---

[1]The toxicity studies were conducted at the DVGW Research Department of the Engler-Bunte-Institute

Tab. 1 Comparison of the mean biodegradation rates

Type and level of contamination	Test characteristics	Degradation rate (mg HC/(kg s.×d))	Test duration (d)
n-Decan (7.4 g/kg soil)	sandy, gravelly undisturbed large sample, artificially spiked	90	54
Toluene (2 g/kg soil)	sandy disturbed sample, artificially spiked	28	47
Gasworks-HC (20 g/kg soil)	sandy, gravelly disturbed sample	20.8	55
Gasworks-HC (20 g/kg soil)	sandy, gravelly undisturbed large sample	19.7	55

Monitoring and Assessing the Progress of Bioremediation in the Field Experiment

Before beginning remediation in the field experiment information was sought about the kind, degree and distribution of contamination by drilling and extracting 3 cores from the site. Mixed samples were taken from the cores at 1 m intervals for analysis. The degree of contamination was quantified in numerous ways of which the principle were: determining the amount of extractable lipophilic organics, determining the hydrocarbons by infrared spectroscopy (IR-HC) and gas chromatographic analyses of the 16 polycyclic aromatic hydrocarbons (PAH) contained in the EPA list.

If the initial amount of contamination is calculated from these data one is faced with a number of problems. Apart from the unevenness of contaminant distribution there is also the problem that the results of its chemical quantification vary according to the method of analysis: while the determination of lipophilic organics resulted in a total contamination of 5000 kg, IR-HC measurements indicated only 2000 kg. The total burden of PHA as determined by gas chromatography was 600 kg. Since contamination specifically associated with gasworks consists of a very large number of individual substances (about 10000, according to *Lauer, 1989*) it is relatively difficult to find a suitable parameter with which to satisfactorily encompass and quantify it. Detailed investigations showed the non-specific parameter, lipophilic organics to be the most suitable (*Hopfstock, 1992*). Thus, this was chosen as our main measure of contamination and biodegradation.

Fig. 3a shows the degree of contamination plotted against soil depth. Lipophilic organics were chosen as the measure of contamination. Maximum contamination was found at a depth of 5 to 7 m. In the 3 cores extracted from the site, maximum contamination varied from 27 to 37 g lipophilic organics/kg soil < 2 mm particle size.

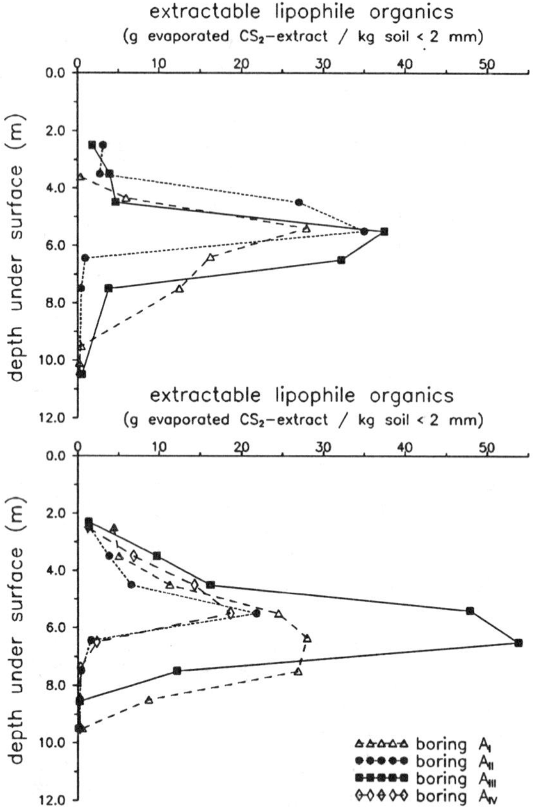

Fig. 3 Profiles of contamination: Because of the heterogenous distribution of hydrocarbons an exact quantification of the starting load is impossible
a. Contamination profile June '90
b. Contamination profile Jan. '92

After one year of biological remediation, a second set of cores was extracted from the
site. In order to minimize the influence of uneven contaminant distribution, the cores
were taken in close proximity to where the original cores were extracted. In the second
set of cores a considerably greater range of contamination concentration was found.
While maximum contamination in the core from the sampling site A_{II} was found to
have decreased by 50 %, in the core from site A_{III} it had increased by about 50 %.
Apparently, the distance of 0.5 - 1 m (which for technical reasons could not be reduced
any further) between the first and second sampling point was too great, owing to the
uneven distribution of contaminants, to enable us to follow the remediation process in
this way. Thus, other means of monitoring the remediation process had to be found:

- Hydrocarbon degradation as determined by measurements of O_2-consumption and
  CO_2-production (on-line biomonitoring).

- Determination of hydrocarbon emissions in the large-scale elution experiments.

A very important criterion for assessing the course of remediation is the determination of soil respiration, as a measure of the **biological degradation rate**. Measurements were made of the O_2- and CO_2-content in the spent air, from which it was possible to deduce the amount of organic substance which had been mineralized. Depending on the amount of newly created biomass, 1.5 - 3 kg of oxygen are required in order to mineralize 1 kg of aromatic hydrocarbons. Under in situ-conditions, an initial, but not a continual increase in biomass is to be expected. Thus, a factor of 3 was employed for estimating the amount of mineralized hydrocarbons.

On account of biotic and abiotic processes not concerned with the degradation of hydrocarbons, but which nevertheless contribute to oxygen consumption (e.g. nitrification), it was also necessary to measure the production of carbon dioxide *(Würdemann, 1990)*. If mainly hydrocarbons are completely mineralized the respiratory quotient (the ratio of CO_2-production to O_2-consumption) will be between 0.6 and 0.85. On average the value obtained in the field experiment was 0.7.

Since microorganisms degrade possible substrates according to their bioavailability and usefulness, it is conceivable that a significant amount of O_2 might be consumed in the oxidation of humic substances which are naturally present in the soil. However, on-site measurements revealed a close correlation between the degree of contamination and the level of biodegradation activity, suggesting that the degradation of organic constituents other than contaminants is negligible. An estimation of the influence of processes not involved with the degradation of organic material suggested an error of about 10 %.

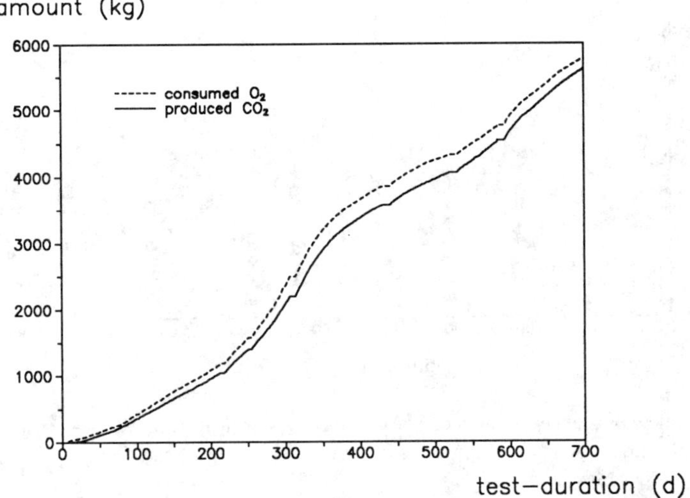

Fig. 4 Course of the biological remediation: amount of consumed O_2 and produced CO_2 during two years of biological remediation

The data obtained from monitoring O_2 and CO_2 content in the spent air are presented in the curves shown in Fig. 4, in which O_2-consumption and CO_2-production are plotted against time. During the course of two years remediation about 5 500 kg O_2 and 5 300 kg CO_2 were consumed and produced, respectively, which indicates a mineralization of about 2 000 kg of organic material.

Apart from providing information on the amount of bioremediation achieved, monitoring O_2 and CO_2 also provided information on the effects of various measures taken to optimize degradation. During the course of the first year of the experimental operation it was possible to increase the rate of degradation considerably. This increase in bioactivity is reflected in the increased slope of both the O_2-consumption and the CO_2-production curve. At the beginning of the operation an O_2-consumption of 4 kg/d was measured. During the summer of the first year this was increased to 17 kg/d by optimising soil aeration and nutrient supply. Parallel to this development an increase in soil temperature was registered. This was due to the intentional preheating of injected air and process water and to metabolic heat produced by the degrading microorganisms. During the cold winter months there was a marked drop in activity, which was accounted for by the drop in temperature and by a reduction in the bioavailability of contaminant carbon sources. In the spring and summer months activity increased again (up to 10 kg O_2/d) with increasing ambient temperatures and the temperature of the process water.

Besides biomonitoring, the determination of **contaminant emissions** is also a very important parameter for assessing the course and progress of the remediation process. It provides an indication of the hazard potential which the contaminated site still possesses.

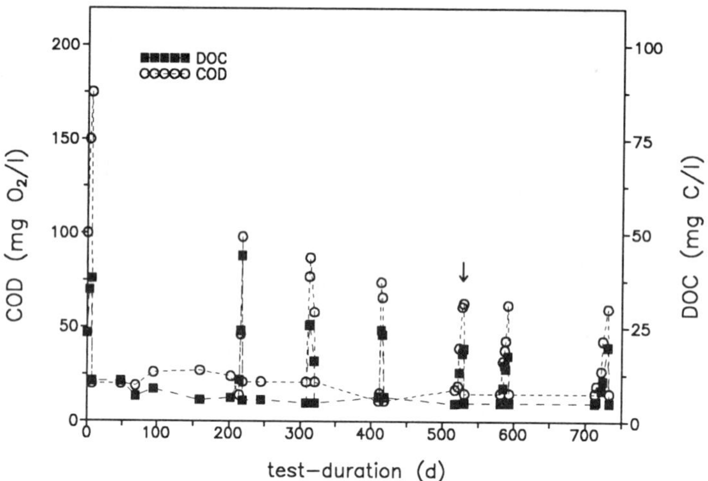

Fig. 5 COD and DOC concentration of water samples taken during two years of a field experiment. Arrow: Beginning of active preheating of the flushing water

Fig. 5 shows the data for chemical oxygen demand (COD) and for dissolved organic carbon (DOC) in samples of water taken from the site during the course of bioremediation. The 7 peaks obtained during the large-scale elution experiments are of particular interest. Because the water used for flushing was repeatedly passed through the body of contaminated soil (10 - 20 fold flushing of soil pore water) it became enriched with contaminants. During the 2 years of bioremediation the thus obtained COD peaks were reduced by about 70 %, while the DOC peaks dropped to about 50 % of there original value. Over the same period there was a parallel decrease of 80 % in the concentration of of PAH (initial maximum concentration 2.5 mg/l) (Fig. 6). The success of the biomediation operation becomes even more evident when one considers the data on toxicity. The bioluminescence test for toxicity indicated 80 % detoxification.[2] Although in places the soil still contained extremely high concentrations of lipophilic organics, the hazard potential of the contaminated site had been considerable reduced.

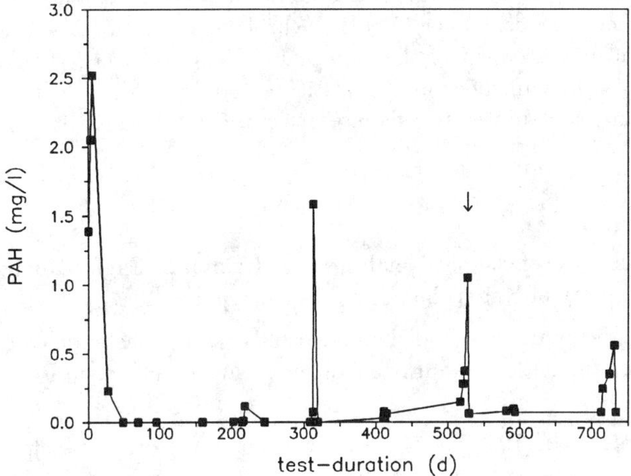

Fig. 6 PAH concentration of water samples taken during two years of the field experiment. Arrow: Beginning of active preheating of the flushing water

Closing Remarks

Detailed reports will later be submitted on the laboratory experiments with gasworks soil *(Świniański)* and on the results of the field experiment *(Würdemann, Lund)*. They will contain proposals for dimensioning and optimising the operation in situ, and for hazard assessment via emissions of contaminants in the water phase. The proposals for optimising and dimensioning have to be justified technically and ecologically. Since the hazard assessment of contaminated sites via chemical analysis of soil samples is becoming questionable, evaluation based on the theory and monitoring of emissions will gain importance. Data will be obtained from large sample cores taken by ground freezing technique and subjected to elution tests in the laboratory. Our aim is to develop a method of assessment for contaminated sites prior to, during and following in situ remediation.

The basis of assessment should be the geotechnical risk, i.e. the sum of products of damage sums (subdivided into human health and financial losses) and their respective probabilities, summarized over relevant damage mechanisms. For contaminated sites, this requires following relevant emission paths as far as potentially exposed objects and

---

[2]see 1

persons and the respective cumulative toxicological assessment *(Gudehus, under preparation)*. In the case of high geotechnical risk an observational method has to be applied, as outlined in recent geotechnical codes of practice.

The financial and managerial support provided by the BMFT and UBA (federal) and by the PWAB and KfK Karlsruhe (state) is gratefully acknowledged. We also thank the City of Karlsruhe for conceptional and technical support, especially Prof. Dr.-Ing. habil. D. Maier for his scientific advice concerning the evaluation of chemical analyses. We are further obliged to the contractor, Bauer GmbH., Schrobenhausen, for his contribution to the construction and management of the field plant. We owe special thanks to Dr.-Ing. N.C. Lund, who, former member of our institute (now with Trischler & Partner Consultants in Potsdam) contributed to this research project substantially.

Literature

Hopfstock, H. (1992): Results of chemical analyses of soil samples. Engler-Bunte-Institut Abt. Gastechnik. University of Karlsruhe. Unpublished report.

Lauer, K.-H. (1989): Herkunft von Boden und Grundwasser belastenden Stoffen bei der Stadtgaserzeugung. In: Altlasten auf ehemaligen Gaswerksgeländen. Report of the DVGW. Gas No. 45.

Lund, N.-Ch., Gudehus. G. (1990): Laborversuche an ungestörten Großproben zur biologischen in situ-Sanierung kohlenwasserstoffbelasteter Böden. In: F. Arendt, M. Hinsenveld und W.J. van den Brinks (Hrsg.): Altlastensanierung '90. Kluwer Academic Publishers.

Lund, N.-Ch. (1991): Beitrag zur biologischen in situ-Reinigung kohlenwasserstoffbelasteter körniger Böden. Dissertation. Report of the Institute of Soil Mechanics and Rock Mechanics, University of Karlsruhe, Report-No. 119.

Stieber, M. (1991): Ökophysiologische Untersuchungen. In: In situ-Sanierung kohlenwasserstoffbelasteter Böden. Unpublished report: 3. Report of the BMFT research projekt "In situ-Sanierung kohlenwasserstoffbelasteter Böden". No. 1460635A5

Würdemann, H. (1990): Schnelle Beurteilung der aktuellen Sanierungsleistung von Bodenregenerationsmieten durch die in situ-Erfassung der biologischen Atmungsaktivität. Master thesis at the Institute of Microbiology. University of Braunschweig.

USE OF ION EXCHANGE AND UV-LIGHT FOR REMOVAL OF COMPLEXED CYANIDES FROM GROUNDWATER

B. AKSAY[1] and R. KAMPF[2]

[1] Heidemij Realisatie BV, P.O.B. 660, 5140 AR Waalwijk, The Netherlands.
[2] Waterboard Uitwaterende Sluizen, Lingerzijde 39, 1135 AN Edam, The Netherlands.

1. ABSTRACT

The soil and groundwater of the site of the sewage treatment plant Beverwijk of the Waterboard Uitwaterende Sluizen are polluted with, mainly, cyanides. The source of the pollution with cyanides had been a dump of a chemical works that made complexed cyanides out of wastes of coal gasification plants.

For the extention of the plant the sites of the new aeration tanks, settling tanks and other parts of the extention polluted ground has been excavated. During the construction of the extension the high groundwater level at the site has to be lowered. For this aim a number of deep-wells will be installed. The pits will also be pumped dry directly. Concentration of cyanides in groundwater varies from 100 up to 30,000 μg/l. The COD and TKN of the groundwater are ca. 150 and 25 mg/l respectively. The turbidity is rather high because of clay particles.

The Waterboard and Heidemij have carried out investigations and pilot plant studies at a 1 m^3/h scale. The aim of the study was to assess the most cost-effective method to meet the effluent standard for cyanides of 250 μg/l during the building period of approx. two years.

It pointed out that concentrating the complexed cyanides on a resin was very effective because of the entrapment of the cyanides on the specially developed resin. The ion exchange process gives a highly concentrated regenerate. The flow to be handled in the photo-chemical reactor is less than 1 % of the total groundwater flow. This saves both peroxide and electricity for the UV-lamps for the conversion of complexed cyanides into harmless components.

To prevent fouling of the resin a rather extensive pre-treatment was incorporated in the process: pre-treatment was done by using aeration, coagulation, flocculation processes and rapid sand-filtration in order to reduce COD, Fe and turbidity. For enhanced floc entrapment activated sludge from the sewage treatment plant was used.

Scale-up to 700 m^3/d is planned to be in action in the first half of 1993.

2. INTRODUCTION

The soil and groundwater of the site of the sewage treatment plant Beverwijk of the Waterboard Uitwaterende Sluizen are polluted with, mainly, cyanides. The source of the pollution with cyanides had been a dump of a chemical works that made complexed cyanides out of wastes of coal gasification plants.

For the extension of the plant the sites of the new aeration tanks, settling tanks and other parts of the extension polluted ground has been excavated.

F. Arendt, G.J. Annokkée, R. Bosman and W.J. van den Brink (eds.), Contaminated Soil '93, 1057–1064.

1058

A complete clean-up of the site was not feasible, the sewage treatment plant is situated in a harbour area with small industry.

The main aim was to deal carefully with the excavated soil and and to remove cyanides from the groundwater. During the construction of the extension the high groundwater level at the site has to be lowered. For this aim a number of deep-wells will be installed. The pits will also be pumped dry directly.

The groundwater has to be treated during the whole building period of two years. The volume of water is limited to 700 m^3/day. The maximum values in the effluent may not exceed in any grabsample:

-     cyanides   250 µg/l

and:

-     IR-analyses (mineral oil)   10 mg/l
-     benzene   10 µg/l
-     toluene   25 µg/l
-     ethylbenzene   25 µg/l
-     xylenes   25 µg/l
-     PAK-Borneff   25 µg/l
-     Naftalene   25 µg/l

3. GROUNDWATER

Beverwijk is situated near the coast. The soil consists of an alternation of clay and peat sediments, covered by andropogenic deposits. Mineralisation of peat through lowering of the groundwater level in the last decades led to a high COD (150 mg O_2/l) and high levels of TKN (25 mg/l) and P (3 mg P/l) in the groundwater. The decomposition of clay particles caused a high turbidity of the groundwater, unfavourable for UV-processes. Above this the groundwater has a high alkalinity (25 mmol/l).

The concentration of cyanides are up to 30.000 µg/l, on an average 2.500 µg/l. During the project there has been many discussions on the analysis of cyanides. Total cyanides has been analyzed by EPA 335.3, the sum of total cyanide and thiocyanates. Four samples have been analyzed by four laboratories, the results are summarized in table 1. (Sample 1: groundwater, Sample 2: groundwater, Sample 3: dilution (5x) of sample 2, Sample 4: dilution (100x) of sample 2). (See table 1).

TABLE 1. Analysis of total cyanide (EPA 353.3), concentrations in µg/l

	lab 1	lab 2	lab 3	lab 4
sample 1	24,000	25,800	21,500	21,000
sample 2	1,500	1,070	920	930
sample 3	200	210	190	190
sample 4	17	12	10	10

4. REMOVAL OF CYANIDES FROM GROUND WATER

For treatment of groundwater with cyanides several methods are available, depending on the type of groundwater, other components in the groundwater and most of all the type of cyanide. Most of all the methods however have a low efficiency for comlexed cyanides.

4.1. Pre-studies

Flocculation with chemicals pointed out to be expensive (over $ 4.-/m^3, chemical costs alone) because of the high alkalinity. Very high dosages were needed for pH changes.

Oxidation of the groundwater by H_2O_2 or O_3, followed by UV-light treatment gave good results on laboratory scale. But again because of the composition of the groundwater the costs would be in the same magnitude as a flocculation processes. The organics in the groundwater required a high dosage of oxidizer, the turbidity influenced the efficiency of the UV-lamps negatively. Total costs of the groundwater treatment would be well over $ 3,000,000.--.

Eventually it was concluded that cyanides were nearly alone in the form of complexed cyanides. Biological treatment had hardly influence on the cyanide concentration.

4.2. Pilot-plant studies

The Waterboard Uitwaterende Sluizen commissioned Heidemij for assessing the possibilities of the new use of a resin for entrapment of cyanides. With several tests Heidemij had already showed that photochemical processes a capable of treating complexed cyanides.

The process have been investigated on a 1 m^3/h scale on the Sewage Treatment Plant of Beverwijk for several months. With the experiments insight have been obtained under practical circumstances in: the efficiency, the choice of unit-operations, the consumption of chemicals, etc. The studies had to result in a design for a 700 m^3/d plant.

The pilot-plant consisted of two parts:

a. -   pre-treatment of the groundwater;

b. -   removal of cyanides by ion exchange and oxidation of the cyanides by UV/H_2O_2 treatment.

4.2.1. Pre-treatment. The main aim of the pre-treatment was to:

- remove suspended and colloidal particles;
- reduce the concentration of organics;
- oxidize S^{2-}, $Fe2^+$;
- remove oil.

This was needed to enhance the effectiveness of the ion exchange, prevent fouling of the resin and minimalize the consumption of chemicals and energy.

Two different schemes have been tested. At first the co-metabolism with sewage have been tried out, figure 1a. The basic idea was: could it be effective to treat the groundwater in a very low loaded activated sludge plant. Sewage from the main plant was added to grow activated sludge (0.2 m^3/h sewage and 1 m^3/h ground water). It was thought that biological removal of cyanides would occur. In fact the removal of cyanides was minor, the reduction of COD of the groundwater was only 10 %.

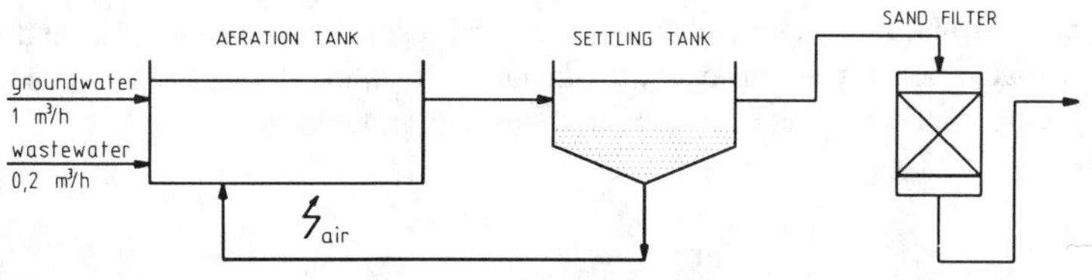

FIGURE 1a.   Pre-treatment of groundwater with addition of waste water

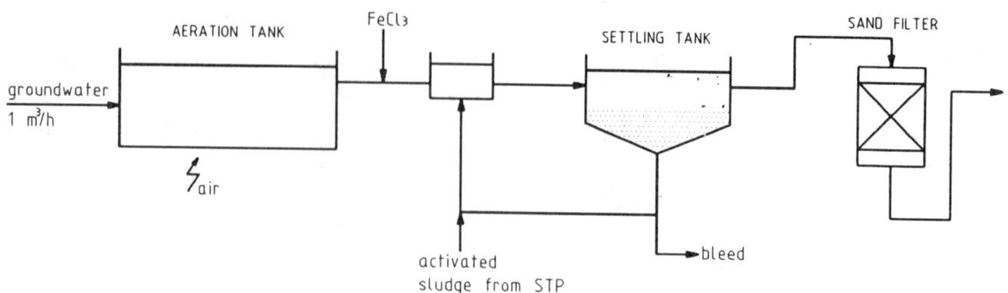

FIGURE 1b.   Pre-treatment of groundwater with entrapment of Fe-flocs with activated sludge from the sewage treatment plant (STP)

After sand filtration the groundwater had become very clear however, the activated sludge did actually bind the fine particles in the groundwater very effective. This observation led to the set up in figure 1b. The aeration tank was used for oxidation of the ground water, a dosage of $FeCl_3$ was installed at the overflow of the aeration tank. In the settling tank a blanket of activated sludge from the sewage treatment plant was maintained to catch small particals. Overflow of sludge particles were removed in the sand filter. Table 2 gives a summary of the results obtained.

TABLE 2. Results of pre-treatment: scheme b in figure 1.

	Influent Concentration (mg/l)	Removal Efficiency (%)
COD	198	60
$Fe2^+$	12	80
TKN	50	10

The adsorption of cyanides to the activated sludge was very low, less than 0.1 % of the cyanides in the ground water.

It was concluded that scheme b in figure 1, as described above would be cost-effective combination of Unit-operations for pre-treatment of the groundwater on the site in Beverwijk. The method gave a clear ground water free of particles suitable for an ion exchange process.

5. REMOVAL OF CYANIDES BY ION EXCHANGE AND UV-LIGHT

The concentration of cyanides by ion exchange on a resin followed by regeneration and UV-treatment has considerable advantages on direct treatment of the whole flow of pre-treated groundwater. The outline of the investigated method is outlined in figure 2.

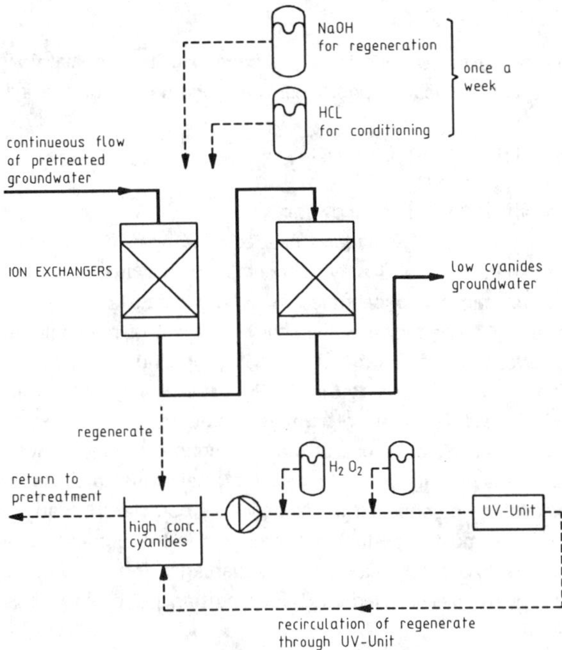

FIGURE 2. Removal of cyanides by ion exchange and UV-light

5.1 Ion exchange

The pre-treated groundwater is percolated through a bed of granular ion exchange resin. The cyanides are adsorbed at the resin by exchange against chloride ions. This will continue until all exchange sites are occupied. Regeneration takes place with NaOH to free the cyanides from the resin. For conditioning of the resin with chlorine ions HCl is used.

The efficiency of the process was higher than expected, namely 99 %. Table 4.

TABLE 4. Removal of cyanides from pre-treated groundwater by ion exchange

	Influent concentration (μg/l)	Effluent concentration (μg/l)
Run 1	760	2.7
Run 2	1940	< 5
Run 3	890	9.8
Run 4	1000	8.4

To test the capacity of the resin an experiment has been carried out with groundwater with 30,000 μg cyanide/l, without pre-treatment. As much as 40 g cyanide/l had been adsorbed in the ion-exchanging resin. After regeneration still the resin contained 5.5 g cyanide/l. Only 85 % of the cyanides could be regenerated. the test was run with crude groundwater, the resin was rather fouled. The concentration in the regenerate was 7.7 g/l. Two thousand times more than the already rather high concentration of cyanides in the groundwater.

5.2 <u>Photochemical degradation</u>

The flow of concentrate to be treated in the UV-unit is only a fraction of the original flow of ground water. This gives considerable advantages on photochemical degradation, as the following example shows:
- maximum adsorption capacity of the resin: 30 g/l resin;
- CN- in groundwater: 2,500 μg/l = 2,5 g/m^3;
- > capacity per l resin is well over 10,000 l of groundwater.

For regeneration of 1 l resin about 1.5 l NaOH-solution is needed. The effectiveness of the resin will lesser after a number of regeneration. But it may be concluded that at least on cyanides from 1,000 to over 10,000 l groundwater can be concentrated in 1 l regenerate.

In the UV/H$_2$O$_2$ process the pretreated groundwater is lead through a set of medium-pressure mercury-lamps. The UV-light has a wavelength of 185 nm UV-light is a continuous flow of energy quants, photons. At short wavelengths as UV-light, energy in the light will be higher than at longer wavelengths, as a consequence cyanides will fall apart in harmless components when radiated with UV-light. H$_2$O$_2$ is added to increase the effectiveness of the treatment, by radiation with UV-light hydroxyl radicals will be formed. In a propagation of hydroxyl radicals other radicals will be formed. These so called chain reactions are responsible for oxidation of complexed cyanides. The formed CN$^-$-ions are directly oxidized. The most important end products in the process are CO$_2$, H$_2$O and N$_2$. The use of H$_2$O$_2$ instead of ozone (O$_3$) has several advantages for use groundwater treatment, where only limited amount of oxidizer is used. H$_2$O$_2$ can be transported in drums, only simple safety precautions are necessary.

The regeneration of the resin takes place with intervals of weeks. The UV-light treatment of the regenerate is therefore a discontinuous process. A batch of groundwater with a high concentration of cyanides is pumped through the UV/H$_2$O$_2$ system until the cyanide concentration has dropped enough to return the regenerate to the pre-treatment step. In table 5 an example is given.

TABLE 5. Results obtained with UV/H$_2$O$_2$ treatment of regenerate

Time (min)	Cyanide concentration (μg/l)
0	28,000
5	24,000
15	19,000
30	12,000
60	6,000
120	540

6. FULL SCALE DESIGN

The Waterboard Uitwaterende Sluizen asked Heidemij to scale up the 1 m³/h design to a 700 m³/day scale for treatment of groundwater with on an average 2500 µg/l cyanide during the extension of the sewage treatment plant of Beverwijk for 2 years from 1993 through 1995. One of the stormwaterbasins of the plant 700 m³ will be used as a storage and balancing tank.

The process is a combination of the following unit-operations:

pre-treatment: -   aeration;
          -     coagulation/flocculation;
          -     removal of fines with activated sludge;
          -     sedimentation;
          -     sand filtration;

cyanide removal: -   cyanide removal with ion exchange;
          -     oxidation of cyanides in regenerate of the ion exchange process with UV/H_2O_2.

The process scheme of the full scale plant is given in figure 3. The whole plant will be placed in a 11 * 25 m² shed.

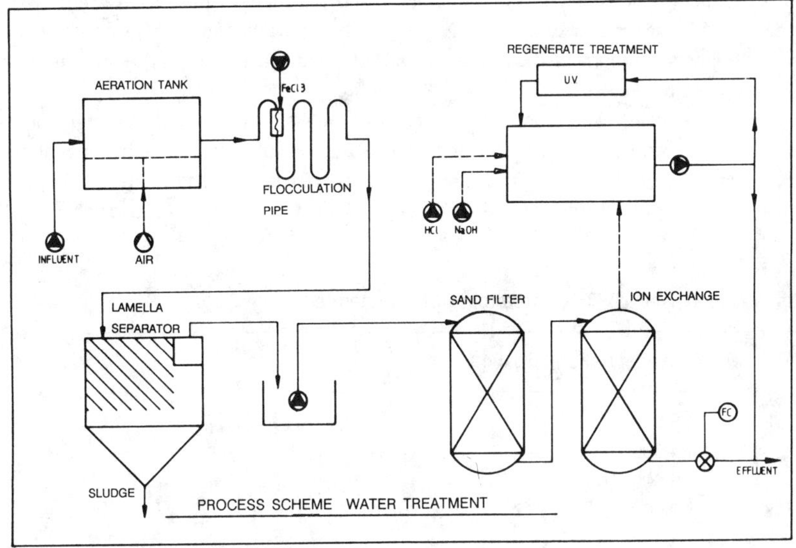

FIGURE 3. Process scheme of the Beverwijk full scale plant

The groundwater will be aerated for oxidation of reduced components in the groundwater. Coagulation is carried out with 10 mg Fe^{3+}/l. For flocculation and removal of suspended particles (mainly clay fines) the flocculated groundwater is brought on contact with a small stream of activated sludge from the sewage treatment plant. For settling of the groundwater/activated sludge mixture a lamella separator will be used. The dual media filtration will have a empty bed residence time of 20 minutes.

The cyanides will be removed in two anion ion exchange filters with a total capacity of 90 kg CN⁻ binding capacity. The regeneration process will be controlled by hand. The oxidation of cyanides will take place in a batch process.

H$_2$O$_2$ and a complexing agent will be added in the UV/H$_2$O$_2$ process. The quality of the groundwater and of the treated water will be monitored by the Waterboard, 24 hour composite sample will be taken.

The costs for the treatment of 500,000 m^3 groundwater in two years time will be about three Dutch Guilder (US $ 1.75) per m^3.

7. CONCLUSIONS

From laboratory work and 1 m^3/h trials it can be concluded that groundwater polluted with complexed cyanides, that has to be discharged up during the construction period of the extension of the Beverwijk sewage treatment plant can be treated in a cost-effective way with the ion exchange process combined with UV/H$_2$O$_2$ treatment. A pre-treatment to make the groundwater as clear as possible is essential to prevent fouling of the resin.
The removal efficiency of cyanides by ion exchange is over 99 % and of the oxidation process more than 90 %.

IN SITU BIOREMEDIATION TECHNIQUES OF A SITE CONTAMINATED WITH PAH
- APPLYING NITRATE AS AN ALTERNATIVE OXYGEN SOURCE -
ON LABORATORY AND PILOT PLANT SCALE

H.B.R.J. VAN VREE*, L.G.C.M. URLINGS*, J.G. CUPERUS*, P. GELDNER**

* TAUW INFRA CONSULT B.V., P.O. BOX 479, 7400 AL DEVENTER, THE NETHERLANDS
** GELDNER INGENIEURBERATUNG, EBERTSTRASSE 49, D-7500 KARLSRUHE 1

1. ABSTRACT

These days in situ remedial technologies are emerging from the experimental stage towards a more proven technology. Until now in situ biorestoration was mainly focused on sites contaminated with volatile organic
contaminants. The application of bioremediation techniques, concerning
Polycyclic Aromatic Hydrocarbons (PAH), has been restricted for sometime,
due to unfavourable technological and microbial reasons. This paper
describes an application of in situ bioremediation of a saturated zone
contaminated by PAH with the use of indegious microorganisms with nitrate
as an electron-acceptor and the application of surfactants.

2. INTRODUCTION

Talking about environmental issues it seems that soil contamination is
more the rule than the exception. For the Dutch situation it goes without
saying that many contaminated sites will have to be remediated in the next
decennia. Table 1.1 gives an overview of an inventory of contaminated
industrial sites (Zeilmaker 1992). The Dutch government has set-up a
special commission (Commission Oele) to tackle the environmental problems
concerning these (former) industrial sites.
TAUW Infra Consult B.V. estimates that (partial) in situ remediation is
applicable in roughly 15% of these cases. Most of these sites are contaminated with gasoline, mineral oil and volatile solvents. Although the
PAH contaminant occurs less frequently hundreds of sites could be
(partially) remediated using in situ techniques.
Gasworks and wood preservation plants are the main sites contaminated with
PAH. Until now remedial action here was only focused on excavation.
The numerous contaminated sites can only be effectively remediated if a
conceptual framework is developed wherein infrastructure, specific site and
political characteristics will lead to a well chosen strategy.

F. Arendt, G.J. Annokkée, R. Bosman and W.J. van den Brink (eds.), Contaminated Soil '93, 1065–1073.
© 1993 *Kluwer Academic Publishers. Printed in the Netherlands.*

TABLE 1.1. Overview of the Extent of Soil Pollution in the Netherlands
         (Zeilmaker 1992)

Number of sites	Type of contaminated site
230	Gasworks
3,300	Landfills
2,100	Scrapyards
430	Military sites
80,000	Former industrial estates
24,000	Present industrial estates

2. REMEDIAL STRATEGIES

For a long time the excavation of large quantities of contaminated soil was the only strategy for soil remediation. Although excavation of the polluted soil is a very effective method, the soil still has to be treated and disposed of, which leads to additional expense. Residual contamination in the soil and/or groundwater may cause environmental pollution again. Excavation can be difficult or even impossible, particularly in city centres and industrial estates. More emphasis should be placed on the development of new innovative remedial techniques (Spuij et. al., 1991). In situ techniques may play an important role here. The main advantage of in situ techniques is that the contaminated soil does not have to be excavated whilst the soil is being treated.

In situ biorestoration has been confined to volatile organic contaminants for some time. The application of bioremediation concerning polyaromatic hydrocarbons (PAH) has been underestimated due to unfavourable technological and microbial conditions. This paper discusses an in situ remediation of PAH.

3. PROCESSING IN SITU BIORESTORATION

3.1. In situ biorestoration

In situ biorestoration of the saturated zone is mainly water based. The main objective is to optimise all relevant biotic factors in order to promote the biodegradation of contaminants by the subsurface microorganisms (Brickel, 1991).
It is important that a suitable oxygen source, nutrients, a "C" source and appropriate microorganisms are present at all times. The biodegradation process will be inhibited if one of these factors is absent. A general process scheme is depicted in Figure 3.1.

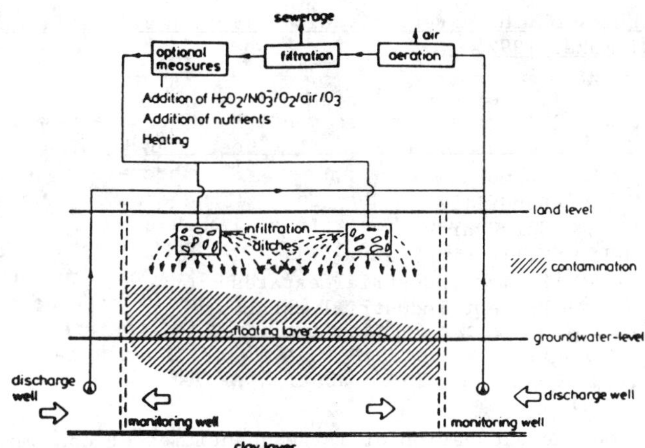

FIGURE 3.1. General Process Scheme of an In Situ Biorestoration (Staps, 1990)

Nutrients and an appropriate oxygen source (e.g. "air" oxygen, nitrate) are added to the infiltrated water, hereafter trace elements (e.g. Cu, Mg, Mo) can also be added. The water is infiltrated with tubes or ditches in the subsoil. Special geohydrological measures will be taken on site to prevent the contaminants migrating beyond the site. This can be established by installing an underground hydrological screen. The infiltrated groundwater stimulates the biological process and partly dissolves the adsorbed contaminants. The groundwater pumped up has to be treated, for example in a bioreactor. The effluent from the bioreactor must be post-treated so that the water can be reinfiltrated.

With regard to PAH contaminants and bioremediation of the saturated zone, the adsorption of the contaminants and the potential iron oxidation should be taken into account.
The adsorption of PAH to the organic matter in the soil is quite substantial. The k_{oc} values for PAH begin at approximately 10^3 dm³/kg for naphthalene and go up to 10^6 dm³/kg for 4 ring PAH. In most types of soil the retardation of the PAH contamination is so high that conventional "pump and treat" techniques would take between 10 to 100 years to remediate the site and in some cases even longer.
When molecular oxygen or hydrogenperoxide is applied, as an electron-acceptor, to the microorganisms, clogging in the saturated zone can pose a serious problem due to the oxidation of iron and gas production. This is undesirable from an engineering point of view and it can even make in situ remediation impossible. To prevent this, the application of nitrate as the main electron-acceptor can be an alternative.
To optimize the in situ bioremediation process one can:-
-look for and apply alternative oxygen sources;
-improve the bioavailability;
-add a special mixture of nutrients;
-add externally grown microorganisms.
This study concentrates mainly on the first two points.

3.2. Biodegradation of PAH

General

The degradation of PAH under aerobic conditions has been widely investigated. A variety of microorganisms have been found which possess the ability to degrade PAH (e.g. Alcaligenes, Flavobacterium, Micrococcus, Mycobacterium, Pseudomonas, and the fungus Phanerochaete crysosporium (Mueller, 1989)). The biodegradation rate of PAH decreases when the molecular weight increases. A low molecular weight PAH, such as naphthalene usually degrades quicker than a 3-ring PAH such as acenaphtene. A characteristic metabolite in the biodegradation pathway of naphthalene is 1,2-dihydroxynaphthalene. Once oxygen has been built-in, the biotransformation of the PAH molecule is not restricted any more. Much research has been carried out into the aerobic biodegradation of PAH derivatives (Mueller, 1989).

The biotransformation of PAH under anoxic or anaerobic conditions has been misunderstood for a long time. Recent research shows that under denitrification conditions low molecular PAH is amenable to microbial degradation in soil/water systems (Mihelcic 1988a, 1988b, 1991). Acclimation periods for the biodegradation of naphthalene and acenaptene took two days under aerobic conditions and 2 weeks under anaerobic conditions. Naphthol (a substituted hydroxy derivative from naphthalene) was also looked into; due to the hydroxy group the biological activity was quicker than for naphthalene alone. The important conclusion which can be drawn is that the biotransformation of PAH is feasible under denitrification conditions.

Bioavailability

Bioavailability is one of the factors which effect an optimal biodegradation process. Subsurface microorganisms must have the opportunity to come into contact with contaminants in the soil. Improving the bioavailability can speed up the in situ biorestoration process. The addition of surfactants can influence the sorption/description process and improve the bioavailability (Aronstein, 1991). Research was carried out recently which tested the desorption effectiveness of several non-ionic and anionic surfactants on the solubilization of anthracene, phenanthrene and pyrene (Vigon, 1989; Liu, 1991). It was concluded that too high surfactant concentrations may even inhibit the biodegradation process.
Factors which influence the effectiveness of the surfactant (derived from Aronstein 1991) are:-
-the properties of the sorbed compounds;
-characteristics of the soil (organic matter);
-type of surfactant applied;
-concentration of the surfactant applied;
-biodegradation of surfactants;
-sort of microbial community in the subsoil.
Stimulation of the biodegradation process can be enhanced by applying low concentrations of certain types of surfactants.

Monitoring

To determine the progress of an in situ biorestoration process one should select certain types of parameters which indicate in situ activity. Table 3.1 gives an overview of relevant parameters.

TABLE 3.1. Overview of Relevant Parameters for the Determination of In Situ Biorestoration Processes

Objective Approach	Explanation/Examples	Usefulness/Reliability
1 enumeration of microbial population	for indirect evidence of biodegradation; Perform Total Plate Counts, Special Plate Counts	little relevance, further evidence of activity
2 stimulation of microbial activity	activity measured related to soil micro-organisms; Oxygen Uptake Rate	supportive elements for proving in situ biorestoration
	Nitrate Consumption Rate monitoring metabolism of m.o. enumeration of possible metabolites (GC/MS Analysis)	direct proof of microbial activity
3 interpretation of consumption of electron-acceptor	O_2 depletion/NO_3 consumption rate; enables the conversion factors to be calculated	confirmatory
4 interpretation of changes in the contaminants in the soil and groundwater	set-up mass balances, leading to conversion rates	influenced by heterogeneity of contaminants in the soil
5 interpretation of essential environmental parameters	monitor pH redox, EC and alkalinity	additional proof of in situ activity

One application of in situ bioremediation will be described for soil contaminated with PAH.

4. IN SITU BIOREMEDIATION OF PAH CONTAMINATED SOIL ON LABORATORY AND PILOT PLANT SCALE

4.1. Introduction

A former tar distilling production site (1.5 hectares) appeared to be heavily contaminated with PAH, phenols and petroleum hydrocarbons. The site is now part of a pharmaceutical plant and is situated in a densely populated urban area. The contaminants are confined by a marly layer approximately 6 m below ground surface. The contamination has a very heterogenic character. PAH, oil (5-10,000 mg/kg d.m.) and phenolic compounds are also present. The PAH concentration varies between 50 to more than 3,000 mg/kg.

4.2. Remedial strategy

Based on literature and previous experience the conventional "pump and treat" method would simply not be effective as this method could take more than 1,000 years! High retardation factors of the individual PAH compounds will prolong the remediation. Alternatives had to be found.

First of all, an intensive laboratory study was set-up. It was recommended to continue the project at field level (on pilot plant scale). Intensive monitoring was necessary. The pilot plant study is now coming to an end. Final sampling will soon be reported. Thereafter, it will be decided whether or not a full scale in situ remediation is financially, technically and politically feasible.

4.3. Feasibility study on laboratory scale

The objectives of the laboratory study were to:-
-investigate biodegradation;
-investigate the improvement of bioavailability;
-selection of a suitable alternative oxygen source.

TABLE 4.1. Results of the Laboratory Study

Stage/Objective		Description/Approach	Results
1	Biodegradability	1.1 Batch experiments	- PAH highly biodegradable (73-99.4%) - high biodegradation rate (> 98%), particularly for marl samples
		1.2 Column experiments	- confirmation of PAH biodegradability, 45% of PAH removed in 24 days, oxygen consumption 5-8 mg O_2/kg/day - high bacterial number 10^5-10^6 CFU/ml
2	Selection of an alternative electron-acceptor	2.1 Microbial activity	- Nitrate Consumption Rate (NCR) $\approx$ 200 mg O_2/kg/day
		2.2 Batch experiments	- oxygen, nitrate and hydrogenperoxide can be used as an "oxygen source", > 95% was removed in 29 days (C_0=104. mg/kg d.m., C_t 3-4 mg/kg d.m.)
		2.3 Column experiments	- column 1 (oxygen): 89% removal 16 EPA PAH column 2 (nitrate): 80% removal 16 EPA PAH After running for 35 days the removal rate was 2-3 mg PAH/kg/day
3	Improving bioavailability	3.1 Screening suitable surfactants	- based on literature 5 surfactants were selected, laboratory experiments showed that two types of surfactants were applicable, K_d was reduced from 41,600-70 (dm^3/kg)
		3.2 Testing surfactants under semi-field conditions	- column experiments to test the leaching effect of surfactants:[PAH] increased from 11 μg/l up to more than 20,000 μg/l, i.e. factor 2,000-3,000
		3.3 Screening eco-toxicity	- closed bottle tests and plate counts showed that certain surfactants did not have an inhibiting effect on microbial growth

The results are summarized in Table 4.1. Based on the results of the feasibility study TAUW Infra Consult B.V. recommended that the project be continued and drew up an intensive monitoring programme for a pilot plant scale. Due to the high oxidation capacity of nitrate, it was decided to use nitrate as an alternative oxygen source.

4.4. <u>Feasibility study on pilot plant scale</u>

Based on the results of the laboratory experiments a pilot plant was designed. At the site an area of 10 x 10 m was equipped with infiltration and extraction lances. Extracted groundwater was treated biologically by means of a biocontactor. The produced effluent was post-treated (sand filter/active carbon) and prepared for reinfiltration by adding nutrients, trace elements and nitrate (oxygen source).

Hydraulic measures were taken at the site to prevent the contaminants migrating from the isolated area. An intensive microbial monitoring programme was set-up to control and follow the impact of the remedial action. The monitoring focused on:-
-enumeration bacterial numbers;
-estimation of bacterial activity;
-estimation of nitrate consumption rate;
-interpretation of environmental parameters;
-interpretation of changes in the contaminants present in the soil and groundwater.
In the first and second stages of the pilot plant study the emphasis was put on pump experiments and groundwater extraction.
The third stage of the study consisted of:-
-starting up the nitrate infiltration (stage 3.0);
-nitrate infiltration combined with passive surfactant infiltration (stage 3.1);
-nitrate infiltration combined with active surfactant infiltration (stage 3.2).
To date the results show an increase in the use of nitrate by denitrifying organisms in the soil. The consumption rates can be found in Table 4.2. The infiltrated water (0.5-1.0 m^3/hour) was supplied with 150-200 mg NO_3/1. Samples from the groundwater filters indicated the consumption of nitrate and a low but non-toxic, nitrite accumulation (Figure 4.1). Large numbers of microorganisms were present in the soil (10^5-10^6 CFU/g soil) and groundwater (10^4-10^6 CFU/ml)

TABLE 4.2. Results of Denitrification Process

Stage	Total Amount Denitrified (kg-N)	Nitrite Accumulation (%)	Nitrate Consumption (infiltrated water) (mg-N/1)
3.0	1.42	61	6
3.1	13.32	11	41
3.2	2.13	30	2

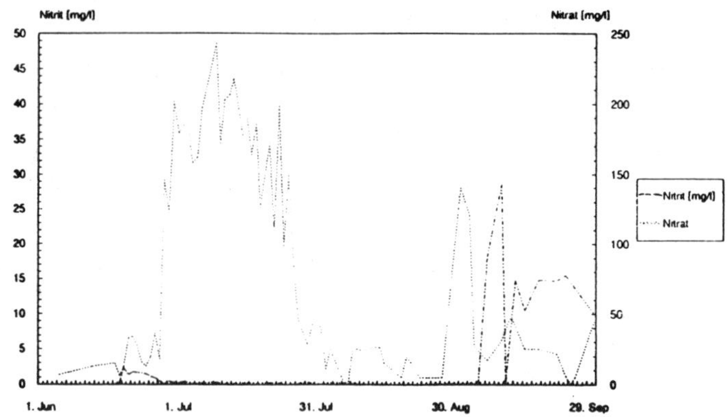

FIGURE 4.1. Nitrate Consumption/Nitrite Accumulation in One of the
        Groundwater Filters

The extracted groundwater showed a low contamination degree (1000-1500 μg/l
PAH). To improve the leaching and bioavailability of PAH in the subsoil, a
mixture of selected surfactants (laboratory study) was injected into the
recirculating groundwater during two stages (3.1 and 3.2). The two
surfactants applied have a non-ionogenic character (ethoxylated
oxoalcohol). The applied concentration in the infiltrated water was 0.5 and
0.2% v/v respectively; both are readily biodegradable (OECD test). Based on
the total amount of nitrate consumed it was estimated that, in theory,
approximately 16 kg of naphthalene could be biodegraded. Approximately
2,600 m^3 groundwater was withdrawn. Roughly 30 kg of contaminants were
removed from the subsoil by groundwater extraction. The addition of surfactants improved the solubility of the PAH and oil contaminants by groundwater extraction. Soil samples were taken on several occasions. Mixed
samples taken from the saturated zone (3-5 m below ground surface) showed
a rate of removal for the phenol concentration, ranging from 60 to >94%
(n=8). Initial concentrations ranged from 4 to 230 mg/kg. Final concentrations ranged from 0.1 to 75 mg/kg (phenol, distilled).
The rate of removal for PAH ranged from 50 to 93%. Originally the concentrations ranged from 15 to 1200 mg/kg PAH (16 EPA). Final values ranged
from <1 to 500 mg/kg. Due to the heterogeneity of the soil not all the
samples gave results which could be interpreted.
Seven of the nine borings carried out for mineral oil gave results which
could be interpreted. The average rate of removal ranged from 68 to >90%
mg/kg. Initial values ranged from 35 to 2900 mg/kg mineral oil compounds.
Final samples showed a contamination concentration of < 20 to 1400 mg/kg.

5. CONCLUSIONS AND RECOMMENDATIONS

In situ biorestoration of sites contaminated with PAH has been restricted for sometime due to unfavourable technical and microbial
conditions.

The case presented in this paper shows that stimulation of the bio-degradation is also possible if an alternative oxygen source, such as nitrate, is used. The application of surfactants is feasible but still expensive. Although biodegradation can be stimulated the biodegradation of PAH remains a relatively slow process.

These types of applications can be seen as part of the supplementary measures to be integrated into an overall remedial plan of action.

A suitable remedial option could consist of:-

- preventing the pollution from migrating into the surrounding soil and groundwater by containing the groundwater using (hydraulic) screens;
- implementation of an in situ biorestoration system to remediate the site on a long term basis.

To reinforce the application of in situ biorestoration processes further applied scientific research should aim primarily at:-

- monitoring the in situ bioactivity by direct indicators e.g. enzymatic activity of m.o., (application of DNA or RNA genetic technique to assess the distribution of potential expression);
- improvement of integrated (input friendly) models to optimise biodegradation and extraction during remediation.

6. REFERENCES

[1] Spuij F., S. Coffa, C. Pijls, L.G.C.M. Urlings.
In Situ Soil Vapour Extraction of Contaminated Soil.
Third Forum on Innovative Waste Treatment Technologies.
11-13 June, 1991, Dallas, Texas.

[2] Madsen, E.L.
Determining In Situ Biodegradation.
Environ. Sci. Technol., vol. 25, no. 10, 1991.

[3] Thomas J.M., C.H. Ward.
In Situ Biorestoration of Organic Contaminants in the Subsurface.
Environ. Sci. Technol., vol. 23, no. 7, 1989.

[4] Staps, J.J.
International Evaluation of In Situ Biorestoration of Soil and Groundwater.
National Institute of Public Health and Environmental Protection
Report 738708006, 1990.

[5] Zeilmaker, D.A.
Milieucompartiment Bodem. Milieubeheer in de industrie. January 1992.

[6] Brickel M.E., J.W. van Vliet.
Optimum Use of Soil and Groundwater Remediation Methods for Petroleum Products. Petrochem, no. 7, 1991.

[7] Mueller, J.G. et. al.
Cresote Contaminated Sites. Environ. Sci. Technol. vol. 23, no. 10, 1989.

[8] Mihelcic J.R., R.G. Luthy.
Microbial Degradation of Acenaphthene and Naphthalene Under Denitrification Conditions in Soil Water Systems.
Applied Environ. Microbiol. vol. 54, 1988a (p1988-1198).

[9] Mihelcic J.R., R.G. Luthy.
Sorption and Microbial Degradation of Naphthalene in Soil/Water Suspension Under Denitrification Conditions.
Environ. Sci. Technol., vol. 25, no. 1, 1991.

[10] Mihelcic J.R., R.G. Luthy.
Degradation of Polycyclic Aromatic Hydrocarbons Under Various Redox Conditions in Soil/Water Systems.
Environ. Sci. Technol., vol. 54, no. 4, 1988b (p 1182-1187).

CLEANING UP GOLDBEKHOF IN HAMBURG

Klaus Marg

ENVIRONMENTAL AUTHORITY, HAMBURG, Amelungstraße 3, D-2000 Hamburg 36, Germany

In Hamburg-Winterhude the Goldbekhaus Arts Centre is accommodated in the buildings of a former disinfectant factory. The soil, groundwater and parts of buildings now standing empty, but which are to be used in the future by the Centre, are badly contaminated with chemicals. The soil and buildings therefore need to be cleaned up. This involves replacing the soil using a special new method while retaining the buildings. The soil is transported to another site in Hamburg for biological cleaning.

1. PREVIOUS HISTORY

Between 1889 and 1963 the Schülke & Mayr disinfectant factory operated on the site of Moorfuhrtweg 9 in Hamburg-Winterhude. One of its products made here was the preparation LYSOL, which was used extremely successfully in combating the cholera epidemic of 1892 in Hamburg. The disinfectant SAGROTAN used in many private households, medical practices and hospitals was developed and produced by the company at a later stage.

The company shifted its production to a new building outside the city in 1963. The city of Hamburg bought the former factory and let the buildings to small craftsman's establishments and artists in the following years.

It was not until 1980/81 that the Goldbekhaus Arts Centre was established in the north wing of the former factory, and subsequently in the old administration building. In 1985 it was intended to make further rooms available to the Centre in the neighbouring buildings. However that same year substantial contamination of the soil, groundwater and some parts of the buildings with the raw materials that used to be used for manufacturing the disinfectant was discovered.

In the following years the extent and type of contamination was investigated in greater detail, and an overall concept for implementing the clean-up developed. Various methods of cleaning up the site were checked out, and in 1991, after the planning had been completed, Keller Grundbau GmbH and HANSATEC GmbH were finally awarded the contract. The cost of about 22 million deutschmarks was borne by the city of Hamburg.

F. Arendt, G.J. Annokkée, R. Bosman and W.J. van den Brink (eds.), Contaminated Soil '93, 1075–1081.
© 1993 *Kluwer Academic Publishers. Printed in the Netherlands.*

2. SOIL CONDITIONS

The soil is very heterogeneous in the contaminated area. Different types of soil arise in close proximity and changing stratification. Under a mostly sandy fill, the site generally has a shallow layer of peat, which has fine to coarse sands below it. Below this again is a peat, sapropel or calcareous sapropel stratum. The base of these soft strata is formed by sands again, with an underlay of glacial marl. The individual strata are not present continuously and are of very different thicknesses and lie at a wide variety of depths. The groundwater is in direct contact with the navigable Goldbek Canal adjoining the site. The water table is about 1 m below ground level.

3. TYPE AND LEVEL OF CONTAMINATION

Over the decades of production chemicals percolated underground. Today they are to be found in the soil, groundwater and even in sections of the north and east wings of the former factory building. They consist of phenol and various cresol and xylenol isomers.

In the yard area and under the north and east wings the contaminants have already percolated though several strata (sandy fill, peats and sapropels, sands, glacial marl) and contaminated the groundwater. Since there is virtually no groundwater flow in the vicinity of the site the contamination has not yet spread out any further horizontally. The soil clean-up and groundwater treatment will cover virtually the same area.

The contaminant contents (total of phenol, cresols and xylenols) are up to 63,000 mg/kg of solids in the soil and 20,000 mg/l in the groundwater. The contaminant content (clean-up limit) in the soil of 20 mg of phenolic substances/kg solids specified as tolerable for this case is greatly exceeded over large areas of the site. About 10,500 m^3 of soil and groundwater down to a depth of 16 m therefore has to be cleaned up to protect the ground water.

Extent of the soil contamination

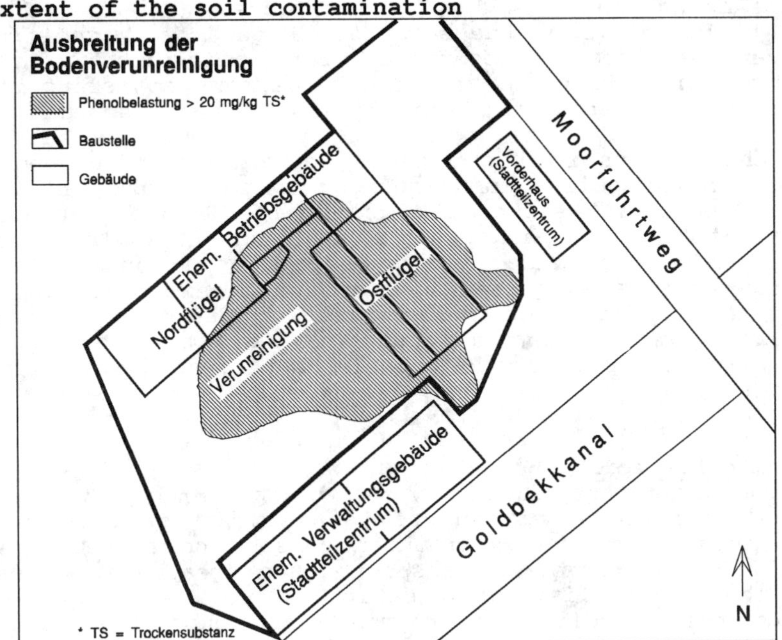

Fig. 1: Extent of the soil contamination

Phenolbelastung = Phenolic contaminants, > 20 mg/kg of solids
Baustelle = Site, Gebäude = Buildings, Ehem. Betriebsgebäude =
Former factory building, Nordflügel = North wing, Ostflügel =
East wing, Ehem. Verwaltungsgebäude = Former administration
building, (Arts Centre), Vorderhaus = Reception building (Arts
Centre), Goldbekkanal = Goldbek Canal

4. CLEAN-UP METHOD

The very heterogeneous soil conditions, which present great
problems for conventional methods, and the very strong smell
of the phenol as well as the cresols and xylenols (threshold
only about 0.5 to 5 ppm) were particular important when it
came to selecting a method for cleaning the site up. Moreover
the clean-up had to be carried out in such a way that the
former factory buildings, which determine the character of the
city in the area and whose historical importance makes them
worthy of classification as a monument, are retained. A method
therefore had to be found that:

a) allows the buildings to be retained,
b) ensures the soil is substantially cleaned up
c) can be implemented in such a way that there are no smelly
   emissions.

In 1988 the method being used today was successfully tested in extensive field trials on the site. The method developed from the proven soilcrete technique for underpinning the foundations of buildings and forming diaphragms satisfies the above-mentioned requirements.

Tests of the process of binding the contaminants found were carried out on the soil columns formed in the field test. The tests showed that with a residual content of 20 mg of phenolic substances (total of phenol, cresols and xylenols/kg of solids) there are no far-reaching negative effects on the surrounding groundwater emanating from the clean-up area. The reason for this is the very small amount of contaminants being carried out from the contaminated area and the very good biological degradability of the contaminants at low concentrations.

4.1 Replacing the soil

Since the contaminants have a very strong smell, extensive precautions have to be taken during the clean-up to prevent any smell nuisance in the surrounding area. The contaminated soil is therefore conveyed, transported and treated in closed systems. The possibility of creating a smell in or even endangering the immediate environment (housing, Arts Centre restaurant with beer garden) as a result of the clean-up work is therefore largely ruled out.

Schematic of the soil replacement

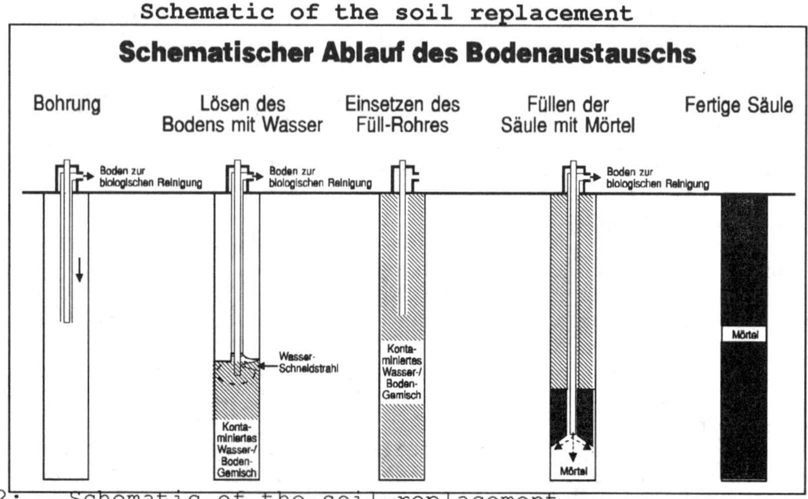

Fig. 2:   Schematic of the soil replacement

Bohrung = Borehole, Lösen des Bodens .... = Loosening the soil with water, Einsetzen der Füll-Rohres = Inserting the injection tube, Füllen der Säule mit... = Injecting the column with grout, Fertige Säule = Finished column, Boden zur biolog.... = Soil to biological treatment, Wasser-Schneidstrahl = Water jet, Kontaminierten ... = Mixture of contaminated water and soil, Mörtel = Cement grout

The contaminated soil is loosened and liquefied underground with a water jet. This is achieved by first lowering a drill rod to a specified depth and then raising it again slowly. While it is being withdrawn the drill rod is turned, and a jet of water emitted from the drill head under a high pressure of 400 to 600 bar that loosens and liquefies the surrounding soil up to a distance of about 0.80 m.

Once the drill rod has been fully retracted the loosened and liquefied soil is replaced, using the contractor method, with clean cement grout that goes off underground. To achieve this an injection tube is lowered to the base of the eroded column of soil and grout injected to displace the liquefied soil with a lower relative density. The soil comes out of the preventer above ground and is carried away in closed pipelines.

The soil is even being replaced under the buildings and their foundations without jeopardising their stability. The columns of soil cleaned up in this way can be set out in rows to cover an area of virtually any shape.

Measurements in columns of soil already completed have shown that the soil replacement is so complete that values well below the clean-up limit are achieved.

4.2 **Treatment of soil**
Once the coarser elements (large stones, pieces of timber etc) have been separated, the mixture of contaminated soil and water is loaded into tank barges on the Goldbek Canal and transported along the Alster and Elbe to Nippoldweg (near Köhlbrandbrücke) in the Port of Hamburg. At this location the mixture is separated into its constituents (sand/gravel, peat/sapropel/silt, water). The solids are treated in batches in bio-beds.

The water is purified biologically in an effluent treatment system and then discharged into the public sewer. The contaminants are largely dissolved in the water, since the contaminated soil is thoroughly mixed with the water during the soil replacement procedure.

The cleaned sand and gravel are returned to the Moorfuhrtweg site, where they are added to the grout for re-injection into the ground. The cleaned peat, the sapropel and silt find other uses, eg in the construction industry. Since the soil replacement is completed before the cleaning of the sands and gravels, residues that cannot be added to the grout at Moorfuhrtweg are also passed on to the construction industry. However experience to date has shown that slight contamination of the fill near the surface with PAHs, which was detected during the clean-up, leads to these substances being concentrated in the peat/sapropel/silt fraction, which can therefore no longer be used without restriction.

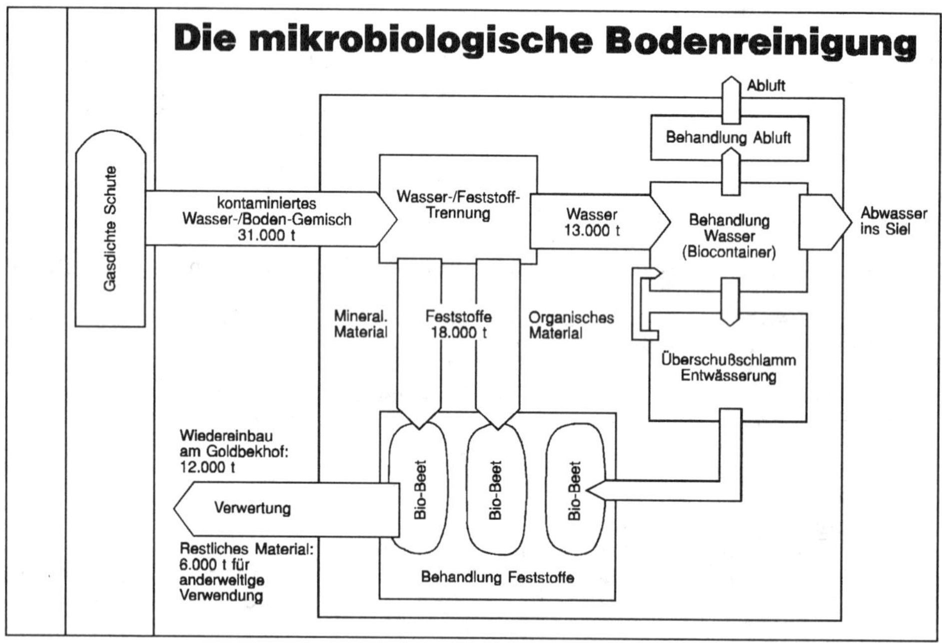

Fig. 3: Microbiological cleaning of the soil

Abluft = Exhaust, Behandlung Abluft = Treatment of exhaust,
Behandlung Wasser..... = Water treatment, (biocontainer),
Abwasser ins Siel = Effluent discharged into sewer,
Überschußschlamm... = Dewatering of excess silt, Wasser... =
Water, 13,000 t, Organisches Material = Organic material,
Wasser/Feststoff... = Separation of water and solids, Mineral
Material = Mineral material, Feststoffe .... = Solids, 18,000
t, Bio-Beet = Bio-bed, Behandlung Feststoffe = Treatment of
solids, kontaminiertes Wasser/.. = Mixture of contaminated
water and soil, 31,000 t, Wiedereinbau ... = Backfilling at
Goldbekhof:, 12,000 t, Verwertung = Recycling, Restliches
Material... = Remainder of material:, 6000 t for other uses,
Gasdichte Schute = Gastight barge

5. CLEANING UP AND RENOVATING THE BUILDINGS

The contaminants in the buildings have penetrated into the flooring and wall-coverings etc. The elements of the buildings affected will be removed and disposed of or treated once the soil clean-up has been completed. At the end of the complete process the buildings are to be renovated and made available to the Arts Centre or let to small businesses and artists. The historic monument protection requirements will have to taken into account in the renovation and extension of the buildings.

6. CLEAN-UP SCHEDULE

The clean-up work was started in January 1992. Operations included relocating supplies and drainage, clearing the buildings of rubbish and preparing them for the clean-up (removing partitions; lowering the cellar floor; demolishing extensions etc). The actual replacement of the soil was commenced in August 1992. Despite initial technical and logistics problems the work should be completed on schedule in 1993. The site will then be landscaped and the buildings cleaned up and renovated.

7. PUBLIC INVOLVEMENT

The "Goldbekhof Clean-up Working Party" was formed under the joint chairship of the Environmental Authority and Goldbekhaus e.V., the association responsible. The working party includes representatives of the district government, the Hamburg North District Authority and the environmental consultants ÖKOPOL, who advise the association and other people affected, as well as the contractors. At the monthly meetings of the working party information is exchanged and queries about the clean-up discussed. Moreover interested citizens have the opportunity during regular public question hours to inquire about the progress of the clean-up work. The public relations exercises have led to a high level of acceptance of the clean-up work and associated inconvenience and restrictions by those affected.

IN SITU AQUIFER REMEDIATION FROM VOLATILE OR BIODEGRADABLE ORGANIC COMPOUNDS, PESTICIDES, AND NITRATE USING THE UVB TECHNIQUE

B. Herrling[*], E.J. Alesi[**], G. Bott-Breuning[***], S. Diekmann[****]

[*]   Inst. of Hydromechanics, Univ. of Karlsruhe, Kaiserstr. 12, D-7500 Karlsruhe, Germany
[**]   IEG Technologies Corp., 1833 D Crossbeam Dr, Charlotte, NC 28217, USA
[***]   GfS mbH, Dettlinger Str. 146, D-7312 Kirchheim/Teck, Germany
[****] Mobitec GmbH, Wagenstieg 5, D-3400 Göttingen, Germany

1. ABSTRACT

Vertical circulation flows around special screened wells have been used for physical (stripping) or biological in situ groundwater remediation. Further, they can be combined with any appropriate on-site remediation when extraction from and infiltration into the same well is realized. Different remediation techniques are explained and discussed in dependency on the groundwater contamination: volatile chlorinated hydrocarbons, BTEX, pesticides, nitrate, and dissolved heavy metals. The effect of the UVB technique is explained with the remediation of a spill with aromatic hydrocarbons and mineral oil including measured field data.

2. INTRODUCTION

Vertical circulation flows around wells with two screen sections in one aquifer, so called "Groundwater Circulation Wells" (German: Grundwasser-Zirkulations-Brunnen, abbr.: GZB), have become increasingly important for aquifer remediation. Originally this idea was used only for in situ remediation of aquifers from strippable contaminants in so called "Vacuum Vaporizer Wells" (German: Unterdruck-Verdampfer-Brunnen, abbr.: UVB)[1,2]. For both, the GZB and the UVB, special wells with two screen sections are employed, one at the aquifer bottom and one at the groundwater surface or below an aquitard. One well should be used to remediate only one aquifer (phreatic or confined) and should not connect different aquifers. Within the well the groundwater is moved vertically. The contaminated water enters the well at the bottom and the stripped or treated water leaves at the top or vice versa. In the vicinity of the well an area of vertical flow circulation is created.

As all the groundwater, being located in the sphere of influence or in the capture zone of a well or well field will flow at least once through the well casing, any conceivable treatment technology can be used to treat or clean the groundwater when flowing vertical through the well. The ways of treatment include a simple addition of nutrients and/or electron acceptors which will stimulate in situ biodegradation in the aquifer. The paper will clarify the wide spectrum of application areas where vertical circulation flows of a UVB or GZB can be used.

The vertical circulation flow around the UVB and GZB has been a matter of continuing numerical investigation. This paper discusses the different remediation techniques and some measured field data. Numerical results and diagrams for dimensioning UVB/GZB installations have been published elsewhere [1,2,3]. All the technologies described in this paper have been patented by IEG mbH, D-7410 Reutlingen, Germany.

3. PHYSICAL REMEDIATION BY IN SITU STRIPPING

At numerous sites in Germany and more recently in the United States the UVB technique has been used for in situ groundwater remediation where the underground is contaminated by strippable substances, e.g. volatile chlorinated hydrocarbons, BTEX. As an alternative to conventional hydraulic redevelopment measures (pumping, off-site cleaning, and infiltration of groundwater), the contaminated groundwater is stripped by air in a below atmospheric pressure field in the UVB. In case of a contamination heavier than water (DNAPL) an upward operating UVB (Fig. 1a) is used while for lighter compounds (LNAPL) the well works downward (Fig. 1b).

F. Arendt, G.J. Annokkée, R. Bosman and W.J. van den Brink (eds.), Contaminated Soil '93, 1083–1092.
© 1993 *Kluwer Academic Publishers. Printed in the Netherlands.*

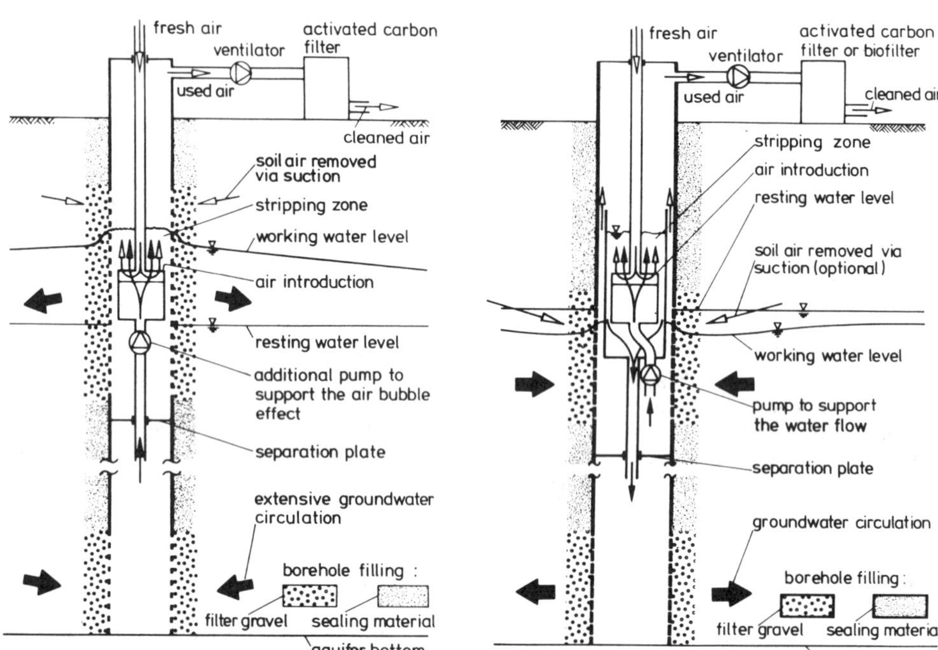

Fig. 1   In situ stripping using a vacuum vaporizer well (UVB) operating (a) upward for DNAPL (e.g. PCE, TCE), and (b) downward for LNAPL (e.g. BTEX, phenol, kerosene)

Often the well is used for vapor extraction at the same time as demonstrated in Fig. 1. The contaminated air is cleaned employing activated carbon or in case of suitable contaminants by using biofilters.

The upper, closed part of the well is maintained at below atmospheric pressure by a ventilator. This lifts the water level within the well casing. The fresh air for the upper part of the well casing is introduced through a fresh air pipe: the upper end is open to the atmosphere, and the lower end terminates in a pinhole plate. The height of the pinhole plate is adjusted such that the water pressure is lower there than the atmospheric pressure. Therefore, the fresh air is drawn into the system. The reach between the pinhole plate and the water surface in the well casing is the stripping zone, in which an air bubble flow develops. The rising air bubbles produce a pump effect, which moves the water up and causes a suction effect at the well bottom, Fig. 1a. In recent wells, a separating plate and an additional pump (Figure 1) are used to reinforce the pumping effect of the air bubbles. Additionally, soil air is drawn from the surrounding contaminated unsaturated zone at many sites. Stripped air and possibly soil air are transported through the ventilator and across activated carbon, onto which the contamination is adsorbed. Thus, only clean air escapes into the atmosphere.

The cleaning effect of the well is based on reduced pressure, which reinforces the escape of volatile contamination out of the water, and as a result of the air intermixing, onto the considerable surface area of the air bubbles and onto the high concentration gradient between water and clean air. In this sense, the permanent vibration caused by the air bubbles is beneficial to the escape process of the contamination. This vibration is transmitted as compression and shear waves into sediment and fluid, and presumably influences the mobility of the contaminants, even outside the well.

The upward-streaming, stripped groundwater leaves the well casing through the upper screen section in the reach of the groundwater surface, which is lifted in a phreatic aquifer by the previously explained pump processes and the below-atmospheric pressure (Fig.1a). In case of a downward operating UVB (Fig.1b) a phreatic groundwater surface is slightly lowered by the pumping process. The amount is small, however, caused by the below-atmospheric pressure which have an opposite effect. The groundwater then returns in an extensive circulation to the well bottom or the upper screen respectively. In this way, the groundwater surrounding the well is also remediated. The artificial groundwater circulation determines the sphere of influence of a well and is overlapped with the natural groundwater flow.

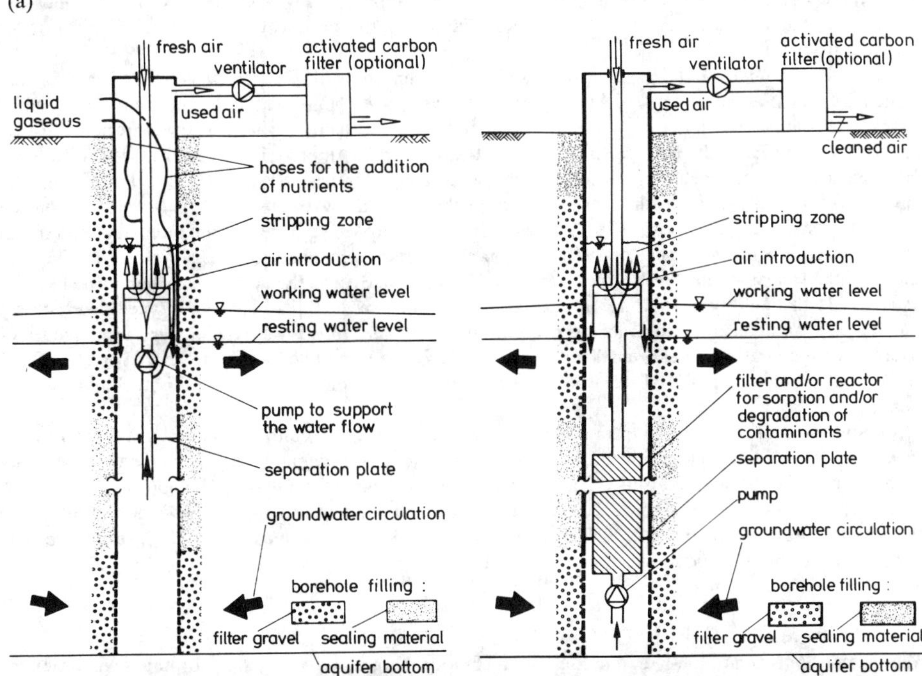

Fig. 2   In situ bioremediation by a vacuum vaporizer well (UVB) using (a) the aquifer as a bioreactor (e.g. biodegradable hydrocarbons), and (b) special reactors within the well casing (e.g. pesticides, nitrate)

4. IN SITU BIOREMEDIATION

When the groundwater is contaminated by other than strippable substances, which are suitable for bioreclamation, the extensive circulation flow around the UVB (Fig.2a) can be used to bring nutrients and/or electron acceptors to all places in the saturated zone where they are needed. Any gaseous or liquid substance, soluble in water, can be added in measured quantities while the groundwater passes the well casing. In this case the aquifer itself is used as a bioreactor. The two phase flow of gas and water in the stripping zone ensures that the gaseous or liquid addition is mixed with the groundwater in an optimal way. Incomplete mixing could produce unwelcome effects in the downstream flow field depending on the respective biodegradation processes.

For in situ bioremediation of e.g. BTEX, phenol essentially oxygen is needed. The in situ stripping process with air of a UVB provides oxygen saturation. While the groundwater quantity captured by a well remains relatively constant, the quantity of circulating water around the UVB can be extremely enhanced by a stronger pump in the well casing. Thus the amount of oxygen supplied for in situ bioremediation can be considerably increased with nearly no additional costs. Carbon dioxide as a by-product is removed from the groundwater by the same stripping process, so the pH-value cannot drop producing new problems otherwise. As the mentioned hydrocarbons are generally lighter than water a downward operating UVB is used: the contaminated water enters the well through the upper screen, Fig. 1b. The contaminated air is cleaned by using e.g. biofilters. A UVB technique which combines vapor extraction with the stripping

1086

process is generally used for those sites; thus lots of oxygen are supplied to the vapor zone and capillary fringe for biodegrading processes. The just explained technique (Fig.1b) combines both the physical stripping of volatile organics and the biological degradation processes enhanced by the added oxygen as electron acceptor. If necessary, further nutrients could be added by using the hoses as demonstrated in Fig. 2a.

On the other hand a special bioreactor or other technologies have been installed within the well casing between the lower and upper screen in case of special contaminants, Fig. 2b. This in situ techniques can be directly combined with the in situ stripping for eliminating the produced biogas (e.g. CO_2, N_2) or for introducing oxygen.

A specially designed well based upon the UVB technology of Fig. 2b was utilized for in situ bioremediation of triazine pesticides. At a polluted site they have contaminated the underground by small leakages at a storage facility over a period of many years. The contaminated water, entering the well through the lower screen, is pumped through an activated carbon filter which is installed within the well casing. Before leaving the well the groundwater is in situ stripped by the standard UVB technique (Fig.2b) supplying oxygen saturation. On the activated carbon the pesticides are removed by adsorption when the well operation started. The exposure to triazines over a relatively long period of time induces the adaption of natural groundwater bacteria to these contaminants. Certain strains of bacteria develop the ability to break down the triazine molecule. Such bacteria are also adsorbed on to the activated carbon particles. The filter is flowed through by circulating groundwater containing oxygen. The bacteria proceed to multiply within the filter and consume the triazines present beside oxygen. In this way, a biological remediation of the triazine contamination is achieved. Thus the activated carbon unit need not be renewed for a very long time. At the mentioned polluted site the measured field data demonstrate that this technique works successfully[4].

Groundwater contaminated by nitrate can be in situ cleaned in containers installed in the well casing between the two screen sections of a UVB, Fig. 2b. In these containers nitrate is eliminated from the passing groundwater in a rapid and efficient process which involves catalytic reduction by immobilized enzymes. The reduction is driven by an electrical current, and results in complete conversion of nitrate to N_2 without residues; the dissolved N_2 is finally removed via in situ stripping using the UVB technique. The development of the technique has been finished in the laboratory scale and is prepared for field applications at the moment [5].

5. ON-SITE REMEDIATION USING A GZB

When any appropriate on-site remediation technique should be used, e.g. for the elimination of dissolved heavy metals from the groundwater, the vertical circulation flow of a GZB can be utilized to advantage: The groundwater entering the well is pumped above ground, treated, and infiltrated in the same well using the other well screen.

Furthermore, following these ideas, the GZB can be also used as a pump or infiltration well for standard on-site remediations. In this case a partial discharge withdrawal or infiltration is taken from or added to the total discharge through the well casing, see Fig.2 in [3]. It is possible to extract or to infiltrate water without any change of head at the well top for a special ratio between the total well discharge and the extracted or infiltrated quantity of water [3], which has been numericly investigated for confined aquifer conditions. Of course, in some distance from the well a smooth deviation of the groundwater head at the aquifer top from the resting position will be found. But in case of low well capacity the pumping operation is continuous for a much higher rate and the infiltration can be realized for a much greater quantity even for a low distance between the surface and the groundwater table. Further, during the pump operation the turbulent mixing of air and water in the filter gravel of the well, which often causes unwanted precipitation, can be avoided.

6. IN SITU TREATMENT WALL BY USE OF VERTICAL CIRCULATION FLOWS

In case that a UVB or a GZB is situated in a plume of contamination, the polluted upstream groundwater (1) is captured by the well, (2) is treated by suitable in situ (or on-site) measures, (3) circulates vertically in the surrounding of the well, and (4) (the same quantity of treated water as captured) leaves the well area flowing downstream. For a wide plume several UVBs or GZBs are arranged in one line normal to the natural groundwater flow at a maximum distance that no water can pass the well line without having been treated. The natural flow field is only locally influenced because no groundwater is extracted. Thus one well or especially a line of several wells represent a treatment wall with some local cleaning process technique: No contamination which comes from upstream can pass the well or the line of wells.

7. UVB REMEDIATION OF AN AROMATIC HYDROCARBON (MAINLY BTEX) AND MINERAL OIL SPILL

7.1 Case description

In Berlin, Germany, the above explained UVB technique, which combines in situ stripping and in situ bioremediation (Fig. 1b), is successfully used to clean a field site from BTEX (benzene, toluene, ethyl-benzene, xylene),styrene, and some mineral oil. The spill was caused by a leaking storage tank of a former retail gasoline station. Some removable soil above the groundwater table was excavated and transported to a landfill. The measured concentration of BTEX and styrene in the groundwater was 280 mg/L and of mineral oil 1 mg/L. The originally hypothetical contamination of chlorinated hydrocarbon was found to be of minor significance.

7.2 Geological and hydrogeological situation

Underneath approximately 1.2 m of fill material consisting of fine sand and fragmented bricks lie 2.7 m of medium to fine sand with recognizable graded bedding. From 3.9 m to about 10.6 m below the soil surface sandy gravel dominates, however, clayey silt as well as fine organic rich layers and occasional stones form intermittent horizontal layers. At around 10.6 m gray clayey marls make up an impermeable layer.

The unconfined aquifer has a thickness of about 6.5 m; the groundwater table lies at around 4 m below the ground surface. With a hydraulic gradient of 0.08% and a hydraulic conductivity of 3.1×10^{-3} m/sec the groundwater flows at a rate of 1.1 m/day to the SW. Pore space in the sandy gravel is 0.2.

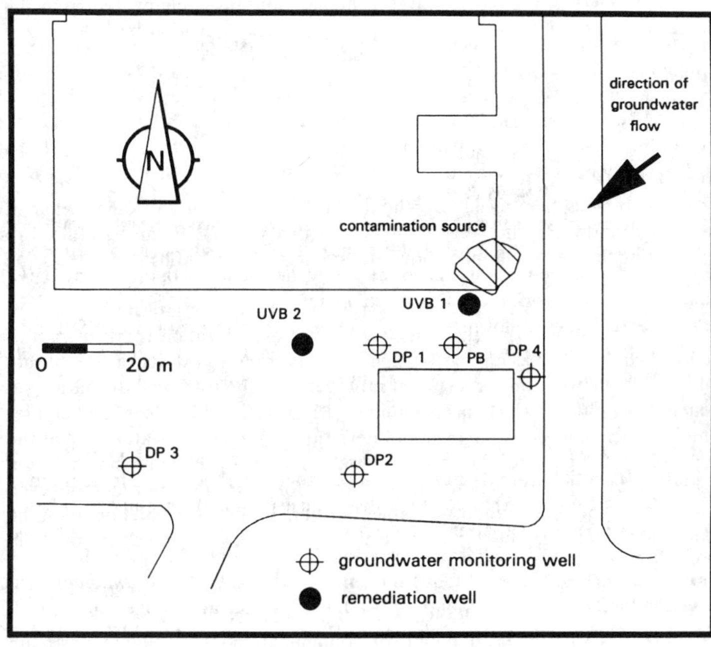

Fig. 3 Site map of the remediation

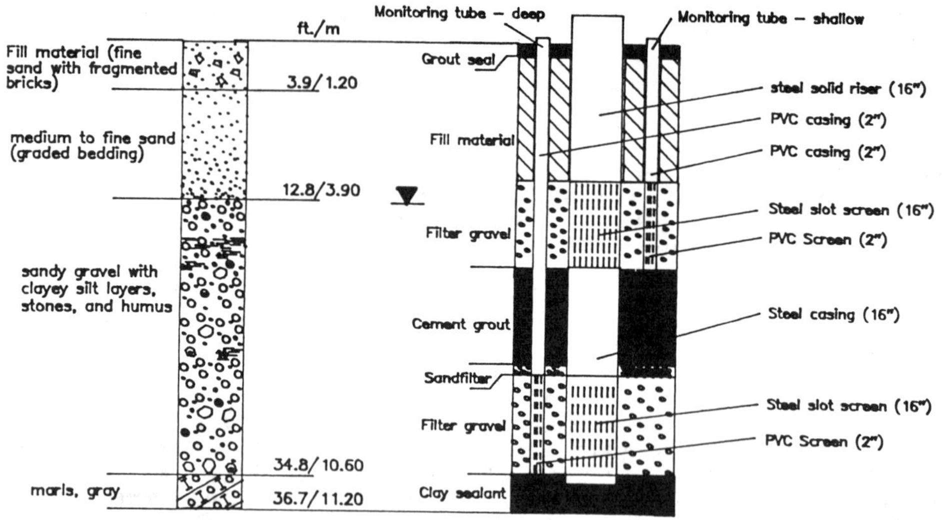

Fig. 4 Geological profile and schematic well construction of UVB 1

7.3 Remediation measures

For remediating the site (groundwater and vadose zone) two UVB were installed (site map see Fig. 3). The source area of the contamination was suspected just upstream of UVB 1. Both UVB were drilled with a borehole diameter of 0.8 m, with a depth of 10.9 m and 11.2 m respectively, and the well diameter of both is 0.4 m. Because phase of LNAPL was found on the groundwater surface being lighter than water both UVB operate upward (see Fig. 1b). The well construction of UVB 1 is demonstrated in Fig. 4, that of UVB 2 is corresponding. The screened sections are from 3.1 m - 5.3 m and 9.0 m - 10.5 m (UVB 1) and from 3.15 m - 5.35 m and 8.2 m - 10.2 m (UVB 2) below ground surface.

The remediation started with a five-day test run of UVB 1 in October 1990 which removed about 88 kg aromatic and mineral oil hydrocarbons in the off-gas of the stripping process; the contaminants were adsorbed on activated carbon. After a four months stoppage caused by the client the operation of UVB 1 started on March 11, 1991 and was interrupted for technical modifications from March 29 to June 19, 1991. The operation of UVB 2, which is situated 40 m downstream of UVB 1, started on October 22, 1991. Since the modification of UVB 1 (June 1991) the stripping reactor has been placed on top of the well to get a better control of the biomass production. UVB 2 has been arranged in the same way.

7.4 Results of the UVB remediation

Until May 4, 1992 about 164 kg aromatic (AHC) and mineral oil (MHC) hydrocarbons have been removed by the off-gas of UVB 1 from the groundwater and vadose zone. Fig. 5 demonstrates the daily total contaminant rate. Starting March 1991 the AHC concentrations, measured in the groundwater before and after the stripping unit of UVB 1 are presented in Fig. 6.

The measured contaminant removal of UVB 2 has been very small with about 10 g (all in the first 14 days) until May 1992. Contrary to expectations the contaminant concentration in the groundwater entering UVB 2 remained below detection limits during the operation. As the total circulation zone around UVB 2 is fully saturated with oxygen in consequence of the stripping process (see Fig.7b) the AHC and MHC contamination could be biodegraded in situ within the aquifer before the groundwater reached the well. The latter seems to be realistic because at monitoring well DP1 (see Fig. 3), situated about 15 m upstream of UVB 2, AHC concentrations of about 3 mg/L have been measured in October 1991 which vanished totally in the following period.

Strong biodegradation processes have been further involved at UVB 1. Visually this could be ascertained by an approximate 2 cm coating of biomass in all pipes and on the inside surface of the

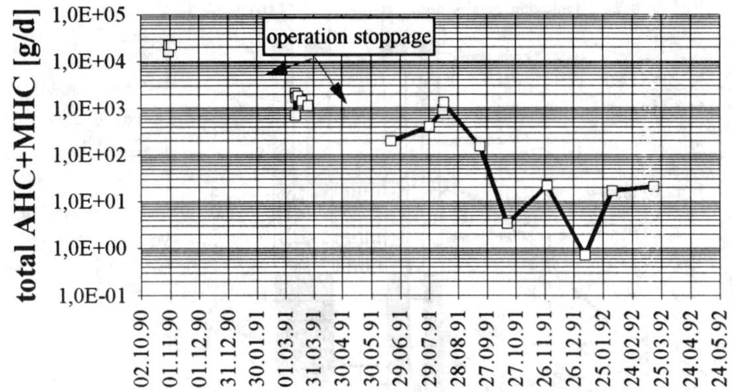

Fig. 5 Daily contaminant rate in the off-gas of UVB 1

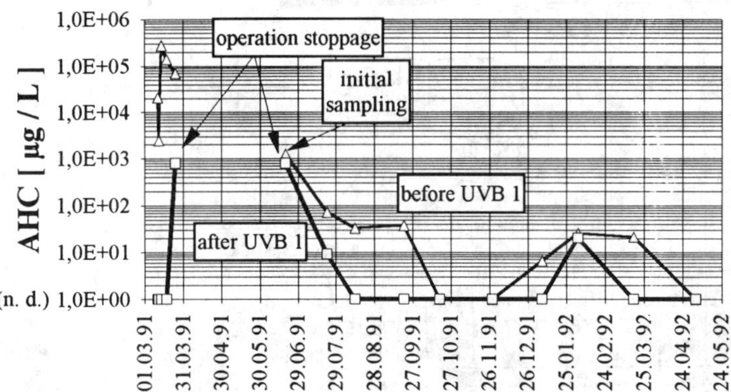

Fig. 6 AHC concentrations in the groundwater before and after the stripping unit of UVB 1

stripping reactor casing; the reactor itself was covered with a thin biocoating. Further, within the well casing lots of biomass were found. Neverless the pump rate of 6 m³/h could be kept up without problems for all the time.

The strong biodegradation was further detected within UVB 1 and in the groundwater circulation zone around it by analyzeing oxygen concentrations and pH values (see Fig. 7 and 8). Before the remediation started or at the end of a stoppage nearly anaerobic conditions and pH values of about 8.1 to 8.2 were found in the groundwater zone at the site. Within the stripping reactor oxygen saturation is always achieved (about 10 mg/L). The groundwater leaving UVB 1, measured at the deep monitoring tube, has a much smaller oxygen content, see Fig. 7a. Because of the lack of contaminants this oxygen consumption cannot be detected in UVB 2, see Fig. 7b. Further oxygen is used up in the circulation zone around UVB 1 where the aquifer itself can be regarded as a bioreactor, see Fig. 7a (shallow monitoring tube). Of course, other processes may use up oxygen too and it should be mentioned that about 30 % of nearly anoxic upstream groundwater is always entering the circulation zone and thus will lower the oxygen content of the water flowing into the well.

Another characteristic should be emphasized: The pH value upon passage through the stripping reactor of UVB 1 increased (see Fig. 8a); this means the produced biogas CO_2 is stripped out of the groundwater.

(a)

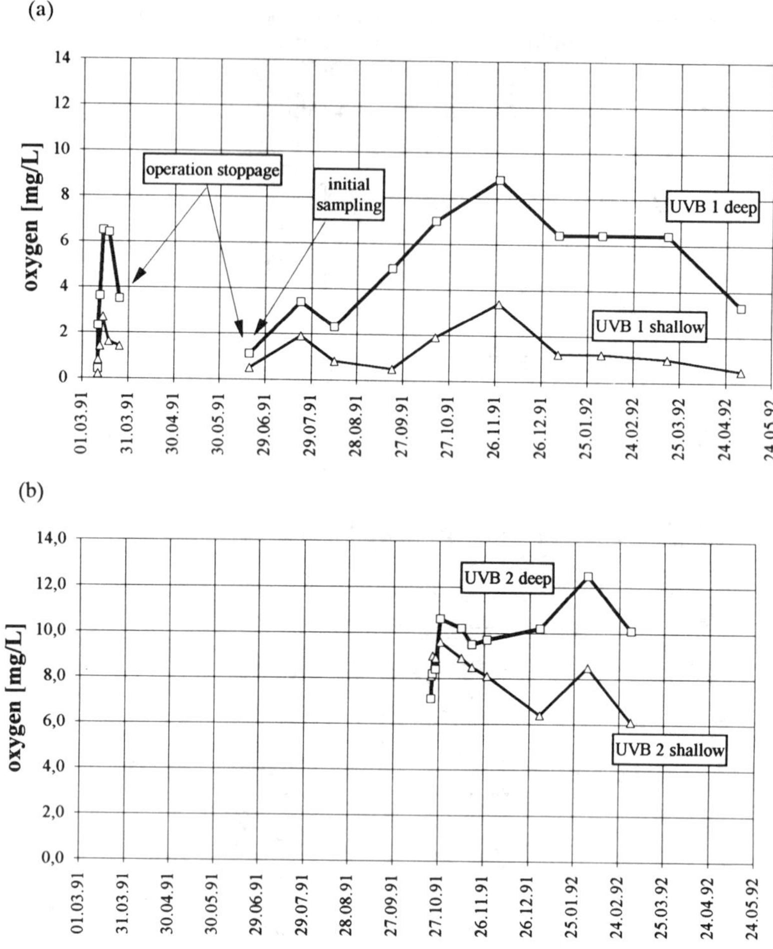

(b)

Fig. 7 O_2 concentrations in the shallow and deep monitoring tube of (a) UVB 1 and (b) UVB 2

As biodegradation is much lower at UVB 2, this phenomenon cannot be ascertained there in the same way (see Fig. 8b). Another point also suggests that biogas is produced in the aquifer: At monitoring well DP1 25m downstream of UVB 1 the pH value decreased to values of 7.3 to 7.4 after January 1992. Further, nearly no oxygen can be measured at DP1 although the outflow of UVB 1 delivers an increasing amount.

Finally, from Fig. 7a and 7b it can be seen that the oxygen consumption in the circulation zone of both remediation wells increases with remediation time because of the growing difference of the out- and inflowing oxygen concentrations. This means that the biodegradation increases in spite of the decrease of contaminant concentration in the groundwater (see Fig. 6).

7.5 Concluding remarks

The presented case study demonstrates that aromatic hydrocarbons and mineral oil can be successfully stripped and biodegraded at the same time. The lack of financial resources during a commercial remediation and the complexity of a field application do not allow a precise balance and an investigation about the fate of all the contaminants. Some microbiological investigations were carried out (e.g. biodegradation tests in the laboratory) but are unfortunately incomplete for lack of financial resources.

Otherwise, the role of biodegradation can be clearly ascertained: The production of biomass and biogas (CO_2) could be determined as well as the fact that bioclogging was not a limiting factor and that the oxygen consumption in the aquifer increases with time though the measured contaminant concentrations at the well decreases. The latter indicates that the aquifer in the circulation zone becomes increasingly a bioreactor. This process can be strengthened by enlargeing the well discharge with a stronger pump. Besides offering additional remediation a UVB further downstream can help to strip out the biogas and to lift pH values.

(a)

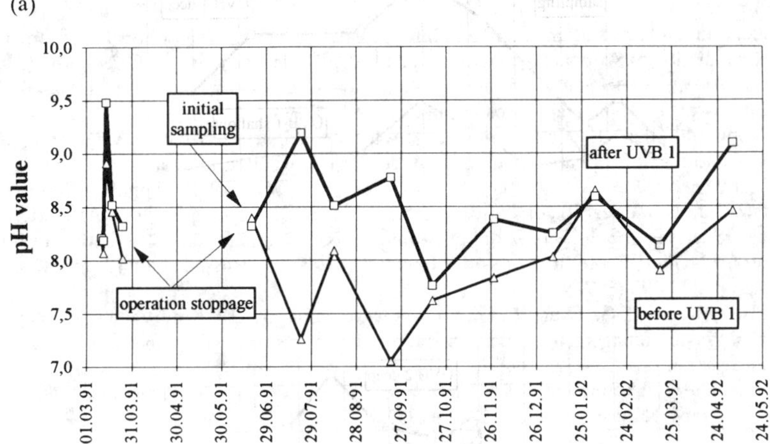

(b)

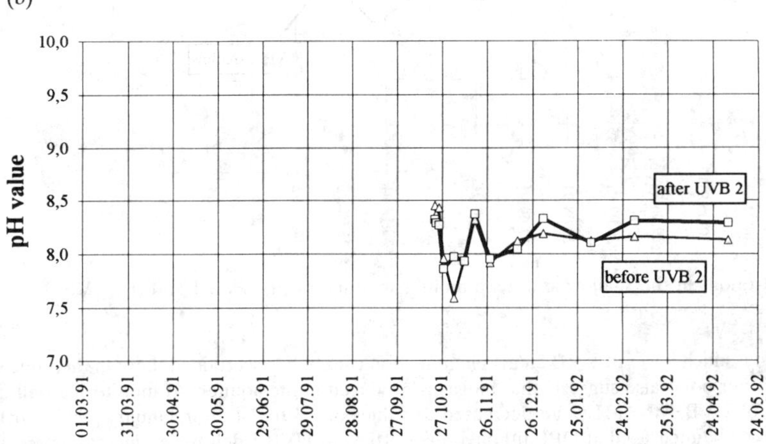

Fig. 8 pH value before and after the stripping unit of (a) UVB 1 and (b) UVB 2

8. ADVANTAGES OF A UVB OR GZB

The presented technologies and in particular the vertical circulation flow offer a lot of advantages, especially when compared with a standard pump and treat system. Beside those advantages explained above the following is mentioned: (1) for wells operating upward no lowering of the heads appears; (2) even at low well capacity, remediation operation is continuous; (3) less permeable, horizontal layers are penetrated vertically; (4) the remediation of the groundwater takes place down to the aquifer bottom; (5) the groundwater flow regime is influenced only locally in case of no partial discharge withdrawal or infiltration; (6) the presented in situ technology avoids waste water and groundwater extraction.

9. ACKNOWLEDGEMENTS

The first author gratefully acknowledges IEG mbH, D-7410 Reutlingen (Germany), for financially supporting these investigations. In particular B. Bernhardt, IEG mbH, D-7410 Reutlingen, inventor and patent holder of the GZB and UVB method; and further P. Brinnel, PROTEC GmbH, D-6370 Oberursel; W. Buermann, University of Karlsruhe, D-7500 Karlsruhe; W. Kaess, D-7801 Umkirch; H.J. Lochte, UTB mbH, D-4020 Mettmann; and J. Stamm, University of Karlsruhe, are thanked for many helpful discussions and contributions to realize the research work.

10. REFERENCES

1. Herrling, B., Buermann, W., and Stamm, J.: Hydraulic Circulation System for In Situ Bioreclamation and/or In Situ Remediation of Strippable Contamination, In: R.E. Hinchee and R.F. Olfenbuttel (Eds.), In Situ Bioreclamation, Applications and Investigations for Hydrocarbon and Contaminated Site Remediation, Butterworth - Heinemann, Boston, pp. 173-195 (1991).

2. Herrling, B., Stamm, J., Alesi, E.J., and Brinnel, P.: Vacuum Vaporizer Wells (UVB) for In Situ Remediation of Volatile and Strippable Contaminants in the Unsaturated and Saturated Zone, Proc. Symp. on Soil Venting, April 29 -May 1, 1991, Houston/Texas, EPA/600/R-92/174,pp.203-228 (1992).

3. Herrling, B., Stamm, J.: Numerical Results of Calculated 3D Vertical Circulation Flows Around Wells with Two Screen Sections for In Situ Aquifer Remediation, Computational Methods in Water Resources IX, Vol.1: Numerical Methods in Water Resources, Eds. T.F. Russell et al., Elsevier Applied Science, London, pp. 483-492 (1992).

4. Buermann, W., Bott-Breuning, G., Krug, R.: Groundwater Remediation Using the Vacuum-Vaporizer-Well. In: Industrial Waste Management (Proc. Envirotech Vienna 1992), W.Pillmann (Ed.), A.Riegelnik Printers, Vienna/ Austria, pp. 723-732, 1992.

5. Mellor, R.B., Ronnenberg, J., Campbell, W.H., Diekmann, S.: Reduction of Nitrate and Nitrite in Water by Immobilized Enzymes, Nature 355, pp. 717-719, 1992.

TRENDS IN IMPLEMENTING INNOVATIVE TECHNOLOGIES IN THE U.S.

WALTER W. KOVALICK, JR., PH.D.

Technology Innovation Office, U.S. Environmental Protection Agency
Washington, D.C.

1. INTRODUCTION

The remediation of hazardous waste sites in the United States presents an important opportunity for the development of new technologies.
Several different cleanup programs represent a future market worth over
a hundred billion dollars. In addition to the Superfund program for
abandoned waste sites, these programs include the corrective action for
active industrial property (pursuant to the Resource Conservation and
Recovery Act (RCRA)), leaking underground storage tank programs, various
state remediation programs, and cleanup efforts for federal facilities.

The United States Environmental Protection Agency (EPA) has committed
itself to encouraging the use of new or innovative technologies capable
of treating contaminated soils and ground water more efficiently, less
expensively, and in a manner more acceptable to the public. EPA recognizes
that new and improved approaches come with an associated risk which serves
as a serious impediment affecting decisions by government, private parties,
and engineering firms. EPA's Office of Solid Waste and Emergency Response
seeks to create an atmosphere which recognizes that reasonable risk-taking,
which is protective of human health and the environment, is necessary
to achieve our objective. We believe it is necessary to develop the
technologies necessary to fulfill the long-term needs of our hazardous
waste cleanup programs.

The status of completion of the different site remediation programs
varies widely, and site assessments are still generally more common than
actual field remediation. Up to this time, most of the progress has been
made in the Superfund program, which consequently provides much of our
operating experience and available data. The discussion which follows
concerns the status of technology use in this program, and an identification of technology development needs with particular attention to Agency
efforts in the area of bioremediation.

2. OVERVIEW OF SUPERFUND REMEDY SELECTION

Our operating definition of innovative technologies includes those
which are alternatives to conventional remedies of incineration and solidi-
fication/stabilization for control of the source of contamination (e.g.,
soil, sludge, and sediment) and above-ground treatment of ground water.
This definition is based on the assumption that adequate cost and performance data are available to select these conventional remedies. The
definition is not meant to imply that no new and important developments
are taking place for these conventional methods of treatment.

F. Arendt, G.J. Annokkée, R. Bosman and W.J. van den Brink (eds.), Contaminated Soil '93, 1093–1100.
© 1993 *Kluwer Academic Publishers. Printed in the Netherlands.*

Hazardous sites are assessed and ranked before being placed on the
Superfund National Priorities List (NPL). For the 1275 locations that
have been listed on the NPL, cleanup decisions have been made for 712
sites through fiscal year (FY) 1991. These decisions were greatly
influenced by the 1986 amendments to the Superfund statute which provide
a preference for treatment rather than containment or land disposal of
waste. In the past four years, 70 percent of all Records of Decision
(RODs, a formal document required for decision purposes at remedial cleanup
sites) for source control involve a provision for treatment of some portion
of the waste. The historical remedy selection pattern given in Figure
1 shows the progress which has been made in selecting treatment at contaminated sites.

With the priority for permanent remedies came an Agency commitment
to encourage the development of new technologies as alternatives to the
conventional approaches. EPA has made considerable progress in selecting
innovative treatment methods and the selection of these processes surpassed
conventional technologies for the first time in FY 1991. Figure 2 illustrates the selection histories for innovative and conventional technologies. The graph also illustrates the impact of the 1986 Superfund Amendments.

Figure 3 provides an overview of the selection of treatment technologies. The figure, which illustrates the number of times each technology
has been selected, shows that incineration is the most frequently selected
form of source control. Both off-site and on-site incineration, combined,
account for 30% of all treatment processes.

Figure 4 provides a more detailed look at the selection trend for
incineration. Since FY 1988, the total number of RODs, specifying incineration has remained essentially constant while there has been a shift
in the selection frequencies for off-site and on-site applications. This
shift can be accounted for, only in part, by an increased acceptance of
preprocessing technologies which separate and concentrate contaminants
prior to subsequent treatment. Although more data would be useful to
confirm this trend, there is certainly an apparent pattern away from the
use of on-site incineration at Superfund sites.

3. SELECTION OF INNOVATIVE TECHNOLOGIES

The selection histories of the four most popular innovative technologies are compared in Figure 5. The figure illustrates the selection
pattern of the four most popular innovative technologies. It clearly
shows that selection of soil vapor extraction (SVE) is increasing faster
than the other processes. Indeed, its growth also exceeds that of any
conventional technology. For the last three years, SVE has constituted
nearly half of the total number of selected innovative technologies.
There is some question as to whether SVE should be considered "innovative"
since it is commonly used for petroleum contamination from leaking underground storage tanks, and it has been selected 84 times for Superfund
remediation. However, despite the frequency of selection, only one Superfund site remediation has been completed using this process and questions
remain regarding its application and effectiveness for the wide variety
of contaminants for which it is being applied. In addition, it is still
necessary to develop a verified means to characterize performance and
to determine when cleanup goals have been met.

Preprocessing technologies such as soil washing and thermal desorption are being regularly chosen at fewer than ten sites per year (Figure 5). However, the potential market for these technologies is believed to be relatively large. Soil washing may be used as a volume reduction step for both metals and semi-volatile organics. Thermal desorption has potentially broad use for volatile and semi-volatile organics and for the separation of metals from organics when they are mixed together. As volume reduction or separation processes, these technologies are primarily applicable for use in treatment trains. It is possible they are not being more widely accepted because we have not fully exposed the potential of these trains. The most frequently selected trains are soil washing followed by bioremediation (five sites) and thermal desorption followed by incineration of residuals (six sites) or solidification/stabilization of residuals (five sites).

4. SELECTION AND EVALUATION OF BIOREMEDIATION

The combined selection trend for ex situ and in situ bioremediation (including nine in situ groundwater applications) is also shown in Figure 5. In situ or ex situ bioremediation comprise seven percent (36 sites) of the treatment remedies for source control. EPA has placed a high priority on developing the full potential of bioremediation to treat hazardous waste, in addition to petroleum spills and leaks, through a diversity of demonstration and technology transfer efforts. These efforts primarily fall under the Bioremediation Field Initiative, which will briefly be described due to its potential implications for future technology use. The initiative was established to provide EPA and state project managers, consulting engineers, and industry with timely information on new developments in bioremediation at hazardous waste sites including: evaluations of performance at selected full-scale field applications; technical assistance to EPA site managers; and a database on field applications of bioremediation.

Currently in the field initiative, EPA is evaluating 7 sites to determine their effectiveness, operational reliability, and costs. Four of these evaluations are examining biotreatment of petroleum contaminants, 2 are of wood treatment sites contaminated with creosote and pentachlorophenol (PCP) and one is a solvent site. Other waste types are being considered for evaluation, including: munitions, pesticides, and coal tars. Treatment methods being evaluated are both in situ and ex situ for soil, including land treatment and bioventing.

In situ ground water treatment evaluations include use of denitrification in treating petroleum contamination, hydrogen peroxide addition for treatment of a creosote and anaerobic plus subsequent aerobic biodegradation for trichloroethylene (TCE). Sites included in the evaluation program include: Eielson Air Force Base, Alaska; Hill Air Force Base, Utah; Park City Pipeline, Kansas; Public Service and Electric, Denver, Colorado; Champion Superfund Site, Libby, Montana; Brookhaven, Mississippi; and Allied Signal Superfund site, Michigan.

Preliminary results are now being reported. They are published along with other information for the user and research community in the bulletin <u>Bioremediation in the Field</u>, produced quarterly by EPA.

5. TECHNOLOGY DEVELOPMENT NEEDS

A summary of ground water remedies is given in Figure 6. Only in situ remedies are considered innovative, although it is recognized that there are important new approaches to above-ground aqueous treatment. In most cases, ground water is pumped to the surface where it is usually treated by established methods. In situ remedies make up less than two percent of the selected technologies to treat ground water (Figure 6), although nearly 80 percent of Superfund sites with RODs require groundwater remediation. Of the 210 innovative technologies selected overall, 12 address ground water in situ. The lack of available in situ groundwater treatment technologies is responsible for their limited application for site remediation. Data on successful applications will be necessary to broaden their acceptance and use. Of the nine Superfund sites for which in situ bioremediation has been selected or used, five are addressing nonchlorinated volatiles (i.e., benzene, toluene, ethylbenzene, and xylene), five are addressing semi-volatiles, and one is targeting chlorinated volatiles. Both volatile and semi-volatile organics are addressed at some of these sites.

Relatively few innovative technologies are being chosen either as pretreatment or as alternatives to conventional solidification/stabilization for metals. Figure 7 shows that a total of 16 innovative technologies have been chosen to treat metals while solidification/stabilization is the preferred remedy at 128 sites (Figure 3). Although 77% of NPL sites with RODs have some metals (including arsenic) contamination, few new technologies are being applied for these constituents. Although solidification/stabilization may be viewed as an effective and relatively inexpensive remedy, it may not eliminate long-term liability concerns and monitoring requirements at sites.

While the remedy selection trends are generally encouraging, relatively few innovative projects have been completed. Figure 8 illustrates this point by showing how few of the technologies have moved to the completion phase as of October 1992 -- only five percent. The figure reveals a lack of actual full-scale cleanup experience for all technologies. It is important to note, however, that there has been a large increase in the number of projects that are being installed or are operational. One year earlier, only 14 projects were in this category. Thus, there will soon be a large increase in the number of operating technologies which will present an opportunity for observation and analysis of actual field experience.

As a final observation, potentially responsible parties (PRPs) are responsible for implementing many innovative projects at Superfund sites. The private sector, rather than the federal government, is the key client for innovative technology services. Approximately two-thirds of the 156 projects currently at the design stage are being managed directly by PRPs through legal agreements with EPA to remediate the sites. Of the remaining sites, EPA is negotiating cleanup agreements in 10 cases and an additional 8 projects are at federal facilities.

The author gratefully acknowledges the assistance of John Kingscott in the preparation of this paper.

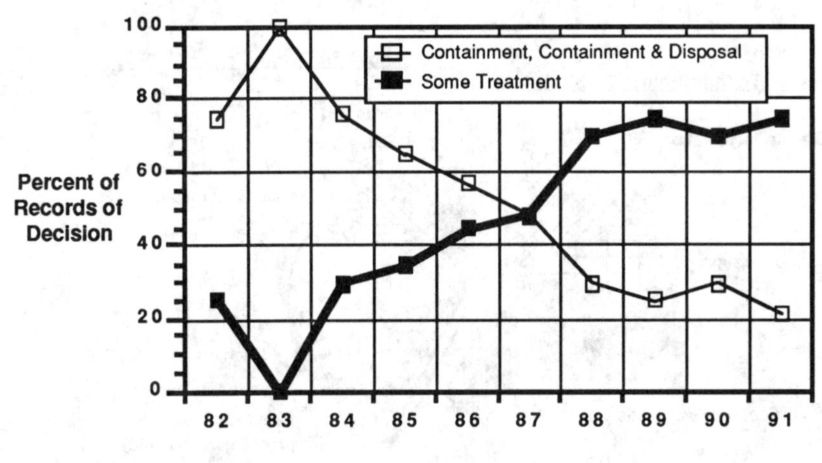

Figure 1.
Superfund Remedial Actions:
Treatment Versus Disposal Decisions for Source Control

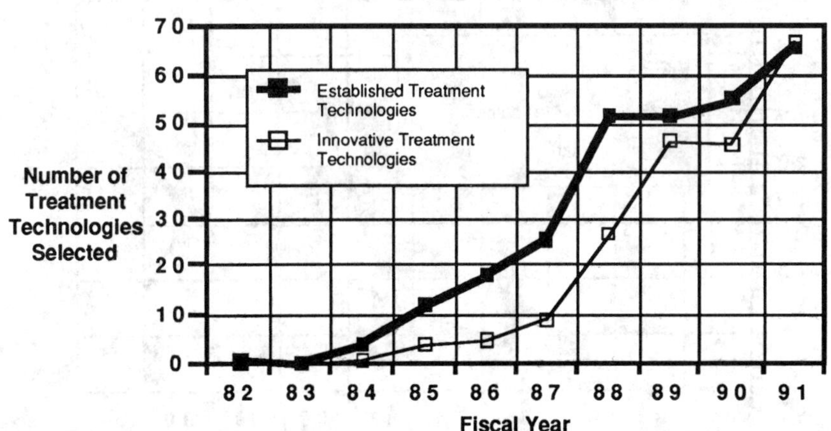

Figure 2.
Superfund Remedial Actions:
Number of Established Versus Innovative Treatment Technologies

Note: Data for innovative technologies are derived from Fiscal Years 1982 – 1991 Records of Decision (RODs) and anticipated design and construction activities as of February 1992. More than one technology per site may be used.

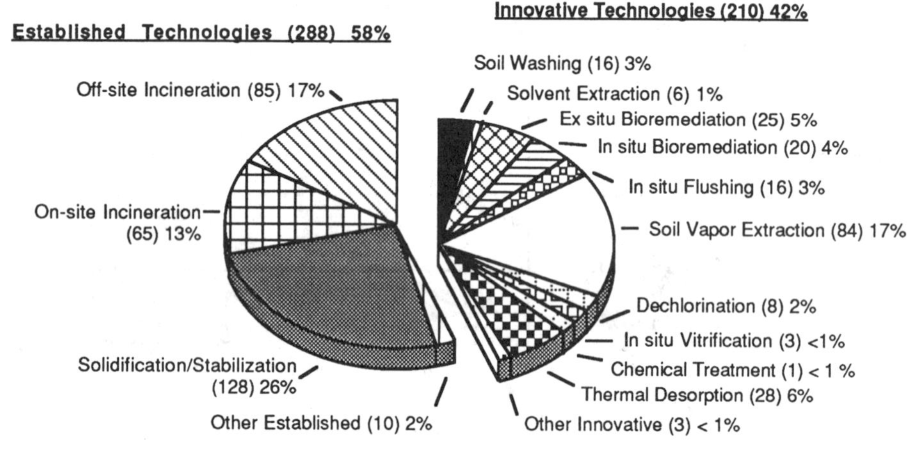
Figure 3.
Superfund Remedial Actions:
Summary of Alternative Treatment Technologies Through Fiscal Year 1991
(Total Number of Technologies = 498)
Established Technologies (288) 58%
Innovative Technologies (210) 42%
Off-site Incineration (85) 17%
On-site Incineration (65) 13%
Solidification/Stabilization (128) 26%
Other Established (10) 2%
Soil Washing (16) 3%
Solvent Extraction (6) 1%
Ex situ Bioremediation (25) 5%
In situ Bioremediation (20) 4%
In situ Flushing (16) 3%
Soil Vapor Extraction (84) 17%
Dechlorination (8) 2%
In situ Vitrification (3) <1%
Chemical Treatment (1) < 1 %
Thermal Desorption (28) 6%
Other Innovative (3) < 1%

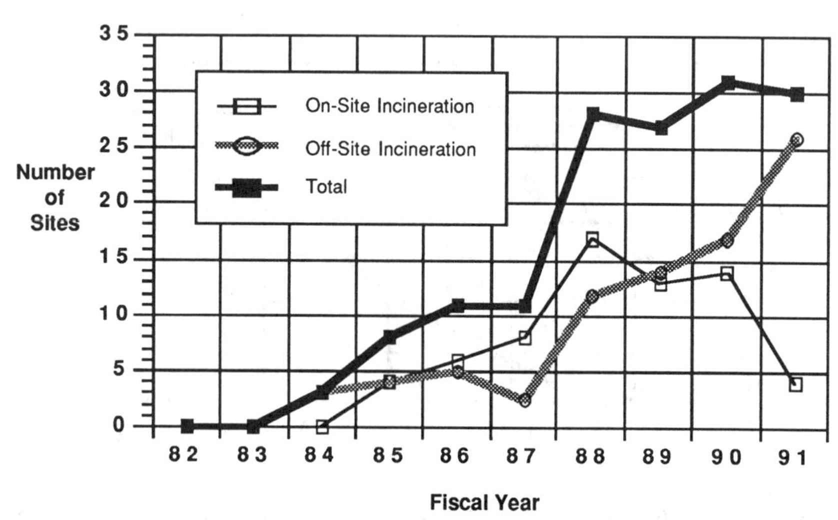
Figure 4.
On-Site and Off-Site Incineration Selected for NPL Sites
3 5
3 0
2 5
2 0
1 5
1 0
5
0
Number
of
Sites
On-Site Incineration
Off-Site Incineration
Total
8 2   8 3   8 4   8 5   8 6   8 7   8 8   8 9   9 0   9 1
Fiscal Year

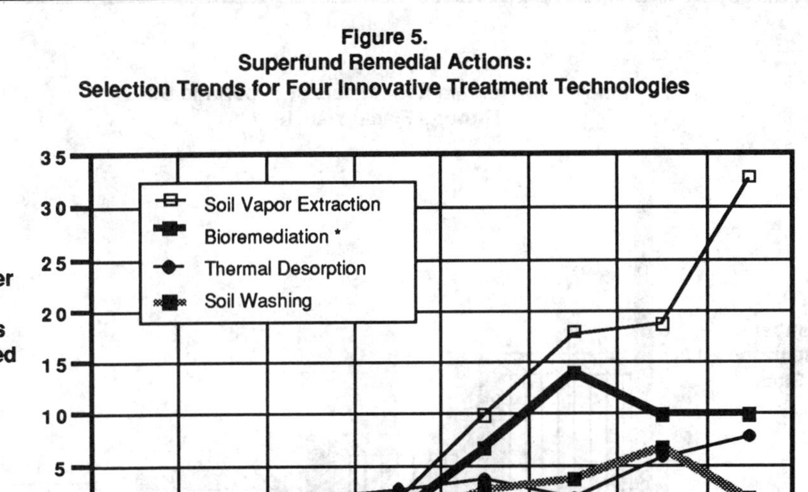

Figure 5.
Superfund Remedial Actions:
Selection Trends for Four Innovative Treatment Technologies

* Includes sites using ex situ and in situ bioremediation of source material and ground water.

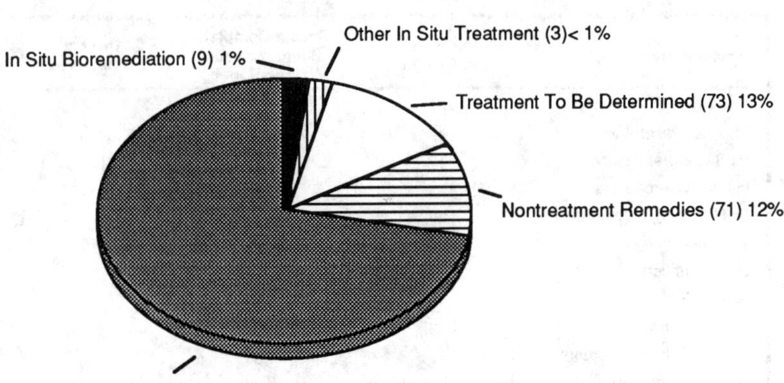

Figure 6.
Superfund Remedial Actions:
Summary of Groundwater Remedies Through Fiscal Year 1991

Note: Data derived from Fiscal Years 1982 – 1991 Records of Decision (RODs) and anticipated design and construction activities.
* Includes in situ groundwater treatment.

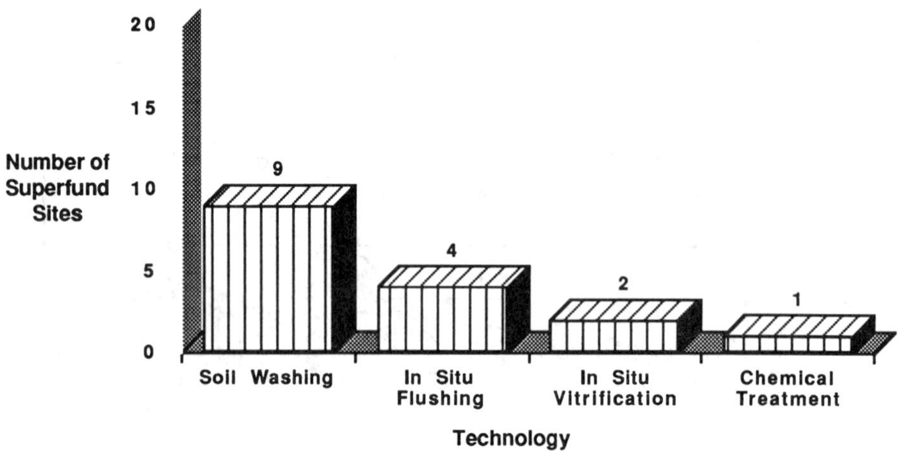

Figure 7
Innovative Treatment for Metals at Superfund Sites
Through Fiscal Year 1991

Note: More than one treatment technology may be used at some sites.

Figure 8.
Project Status of Innovative Treatment Technologies at NPL Sites
as of October 1992

Technology	Predesign/ In Design	Design Complete/ Being Installed/ Operational	Project Completed	Total
Soil Vapor Extraction	62	18	3	83
Thermal Desorption	19	4	4	27
Ex Situ Bioremediation	17	7	1	25
In Situ Bioremediation*	14	5	1	20
Soil Washing	16	2	0	18
In Situ Flushing	12	5	0	17
Dechlorination	5	1	1	7
Solvent Extraction	5	1	0	6
In Situ Vitrification	3	0	0	3
Other Innovative Treatment	3	0	0	3
Chemical Treatment	0	0	1	1
TOTAL	**156** (74%)	**43** (21%)	**11** (5%)	**210**

Note: Data derived from Fiscal Years 1982 – 1991 Records of Decision (RODs) and anticipated design and construction activities.
   *   Includes in situ groundwater treatment.

EXPERIENCE WITH DIFFERENT CLEAN-UP TECHNOLOGIES AT AN ABANDONED INDUSTRIAL SITE

DR. ECKART HILMER

L.U.B. Lurgi-Umwelt-Beteiligungsgesellschaft mbH
Emil-v.-Behring-Straße 2, 6000 Frankfurt 50

Abstract

One of the largest site remediation projects in the Federal Republic of Germany is currently being carried out at the 1 million m^2 site of the former Vereinigte Deutsche Metallwerke AG (VDM) in Frankfurt-Heddernheim. The site is contaminated with chlorinated hydrocarbons, polycyclical hydrocarbons, mineral oils and heavy metals.

The cleanup of this site proceeds in two phases. The first phase involves the removal of highly volatile VOCs from the subsoil by vacuum extraction. The cleanup objective is a residual VOC concentration of < 1 mg/m^3 compared to an average initial concentration of 100 mg/m^3.

In parallel with soil vapour extraction, the contaminated groundwater is cleaned of VOCs. Original concentration 1.2 mg/l Σ VOC. Treatment objective < 3 µg/l.

The second phase involves the cleanup of the soil by wet-mechanical methods. The contaminated soil exhibits up to PAH concentrations of 800 mg/kg. The cleanup objective is < 10 mg/kg Σ PAH.

Introduction

Remediation of the 1 million m^2 site of the former Vereinigte Deutsche Metallwerke (VDM) in Frankfurt-Heddernheim is one of the largest and most elaborate site cleanup projects in progress in the Federal Republic of Germany. The soil at this abandoned industrial site is contaminated with chlorinated hydrocarbons (VOCs), mineral oils, polycyclical aromatic hydrocarbons (PAHs) and heavy metals while the groundwater is laced with VOCs.

Over a period of 150 years, copper-containing alloys for the manufacture of semis were produced at this site, last by VDM. In 1982, the plant was closed down for economic reasons. Demolition of the plant began in 1986. After demolition, the entire site was levelled with some deep foundations remaining in the ground.

In 1987, a site development plan was agreed which envisaged the construction of industrial and commercial buildings, administrative buildings, blocks of flats and terraced housing. Moreover, a hotel complex, a shopping mall and a kindergarden are to be established or have already been established at this site.

Site exploration programmes conducted from 1986 identified severe contamination of both the groundwater and the soil which called for immediate remedial action in the interest of public health. By 1990 the full extent of contamination had been established and a remediation concept developed.

F. Arendt, G.J. Annokkée, R. Bosman and W.J. van den Brink (eds.), Contaminated Soil '93, 1101–1107.
© 1993 Kluwer Academic Publishers. Printed in the Netherlands.

The two-phase remediation concept agreed with the responsible authorities provides for soil vapour extraction and cleanup in a first step followed by soil washing. In parallel with these remediation steps, the groundwater is treated at identified source areas on the site and along the site boundaries using a well gallery extending over 1 km. The treated groundwater is recharged downstream of the well gallery.

The site is located on the central terrace of the Main River. The stratification profile is as follows: a backfill layer, 2 to 3 m thick, composed of building rubble and soil mixed with slag and gravel followed by a 0.75 to 1.50 m layer of loess loam consisting mainly of stiff siltic clay originating from the Holocene age. This siltic clay has a permeability coefficient of $K_f = 3 \cdot 10^{-8}$ m/s. Below the silt layer is an aquifer composed of sandy gravel interspersed with clay lenses stemming from the Tertiary age. The groundwater table is 3 to 4 m below surface.

Soil Vapour Extraction

Soil vapour extraction is carried out all over the site to prevent VOCs from escaping to atmosphere during construction work later on or during excavation of the contaminated soil layers for cleanup.

The VOCs identified at the Heddernheim site mainly comprise:

    cis-1.2-dichloroethene
      1.1.1-trichloroethane
         trichloroethene (TRI)
    and   tetrachloroethene (PER)

The VOC concentration is generally expressed as cumulative parameter Σ VOC. The soil vapours exhibit maximum Σ VOC levels of up 1400 mg/m3. After determining the radius of influence of the soil vapour extraction system, wells were drilled in the affected areas to install the suction lances. Moreover, wells, 620 mm in dia. were sunk for simultaneous groundwater pumping. The lances are up to 7 m deep, the combined groundwater/soil vapour extraction wells 10 m.

To boost the effectiveness of the soil vapour extraction measures, the entire surface of the site was sealed with a 20 cm mineral layer with a grass cover on top.

The soil vapours are extracted via the lances and wells using vacuum pumps. The delivery rates of the individual pumps are adjusted to suit the specific soil conditions, i.e. the soil permeability has a direct influence on the operation of the system. The permeability coefficients of the backfill at the Heddernheim site are in the range of abt. 10^{-5} to 10^{-6} m/s.

The extracted soil vapours are routed through a water separator where they are freed from entrained water droplets. Actual treatment is accomplished in two series-connected activated carbon filters.

Two vacuum pumps and water separators each are accommodated in a common container. 12 containers are in operation. Via a header system, the extracted soil vapours are transferred to a central activated carbon unit where they are cleaned to the prescribed Σ VOC level of < 1 mg/m^3.

Each activated carbon unit consists of 2 series-connected adsorbers so that one adsorber is always on duty while the other is available as standby and can be directly connected when the breakthrough level is attained in the duty adsorber. The exhausted carbon is reactivated for reuse in Lurgi's own activated carbon regeneration plant.

In some areas of the Heddernheim site, soil vapour extraction has already been successfully completed. One example for successful soil vapour extraction is unit D-LL-9 where a total of 38.8 kg Σ VOC was extracted from the unsaturated soil zone over a period of 379 days. At the beginning 500 g Σ VOC were extracted per day.

As cleanup of the soil progressed, increasing amounts of vinyl chloride of the order of 40 mg/m^3 were determined in the soil vapours. As VC has a low sorption equilibrium on activated carbon, the maximum run times for the activated carbon adsorbers would be reduced respecitvely.

For this reason, a UV system was installed upstream of the adsorbers. UV radiation permits the VC inlet concentration to be reduced by 90 % and hence increases adsorber run times appreciably.

The soil vapours extracted from the well by the vacuum pumps are passed through a water separator from where they go to the UV reactor. In this reactor the VOCs will be oxidised by UV radiation.

Groundwater Rehabilitation

In parallel with soil vapour extraction and cleanup, groundwater treatment was tackled. Emergency measures for groundwater rehabilitation at some selected points across the site have been in progess for 4 years. At the beginning, the contamination level was extremely high, e.g. 133 mg/l Σ VOC at well PP4. The average Σ VOC concentration of the ground water is 1.2 mg/l.

Since 1991 a hydraulic barrier has been in operation to prevent contaminated groundwater from migrating to non-polluted aquifers. Over a distance of 1 km, barrier wells were drilled at 50 m intervals. Via these wells the contaminated groundwater is pumped to a central groundwater treatment systems. The treated groundwater is recharged to the aquifer area downstream of the wells via infiltration trenches.

The central groundwater treatment system is designed for a maximum throughput of 180 m^3/h.

The groundwater exhibits the following concentrations of dissolved contaminants:

Fe   4.0 mg/l
Mn   3.0 mg/l
VOC 1.2 mg/l

Prior to admission to the treatment system, the groundwater is subjected to compressed air aeration in an in-line mixer. This causes the iron which is present in dissolved form to be oxidized and precipitated as ferrous hydroxide flocs. The latter are separated from the water in a downstream filter. Dissolved manganese in water is not amenable to oxidation by air oxygen. It requires a more powerful oxidant like ozone or potassium permanganate. In Heddernheim, potassium permanganate has been selected as oxidant due to its ease of handling. After Fe removal, the water is inocculated and intimately mixed with potassium permanganate solution which is prepared in the in-plant dissolving station. The manganate is oxidized to manganese hydroxide which is separated from the water in a second identical filter unit. In a final step, the water is cleaned of VOCs in a granular activated carbon filter.

The metal hydroxides retained on the filter media are removed from the filters by backwashing. The backwash effluent is sent to one of two settling tanks which are alternately filled with backwash effluent. In these tanks, the metal hydroxide flocs settle out, the settling process being promoted by adding a polyelectrolyte to the tank influent. The supernatant is returned to the ferrous hydroxide filters. The settled solids are largely dewatered and routed to disposal.

The activated carbon filters also require occasional backwashing, especially after replacement of the activated carbon to remove carbon fines and dust deriving from abrasive action during transport. The resultant backwash effluent is likewise directed to the settling tanks.

The treatment objective for the groundwater is a residual Σ VOC concentration of < 3 µg/l.

Apart from groundwater rehabilitation by the hydraulic barrier and the central water treatment system, individual particularly badly afflicted groundwater tables are subjected to decentralized treatment. As the groundwater, too is laced with vinyl chloride in some areas of the site, a system has been selected for treatment using peroxide and UV radiation. The Σ VOC load of the groundwater in these badly afflicted areas is between 1 and 7 mg/l.

The process operates on the atmospheric chemical wet oxidation principle. It makes use of the capability of UV radiation to decompose hydrogen peroxide photolytically into hydroxyl radicals.

The hydroxyl radicals have a very high oxidation potential and react with unsaturated VOCs by aggregating to the free double bond, thus activating the VOCs for oxidation.

The overall decentralized water treatment system comprises the following unit operations:

    Well with pump
    Quartz filter unit
    Storage tank
    Oxidation system
    Activated carbon adsorber

The decentralized groundwater treatment system is designed for a throughput of 20 m³/h at a Σ VOC load of max. 35,000 µg/l. So far no VOC levels larger than 1 µg/l have been determined in treated effluent.

By installing an oxidation system upstream of the activated carbon adsorber, the adsorber run time could be increased many times over that achievable without previous wet oxidation.

Soil Cleanup

After completion of soil vapour extraction, the subsoil and backfilling materials which are contaminated with mineral oils, PAHs and heavy metals, are cleaned in a Lurgi Deconterra soil washing plant. In determining the intervention threshholds, the soil was classified into a top layer, 1 m thick, and subsoil located below - 1 m.

Intervention Threshholds

Contaminant	down to 1 m	below 1 m	
Σ PAH (to EPA)	10 mg/kg	50 mg/kg	
Mineral oil-borne HCs	500 mg/kg	1000 mg/kg	
Σ BTX	5.5 mg/kg	20 mg/kg	
Cr	50 mg/kg	0.5 µg/kg	
Cu	50 mg/kg	5 µg/l	
Zn	150 mg/kg	5 µg/l	leachate
Pb	100 mg/kg	1 µg/l	

If the above threshholds are overridden, the soil has to be treated in a soil washing system.

The contaminated soil is hauled by truck from the excavation site to the raw soil stockpile located at a distance of about 500 m. This stockpile has an area of 21,000 m^2 and a storage volume of 150,000 m^3. The stockpile base is sealed by means of a multi-barrier system with monitoring drains and a leachate drainage and collection system installed underneath.

The soil washing plant is designed for a throughput of 25 t/d wet material. It is operated in 2 shifts with 5 operators being required per shift. Up to February 1993, a total of 28,000 tons of soil and backfilling material had been cleaned. Of the 28,000 t clean soil, 23,100 t were backfilled at the original site and/or stored on a clean soil stockpile. 4,900 tons of highly contaminated cleaning residues were dumped on a controlled landfill.

The soil recovery rate of the soil washing plant is all the more impressive as 70 % of the raw soil is accounted for by siltic clay.

The soil washing plant operates on the Lurgi Deconterra process. It consists of individual modules which are configured to suit the soil charateristics and the specific application.

The raw soil is continuously reclaimed from the stockpile and transported to the feed hopper by a wheel loader.

From the hopper, the soil feeds continuously onto a roller bar grizzly where it is screened at a cutoff level of 150 mm. The oversize plus 150 mm is crushed in a jaw crusher and once again united with the size fraction minus 150 mm. Tramp iron is removed on the conveyor belt by a magnetic separator.

Attrition Scrubbing

In a first key process step, the raw soil is subjected to attrition scrubbing. Depending on the soil to be treated, the attrition drum is operated at specific energy inputs of up to 10 kWh/t raw soil. This presupposes that the operating conditions are controlled within a narrow range. Attrition scrubbing results in the almost complete disintegration of the soil minerals and components down to the micron range. Another important aspect is that the clay agglomerates are largely broken up, thus increasing the surface area avaialable for adsorption of the contaminants liberated from the coarser fractions.

Classification

After attrition scrubbing, the soil is classified applying cutoff levels suited to the requirements of the downstream sorting stages and the specific soil characteristics.

The upper screen deck has a mesh size of 20 mm. The fraction plus 20 mm is routed to a downstream impact crusher, the discharge of which is returned to the first attrition scrubber stage. If necessary, wood debris can be removed from this circuit by hand.

The bottom screen deck has a mesh size of 1 mm. The screen overflow minus 20 mm to plus 1 mm is directed to the gravity sorting system where the highly contaminated light-gravity constituents are separated out.

A bottom screen mesh of 1 mm has been selected for the following reasons:

1. With soils containing fibrous components, a cutoff level of 1 mm is just about the limit that can be reliably handled by commercial screens in a continuous operating mode.

2. Particle sizes below 1 mm have an adverse effect on gravimetric sorting.

3. Particles coarser than 1 mm interfere with the subsequent selective flotation process as they tend to be sedimented.

The size fraction minus 1 mm makes the highest demands on the sorting and classification process and hence requires sophisticated process technology.

This fines fraction exhibits specific surface areas from 4000 cm^2/g up to 16,000 cm^2/g. The adsorptive capacity is known to rise with decreasing particle size and increasing pore volume. However, as the desorbability decreases in the same proportion, desorption of the contaminants by physical methods is no longer possible. Therefore, the fines fraction has to be separated and discharged as residue.

In the Deconterra process, fines separation, i.e. desliming, is accomplished in hydrocyclones applying a d_{80} mesh of separation. The cyclone overflows having a d_{80} value of 0.015 to 0.02 are discharged as slurry. This fraction is routed to the treatment residue without further processing.

Treatment of Coarse Fraction

The coarse fraction is further processed in a jig. Here the material is jigged up and down by a stream of pulsating water while being slowly moved in horizontal direction. As a result, the heavy material is separated from the light material, each fraction being drawn off separately. These two fractions can be further sorted into one light and one heavy fraction each by corresponding sorting equipment. The heavy fraction, which is free from wood, coal and porous components constitutes the coarse soil yield.

Fines Treatment

If required the cyclone overflows are routed to a second attrition scrubber where they are washed in the absence of the highly contaminated fines fraction. This not only forestalls re-contamination but is also beneficial to the subsequent sorting process as the fraction plus 0.8 mm is further reduced.

As already mentioned, the principal contaminant sources such as coal, wood and roots have to be removed from the fines fraction as completely as possible. This is best achieved by selective froth floatation. In the Lurgi Deconterra plant, the fraction minus 1 mm is processed through a three-stage pneumatic floatation unit.

The floatation tanks are arranged in series so that the effluent of the first tank can be further treated in the second tank and so on.

As floatation agents, rapeseed oil, an ecologically neutral collector, and a biodegradable foaming agent are added at ratios of 120 g/t and 40 g/t respectively. The contaminants collected in a layer of froth on the liquid surface are routed to a thickener together with the cyclone overflow before being discharged as residue.

Residue treatment

The fines fractions are flocculated and thickened. The thickener underflow is then dewatered on an plate-and-frame press. Dewatering is carried out batchwise. The resultant filter cake normally has a residual moisture of < 20 wt. %.

As 90 to 95 % of the contaminants originally contained in the raw soil are transferred to the treatment residue, the latter is stored in special containers and dumped on a controlled landfill.

In conclusion, the effectiveness of the Deconterra process can be demonstrated by reference to PAH. The average PAH level of the cleaned soil is 9 mg/kg.

The total soil volume to be cleaned in the soil washing plant is estimated at some 190,000 t.

Remediation of the Heddernheim site will probably be completed by 1996 with operation of the central groundwater treatment system being continued till 1998.

SOIL WASHING, FROM CHARACTERIZATION TO TAILOR-MADE FLOW DIAGRAMS, RESULTS OF FULL-SCALE INSTALLATIONS

M. PRUIJN and E. GROENENDIJK

Heidemij Realisatie BV, P.O.BOX 660, 5140 AR Waalwijk, The Netherlands

ABSTRACT.
A well known method for the ex-situ treatment of contaminated soil is soil washing, also known as extraction or classification. For successful development and operating a characterization of the contaminated soil with respect to the contaminants and the soil matrix is essential. A difference in soil- and contaminant characteristics can lead to an adjustment of the process or to a revised process lay-out. These tailor-made full-scale flow diagrams are presented for four case studies. The characterization results are included for each project. The results of the projects in terms of removal efficiency and concentration of contaminant in the sand product are also presented. The case studies show that tailor-made flow diagrams lead to good performances of the full-scale soil washing installation.

1. INTRODUCTION

The principle of soil washing technology is to concentrate the contaminants in a small residual fraction by separation. The soil is mixed with water to form a slurry and then several separation/classification-technologies are used to remove the contaminants. Examples are screening, hydrocycloning, gravity separation and froth flotation. The sand product can be reused, the residue has to be landfilled or treated. Soil washing is applicable to a wide variety of contaminations such as heavy metals, Polycyclic Aromatic Hydrocarbons (PAH), mineral oils, pesticides, cyanides and others.

The objectives of soil washing are two fold. The first is to decrease the level of the concentration of contaminant. The second is to decrease the quantity of the concentrated residue. Often these two objectives interfere. Adding an extra separation step will create more residue, not adding a step means a higher concentration of contaminant in the product.

In designing a soil washing process it is important to understand in what form the contaminant is present within the contaminated soil. This <u>characterization</u> is the key to find an optimum in the two objectives; the lowest concentration of contaminant in the product and a minimal quantity of residue.

2. CHARACTERIZATION

The characterizing of contaminated soil consists of a characterization of the contaminant and a characterization of the soil. In the dutch legislation the standards for the contaminants are made referring to their chemical form (like benzo(a)pyrene) or to their absolute concentration (like total lead). The physical form in which the contaminant occurs within the soil is still not clear when the concentration is known. Determination of this physical form is the objective of the characterization.

1109

F. Arendt, G.J. Annokkée, R. Bosman and W.J. van den Brink (eds.), Contaminated Soil '93, 1109–1118.
© 1993 *Kluwer Academic Publishers. Printed in the Netherlands.*

2.1. Contaminant characterization

In general there are four main physical forms in which a contaminant can occur in the soil.

- Particles that have a high concentration of contaminant. These particles can be pure contaminant, like lead dust or pesticide pellets, or not pure, like tar particles or slags.

- Coatings of contaminants on the sand particles.
An example is the lead coating that often is found on shooting range sand.

- Water soluble components.
Since soil contains 10-20 % of water, the pollution of the water will also pollute the soil. Phenol is a good example for this type of contaminant, but also some metal salts are soluble enough to give significant soil contamination.

- Contaminants that adsorb to the soil.

Adsorption is possible whenever the contaminant is (slightly) water soluble. From the water phase the contaminant adsorbs to a place with a higher binding energy. In this way the contaminant can migrate from one soil fraction to another. Almost all contaminants show this adsorption behaviour. Very often adsorption takes place onto the fine clay fraction and onto organic material.

The actual contamination in a soil is often a combination of more characterization types. All kinds of contaminants occur in the soil, and every contaminant can have a different occurrence.

The first step for a proper characterization consists of evaluation of the historical information available concerning the source of the contamination. Examples for common sources are:
- chemical industry calamities and spillage,
- gasoline station spillage,
- use of slag material in road construction,
- hunting,
- waste dump.

After the initial contamination of the soil, the physical occurrence of the contaminant can and will change because of the adsorption processes.

The above mentioned historical analysis gives a prediction in what physical form the contaminants are present but needs to be confirmed by soil characterization. This physical and chemical analysis includes wet screening, Scanning Electron Microscopy (SEM), stereomicroscope, density separation and common chemical analyses such as GC, HPLC, AAS etc.

As an example the physical and chemical form of some contaminations are related to the four characterization types in table 1.

TABLE 1. Some examples of different physical occurrence of the same contaminant

	Particles	Coating	Water	Adsorbed
Lead	dust	shooting range sand	leadchloride	at clay
Benzo(a)pyrene	slags	tars	low conc.	at clay
Arsenic	fungicide powder	--	as complexes	organic mat.

2.2. Soil characterization

The characterization of the soil is very straight forward. Important factors are the particle size distribution, the organic matter content and the calcium carbonate content.

This physical characterization is important for the amount of residue that the process will generate. Also a large content of certain fractions like the sludge, the organic matter or the oversize can cause operating problems in the process.

2.3 Conclusion

A combination of the soil- and the contaminant characterization is the best starting point for developing a successful soil washing process.

The effort that is put into the characterization phase will result in lower concentrations contaminant in the clean sand product, lower production costs and lower quantities of residue. This way the environmental and the commercial efficiencies are optimized.

Heidemij has a large experience in translating the characterization to a full-scale soil washing process. The translation path is visualised in figure 1.

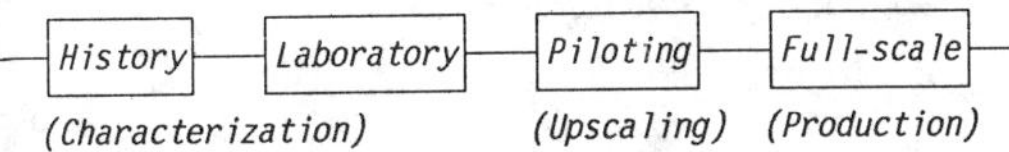

FIG. 1. Linking characterization and full-scale soil washing

3. CASE STUDIES

3.1. Introduction

In each case study presented, the characterization results as well as the actual process used are explained. The projects are realised with a capacity that varies between 15 and 40 tonnes per hour. *The product sand in all projects was reused.* The residual sludge and oversize are landfilled or treated with other technologies.

Special attention is paid to the contamination level of the soil before and after soil washing. The efficiency for each component is presented. These efficiencies are, of course, only valid in relation with the described project.

3.2. Case studies

3.2.1. Case study 1.

Characterization. In this case study the soil was contaminated with dredged sediment that was landfilled. The sediment was contaminated with cadmium coming from industrial activities, metallurgical as well as phosphate processing. The concentration of cadmium was 8-18 ppm.

The relation between particle size and contaminant concentration is shown in Fig 2. This figure was assembled from concentration analysis data derived from samples with a different content of fines.

1112

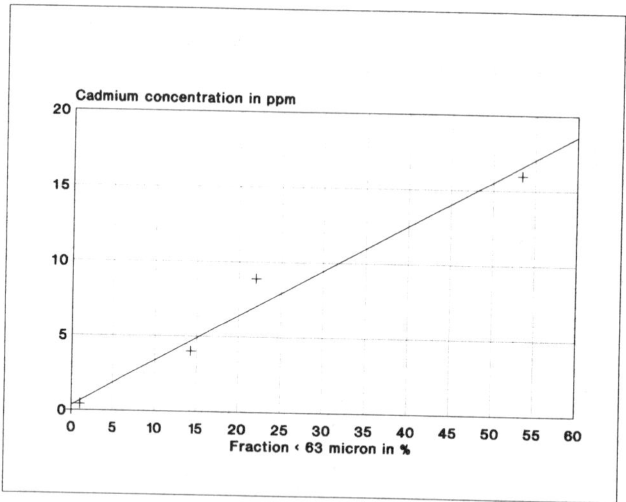

FIG. 2. Relation fines content and cadmium concentration for case study 1.

There is a clear relation between the content of fines (defined as fraction smaller than 63 μm) and cadmium concentration. In this case the contaminant appears to be associated only with the fines.
The cadmium might be readsorbed to the fine particles.

Process. The soil was wet screened to separate the coarse particles (clay lumps, gravel, waste material). After wet screening the fines were separated from the sand fraction using hydrocyclones. The fines were removed almost completely. The sand product was reused at a residual cadmium concentration of 0.4 - 0.8 ppm (removal efficiency approx. 95%).
Figure 3 shows the schematic flow diagram.

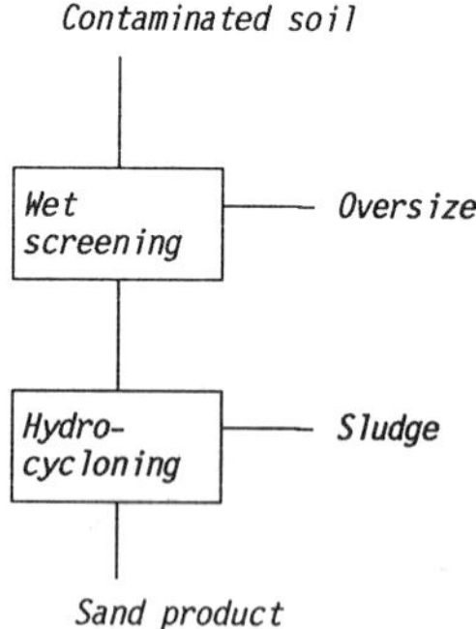

FIG. 3. Schematic flow diagram case study 1.

3.2.2. <u>Case study 2</u>.

<u>Characterization</u>. Shooting at clay doves with lead pellets (usually 1.1-1.3 mm) contaminated surface soils over a large area. The top layer of the soil was contaminated with lead shot, but the pellets also leached out. The leachate readsorbed to the soil, specifically to the clay fraction and the organic matter. The pellets also disintegrated due to mechanical forces and corrosion. Therefore also lead slices with a diameter smaller then 1.0 mm were found. Figure 4 shows the relation between lead content and particle size. In the plus 1,000 μm fractions the lead content varied between 5,000 ppm and 200,000 ppm. Higher concentrations than 5,000 ppm are not presented in the graph.

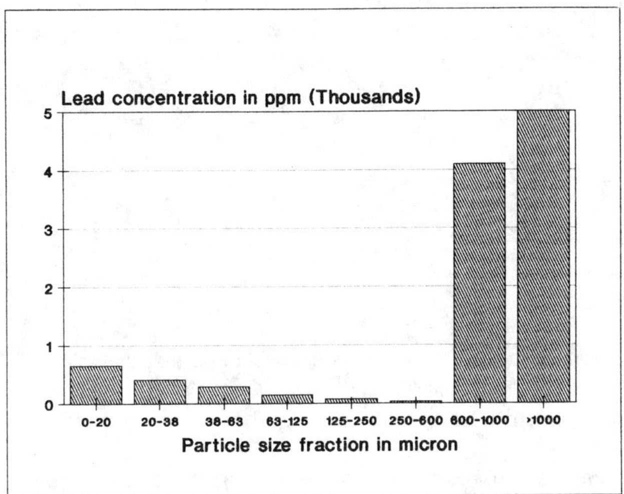

FIG.4.   Distribution of lead over the size fractions.

The lowest concentration of lead is found in the 63-600 μm fraction, which is also the main soil size fraction.

<u>Process</u>. The soil with an input lead concentration of 1,000-5,000 ppm was first wet screened to remove oversize material. In a next screening step the soil was screened to remove the lead pellets. The clay fraction was separated by use of hydrocyclones. The remaining soil was scrubbed and leached with slightly acidic solution and finally dewatered. The sand product contained 50-60 ppm lead (removal efficiency approx. 98%).
The schematic flow diagram is presented in figure 5.

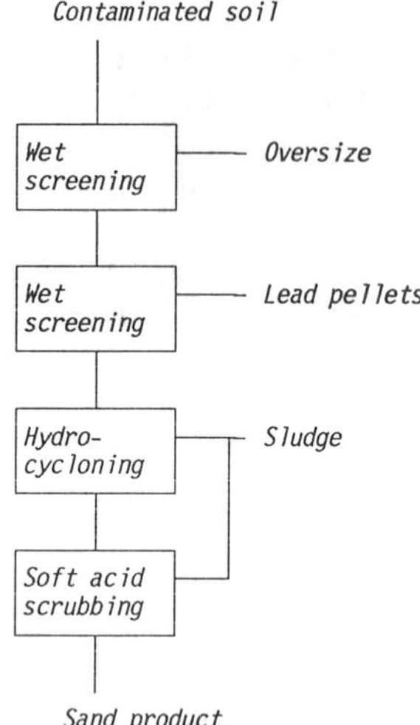

FIG. 5. Schematic flow diagram case study 2.

3.2.3. Case study 3.

<u>Characterization</u>. The historical analysis of this case study showed that the soil was contaminated with slag residue produced from metallurgical industrial activities. The slag material was easily visible in the soil and appeared to be lighter than sand. They were also disintegrated and leached out to the groundwater.

Soil samples were wet screened and in the different size fractions the contamination-content was analyzed. There turned out to be a clear correlation between the three contaminants. This indicates that the characterization of the three different contaminants are the same, except for the fines, where the relation was different probably due to different solubilities of the metal salts. The results are shown in figure 6.

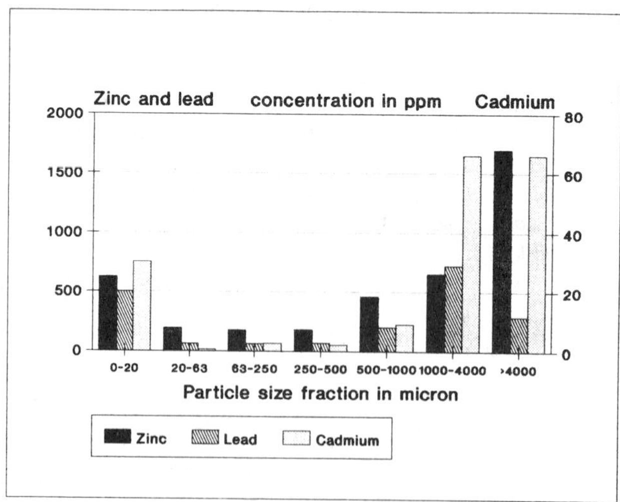

FIG. 6. Relation cadmium, lead and zinc content as a function of particle size.

The sand fraction (38-2,000 μm) was analyzed with Scanning Electron Microscopical analysis to identify the physical form of the heavy metal contaminants present. The results of this analyses are shown in table 2.

TABLE 2. Elemental composition of contaminants as identified by SEM-analysis.

Elemental analysis	Mineralogical description
Zn, Si, (Fe, K, Al)	Zn on silicate matrix
Zn, C, (Ca, Al, Si, Fe, Pb)	Zn on organic matrix
Fe, Zn, (Mn, Pb, Si)	Zn/Fe-compound
Zn, O, Si	pure ZnO with Si
Zn, Fe, P, O, (Pb, Al)	metallic Zn on iron-phosphate
Zn, Pb, Al, Si	Zn/Pb on Al/Si-matrix
Zn, S, Si, O	Zn-sulfide on SiO_2 matrix
Pb, S	Pb-sulfide
Pb, S, Ca, P, O	Pb-sulfide on Ca-phosphate
Sn	Sn-compound
Mn, Zn	Mn with Zn
Fe, S, (Al, Ni)	Fe_2S (pyrite)
Zn, S	Zn-sulfide
Fe, Zn	Zn/Fe-compound
W, Pb	W/Pb-compound
Mn, Si, (Zn, Pb, Fe)	Mn-compound on Si-matrix
Fe, S, Si, O	Fe_2S (pyrite) on quartz
Fe, Pb, Si (Zn)	Fe-compound

It was frequently found that zinc and lead contaminants were present in the form of different discrete particles in a "free" form or attached to organic (slag) material. From the mineralogical analysis it appeared that a great variety of zinc and lead containing compounds were present.

The presence of non discrete zinc and lead within slag particles could not be detected by the SEM analysis technique but chemical analysis showed that these slag particles also contained high concentrations zinc and lead in their structure.

The great variety of mineralogical forms in which zinc and lead were found is typical for thermal processes that produce slag materials and can be directly related to the original source of the contamination: pyro metallurgical industrial processes.

<u>Process</u>. The soil was screened to remove oversize material and large slag particles. Hydrocyclones were used to remove the fine particles. The low density slag particles were removed by gravity separation, other slag particles were removed by froth flotation. The concentration of the contaminants are listed in table 3. Figure 7 shows the flow diagram.

TABLE 3. Contamination levels of input and output of case study 3.

	Lead	Cadmium (ppm)	Zinc
Input	140-611	5-17	640-2822
Output	60	1.3	200
Efficiency (%)	57-90	74-92	69-93

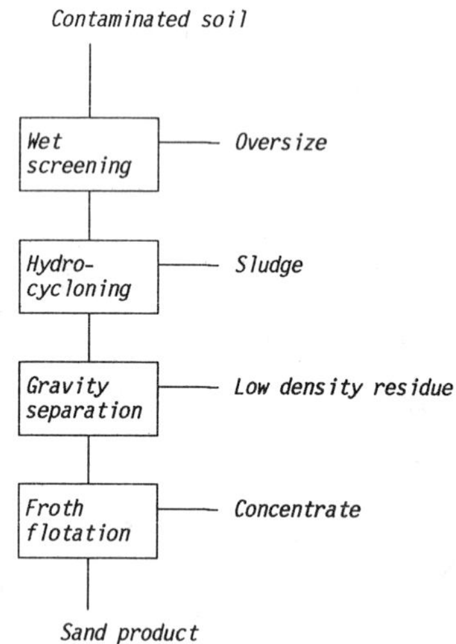

FIG. 7. Schematic flow diagram case study 3.

3.2.4. Case study 4.

Characterization. The soil in this case study is contaminated with coal ashes containing Polycyclic Aromatic Hydrocarbons (PAH). There is a side contamination of copper, lead and zinc due to traffic emissions and other urban activities. This case study focuses on the PAH occurrence and process performance because PAH was the critical contaminant. In The Netherlands it is common to use a selection of ten types of PAH as a total PAH content. This so called 10-PAH-content is used in this study. The PAH content averaged 85 ppm.

Figure 8 shows the distribution of the PAH over the size fractions. The size distribution of the sand is also presented.

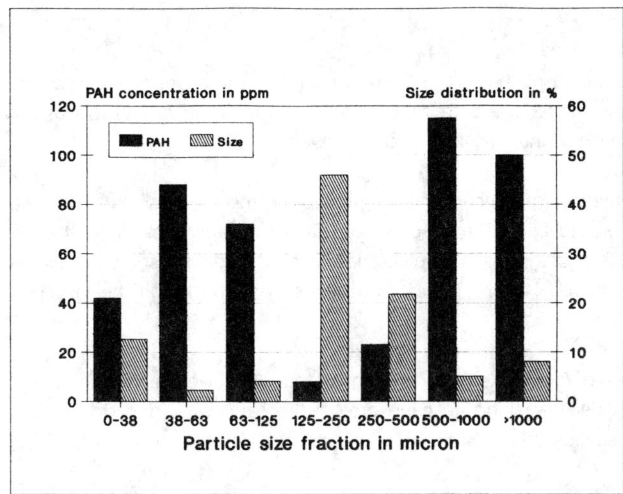

FIG. 8. Distribution of total PAH over the size fractions for case study 4.

The PAH concentration is the lowest in the main size fractions. The readsorbtion to the fines is not very large, but still important.

<u>Process</u>. The coarse particles were removed by wet screening. Special attention was paid to the hydrocycloning because of the high efficiency required. It was necessary to remove practically all the fines. The PAH containing low density particles were removed in gravity separation step. Other PAH containing particles were removed by froth flotation. Figure 9 shows the flow diagram of the used process. The concentration of the PAH is also given after each treatment step.

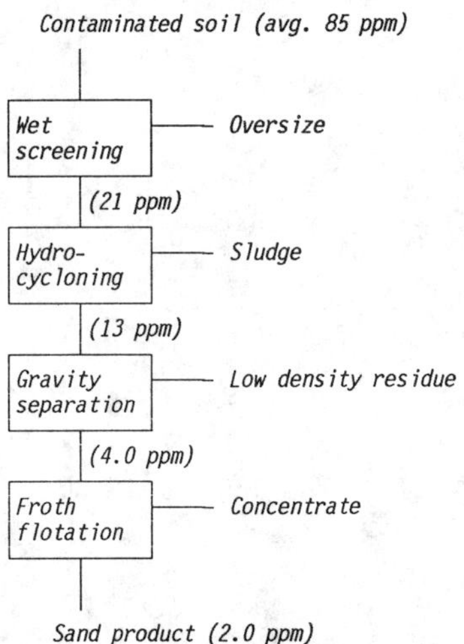

FIG. 9. Schematic flow diagram case study 4 with total PAH concentration.
The overall removal efficiency for PAH was 98%. The side contaminants were also removed to satisfactory levels.

4. CONCLUSIONS

Based on a good characterization of the soil and the contaminant it is possible to design optimized tailor-made flow diagrams. The characterization starts with a historical analysis and is completed with an extensive chemical and physical characterization.

For the case studies presented it was shown that differences in the physical occurrence of the soil and the contaminants resulted into different optimized flow diagrams. The case studies presented indeed have different flow diagrams. The difference in flow diagrams sometimes meant little adjustments in the process, in other cases it was also necessary to redesign the installation completely.

For the case studies presented removal efficiencies higher than 90% were achieved. These removal efficiencies are typical for soil washing. In each project the sand was treated to a contaminant level that made reuse possible.

The fundamental approach from characterization to tailor-made flow diagrams lead to optimum results in the realisation of soil washing projects. Using this concept of tailor-made flow diagrams, Heidemij has successfully treated more than 200,000 tonnes of contaminated soil.

DOCUMENTATION OF AN ENVIRONMENTALLY SOUND SOIL REMEDIATION

Dr. Christoph Munz and Dr. André Bachmann

MBT Environmental Engineering Ltd., 8048 Zürich, Switzerland

1. ABSTRACT

This article describes the overall approach and measures implemented to remediate an accidentally contaminated industrial site with pesticides and mercury. The individual elements, from problem quantification, over chemical risk assessment for defining clean-up levels, to the actual soil clean-up operation with a soil washing process, and their technical, scientific and time interrelationships are discussed.

2. INTRODUCTION

During the fire of a chemical warehouse, approximately 9000 kg of pesticides (mainly phosphoric acid esters) as well as 130 kg of mercury as organic compounds seeped into the ground with part of the fire-fighting water. The contamination in the soil is detectable up to depths of 11 meters; however, the severe contamination is generally only present up to 6 meters depth. *Capel et al.(1988)* provide a summary of the stored chemicals, their quantities, properties, and the observed environmental contamination after the accident.

3. OBJECTIVES

For the situation described above the overall goal is to develop and implement an approach leading to an integral solution that will guarantee the industrial reutilization of the area without any restrictions. The remediation philosophy is neither one aiming for a minimum nor a maximum solution, but for a solution that is technically, politically, and economically acceptable as well as environmentally sound. Based on the 5-year practical experience, the various elements involved in such a soil clean-up operation, their timing and their technical/scientific interrelationships will be presented and discussed.

F. Arendt, G.J. Annokkée, R. Bosman and W.J. van den Brink (eds.), Contaminated Soil '93, 1119–1126.
© 1993 Kluwer Academic Publishers. Printed in the Netherlands.

4. IMMEDIATE MEASURES

Various immediate measures were implemeneted to prevent further spreading of contaminants in soil and groundwater:

4.1 Groundwater protection

Decreasing the groundwater level at and around the area most affected by contamination and temporary shutdown of the closest drinking water pumping stations

4.2 Sealing of the ground

All surfaces not yet sealed at the contaminated site were sealed with an impermeable asphalt layer to prevent transport of contaminants into greater depths by rain.

4.3 Coverage

The former warehouse was covered with a tent

4.4.Drainage

Construction of a drainage ditch around the contaminated site to avoid lateral water flow through into the contaminated area.

5. RISK ANALYSIS

The objective of a risk analysis is to quantify the degree and extent of contamination and to estimate the risk or hazard potential to the environment, that is, for humans, animals, and plants. The decision whether or not a clean-up is necessary and to what degree, i.e. how clean is clean?, is based on the risk analysis. In the present case, the risk analysis required approx. 2 1/2 years and included the following aspects:

5.1 Extent and distribution of the soil contamination

By various means over 1000 soil samples were collected and analyzed. With these data 3 dimensional soil contamination charts were produced for the contaminated area of approximately 10'000 m². At the same time the main contaminants (indicator compounds) could be defined based on their physical-chemical and toxicological properties. The mass distribution of the two main contaminant types, total pesticides and mercury, in soil as a function of depth is shown in Figure 1. From Figure 1 it is apparent that the mercuric compounds were not transported as deeply as the purely organic pesticides. While approximately 75% of the mercury is present in the first upper meter, almost 50% of the organic pesticides are present at depths larger than 1 meter.

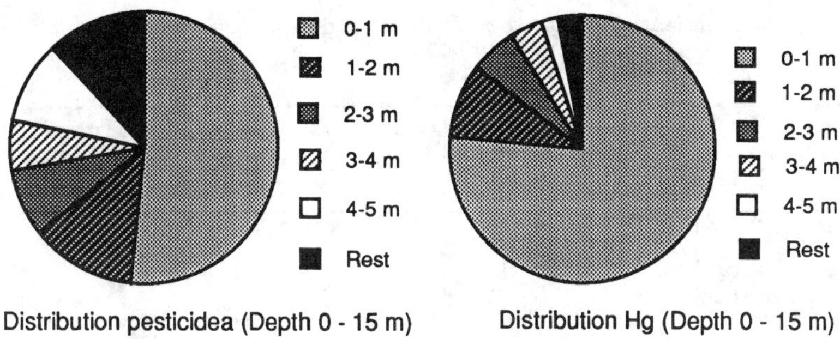

FIGURE 1. Mass distribution of the main contaminant types, mercury and total pesticides, in soil as a function of depth.

5.2 Environmental behavior of the main contaminants

Based on the compounds' properties and quantities present in the soil, three relevant contaminant classes could be defined: Phosphoric acid esters (Thiometon and Disulfoton), Oxadixyl, and organic mercury compounds. The environmental behavior and fate of these contaminants was investigated in depth.

5.2.1 Mercury compounds. The behavior of the organic mercury compounds was investigated with actual soil in laboratory experiments at conditions with various redoxpotentials. The results indicate that the production of methyl-mercury could be excluded under all possible conditions. Further it was shown that the original mercury compounds, ethoxy-ethyl-Hg-hydroxide and phenyl-Hg-acetate, had been transformed to a form that exhibits very strong sorption characteristics towards the soil particles, and hence, did not represent a siginicant risk to groundwater.

5.2.2 Pesticides. *Wanner et al.(1989)* describe the chemodynamic behavior of thiometon and disulfoton in aqueous media. The required data to assess the behavior of the organic pesticides in the unsaturated (moist) soil was obtained from further laboratory investigations (*Rich et al.,1993*). Table 1 summarizes the most important data obtained regarding the sorption properties (soil-water partition coefficient, K_d (l/kg)) and abiotic degradation (half-life, $t_{1/2}$) of the main pesticidal contaminants. The values shown in Table 1 are the corrected values applicable to the average field conditions.

TABLE 1.   Environmental behavior of pesticides in the unsaturated soil at field conditions (10°C)

Compound	K_d ($l \cdot kg^{-1}$)	abiotic degradation $t_{1/2}$ (years)	Solubility ($mg \cdot l^{-1}$)
Thiometon	0.9	0.8	200
Disulfoton	2.4	2.0	25
Oxadixyl	0.5	10	3400

5.3 Toxicological risk assessment

This quantifies the extent of potential risk to man through direct (inhalation, eating) or indirect (via food, drinking water) ingestion of contaminants (Steigmeier und Bachmann 1990)

5.4 Hydrogeology, modeling

Contaminant transport was estimated with two models. A first model was developed to simulate contaminant transport in the unsaturated soil zone. A groundwater contaminant transport model was calibrated and validated under consideration of the local geological and hydrogeological situation, various scenarios of regional and local groundwater use, and the contaminant properties. Figure 2 shows, as an example, the extent of groundwater contamination of pesticides for a "worst-case" hydrogeological scenario.

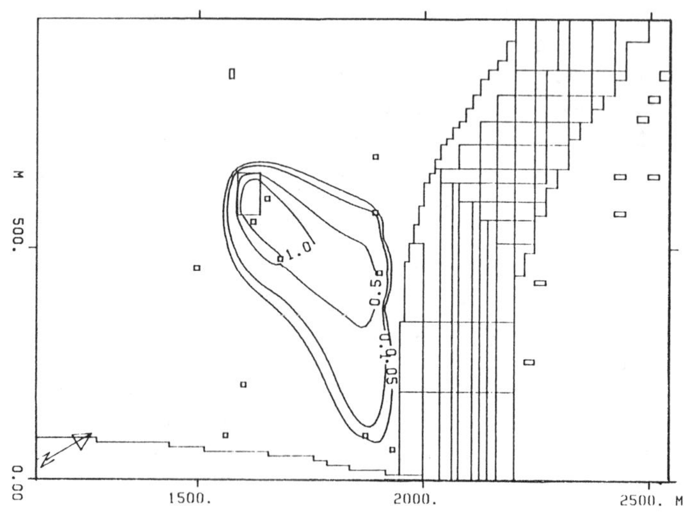

FIGURE 2: Calculated groundwater contamination (total pesticides) for a "worst case" hydrogeologic scenario (concentration values in µg/l).

<u>5.5 Residual soil contamination - how clean is clean?</u>

The risk analysis conducted showed that the largest potential risk was groundwater contamination by pesticides. The residual mass of contaminants in the soil after clean-up can be estimated by combining the expected removal efficiency with a given soil treatment process and various structural measures limiting the water infiltration potential. This is the basis to meet the regulatory requirements (outside the industrial area groundwater is considered drinking water).

6. SOIL REMEDIATION PROCESSES

Parallel to the chemical risk analysis various in-situ and on-site soil remediation processes, thermal treatment methods, and chemical immobilization methods were evaluated experimentally. A wet mechanical process with a so-called flotation step was considered the most effective - from a technical, energetical, time for implementation, and cost-effectiveness point of view - and hence, optimal process.

The soil treatment plant with a throughput capacity of 15 tons per hour was tested and optimized with various categories of contaminated soil during a 4-month pilot phase. The clean up operation started in July 1990.

7. THE CLEAN-UP PROCESS IN THE SOIL WASHING PLANT

The contaminated soil is excavated and sampled, stored in airtight containers (due to the odor), analyzed, and transported to the soil remediation plant. A schematic diagramm of the flow sheet is shown in Figure 3. In the plant, gravel is first separated by sieving. Gravel, being relatively easy to treat, is washed in a countercurrent drum washer with process water. The sand and fines are fed with the washing water into the hydrocyclone to separate the fines. The fines, having a large specific surface area, contain the majority of the contaminants. This fraction is thickened in a static thickener, and subsequently filled in barrels for disposal as toxic waste. After the hydrocyclone, the sand passes through a series of attrition mills, where the grains are rubbed against each other to release the contaminants adsorbed on their surface. Then, the contaminants are separated from the sand in the flotation cells with the aid of surfactants. The flotation foam is thickened with the fines, while the sand is rinsed with clean water. Approximately 90 % to 95 % of the contaminants are removed from the sand and gravel with this method.

The process water of the various steps is collected, subjected to oxidative treatment (*Munz et al.,1992, Egli et al.,1992*), and recirculated. Only the water which leaves the plant with the gravel, the sand, and the concentrates needs to be replaced.
The gravel and the sand (approx. 95 % of the soil material) are reincorporated at the site after passing the appropriate quality control.

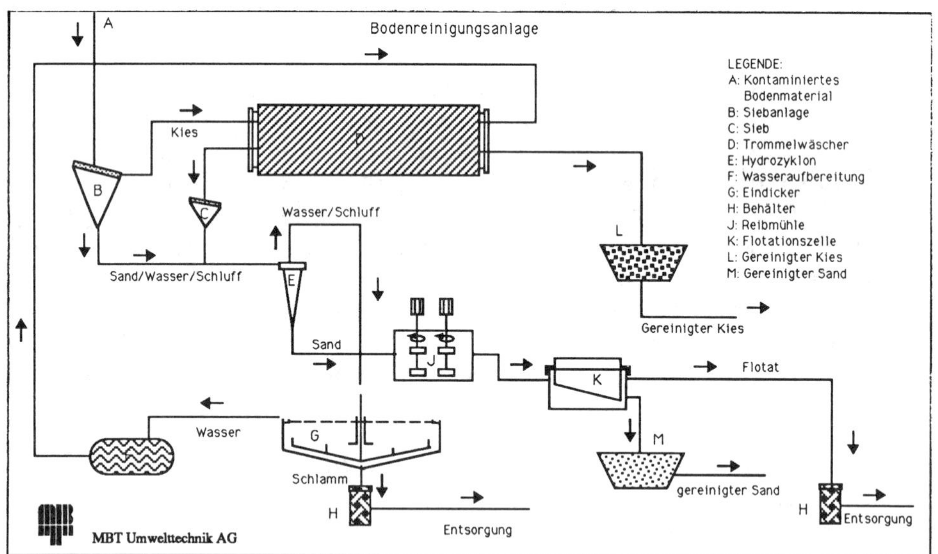

FIGURE 3. Schematic diagram of the soil treatment plant.

8. DISPOSAL
The highly mercury-contaminated residues are solidified and deposited at a former salt mine in Herfa-Neurode, Germany. The non-mercury-containing residues with high pesticide contents could not be incinerated in municipal sludge incineration facility due to technical reasons. Finally all residues had to be disposed off at Herfa Neurode.

9. LOGISTICS
The entire contaminated site was divided into three-dimensional modules and input into a computer data bank. The excavation occurs containerwise, and a composite soil sample is taken and

analyzed for each container. Each container is labeled with the date, location and depth of excavation, and a color code. The color code describes the type and degree of contamination of each container, such that at a later stage each soil container can be treated at the corresponding optimal conditions in the plant. All activities and the entire data are stored in a central computer and continuously updated, so that the whereabouts of every soil container can be determined at all times (excavation, storage 1, plant, storage 2, reincorporation, etc). This guarantees a flawless coordination of the various activities (excavation, soil washing, analytics, disposal, and reincorporation). The clean-up efficiency of the plant and the material flow analysis (including contaminants) of the overall system are also recorded automatically with the logistics program developed by the project team.

10. ANALYTICAL INVESTIGATIONS

The chemical analyses (pesticides, mercury) are conducted on site within four days. As a precautionary measure, control analyses are performed regularly by an external laboratory. Soil which conforms to the requirements regarding contaminant content for reincorporation, undergoes an additional acute toxicity test (*Gälli et al.,1992*). This test indicates possible contaminations not detected by the chemical analyses.

11. SAFETY MONITORING

All activities during the excavation and in the soil remediation plant are monitored by the safety team who perform air and dust measurements regularly. All employees involved in this project undergo regular medical examination.

12. COSTS

It is estimated that a total of 40'000 tons of soil were excavated of which approx. 15'000 tons had to be treated. The costs for the _entire_ clean-up operation will be on the order 60 million Swiss Francs. The process engineering costs by itself are about 300 Swiss Francs per ton of treated soil.

13. OUTLOOK

The soil remediation plant started operating in July 1990. In September 1992 the last batch of soil was treated in the plant. Hence, the remediation was succesfully concluded such that as of January 1st, 1993 the area is again available for industrial use.

ACKNOWLEDGEMENTS
This paper is being presented in the name of all employees of MBT Environmental Engineering Ltd., Switzerland that participated and led to the successful conclusion of this project.

REFERENCES
Capel, P.D., et al. (1988). Accidental Input of Pesticides into the Rhine River. <u>Environ. Sci. & Technol.</u>, 22(9): 992-996.

Egli, S., et al. (1992, February). Oxidative Treatment of Process Water in a Soil Decontamination Plant: II. Pilot Plant and Large Scale Experiences. Proceedings of the 2nd Intl. Symposium <u>"Chemical Oxidation, Technology for the Nineties"</u>, Nashville TN, in press, Technomic Publishing Co. Inc.

Gälli, R., et al. (1992, in press). Evaluation and Application of Aquatic Toxicity Tests: Use of the Microtox Test for the Prediction of Toxicity Based upon Concentrations of Contaminants in Soil. <u>Hydrobiologia.</u>

Munz, C., et al. (1992, February). Oxidative Treatment of Process Water in a Soil Decontamination Plant: I. Laboratory Studies. Proceedings of the 2nd Intl. Symposium <u>"Chemical Oxidation, Technology for the Nineties"</u>, Nashville TN, in press, Technomic Publishing Co. Inc.

Rich, H.W., et al. (1993). Abiotic Degradation of various Pesticides in a Contaminated Soil. Manuscript in preparation.

Steigmeier, M. und A. Bachmann, (1990). Toxikologische Betrachtungen zur Einhaltung von Grenzwerten bei Bodensanierungen. In: F. Arendt, M. Hinsenvelt und W.J. van den Brink (Eds.), <u>Altlastensanierung '90,</u> (I): 289-295, Kluwer Academic Publishers

Wanner, O., et al. (1989). Behavior of the Insecticides Disulfoton and Thiometon in the Rhine River: A Chemodynamic Study. <u>Environ. Sci. & Technol.</u>, 23(10): 1232-1242.

SOIL DECONTAMINATION CENTRE BOCHUM

Dr. Ulrich Törk

BSR-Bodensanierung und Recycling GmbH, D 4630 Bochum

## 1.	ABSTRACT

The Soil Decontamination Centre Bochum shall contaminated Soils and other bulk materials, which otherwise are to be deposited, decontaminate. The decontaminated materials are to be reused in remedial-action-projects.

This article describes the soil decontamination centre Bochum and the process of developing the application for the official approval.

The application for the official approval for the soil decontamination centre Bochum will be applied to the Regierungspräsident Arnsberg at the end of 1992.

## 2.	TASK OF THE SOIL DECONTAMINATION CENTRE

The BSR-Bodensanierung und Recycling GmbH has set with the soil decontamination centre the goal, to clean contaminated bulk materials, so that these materials can be recycled at the original site or sometimes to another. The sites are to be used after the soil decontamination for several purposes.

The centre will be installed in the Ruhr-area, the commuter belt is 40 kilometres around the location of the centre. In the commuter belt we find soils with a very high rate of silt. Because the contaminating agents in the soil mostly descend from coal mining and processing, a thermal treatment process is the best method to deal with these contaminated soils. So an thermal treating plant is planned in the application for the official approval.

We think, that we need a strategy using more treating modules in the soil decontamination centre, in order to decontaminate the soil on an high ecological level. So the centre shall be supplemented in the future with an biological and an extracting technique for soil decontamination. These techniques are not included in the present application. When these techniques are accessible, even sorted out soils with suitable structures and contamination's or particularly decontaminated materials will be cleaned in a proper way. Transport of soil to other plants and the joined emissions and risks will be avoided. Additional the recycling principle can be better kept.

The application for the official approval includes the infrastructure, temporary storage and the thermal treatment plant. The transport of soil to and from the centre will not be done by the centre. The standards of the German ´TA Abfall´ and ´TA Luft´

F. Arendt, G.J. Annokkée, R. Bosman and W.J. van den Brink (eds.), Contaminated Soil '93, 1127–1132.
© 1993 *Kluwer Academic Publishers. Printed in the Netherlands.*

are fulfilled, the thermal treatment plan additional fulfils the standards of the German '17. BImSchV'.

The amount and origin of the contaminated soils depend on the sites, which are to be remedied. Criteria for the acceptation of contaminated soil are fixed in order to guarantee the best operation of the thermal treatment plant and the decontamination efficiency.

3.   DEMAND-ANALYSIS

In the Ruhr area is a lot of operation and production places located, which have been shut down because of economic changes. These places are often contaminated sites as a result of their operation. These sites are to be remedied before they can be used for the settlement of new companies, for house building or other urban development.

Among the contaminated soil even materials from accidents, which led to soil contamination and other mineral bulk materials as materials from road-sweeping machines, used sand from pavements, sand from sand mold casting and mineral materials from rebuilding of sewerage systems should be decontaminated.

Table 1 shows the potential amount of contaminated soil and other bulk materials. In the first column we find the types of materials, in the second column is the potential amount for the area of the town of Bochum an in the third column the amount for the whole commuter belt of 40 kilometres around the plant. As we intend to get the official approval for the thermal treatment plant for 15 years, we have calculated the amount of contaminated soil per year by dividing the total amount through 15 years.

4.   SELECTION OF THE SITE OF THE SOIL DECONTAMINATION
          CENTRE

For the soil decontamination centre an comparing investigation for the best site, was carried out. It included all trade and industrial areas in Bochum.

The first step in the selection process was to find out the possible sites. These sites must be large enough, when the soil decontamination centre is to be installed. Considering protection from noise and pollutants the soil decontamination centre can only be build in an industrial area. Only the five candidates shown in Table 2 fulfil these conditions and were taken for the comparing and assessing selection process.

After fixing the five possible sites followed an detailed investigation of the sites and their neighbourhood. This investigation led to an assessing comparison, from which the recommendation of the best site was derived.

Table 1: Potential amount of contaminated soil and bulk materials

<u>Bulk Materials</u>	<u>Town of Bochum [t/a]</u>	<u>Total commuter belt [t/ a]</u>
Contaminated soil	220.500	1.070.000
Soils from accidents	200	1.500
Bulk materials from road-sweeping machines	2.800	116.000
Bulk materials	-	3.700
Mold casting sand	4.200	283.000
Soils from building-actions	10.200	585.000
Materials from reconstruction of sewerage systems	4.200	232.000
Pavement sand	2.500	146.000
Sandblast materials	1.700	97.000
Dragged materials	2.500	146.000
<u>Sum of bulk materials</u>	<u>250.000</u>	<u>2.680.200</u>

Table 2: Candidates for the comparing site investigation

Site	Physical planning (GEP+)	Total area	Un-used area	Owner
Schmaler Hellweg	GI	15,3 ha	11,6 ha	Stadt, Privat
Lothringen IV	GI	24,3 ha	21,0 ha	EBV, Stadt
Rombacher Hütte 2	GI	11,4 ha	11,4 ha	Stadt
Engelsburg	GI	15,2 ha	9,1 ha	Krupp
Obere Stahlindustrie 2	GI	37,3 ha	20 ha	LEG

+ Industrial area

The site "Obere Stahlindustrie 2" came out to be the best site for the installation of the soil decontamination centre in the town of Bochum:

- comparatively good, hardly to see in the middle of an connected big industrial area,

- biggest distance to the next residential area,

- good regional traffic connections, direct connection to an noise-protected motor way feeder road. This leads to a low noise pollution for the residential areas in the neighbourhood,

- comparatively lowest pollution load,

- no nature or water conservation area in the neighbourhood,

- only at this site one can follow the principle of physical planning "extension is better than new building". Using one of the other four sites one has to consume ökological valuable at this time in part agricultural or for recovery used green areas,

- in the direct neighbourhood are supplementary lines of business.

5.   TECHNICAL AND ORGANISATIONAL STRUCTURE OF THE SOIL
          DECONTAMINATION CENTRE

In the soil decontamination centre contaminated soils and bulk materials from the Ruhr area shall be treated under ökolgical and ökonomical best conditions. Therefor an technical and organisational structure is necessary, in which different soil decontamination techniques, which are useful for different soil- and pollution-structures, are connected to an central installation. This principle is shown in figure 1.

The organisation of the soil decontamination centre fulfils the rules of the "TA Abfall". The organisational unit ´supervision´ summarises the ´Input- and Output-Supervision´ and the ´Laboratory´.

The organisational unit ´Central technique´ summarises the ´supply and draining´ with energy, water and telecommunications, ´Extra installations´ and the ´Temporary storages´ for contaminated and decontaminated soils.

The storage for contaminated soil will be installed in an former rolling-mill-hall. The supplied soils and bulk materials are stored in closed boxes. These boxes are equipped with an double and controllable ground-sealing to prevent ground-pollution. The exhaust air from these boxes is cleaned by and adsorptive purification equipment. This store has an capacity of about 10,000 m³.

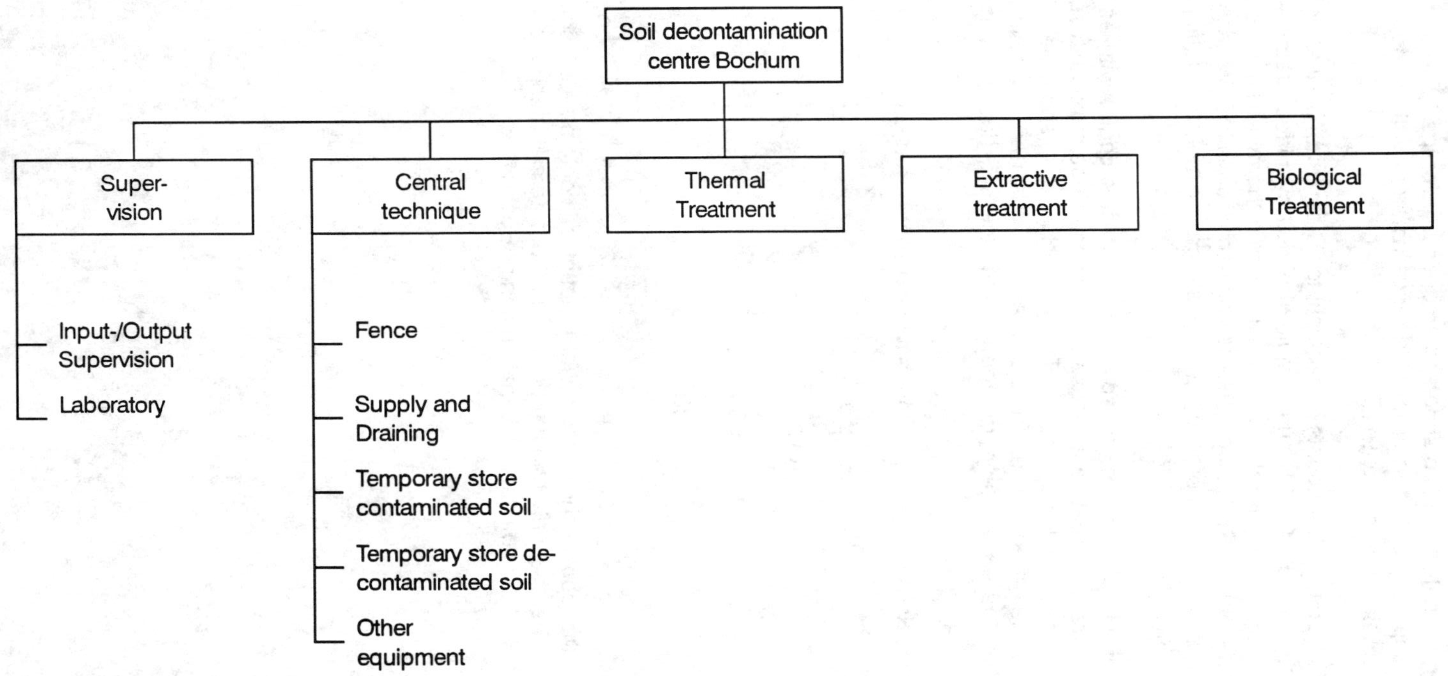

Figure 1: Structure of the soil decontamination centre Bochum

The storage for decontaminated soil is planned as an open air storage. The storage will be done even in separate boxes in order to arrange different soils in separate boxes. The capacity of this store is about 12,500 m³.

In the Ruhr area we find soils with amounts of silt until to 80 percent. They are often contaminated with BTX-Aromates, PAH and complex bonded cyanides. Because of these reasons is in the application for the official approval an thermal treatment unit according to the Ecotechniek/ RUT Ruhrkohle Umwelttechnik GmbH process included. this plant has a new engineered exhaust gas cleaning unit and fulfils the standards of the '17. BImSchV'.

For the future is planned to supplement the soil decontamination centre with an micro biological and an extractive treatment. These techniques will be applied in separate procedures for the official approval. These Techniques shall be used depending on their special opportunities for the treatment of sorted out soils and bulk materials. In table 3 are the planned capacities for the soil decontamination techniques shown.

Table 3: Treatment capacities in the soil decontamination centre Bochum

Treatment technique	Capacity [t/ a]
Thermal treatment	150.000
Extractive treatment	100.000
Biological treatment	25.000

The comparison with the above shown potential amount of contaminated soil an other bulk materials shows, that only about 9 percent of the potential amount can be treated with the capacity of the soil decontamination centre Bochum.

INVESTIGATION AND REMEDIATION OF AN ARSENIC AND COPPER CONTAMINATION

Dr. Michael Altmayer
Dr. Reinhard Röder
Bayer. Landesamt für Wasserwirtschaft
Lazarettstr. 67
D-8000 München 19

Dr. Johann Rietzler
Dr. Rietzler & Heidrich GmbH
Rüberstr. 4

D-8500 Nürnberg 50

HISTORICAL INVESTIGATION

While investigating a mineral oil contamination in a northern bavarian
town high concentrations of arsenic in groundwater were discoverd.
Through the examination of historical records it was ascertained that up
to 1915 a local firm produced paint, to impregnate ships, on the basis
of the so-called "Schweinfurter Grün" (copper-arsenite-acetate). The
company deposited their misproduction in the ground in various places
inside and outside the town.

INITIAL INVESTIGATIONS AND IMMEDIATE ACTIONS

The initial investigations and measurements of the extent of damage revealed were carried out. In the area of the former factory arsenic and
copper contaminations in the soil were found in the magnitude up to
10 000 mg/kg. The contamination reaches from the surface to the
aquiclude in ca. 6-8 m depth.

The first reaction was to close down all house wells still in use. The
next important step was to tear down the old factory storage sheds located in the midst of inhabited houses. From the ruins and in the directly
affected soil concentrations up to 17 000 mg/kg arsenic and copper were
measured.

DETAIL INVESTIGATION

In the course of the detail investigation 14 monitoring wells were drilled in order to determine the extent of damage to the groundwater.

Through the pumptests on the monitoring wells the geological and hydrogeological data (direction of groundwater flow, groundwater gradient,
groundwater thickness and aquifer yield etc.) were determinated. In the
groundwater samples arsenic concentrations up to 5 600 mg/l (!) and copper concentrations up to 0.7 mg/l were found. The immense pollutant input into the 4 m thick aquifer caused a decrease in the permeability.

REMEDIATION

The remediation will be executed in stages. The first step will be a
groundwater remediation with oxidation, flocculation, precipitation and
filtration that will be in operation in 1993 (Fig.).

F. Arendt, G.J. Annokkée, R. Bosman and W.J. van den Brink (eds.), Contaminated Soil '93, 1133–1134.
© 1993 Kluwer Academic Publishers. Printed in the Netherlands.

1134

In the second step the soil remediation will be started. The undeveloped
or low-value developed areas should be excavated down to the basis of
the quaternary unconsolidated rock and cleaned with a method which has
still to be found out. In consideration is an electrolytic process, a
wash process or a combination of both. Corresponding pilot tests have
already been conducted with promising results. The soil remediation
plant should be located next to the groundwater remediation unit so the
resulting highly contaminated process water can be easily treated.

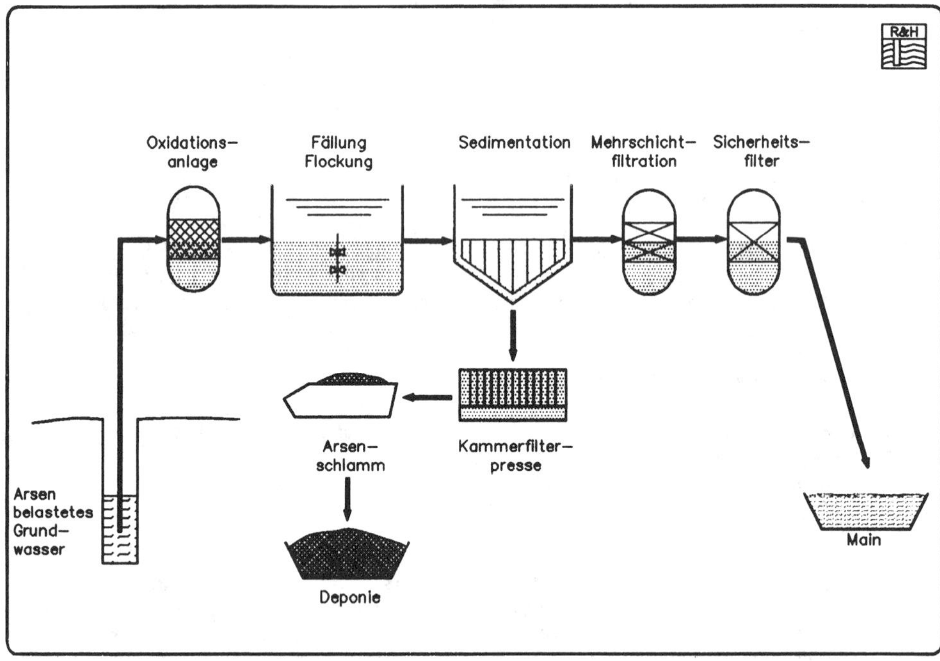

Fig.: Flowchart of the planned groundwater remediation plant

PILOT AND LARGE SCALE EXPERIENCES IN THE IN-SITU BIOREMEDIATION OF AN REFINERY-SITE POLLUTED WITH HYDROCARBONS

Dr.-Ing. Gerhard Battermann, Ricardo Fried,
Matthias Meier-Löhr, Dr. Peter Werner[*]

Technologieberatung Grundwasser und Umwelt GmbH,
Kurfürstenstraße 87 a, 5400 Koblenz

[*] DVGW-Forschungsstelle am Engler-Bunte-Institut der
Universität Karlsruhe, Richard-Willstätter-Weg 5,
7500 Karlsruhe

## 1.	INTRODUCTION

About 25 years ago a restoration of a hydrocarbon
spill was successfully terminated when the free phase was
removed from the groundwater surface [1]. The risk assessment for the groundwater after such activities was still
very high due to the remaining hydrocarbons attached to the
soil. A permanent dissolution of the pollutants into the
groundwater cannot be excluded. Today the authorities raised
the requirements and standards for restoration strategies in
order to reduce the risk assessment as much as possible.

The concept, the preparation and the beginning of the
in-situ bioremediation at the former refinery site are described and discussed in the paper presented.

## 2.	HYDROGEOLOGICAL CONDITIONS AND GROUNDWATER CONTAMINATION

The subsurface in the not filled-up parts consist of
flood originated loam of 0,5 to 1 m thickness. Underneath
there is nearly about 20 meters a mixture of pleistocene
sands, gravely sand or sandy gravel intersected in different
depths with silt layers of small areal extent.

The upper aquifer is formed by the pleistocene layers.
The permeability is in the range of 7E-4 to 2E-3 m/s. An
lower aquifer is composed of about 100 meters of pliocene
sands intersected with clay, silt and gravel lenses. The two
aquifers are separeted by an hydraulically effective
aquiclude in an depth of approximatly 20 m.

The contamination of the soil and the groundwater predominantly was caused by a leakage of a subsurface supply
pipe. The discharged mixture of hydrocarbons was a intermediate of gasolineproduction consisting mainly in BTEX-aromates. The solubility in water is relatively high in a range
of about 250 mg/l. Hydraulic measures were done immediately
after realizing the spill in 1974 in order to prevent spreading out of the environmental pollution.

The groundwater was discharged by a changing number of
wells with removal of free oil. The measures are continued
till the beginning of the bioremediation in 1991.

F. Arendt, G.J. Annokkée, R. Bosman and W.J. van den Brink (eds.), Contaminated Soil '93, 1135–1144.
© 1993 *Kluwer Academic Publishers. Printed in the Netherlands.*

The free oil on the groundwater surface was nearly completly removed in 1986, when the planning for the bioremediation started. About 2000 t of oil are recovered.

Only the immobile residual concentrations of oil in the pore space are remained in the subsurface. They cover in a depth between 6 to 10 m only in the saturated zone an area of about 7 ha. By mean concentrations of 1 to 1,5 g/kg the total mass of oil is calculated to about 500 tons in an volume of 500.000 tons of polluted soil. Originated by this immobile saturations in the soil in the groundwater concentrations from 10 to 20 mg/l of hydrocarbons are observed.

3. REMEDIATION STRATEGY

A remediation strategy had to be developped in order to reduce the costs of the permanent pump operations. Moreover an clean-up method with an acceptable expenditure of time had to be found for a reuse of the area for other purpose. Some conventional remediation techniques could be excluded due to the widespread contamination located deep below surface. The most efficient and most economic solution seemed to be a in-situ bioremediation, activating the almost at the site proved hydrocarbon degrading microflora, which mineralizes the contaminants attached to the soil and diluted in the groundwater. Experiences in enhanced bioremediation of a hydrocarbon polluted area of a previous large scale case were available [2] and were helpful in the development of the clean-up strategy.

In a preinvestigation study it could be proved, that the hydrocarbons (mainly BTX-compounds) are fairly biodegradable under aerobic and denitrifying conditions by the indigenous microflora [5].
As further optimum requirements are to mention the high hydraulic conductivity in the subsurface and the absence of hydrocarbons in free phase on the groundwater surface.

The clean-up stragety elaborated consists in the following action steps:
The intensive investigations using groundwater models [a.g. 8, 14] ended up in a conception of hydraulic activities covering a complete and effective groundwater transport in all contaminated areas. By 16 wells about 520 m^3/h groundwater are extracted, 400 m^3/h are reinjected via infiltration ditches, the mean residence time in the remediation area is in the range of about 40 - 60 days. A limitation against spreading of the pollution is realized by a netto discharge of 100 m^3/h and an protective infiltration of 20 m^3/h (fig. 1 and 2).

The enhancement of the bioremediation in the groundwater consists in the increase of the missing electronacceptors and the addition of nutrients as there are N- and P-compounds. Due to the large amount of contaminants at the site the dissolved oxygen in the water reinjection is not sufficient, and the use of nitrates as an additional and effectful electronacceptor was necessary.

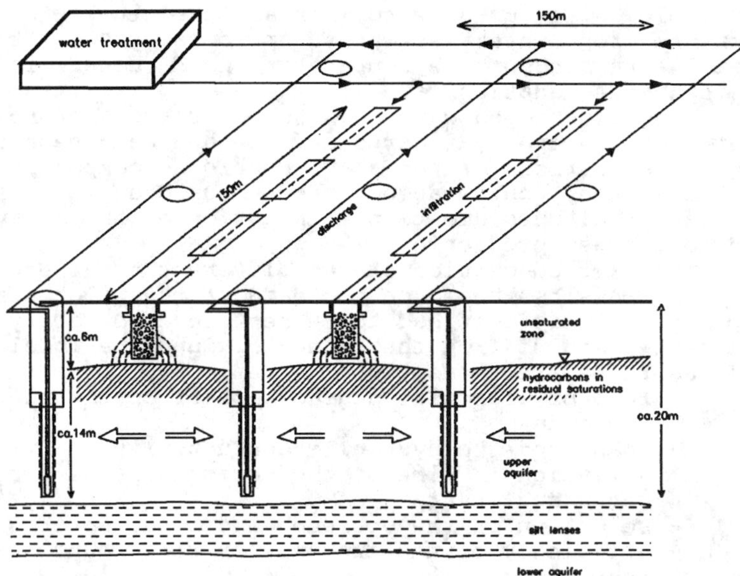

Figure 1 : Schematic of installations for large scale remediation

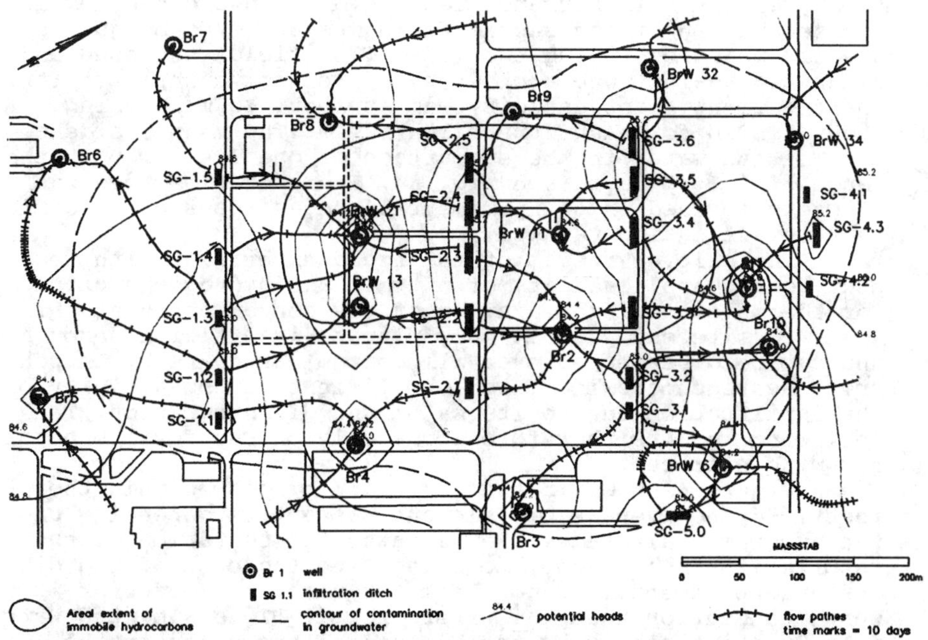

Figure 2 : Flow conditions of the groundwater

As complementary measure to increase the kinetik of degradation the groundwater was warmed up to 18 - 20°C. It is known that the raising of the temperature has a beneficial effect on the biodegradation rate.

The discharged water has to be treated before reinfiltration to remove hydrocarbons, iron and manganese as well. The concentrations of the above mentioned compounds are so high that they would cause a tremendous head loss of the infiltration ditches due to precipitation of the heavy metals and to biomass production.

The effective conception of the different treatment steps was only possible based on the data of the pilot scale field experiment. This test had to be performed to verify and evaluate the results from the laboratory and the modelling system as there are to mention:

- rate of microbial degradation of contaminants at the site
- change of the hydraulic conductivity
- variation in time of the decisive parameters in water quality
- selection of the optimum electronacceptors
- efficiency of the increase of the groundwater temperature

4.   PILOT SCALE FIELD TEST

The pilot plant was constructed in a way to enable a parallel and separated test for two different electronacceptors (Testfield I: Nitrate; Testfield II: Hydrogenperoxide). The separation of the second field was achieved by just hydraulic measures, using two inner flow fields enclosed and separated against one another by a protective infiltration. The treatment plant capacity was at about 20 m^3/h using about 5 m^3/h for each inner flow field. The mean residence time of the water in the subsurface of one field was in the range of 3 - 5 days (see fig. 3). The operation time was 27 weeks devided in 6 experimental stages focusing on different investigation targets.

The allover result of the field experiment with respect to the applicability of nitrate and hydrogenperoxide for the in-situ bioremediation of the contaminants at this specific site showed that only nitrate is efficient enough and can be distributed over all contaminated areas. In contrast hydrogenperoxide is only efficient in the vicinity of the infiltration due to its catalytic disintegration. The following results therefore focus only on the results obtained with nitrate.

Figure 4 shows in the direction of flow within the testfield the average of relevant parameters depending on the experimental stages. For a better presentation of the results it was necessary to select the sequence of the different experimental stages back- or foreward. Moreover it is worth to mention that the water of well B1 is a mixed water only 50 % influenced by the biological processes in the testfield. Therefore the concentration of hydrocarbons, iron and manganese is higher and that of nitrate and nitrite is lower.

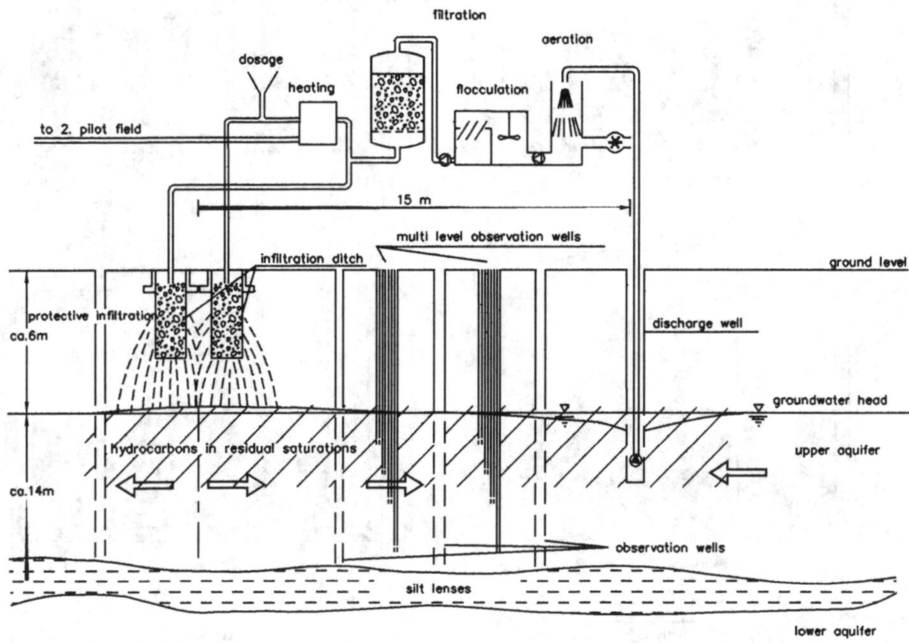

Figure 3 : Schematic cross-section of pilot plant

When starting the test runs (experimental stage 2-tracer experiment) the environmental conditions were reducing due to the high content of dissolved aromatic hydrocarbons resulting in high concentrations of iron and manganese and a very low redoxpotential. Moreover methan and ammonia could be proved in the water in the range of mg/l.
The beginning of experimental stage 3 is defined with the addition of nitrate which immediately changed the environmental conditions. Dissolved reduced iron is no more detectable due to the increase of the redoxpotential. Regarding manganese the concentration was reduced to half.

Of a special importance in balancing the application of nitrate in this case is the evaluation of the hydrocarboncontent. Soon after the addition of nitrate the flushing water was almost free of hydrocarbons, which means that the degradation rate is higher than the solubilization rate of the immobilized residuals causing up to 20 mg/l hydrocarbons before the measure.

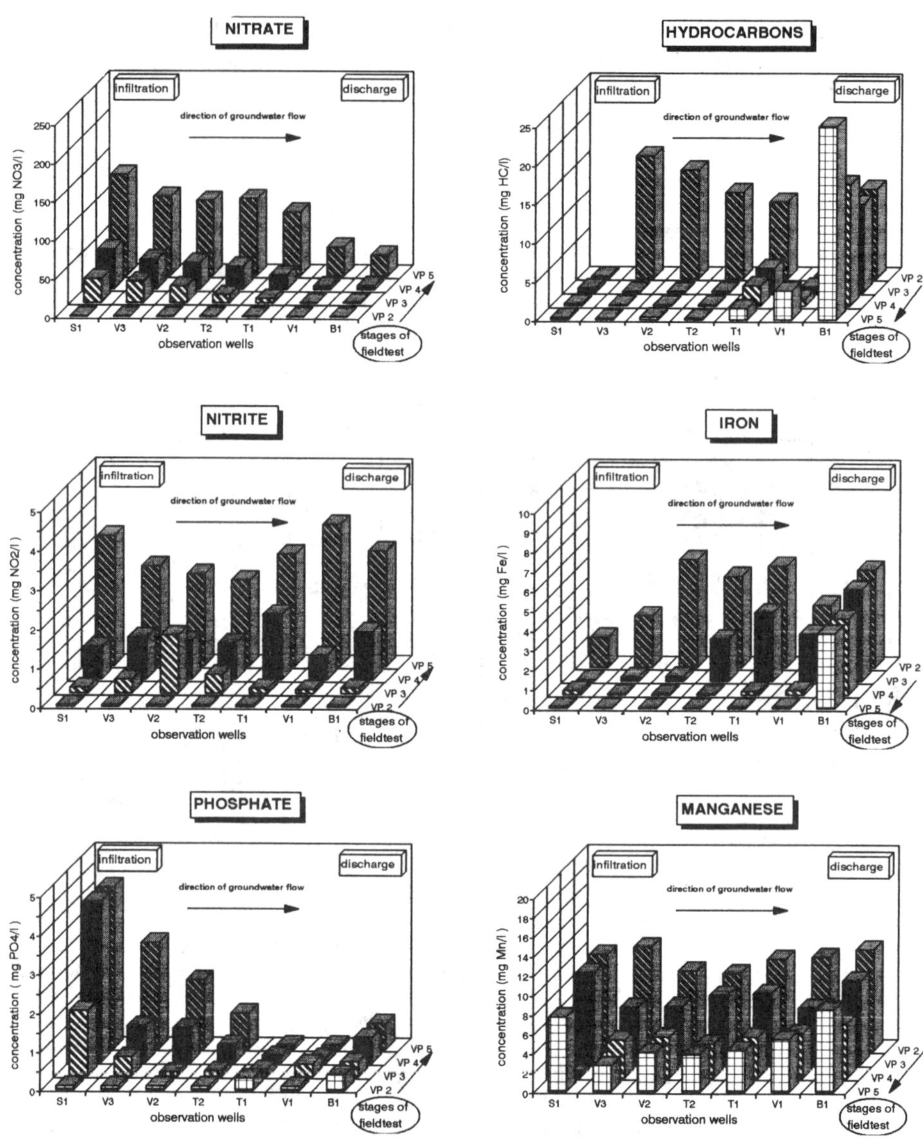

Fig.4 :   averages of chemical parameters in groundwater during the pilot scale field test

The laboratory experiences [10] showing the metabolization of the hydrocarbons proved at the site by denitrification could clearly be verified under field conditions as well. Along the path of flow the hydrocarbons were biodegraded, nitrate was consumed and nitrite as an intermediate was formed. The nitrite formation is in the range of 1 mg/l during the experimental stages 3 and 4. Even the increase of the nitrate concentration (up to 200 mg/l) at the end of the experiment caused only a slight increase of nitrite up to 4 mg/l.

It was tried to maintain optimum environmental conditions for the bacteria in the subsurface concerning the N- and P-compounds. To provide the microflora with sufficient phosphates created problems due to precipitation reactions.

According the mass balance based on nitrate consumption the degradation rate was between 1 to 2 g of hydrocarbons per cubicmeter soil and day. This rate could be doubled by raising the groundwater temperature from 12°C to 20°C. These data were calculated with the aid of models and are only valid under the defined and special conditions of the testfield. Decisive for the scaling up of the experiment is among others the availability, composition and concentration of the hydrocarbons at the site. Moreover the absence of inhibiting compounds must be assured.
A rough calculation based on the data provided from the testfield shows that for the entire clean-up of the area about 5 years operation time is necessary. In this time it is expected, that by the in-situ measures almost all bioavailable hydrocarbons can be removed. The experiences [2] showed that the remaining hydrocarbons are as a rule less soluble and cause only negligible concentrations in the groundwater. That means a decisive reduction of the risk assessment for the groundwater.

5. LARGE SCALE REMEDIATION

The large scale plant is operating since beginning 1991 with a capacity of 550 m^3/h circulation water. The hydraulic conditions correspond in a wide range the groundwater flow model calculations according to Fig. 2. To verify the model based assumptions a tracer test was performed over the entire area. It could be shown that the flow conditions and the efficiency of the hydraulic measures are like assumed. The detection of uranin, which was used as a tracer, at the partly depth differentiated observation wells in the upper groundwater layer enables a realistic prediction of the overall spreading of nitrate in the circulating groundwater flow.

The next step was the increase of nitrate and inorganic growth factors in the water to be reinjected to the subsurface in order to enhance the bioactivity.
Due to the decrease of recharge capacity of the infiltration ditches after 6 month of operation the process of remediation was slightly retarded. The cause of the capacity loss

could not be determined, but biological processes could be excluded. The problem was overcome by drilling 9 m deep holes in the infiltration ditches and filling them up with coarse gravel. To exclude one of the possible causes the P-compound added to the system was changed to a more stable one. In the following year of operation no decrease of the infiltration capacity was observed.

The nitrate and growthfactor addition was optimized and controled according to the consumption in the system. Furthermore the hydraulic system was brought to an optimum and recharge water was warmed up until 25°C.

A nitrate consumption of 100 mg/l in a mean residence time of 50 days could reached at an average groundwater temperature of 18°C. That means a degradation rate of about 1,5 g hydrocarbons per cubicmeter soil and day.

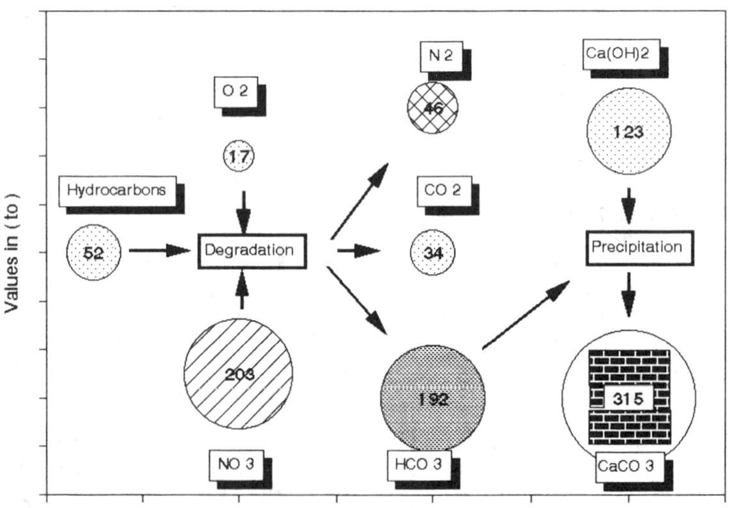

Fig 5 : Balance of bioremediation (Jan - Aug 1992)

Figure 5 shows a mass balance of the overall degradation processes in the subsurface based on the measured parameters oxygen, nitrate, carbondioxide and hydrogencarbonate. Over the time period between January to August 1992 a mineralization of about 50 t hydrocarbons by consuming 17 t of oxygen and 200 t of nitrate was calculated. Inorganic carbon is formed by the denitrifying processes and can be proved as hydrogencarbonate in the flushing water after groundwater passage and is precipitated as limesludge by flocculation in the water treatment process.

Next to the biodegradation the stripping of volatile dissolved hydrocarbons contributes to the clean-up with 30 t/a. The concentration of the hydrocarbons increase slightly in some wells and as an average about 8 mg/l dissolved hydrocarbon can be detected. The explanation for this unexpected increase is a mobilizing effect probably caused by the formation of biosurfactants, which could even be detected in some wells.

The raise of the redoxpotential in the groundwater caused as expected a decrease of the content of iron and manganese having nowadays concentrations of 1 mg/l for iron and 2 mg/l for manganese respectively in the influent of the treatment plant.
The operation of the large scale remediation confirm up to now that the experiences of the pilot plant cannot be transferred completely due to the multifactorial interference in the system. But nevertheless they provided the necessary and valuable background the large scale remediation is based on.

It will be expected that the removal of the dissolved hydrocarbons in the groundwater at the site will be reached in about 5 years of operation as it was scheduled. Optimizing the system to shorten the operation time and to reduce the costs as well will be the target.

The further experiences of the project will be reported as soon as they are available.

6.   REFERENCES

1. Arbeitskreis "Wasser und Mineralöl" (1970). Beurteilung und Behandlung von Mineralölunfällen auf dem Lande im Hinblick auf den Gewässerschutz. Bundesministerium des Innern, Bonn, 2. Auflage, Dezember 1970.

2. Battermann, G.; Werner, P. (1984). Beseitigung einer Untergrundkontamination mit Kohlenwasserstoffen durch mikrobiellen Abbau. gwf-wasser/abwasser 125, H. 8, S. 366 - 373.

3. Battermann, G. (1987). Einsatz mikrobiologischer Verfahren zur Sanierung von Untergrundkontaminationen. VDI-Berichte Nr. 628, 1987, S. 287 - 300. VDI-Verlag Düsseldorf.

4. Battermann, G. (1987). Hydraulisch geologische Voraussetzungen biologischer Reinigungsprozesse im Untergrund. Biotechnische In-Situ Sanierung. Fischer Stuttgart.

5. DVGW-Forschungsstelle am Engler-Bunte-Institut der Universität Karlsruhe (1986). Chemische und mikrobiologische Untersuchungen von Boden- und Wasserproben. Karlsruhe.

1144

6. DVGW-Forschungsstelle am Engler-Bunte-Institut der Universität Karlsruhe (1988). Untersuchungen zum Versuchsbetrieb. Karlsruhe.

7. DVWK-Schriften 98 (1991). Sanierungsverfahren für Grundwasserschadensfälle und Altlasten - Anwendbarkeit und Beurteilung (spez. Kap. 4.3). Verlag Paul Parey, Hamburg und Berlin.

8. Fried, R.; Horalek, U.; Zipfel, K. (1990). Abschirmung eines Deponiestandortes. BBR 6/90, S. 323 - 328.

9. Hessisches Landesamt für Bodenforschung Wiesbaden (1980). Erläuterungen zur geologischen Karte von Hessen, Blatt Nr. 5917 Kelsterbach. Wiesbaden.

10. Major, D.W.; Mayfield C.I.; Barker J.F. (1988). Biotransformation of Benzene by Denitrifikation in Aquifer Sand. Groundwater Vol. 26, No. 1, S. 8 - 14.

11. Riss, A. (1988). Mikrobiologische Untersuchungen über wesentliche Faktoren des subterratrischen Altlastabbaus beim Einsatz von Nitrat als terminaler Elektronenakzeptor. Dissertation Universität Saarbrücken.

12. Riss, A.; Barenschee E.R.; Helmling O.; Ripper P. (1991). Einsatz von Wasserstoffperoxid zum mikrobiologischen Kohlenwasserstoffabbau, Labor- und Feldversuche in-situ. gwf-Wasser/Abwasser 132 (1991), Heft 3, S. 115 - 126.

13. Tangermann, H. (1989). Beiträge zur Umweltgeologie. Jber. Mitt. oberrhein. geol. Ver. N.F 70. S. 141 - 159, Stgt.

14. Zenz, Th. und Zipfel, K. (1989). Grundwasserschutz durch Einsatz numerischer Modelle zum Stofftransport. bbr 6/89, S. 336 - 341.

KORNHARPEN EXPERIMENTAL LANDFILL - BALANCE-SHEET OF A RESEARCH
PROJECT

PETRA BECKEFELD
HEITKAMP UMWELTTECHNIK GMBH, HEINRICHSTR. 67, D - 4630 BOCHUM

From 1987 to 1991 a comprehensively scientifically monitored,
full-scale trial of immobilisation and disposal of contaminated
soils using Heitkamp Umweltechnik GmbH's patented
immobilisation technique was carried out on the Bochum-
Kornharpen central landfill. The provisional final report on
the monitoring was submitted by the Technical University of
Brunswick in March 1991.

During this trial about 10,000 m^3 of soils contaminated
specifically with tar - silt or very silty sands and gravels,
interspersed with building rubble - were treated in a mixing
plant installed on the trial site. The mixed material was
deposited and compacted on a sealed surface in layers at the
normal soil moisture content. In two separate areas, monolithic
landfill mounds were formed using different immobilisation
mixes, and exposed to weathering.

The scientific monitoring included checking the suitability of
the immobilisation technique, the quality of treatment being
achieved by the plant and of the mixed material backfilled, as
well as long-term monitoring of the monolithic mounds,
particularly in relation to their elution characteristics and
stability.

The mechanical properties of the immobilisation product were
optimised in the process of checking the suitability. For the
second mound for instance a coefficient of permeability $k_f = 6
\times 10^{-11}$ m/s and uniaxial compressive strengths $q_u = 5.47$ MN/m^2
(after 14 days), $q_u = 7.93$ MN/m^2 (after 28 days) and $q_u = 11.3$
MN/m^2 (after 56 days) were measured. There was no decomposition
of the specimens used in the ENDELL test.

Statistical evaluation of the records of the electronically
controlled mixing plant, which was undertaken in the course of
monitoring quality, showed good compliance with the specified
mixes and the mixing times. Testing of specimens of material
which had been treated and was intended for backfilling yielded
results at least as good, and sometimes even better, than those
from checking the suitability. This showed that the good
laboratory results achieved by this immobilisation technology
are transferable to full scale applications.
The immobilisation of contaminants achieved is shown for the
example of the total of the 6 polycyclic aromatic hydrocarbons
(PAH) to TVO. Whereas the average eluate concentration of the

F. Arendt, G.J. Annokkée, R. Bosman and W.J. van den Brink (eds.), Contaminated Soil '93, 1145–1146.

contaminated soil was 54 µg/l (maximum value: 895 µg/l), in the mixed material only 1.9 µg/l could still be detected. In specimens from surface and circulating water (leachate) collected, maximum PAH contaminations of 0.75 µg/l were measured during the two-year observation phase. After two years the concentrations were no longer measurable.

The following conclusions can be drawn from this large-scale trial:

1. Virtually complete immobilisation of PAH, phenols and heavy metals can be demonstrated when a suitable technology is used.

2. Since extensive laboratory tests and observation of the landfill mounds gave no indication of the product's properties being impaired with time, long-term stability may be assumed.

3. The methods developed for testing the mechanical properties of immobilisation products as part of a research project are suitable in practice for determining and evaluating the quality of the products. However different tests are proposed for evaluating the leaching characteristics.

4. A quality assurance system, consisting of a check on suitability and monitoring of quality, is absolutely essential in providing evidence of immobilisation. Long-term control is to be provided for large projects in particular.

Thus the immobilisation technique used in this trial represents a reliable yet cost-effective solution to the problem of remedial action for contaminated sites. There are a variety of possible applications handling large-scale contamination in the new states of the Federal Republic of Germany in particular, as well as industrial sites which it is planned to reuse in the old states. A report on its adaptation to clean up the sites of old mines in the Ruhr has already been given in another paper.

REFERENCES
Beckefeld, P (1991). Schadstoffaustrag aus abgebundenen Reststoffen der Rauchgasreinigung von Kraftwerken - Entwicklung eines Testverfahrens [Extraction of contaminants from bound residues from the cleaning of power station flue gases] - dissertation. Mitteilung des Instituts für Grundbau und Bodenmechanik, Technical University of Braunschweig, Vol 33, Braunschweig 1991.

Beckefeld, P (1992). Schadstoffeinbindung durch Verfestigung am Beispiel der Versuchsdeponie Kornharpen [Immobilisation of contaminants taking the Kornharpen experimental landfill as an example]. Loseblattsammlung Altlasten, published by WEKA-Verlag.

TECHNICAL FACTORS IN THE DESIGN OF A PROGRAMME FOR EX-SITU AND IN-SITU BIOREMEDIATION OF A FORMER OIL DISTRIBUTION TERMINAL.

R J F BEWLEY and J G ALEXANDER

Dames & Moore International, Blackfriars House, St Mary's Parsonage, Manchester, M3 2JA, England

This paper discusses the key issues relating to the design of remedial works for clean-up of a former Lube Oil Terminal site in England, using both ex-situ and in-situ bioremediation.

The former terminal is located in an industrial park between an active oil distribution terminal and a canal and covers approximately 3.5 hectares. Site investigations of the area established the presence of significant soil contamination by petroleum hydrocarbons (up to 126,000 mg/kg) in the unsaturated zone, consisting of brick and concrete rubble overlying silty clay.

Contamination also existed in the saturated zone where it tended to be associated with groundwater within the underlying sand/gravel aquifer. Such contamination tended to consist of hydrocarbons (up to 1300 mg/l) trapped below the overlying clay in an emulsified, partially degraded state rater than occurring as floating product. This originated from past activities on the terminal site, continuing vertical migration from the unsaturated zone where the clay is breached, and past and ongoing lateral migration in groundwater from the active oil distribution terminal across the site towards the canal. The groundwater migrating across the active site is fed by an artificial lake located in a park immediately upgradient of the site.

A scheme was devised to address contamination both in the unsaturated and saturated zone. The former involves excavation of approximately 11,500m^3 of soil to 1.7m depth from the contaminated areas of the unsaturated zone, processing and crushing the material to reduce it to optimal particle size and placement in three treatment beds constructed both on-site and in the adjacent area. The beds were underlain by a synthetic liner to provide an impermeable membrane, over which a layer of sand was placed to collect leachate and minimise the risk of damage to the liner. A nutrient mixture based on previous treatment of petroleum hydrocarbons, (Bewley et al, 1990) was formulated to provide optimal conditions for microbial activity, and is applied to the material using a boom spray. Rotovation (spading) of the beds is then undertaken over a period of approximately 9 weeks until target concentrations are achieved. Reinstatement of the material can then take place.

Clean-up targets for the soil were based upon the risks of migration of residual hydrocarbons in groundwater to the adjacent canal. As the site is to be used for industrial purposes, exposure of individuals to residual concentrations of hydrocarbons

F. Arendt, G.J. Annokkée, R. Bosman an[d] ... [s]il '93, 1147–1148.
© 1993 Kluwer Academic Publishers. P[rinted in the Netherlands.]

in soil is considered to be minimal. Leach tests were performed to assess the potential mobility of the petroleum hydrocarbons, based upon DIN 38,414 (1984), but using a 1:5 soil:water ratio. For total petroleum hydrocarbon (TPH) concentrations in soil below 1500mg/kg, less than 40mg/l TPH was present in the leachate (this being the current consent for discharge of surface water run-off into the canal). The target concentrations that were negotiated with the regulatory body, the National Rivers Authority (NRA) are to achieve TPH concentrations no greater than 1500mg/kg in 80% of soil samples within a given volume of treated material. Alternatively the material is deemed acceptable for reinstatement if 80% of samples do not exceed 2500mg/kg TPH, provided that the TPH concentrations in a leach test performed on such samples does not exceed 10 mg/l using a 1:10 soil:water ratio.

For addressing contamination in the saturated zone, contaminated groundwater is abstracted from four wells located at strategic points across the site. A mixture of nutrients and hydrogen peroxide is added to the abstracted groundwater which is pumped via a system of reticulation pipework to a 275m long, 5m deep infiltration trench constructed along the boundary between the Lube site and active distribution terminal. The Wellhead Protection Areas (WHPA) groundwater flow model developed by the USEPA was used to model groundwater flow assuming a 1:200 hydraulic gradient across the site, an aquifer transmissivity of $60m^2$/day and a porosity of 25%. It was calculated that from a $300m^3$/day total abstraction rate the re-circulation time from injection trench to abstraction borehole will be approximately 50 days. Following active clean-up, the trench will remain as a cut-off system to intercept potentially contaminated groundwater migrating from the up-gradient active distribution site. A clean-up target of 25 mg/l dissolved TPH was negotiated for the groundwater remediation. This figure was based upon the results of a model developed to estimate future concentrations of petroleum hydrocarbons in groundwater following completion of the active remediation programme. This model incorporated the effects of advective transport, sorption/desorption on soil particles, contaminated leachate (recharge) and degradation. The model was run for different values of the partition coefficient Kd. (Kd = concentration sorbed onto soil particles/concentration dissolved in groundwater). Residual concentrations of 5mg/l TPH were estimated for Kd values of 0.5, 1.0 and 2.0 after approximately 2.7, 4.7 and 7.5 years respectively.

Following acceptance of these criteria by the regulatory authorities implementation of the remedial works has commenced and is expected to be completed over a 15 to 18 month period.

REFERENCES

Bewley, R J F (1990), Land Deg. & Rehabil., 2, 1-11.
Deutsche, Norman DIN 38 414 Part 4, Sludge and Sediments (group S). Determination of leachability by water (S4), Beuth Verlag GmbH, Berlin (1984).
Dibble, J T & Bartha, R (1979), Appl. Environ. Microbiol., 37,729-739.
Walker, J D, Colwell, R R and Petrakis, L (1976). Can. J. Microbiol., 22, 1209-1213.

ENHANCED BIOTECHNOLOGICAL REMEDIATION OF OIL POLLUTED SOILS WITH THE CUM BAC® SYSTEM BY OPTIMIZATION AND MODELLING OF THE LIMITING FACTORS.

E. TEN BRUMMELER

Heidemij Realisatie BV, P.O.B. 660, 5140 AR Waalwijk, The Netherlands.

1. INTRODUCTION

Biotechnological remediation in so called 'landfarms' is an attractive method for oil polluted soils. Landfarming is a simple technology and can therefore be carried out at low costs. During the biotechnological remediation the soil keeps its structure and can be reused for most purposes afterwards.

Nowadays (conventional) landfarming is a widely applied method for remediation of oil polluted soil. However, the factors that determine the rate and the level of the aerobic microbiological degradation of the pollutants in landfarm systems are poorly controlled. A major draw back of "open" landfarming is the possible emission of volatile compounds, nutrients and leachate to the environment.

Since 1985 Heidemij is treating oil polluted soil in the improved landfarming system Cum-Bac® on full scale. The actual capacity of our four remediation sites amounts to 85,000 ton per year. In the Cum-Bac® system soil is treated in a type of green house, equipped with a leachate recycle and spraying system, water drains and air drains. In the green house the temperature of the soil normally is at least 5 °C higher than non covered soil. In general the soil is not inoculated, but the bacterial population present in the soil is taken advantage of during the remediation. Just at high levels of mineral oil (> 10,000 ppm) inoculation is essential.

Depending on the type and concentration of pollutant, air suction is applied to remove volatile hydrocarbons and to supply oxygen to the process. The air is subsequently treated in a bio-filter, where the volatile compounds are degraded by specialized organisms. As one of the main limiting factors in the biodegradation of oil pollutants in soil is the availability and way of addition of nutrients to the microorganisms, the nutrients (nitrogen, phosphorous) are supplied in dissolved form. This way of nutrient addition prevents the slow release of the nutrients from solid fertilizer. The slow release of nutrients can limit the rate of degradation of mineral oil because the actual need for nutrients may be higher than is released from the solid fertilizer.

2. OPTIMIZATION AND MODELLING OF LIMITING FACTORS

In the Cum-Bac® system the limiting factors for biodegradation of hydrocarbons (varying from C8 to C40) present in mineral oil and refined products, i.e. temperature, moisture, nutrients, and oxygen are optimized. Depending on the type of soil and type and concentration of pollutant the retention time needed to reach the level of 100 ppm oil amounts to 7 months for sandy soils and 6-12 months for clay soils. Although a body of practical experience exists with landfarming, precise prediction of the exact time needed for remediation is hardly possible due to the lack of fundamental knowledge of the occurring processes during biotechnological soil remediation.

It is tried to improve the system by creating a simulation model of the microbiological, physical and chemical processes that are responsible for the degradation of oil.

F. Arendt, G.J. Annokkée, R. Bosman and W.J. van den Brink (eds.), Contaminated Soil '93, 1149–1150.
© 1993 *Kluwer Academic Publishers. Printed in the Netherlands.*

1150

One dimensional deterministic transport simulation models can be used for simulating different treatment scenarios in landfarming. These models were developed by the Amsterdam University, Department of Physical Geography and Soil Science, predicting the positive effect of leachate recycle and the increased and temperature level of the greenhouse on the rate of biodegradation.

3. RESULTS

The results of the simulations were used to develop an ideal landfarming scenario. This scenario was applied to remediation of a sandy soil. The scenario includes the following features:
- mean soil temperature during remediation is 25 °C,
- nutrients are supplied in dissolved shape,
- soil is kept at optimum moisture content by sprinkling of water,
- aeration is provided by suction of the drains,
- soil is homogenized frequently.

In Fig. 1 The course of mineral oil concentration is shown during remediation of the sandy soil with an ideal scenario. The amount of soil amounted to 1,500 tons. As reference also the results of a conventional landfarming of this type of soil is shown. Course of the mineral oil concentration in conventional and enhanced biotechnological remediation with the Cum-Bac® process .

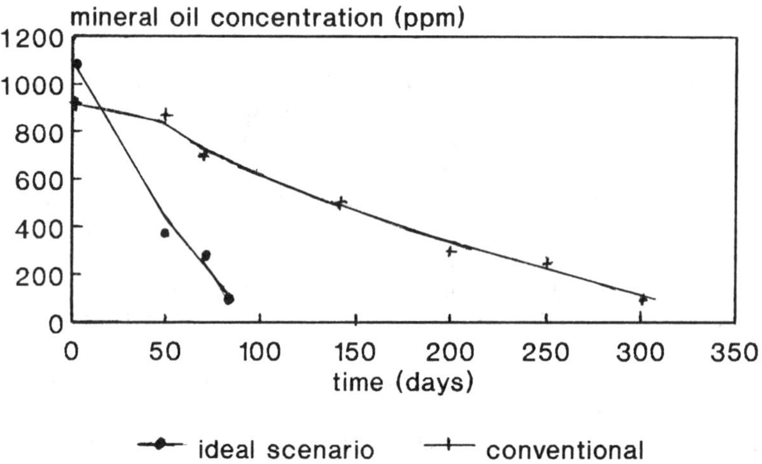

FIG. 1. Course of the mineral oil concentrations of enhanced landfarming with the Cum-Bac® process and of conventional landfarming.

4. CONCLUSION

The conventional landfarming includes outdoor remediation (lower temperature) application of solid fertilizer, no sprinkling of water, and no aeration. It can be concluded from Fig. 1 that the remediation process can be accelerated by application of an ideal remediation scenario, that partially has been developed from simulations with a deterministic transport model. The time needed to remediate sandy soils biotechnologically from 1,000 ppm mineral oil to less than 100 ppm can be shortened by a factor 2. For other types of soils, such as light clay soils, similar results are observed.

EBIOX BIOREMEDIATION OF CONTAMINATED SOIL MATERIAL
USING VACUUM HEAP TECHNOLOGIES

DANIEL ROLF EIERMANN

PROJECT COORDINATOR EBIOX AG, SURSEE, SWITZERLAND

1. ABSTRACT

This enhanced off-site multi-layer bioremediation technique will be applied when large volumes of soil (up to several thousands of cubic metres at a time) are to be microbiologically treated. The microbiological degradation of contaminants is significantly accelerated with the vacuum heap technology. It will only be applied for soil volumes of at least 1'000 m^3. The bioremediation is going to be extremely competitive by optimizing the milieu conditions in the specific soil ecosystem. Various different mineral oil pollutant products can be treated (e.g. BTEXs, fuels, kerosene, diesel oil, lubricants, crude oils and in particular, polyaromatic hydrocarbons PAH's).

2. THE PRINCIPLE OF EBIOX BIOREMEDIATION

Ebiox Bioremediation represents enhanced natural decontamination processes by means of biostimulation of contaminant adapted micro-organisms. This yields not only in an accelerated natural pollutant degradation but in improved economics as well and generates, furthermore, an environmentally beneficial ecosystem.

If such a natural degradation shall be spontaneously initiated, various milieu limiting parameters must be optimized:

* oxygen content
* nutritive substances (macro-nutrients)
* trace elements (micro-nutrients)
* moisture content
* pH/Eh-values
* temperature
* soil structure
* organic carbon content
* bioavailability of pollutants

Only the optimal interaction between natural micro-organisms on the one hand and carefully limiting factor-tuned nutritive substances (macro-nutrients) and trace elements (micro-nutrients) on the other guarantee a best possible and above all a fast pollutant degradation process.

The microbiologically remediated soil material represents a natural-intact ecosystem after bioremediation completion.

Large volumes of contaminated soil are mechanically mixed (rubble recycling plants), homogenized and irrigated with the appropriate nutrient composition and trace elements. The heap is then set up in layers (see also figure 1).

F. Arendt, G.J. Annokkée, R. Bosman and W.J. van den Brink (eds.), Contaminated Soil '93, 1151–1154.
© 1993 *Kluwer Academic Publishers. Printed in the Netherlands.*

In between those, a perforated plastic piping is installed which is
connected with manifolds to a vacuum blower system.
By pulling a vacuum through the perforated pipes, outside air can be
uniformly drawn through the soil to provide sufficient oxygen support for
the aeobic micro-orga-nisms. Permanent low pressure conditions preclude air
emissions to the atmosphere.

All volatile compounds are, therefore, sucked out initially through the
vacuum system and purified via a biofilter. Seepage waters are captured in
a drainage system and pumped into the bioplant system. The contaminants are
completely degraded into the end products water and carbon dioxide.

A sector-controlled irrigation system lies on top of the vacuum heap. This
guarantees nutritive substances and trace elements supply on the one hand
and also most optimal moisture content conditions on the other hand. The
entire heap is covered with a black plastic liner as a precaution to
prevent moisture evaporation and loss of passive solar heat.

After treatment in the bioplant system, the purified, oxygen saturated and
with nutritive substances and trace elements laden waters are resprayed
onto the heap again. Therefore, all heavier petroleum fractions undergo a
double effect bioremediation.

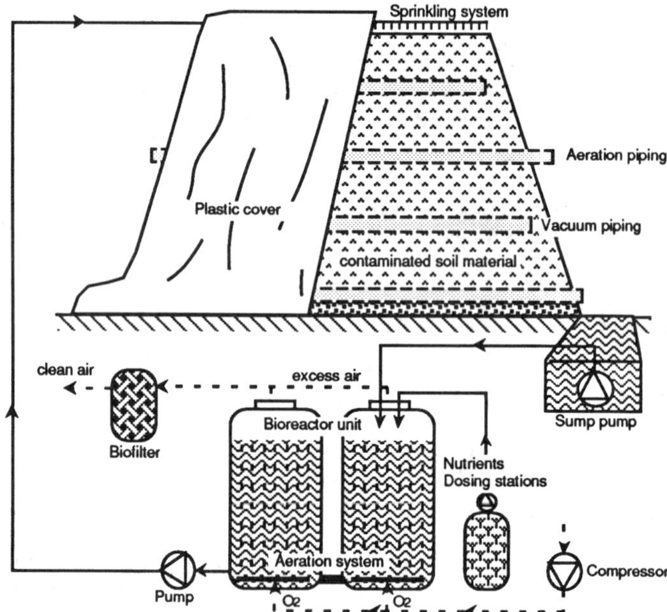

Figure 1:   System flow chart of vacuum heap technology

3. A VACUUM HEAP TECHNOLOGY CASE HISTORY

3.1.   Introduction

Ebiox Bioremediation treated microbiologically 1'092 m^3 oil contaminated soil during a giant building project.The average hydrocarbon load after excavation defined as project start-up concentration was 1'419 mg/kg.

3.2.   Bioremediation targets

Threshold values to be achieved were defined with a hydrocarbon concentration of less than 100 mg/kg to allow a subsequent reuse of the bioremediated soil as back-fill material for the building project.

3.3.   Project-Monitoring

Daily bioplant system controls were maintained by specialised personnel during the whole bioremediation.

A fortnightly monitoring programme including competent and thorough sampling procedures and analytical methodologies were carried out to follow the bioremediation progress. Furthermore, complete weekly system checks with adequate maintenance were carried out.

VOLUME: 1'092 m^3 soil

* 25 sample taking series with rotary auger device
  (drilling depth 3 m)
* Sample preparation, storage of reference samples
* 36 microbiological soil investigations (plate counts, ammonia, nitrate, ortho- and total phosphate)
* 138 total petroleum hydrocarbon analyses (TPH) DIN H 18, all as duplicate measurements against dry substance
* 9 start-up analyses (neutral Laboratory)
* 18 final analyses (neutral Laboratory)
* Weekly field analytical investigations (ammonia, nitrate, ortho- and total phosphate, dissolved oxygen content, temperature, pH/Eh)

3.4.   Results

The relevant data was plotted as a project graph to give a better picture of the bioremediation process (see figure 2). The initial hydrocarbon load was defined prior to the vacuum heap's civil works.

The average hydrocarbon concentration determined in a neutral laboratory was 1'419 mg/kg prior to project start-up (start-up analyses). 5 sampling series in regular intervals were analysed in the Ebiox laboratory to document a statistically well supported bioremediation progress after completion of vacuum heap's civil works.

The last sampling series was analysed again in the neutral laboratory (final analyses) to get fastest authorities approval for the subsequent filling of the bioremediated soil in the building project.

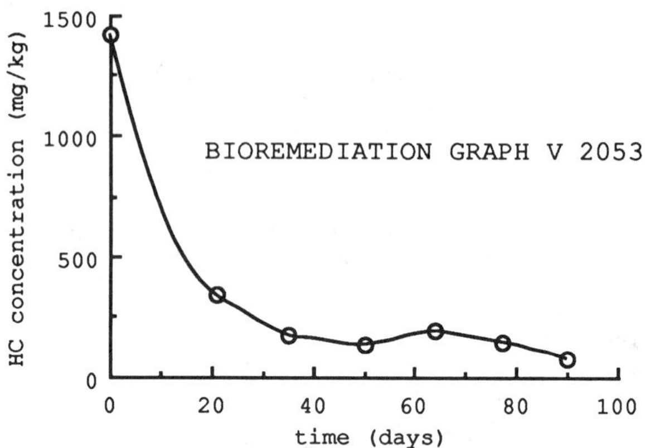

Figure 2:   Bioremediation graph project V 2053 plotted time vs
             hydrocarbon concentration. Start-up- and final analyses
             carried out by neutral laboratory, project analyses carried
             out by Ebiox laboratory.

The last sampling series (final analyses in neutral laboratory) yielded an
average hydrocarbon concentration of only 79 mg/kg after 90 days treatment
with Ebiox Bioremediation which represents a degradation performance of
94.4 %.

3.4.   <u>Conclusions</u>

The analytical data carried out clearly show that the average
initial hydrocarbon concentration of 1'419 mg/kg in the contaminated soil
material was significantly reduced to 79 mg/kg with Ebiox Bioremediation
within three months. The achieved soil quality nearly represents "clean
excavated soil" which is 50 mg/kg. As the natural degradation process still
continues for several weeks after project completion, the soil quality
"clean excavated soil" is achieved at the time of back-fill for the giant
building project. Several different benefits using Ebiox Bioremediation
(natural-intact soil, unchanged soil structure, no addition of filling
materials) remarkably demonstrate the various application possibilities in
ongoing building projects

Ebiox Bioremediation of contaminated soil using vacuum heap technologies
takes maximum advantage of large scale treatment of excavated soil
(thousands of cubic metres at a time) and is particularly well-suited for
projects with accelerated completion deadlines, confined working space and
air emission restrictions.

REMEDIATION OF AN ACIDIC-TAR-POND AND CONTAMINATIONS OF THE SURROUNDING SOIL WITH PAH, PCB, BTX AND CHLORINATED HYDROCARBONS BY IMMOBILIZATION

THOMAS ERTEL, DIPLOM-GEOLOGE

UMWELTWIRTSCHAFT GmbH, JUL.-HÖLDER-STR. 39, 7000 STUTTGART 70

1. SITUATION

During the enlargement of a disposal site, an acidic-tar-pond situated in a depression at the margin of the disposal, had to be remediated. Besides acidic tars from oil-recycling other liquid organic wastes partly in drums have been deposited. After the deposition, the depression was of liquid to pasty consistency and surrounded by an extensive zone of contaminated soils with contents of up to 0,7 g/kg PCB and 2,5 g/kg PAH.

2. CONCEPTS OF REMEDIATION

After the separation of the different substances and grades of contamination, the remediation elapsed as follows (see figure).

A continuous observation and analytic control of the dredging operations and the recovered material provided a good classification of the dredged materials.

Due to the fact that the bulk of the contaminants made up of tary lumps and nodules, the solidification and longtime immobilization was the most difficult part of the on-site treatment.

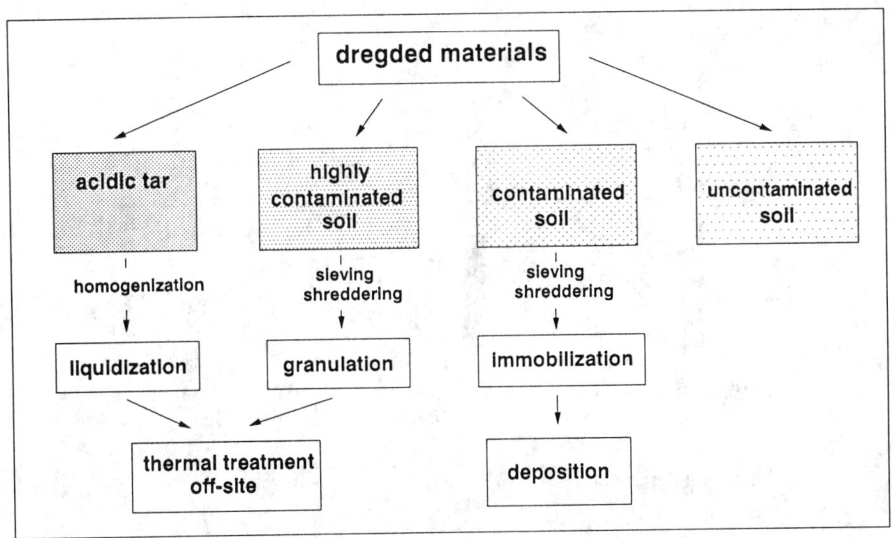

F. Arendt, G.J. Annokkée, R. Bosman and W.J. van den Brink (eds.), Contaminated Soil '93, 1155–1156.
© 1993 *Kluwer Academic Publishers. Printed in the Netherlands.*

3.   IMMOBILIZATION

The first step of the treatment was the homogenization and investigation of the dredged material by shredding and sieving. The following solidification took place in a mixer by adding a mixture mainly consisting of cement, bentonite, water-glass and lime. To crack the tary lumps hot water, partly as water-vapour, was added. After a final solidification and shredding, solid samples and eluates of the material were analyzed. Even if the samples for the preparation of eluates had to be pulverized, the analytic results remain far below the acceptable levels (see figure). Solidification enabled an effective immobilization of the contaminants. Especially for complex mixtures of contaminants, where there are no appropriate and cheap remedial action techniques, immobilization proves to be a very effective method.

Contaminant levels in solidificated material and its eluates

	PCB Original [mg/kg]	PCB Eluat [mg/l]	Phenole Original [mg/kg]	Phenole Eluat [mg/l]	KW Original [mg/kg]	KW Eluat [mg/l]
sample 1	5,5	n.n.	5,8	0,4	6243	<0,1
sample 2	2,55	n.n.	2,4	0,33	24800	<0,1
sample 3	4,7	n.n.	3,6	0,2	8700	0,19

Original [mg/kg]
Eluat [mg/l]

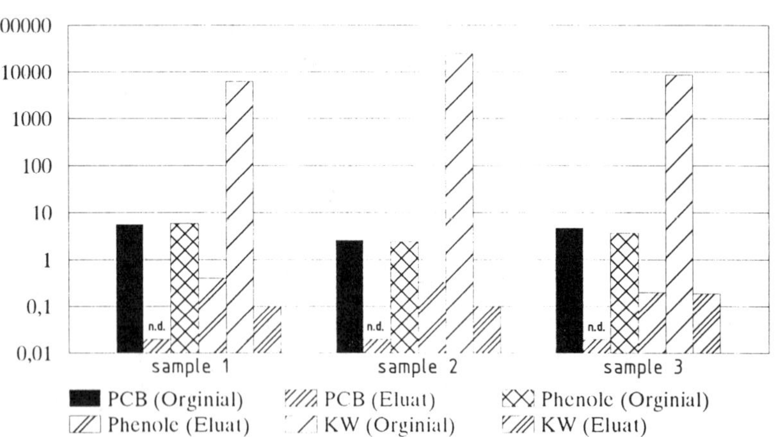

n.d. not detectable

EXTRACTION OF HEAVY METALS FROM CONTAMINATED MATERIALS BY AMINO ACIDS

Fischer, K.[1,2]; Rainer, C.[1]; Bieniek, D.[1] and Kettrup. A.[1,2]

[1]GSF-Forschungszentrum, Institut f. Ökologische Chemie, D-8042 Neuherberg
[2]Lehrstuhl für Ökologische Chemie, TU München, D-8050 Freising-Weihenstephan

1. INTRODUCTION

The remediation of soil and abandoned waste sites which are contaminated with heavy metals requires major financial and technological investments. Although a number of methods have been developed and used for this purpose, they are frequently not environmentally adequate and can therefore lead to ecologically problematic consequences. Starting points for the development of an environmentally acceptable remedial action technique may be found by studying the influence of natural substances, esp. organic complexing agents, on the mobilization of metals in soils. The mobilization process is influenced by numerous parameters, such as soil chemical conditions, the strength of the metal binding in the soil matrix as well as the stability, solubility and sorption behaviour of the formed complex. For an assessment of the ability of amino acids to extract heavy metals from typical soil components, bentonite and peat were purposely contaminated with metals (target concentrations: 2 and 20 times higher than the guideline values established by the German sewage sludge decree [KSV]). Afterwards, the contaminated materials were suspended in solutions of amino acids and the release of metals under different experimental conditions (e.g. variation of pH) was maesured in fixed periods.

2. EXPERIMENTAL

<u>Adsorbents</u>: a.) Bentonite, a smectite rich mud stone (approx. 95% montmorillonite). Cation exchange capacity (CEC): 39.25 meq/100g.
b.) Peat, sampled from the upper 30cm of a Bavarian fen soil, dried at 60°, 0.56 mm sieved, 49% organic matter, 2.4 % carbonate content, CEC 82.7 meq/100g.
<u>Metal salts</u>: Analytical grade divalent acetates of Cu, Cd, Ni, Pb and Zn. Adsorption pH buffered at 4.65 (acetate buffer).
<u>Leaching solutions</u>: 5% (w/w) solution of glycine, 5 or 1% solution of cysteine and penicillamine .Desorption pH buffered at 7.0 with 0.1m HEPES and at 4.5 with 0.05m MES.
<u>Procedure</u>: The experiments were carried out in centrifuge tubes, clamped into a rotation apparatus. The solid:liquid ratio was 1:10.
<u>Evaluation</u>: The amount of adsorbed/ desorbed metal ions was calculated from the differences between the initial and final concentrations in the liquid phase, analyzed by ICP-AES.

3.RESULTS

3.1 <u>Leaching of metals from bentonite and peat by glycine at neutral pH</u>

High contaminated (approx. 20-fold KSV value, Pb 31-fold) bentonite samples were suspended in glycine solutions and rotated (Fig.1). After eight days, the following leaching rates were obtained: Cu 94.7%, Ni 84.9%, Zn 69.9%, Cd 61.3% and Pb 9.4%.
Experiments with metal loaded peat, conducted under analogous conditions, resulted in about 50% decrease of the desorption of Cu, Ni and Zn. Less than 10% of Cd and Pb could be removed from the organic material (Fig.2). Due to the peat derived protons their concentration in the test suspensions increased slightly. The influence of the pH value on the leaching rate was evidenced in the two copper test series.

3.2 <u>Influence of the experimental conditions on the leaching of Cd</u>

The effect of glycine upon the removal of Cd indicates a significant influence by the pH value. Although less pronounced, an influence of the metal concentration on the desorption rate

F. Arendt, G.J. Annokkée, R. Bosman and W.J. van den Brink (eds.), Contaminated Soil '93, 1157–1158.

1158

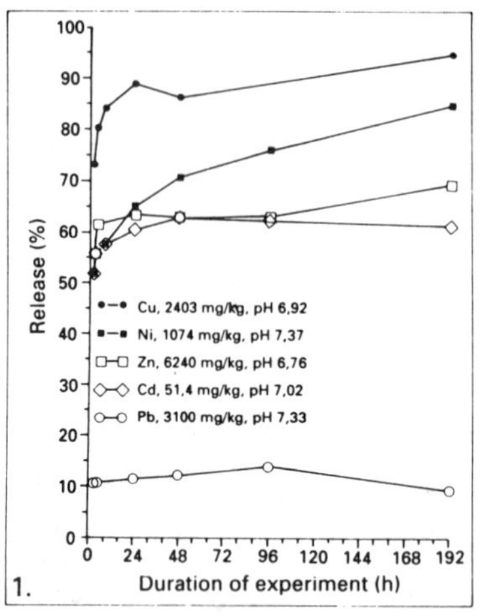

Desorption of heavy metals from bentonite by means of glycine

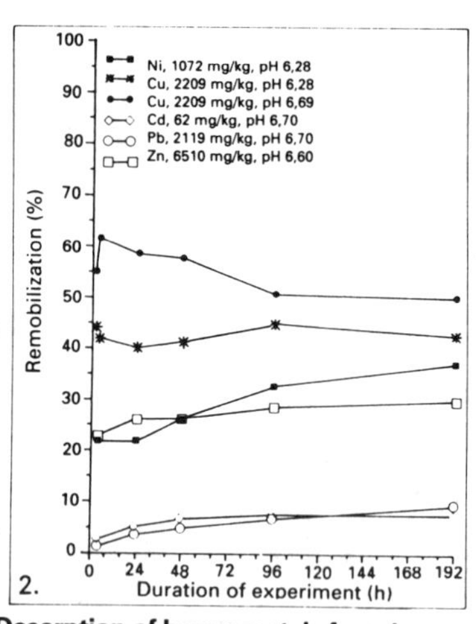

Desorption of heavy metals from humus

can also be seen. At neutral pH, approx. seven times more Cd is released from bentonite than at pH 4.5. The extractable portion increases with increased Cd content.

3.3 Removal of Zn from bentonite by means of amino acids

Corresponding to increases in the formation constants of the amino acid complexes at pH 7, the release of Zn from bentonite increases in the sequence glycine < cysteine < penicillamine. At pH 4.5, the leaching rates decrease and differences in leaching efficiencies are petering out.

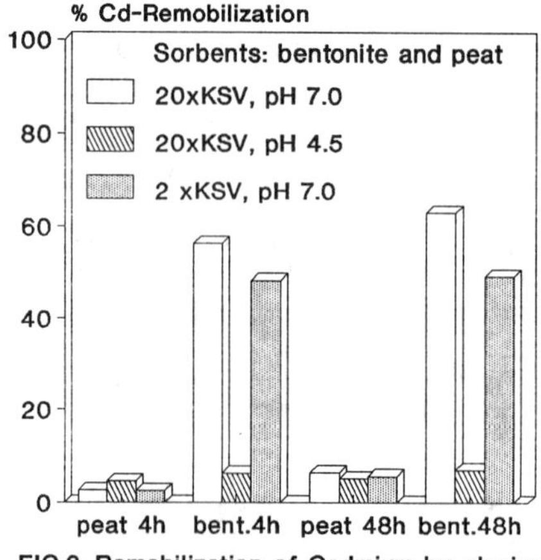

FIG.3: Remobilization of Cadmium by glycine

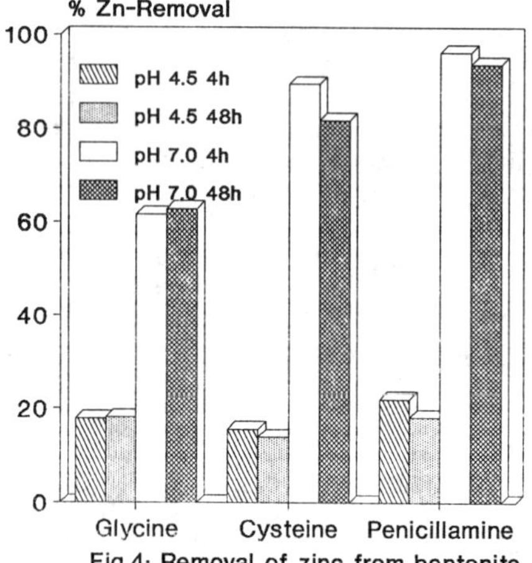

Fig.4: Removal of zinc from bentonite

4. CONCLUSIONS

a.) Under analogous experimental conditions, glycine leads to a higher removal of metals from bentonite than from peat.

b.) Desorption at neutral pH is more effective than at acidic pH.

c.) In most cases, the desorption rate increases with the increase of the metal content.

d.) At least for the remobilization of Zn, the leaching efficiency increases following the order glycine < cysteine < penicillamine.

e.) If equal degrees of contamination are used as a basis for the comparison of the extractions obtained, the following order of remobilization rates, as effected by glycine, could be established: **Cu > Ni > Zn > Cd > Pb**

SEPARATION OF URANIUM FROM POLLUTED WATER WITH POLYACRYLAMID-OXIMES

Ulrich Gohlke   FhG-Einrichtung für Angewandte
                         Polymerforschung
                         Kantstraße 55
                         O-1530 Teltow-Seehof

Andreas Otto   HeGo Biotec GmbH
                         Uhlandst. 16
                         O-1530 Teltow-Seehof

Jürgen Meyer   Wismut GmbH i. G.
                         Jagdschänkenstr. 29
                         O-9030 Chemnitz

Aminolysis of polyacrylonitrile with hydroxylamine leads to a polymer (named GoPur[R] 3000) with hydroxamic acid and amidoxime groups. These groups permit an effective complexation of uranylions. The polyampholyte is insoluble in a large pH-range around the isoelectric point and sedimentates together with the bounded uranium salt. The insolubility is caused by aggregation of polymer molecules due to polymer salt bridging or under the influence of polyvalent inorganic salts like sulfate, carbonate or by alkaline earth metals. These aggregation was measured by light scattering. The molecular weights of these aggregates in solved state are in the range of 1010 Dalton.

The aggregates formed after addition of GoPur[R] 3000 to the contaminated water can be effectively floated. After floating the polymeraggregates are recovered from the flotage for reuse.
In addition to this a concentrated solution of uranium salts results as a product of this recycling process which subsequetly can be used for the recovery of uranium as yellow cake in a commonly used manner.

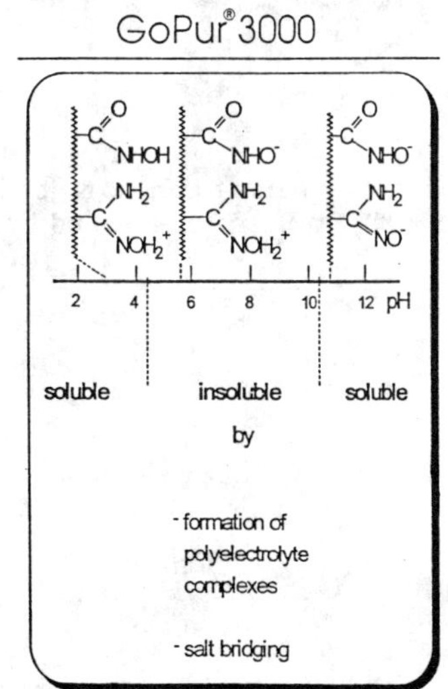

F. Arendt, G.J. Annokkée, R. Bosman and W.J. van den Brink (eds.), Contaminated Soil '93, 1159–1160.

The principle of these cleaning process is described by the following scheme

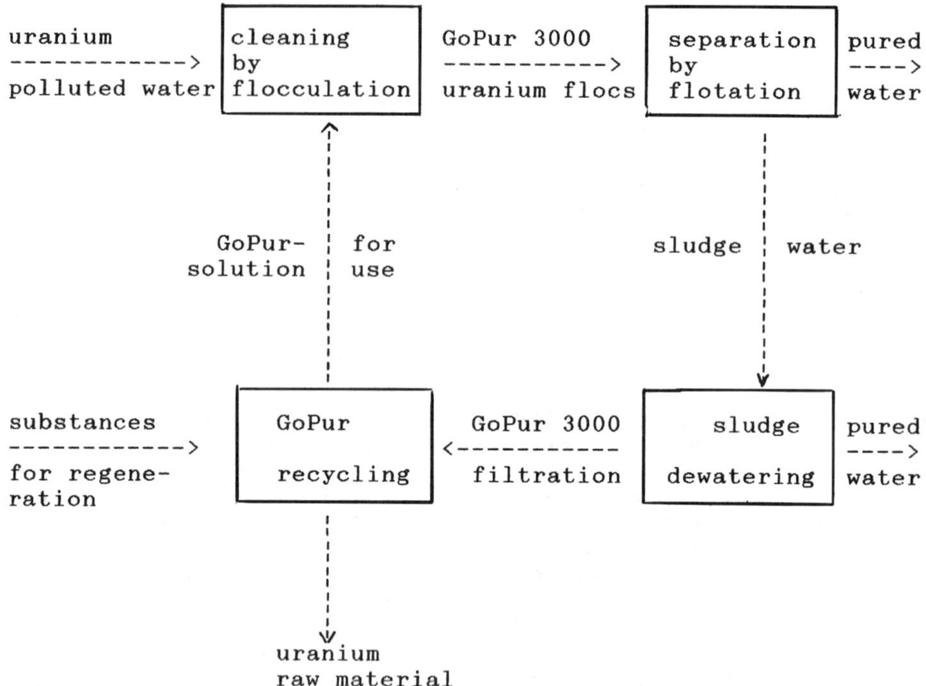

A technical variation of this process was proofed together with WISMUT GmbH successfully.

DEVELOPMENT OF A FULL SCALE SOIL WASHING SYSTEM FOR THE KING OF PRUSSIA SUPERFUND SITE, NEW JERSEY, USA - A CASE STUDY

E. GROENENDIJK [1] and M. PRUIJN [2]

[1] Alternative Remedial Technologies (ART) Inc., 14497 North Dale Mabry Hwy, Suite 115, Tampa, Florida 33618, USA

[2] Heidemij Realisatie BV, P.O. Box 660, 5140 AR Waalwijk, The Netherlands

1. INTRODUCTION

For a number of Superfund sites in the US, Soil Washing has been selected as the appropiate remedial technology for clean-up of these sites. The design of the "Soil Washing" process however is not "universal" but is largely dependent on the type of soil and nature of the contaminants present. Since each contaminated soil is basically unique, the general arrangement of the soil washing process and choice of treatment technologies: physical, chemical and/or biological, has to be made based on the soil and contaminant characteristics. Treatability and pilot-plant testing is generally required to provide the neccesary data for the design of the soil washing process.

In the development of a soil washing process for heavy metal contaminated soils of the King of Prussia (KOP) site, this fundamental step by step approach was followed in designing a process that would effectively remove the contaminants and could deal with the expected variability in contaminated soils composition.

2. SOIL CHARACTERISATION

Understanding of the soil matrix/contaminants relationship (for both metals and/or organic contaminants) is essential for a succesfull soil washing process design.
A characterisation of contaminated soils generally includes the determination of the soil particle size distribution, chemical analysis of the different size fractions and optical or Scanning Electron Microscopy for determination of the physical occurrence of the contaminants.
An example of the distribution of the heavy metal contaminants Chromium, Nickel and Copper as a function of the soil particle size in soils of the KOP-site is presented in figure 1.
Scanning Electron Microscopical analysis showed that the contaminants were present as trace metals within a matrix of Calcium, Magnesium and Aluminum sulfate and carbonate compounds.

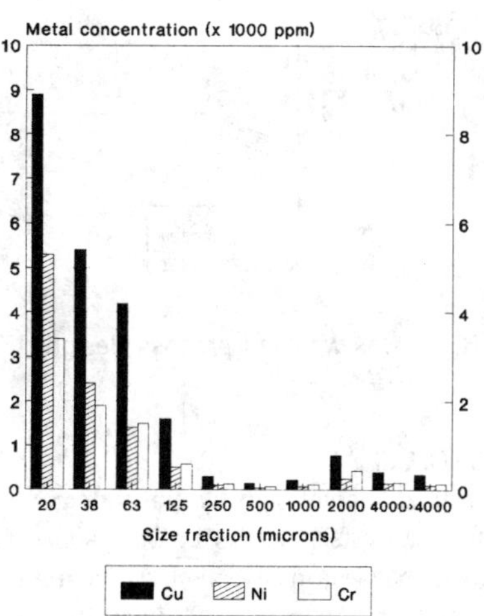

FIGURE 1: Chromium, Nickel and Copper concentration as function of soil particle size (KOP-soils)

F. Arendt, G.J. Annokkée, R. Bosman and W.J. van den Brink (eds.), Contaminated Soil '93, 1161–1162.

2. BENCH SCALE UNIT PROCESS TESTING AND LABORATORY PROCESS SIMULATION

Based upon the results of the soils characterisation, potential applicable technologies as wet screening, hydrocycloning, froth flotation and gravity separation were selected for bench scale unit operations testing. As a next step a series bench scale process simulation runs were performed, simulating the effect of a continuous process.

3. PROCESS DESIGN AND DEMONSTRATION RUN

Based upon the results of the characterisation, bench scale process testing and process simulation runs, a final selection of treatment technologies was made and configured into a continuous process, as schematically shown in figure 2.

To confirm and demonstrate the effectiveness of the process, a demonstration (pilot) run was performed with a blend of about 200 tonnes of contaminated soils excavated from the KOP site and shipped over to the Heidemij Soil Washing facility at Moerdijk, The Netherlands. For this pilot run the Moerdijk plant was configured exactly as proposed in the "process design". In the demonstration run the effectiveness of the soil washing process design was confirmed and EPA-approval was obtained for implementation of the project in the field.

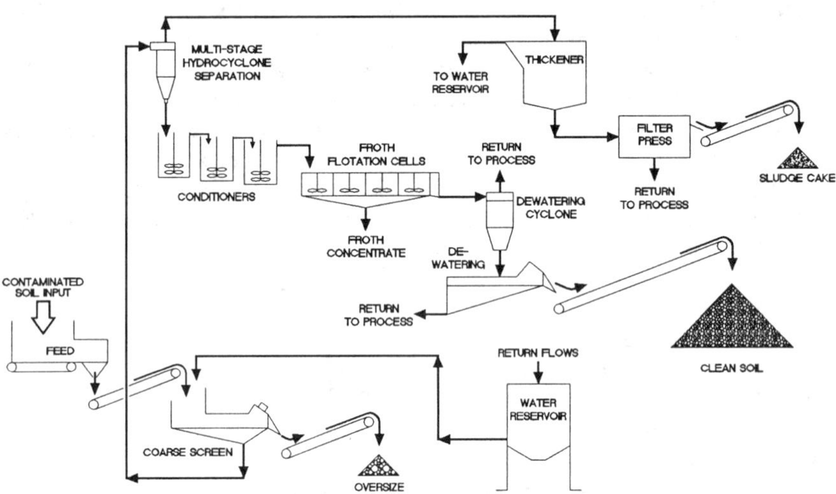

FIGURE 2: Soil washing process design (KOP-soils)

4. CONCLUSION

The fundamental step by step approach as followed in the design of a soil wash process for the KOP-site was found to be very succesfull. As a result a soil washing process could be designed that was fully optimized for the treatment of KOP-heavy metal contaminated soils. The pilot run at the Heidemij Soil Wash facility at Moerdijk confirmed the effectiveness of the process and EPA-approval was obtained for project implementation in the field. A mobile 25 t/hr capacity mobile soil wash plant was designed and built for on site erection. The modular plant design concept allows for easy transportability of the plant and allows for changes in process lay-out in future projects.

IN-SITU BIOREMEDIATION OF HYDROCARBON IN SOIL: PILOT TESTS AND FIELD EXPERIMENTS

Katalin Gruiz

Technical University of Budapest Department of Agricultural Chemical Technology
1111 Budapest, Gellért tér 4. HUNGARY

1. ABSTRACT

Laboratory and field experiments were carried out for bioremediation of soils polluted by fuel oil and motor oil. Bioventing was combined with the application of selected bacteria and dissolved nutrients. In the field experiments soil gas was evacuated by ventilators from the permeable boreholes.The process was followed by both soil and gas analysis. The biodegradation of the oil pollutant and of the microbial activity was measured by the oil and cell concentration in the soil. In two months the oil content decreased considerably, the cell number increased by one order or more. The evacuated gas was tested for CO_2, O_2 and volatilized hydrocarbon content. The CO_2 level proves the presence of biodegradation: a permanent high value - about ten times the normal - could be measured for two months, to slowly decrease in the third month. Volatilized hydrocarbon content was the highest in the first two days, after a continuous decrease it became unmeasurable for the third month.

Selective biodegradation of hydrocarbon mixtures (oily wastes) was investigated as well: gas chromatographic oil analysis showed the changes in the oil composition. The appropriate micro flora worked in an ideal commensalism, and as a result, all of the hydrocarbon components degraded at about the same rate.

2. LABORATORY PILOT TESTS

The in situ oil biodegradation in soil was modelled in a 15 cu.dm tank with air exhaust. The bottom of the tank was perforated, and there was a connection for the water jet pump on the top. Experiments were carried out both with soil contaminated artificially, and with soil undergoing pollution for years.

The necessary amount of nitrogen source was calculated by means of the following equation.

$$C_7H_{12} + 5\,O_2 + NH_3 = C_5H_7N + 2CO_2 + 4H_2O.$$

Molecular weight   96   16   14 (N)   113 (biomass)

C_7H_{12} is a hypothetical hydrocarbon, C_5H_7N is empirical formula of the microbial cell. Twice the calculated quantity of N-source was added in the form of ammonium nitrate in five steps. The amount of the P was one fifth of the amount of N in grams.

Experiment I.: The soil was artificially polluted with an 1:1 mixture of 20 000 ppm (2%) motor oil and Diesel oil. A loose garden soil was used in the experiment, easily transmitting air. The starter culture was a mixture of ten bacterial strains, isolated from a naturally adapted mixed microflora from the shore of an oil-polluted open channel. The inoculum contained $8*10^9$ cells in one cu.cm. 100 cu.cm of this mixture was injected into the soil.

Changes in the oil content and the cell concentration of the soil are shown in Table 1. Oil content was determined after Soxhlet extraction of air dried soil samples.

Table 1: Decrease in the oil content [percentage] and cell concentration [cell/g soil] as a result of
         aeration and supplementation

weeks	control		only aeration		aeration microbes nutrients	
	oil	cell	oil	cell	oil	cell
2	100	$3.1*10^6$	66	$2.0*10^7$	52	$8.2*10^7$
4	100	$4.6*10^6$	43	$3.6*10^7$	36	$9.5*10^7$
6	89	$3.9*10^6$	38	$3.9*10^7$	25	$9.5*10^7$
10	90	$3.5*10^6$	19	$6.0*10^7$	8	$9.2*10^7$
20	82	$5.2*10^6$	3.5	$8.1*10^7$	1.5	$1.1*10^8$

F. Arendt, G.J. Annokkée, R. Bosman and W.J. van den Brink (eds.), Contaminated Soil '93, 1163–1164.

The results show that microbial activity and oil biodegradation could be much increased by aeration. Further increase occurred as a result of adding microbes and nutrients to the soil. Composition of the residual oil was investigated by capillary gaschromatography. Chromatogram of the residual oil is similar to that of the undegraded control. This means that there is no considerable scale of selection -- quicker biodegradation of easily degradable hydrocarbon components.

Experiment II: Fuel oil pollution for several years was found at the coal mining site of Tatabánya, Hungary. The polluted soil was put into the model-equipment. The average oil content was 60 000 ppm (60g/kg). The polluted sandy-gritty soil was aerated, starter culture and nutrients were added in the same way as in, Experiment I. The quantitative results were very bad, the oil content hardly decreased: oil content was 80-82 percentage after 10 weeks in the treated tanks and 96 percentage in the control.

The gaschromatographic analysis gave the explanation for this extremely slow biodegradation. The easily degradable hydrocarbon components, e.g., normal paraffines had been consumed by the spontaneously formed soil microflora, but the isomeric, the cyclic and the long-chain components got enriched in the soil. Biodegradation of these components is slower, and needs special microflora.

3. FIELD EXPERIMENTS

An area 5 m by 10 m in the territory of the Tatabánya Mines, Hungary was polluted 4 meters deep with heating oil rather inhomogeneous. Oil content of the soil ranged from 1000 to 20 000 ppm. Bioremediation was carried out by a bioventing-type method developed jointly by the Technical University of Budapest and the Tatabánya Mines.

The technology: Three rows of boreholes were drilled on either side of the contaminated area and one row in between. Soil gas was evacuated by ventilators through the middle row of boreholes. Lateral boreholes served for intaking atmospheric air to the underground area. Trials were made to increase pressure by compressors in lateral boreholes. The tubes in the boreholes were impermeable in the upper region (1,5m), but permeable deeper down, at 1,5 to 5 meters. The underground air-flow was thanks to this construction- nearly linear between the two rows of boreholes.

Nutrients (N, P) and oil degrading microbes were added across the boreholes in three cases during the 3 months passed. Field experiments are still going on.

Results of the field experiments. Oil content of the soil at different depths is shown in **Table 2**. Samples were taken near the boreholes, at 30-50 cm distances .

Table 2: Oil content [ppm] of the soil near boreholes C_1, C_4 and C_5, after 2 and 3 months

depth	C_1			C_4			C_5		
cm	start	2 mo.	3 mo.	start	2 mo.	3 mo.	start	2 mo.	3 mo.
0-50	3600	420	280		520	330	650	320	260
50-100	8600	1500	1500	19500	210	100		310	100
100-150	12700	3360	2300		540	330		280	270
150-200	8800	1790	980	18700	840	250	960	370	170
200-250	6000	1010	690	15000	460	430	860	110	180
300-350	3300	990		400			240	80	
350-400	9600	800		100			650		

The oil content decreased considerably in three months. The characteristic control value of the area is 170-380 ppm oil. The bacterial cell number increased during the process. After 2 months, cell number was 5-50 times higher, than those at the beginning: $10^5 \Rightarrow 10^6\text{-}10^7$ cell/g soil.

CO_2 **content of the evacuated gas** increased from an average 0,5 % by vol. to 3-4 % by vol. during aeration. This increased value is characteristic of bioventing in the first 3 months. After a slight maximum (at 2 months) a slow decrease appears.

Comparing CO_2 values in the two alternative cases, evacuation by venting is much more effective than injection of air by compressors. In an other experiment one row of boreholes served for evacuation the other row for air jet. The absolute CO_2 quantity did not increase with the higher air flow.

The volatilized hydrocarbon content of the vent gas had a relatively high value in the first 48 hours with a maximum of 0,04 % by vol. In the first two months it changed from 0,01 to 0,001 % by vol. After 3 months there was no measurable volatile hydrocarbon quantity in the evacuated gas.

THE PREUSSAG ANLAGENBAU APPROACH TO SOIL WASHING:
OPTIMIZED EXTRACTION OF INORGANIC AND ORGANIC CONTAMINANTS

Dr. Frank Härig
c/o Preussag Anlagenbau GmbH
Kaiserstraße 4
W-2300 Kiel 14
Germany

INTRODUCTION

Soil washing has been chosen for approximately 60 % of all ongoing soil remediation projects in Germany. This illustrates that physico-chemical processing is recognized as a useful and efficient approach towards soil sanitation, as are biological and thermal techniques.

The range of contaminants removable by means of soil washing includes heavy metals, cyanides, mineral oil and its derivatives, as well as aromatic hydrocarbons and chlorinated hydrocarbon compounds.

Soil washing services are being offered by an increasing number of contractors. Nevertheless, there is only a small number of soil washing techniques which have proved reliable on a broad basis of applicational experience.

In the field of soil remediation, Preussag Anlagenbau GmbH cooperates with the Dutch company Heijmans Milieutechniek. Experience has grown in the course of cleaning more than 300000 t of contaminated soil.

In Germany Preussag Anlagenbau GmbH runs a stationary soil washing plant in Kiel.

F. Arendt, G.J. Annokkée, R. Bosman and W.J. van den Brink (eds.), Contaminated Soil '93, 1165–1169.
© *1993 Kluwer Academic Publishers. Printed in the Netherlands.*

GENERAL DESCRIPTION

The Preussag Anlagenbau soil remediation technique can be characterized as a physico-chemical washing process with water as extractant, where mineral acids, alkaline solutions, oxidizing agents or surfactants are added according to the type of contaminants concerned. The process is particularly effective for degrading or removing cyanides, heavy metals and hydrocarbons from soil containing fines, i. e. silt and clay, up to 25 %.

Mechanical energy is applied to remove contaminants from coarse soil components and to transfer them into the washing fluid. As a next step, contaminants are readsorbed by suspended fine material. There is a considerable adsorptive capacity in clay and silt, which results from a specific surface exceeding 4000 cm²/g. The fine fraction is separated and treated further, or it must be disposed of.

Fig. I illustrates the washing process which consists of 3 stages:

> (1) treatment of gravel
> (2) treatment of sand material
> (3) purification of washing fluid

As a first step, coase material > 40 mm is separated by means of a roller sieve. Iron particles are removed magnetically.

The soil is then conveyed to a sieve equipped with spray nozzles. The sieve retains the 8 - 40 mm fraction.

The undersize < 8 mm, which is the bulk fraction of the soil undergoes intensive wet scrubbing in an attrition mill. Water is used as a washing liquid. Friction and shearing abrase contaminants from the grains' surface and transfer them into the washing liquid. Chemicals are added to intensify the desorption process.

In the next procedural step hydrocyclones separate the cleaned sand fraction > 63 μm from the contaminated washing liquid. The sand is flushed with purified process water in an upstream separator, which also removes light particles such as straw, grass, roots, and other sorts of particulate carbon. The sand fraction is then dewatered and delivered as a reusable material.

Depending on the granulometry of soil suitable for cleaning by washing, the fine fraction < 63 μm amounts between 5 and 25 % (dry weight). These fine solids are precipitated in a lamella separator. In general the sedimentation process requires flocculants to be added. The precipitate is concentrated and dewatered in a belt-screening machine to yield a residue sufficiently solid for disposal or further processing. Being cleared from fine particulates the process water may still contain dissolved or emulsified contaminants and perhaps some particulate impurities. Further purification of the washing water is attained by alkaline precipitation of dissolved metals or flocculation of emulsified organics, followed by the flotation of precipitates and flocks, and substances skimming them off.

The water is then recirculated and used for the washing process.

Experiences

Using the process described here, a broad basis of experience has been acquired in remediating soil with most different types of contaminants. The soil originated from sites such as former gas works, galvanizing plants, fuel depots and mineral oil refineries.

The soil washing process is applicable to a broad scale of organic and inorganic contamination. The technology proved most effective in removing contaminants as follows:

1168

- mineral oil and fuel
- heavy metals (e.g. lead, cadmium, nickel, chromium, arsenic)
- anorganic compounds (e.g. soluble and complex cyanides, fluorine compounds)
- aromatic compounds (benzene, toluene, xylene)
- polycyclic aromatic hydrocarbons
- halogenated hydrocarbons
- insecticides, herbicides, fungicides

THE PREUSSAG ANLAGENBAU CONCEPT

Beginning in 1991 the Heijmans soil washing technique has also been applied in the Federal Republic of Germany. Preussag Anlagenbau GmbH has obtained the licence for building the plants required to run the technique. The license includes an option for all countries bordering the Baltic sea.

Considering the experience accumulated "on the job", a new plant was designed and built. It is a containerized semi-mobile plant intended for transportation, and for soil remediation on site. The capacity is approximately 10 t/hr. The plant has successfully been working since 1991. At present it operates in Kiel in a stationary mode. Formal allowance according to BImSchG (Federal Law of Immission Limitation) has been obtained for washing in Kiel soils from other places.

LINES OF FUTURE DEVELOPMENT

Soil washing produce secondary wastes, which remain to be disposed of at special sites. A subsequent treatment of these wastes by thermal, biological or chemical means is possible, and appears necessary for economical and ecological reasons.

The Preussag Anlagenbau activities in soil washing aim at developping complementary techniques for

- disintegrating organic contaminants in the fines residue
- precipitating residual heavy metal contamination
- immobilizing contaminants by binding them to inert material
- thermal treatment

Such additional steps are intended to reduce the amount of secondary wastes to be disposed of. If applicable, inert material might be reused, e.g. for construction purposes. At least most of the residues from soil washing should be rendered less toxic, and thus more easily disposable at lower-risk sites.

At present there are cooperative efforts with experienced partners for developping technologies for subsequent decontamination steps, and integrating them into the plants already operating.

<u>SUMMARY</u>

- The Preussag Anlagenbau technologie provides an approved method for remediating contaminated soil in Germany.

- There is no wastewater from soil washing, since the washing water is recycled completely. Watertight fundamentation makes sure that contaminated washing fluid will not be emitted to the environment.

- Multiple options in adjusting the technique allow to remediate in a single run soil containing different types of contaminants.

- In general decontamination ranges between 85 and 97 % of initial levels, with best results of 99 %.

DEVELOPMENTS AND OPERATING EXPERIENCE IN SOILCLEANING, A PROGRESS REPORT ON THE DEVELOPMENTS IN CLEANING SOILS WITH CHLORINATED HYDROCARBONS, AND OPERATING EXPERIENCE IN CLEANING CONTAMINATED GROUNDWATER WITH BIOROTORS.

Ir. H.J. van Hasselt

NBM Bodemsanering B.V.
P.O. Box 16032, 2500 BA The Hague Netherlands
Phone 31 70 3814331 / Fax 31 70 3476983

NBM Bodemsanering B.V., subsidiary of NBM Amstelland N.V., one of the biggest building and construction groups in The Netherlands, has developed an indirect heated thermal treatment system and became one of the leading firms in The Netherlands in the cleaning of soil and the execution of projects for soil pollution.

As a first area of interest, NBM elected to clean soil containing pollutants which are hard to break down. In the first phase, NBM concentrated on cyanide, aromatic hydrocarbons, and oily type pollutants.

NBM developed an indirect heated thermal system using a dryer and a rotary tube-furnace.
After laboratory experiments and a pilot plant phase, the first production plant was built in 1986. Up to june 1992, over 400.000 tons of soil, containing cyanides, aromatic hydrocarbons and oily type pollutants were cleaned. Typical cleaning results are concentrations of less than 1 mg/kg total PAH's and less than 1 mg/kg total cyanide, which means cleaning efficiencies of > 99.8 - 99.999 % .

As soon as the cleaning of soil contaning non-halogenated hydrocarbons was a normal daily practice, NBM carried out further research.
Based on the results of laboratory work at the technical university of Delft, NBM decided did practical tests at the production plant in Schiedam to prove that it is possible to clean soil contaminated with halogenated hydrocarbons, and that safe incineration of the offgases is possible.
The practical experiments were carried out in several phases.
First, in 1989, the incineration conditions were tested. While running the plant with an input of sand contaminated with fuel oil a known amount of a mixture of trichloroethene and tetrachloromethane was injected into the incinerator, and stack emissions were measured. The destruction efficiency was higher than 99,999 % .
Further experiments were carried out in 1990 and 1991. Soil containing a.o. aldrin, dieldrin, and lindane was processed during test runs. In these tests, the rest concentrations of contaminants, were below the detection limit of 10 µg/kg dry solids.

F. Arendt, G.J. Annokkée, R. Bosman and W.J. van den Brink (eds.), Contaminated Soil '93, 1171–1172.
© *1993 Kluwer Academic Publishers. Printed in the Netherlands.*

The stack emissions showed variable results for the several test runs. Partly, they were below the emission standard (0,10 ng/m³) set by the dutch authorities, partly they were above these standards. Values ranging from 0,074 up to 2,01 ng/m³ were measured in the first series of test runs.

Also in the second series of tests, the rest concentrations of contaminants in the soil were below the detection limit. The stack emissions of dioxins varied between 0.2 and 3 ng/m³. No other contaminants were detected in the stack emissions above the dutch standards.

Based on the results of the tests, NBM Bodemsanering decided to build a flue gas cleaning system. The flue gas cleaning system will be of the dry adsorption type, and will be installed in 1993.

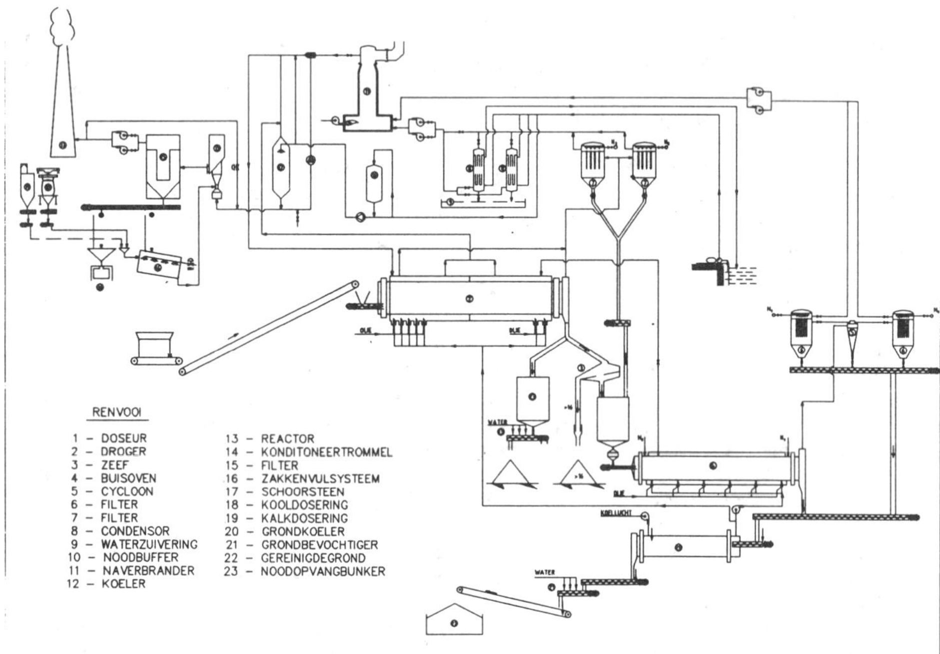

NBM Bodemsanering also carries out remediation projects on site. In Holland, very often groundwater remediation is necessary. For projects that take a period of approx. 1 year and longer, the water purification with biorotors proved to be very succesfull. The method is also cost effective because of low maintenance cost and low consumption of energy and additives.

Groundwater contaminated with cyanide, aromatic hydrocarbons and polyaromatic hydrocarbons can succesfully be treated to values below dutch standards. With an input concentration of 60 mg/l of aromatic hydrocarbons, output values < 1 mg/l have been acheived with a single biorotor. At the conference, data from several projects will be presented.

3D VERTICAL CIRCULATION FLOWS AROUND GROUNDWATER CIRCULATION WELLS (GZB) FOR AQUIFER REMEDIATION: NUMERICAL RESULTS

B. Herrling, J. Stamm

Institute of Hydromechanics, University of Karlsruhe
Kaiserstrasse 12, D-7500 Karlsruhe, Germany

Three-dimensional vertical circulation flows around wells with two screen sections, so called "groundwater circulation wells" (GZB), are an important subject of numerical investigation. Often these flow systems are situated at several locations beside one another and can be located within a natural groundwater flow field. The circulation flow around the wells is initiated by vertical pumping measures within the respective well casing. The two screen sections are placed at the bottom and top of an aquifer.

At numerous locations in Germany and recently in the United States among other countries, this technology (patented by IEG mbH, D-7410 Reutlingen) is used for physical and/or recently for biological in situ groundwater remediation [1,2,3,4]. This includes in situ stripping of volatile organic compounds in the vacuum zone of the upper well casing (vacuum vaporizer well; abbr.:UVB) or in situ biodegradation by controlled addition of nutrients and/or electron acceptors as the groundwater passes through the well casing. Moreover, partial discharge withdrawal and infiltration from the total discharge through the well casing with two screen sections have been investigated numericly and are used in practice: The possibilities of no lowering of groundwater levels near the well and continuous groundwater extraction in case of low well capacity make this technique to an interesting alternative to conventional pumping; further this idea can be used as well for infiltration of treated groundwater without any raising of the groundwater table at the infiltration well [2,5].

Figure 1 clarifies the resulting vertical cirkulation flow by streamlines plotted in a vertikal plane section parallel to the natural groundwater flow. In a highly contaminated source area sometimes it may be useful to arrange a grid of UVB/GZB installations to get an extreme strong flow between the wells. In figure 1 the flow for two wells behind each other is demonstrated. Figure 2 describes the curved separating stream surface of the well's capture zone and the treated water bodies of the circulation flow around the well and of the flow moving downstream, demonstrated for three wells which mutually affect one another. A line of wells situated normal to the natural groundwater flow can be employed as an in situ "treatment wall" cleaning a passing wide contaminated plume.

Using dimensionless diagrams, the sphere of influence of a well is described and discussed as well as the upstream cross section of the capture zone, the stagnation point distance, the maximum well distance such that no groundwater can pass between the wells without having been treated, and the dilution ratio between the upstream water flowing into the capture zone and the circulating groundwater around the well [1-5].

Concerning the partial discharge withdrawal or infiltration, the extracted or infiltrated quantity of water from the total pumping amount has been calculated and presented using dimensionless diagrams for the condition of no changing groundwater head at the well top. Additionally, the small amount of maximum head lowering or raising and the distance from the well axis, where this value appears, have been presented in diagrams [5].

REFERENCES

[1] Herrling, B., Buermann, W., and Stamm, J.: Hydraulic Circulation System for In Situ Bioremediation and/or In Situ Remediation of Strippable Contamination, In: R.E. Hinchee and R.F. Olfenbuttel (Eds.), In Situ Bioreclamation, Applications and Investigations for Hydrocarbon and Contaminated Site Remediation, Butterworth-Heinemann, Boston, pp. 173-195 (1991).

F. Arendt, G.J. Annokkée, R. Bosman and W.J. van den Brink (eds.), Contaminated Soil '93, 1173–1174.
© 1993 *Kluwer Academic Publishers. Printed in the Netherlands.*

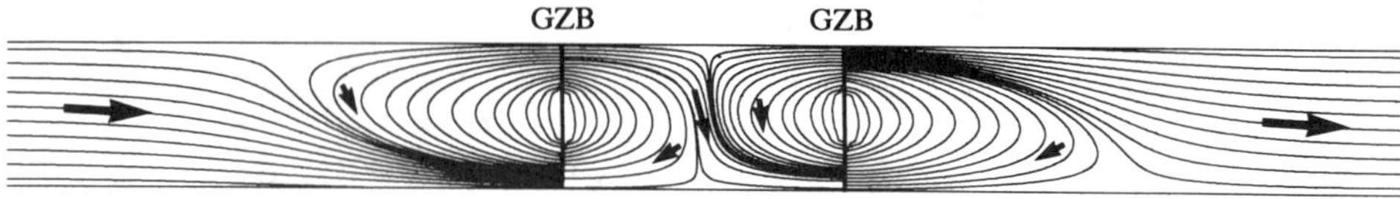

FIGURE 1   Streamlines in a vertical longitudinal section parallel to the natural groundwater flow field demonstrating the circulation flow around two UVB or GZB in one line one behind the other

(a)   (b)

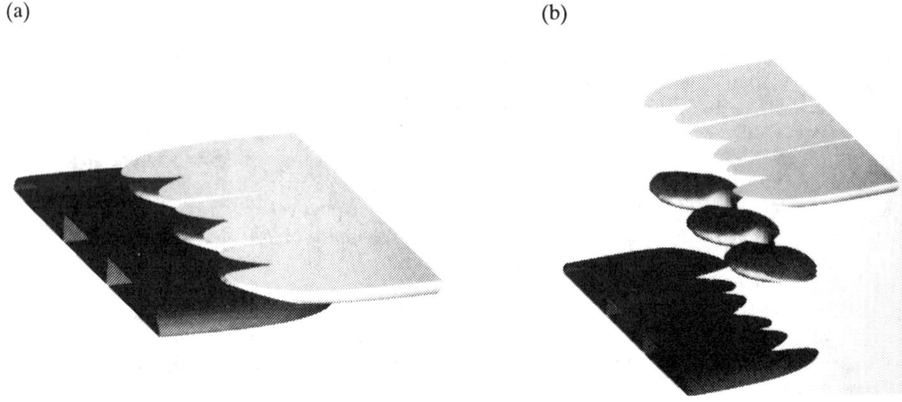

FIGURE 2   Separating stream surfaces of the different water bodies in the outside flow around three UVB or GZB situated in one line normal to the natural groundwater flow: captured, circulating, and flowing downstream water in (a) a real situation and (b) water bodies separated for clarification (copies of a computer screen photo)

[2] Herrling, B., Stamm, J.: Groundwater Circulation Wells (GZB) for In Situ and On-Site Aquifer Remediation, In: Proc. SPECTRUM '92, Int. Topical Meeting, Nuclear and Hazardous Waste Management, August 23-27,1992, Boise/Idaho, publ. by Am. Nucl. Soc., pp. 1188-1193 (1992).

[3] Herrling, B., Stamm, J., Alesi, E.J., and Brinnel, P.: Vacuum Vaporizer Wells (UVB) for In Situ Remediation of Volatile and Strippable Contaminants in the Unsaturated and Saturated Zone, Proc. Symp. on Soil Venting, April 29-May 1, 1991, Houston/Texas, EPA/600/R-92/174, pp. 203-228 (1992).

[4] Herrling, B., Stamm, J.: Vertical Circulation Systems for In Situ Bioreclamation and In Situ Remediation of Strippable Contaminants, in: Proc. Int. Symp. on Soil Decontamination Using Biological Processes, December 6-9,1992, Karlsruhe/Germany, in press (1992).

[5] Herrling, B., Stamm, J.: Numerical Results of Calculated 3D Vertical Circulation Flows Around Wells with Two Screen Sections for In Situ Aquifer Remediation, in: Computational Methods in Water Resources IX, Vol.1: Numerical Methods in Water Resources , Eds. T.F. Russell et al., Elsevier Applied Sciences, London, pp. 483-492 (1992).

Concept of a redevelopment of a former gasworks plant area

Jaar, M., Krebs, H.; Kilian, U.; Sauer, I.; Loock, R.;Mauck, P.

Holsteiner Gas-Gesellschaft mbH, Kaiser Wilhelm Strasse 115,
2000 Hamburg. Germany.

1. Starting Situation and Problematic

Analyses of soil, air and water in the area of a former gasworks plant situated in Hamburg showed contamination of different areas and in different layers which was possibly caused during the gas production procedure or by neighbouring factories. The contamination consists of organic compounds such as stationary PAH, BTEX, petrol components and non-organic compounds such as cyanides.

2. Concept of Redevelopment

To redevelop former gasworks sites generally different ways of treatment have to be used, considering the variety of pollutants.
In this special case the following possibilities are discussed:

combined physical chemical treatment; thermal treatment; disposal

2.1 Objective

The primary target of the redevelopment is to avoid the spread of contamination on the waterway. As a second step the first aquifer should be purificated, but because of the extent of the contamination a cooperation with neighbours will be necessary to succeed.

In order to prevent further wash of pollutants towards the aquifer a complete sealing of the surface is planed.

To support the ground water purification measures for soil and soil air treatment will be undertaken.

Priority of the first step is the treatment of soil and will be accompanied by de-watering and de-gasing measures, with treatment of water and air. For the hydraulic measures following procedures will be taken into consideration:

1) Complete removal of the soil pollution
2) Removal of the desolved contamination

The contamination of the gaswork consists of compounds with different solubility and some of them were found in high concentrations in some areas. An exclusive hydraulic treatment would take a long time period, so that the second procedure seems to be more practicabal. The not yet desolved pollutants have to be screened against any water flow and in a second step the polluted soil will be treated off-site.

2.2 Methods of Redevelopment

The water phase of the encapseld area and especially the high polluted part of the first aquifer will be treated in the following procedure (Fig.1):
oil seperator - precipitation/flocculation - chemical oxidation - adsorption on activated carbon - biological nitrification

After the treatment the water will be re-infiltrated into the contaminated ground to wash the soluble and volitle pollutants. This measure also supports the biodegradation of pollutants by importing oxygen into the ground.

F. Arendt, G.J. Annokkée, R. Bosman and W.J. van den Brink (eds.), Contaminated Soil '93, 1175–1176.
© 1993 Kluwer Academic Publishers. Printed in the Netherlands.

1176

2.2.1 Soil Treatment in the tar contaminated area

The tar containing holes and their environment have to be encapseld, de-watered and de-gased in the first step (Fig.2). In the following process of soil excavation there still has to be a precautional air exchange in the working area, with biological and activated carbon filter.

The thermal treatment of the contaminated soil seems practicable, but cost intensive, because suitable plants can only be found in foreign countries at the moment. To proceed further steps of the redevelopment the excaved tar polluted soil will be stored in containments till an accepticle way of treatment is found. Soil of the tare hole environment, which seems to be less contaminated, will be treated by a less expensive washing process.

2.2.2 Soil Treatment of the cyanide contaminated area

One area is contaminated with cyanide compounds in different layers and depths. The concept provides a plan for excavating the whole area, but to reduce the amount of contaminated soil which has to be washed off-site by classifying the ground during the excavation and maybe by separating fine (contaminated) and coarse grained sands. The treatment will probably be a combined chemical physical washing process. The remainings have to be deposited.

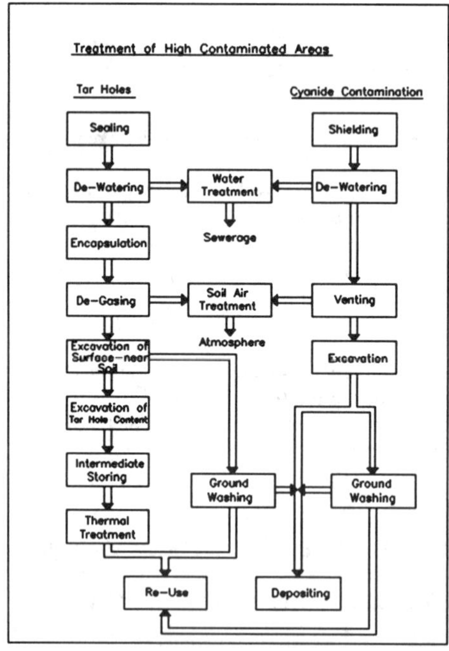

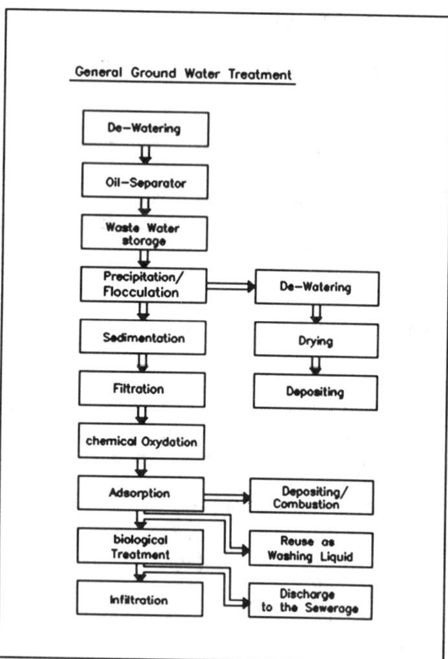

Fig.1: Treatment of high contaminated areas

Fig.2: Ground water treatment

TESTING AND EVALUATION OF SOIL AND GROUNDWATER REMEDIATION PROCESS: ACT*DE*CON^{SM} AND MAG*SEP^{TM}*

Donald O. Johnson, Argonne National Laboratory
Dornald E. Edgar, Argonne National Laboratory
Michael L. Wilkey, Argonne National Laboratory
Michael J. Dunn, Bradtec - US, Inc.
Michael P. Marlone, Program Manager, U.S. Department of Energy

BACKGROUND

Low-level radioactive, hazardous, and mixed contamination remediation is a major concern within the U.S. Department of Energy (DOE) complex at sites located across the United States. Soils, fly ash, waste sludges, surface and groundwater, and other materials must be treated so that they no longer pose a threat to human health and the environment. The contaminants tend to be primarily inorganic in nature and are locally present in significant volumes.

Throughout its Office of Technology Development (OTD), DOE is supporting research to develop and evaluate processes for the treatment of hazardous, mixed, and radioactive wastes in soil and groundwater. To perform this research, evaluation, and demonstration process more effectively, DOE takes full account of innovative treatment technologies being developed in the private sector. Two such technologies are ACT*DE*CON^{SM} and MAG*SEP^{TM}*, processes developed by Bradtec, Ltd. (UK), and licensed to Rust Remedial Services, Inc.

Argonne National Laboratory (ANL), a DOE facility, is providing OTD with credible, defensible and timely information about the performance and costs of selected innovative and promising treatment processes for specific contaminated soil and water systems as found at sites located within the DOE complex. This support activity consists of a number of integrated tasks focused toward the goal of evaluating remediation technologies for immediate demonstration and implementation to facilitate attainment of the stated 30-yr goal of total environmental compliance within the DOE complex.

Argonne has established a collaborative program integrating small business technology development (Bradtec) with private sector interest in technology application (Rust Remedial Services) under the umbrella of National Laboratory scientific and engineering expertise.

APPROACH

Bradtec has developed the ACT*DE*CON^{SM} process for use in soil washing to chemically dissolve and remove metal and actinide contamination from soils and sediments. The process uses benign chemicals, thus eliminating concern over the potential for residual chemicals causing further contamination. This dissolution solvent is then treated with Bradtec's MAG*SEP^{TM} process to remove the dissolved contaminants. The contaminants can be recovered for recycle or stabilized for disposal.

An in-situ ACT*DE*CON^{SM} process is currently under investigation for the treatment of soils and sediment from the DOE Mound plant in Miamisburg, Ohio, a former plutonium-processing facility.

In Phase 1 of this study, ANL evaluated the ACT*DE*CON^{SM} process on soil materials spiked with Pu to levels similar to those in the abandoned Miami-Erie canal at the Mound site. Because of the promising results of these laboratory tests, bench-scale testing is being conducted in Phase 2. If this bench-scale performance testing proves successful, ANL will proceed to pilot-scale performance tests and process design and cost estimation for full-scale implementation.

---

* Work supported by the U.S. Department of Energy, Assistant Secretary for Environmental Restoration and Waste management under contract W-31-109-Eng-38

F. Arendt, G.J. Annokkée, R. Bosman and W.J. van den Brink (eds.), Contaminated Soil '93, 1177.

THERMAL SOIL CLEANING PLANT
HOCHTIEF

Dipl.-Ing. H. Kimmel
HOCHTIEF Aktiengesellschaft
vorm. Gebr. Helfmann
Rellinghauserstraße 53-57
D-4300 Essen 1

Base: Herne/Westfalen/Germany

Capacity: 7t/hour or 35 000 t/year

start of operation: June 1992

PROCESS DIAGRAMM

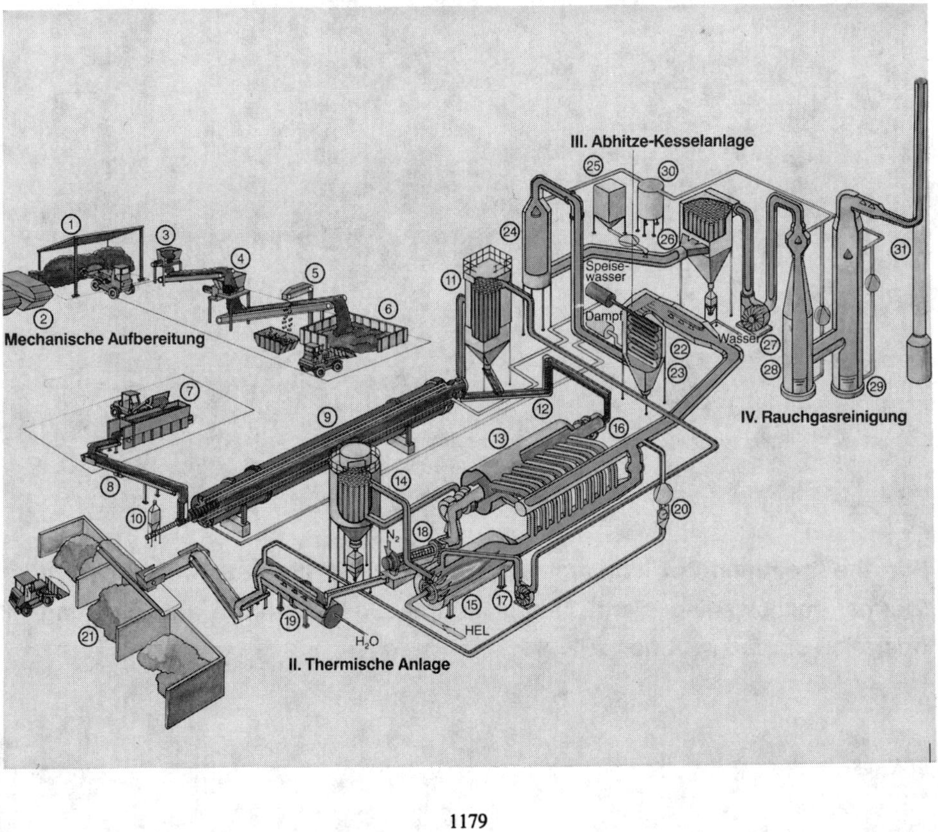

F. Arendt, G.J. Annokkée, R. Bosman and W.J. van den Brink (eds.), Contaminated Soil '93, 1179–1180.
© 1993 Kluwer Academic Publishers. Printed in the Netherlands.

THERMAL SOIL CLEANING PLANT HOCHTIEF

Base: Herne/Westfalen/Germany
Capacity: 7t/hour or 35 000 t/year
start of operation: June 1992

For the treatment of contaminated soil HOCHTIEF has developed a thermal soil cleaning plant. The plant was built in Herne and cleaning operation started in June 1992.

In-Situ Bioremediation of Underground Storage Tank Sites Using Different IEG-Circulation Methods Combining Physiological and Biological Treatment Techniques.

Ligner T[1], Brinnel P[2], Lehr C[2], Blömer R[3], Adam R[3]

1. IMA GmbH, Teichmannstr. 16, O-7125 Liebertwolkwitz, Germany
2. Protec GmbH, An den drei Hasen 21, W-6370 Oberursel, Germany
3. IMA GmbH, Admiral-Rosendahl-Str. 16, W-6073 Neu Isenburg, Germany

At a former truck filling station in Brandenburg-Vorpommern, Germany, three different sites are polluted with diesel oil. Approximately 800 m^3 soil is contaminated with concentrations between 2000 and 8000 mg of hydrocarbons/kg. The concept for the remediation of these sites will be presented.

As oxygen is one of the most limiting factors by in-situ bioremediations, a new combination of soil-aeration and biological treatment has been designed in order to stimulate the underground biological degradation of hydrocarbons.

At the first site, the contamination in the unsaturated zone, in the capillary fringe and in the uppermost section of the aquifer will be treated using a combination of "IEG-circulation-methods". A "soil air circular flow system" (-BLK) is applied to remove contaminants form the unsaturated zone. Additionally the capillary fringe and the groundwater will be aerated using the "coaxial ground water aeration system (-KGB). Thus the microbial activity will be stimulated both in the unsaturated zone and in the ground water.

At the second site the ground water contamination will be treated using a spezial "vacuum vaporizer well" (UVB/GZB). A circular flow regime will be established in the aquifer. The flow path of the contaminated water will be prolonged to ground level. For the biodegradation of the diesel oil it will pass a bioreactor befor it runs back into the underground circle.

At the third site two big concrete tanks are used as biobeds for on-site treatments. To supported biodegradation by passive soil aeration while adding nutrients and water via a circulating nutrient solution, conventional soil air venting with its effects of soil drying should be avoided. To keep the soils moisture, the "double cased capillary screen" (DMF) is used for "soil air venting with passive soil aeraton"(IEG-BLA). The off air is treated using active carbon filters.

F. Arendt, G.J. Annokkée, R. Bosman and W.J. van den Brink (eds.), Contaminated Soil '93, 1181.
© 1993 *Kluwer Academic Publishers. Printed in the Netherlands.*

THE GEUZENHOEK PROJECT - FIELD TESTS OF CHEMICALLY INTERACTIVE SYSTEMS
FOR CONTAINING HIGHLY CONTAMINATED DREDGING SPOIL

M. Loxham [1], E. van den Eede [2] and J. Taat [1]

[1] Delft Geotechnics, P.O. Box 69, 2600 AB Delft, The Netherlands
[2] Ministerie van de Vlaamse Gemeenschap, Departement Leefmilieu en Infra-
structuur, Bestuur Havens, Simon Bolivarlaan 30, 1210 Brussel, Belgium

1. INTRODUCTION

Experience gained in the design and construction of the first generation
of major dredging spoil disposal sites at Rotterdam and Gent has focused
attention on the value of containment systems that chemically interact
with pollutants leaching out of the site. Theoretical studies have been
able to identify significant long and short term advantages in such
technologies, especially in mitigating the consequences of hydrological
failure of the barrier at some point in its lifetime, a major cause of
concern in the design and permitting process. In order to evaluate, at
field scale, these concepts, the Belgian Ministry of Environment and
Public Works have initiated a five year program centred around seven large
pilot containment compartments at their disposal site Geuzenhoek in
Belgium.

2. EXPERIMENTAL

In the seven experiments differing liners were installed in compartments
chosen large enough to enable the barrier technologies to be constructed
using full scale civil engineering equipment. Each compartment is 500
square meters large and has been filled using a common sludge which was
further mixed and blended. The following liner technologies have been
chosen for study under various conditions:

1) Untreated dredged material   2) Glauconite
3) Mix in place clay-sand systems   4) Pre-treated dredged material
5) Peats   6) Biological treatment
7) Vertical screens

The test compartments have been filled with actual contaminated spoil from
the canal between Gent and the Eastern Scheldt. The sites have been carefully monitored by an extensive testing and measuring program from the
beginning following the different exploitation phases encountered in
practice.

The field test have been backed up by a series of laboratory test and
laboratory scale simulations. Mathematical models are used for the evaluation, the collation and the extrapolation to larger scales of the experimental results. The overall methodology of the study is shown in figure 1.

F. Arendt, G.J. Annokkée, R. Bosman and W.J. van den Brink (eds.), Contaminated Soil '93, 1183–1184.
© 1993 Kluwer Academic Publishers. Printed in the Netherlands.

3. RESULTS

The monitoring system was designed around intercepting leachate just below the liner in the unsaturated zone using porous geomembrane mats sloping horizontally to a central collection sump. The system was presimulated for a given stimulus-response behaviour using finite difference contaminant migration codes.The predictions were checked against tracer experiments carried out using a sodium iodide tracer added to the sludge upon filling. The results of the tracer experiment are shown in figure 2. where it can be seen that the presimulation and the experimental results agree quite well.

The results from the filling period also show that the volume of leachate percolating in the first few days is much larger than after some sedimentation and colmatation has taken place. This effect will have significant consequences for the filling regimes of such sites.

The program is on-going but in the 500 days to-date no significant release of heavy metals in the leachate has been detected.

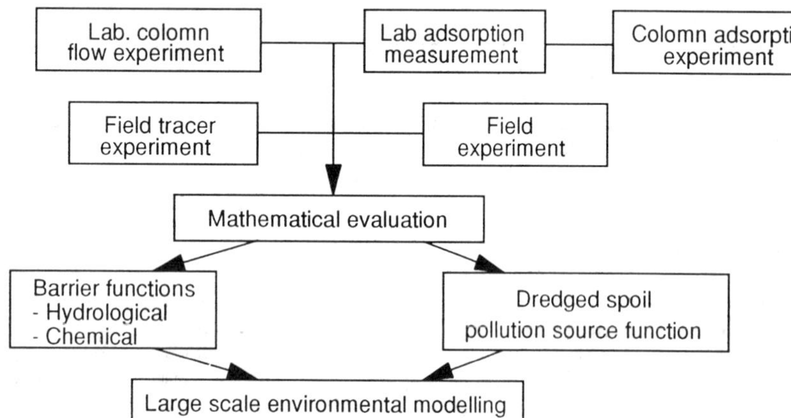

Figuur 1. Schematic overview of the Geuzenhoek project

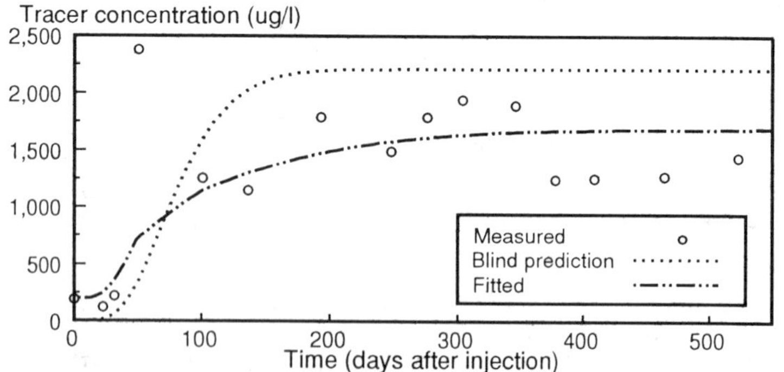

Figuur 2. Tracer experiment in the reference compartment

Treatment of Contaminated Soils by a Combination of Suitable, Proven Technologies

Klaus Mackenbrock
Deutsche Babcock Anlagen GmbH, Abt. 3230 - Recyclingverfahren
Duisburger Straße 375; 4200 Oberhausen 1

A critical evaluation of soil decontamination as practised so far shows that soil remidial action techniques shall in future be determined by a combination of several technologies which have so far only been applied individually. Three reasons lead to this decision:

1. Soils with a complex mixture of contaminations, which have to be treated in major site clearance mainly, cannot be satisfactorily treated by only one technology.

2. Soils to be treated at the moment consist of medium- lor small-scale lots. Type and contamination of these soils are continuously varying and optimum treatment has to be adjusted to each individual soil.

3. All kinds of treatment produce residual fractions which have to be treated further.

Bearing in mind these facts, each contamination process has to be re-evaluated for each individual clean-up activity. Feasibility and limitations of all technologies, as well as advantages and disadvantages of combinations of biological, chemico-physical and thermal methods have to be weighed up, considering ecologic and economic aspects. This will be of decisive importance for the design and operation of so-called soil remidiation centres.

Basically, the following treatment methods have to be considered:

- physico-chemical treatment
washing with or without additives
washing with or without oxidation agents
extraction with non water-soluble agents

- thermal treatment
high-temperature treatment, direct combustion up to 1,200 °C
high-temperature treatment, direct + indirekt combustion at 1,200 - 800 °C
low-temperature treatment, indirect heating up to approx. 450°C (pyrolysis)

- biological treatment
static processes in stacks resp. "bio-pits"
dynamic processes for the treatment of preparated soils and soil suspensions in bio-reactors.

A soil remidiation centre which is to cover a large scale of different types of soils and contaminations, should be equipped with all available methods. For each application, they may then be optimized and/or combined as required. This is the only way to obtain soil treatment with a minimum of residues, which in turn ensures that no contaminated sites are created for the future.

A combination of several technologies leads to a number of criteria for the decision and consequences which have to be considered for day-to-day operation of a remidiation centre.

F. Arendt, G.J. Annokkée, R. Bosman and W.J. van den Brink (eds.), Contaminated Soil '93, 1185–1186.
© 1993 *Kluwer Academic Publishers. Printed in the Netherlands.*

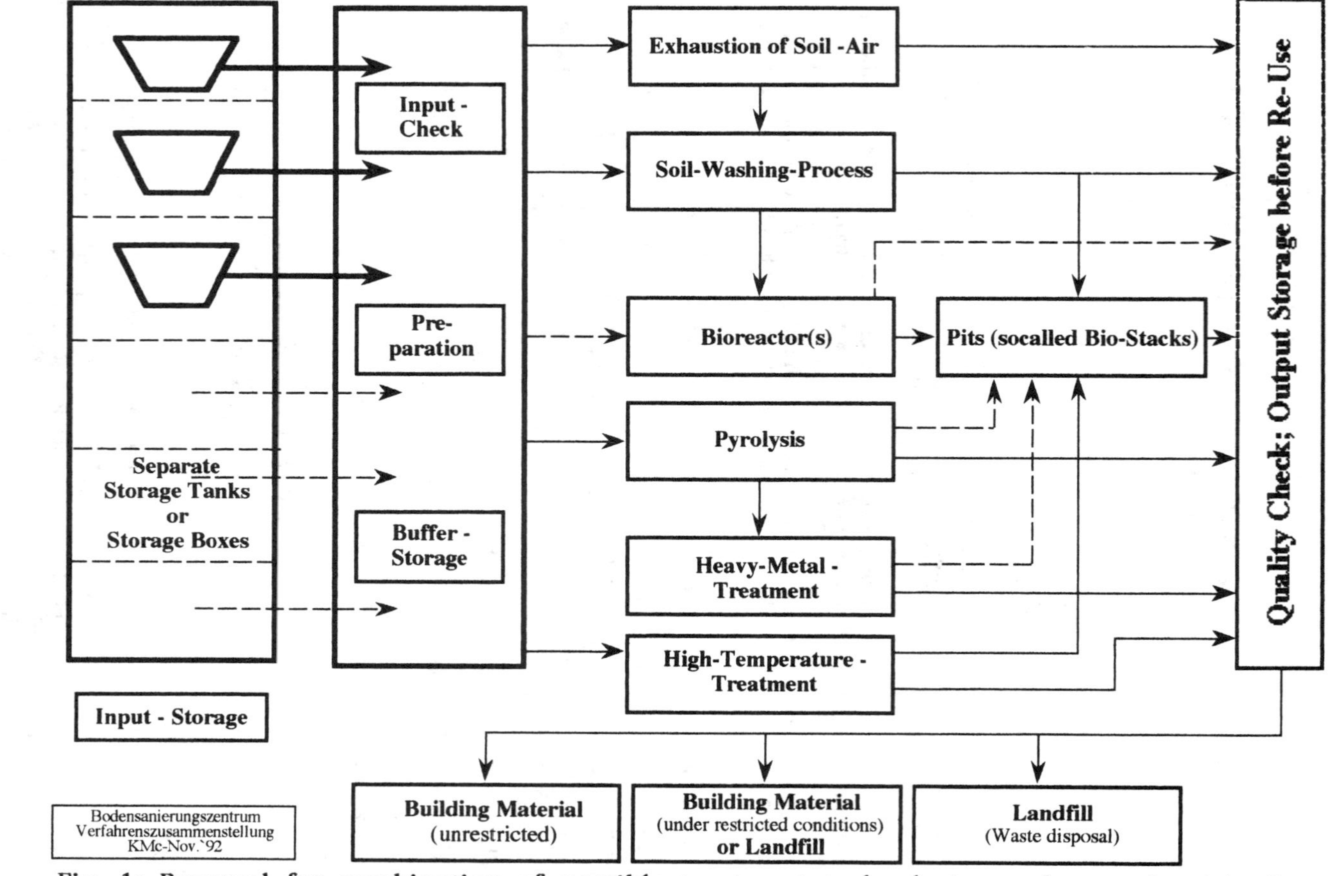

Fig. 1: Proposal for combination of possible treatment technologies and procedure(s) of decontamination in a complex soil remidiation centre.

COST EFFECTIVE BIOLOGICAL GROUNDWATER TREATMENT

E.H. MARSMAN, H.B.R.J. VAN VREE, B.A. BULT, L.G.C.M. URLINGS

TAUW INFRA CONSULT B.V., P.O. BOX 479, 7400 AL DEVENTER, THE NETHERLANDS

1. INTRODUCTION

In industrialized countries, many sites are contaminated with organic compounds. The soil as well as the groundwater and soil vapour are contaminated. For infrastructural reasons, in situ soil remediation would be advantageous. However, the costs of groundwater and soil vapour treatment are considerable. Techniques which are cost effective have yet to be developed. Biological treatment techniques can be used as many of the contaminants are biodegradable. A new aerobic fixed film bioreactor "the BIOPUR®" has just been developed to treat both the soil vapour and the groundwater simultaneously.

2. BACKGROUND

TAUW Infra Consult B.V. has used biological groundwater treatment in fixed film reactors, since 1986. Initially, trickling filters were used followed by Rotating Biological Contactors (RBC). Groundwater heavily contaminated with BTEX, mineral oil, mono-chlorobenzene, phenols, naphthalene and HCH were treated. High removal efficiencies were achieved. Volatilization of organic compounds hardly ever occurred.

3. BIOPUR®

To stimulate the application of in situ soil remediation a new aerobic biofilm reactor was developed to treat soil vapour in combination with groundwater. Reticulated polyurethane (PUR) is used as a growth media for the biomass. The pressure drop in PUR was slight and high biomass concentrations were achieved. Air and water flow co-current through the plug flow bioreactor to prevent the volatile compounds from being stripped.

4. RESULTS

BIOPUR® was applied to groundwater treatment on both pilot scale and full scale, see Table 1. The results showed high removal efficiencies with short hydraulic retention times. Together with a contractor, TAUW has been gaining full scale experience with the BIOPUR® for more than three years, involving nine treatment plants. Less than two percent of the load offered was stripped. Maintenance was confined to reading the meters and re-filling the nutrient solution. Groundwater containing iron, in concentrations of up to 25 mg/L, was treated without any problem.

F. Arendt, G.J. Annokkée, R. Bosman and W.J. van den Brink (eds.), Contaminated Soil '93, 1187–1188.

1188

Table 1. Removal efficiencies on pilot plant and full scale BIOPUR®

site	capacity (m^3/h)	hydraulic retention time (hr)	influent (μg/L)	effluent (μg/L)	efficiency (%)
Raalte 1	15	0.25			
BTEX			650	<0.5	99.9
Raalte 2	4	0.9 -1.2			
BTEX			300-1980	<1	>99.6
mineral oil			40- 330	20-110	>66
Overslag	6	0.87-1			
BTEX			135- 460	<1	>79
mineral oil			50-2300	<100	>84
Utrecht	13	0.5			
BTEX			390-1300	3-5	99
mineral oil			323-1000	<50	>80
Amersfoort	10	0.4			
BTEX			10390	7	>99
mineral oil			1300	<100	>92
Borculo	5	1			
BTEX			420- 2600	0.4-17	>95
mineral oil			6000-18000	40-2800	>55
Olst	0.05	1			
BTEX			7390	295	>96
naphtalene			4180	-41	>99

5. IN SITU REMEDIATION AND ON SITE BIOLOGICAL TREATMENT OF A GASOLINE STATION

The BIOPUR® was used at a gasoline station for the on site treatment of groundwater and soil vapour which were contaminated with BTEX and mineral oil. The bioreactor was coupled to an in situ soil remediation. Every day 5 kg of petrol was removed in the bioreactor, retention times were less than 15 minutes. The exhaust gases hardly contained organic compounds. The groundwater was discharged directly into the surface water. The cost of this remedial technique was 30% less than conventional techniques, such as excavation.

6. TREATMENT COSTS

Table 2 makes a comparison of the treatment costs of four types of groundwater treatment plants (hypothetical case (5 mg BTEX/L, 10 mg mineral oil/L, over a 2 year period).

Table 2: Comparison of the treatment costs.

Type		Costs (DM per m^3 of groundwater)		
	Flow rate:	10 m^3/hour	20 m^3/hour	40 m^3/hour
1	BIOPUR®	0.53	0.44	0.35
2	Rotating Biological Contactor	0.78	0.65	0.59
3	Stripper/activated carbon	0.85	0.54	0.36
4	Stripper/compost filter	1.16	0.87	0.72

7. CONCLUSIONS

The BIOPUR® reactor has proven to be a cost effective method for the full scale treatment of groundwater and/or soil vapour contaminated with hydrocarbons. Removal efficiencies were virtually > 90-95%. The BIOPUR® has good perspectives in cases where the (ground)water and/or soil vapour are contaminated with BTEX, PAH, mineral oil or other biodegradable compounds. The treatment of chlorinated hydrocarbons on pilot scale is currently being investigated.

Cleaning Up a Former Industrial Site with a Soil Washing System

ULRICH NICKEL

ENVIRONMENTAL AUTHORITY, HAMBURG
Amelungstraße 3, D-2000 Hamburg 36, Germany

The site of the former Stülcken shipyard in Hamburg, which had been raised artificially last century to protect against flooding when it began to be used for this purpose, was tested for soil contamination after relocation of the production facilities and demolition of the buildings ready for the new use planned.

The contamination detected in one section, including hydrocarbons, had to be cleaned up before the site could be put to its new use. The volume of soil found to be affected in the preliminary tests was limited to about 8500 m^3 over an area of 1.2 hectares and extending to a depth of 2-5 m.

At the end of 1989 the Hamburg Environmental Authority obtained estimates from three companies offering soil washing systems and one offering a microbiological cleaning process. Other clean-up techniques were ruled out by the boundary conditions.

At the conclusion of these negotiations a contract was awarded to NORDAC, a firm which was being established at the time, with the 2000 high-pressure soil washing system manufactured by Klöckner. The site was cleaned up successfully between May and October 1990.

The limits specified by the authority responsible were as follows:
  soil to be washed: 1000 mg of mineral oil
                     hydrocarbons/kg of solids
  washed soil   : 500   "
                     10 mg of PAH/kg of solids

The clean-up was split into the following phases:

  excavation
  cleaning including disposal of residues
  backfilling.

The earthworks were split into excavation of uncontaminated soil, contaminated soil, and excavated material that could not be washed.

A total of 59,000 m^3 (100%) was excavated.

The uncontaminated soil placed to one side and subsequently backfilled amounted to 25,734 m^3 (44%), the soil that had to be washed 17,707 m^3 (30%).

F. Arendt, G.J. Annokkée, R. Bosman and W.J. van den Brink (eds.), Contaminated Soil '93, 1189–1190.
© 1993 *Kluwer Academic Publishers. Printed in the Netherlands.*

In addition 9,100 m^3 (15%) of sticky soil that was not suitable for the washing process had to be excavated from natural ground and the raised section.

Moreover 6,430 m^3 (11%) of highly contaminated carbon concentrates had to be removed.

The soil washing process used separates the contaminants hydromechanically from the soil structure at 350 bar and transfers them to the wash water, which is purified in the downstream water treatment system.

Once the soil has been washed and the water treated the contaminants are concentrated in the light products, flotation sludge and sediment filter cake dewatered in filter presses.

Each 500 t batch of the washed soil discharged from the system was tested for compliance with the purity limits specified and approved for backfilling once they were achieved. The average residual contamination was found to be 161 mg of mineral oil hydrocarbons/kg and 7.8 mg of PAH/kg.

The flotation sludge and light product residues were incinerated.

As was the sediment filter cake in 1992 due to the contamination with organic material and the heavy metal contents; microbiological treatment was not feasible.

The clayey soil and carbon concentrate fractions that could not be washed were disposed of in different ways. Whereas the only possibility for the concentrates was storage on a disposal site for hazardous waste, the clayey soil is being cleaned microbiologically on a former oil tank site in the Port of Hamburg.

The net cost of the measures amounting to 18.6 million deutschmarks breaks down as follows:

	Million DM
Earthworks	1.6
Washing soil	6.8
Disposal of residues	3.7
Disposal of carbon concentrates	3.0
Microbiological treatment of clayey soil	2.9
Miscellaneous	0.6

Despite sufficient boreholes it was only possible to obtain an inexact estimate of the extent of the soil contamination.

The disposal of residues cannot be specified definitively at the time a contract is awarded, since the concentration and consistency of contaminants are unknown.

It proved possible to make the site available for its new use on time.

Recycling biologically reclaimed soils

Dr. rer. nat. Axel Oberbremer
Dipl.-Geol. Rudolf Petersen jr.

Biodetox Gesellschaft zur biologischen Schadstoffentsorgung,
Feldstrasse 2, 3061 Ahnsen. *Germany.*

Summary
The objective of the biological reclamation of soils contaminated with hydrocarbons is ultimately recycling, to save valuable landfill space. There are various alternatives for recycling reclaimed material, for example road building, noise absorption walls, landfill recultivation and industrial zone backfill. This paper focuses on the BEZ biological waste management centre in Ahnsen to illustrate and describe procedures for recycling.

The Process
In standard practice the target of any biological reclamation of soils contaminated with hydrocarbons should be recycling, i.e. the reintroduction of the material into some use-cycle as a means of saving and protecting valuable landfill space.

The biodetox company operates a permanent plant at Ahnsen, with authority approval, for the biological reprocessing of soils contaminated with hydrocarbons. To achieve the recycling objective for the reclaimed soils the remaining hydrocarbon content must be less than 500 mg/kg TS (0.2 mg/l leachate).

In Ahnsen contaminated soils are treated in a microbiological process combined with a system for drawing off and filtering gases exhausted from the soil. The average duration of reclamation is 3 to 6 months. The hydrocarbon-bearing exhaust gases are passed through a scrubber (FBK-bioreactor) with a biofilter, as is the water collected from the process.

Usage of soils
The first batch (approx. 650 m³) had a remaining hydrocarbon content of average 700 mg/kg TS (range 500 - 1000 mg/kg TS). It was used to recultivate a landfill site. A topsoil cover of thickness 0.3 m was spread and seeded as fallow land.

Over a period of in total 11 month (including winter) the symbiotic action of the microorganisms and flora reduced the hydrocarbon content to approx. 320 mg/kg TS (see table 1).

F. Arendt, G.J. Annokkée, R. Bosman and W.J. van den Brink (eds.), Contaminated Soil '93, 1191–1193.

In the second batch, this time implementing combined microbiology and soil gas extraction, the hydrocarbon content was
reduced within a 5 month period by around 86 % from approx.
1400 mg/kg TS to less than 200 mg/kg TS (see table 2). The
material was then spread liberally.

Toxicity tests (see table 3)
The soil of the second batch was subjected to a series of
toxicity tests after reclamation treatment. To allow comparison and control a contaminated soil (hydrocarbon level 2000
mg/kg TS) and an uncontaminated soil from a mixed woodland
were also tested.

The leachates required for the dehydrogenase activity test
were prepared using the DEV S 4 method and with tap water (pH
4.5).

The reference soil did not inhibit the activity of live slurry
bacteria, but the reclaimed and the contaminated soils did
have an inhibiting effect. However, this inhibition is thought
to be due to other soil constituents, since there is no
correlation between the hydrocarbon content and the degree of
inhibition.

In the bioluminescence test the inhibition of all soils was
less than 20 %, i.e. the luminescent bacteria showed no toxic
effects.

In the cress seed the DEV S 4 and tap water (pH 4.5) leachates
were used for germination tests with cress seeds. It was
observed that both the reclaimed and the untreated soils
produced root length growth than did the reference soil.

	Initial mg/kg TS 08/91	19.11.1991		28.04.1992		01.07.1992	
		mg/kg TS	mg/l	mg/kg TS	mg/l	mg/kg TS	mg/l
MP I		390	0.4	350	0.4	310	< 0.1
MP II	Ø 700	350	0.2	310	0.2	350	0.1
MP III		400	0.2	530	0.2	300	0.13
MP IV		430	< 0.1	420	0.3	335	0.13
Ø		390		400		320	

Tabel 1:   Residual hydrocarbon contents in soils and leachate (DEV S 4)

	X	0 - 0.5 m	0.5 - 1 m
08.05.92	1400	–	–
29.06.92	860	610	1100
16.07.92	530	–	–
12.08.92	590	580	600
15.09.92	500	450	560
15.10.92	195	200	190

Tabel   2: Content of hydrocarbon (mg/kg TS) at BEZ-Ahnsen, Bed 1

	Dehydrogenase activity inhibition test (%)		Luminescent bacteria inhibition test (%)		Cress seed test %-inhibition of root growth	
	I	II	I	II	I	II
biologically re-claimed soil: HC < 300 mg/kg TS	50	60	– 10	– 10	+ 23	+ 240
contaminated soils: HC > 2.000 mg/kg TS	80	50	+ 10	– 1	+ 43	+ 56
Mixed woodland soil= refer. soil no contamination	0	7	– 12	+ 4	0	0

Tabel   3:   Results of toxicity tests

I   leachate as per DEV S4
II   leachate with tap water (pH 4.5)

INHIBITION OF NATURAL MICROBIOLOGICAL LEACHING PROCESSES

Dr. J. Ondruschka [1], Dr. F. Glombitza [2]

KAI e.V., WIP-AG "Laugung", Permoserstr.15, O-7050 Leipzig [1]
DFA GmbH Consulting und Engineering, Jagdschänkenstr.52, O-9030 Chemnitz [2]

Microbial oxidation in pyrite containing wastes and dumps of the Thuringian uranium mining region as well as in low grade ore dumps in the same region after finishing the uranium leaching process produce acidic drainage water with a high concentration of heavy metals and a high hardness.
The reasons for this are well known. They are the microbial oxidation of the sulfide to sulfate in the pyrite connected with the release of iron-II-ions and their oxidation to iron-III-ions as well as the the chemical and microbial oxidation of the uranium-IV-ions to uranium-VI-ions.
It must be tried interrupt this process of reduce or avoid the consequences of the natural leaching processes.
At present, the dumps as a result of uranium mining in Australia and on the Wismut area in Germany were redeveloped covering with earth layers or canvass. The resulting deficiency of oxygen changes the microaerobic into anaerobic conditions and microbiological oxidation and leaching processes are inhibit.
This measure, however, do not solve the original problem of inhibition. For this purpose, compounds or substances were sought which inhibit the growth of the Thiobacillus populations.

The microbial leaching processes induced by Thiobacillus species can be reduced or stopped with growth inhibitors and bactericidal components.
Therefore we tried to find suitable substances and found fluorspar, a biodegradable tenside and Kathon.
In systems with a circulation of water fluoride-containing solution or a tenside can be added. If new dumps are constructed layers with fluorspar could be added.

Inhibition by fluorspar

Fluorides inhibit the growth of Thiobacillus species at low concentrations. The amount depends on the kind of fluorides used. The concentration is 400 mgF / 1, if calcium fluoride is used.
Fluorides are wide spread in nature. Therefore the application of fluorspar suggests itself. It is nearly insoluble in water under normal conditions (16 mg / 1). But it is possible to dissolve calcium fluoride in the form of fluoro-containing metallic complex. Such complex forming elements are iron and aluminium which exist in the aqueous

F. Arendt, G.J. Annokkée, R. Bosman and W.J. van den Brink (eds.), Contaminated Soil '93, 1195–1196.
© 1993 *Kluwer Academic Publishers. Printed in the Netherlands.*

medium in a leaching process.
About 2500 mgF / l is dissolved in the acidic leaching solution (pH: 1,8; iron-III-concentration: 3,3 g / l). The fluoro-complex can be reached completely by shifting the pH-value to a level higher than 2,5.

A lab perkolator is filled with 1 kg uranium ore and layers of fluorspar. The ashes on the end of column to serve precipitate the soluble fluoride by pH shift.
For the pilot scale 2000 t of a low grade ore was blocked after an activ leaching process. The extraction rate from uranium and iron decreased after the blocking nearly to zero. Living Thiobacillus microorganisms could not be determined.

Inhibition by Tensid and Kathon

Different surface active substances were tested and a sodium paraffinsulfonat (E30) was chosen, which acts as an inhibitor if the concentration is greater than 20 mg / l. Because the tenside adheres to the surface of the ore body, a specific amount (needed for the surface of 1 cm^2) was determined to 0,016 mg E30 /cm^2.
For the inhibition of growth of microorganisms by tenside in a test dump (1000 t; 0,25 km^2 ore surface) 40 kg E30 in 160 m^3 water was dissolved and the dump was sprayed with the solution. The drainage water did not contained Thiobacilli or tenside.
After complete spraying the activity of microorganisms reaches a limit, below which growth of microorganisms is not longer possible.

Experiments to inhibit the growth of Thiobacilli were continued with Kathon. This biocidal substance belongs to the family of isothiazolones.
Concentrations of 0,03% Kathon (applied to active surface of ore) are sufficient for the inhibition of growth. All runs of the restoring to life of the Thiobacilli were negative.

Summary

It was tried in a laboratory and pilotplant scale to inhibit the natural microbiological leaching processes in uranium-containing ore.
Fluorides, a sodium paraffinsulfonat and a isothiazolone were successfully used as biocidial substances for blocking microbial oxidation reactions due to chemolitho-autotrophic microorganisms Thiobacillus species.
These methods can used for restoration of environment in mining and in chemical industry.

ADVANCED STAGES OF REMEDIATION ON THE PROPERTY OF THE CHEMICAL FACTORY
MARKTREDWITZ (CFM) 1992 - 1995

KARL-GERDT PEDALL

HPC HARRESS PICKEL CONSULT GMBH

Founded in 1788, the former Chemical Factory Marktredwitz (CFM) in Upper
Franconia was the oldest plant of this type. At an early date, production
concentrated on the manufacture of antimony compounds and was also
especially active in all aspects of mercury chemistry. During this century,
predominantly mercury-containing agrochemicals, also including pesticides
and herbicides, as well as arsenals, were produced.

This spectrum of contaminants was disseminated, in some instances in
extremely high levels, over a long period, from the predominantly
antiquated facilities of the plant, into soil, air and ground water. The
mercury and antimony emissions in particular led to a grave contamination
of the surroundings.

Under section 10, Paragraph 3 of the Federal Hazard Protection Law
(BImSchG), the CFM was shut down in 1985 by the state authorities
(Wunsiedel office). In the following years extensive immediate action and
protective measures were undertaken in order to come to grips with the
remediation.

In 1992 demolition of the massive (in part multiple storied) structure was
performed. In June 1993 the excavation of contaminated soil will be begun,
which will be concluded in mid-1995. At that time the CFM property will be
available for new construction for a different utilization.

In order to dispose of the more than 120,000 tons of demolition rubble and
soil, beside deposition of the highly contaminated material in underground
waste disposal sites, the following alternatives are available, depending
on the degree of contamination:

- Material heavily contaminated with mercury (50 - 5,000 mg/kg) will be
brought to a soil purification facility, which operates on the basis of a
wet sieving and subsequent vacuum distillation step.

- Material contaminated with mercury at a lower level (10 - 50 mg/kg) will
be disposed of in a single purpose landfill. The material purified in the
above mentioned soil cleaning process will also be disposed of in this
landfill.

F. Arendt, G.J. Annokkée, R. Bosman and W.J. van den Brink (eds.), Contaminated Soil '93, 1197–1198.
© 1993 *Kluwer Academic Publishers. Printed in the Netherlands.*

- Material with mercury content below 10 mg/kg can be refilled into the excavation, provided that the levels of contamination with other production-specific substances lie below the applicable "level 2 values" of the Bavarian Contamination Guideline.

Single-purpose landfills, as well as the soil purification unit, are set up especially for the CFM contamination project outside of the city of Marktredwitz.

Basic concept of the excavation, reaching, in places, as deep as 5 m, is a backward-directed benching. This benching is proceding in the flow direction of ground water. In order to minimize the increase of mass of contaminated material the various strata, occuring one above the other, are seperately excavated and stocked.

In addition to an in parts thick and highly contaminated layer of artificial fill, a sequence of layers, consisting of meadow loam, gravel and in places weathered hard rock, must also be excavated.

A mass flow distribution control plan was developped in order to facilitate decision making and supervision of the contamination-oriented disposal process, as well as to document the sucess of remediation. By means of continual representive bulk sampling, sample preparation and on-site rapid analysis conducted without impeding the course of remediation, it is possible to select the correct and most cost-effective alternative of disposal.

Partial securing of the excavation sides by means of a wall of bore piles is required, which also minimizes the entry of ground water, which is contaminated to various degrees, protects neighboring buildings and prevents recontamination of the remediation area.

Water from the excavation site will either be fed into a tributary directly or stored in an storage basin leading to a ground water purification plant, depending on the level of contamination.

As a consequence of the site location being within the city center, all remediation measures are accompanied by comprehensive and continuous measurements to determine immission of mercury vapor or mercury bound within small particulates.

Transfer Of Polycyclic Aromatic Hydrocarbons From Contaminated Soil In Vegetables

M. Preusser/H. Ruholl
Chemisches Laboratorium
Dr. E. Wessling, Altenberge

J. Schwermann
Chemisches Untersuchungsamt
der Stadt Essen, Essen

Summary

The transfer of polycyclic aromatic hydrocarbons (PAH) from contaminated soil in selected types of vegetables was studied within two vegetation periods by outdoor experiments in allotments of the city of Essen. Some types of vegetable showed an increased PAH content. A correlation between the PAH concentrations of the soil and the vegetables was not observed. The experimental results lead to the conclusion that the incorporation of the pollutants mainly takes place from the air.

1. Introduction

Inside several allotments of the city of Essen the soil (0,3-1,0 m under surface) is contaminated with polycyclic aromatic hydrocarbons. The detected PAH concentrations reach up to 120 mg/kg (16 EPA compounds) respectively 6,2 mg/kg benzo[a]pyrene.

Because of this fact an endangering of the allotment holders by the consumption of contaminated vegetables could not be excluded. To evaluate an advanced risk assessment the possible transfer of polycyclic aromatics from soil in vegetables had to be investigated.

2. Experimental Procedure

The experiments were carried out in 9 outdoor patches, which covered a size of 25-30 m², including a not contaminated control patch. To detect the possible influence of the weather the investigations lasted two vegetation periods (1990, 1991).

The following list of the selected types of vegetable took different properties into account, e.g. fat content, fruit vegetables resp. leaf vegetables as well as low and flat rooting vegetables:

carrot, leek, celeriac, dwarf bean, curly kale, red cabbage, endive, spinach

To ensure equal growth conditions in all patches the nutrient content of the soil was analyzed before and furtilizer was applied if necessary. The PAH analyses was performed after usual cleaning and preparation of the vegetables.

F. Arendt, G.J. Annokkée, R. Bosman and W.J. van den Brink (eds.), Contaminated Soil '93, 1199–1200.
© *1993 Kluwer Academic Publishers. Printed in the Netherlands.*

1200

3. Results

The detected PAH concentrations differed quite considerably between the different types of vegetable. Carrot, leek, celeriac and red cabbage only contained small amounts of PAH (n.d.-12 µg/kg) whereas some dwarf bean, endive and especially spinach as well as curly kale samples proved to be contaminated (1,0-170 µg/kg). Within the group of 16 analyzed PAH compounds phenanthrene, fluoranthene and pyrene showed the highest concentrations.

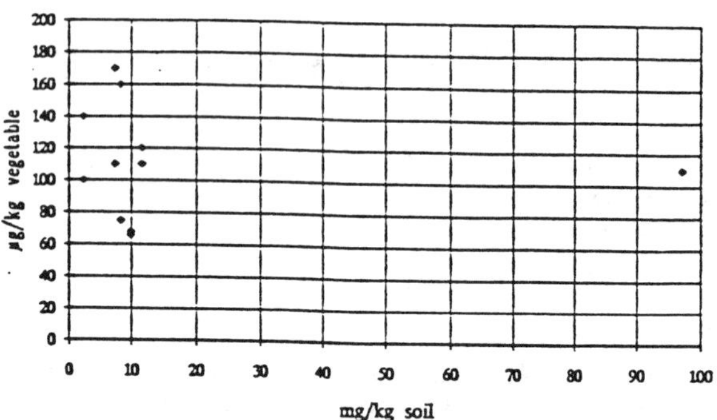

Figure 1   PAH concentration in soil and vegetable (curly kale)

A correlation between the PAH concentration in the soil and the detected PAH contents of the vegetables was not observed (figure 1). Vegetable samples deriving from contaminated patches did not contain a significant higher PAH amount than samples growing on not contaminated soil.

Even types of vegetables with a high fat content (carrot, celeriac) showed lower PAH concentrations than types with lager above ground parts like spinach and curly kale in spite of the direct contact with the soil adsorbed pollutants. This result indicates that the pollutants are mainly incorporated from the air and not from the contaminated soil.

4. References

(1)   Crößmann, G., 1990. PAK-Transfer Boden/Pflanze. Vortrag beim LÖLF-Kolloquium "Schadstoffe im System Boden/Pflanze" am 22.03.1990 in Recklinghausen

(2)   Friege, H. et al., 1989. Belastung von Klärschlämmen und Böden mit organischen Schadstoffen. Korrespondenz Abwasser 36, S. 601-608

(3)   Fritz, W., 1983. Untersuchungen zum Verhalten von Benzo(a)pyren im Boden und zum Übergang aus dem Boden in Erntegüter. Zbl. Mikrobiol. 138, S. 605-616

(4)   Hechler, U.; Storkebaum, G.; Stumpf, G.; 1992. Bestimmung von polycyclischen aromatischen Kohlenwasserstoffen (PAK) in pflanzlichen Lebensmitteln. Lebensmittelchemie 46, S. 25

(5)   Hembrock-Heger, A.; König, W., 1990. Vorkommen und Transfer von polycyclischen aromatischen Kohlenwasserstoffen in Böden und Pflanzen. VDI-Berichte Nr. 837, S. 815 - 830

AIR INJECTION AND IN SITU SOIL VAPOUR EXTRACTION

C.G. PIJLS[**], J.A.M. VAN DER MEER[*], E.A. HESLINGA[*], A. NIJHOF[**]

[*] PROVINCIE LIMBURG, P.O. Box 5700, 6202 MA Maastricht, The Netherlands
[**] TAUW Infra Consult B.V., P.O. Box 479, 7400 AL Deventer, The Netherlands

1. INTRODUCTION

Several applications of soil vapour extraction have been realized. In recent applications soil vapour extraction was used as a means of in situ biological treatment of volatile and non-volatile organic contaminants. This biological degradation is stimulated by supplying the soil with oxygen. The addition of nutrients can further enhance the biological activity of micro-organisms in the soil.

Soil vapour extraction can only be used above the groundwater level (i.e. in the unsaturated zone). In order to achieve remediation by means of soil vapour extraction the groundwater level has to be lowered which can be an expensive business. With the injection of compressed air into the soil, below the groundwater level, oxygen is transferred into the aquifer and biological degradation can be activated without actually having to lower the groundwater level.

2. SITE DESCRIPTION

The site is located in Gennep in the Province of Limburg in the Netherlands. In the described project a gasoline (domestic fuel) spill caused a contamination of more than $20,000$ m^3 soil. The composition of the soil consists of moderate to very coarse sand with a high aerodynamic conductivity. The soil and groundwater are heavily contaminated, mainly with hydrocarbons and traces of aromatics.

Investigations showed that excavating and cleaning the polluted soil would be very expensive (Dfl 8 to 12 million). Laboratory studies demonstrated that air injection, in combination with soil vapour extraction, is a promising cost-effective alternative remedial technique. Under laboratory conditions removal rates of more than 10 mg/kg/day have been achieved. Air injection showed similar biological breakdown compared to soil vapour extraction.

F. Arendt, G.J. Annokkée, R. Bosman and W.J. van den Brink (eds.), Contaminated Soil '93, 1201–1202.
© 1993 *Kluwer Academic Publishers. Printed in the Netherlands.*

3. PILOT PLANT STUDY

A pilot plant was operational for four months on a site some 150 yds^2. The radius of influence of the air injection was determined by monitoring the oxygen levels (see figure below) and by conducting tracer experiments in the air. Tracer experiments in the groundwater proved that the horizontal groundwater flow was negligible due to air injection. Throughout the experiment the contamination did not spread.

The biological activity was measured by monitoring the (in situ) oxygen consumption and the carbon dioxide production. The injection of nutrients caused an increase in oxygen consumption (see figure below).

The groundwater concentrations of aromatics and hydrocarbons, within the radius of influence of the air injection, fell under the detection limit.

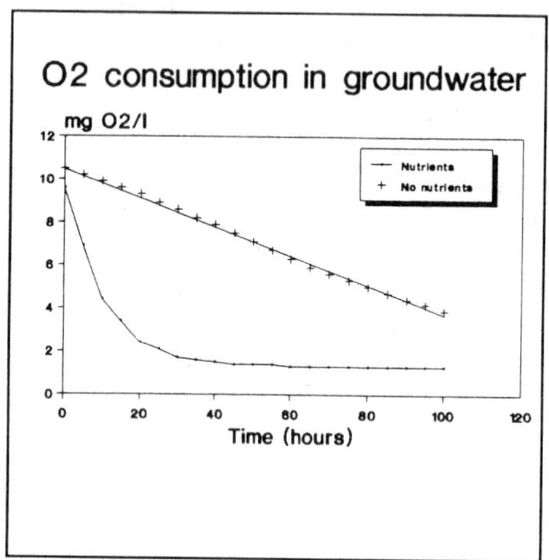

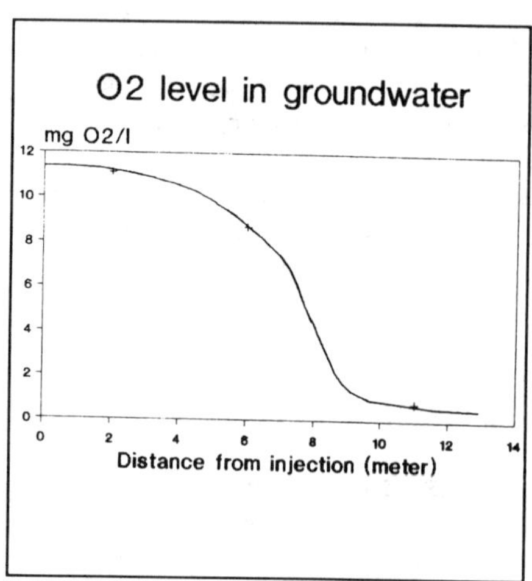

4. FULL SCALE REMEDIATION

A full scale remediation system is currently being set-up. The system consists of 35 air injection filters. The soil vapour will be treated biologically. A groundwater injection and withdrawal system will supply the soil with nutrients. The withdrawal and injection of water and air will be controlled by a Programmable Logic Controller (PLC). The project is scheduled to start at the beginning of 1993. The first results on full scale level are expected in the spring of 1993.

Investigations into the Effect on the Groundwater of Deep Compacting in the Area of a Former Dump

Reichert J.K.; Roemer M.; Tillmanns W.

1) Reason for investigation and description of existing situation

A large hall is to be built on the site of a refilled former excavation site. The non-homogeneous subsurface, in some places comprising loosely deposited soil, is being improved by the vibro-replacement method of deep compaction down to a maximum depth of 16 m over a surface area of 16,000 square metres with some 6,000 vibration points.

The refilled excavation (former dump) used to be a large gravel pit covering an area of several hectares. The in-situ subsoil is made up of Quaternary sands with a primary thickness of approx. 25 m and a k_f value of approx. 3×10^{-3} m/s. The permeability of the underlying silty sand from the Tertiary is much lower (approx. 10^{-6} m/s). The excavation depth ranges from 16 m to 25 m below ground level. In the excavation area the groundwater level is between 8 m and 4 m below ground, so that groundwater flows through most of the former dump. In the sixties and seventies the excavation site was filled predominantly with excavated soil, building rubble and broken-up road surfaces. An examination of the filling constituents revealed that PCA-contaminated materials were also dumped. Analyses of groundwater from the dump indicate a high PCA content. PCA discharges have been detected in the immediate tail of the former dump.

2) Preliminary investigations

Extensive soil, soil air and groundwater analyses were carried out at the envisaged construction site and the surrounding area. Twenty groundwater measuring points were installed for test purposes in the immediate vicinity. The groundwater samples were taken horizontally and the PCA (U.S. EPA) content was determined. An evaluation of the groundwater levels of the last 25 years completed the basis of the analysis.

Authors' addresses:
Univ. Prof. Dr. J.K. Reichert: Chemie der Wassergewinnung und des Gewässer-
Dipl.-Ing. M. Roemer   schutzes, RWTH Aachen,
                      5100 Aachen, Kopernikusstr. 16
Dr. W. Tillmanns, Lecturer: Büro Dr. Tillmanns & Partner GmbH, 5010 Bergheim-
                      Ahe, In den Benden 3

F. Arendt, G.J. Annokkée, R. Bosman and W.J. van den Brink (eds.), Contaminated Soil '93, 1203–1204.
© 1993 *Kluwer Academic Publishers. Printed in the Netherlands.*

3)   Civil engineering

Torpedo and infiltration vibrators were employed with a frequency of 50 Hz.
Power input: up to 150 ampere at 380 volt, plus 10 to 12 t machine weight
Working speed: 20 m/h = 3 min/m column
The consolidation of the final column is more or less equivalent to that of a compactly deposited gravel and sand mixture with a pore volume of 20 % by volume.

4)   Monitoring

In the context of monitoring measures, samples were continuously taken at the available groundwater measuring points on and in the vicinity of the building site at intervals agreed with the supervisory authorities. The samples were analyzed with reference to the contaminants anticipated according to the preliminary investigations. Powerful wells were installed and activated carbon filters used as a means of pumping away and cleaning contaminated groundwater.

5)   Compaction effects

In connection with the building work, the principal factors to be determined were the hydrochemical effects of the vibro-replacement and, on conclusion of the construction work, the effects on the hydrogeological and hydraulic conditions at the building site and in the surrounding area.

6)   Investigation findings

The evidence provided by the continuous groundwater level measurements showed that the vibro-replacement resulted in a groundwater level increase of max. 0.7 m at the building site.

Vibration measurements verified soil shifting caused by the vibro-replacement up to a distance of 280 m from the actual building site. A spherical propagation from the point of energy introduction can be assumed.

The groundwater was closely monitored with samples being taken at weekly intervals. The evaluation indicates that the energy input results in a desorption of the contaminants on the grain surfaces to the aqueous phase. The extent to which the substances are mobilized depends on the amount of energy input and the substance potential. Considerably higher contents were measured inside the dump, for example, than in the tail.

The contaminant concentration changed simultaneously at practically all of the measuring points.

A rapid fall in the contaminant concentrations was ascertained when the compacting work was interrupted. This is attributed to reabsorption on the grain surfaces. On conclusion of the building work, the contaminant concentrations in the groundwater had returned to their initial values.

CLEAN-UP OF CONTAMINATED INDUSTRIAL SITES BY DIRECTED SOIL AIR CIRCULAR FLOW (BLK).

MARC R. SICK, PH.D.; EDUARD ALESI, PH.D; SUSANNE BORCHERT; RAINER KLEIN

GfS -Gesellschaft für Boden und Grundwassersanierung mbH, D-7312 Kirchheim/Teck

1. ABSTRACT:

Directed soil air circular flow (BLK: Boden-Luft-Kreislaufführung) systems are employed for the remediation of soils polluted with volatile contaminants e.g. chlorinated and aromatic hydrocarbons (CHC, BTEX). The remediation efficiency of the BLK was tested on an industrial site by carrying out extensive measurements within four soil borings. After 6 months of operation only 12% of the original contamination remained in the vadose zone. The increase in temperature of the circulating air is the most important factor responsible for the effectiveness of the remediation. Assuming equilibrium conditions in the subsoil, an increase in temperature from 10°C to 20°C leads to a 50% reduction in the amount of time needed to complete a remediation.

2. DESCRIPTION OF THE BLK TECHNOLOGY (Fig. 1):

Two screen sections built into a bore hole are separated into an upper and a lower section, each of which is connected to a ventilator. This allows for the withdrawal of air from either segment individually or from both simultaneously. The air extracted, after passing through a suitable remediation unit (i.e. activated carbon filter), is reinfiltrated into the soil. Horizontal and vertical flow circulations are generated in the soil surrounding the extraction well. The circulation direction is reversible and can be adjusted according to the pollutant distribution in the soil. The BLK, in contrast to conventional venting methods, is capable of generating a directed circulation through the center of the contamination. Air passing through the ventilator is heated, thereby enhancing desorption of contaminants adsorbed onto soil particles. This leads to a more effective remediation of the site. Temperatures of circulating air reach up to 40°C.

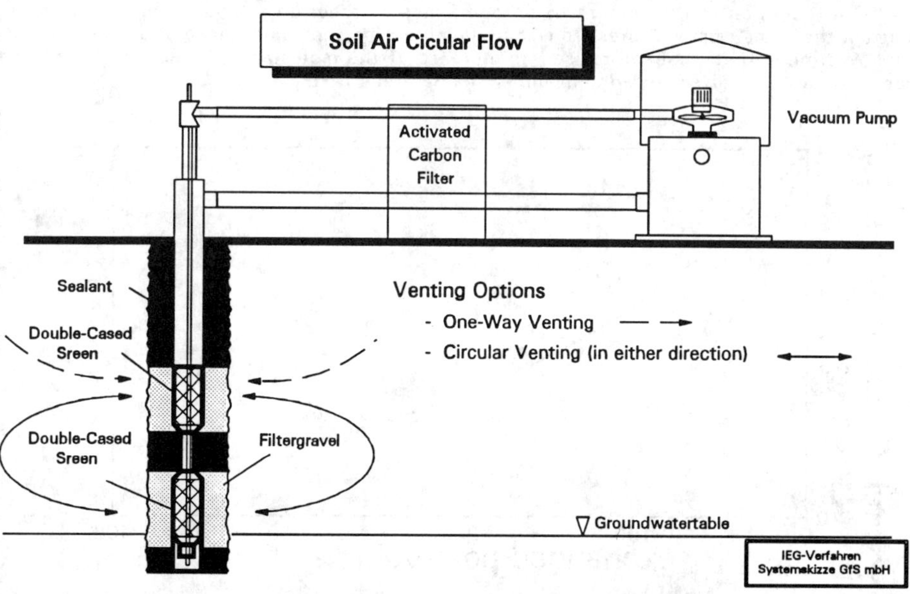

F. Arendt, G.J. Annokkée, R. Bosman and W.J. van den Brink (eds.), Contaminated Soil '93, 1205–1206.
© 1993 Kluwer Academic Publishers. Printed in the Netherlands.

3. FIELD DATA - EVIDENCE FOR A SUCCESSFUL REMEDIATION USING THE BLK:

The efficiency of the BLK has been proven successful at a CHC remediation site using four soil borings and analyzing numerous soil samples. Results showed a definitive contamination reduction in the treated vadose zone. Remediation took place with a directed soil air circulation in the upper section (about 4 m) of the unsaturated zone. The arrows in fig. 2 show the BLK in operation with soil air being drawn into the upper screen section and being injected into the lower screen. In spite of unfavourable soil conditions (high percentage of organic material and low permeability), the contaminant concentrations were reduced to 12% of their original concentrations after only 6 months. By changing the direction of air circulation, the lower section of the vadose zone is in the process of being remediated.

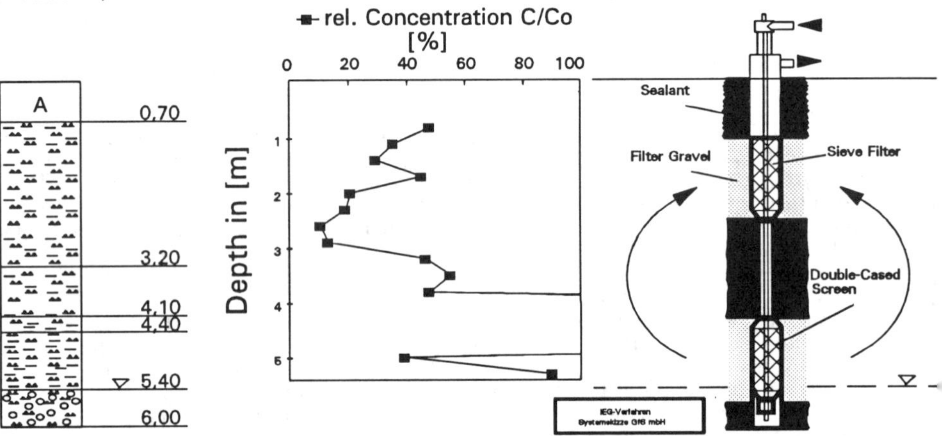

4. EFFECT OF THE TEMPERATURE INCREASE ON THE SUBSOIL:

CHCs are found in the subsoil according to their distribution coefficients - soil (solid phase) - pore water (liquid phase) - and soil air (gas phase). With an increase in temperature the entire distribution shifts towards the gas phase causing a higher percentage of the contamination to be present in the soil air, which can more readily be removed. Fig. 3 depicts the effect of a temperature increase from 10° to 20°C on the subsoil. The rise in temperature leads to a 50% reduction in the amount of pore volumes to theoretically be exchanged in order to attain the same removal rates from the subsoil during the same remediation time. In addition, warming of the subsoil causes an increase of the molecular diffusion constant - an especially positive effect in poor to impermeable areas of the soil.

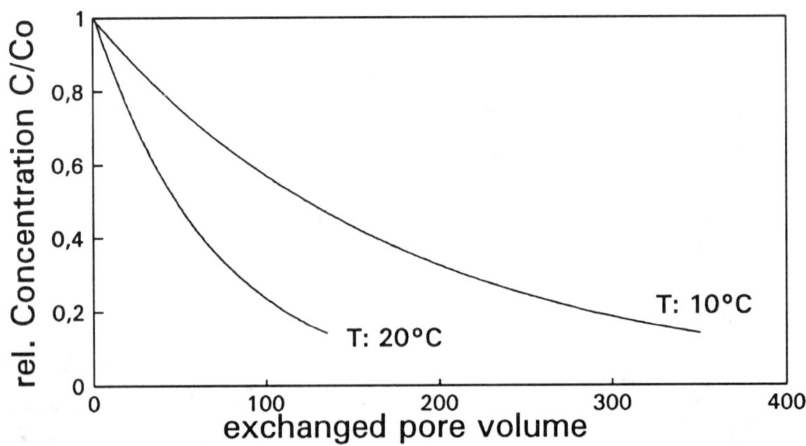

REMEDIATION OF CONTAMINATED WATER BY U.V. OXIDATION

Mr. Rainer Träxler Dipl.-Ing. (FH), Dipl.-Ing. (FH)

SAN Sanierungstechnik für den Umweltschutz GmbH, Nördlinger
Strasse 2, Harburg 8856. *Germany.*

Definition:

Oxidation of contaminants aided by high-energy ultraviolet light. During
this process the substances can be degraded to their basic components CO_2,
H_2O and HCL. As an oxidizer added H_2O_2 or ozone is used in most cases.

Applicability:

So far only for the degradation of water constituents, tests are, however,
carried out presently for the degradation of contaminants in a gaseous
phase or on the surface of solids.

Prerequisite: The molecules must be able to adsorb ultraviolet light. This
applies to almost all unsaturated, aromatic and halogenated hydrocarbons.

Prior to application, this procedure can easily be tested in the laboratory.
This enables us to offer guarantees and to convince the supervisory
authorities. In most cases the approving authorities will demand an on-
site test run.

Principle of Reaction:

The contaminant molecules are excited by U.V. irradiation, what means an
increase of reactivity. Additionally, more aggresive radicals, which react
with the excited contaminant molecules in several phases, are produced
from the added oxidizers. In the case of a complete oxidation process, the
contaminants are mineralized to the harmless end products carbon dioxide,
water and hydrochloric acid.

Reactor Design:

The following applies to all reactors: A radiator emitting ultraviolet
light submerges - protected by a quartz tube - into the contaminated water.
By means of construction details certain reaction conditions are obeyed
within the reactor in order to guarantee an oxidation as complete as
possible. Particularly important are, for example,

* maximum transirradiation of the contaminated water
* high radiation flux with the U.V. area
* short distance between U.V. radiator and contaminant

F. Arendt, G.J. Annokkée, R. Bosman and W.J. van den Brink (eds.), Contaminated Soil '93, 1207–1208.
© *1993 Kluwer Academic Publishers. Printed in the Netherlands.*

Conclusion:

The U.V. oxidation represents a useful procedure for the clean-up of contaminated water. Mostly, the contaminants can be eliminated completely, what makes a further disposal, as it is for example required in the case of filtering techniques, redundant.

Modular design allows an economic process engineering. The costs can be compared with conventional techniques, they might even be lower, especially when taking the costs for disposal or recycling of the charged activated carbon into consideration. Therfore this method represents a cost-saving process alternative and should be contemplated in each case of ground-water remediation.

IN SITU REMOVAL OF A DECANOIC ACID SPILL BELOW A STORAGE TANK AT A LARGE
PETROCHEMICAL SITE IN THE NETHERLANDS

VISSER, W., REE, C.C.D.F. van

DELFT GEOTECHNICS, P.O. BOX 69, 2600 AB DELFT, THE NETHERLANDS

VREEKEN, C.

BMS ENVIRONMENTAL SOIL SYSTEMS, P.O. BOX 238, 2650 AE BERKEL EN
RODENRIJS, THE NETHERLANDS

Due to a punctured bottom plate of a large above ground storage tank
decanoic acid ($C_{10}H_{20}O_2$) was spilled into the soil at a petrochemical site in
the Netherlands. The spill was discovered in 1989. Site investigation using
various field methods showed that the spill was largely present below the
tank foundation. The total amount of spilled product was estimated at 16
tons. From the extent of the ground water contamination it could be concluded that the leakage had began more than seven years before.

Removal of the tank and excavation of the contaminated soil was not feasible
for technical and economical reasons. Therefore the clean-up had to take
place in situ. Removing the product from the soil by pumping was impossible
because of the high viscosity of the product (17 cp). Water flushing was not
possible because the solubility of the product in water was too low (300
mg/l). The applied method had been based on the principle that the disso-
ciation (and the solubility) of a weak acid like decanoic acid depends on
the pH. Flushing with an alkaline solution instead of water can reduce the
time for the removal of decanoic acid considerably. The flushing system is
illustrated in figure 1.

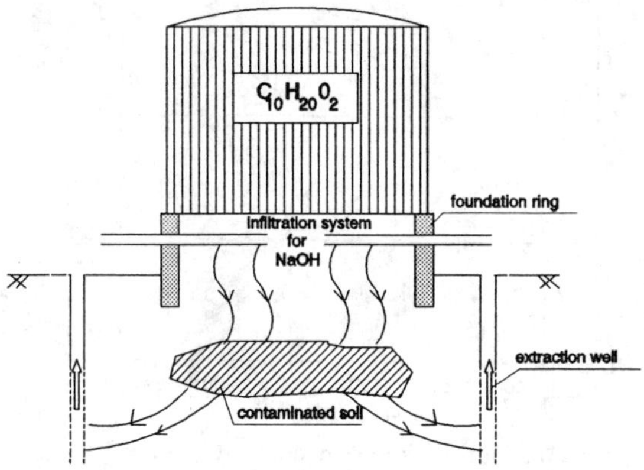

figure 1: Flushing system for the removal of decanoic acid

F. Arendt, G.J. Annokkée, R. Bosman and W.J. van den Brink (eds.), Contaminated Soil '93, 1209–1210.

1210

The reaction between an alkali solution and decanoic acid (HDe) is almost
complete.

$$HDe + OH^- \rightarrow De^- + H_2O$$

$$\frac{[De^-]}{[OH^-].[HDe]} \approx 9 \times 10^8$$

It can be calculated that with 1 m^3 0,25 m NaOH-solution 43.75 kg decanoic
acid can be dissolved.

For the extraction of contaminated ground water (flushing agent) around the
tank 12 extraction wells were placed. To prevent infiltration of contaminated flushing agent into the second aquifer so much ground water was
extracted that a seepage flow was established. This could be checked with
standpipes around the tank placed in the first and second aquifer. It
appears that by extraction of 2 m3/hour a seepage flow could be established.
One vacuum pump in tank pit extracted ground water from the wells and pumped
it to the ground water treatment plant. The ground water treatment plant
consisted of an acid mixing unit and a separation unit (flotation system).
The actual sanitation started the 28th September 1990 and ended the 30th May
1991. In November 1990 the infiltration of alkali solution was stopped for
three weeks because of problems with the water treatment system. Throughout
the rest of the period for 5 days a week a mean quantity of 20 m^3/day of a
0,25 m NaOH solution was infiltrated under the tank and 7 days a week a mean
quantity of 50 m^3/day ground water was extracted from the wells around the
tank. In figure 2 the cumulative amount of decanoic acid is illustrated.
It is shown that the amount of decanoic acid which is removed is 16.8
tons. This amount equals the estimated quantity of leaked product.
It was concluded that the product was completely or almost completely
removed from the soil.

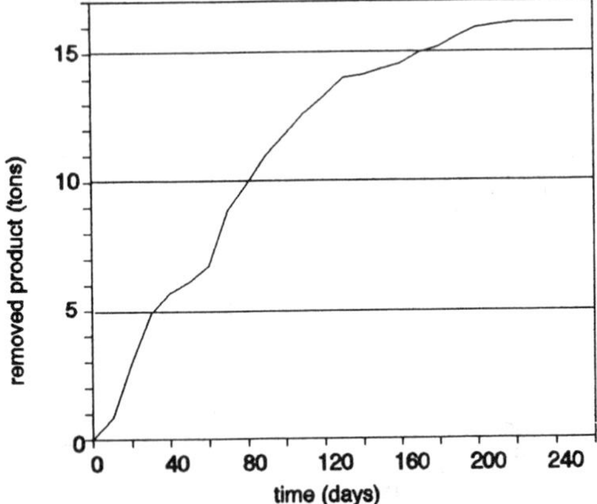

figure 2: Cumulative amount of removed decanoic acid

Decontamination of Polluted soil and waste water by Ozone Treatment

Rolf Voigtländer

TELAB Technik Labor GmbH
Oststraße 7 - 4417 Altenberge - Germany

Industrial development has a crucial influence on natural resources. This also applies to soil and water. Nowadays there is soil and water pollution in many areas which can give rise to many dangers for both man and nature. For this reason, a growing market has developed for purification techniques. Various methods exist for the purification of soils. Incineration is widely used to get rid of organic substances such as PAH and other hydrocarbons. Also so-called soil washes are frequently employed. Ozone treatment constitutes a new process in the purification of soils with which soils polluted with PAH or hydrocarbons can be purified. This process, which has already been patented, can be applied in-situ. It is also possible to treat the soil in reactors or by landfarming.

An initial demonstration project was succesfully concluded in Germany. It deals with the remediation of a gas station grounds. High pollution levels with diesel oil were present in the soil. Today there is no more danger. The purification was carried out in-situ. The plant was located on a car park which could continue to be used.

The purification of water with ozone has a long history. Drinking water is treated with ozone as is also water in swimming-pools where the germicidal effect of ozone can be applied. It is also possible to oxidize organic substances with ozone. The main product given off is carbon dioxide. In this way water with certain pollutants can be purified. In particular water which is difficult to purify biologically is suitable for successful treatment with ozone. For example, water polluted by a pesticidewas successfully purified. Here ozone treatment was less expensive than other current methods.

Water polluted with PAH can be purified very easily. Also ozone treatment is very suitable for successful application of chemical sewage in many cases.
Whether ozone treatment of soil or water is practical can be tested in laboratory experiments. Experiments developed in TELAB Ltd. serve this purpose.

F. Arendt, G.J. Annokkée, R. Bosman and W.J. van den Brink (eds.), Contaminated Soil '93, 1211.
© 1993 *Kluwer Academic Publishers. Printed in the Netherlands.*

IMMOBILIZATION OF SOIL CONTAMINATED WITH LEAD

Margareta Wahlström[1], Esa Mäkelä[1], Pasi Vahanne[2], Jaakko Paatero[3], Bob Talling[3], Martti Keppo[4]

[1] Technical Research Centre of Finland, Chemical Laboratory, Postbox 204, SF-02151 Espoo, Finland
[2] Technical Research Centre of Finland, Road, Traffic & Geotechnical Laboratory
[3] Partek Cement Ltd, SF - 21600 Parainen, Finland
[4] Lohja Rudus Ltd, Pronssitie 1, SF-00400 Helsinki, Finland

The contaminated soil materials studied were taken from the site of an old lead smeltery. The site had been contaminated partly by air emissions from the smeltery and partly by waste slag which was stored at the site. The lead concentrations were 1-2 % in the soil layers (clay, sand, humus, peat) and on the average 9 % in the slag. The total amount of soil and slag to be treated is about 17000 m^3.

Optimal binding systems were developed for the contaminated soils. The contaminated soils were mixed with binders including e.g. portland cement and additives. The optimal binding systems were studied with mortar bar tests, where properties, like the compression strength, frost resistance and swelling were studied. In 1991 a roller compacted concrete pavement using about 500 m^3 stabilized contaminated soil as aggregate was constructed.

After 7 days the compression strength of the laboratory samples was rather high. The compression strength of the field test samples was 19 MPa after 28 days. Also the frost resistance of the immobilized soil was satisfactory after 50 cycles in 30 % NaCl-solution.

The environmental impacts of the immobilized materials were evaluated with a Dutch tank leaching test. In the Dutch test a specimen is immersed in water and at certain time intervals the solution is renewed and analyzed. With this test also the surface waters at the full-scale construction were analyzed.

In table 1 some results from the Dutch leaching tests are presented as the cumulative leached amount of lead per surface area as a function of contact time. One should notice that the soil samples studied with the leaching test contained more lead than the average lead concentration in the contaminated soil to be treated at the site.

F. Arendt, G.J. Annokkée, R. Bosman and W.J. van den Brink (eds.), Contaminated Soil '93, 1213–1214.
© 1993 *Kluwer Academic Publishers. Printed in the Netherlands.*

Table 1. Leached amounts of lead from some stabilized soil mixtures.

soil sample	lead content in the stabilized soil mg Pb/kg	leached amount mg Pb/m^2/64 days
laboratory samples:		
surface soil	22 000	75
slag + soil	120 000	4400
peat	19 000	120
field sample:		
containing slag	-	900

In the beginning of the leaching test the leaching mechanism for lead was diffusion controlled. At the end of the leaching test there were nearly no lead available for leaching. The leached amounts of lead corresponded to the lead content in the samples. The leached amounts were also related to the optimal binding systems. Furthermore the compression strength correlated generally to the leached amount of lead.

From the leaching results the maximum load on the surroundings can be calculated and also compared to other emissions. For example airborne emissions in southern Finland (non polluted area) were in the years 1980-1989 in average 9 mg/m^2/year. This means that in ten years the air emissions are probably larger than the leached amounts from immobilized surface soil.

The results from the leaching tests indicate that immobilized soil contaminated with lead can be disposed on landfills and be used in the construction of storage area or roads on landfills. Slightly contaminated soils can also be used in the construction of parking places in urban areas, which are already polluted by traffic and industrial activities. The disposal site must however be situated above ground water level and the area should not be an important ground water area.

The next step in this case study is to prepare a plan for the clean-up action. Special attention has to be paid to the risk of ground water pollution, when the contaminated soil is moved from the area.

IN SITU-BIORECLAMATION OF GROUNDWATER CONTAMINATED
BY CHLORINATED AND NONCHLORINATED ALIPHATIC AND
AROMATIC HYDROCARBONS - FUNDAMENTAL CONSIDERATIONS
AND RESULTS OF IN SITU-EXPERIMENTS

WICHMANN, K.; CZEKALLA, C.; CRON, K.

CONSULAQUA HAMBURG CONSULTING ENGINEERS
HEIDENKAMPSWEG 77, D-2000 HAMBURG 1

Keywords: Biological Treatment, Chlorinated Hydrocarbons, Groundwater Treatment, in situ-Treatment.

To ensure good results when using in situ-bioreclamation techniques in groundwater treatment in situ-experiments in a limited area of the contaminated site have been proved helpful. They are allowing the verification of preliminary laboratory studies, the optimization of the chosen technology and adaptation to the site-specific conditions.

A successful transfer of results from field studies to the full scale process demands special consideration for the planning of the studies and the chemical, microbiological and hydraulic analyses have to be done with special care. Main tasks are:

- to characterize the transport mechanisms by tracer studies

- to distinguish between degradation and dilution

F. Arendt, G.J. Annokkée, R. Bosman and W.J. van den Brink (eds.), Contaminated Soil '93, 1215–1216.
© 1993 *Kluwer Academic Publishers. Printed in the Netherlands.*

- to quantify the microbial degradation potential

- to avoid the formation of toxic metabolites of biodegradation

- to avoid the accumulation of nutrients in the aquifer.

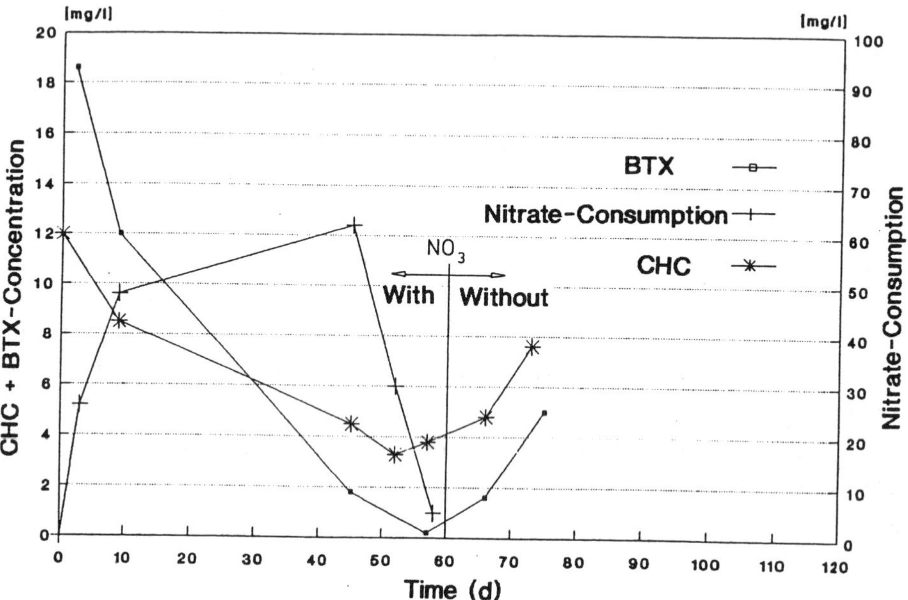

Fig. 1: In situ-elimination of BTX-aromates and volatile chlorinated hydrocarbons by addition of nitrate

These tasks are discussed by presenting results from field studies at sites with different contaminants (chlorinated and nonchlorinated aliphatic and aromatic hydrocarbons).

5. RESEARCH AND DEVELOPMENT, EMERGING TECHNOLOGIES

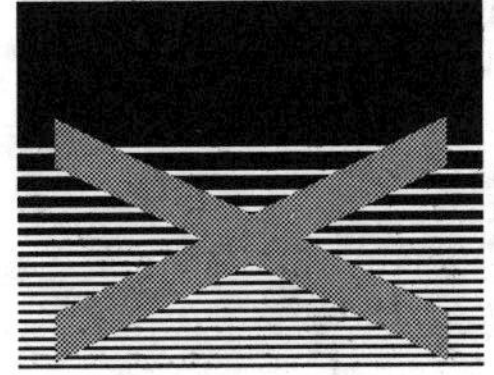

CAPABILITIES AND LIMITATIONS IN THE PERFORMANCE OF MICROBIAL
REMEDIATION PROCESSES

U. STOTTMEISTER

UFZ CENTRE FOR ENVIRONMENTAL RESEARCH LEIPZIG-HALLE
SECTION FOR REMEDIATION RESEARCH

Microbiological processes for degradation of contaminants in different sites
are more economical than other ones (Tab.1). Some environmental harms can be
advantageously remediated by their application (Tab2). The cons, however, are
not to be forgotten for all the pros (Tab.3). This field threatens to be discredited
because expectations cannot be realized. The situation is comparable to the 70ies
with intentions to replace chemical syntheses by microbiological ones.
This should be avoided by
- a realistic assessment of possible applications and the performance potential
  of biological remediation methods,
- enhanced research and development in promising areas free of commercial or
  political compulsions.
In three sections we present
1. Tendencies in r&d focused on increased performance in bioremediation
2. Examples for "non-traditional" solutions
3. Projects and results dealt with in our own labs.

1. TENDENCIES IN RESEARCH AND DEVELOPMENT IN BIOREMEDIATION

The aerobic and anaerobic destruction of harmful substances to carbondioxide
and to methane (+ CO_2) resp. using metabolic activities is the general aim of
bioremediation methods. The physiological potential of microorganisms is
optimized by technical procedures serving this purpose. The technical and
engineering standard is similar to that in the biological waste water treatment 20
years ago. However, a further principally new development in bioremediation
performance with more severe laws and changed requests to energy apply cannot
be expected.
Every site is unique and can only be remediated in a complex interrelation
between the concrete local requirements. General correlations have to be found
out to realize this target.
The knowledge of the general biotechnology going beyond the standards, like
- aeration
- nutrient addition
- warm- up
- inoculation by "performance" strains,
is applicated these days, mostly in lab scale.

1.1. The application of "new" microorganisms and microorganism associations

The physiological potency of fungi is of special interest for degradation
processes (Fritzsche 1992).
New knowledge was acquired by
- application of cosubstrates (f.e. glucose ---> amines), especially of metabolites
  of the degradation pathway (halo- and nitroaromatics, Scheibner et al. 1992)
- use of basidiomycetes (Lambert et al. 1992)

F. Arendt, G.J. Annokkée, R. Bosman and W.J. van den Brink (eds.), Contaminated Soil '93, 1219–1227.
© 1993 Kluwer Academic Publishers. Printed in the Netherlands.

- investigation of radical forming enzymes (Sack et al., 1991).
- application of methanotrophic microorganisms (halogenated solvents, Hazen et al., 1992)

The addition of precultivated bacteria into contaminated soils is under discussion. It must be a future task of research to cultivate adapted biocenoses preserving their performance state in bioreactors.

1.2. The enhanced performance of microorganisms realized by specific variations of their environmental conditions

The desired performance of microorganisms can be increased by changes in their life conditions. There are to apply:
- variations of the redox-potential by nitrate addition (realizing two effects: assimilatoric and dissimilatoric nitrate reduction by different organisms, Bouwer et al., 1992)
- limitations by substrates or nutrients resp.(oxygen, sulfur, nitrogen) (f.e. Knackmuß 1991, Maghon et al. (1992)
- specific changes in the environment f.e. from anaerobic to aerobic conditions with the target to realize in the first step a dehalogenation, followed by an aerobic decomposition in the second (Quensen et al., 1992)
- the use of the specific reaction of biocenoses to changes in the temperature (Wiegel et al., 1992)

The connection between stress situations and the toxicity of pollutants to the microorganisms is an open question today.

1.3. The application of technological specifications used in the general applied microbiology

The use of bioreactors in soil remediation research allows not only the matter balance (Lotter et al., 1992) but an intensification of the degradation processes.
Examples of application are given by
- the sequencing batch reactor (Irvine et al., 1992)
- the slurry reactor (Luyben and Kleigntjens 1992)
- the airlift reactor (Bryniok et al., 1992)
- the sequencing film reactor (Speitel and Leoard, 1992).

1.4. Chemical structure and microbial degradation

The connection between chemical structure and microbial degradation in waste water treatment was investigated by Pitter and Lischke (1989). The relative degradation speed of substituted aromatics was illustrated by graphs (fig. 1).
Knackmuß (1991) gave an excellent summary of structure-degradation relations.
Kaminski et al. (1990) demonstrated the connection between the methanogenic degradation of cresols and the adaption of the coculture to the substrates.

1.5. Bioavailability and degradation

The microbial degradation of organics in aquatic solutions or via aquatic phases can be disturbed by toxicity, steric effects, non-inducible enzymes and others. Many additional effects, however, are present in solid phases like sorption-desorption, diffusion, and other, whose overlapping makes a scientific explanation more difficult.
In dependence on the size and surface, inorganic particles e.g. adsorb organic molecules. The embedding in organic matrices (humic substances) is realized by covalent and electrostatic bondings, but mainly by unspecific hydrophobic interactions. Concentration potentials are responsible for diffusions steps (Rijnaarts et al. 1991). Increased problems arise from metabolites with lower polarity, resulting in higher mobility in soils. Plumes of this metabolites can

contact the saturated zones with lower microbial activities and contaminate the groundwater (f.e. Hexogen: reduction of -NO_2-groups to - NH_2)
An enhanced biodegradation is the result of the substitution of the polar water sphere by unpolar liquids, like proposed by Bryniok et al., 1992. According to Breasson et al. (1991) this effect is based on a changing diffusibility in the organic matrix.

2. "NON-TRADITIONAL" REMEDIATION SOLUTIONS
2.1. Site aeration
Hanert et al (1992) are sceptically evaluating the traditional landfill site:
- reduced gas components in the formed gas (up to 100 Vol % H_2S, 100 % - Ammonia, 80 % methane, and different nitrogen oxides) can cause local harmful situations
- gas forming processes can continue over decades
- the formed site leachates are highly loaded with organic and inorganic contaminants and need an additional treatment.
These negative consequences of the "traditional" landfill site can be avoided by an intermitting aeration with anaerobic-aerobic phases (alternating weekly) resulting in decreased emissions.

2.2. The subhydric site
Harmful substances can be enriched in the natural circuit without any danger for the environment, when there is realized a state free of potentials (f. e. in the black sea). Ripl (1991) facilitated the theoretical consideration by the ETR- model (energy transport model). Against all expectations, such effects of a stable deposit of contaminants can be observed in old landfill sites, existing without any measures for safety.
Obviously states free of potentials can be reached with the property of a "final" disposal. The conditions for them could possibly be realized f.e. in former lignite open-cast minings.

2.3. The microbial immobilization of organic contaminants
The previous strategy of microbiological remediation is focused on a decomposition as complete as possible (mineralisation, methanogenese). All efforts for enhanced degradation aim at on a fast substrate availability (bioavailability) and a quick and complete subtrate utilization. However, Nature provides a second way of removal of organics: The non-bioavailabilty as the end-point of a long-term complex processes with participation of bioprocesses(coal, mineral oil) or the immobilization on or in matrices, which are inert as a whole. Lotter et al. (1992) determined a part of 35 % in matter balances, which was to add to an unspecific solid - matter immobilization.
The role of white- rot - fungi enzymes like laccase is well known in the process of humification forming macromolecules f. e. from phenolics. The amendment of contaminated soils by addition of lignin containing materials supported these processes in the humification with a fixation of low molecular susbtances in a solid matrix. The formation of covalent bonds by the radical forming enzymes can be assumed.
The non-covalent immobilized mostly strong polar low molecular organics are to be moved from the cage structure of the humic substances by diffusion influenced by the concentration potential according to the general matter transport equation. That means a long term giving off of pollutants by new equilibrium states after a supposed remediation. Contrary to all suppositions, the pore water of a highly contaminated sediment from the Leipzig region is nearly nontoxic in the bioluminescence test in the "natural" equilibrium. Probably a faster biodegradation proceeds in comparison to the diffusion step. However after

contact to fresh water (with low microbial activity) a long term giving off of toxic organics (Weißbrodt and Stottmeister 1993) can be detected.
As a conclusion it should be a target of remediation strategies to immobilize harmful organics in macromolecular matrices by covalent bonds in suitable complex matrices. Adsorption or other nonspecific interactions can result in an unwished long term giving off of the pollutants. But this target can be only a part of a remediation strategy in complex with degradation steps and focused on some groups of contaminants.

3. PROJECTS WORKING ON IN THE SECTION FOR REMEDIATION RESEARCH OF THE UFZ
Our work deals with the previously discussed aspects. It is the target to support natural self cleaning processes by technical measures and processes. This will be done interdisciplinarily in the newly founded UFZ research centre.
Some of the research projects , the scientific background and some results will be presented briefly:
1. enhanced degradation in remediation processes
- The induction of degradative and radical forming enzymes:
Monooxigenase systems in bacteria and fungi are to be induced or increased in their activity by a substrate excess and oscillating oxygen stress (fig. 2, Martius et al., 1989).
- The application of new biocenoses:
-- Methanotrophic bacteria show an important but not well-known role in natural processes as in aquifers as well as in landfill sites. The use of methane as cosubtrate for the oxidative dehalogenation in-situ is described by Hazen et al. (1992); the sequencing-film reactor by Speitel and Leonard (1992) was used with the same aim in lab scale. We use the experience from single cell protein synthesis by methanotrophs. The fundamentals were developed of a stable two-component methanotrophic mixed culture (Wendlandt et al., 1990). A quantification is possible for connections between growth rate and methaneconcentration in dependence on the pressure (Wendlandt et al., 1992)
-- The application of stable adapted biocenoses of autotrophic and heterotrophic organisms allows the simultaneous decomposition of N-heterocyclic compounds and nitrification (Mauersberger et al., 1990)
-- The interaction between heterotrophic and autothrophic consortia in bioleaching is described by Ondruschka et al., 1993)
2. Sorption-desorption and deposit
- Pore water from highly contaminated sediments is nearly untoxic in a equilibrium between microbial activity and remobilization. A swirl-up of the sediments acts in resolutions with an asymptotic course (Weißbrodt and Stottmeister, 1993).
- Dumped phenolic waste waters from lignite pyrolysis form humic acid- like compounds after air contact. Aromatics are so fixed that an analytical detection is impossible (Pörschmann and Stottmeister 1993).
- Sieved fine grain fractions of sediments and sediment ashes show the capability in multiphase systems to catalyse the irreversible fixation of PAHs. Preliminary results show a non-extractable part of 30 % of the initial concentration (Remmler et al., 1993).
3. case study "carbonization waste water deposit"
The industrial site in a former lignite open-cast mining with 2 billion cubic meters of highly contaminated phenolic waste water is investigated as case study for an interdisciplinary restoration. The aim is a combination of physico-chemical pretreatment of the water, followed by an intensification of aerobic and anaerobic natural processes and the search for a potential-free disposal. ("pontus euxinos"-relations).

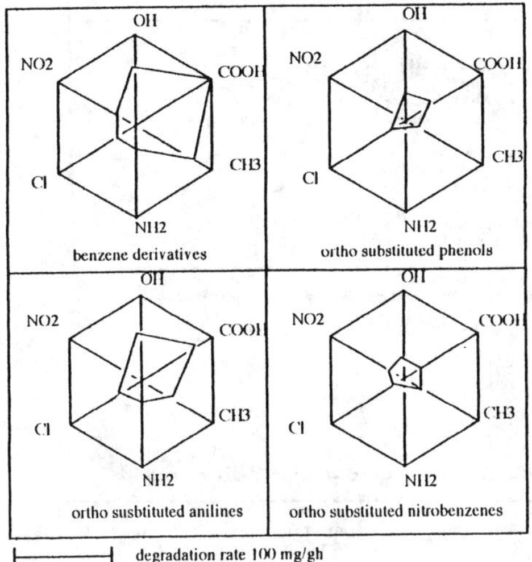

Fig.2 **Auswirkung von aerob- anaerob-Wechseln auf spezifische Leistungsparameter bei Kultivierung von Acinetobacter calcoacaticus EB 104 auf Acetat / Hexan-Modellmedium**

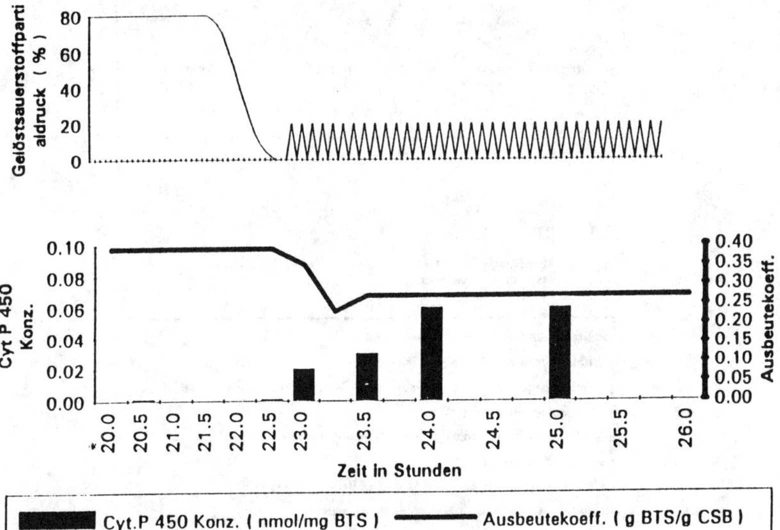

TABLE 1. Approximate Unit Cost for Some Selected Treatments of
Contaminated Soil   (acc. to Taylor and McLean 1992)

	Method	Cost range ($£/m^3$ or tonne)
1.	Landfill	25-100
2.	Thermal destruction .(1)	
	a) Off-site Holland	25-100
	b) Off-site North America	100-500
	c) On-site North America	75-300
3.	Vapour extraction (2)	20-50
4.	Solvent extraction	35-100
5.	Biological treatment	5-40
	Advanced landformig bioreators	50-75

(1) Lower end of range - refinery site soil mainly petroleum hydrocarbons
(2) Higher end of range - coal tar sites, mainly PAH

TABLE 2. Useful application of biological treatment in site remediation

	contaminants	method		remarks
		ex situ off site	1	
		ex situ on site	2	
		in situ	3	
		aerobic	4	
		anoxic	5	
		anaerobic	6	
soils	aliphatic HC	1, 2, 3		
	PAH	1, 2, 3, 4, 5		
	phenolics	1, 2, 4		
	solvents (halogenated))	3, 6		+ stripping
	BTX	1, 2, 4, 5		
	explosives	1, 2, 4, 5		
	biocides	1, 2, 4, 5		
	haloaro-mates	1, 6		
groundwater	solvents	3, 4		+ stripping
	aliphat, HC	3, 4		+ stripping
sediments	PAH	3, 4		

TABLE 3. Disadvantages of biological methods in site remediation
(esp. in situ methods)

- long-term processes (up to some years)

- possible formation of unknown metabolites (e.g. with higher toxicity)

- possible contamination of groundwater by metabolites or microorganisms

- undegradable residues (PAHs)

- low effectivenies (for some compounds lower than 50 %)

- final concentration unsure

- no influencing in the course of the remediation process

- possibility of blockage of soil pores by biomass

- possibility of "blooding" of remediated soils after long-term observation

4. REFERENCES

Bouwer, E.I., Trizinsky, M.A., and Zhang, W. (1992). Influence of redox conditions on organic contaminant biotransformation.
In-situ Bioremediation Symposium '92, Niagara-on-the-Lake, Ont. Can. Sept. 20th-24th 1992, ed. S. Lesage p. 66-67

Brusseau, M.L., Jessup, R.E., Rao, S.C. (1991). Non equilibrium sorption of organic chemicals: Elucidation of rate-limiting processes
Environ. Sci. Technol. 25, 134-142

Bryniok, D., Brunner, W., Eichler, B., Köhler, A., Mackenbroch, K., and Knackmuß, H.J. (1992). Biodegradation of PAH in Airlift-Bioreactors.
Internationales Symposium Soil Decontamination Using Biological Processes, Karlsruhe 6-9 Dec. 1992, Prereprints 639-644

Bryniok, D., Eichler, B., Reinert, C., von Watzdorf, R., and Knackmuß, H.J. (1992) Enhanced bioavailability of polycyclic aromatic hydrocarbons in multiphasic systems.
International Symposium Soil Decontamination Using Biological Processes, Karlsruhe 6-9 Dec. 1992, Prereprints 645-650

Bürger, G., Mauersberger, P. Stottmeister, U. (1990). Untersuchungen zum Abbau organischer Stickstoffverbindungen durch nitrifizierende Mischkulturen.
awt Abwassertechnik, 6/90

Fritzsche, W. (1992). Degradations of xenobiotics by fungi.
International Symposium Soil Decontamination Using Biological Processes, Karlsruhe 6-9 Dec. 1992, Prereprints 31-36

Hanert, H., Harborth, P., Kucklick, M., Lang, E., Rohde, R., Waschke, C., and Wittmaier, M. (1992). Möglichkeiten mikrobieller Stoffumsetzungen bei der Abfallentsorgung.
Schriftenreihe des Zentrums für Abfallforschung der TU Braunschweig, 1992, Heft 7

Hazen, T.C., Dougherty, I.M., Enzien, M., Franck, M.M., Fliermans, C.B., Eddy. C.A. (1992). DOE/SRS Integrated demonstration: In situ bioremediation of soil and groundwater at a chlorinated solvent contaminated site using horizontal wells to inject air and methan.
In-situ Bioremediation Symposium '92, Niagara-on-the-Lake, Ont. Can. Sept. 20th-24th 1992, ed. S. Lesage, p. 22

Irvine, R.L., Earley, J.P., Yocum, P.S., and Kehrberger, G.J. (1992). Slurry Reactors for Assessing the treatability of contaminated soil.
International Symposium Soil Decontamination Using Biological Processes, Karlsruhe 6-9 Dec, 1992, Prereprints 187-194

Kaminski, U., Kuschk, P., and Janke, D. (1990). Degradation of different aromatic compounds by methanogenic consortia from Saale river sediment acclimated to either o-, m- or p-cresols.
J. Basic. Microbiol. 30, 4, 259-265

Knackmuß, H.J. (1991). Chemische Strukturen und biologische Abbaubarkeit.
9. Dechema Fachgespräch Umweltschutz: Mikrobiologische Reinigung von Böden, 1991, Frankfurt/M., 41-67

Lambert, M., Kremer, S., and Anke, H. (1992). Degradation of PAHs by fungi: Comparison of deuteromycetes isolated from polluted and non polluted soil with saprophytic basidiomycetes.
International Symposium Soil Decontamination Using Biological Processes, Karlsruhe 6-9 Dec. 1992, Prereprints 333-341

Lotter, S., Heerenklage, J., and Stegmann, R. (1992). Carbon Balance and Modelling of the Oil Degradation in Soil Bioreactors.
International Symposium Soil Decontamination Using Biological Processes, Karlsruhe 6-9 Dec. 1992, Prereprints 219-227

Luyben, K.Ch.A.M., and Kleijntgens, R.H. (1992) Bioreactor Design for soil decontamination.
International Symposium Soil Decontamination Using Biological Processes, Karlsruhe 6-9 Dec. 1992,
Prereprints 195-204.

Maghon, G., Schaffrath, S., Kneifel, H., und Webb, L. (1991). Bildung von Metaboliten beim Abbau
von Naphthalin unter suboptimalen Sauerstoffbe-dingungen.
9. Dechema Fachgespräch Umweltschutz: Mikrobiologische Reinigung von Böden, 1991, Frankfurt/M.,
259-262

Martius, G., Bock, A., Demme, R., Stottmeister, U. (1990). Substrate mixture and Cytochrome P 450
induction in Acinetobacter calcoaceticus.
Biocatalysis 1990, 4, 75

Ondruschka, J., Seidel, H., Stottmeister, U. (1993). Mobilisation of heary metals in contaminated river
sediments by bacterial leaching.
Poster this symposium

Pitter, P., und Lischke, P. (1989). Biochemische Abbaubarkeit organischer Stoffe.
Acta hydrochim. hydrobiol., 17, 2, 217-226

Pörschmann, J., and Stottmeister, U. (1993). Methodical investigation of interaction between organic
pollutants and humic organic material in coal waste waters
Chromotographia, in press

Quensen, J.F., Tiedge, I.M., Boyd, S.A., Enke, C., Lopshire, R., Giesy, I., Mora, M., Crawford, R., and
Tillitt, D. (1992). Evaluation of the suitability of reductive dechlorination for the bioremediation of PCB-
contaminated soils and sediments.
International Symposium Soil Decontamination Using Biological Processes, Karlsruhe 6-9 Dec. 1992,
Prereprints 91-100

Remmler, M., Kopinke, F.-D., Stottmeister, U. (1993). Investigation on the fixation and release of
organic compounds in river sediments from the Leipzig region
Poster this symposium

Rijnaarts, H.H.M., Bachmann, A., Jumelet, J.C., and Zehnder, A.J.B. (1991). Effect of desorption and
intraparticle mass transfer on the aerobic biominerali-sation of alpha-hexachlorocyclohexane in a
contaminated calcareous soil.
9. Dechema Fachgespräch Umweltschutz, Frankfurt/M. 27.-28. February 1991, 207-210

Ripl, W. (1991). Das Energie-Transport-Reaktionsmodel (ETR-Modell) - ein prozessorales Wasser-
modell als Grundlage für eine reduzierte gesamtökologische Betrachtung.
DGL Deutsche Gesellschaft für Limnologie, Zusammenfassung der Jahrestagung, 531-535

Sack, U., Günter, T., Schade, W., und Fritzsche, W. (1991). Screening von Pilzen auf die Fähigkeit
zum Abbau polycyclischer aromatischer Kohlenwasserstoffe (PAK).
9. Dechema Fachgespräch Umweltschutz, Frankfurt/M., 1991, 234-239

Scheibner, K., Günther, Th., Hofrichter, M., and Fritzsche, W. (1992). Potential of fungi for degrade
anilin and related compounds.
International Symposium Soil Decontamination Using Biological Processes, Karlsruhe 6-9 Dec. 1992,
Prereprints 351-357

Speitel, G.E., and Leonard, J.M. (1992). A sequencing biofilm reactor for the treatment of chlorinated
solvents using methanotrophs.
Water Environment Research 4, 5, 712-719

Taylor, M.R.G., and Mclean, R.A.N. (1992) Overview of clean-up Methods for Contaminatied Sites
J. IWEM 6, 408-415

Weißbrodt, E., Stottmeister, U. (1993). Gefährung der Gewässer im Raum Leipzig durch die
Abwasserbelastung der braunkohleverarbeitenden Industrie (Sedimentologie)
BMFT-Fördernummer BEO 0339419F, Abschlußbericht 1993

Wendlandt, K.-D., Jechorek, M. (1992). <u>The influence of pressure on the mass transfer and growth of bacteria in processes with methane/natural gas</u>
High Pressure and Biotechnology, Eds, C. Balny, R. Heyashi, K. Heremans u. P. Masson
Colloque INSERM/John Libbey Eurotext Ltd, Vol. 224, 465-466

Wendlandt, K.-D., Wand, A., Brühl, E., Karbaum, K., Schurig, K.H., Wagler, D. (1990).<u>Verfahren zur Kultivierung von Mikroorganismen</u>
Patentschrift DD 283738A7 (24.10.90) Akademie der Wissenschaften der DDR

Wiegel, J., Kohring, G.W., Zhang, X., Utkin, I., Dalton, D., He, Z., Wu, Q., and Bedard, D. (1992). <u>Temperature an important factor in the anaerobic trans-formations and degradation of chlorophenols and PCBs.</u>
International Symposium Soil Decontamination Using Biological Processes, Karlsruhe 6-9 Dec. 1992, Prereprints 101-108

DETERMINING AND PREDICTING THE DEGRADATIVE ACTIVITY OF A GEM IN AQUIFER SEDIMENT MICROCOSMS

Egestorff, J., Dwyer, D.F.

Molecular Microbial Ecology Group
Division of Microbiology
National Research Centre for Biotechnology of Germany (GBF)

1. INTRODUCTION

In recent years, a variety of toxic and noxious chemicals have been introduced into the environment. Conventional remediation methods, such as chemical degradation and burning, are often expensive and inadequate. An alternative is to use the natural biodegradative capabilities of microorganisms for degradation and removal of pollutants. Unfortunately, chemicals with novel structures or with substituents which are rarely found in nature (xenobiotics) are often resistent to biodegradation because appropriate degradation pathways have not evolved. It may be possible to use genetically engineered microorganisms (GEMs) with novel and effective degradation pathways for the bioremediation of sites contaminated with chemical pollutants. This report describes experiments which were carried out to assess this possibility by following the fate and degradative activity of a GEM once introduced into microcosms made with aquifer sediment obtained from a sand and gravel aquifer (5 - 8 m depth) located on Cape Cod, Massachusetts, USA.

Aquifer sediment microcosms were developed for these experiments in order to simulate *in situ* conditions; key environmental parameters (temperature, pH, conductivity, etc.) were controlled. Microcosms are useful laboratory systems consisting of an excised part of the environment which are maintained under contained conditions and with vital ecosystem parameters (Wimpenny, 1988). The general usefulness and appropriate procedures for validating microcosms were demonstrated by co-workers from our laboratory who constructed lake sediment microcosms (Wagner-Döbler et al., 1992).

Pseudomonas putida KT2440 (pWWO-EB62) (EB62), was used as the model-GEM. EB62 contains a TOL-plasmid encoded catabolic pathway for degradation of alkylaromatics that has been well described on the genetic and biochemical level (Timmis, 1990). The catabolic pathway was previously modified for the degradation of 4-Ethylbenzoate (4-EB) by experimental broadening of the substrate range (Ramos et al., 1986); 4-EB was used as the target pollutant to be degraded by EB62 within the aquifer microcosms. A GEM requires certain features to function *in situ* as a bioremediation agent which include:

-   the ability to survive and multiply in the target ecosystem, and
-   the ability to degrade the target pollutant by way of the engineered degradative pathway

F. Arendt, G.J. Annokkée, R. Bosman and W.J. van den Brink (eds.), Contaminated Soil '93, 1229–1233.
© 1993 *Kluwer Academic Publishers. Printed in the Netherlands.*

These features were assessed for EB62 in the microcosm experiments in order to obtain data for making predictions about the usefullness of the bacterium *in situ*. Batch- and flow-through microcosms were set up with EB62 and 4-EB to determine survival and degradative activity of EB62. Control microcosms were used to assess the degradative ability of indigenous microorganisms. Concentrations of 4-EB within the microcosms were measured using HPLC. The effects of (i) substrate concentration, (ii) density of EB62, and (iii) temperature on the rates of degradation of 4-EB by EB62 were determined. Data were fit into a first-order kinetic model to determine the half-life of the substrate. The half-life (time in which 50% of the substrate is degraded) can be used to predict the residence time of the chemical in the environment. One goal was to predict the effect of EB62 on this residence time. In addition, the kinetic parameters, V_{max} (maximum velocity) and K_m (Michaelis-constant), were calculated from rates of degradation of 4-EB to determine the maximum rate of degradation and the affinity of EB62 for 4-EB.

2. MATERIALS AND METHODS

Batch microcosms consisted of serum bottles (250 ml) containing 100 g of aquifer sediment and 4-EB. Flow-through microcosms consisted of glass-columns (1600 cm^3) packed with 1500 g sediment through which a solution of 4-EB (200 µM) with a flow-rate of 110 ml h^{-1} was passed (Egestorff, 1992). To both types of microcosms EB62 was added following overnight cultivation of the microorganisms at the *in situ* temperature of the aquifer (11°C) and a starvation period of 2 hours. In addition to 4-EB, synthetic sewage (Nüßlein et al., 1990) was passed through the flow-through microcosms to simulate contamination via disposal of treated sewage.

Following extraction of EB62 from the sediment, the microbe was identified and enumerated by using selective plating and immuno-fluorescence techniques (detailed discription is given by Egestorff, 1992) A monoclonal antibody (Ramos-Gonzales et al., 1992) which recognizes the lipopolysaccharide-O-antigen of EB62 was labelled with a fluorochrome dye (FITC) and used to identify and enumerate EB62 by fluorescent microscopy.

Concentrations of 4-EB obtained over appropriate time intervals were analyzed by non-linear regression (Computerprogram Enzfitter 1.05) to calculate first order rate constants. Kinetic constants, V_{max} and K_m, were determined from a Lineweaver-Burk plot in which the rates for degradation of 4-EB (1/v) were plotted against substrate concentrations (1/S).

3. RESULTS AND DISCUSSION

3.1. Batch microcosms

In control microcosms without EB62, the total number of bacteria remained constant during each experiment (data not shown); which indicated that the indigenous microbial community was viable for at least 50 days. Identification and enumeration experiments revealed that EB62 colonized in the batch microcosm at levels of 10^4 - 10^5 CFU/g dw and survived for a period of at least 50 days following introduction into the microcosm sediment. In the aquifer sediment no indigenous bacteria were present which degraded 4-EB; this allowed for enumeration of EB62 by selective plating.

EB62 not only survived in batch microcosms, it also degraded 4-EB demonstrating expression of the engineered catabolic pathway. Degradation data were fit into a first-order kinetic model to determine rates of degradation of 4-EB. These rates were observed to range between 2.9 - 14.9 μM h^{-1} for substrate concentrations of 18 - 447 μM. The density of EB62 was 10^7 CFU/g dw sediment.

Further experiments assessed the effect of different (i) substrate concentrations, (ii) densities of EB62 and (iii) temperatures on the initial velocities of degradation of 4-EB. The data were used to predict the rate of degradation in the flow-through microcosms. (i) Substrate concentrations of 4-EB of 18 - 450 μM are close to what one could expect for pollutants present in a contaminant plume of an aquifer. In the environment many chemicals can be found in relatively high concentrations such as resulting from oil spills or other cases of human activities (Alexander and Scow, 1992). Rates of degradation of 4-EB at all substrate concentrations were linear, which indicated that the reaction was following first-order kinetics.

(ii) Rates of degradation of 4-EB were assessed for two densities of EB62 (10^6 - 10^7 CFU/g dw sediment). Introduction of a GEM into an aquifer may result in a heterogenous distribution of the microorganism. A higher density of added bacteria would be present inside introduction wells, and lower densities could be present further away. Therefore, the effect of different densities of EB62 on the rates of degradation was determined. Observed rates were linear at both densities ranged from 1.6 μM h^{-1} (10^6 CFU/g dw and 450 μM 4-EB) to 14.9 μM h^{-1} (10^7 CFU/g dw and 450 μM 4-EB).

(iii) Rates of degradation of 4-EB were determined at three temperatures 11, 22 and 30°C because previous studies have shown, that temperature has an effect on the rates of degradation of chemicals (Alexander and Scow, 1989). Observed rates (10^7 CFU/g dw and 450 μM 4-EB) ranged from 14.9 μM h^{-1} (11°C) to 85.9 μM h^{-1} (30°C). The observed rates of degradation of 4-EB were linear in this temperature range. V_{max} and K_m were calculated for rates at each temperature. It was observed that when the temperature went up the maximum velocity went up and the affinity of EB62 for 4-EB went down. At a density of calculated 10^7 CFU/g dw of EB62 and a temperature of 11°C, the calculated half-lifes of 4-EB ranged from 3.6 - 20.6 h (18 - 450 μM) respectively.

Alexander and Scow (1989) proposed that the degradation of a chemical follows first-order kinetics when the initial substrate concentration is relatively low so that the microorganisms do not increase in biomass. Although the half-life of a chemical can be used to predict its residence time, the calculated value does not reflect changes in the rates of degradation which occur during the degradation process. However, when the population density of microorganism which grows at the expense of the chemical degraded by the micobe increases, this will alter the rate of degradation over time. Kinetic models other than the first-order reaction model are be used to describe more precisly the degradation of a chemical when growth occurs. Simkins and Alexander (1984) described the metabolism of benzoate by *Pseudomonas*. spp. at a substrate concentration near K_m with the "Monod-with-growth-model". Further experiments were made to determine the effect of growth by EB62 (10^6 CFU/G dw) on the rates of degradation of 4-EB. As degradation procedes, the rates of degradation of 4-EB were faster than those predicted by first-order kinetic model due to the increasing density of EB62 at relatively high concentration of 4-EB (166 - 450 μM); first-order kinetic models would underestimate the *in situ* true half-life for 4-EB when degraded by EB62. Growth was not observed when concentrations ranged from 18 - 166 μM. The results suggests, that at a population density of 10^7 CFU/g dw and a concentration of 200 μM 4-EB is necessary in order to see growth of the inoculated GEM.

3.2. Flow-through microcosms

Flow-through microcosms were developed and used to determine the fate of EB62 at a density of 10^4-10^5 CFU/g dw under relatively more *in situ* conditions. These had the advantage of simulating ground-water flow. Comparison of the concentration of 4-EB (200 μM) in the inflow of the microcosm and the concentration of 4-EB in the outflow showed, that 31.1% of the incoming 4-EB was degraded. In these microcosms synthetic sewage was added to the inflow. In the microcosms receiving no synthetic sewage degradation of 4-EB was not detected since the density of EB62 was apparently too low to enhance removal of 4-EB.

At a population density <10^5 CFU/g dw and a concentration of 4-EB of 200 μM degradation was not detected in batch microcosms over a period of 6 h; rate of degradation of 4-EB was < 0.05 μM h^{-1} which would result in a total removal of 4-EB of 0.3 μM h^{-1}. Concentration of 4-EB at that level cannot be observed with any statistical significance. In flow-through microcosms, with a retention time of 119 h, the total predicted removal of 4-EB would have been < 6.0 μM, which also cannot be statistically observed.

4. SUMMARY

From the studies it was demonstrated that EB62 survived and degraded 4-EB in aquifer sediment. This was also observed to occur in activated sludge microcosms (Nüßlein et al., 1992). These results indicate that the laboratory strain EB62 adapted to different *in situ* conditions which it managed to function as an bioremediation agent. In addition, another genetically engineered microorganism, *P*. sp. B13 FR1 (pFRC20P), which was designed in our laboratory to simultaneouly degrade 3-chlorobenzoate and 4-methylbenzoate colonized activated sludge microcosms (Nüßlein et al., 1992) and lake sediment microcosms and degraded 3-CB and 4-MB (Pipke et al., 1992).

Within our laboratory group, field studies were done to follow the fate of an introduced bacterium in an aquifer. Co-workers demonstrated that *P*. sp. B13, the parental strain for a variety of GEM, including *P*. sp. B13 FR1 (pFRC20P), survived in a sewage-contaminated aquifer for at least 1 year. These studies, as well as the results obtained in our work with batch- and flow-through microcosms, suggests that GEMs may be useful as bioremediation agents to effect clean up of contaminated sites when alternative methods are not available.

ACKNOWLEDGMENT

This work was in part supported by the German Ministery of Research and Technology (BMFT Vorhaben 0319-433A) and the Biotechnology Action Program grant from the European Community (BAP-O417-D).

4. LITERATURE

Egestorff, J. (1992). Survival and Degradative Activity of a Modell-GEM in Aquifer Sediment Microcosms. (unpublished doctoral dissertation)

Krumme, M.L., Smith, R.L., Thieme, M., Egestorff, J., Tiedje, J.M., Timmis, K.N., Dwyer, D.F. (1992). Aquifer Bioremediation: Use of Field Tests and MIcrocosms to predict Fate of Introduced Microorganisms (submitted to Science)

Nüßlein, K., Maris, D., Timmis, K.N., Dwyer, D.F. (1992). Expression and Transfer of Engineered Catabolic Pathways harbored by Pseudomonas sp. introduced into Activated Sludge Microcosms. Appl. Environ. Microbiol. 58:3380-3386

Pipke, R., Wagner-Döbler, I., Timmis, K.N., Dwyer, D.F. (1992). Survival and Function of a Genetically Engineered Pseudomonad in Aquatic Sediment Microcosms. Appl. Enivron. Microbiol. 58:1259-1265

Ramos, J.L., Stolu, A., Reinike, W., Timmis, K.N. (1986), Altered Effector Specificities in Regulators of Gene Expression: Tol Plasmid *xyl*S Mutants and their Use to Engineer Expansion of the Range of Aromatics Degraded by Bacteria. Proc. Natl. Sci. 83:8467-8471

Simkins, S., Alexander, M (1984). Models for Mineralization with Variables of Substrate Concentrations and Population Density. Appl. Environ. Microbiol. 47:1299-1306

Timmis, K.N. (1990). Genetic Construction of Novel Metabolic Pathways: Degradation of Xenobiotica. In: G. Hauska and G. Thauer (Eds.). <u>The Molecular Basis of Bacterial Metabolism,</u> 41. Colloquium Mosbach, 69-83, Berlin-Heidelberg, Springer-Verlag

Wagner-Döbler, I., Pipke, R., Timmis, K.N., Dwyer, D.F. (1992). Evaluation of Aquatic Sediment Microcosms and Their Use in Assessing Possible Effects of Introduced Microorganisms on Ecosystem Parameters. Appl. Environ. Micrbiol. 58:1249-1258

Wimpenny, J.W.T. (1988). Models and Microcosms. In: <u>Handbook of Laboratory Model Systems for Microbial Ecosystems.</u> Vol 1, 1-17, Boca Raton, Fl., CRC Press

CARBON BALANCE OF A PAH-CONTAMINATED SOIL DURING BIODEGRADATION AS A
RESULT OF THE ADDITION OF COMPOST

S Lotter[1+], A Brumm[1], J Bundt[1], J Herrenklage[1],
A Paschke[2+], H Steinhart[2], R Stegmann[1]

[1]   Waste Management Department of the Technical University of Hamburg-
        Harburg
[1+]   Current address: TÜV Norddeutschland, Hamburg
[2]   Institute of Biochemistry and Food Chemistry of the University of
        Hamburg
[2+]   Current address: Waste Management Department of the Technical
        University of Hamburg-Harburg

1   ABSTRACT

A soil from a site contaminated with polycyclic aromatic hydrocarbons (PAH)
was mixed with compost in the ratio of 2:1. The degradation experiments were
conducted in glass vessels at 30°C over a period of 84 days. The PAH/carbon
balance contains the following parameters: contaminant (PAH), PAH after
saponification, mineralisation (CO_2 production) and biomass. The
contamination was analysed using gravimetric and chromatographic methods.
The 16 EPA PAH represented only 11% of the total PAH measurable using high-
resolution gas chromatography (HRGC). For balancing, the PAH content
analysed by means of HRGC at the start of the experiment was taken as 100%.
By the end of the experiment the PAH/carbon had distributed itself as
follows: 6% extractable PAH, 35% PAH measurable after saponification, 26%
mineralisation. The discrepancy in the balance is 32%. This can be explained
by the formation of bound residues and humus. Almost no discrepancy is
evident when the contamination is determined gravimetrically. In the total
carbon balance the ratio of humus/bound residue in the carbon compartment
was 60%.

2   INTRODUCTION

The production of carbon balances during the biological clean-up of the soil
is interesting in two respects. Firstly the progress of the clean-up
operation can be evaluated with this tool; moreover the balancing process
yields information about the degradation of the contaminant in the soil. The
degradation processes in the contaminated soils have not be described in
great detail to date.

In this investigation a method of balancing the carbon, which was developed
for the example of synthetic oil contamination (Lotter et al, 1992), was
applied to real contamination. The contamination of the soil involved
polycyclic aromatic hydrocarbons (PAH). Carbon balances are normally
produced in research with substances marked with C^{14} (Herrenklage et al,
1992). This approach is generally ruled out for real cases of contamination,
since the range of contaminants and the forms in which they are bound in the
soil cannot be simulated.

F. Arendt, G.J. Annokkée, R. Bosman and W.J. van den Brink (eds.), Contaminated Soil '93, 1235–1246.
© 1993 Kluwer Academic Publishers. Printed in the Netherlands.

The following parameters are included in the balance: PAH content (extractable), PAH content (after saponification), mineralisation and biomass. The total carbon balance is also produced (total carbon of the solid).

3   MATERIAL AND METHODS

The soil contaminated with PAH comes from a site on which tar and similar products, mineral oils and train-oils were traded in the 30s and 40s. The contamination of the soil was caused by war damage. The material was to be described as fill, and consisted of sandy and silty components, which were interspersed with slags and building rubble. The soil was sieved to a grain size of 2 mm for the degradation experiments.

1 part of compost solids was added to every 2 of the contaminated soil. Compost and PAH contamination were tested in separate evaluations as controls. In the case of both synthetic contamination with oils, and contamination with PAH, the addition of compost led to an acceleration of the degradation of the contaminant and to increased mineralisation (Stegmann et al, 1990, Kästner, 1991).

The degradation experiments were carried out at 30°C in 1.5- litre glass vessels with 75 g of soil solids, which was moistened to a degree of saturation of 60%. The supply of oxygen within the vessels was checked for adequacy. The experiments were carried out over a period of 84 days. The mineralisation was measured by absorbing the CO_2 evolved in NaOH (in a beaker). The analytical methods of determining the contamination are assembled in Table 1.

Table 1: Analytical methods of characterising the PAH contamination

Method	Extracting agent	Extraction	Apparatus
Gravimetry	DM	Sox	
HRGC	DM	Sox	MEGA 5300 (Carlo Erba)
HPLC	EA, DM	US	L-6200 (Merk/Hitachi)
Organic analysis	EA	Sox	C/H/N analyser (C Erba)

Acronyms:
HRGC: high-resolution gas chromatography   DM: dichloromethane
HPLC: high-pressure liquid chromatography   EA: ethyl acetate
Sox: Soxhlet   US: ultrasound

For the gravimetric determination of the contamination the extract was confined in a rotary evaporator. The determination of the total of the PAH and the 16 EPA PAH with HRGC was carried out using the method after Paschke et al, 1992. The HPLC analysis was carried out after Bundt et al, 1993. For the saponification the sediment from the ultrasonic extraction for the HPLC was mixed with methanol and NaOH (2m) in a Hungate tube, and stood in a hot water bath for 1 hour (Eschenbach et al, 1991). After centrifuging, the liquid separated was drawn off and injected into the HPLC apparatus.

The biomass was determined using the substrate induced respiration (SIR) method after Anderson and Domsch, 1978. The total carbon was measured in a Dohrmann analyser, after the solids samples had been dried for 48 h over P_2O_5 and diluted with sand. The carbonate content (inorganic carbon) was determined with a Scheibler apparatus.

4   EXPERIMENTAL RESULTS

4.1   <u>Reducing the PAH concentration</u>

A knowledge of the original contaminant concentration is indispensable in producing the carbon balance of the PAH degradation. Different determination methods were therefore used to cope with the complex contamination. The result of the measurements is reproduced in Table 2.

The table gives the carbon content from the particular measurement. This makes it easier to compare the results; and the carbon value is needed for the balance. The gravimetric analysis allowed calculation of the difference between the results for the mixture of soil and compost and for the uncontaminated compost. The carbon content was determined using the result of the organic analysis (81.4% by weight of C). With the chromatographic methods (HRGC and HPLC) calculation was performed with the average carbon content of the 16 EPA PAH (94.8% by weight C).
Gravimetry gave the largest value at 33084 mg of C/kg of solids. The total of the PAH (HRGC) was 20721 mg of C/kg of solids. The fraction constituted by the 16 EPA PAH here was only 11%. The results of the HRGC and the HPLC showed good agreement for the EPA PAH when dichloromethane was used as the extracting agent. With ethyl acetate the yield was significantly lower.

Table 2: Determination of the contaminant concentrations of the mixture of soil and compost at the start of the experiment with different analytical methods

Analysis Extraction	GRAV DM	HRGC DM	HRGC-EPA DM	HPLC-EPA DM	HPLC-EPA EA
mgC/kg sol.	33085	20721	2361	2546	1250

<u>Acronyms</u>: see Table 1
GRAV: gravimetry   EPA: 16 EPA PAH

Figure 1 shows the relative decrease in the contaminant concentrations relative to the initial value. Despite the considerable difference in the concentrations, the four chromatographic methods exhibit a comparable variation with time (exponential decrease). Only the gravimetric method led to completely different results. In this case the measurement at the end of the experiment was about 75% of the initial concentration.

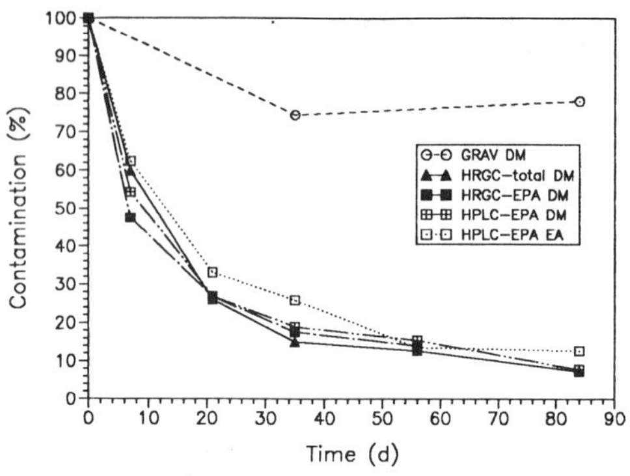

Kontamination	=	Contamination
gesamt	=	total
Zeit (d)	=	Time (d)

Figure 1: Percentage reduction in the PAH contamination relative to the initial measurement for each method (see Tables 1 and 2 for acronyms)

After the saponification the 16 EPA PAH were analysed using HPLC. Figure 2 plots the results of this measurement against time. The "total" in the figure is calculated by adding the results of the extraction and saponification.

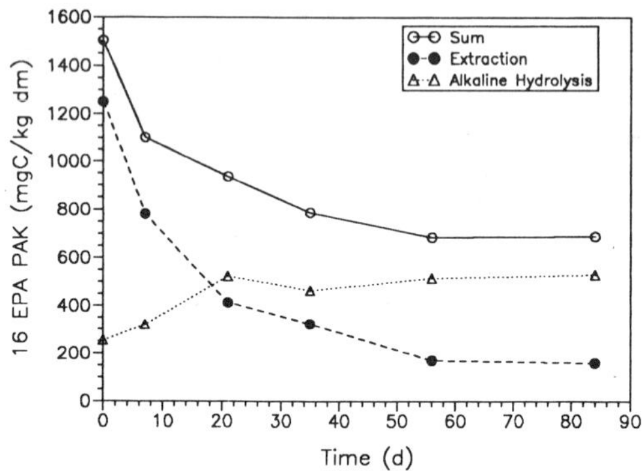

6 EPA ...	=	16 EPA PAH (mg C/kg solids)
Summe	=	Total
Extraktion	=	Extraction
Verseifung	=	Saponification
Zeit (d)	=	Time (d)

Figure 2: Determination of the 16 EPA PAH by extraction and saponification with ethyl acetate

The extractable PAH exhibited the exponential decrease already described. However the PAH concentration measurable after saponification doubled from 254 to 529 mg of C/kg of solids. The total line is accordingly significantly flatter than the curve of the extractable PAH.

4.2 CO_2 production and biomass

The mineralisation of the contamination is manifested in the CO_2 production. Figure 3 shows the cumulative CO_2-C curves of the mixture of soil and compost (PAH + comp.), the PAH contamination (PAH) and the compost (comp.). The total of PAH and comp. is also plotted. To enable comparison of the results, the PAH contamination without addition of compost was related to 2/3 kg of solids and the compost to 1/3 kg of solids (corresponding to the proportions in PAH + comp.).

The variation in the compost's endogenous CO_2 production was virtually linear. The PAH-contaminated soil without added compost exhibited increased production of CO_2 at the start of the experiment, which reduced as the experiment continued. On the other hand sharply increased production of CO_2 was evident in the PAH + comp. mixture, which was reflected in the steepness of the total curve over the first 45 days of the experiment. The TOTAL curve in Figure 3 reproduces the result of adding the control evaluations of the PAH-contaminated soil and of the compost. The increased mineralisation resulting from the addition of the compost is elucidated by comparing the PAH+comp. with the TOTAL curve. 45 days into the experiment the rate of CO_2 production (mg of CO_2/(d x kg of solids) was equal to that of the PAH-contaminated soil plus that of the compost fraction.

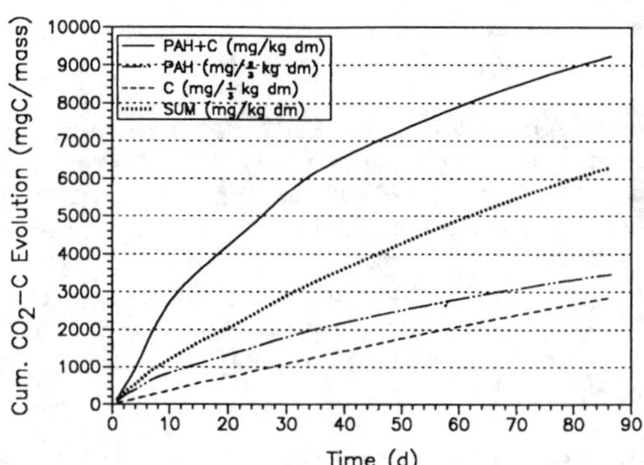

kum. CO_2-C ...	=	Cumulative production of CO_2-C (mg of C/unit weight)
PAK+K....	=	PAH+comp. (mg/kg of sol.)
PAK......	=	PAH (mg/2/3kg of sol.)
comp. ...	=	comp. (mg/1/3 of sol.)
TOTAL....	=	TOTAL (mg/kg of sol.)
Zeit (d)	=	Time (d)

Figure 3:   Cumulative production of CO_2 by the mixture of PAH-contaminated soil and compost (PAH+comp.), the PAH contamination (PAH) and the compost (comp.)

The results of the determination of the biomass using the SIR method are entered in Table 3. The sharply increased value in the PAH evaluation 7 days into the experiment seems to be atypical. The values of the compost experiment (comp.) fluctuate more than had been the case in other series of our own experiments.

Table 3:   Biomass contents according to the SIR method

Time (d)	PAH+comp. (mg C/kg solids)	PAH (mg C/kg solids)	comp. (mg C/kg solids)
0	873	273	n.d.
7	1287	1099	669
21	470	247	434
35	497	267	647
56	655	151	847
84	403	n.d.	n.d.

Acronyms:
PAH: PAH-contaminated soil   comp.: compost
n.d.: not determined

4.3   Carbon balances of the contamination

The carbon balance of the contamination was calculated with two different methods of analysing the contamination. With the first the total of the PAH is selected from the HRGC analysis and in the second balance the gravimetric determination is chosen.

The carbon balance based on the HRGC measurements is shown in Figure 4. The initial contamination involved, which was taken as 100%, was 24900 mg of C/kg of solids. Of this 20700 mg of C/kg of solids could be extracted immediately, and 4200 mg of C/kg of solids after the saponification. The proportion of PAH after the saponification was extrapolated from the EPA PAH and the extraction with ethyl acetate, assuming that the total PAH behaved like the 16 EPA PAH. The mineralisation was calculated from the difference between the CO_2 produced by the PAH + comp. mixture and the control evaluation (compost).

The extractable PAH content decreased to 6% of the initial contamination by the end of the experiment. The PAH concentration after saponification doubled from 17% at the start of the experiment to 35% at the end. The mineralisation (CO_2 production) was 26% after 84 days. With values between 0.2 and 1.1% the biomass contents remained low. The total C in Figure 4 represents the addition of the four balanced parameters. This total was 68% at the end of the experiment. The 70% mark was reached after only three weeks, and was approximately maintained throughout the rest of the experiment.

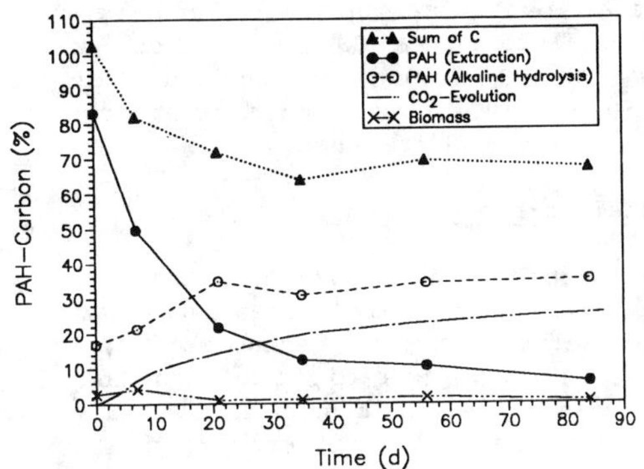

PAK-Kohlenstoff (%)	=	PAH carbon (%)
Summe C	=	Total C
PAK (Extraktion)	=	PAH (extraction)
PAK (Verseifung)	=	PAH (saponification)
CO_2-Austrag	=	CO_2 production
Biomasse	=	Biomass
Zeit	=	Time

Figure 4: Carbon balance of the mixture of PAH-contaminated soil and compost (contamination determined from the total of the HRGC PAH)

Determining the contamination from the gravimetric measurements gave the carbon balance shown in Figure 5. This described the degradation of the contaminant with the parameters gravimetry, mineralisation and biomass. For the mineralisation and the biomass, the same absolute values as in the balance calculated beforehand were used.

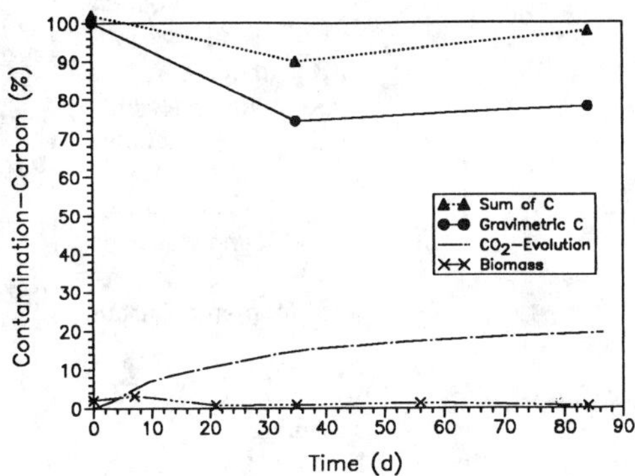

Kontamination.....	=	Contamination carbon (%)
Zeit (d)	=	Time (d)
Summe C	=	Total C
Gravimetrie C	=	Gravimetry C
CO_2-Austrag	=	CO_2 production
Biomass	=	Biomass

Figure 5: Carbon balance of the mixture of PAH-contaminated soil and compost (contamination measured gravimetrically)

The initial contamination was 33085 mg of C/kg of solids. The compost fraction of 1500 mg of C/kg of solids was subtracted from the result of measuring the PAH + comp mix. The total of the three balance parameters lay between 90 and 100%. Thus the discrepancy in the balance was significantly less than in the HRGC balance.

4.4   Total carbon balance

The total carbon balance comprises the carbon compartments shown graphically in Figure 6. A basic distinction was made between 5 carbon compartments: inorganic carbon, humus, contamination, biomass and mineralisation. The contamination (CON) was divided into the compartments EPA PAH, remaining PAH, PAH measurable after saponification and metabolites. The TIC, EPA, BIO and CO_2 were measured directly from these compartments. The remaining compartments were determined by calculation. The total contamination was taken as equal to the gravimetric measurement (GRAV). PAH results from the difference between the total of the HRGC measurement and EPA. The PAH measured after the saponification were again extrapolated from the results of the 16 EPA PAH. The metabolites were calculated as MET = GRAV - HRGC - SAPO.

The solid carbon was analysed in the form of the TC (total carbon) and TIC. The TOC (total organic carbon) was calculated as TC - TIC. HUM/BR was determined from the balance equation TOC = HUM/BR + CON + BIO. The HUM/BR compartment was assembled in this form, since without more extensive investigations no distinction could be made between humus and bound residues.

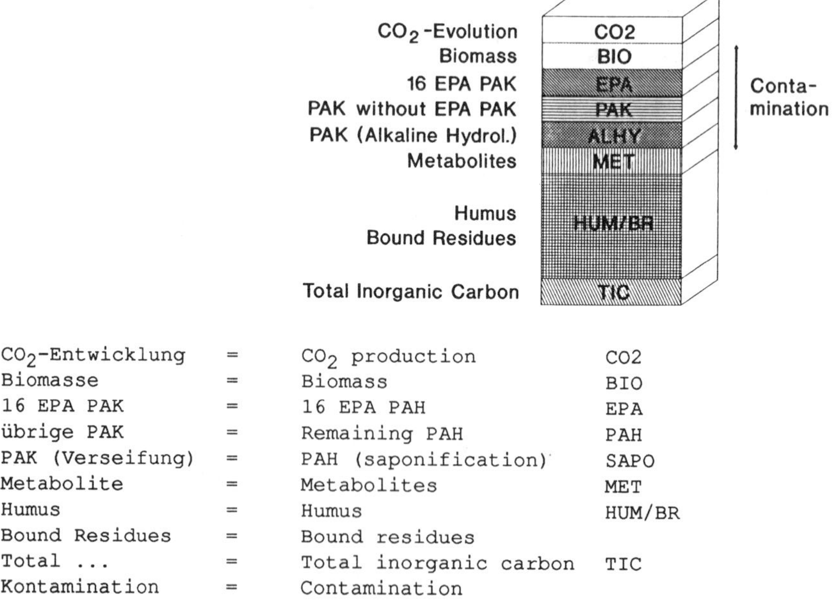

CO_2-Entwicklung	=	CO_2 production	CO2
Biomasse	=	Biomass	BIO
16 EPA PAK	=	16 EPA PAH	EPA
übrige PAK	=	Remaining PAH	PAH
PAK (Verseifung)	=	PAH (saponification)	SAPO
Metabolite	=	Metabolites	MET
Humus	=	Humus	HUM/BR
Bound Residues	=	Bound residues	
Total ...	=	Total inorganic carbon	TIC
Kontamination	=	Contamination	

Figure 6:   Carbon compartments of the total carbon balance

Table 4 gives the percentage distribution of the total carbon amongst the compartments described for days 0, 35 and 84 of the experiment. The total of all compartments was taken as 100% for each day of the experiment. The absolute values lay between 97 and 105 g of C/kg of solids.
The inorganic carbon constituted between 2.9 and 5.5% of the total carbon. The biomass fraction of between 0.4 and 0.9% was low again. At the end of the experiment the mineralisation was 9.3%. The metabolite fraction and the PAH after saponification increased in the course of the experiment. By contrast the PAH and EPA PAH decreased significantly. The HUM/BR compartment maintained a constant percentage of about 60%.

Table 4:   Total carbon balance - percentages of the carbon compartments

Time (d)	TIC (%)	HUM/BR (%)	MET (%)	SAPO (%)	PAH (%)	EPA (%)	BIO (%)	CO2 (%)
0	2.9	60.9	10.0	4.3	18.6	2.4	0.9	0.0
35	5.6	61.1	15.4	7.9	2.7	0.4	0.5	6.3
84	3.8	58.7	17.3	8.9	1.6	0.0	0.4	9.3

Acronyms:
TIC: total inorganic carbon   HUM/BR: humus/bound residues
MET: metabolites   SAPO: saponification
PAH: total of all PAH without EPA   EPA: 16 EPA PAH
BIO: biomass   CO2: cumulative CO_2 production

5   DISCUSSION AND CONCLUSIONS

The description of the contamination provided by the 16 EPA PAH is quantitatively inadequate, since these only represent 11% of the HRGC total. Moreover the extraction method and the extracting agent play an important role. In our investigations the highest yields were achieved with dichloromethane and Soxhlet extraction. In all probability the behaviour of the 16 EPA PAH and the total PAH (total HRGC) during the degradation in the soil is comparable, since the relative decreases follow the same pattern.

It is possible that metabolites from the degradation of the PAH are determined quantitatively by the gravimetric measurement. This conclusion can be drawn from the high values measured gravimetrically over the entire duration of the experiment, and from the balancing with the gravimetric values.

The PAH contents after saponification show that a not insignificant proportion of the PAH is masked by the organic processes in the soil. At the end of the experiment this amounts to 3.3 to 5.5 times the extractable PAH. Because of the lower yield achieved with ethyl acetate extraction, the PAH content after saponification may however be overestimated. The obvious assumption is that the polar products of the degradation of the PAH interact even more strongly with the organic processes and escape being extracted.
The stimulation of the mineralisation produced by adding compost can be read off from the CO_2 production. On the one hand a sharp peak in the rate of CO_2 production (CO_2 produced per unit time) is evident at the start of the

experiment, which would not necessary be expected for PAH contamination. On the other the total CO_2 production is significantly higher than that without the addition of compost. 45 days into the experiment the mineralisation in the mixture of PAH-contaminated soil and compost has stabilised to the extent that the rate of CO_2 production agrees with the total of the rates determined separately in the soil contaminated with PAH and compost.

The use of the SIR method of determining the biomass in a soil contaminated with PAH is not based on any certain knowledge of the reliability of this method under the given conditions. The only investigations available are for soils with synthetic oil contamination (Fricke, 1992). These involved estimating the order of magnitude of the biomass present in the soil. It can be concluded from the results that the biomass does not play any quantitatively decisive role in the carbon balance.

The mineralisation of the PAH in the carbon balance was based on the PAH determination using HRGC, and at 26% at the end of the experiment, was of the same order of magnitude as in other investigations (Heerenklage et al, 1992, Bumpus, 1989, Qiu and McFarland, 1991). However these investigations were carried out with PAH marked with C^{14} and under different incubation conditions. An allowance for the PAH measurable after saponification had not yet been represented in balances. In the investigation being presented they still amounted to 35% at the end of the experiment. The degradation of the PAH led to a stable soil system after 84 days. This fact is indicated by the small changes in the balance parameters with time over the last 40 days. At the end of the experiment a discrepancy in the balance of 32% is evident. This is caused by the formation of metabolites, bound residues and humus.

Determining the PAH concentration by gravimetric measurement gives a different picture of the carbon balance. The balance total was close to 100% over the entire experimental period. This outcome may be a result of determining the extractable metabolites from the degradation of the PAH by means of gravimetry. The balance would not look distinctly different if it included the saponifiable fraction of the PAH. Gravimetric determination of contamination is certainly not yet very satisfactory from the analytical viewpoint. However from the aspect under consideration it does represent an interesting starting point, which should be investigated in even greater detail.

A series of assumptions and simplifications were made in producing the total carbon balance; for instance that the contamination is determined quantitatively by the gravimetric measurement. Moreover the analysis of the solids in the soil samples is subject to unavoidable fluctuations. Given these limitations the result obtained for the overall carbon balance is to be categorised as very satisfactory. It shows the quantitative importance of the humus/bound residues carbon compartment (about 60% of the total carbon). More extensive study of this compartment, to enable explanation of the form of degradation of the contaminant, is of great importance for the acceptance of biological clean-up techniques.

The balancing methods presented lead to results that allow more to be said about the degradation of the PAH in the soil. On real contaminated sites the results of balancing the carbon depend crucially on the determination of the initial concentration of the contaminants. The use of various methods depending on the type of contamination is advisable. The easily measured degree of mineralisation represents an important parameter of the degradation of the contaminant. The formation of humus and bound residues requires the development of powerful methods of characterising these carbon compartments.

ACKNOWLEDGEMENTS

We would like to thank G Schäfer and M Breuer-Jammeli, of Biotechnology Department II of the Technical University of Hamburg-Harburg, for their help with the extraction and saponification of the soil samples for the HPLC analysis. The work presented was financed by funds from the German Research Association for special research area 188 "Cleaning contaminated soils", Hamburg.

REFERENCES

Anderson J P E and Domsch K H (1978): A physiological method for the quantitative measurement of microbial biomass in soils. Soil Biol. Biochem. 10, 215-221.

Bumpus J A (1989): Biodegradation of polycyclic aromatic hydrocarbons by Phanerochaete Chrysosporium. Appl. Environ. Microbiol. 55, 154-158.

Bundt J, Herbel W, Steinhart H (1993): Bestimmung von aromatischen Kohlewasserstoffen aus Mineralölen und deren Altlasten [Determination of aromatic hydrocarbons from mineral oils and sites contaminated with these]. In InCom '93 (Tagesband), 1 - 5 March 1993, Düsseldorf.

Eschenbach A, Gehlen P and Bierl R (1991): Untersuchungen zum Einfluß von Fluoranthen und Benzo(a)pyren auf Bodenmikroorganismen und zum Abbau dieser Substanzen [Investigations of the effect of fluoranthene and benz(a)pyrene on soil microorganisms and on the degradation of these substances. Mitteilungen d. Dt. Bodenkundlichen Gesellschaft 63, 91-94.

Fricke, K (1992): Untersuchungen zur Bestimmung der Biomasse in kontaminierten Böden mit der SIR-Methode [Investigation of determination of the biomass in contaminated soils using the SIR method]. Unpublished internal report. SFB 188, Waste Management Department of the Technical University of Hamburg-Harburg.

Herrenklage J, Breuer-Jammeli M, Kästner M, Lotter S, Stegmann R and Mahro B (1992): Evaluation of the degradation of anthracene and hexadecane in soil/compost mixtures. In press.

Kästner M (1991): Mündliche Mitteilung [verbal report].

Lotter S, Heerenklage J and Stegmann R (1992): Carbon balance and modelling of the oil degradation in soil bioreactors. In Soil contamination using biological processes (proceedings) (DECHEMA International Smposium), 6 - 9 December 1992, Karlsruhe.

Paschke A, Herbel W, Steinhart H, Franke S and Francke W (1992): Determination of polycyclic aromatic hydrocarbons in lubricating oil. Journal of High Resolution Chromatography, in press.

Qiu X and McFarland M J (1991): Bound residue formation in PAH contaminated soil composting using Phanerochaete Chrysosporium. Hazard. Waste Matter. 8, 115-126.

Stegmann R, Lotter S and Heerenklage J (1991): Biological treatment of oil-contaminated soils in bioreactors. In On-site bioreclamation - processes for xenobiotic and hydrocarbon treatment (R E Hinchee and R F Olfenbuttel), 188-208. Butterworth-Heinemann, Stoneham.

MODELLING AND OPTIMIZATION OF IN SITU SOIL REMEDIATION

J.G. CUPERUS*, L.G.C.M. URLINGS*, W. RULKENS**, M.G. KEIZER ***

*   TAUW INFRA CONSULT B.V., P.O. BOX 479, 7400 AL DEVENTER, THE
    NETHERLANDS
** AGRICULTURAL UNIVERSITY WAGENINGEN, DEPARTMENT OF ENVIRONMENTAL
    TECHNOLOGY, P.O. BOX 8129, 6700 EV, WAGENINGEN, THE NETHERLANDS
*** AGRICULTURAL UNIVERSITY WAGENINGEN, DEPARTMENT OF SOIL SCIENCE AND
    PLANT NUTRITION, P.O. BOX 8129, 6700 EV, WAGENINGEN, THE NETHERLANDS

In the Netherlands it is expected that far more than 100,000 sites will be considered for soil remediation. Many of which are industrial sites currently in use. These sites are often unsuitable for conventional remediation, and excavation of the contaminated soil is thus not feasible. In situ remediation is an attractive alternative for these type of sites, but also provides an interesting option for abandoned sites. Rough estimates show that at least 16% of contaminated sites in the Netherlands can be considered for in situ remediation, based on the nature of the contamination and the soil structure.

Due to an expected increase in the in situ remediation market, there will also be an increase demand for better control of the in situ process. Bottlenecks can be formulated as:-
-       the process is relatively slow and ineffective if improperly applied;
-       the duration is difficult to assess;
-       the control of the process is limited.

A project was started in which mainly the last two points were considered. A general user-friendly model will be developed for the in situ remediation of the saturated zone which can easily be coupled to laboratory experiments and field investigations. The model will consist of the following key elements:-
-       sorption and speciation;
-       mass transfer;
-       biodegradation;
-       flow patterns of air and/or water in heterogeneous media;
-       geostatistics.

The first three points will be modelled and validated on column scale. Hereafter, coupling with flow patterns in the soil will be established by means of geohydrological models.

Important for a successful application, per site, is the determination of the data available and the expected increase in accuracy by assessing the course of a remediation using this model. A major problem is posed by heterogeneity in things like soil structure and the distribution of contaminants.

F. Arendt, G.J. Annokkée, R. Bosman and W.J. van den Brink (eds.), Contaminated Soil '93, 1247–1248.
© 1993 Kluwer Academic Publishers. Printed in the Netherlands.

Different means will be pursued in an effort to enhance the process of remediation. The model will also be applied to assess the performance of e.g. surfactants during remediation, based on data obtained from laboratory research. Results of e.g. extraction with detergents will be evaluated using the model, and subsequently by upscaling to pilot scale.

MECHANISMS OF MICROBIAL DEGRADATION OF POLYCYCLIC AROMATIC HYDROCARBONS (PAH) IN SOIL-COMPOST MIXTURES

B.Mahro and M.Kästner

Arbeitsbereich Biotechnologie II, Technische Universität Hamburg-Harburg; Denickestr.15, 21 Hamburg 90 (FRG)

The supplementation of contaminated soils with ripened compost can help to support the clean up especially of oil contaminated soil sites (Stegmann et al.,1991). We followed this line by investigating the effect of compost if it was mixed into soil which had been contaminated with more recalcitrant xenobiotics like polycyclic aromatic hydrocarbons (PAH). We found, as it can be seen in Fig.1, that also with PAH the degradation process in soils can be improved substantially if the soil is supplemented with ripe compost (Fig.1 ; Kästner et al., 1992).

The stimulating effect of compost on the microbial PAH-degradation in soil was even greater than the effect which can be obtained by the addition of laboratory grown cultures being able to grow on PAH as sole carbon and energy source (Kästner et al., 1992). But analyzing the composition of the microbial population of such artificial PAH/soil/compost mixtures we were not able to find any bacterial isolate which was able to mineralize PAH as sole carbon and energy source, though such isolates could be frequently found on soil sites which had been contaminated with PAH for a long time (Kästner et al., 1991). These findings suggested to ask by which mechanism PAH might be degraded in soil/compost mixtures.

The microbial degradation of polynuclear aromatic hydrocarbons is well examined with pure cultures of bacteria and fungi in liquid systems (for review see Cerniglia and Heitkamp, 1989; Mahro and Kästner,1993). Several bacterial strains were described which are able to grow on PAH with two, three or four condensed aromatic rings as a sole source

1249

F. Arendt, G.J. Annokkée, R. Bosman and W.J. van den Brink (eds.), Contaminated Soil '93, 1249–1256.
© 1993 *Kluwer Academic Publishers. Printed in the Netherlands.*

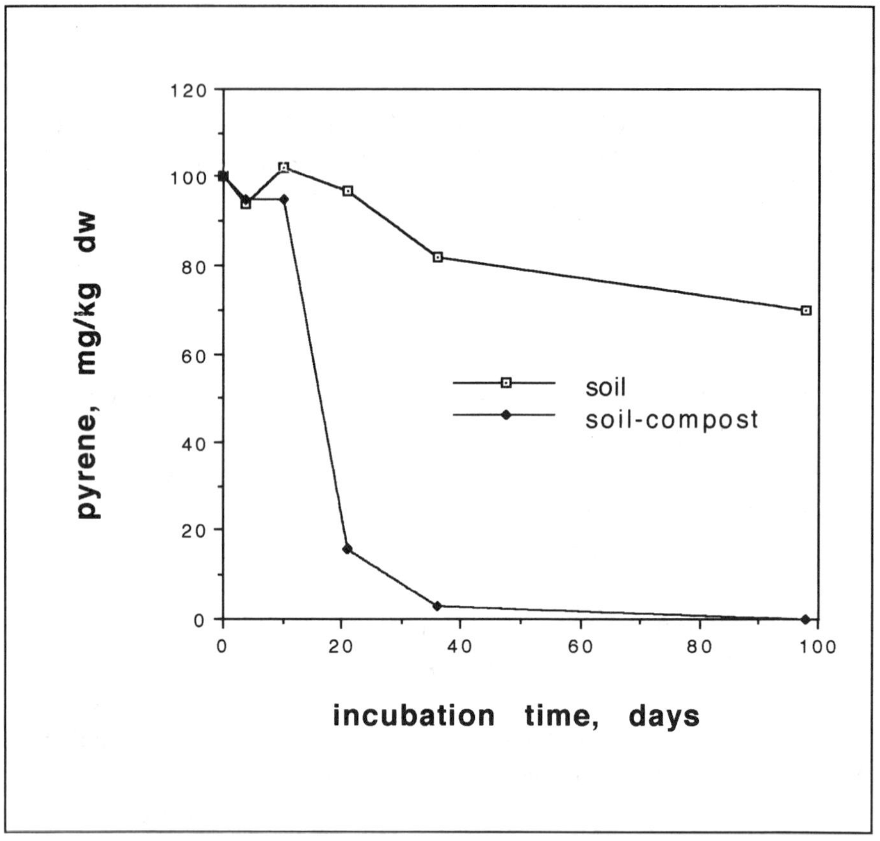

Fig.1: Effect of compost supplementation on the degradation of pyrene in soil; PAH were extrac
by ethylacetate including a subsequent humic acid extraction (modified from Eschenbach et
1991) and analyzed by HPLC.

of carbon and energy. This type of degradation mostly leads to <u>complete mineralization</u> of the PAH and is accompanied simultaneously by the formation of microbial biomass. A second group of microorganisms is not able to mineralize PAH completely but is able to convert them into partially oxidized products (<u>cometabolic degradation</u>). These compounds, mostly phenolic or carboxylic PAH-derivatives, tend to accumulate within the culture and were usually not metabolized any further by the same microorganism. A third type of degradation which has to be taken into account is the degradation of PAH by <u>unspecific radical oxidation</u> reactions due to the activity of ligninolytic enzymes of white rot fungi or other ligninolytic microorganisms like actinomycetes. This type of oxidation produces intermediate aryl radicals, arene oxides or chinoid products but (with the fungi cultures) a partial mineralization was also observed.

In general it is assumed that microorganisms can use the same degradation mechanisms observed with pure liquid cultures in soil as well. But one must be aware of two important differences. The first is that microbial transformation processes are carried out in non-sterile biotopes. All oxidized PAH-derivatives, secreted or transformed by one member of the microflora may be used by other members of the soil microflora for further metabolic transformations. The probability that such cooperative degradation actually occurs is rather high in soil/compost mixtures since it contains both, a higher number of active microorganisms and a larger variety of microorganisms.

Another evident difference to liquid cultures is that soil/compost mixtures contain a hig number of humic and organic matrix surfaces which make it rather diffficult to distinguisl analytically between PAH-disappearance caused by biodegradation and PAH-disappearance which is caused by fixation processes. Three different fixation pathways must be considered. The first is that the PAH can become strongly <u>adsorbed to mineral or organic surfaces</u> in the soil which, in turn, makes them unaccessible to analysis. That this phenomen actually occurs has been reported repeatedly and it was tried minimize the problem by special PAH-soil extraction techniques (Eschenbach et al., 1991; own unpublished results). The second possibility is to assume the formation of a covalent binding of the oxidized PAH or their metabolic derivatives to humic matter in the soil. The polymerisation of oxidized PAH or PAH-metabolites is another possible mechanism to explain the analytical disappearance of PAH in soil. The role of the latter two disappearance pathways is supported by findings which show that oxidized phenolic products or arylradicals are able to trigger <u>oxidative coupling reactions</u> in vitro. Oxidative coupling reactions have been shown between both, the oxidized xenobiotics themselves (polymerisation) and between the xenobiotic and model compounds, mimicking molecules of the organic soil matrix (Bollag et al., 1980; Bollag and Loll, 1983; Dec and Bollag, 1987).

Oxidative coupling activities may also contribute substantially to the disappearance of PAH in compost stimulated soil. Compost contains - compared with soil - a much larger number of microorganisms (mainly streptomycetes and fungi) which are able to transform solid water-unsoluble plant polymers like cellulose, hemicellulose or lignin. To transform such macromolecular molecules these microorganims excrete a whole set of different enzymes like cellulases, ligninases, H_2O_2-generating enzymes, peroxidases, laccases or tyrosinases. Since several of these exoenzymes are known to participate in oxidative coupling processes in soils (Claus and Filip,1991) one can assume that the stimulation effect of compost is therefore to a large extent caused by stimulation of the oxidative coupling activities in the soil-compost mixture. Experiments, which we carried out with radiolabelled anthracene,

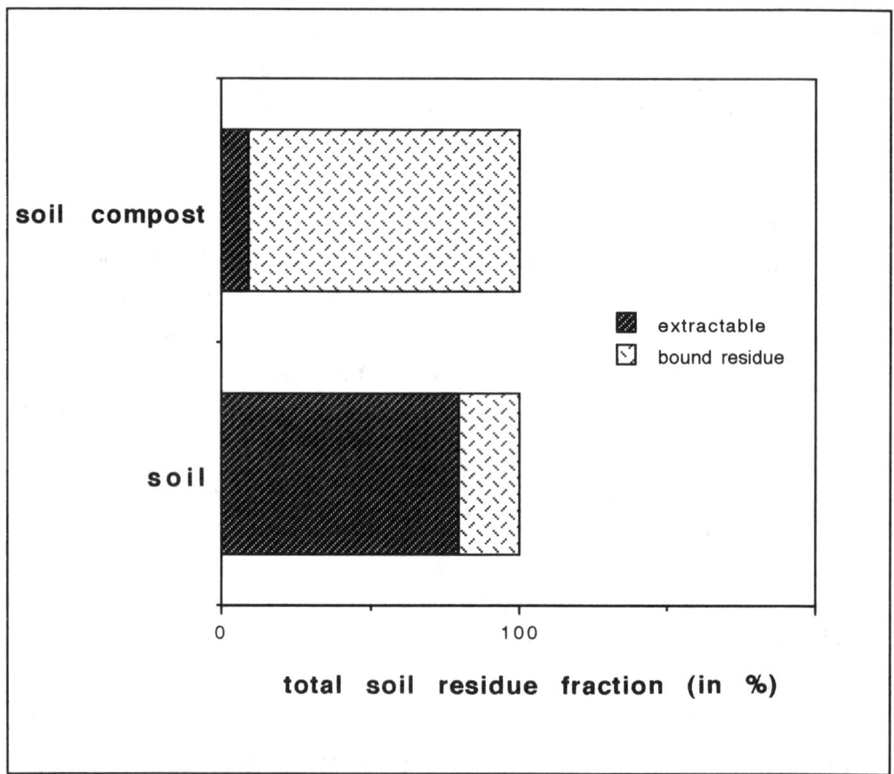

Fig. 2 : Comparison of the relative amount of bound and extractable ^{14}C-labelled anthracene residues in soil with or without compost supplementation (Experimental details at Heerenklage et al., 1992)

support this assumption (Heerenklage et al, 1992). It was found that - if one compared the label-radioactivities which still could be extracted from soil after 100 days of incubation with those which remained unextractable ("bound residues") - the bound residue fraction added up to about 9% in the compost-free bioreactor but reached about 80% of the whole residue fraction if the contaminated soil had been supplemented with compost (Fig.2). The results were not much different if organic solvent or more rigorous extraction techniques like humic acid saponification (Eschenbach et al., 1991) were applied.

We assume therefore that many PAH-xenobiotics became involved in the soil-humification process, especially since enzymatic activities were added to the soil which stimulated oxidative coupling reactions. However at the same time the supplementation of PAH-contaminated soil with ripe compost led also to an excessive stimulation of CO_2 formation. Most of this CO_2 originated from the respiratory degradation of the abundant organic material. But based on labelling experiments with anthracene, it could be shown, that parts of the carbon dioxide flow were also fed by micobial anthracene degradation (Heerenklage et al., 1992). This means that the compost microflora must be able to cleave and degrade at least parts of the PAH-ring structure, either directly or after its binding to the organic soil matrix ("delayed mineralization" in Fig.3). For compost supplemented soil it is not known yet whether the ring cleavage reaction takes place intra- or extracellularly and which enzymes carry out this ring cleavage reaction. Figure 3 summarizes our current view on the mechanisms of PAH-degradation in soil and soil compost-mixtures.

The given model on PAH-depletion pathways in soil-compost mixtures implies a different view on the effect of compost stimulated soil remediation techniques. One has to be aware of the fact that large parts of the PAH-contamination might become efficiently inactivated but not completely destroyed. Nevertheless the compost-based biogenic stimulation of the formation of bound residues might be sufficient for most remediation purposes. Since the humic matrix can be considered as a continously growing polymer it is rather unlikely that the assimilated aromatic compound will ever occur again in its free, original structure. However, extensive analytical and toxicological efforts should be taken to make sure that the xenobiotics stay inactive also on the long run.

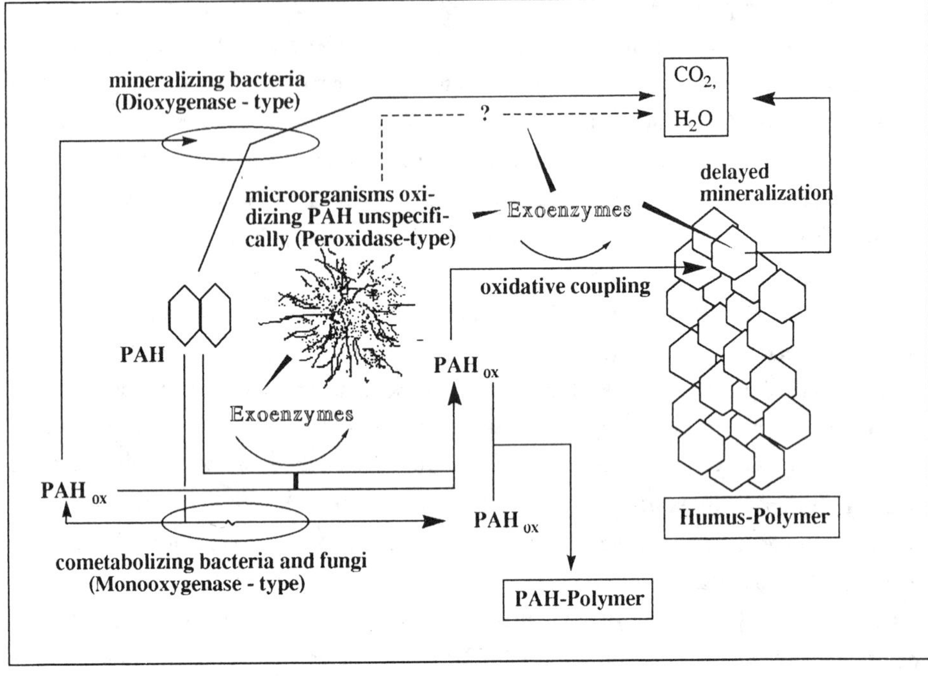

Fig.3: Depletion pathways for polycyclic aromatic hydrocarbons in soil-compost mixtures

ACKNOWLEDGEMENT:

This work was supported by a grant of the Deutsche Forschungsgemeinschaft for the "Sonderforschungsbereich 188: Remediation of contaminated soils" (Project B1). We also appreciate the technical assistance of M.Breuer-Jammali and the very fruitful cooperation with J.Heerenklage, S.Lotter and Prof.Dr. R. Stegmann (TUHH).

REFERENCES:

Bollag J.-M., Loll, M. J. 1983: Incorporation of xenobiotics into soil humus. Experientia, 39, 1221-1231.

Bollag, J.M., Liu, S.Y., Minard, R.D. 1980: Cross coupling of phenolic humus constituents and 2,4-dichlorphenol. Soil Sci. Am. J. 44, 52-56.

Cerniglia, C.E., Heitkamp, M.A. 1989: Microbial degradation of polycyclic aromatic hydrocarbons (PAH) in the aquatic environment. In: Varanasi, U. (ed): Metabolism of polycyclic aromatic hydrocarbons in the aquatic environment. CRC Press Boca Raton. S. 41-68.

Claus, H., Filip, Z. 1991: Phenoloxidierende und andere Enzyme als Mittel zur Umwandlung organischer Schadstoffe im Boden und Grundwasserbereich. Forum Städte-Hygiene 42, 214-223.

Dec,J., Bollag, J.M. 1987: Microbial release and degradation of catechol and chlorophenols bound to synthetic humic acid. Soil Sci. Am. J. 52, 1366-1371.

Heerenklage, J., Breuer-Jammali, M., Kästner, M., Lotter, S., Stegmann, R. und Mahro, B. 1993: Balance of anthracene and hexadecane degradation in soil-compost mixtures. Soil Biol. Biochem., submitted.

Kästner M., Breuer M., Mahro B. 1991: Bakterien-Isolate aus unterschiedlichen Altlastenstandorten zeigen ein vergleichbares Abbauprofil für PAK und Ölkomponenten. GWF Wasser-Abwasser 132, 253-255.

Kästner, M., Schaefer, G., Breuer-Jammali, M., Mahro, B. 1992: Comparison of the microbial degradation potential for PAK in liquid medium, soil and soil compost. Dechema Biotechnology Conferences, Vol.5, S.1043-1046.

Mahro, B., Kästner M. 1993: Der mikrobielle abbau polyzyklischer aromatischer Kohlenwasserstoffe (PAK) in Böden und Sedimenten: Mineralisierung, Metabolitenbildung und Entstehung gebundener Rückstände. Bioengineering 9: in press

Stegmann, R., Lotter, S., Heerenklage, J. 1991: Biological treatment of oil-contaminated soils in bioreactors. In: Hinchee, R.E., Olfenbüttel, R.F. (eds) On site bioreclamation: Processes for xenobiotic and hydrocarbon treatment. Butterworth-Heinemann, Boston. pp 188-208.

THE VACUUM-VAPORIZER-WELL (UVB): BASICS, KARLSRUHE TESTFIELD MEASURING RESULTS

W. Bürmann (Speaker)[1] und H. Wagner[2]

[1]Institute for Hydromechanics, University of Karlsruhe, D-7500 Karlsruhe 1

[2]MG Geocontrol GmbH, D-6000 Frankfurt/Main 1

INTRODUCTION

Worldwide, not only in the industrialized countries, the number of known groundwater and soil air contaminations by hydrocarbons, BTX, pesticides, nitrates, etc., increases. Efficient remediation techniques at low costs are needed.

The Vacuum-Vaporizer-Well (UVB) technology (German: Unterdruck-Verdampfer-Brunnen (UVB). Inventor: B. Bernhardt. Patents: IEG mbH, D-7410 Reutlingen) is a new method for the in-situ remediation of groundwater and soil air. Initially developed to clean groundwater contaminated by chlorinated hydrocarbons (CHC), in between this technology is applied also in biological groundwater remediation of pesticides, for example [1,2].

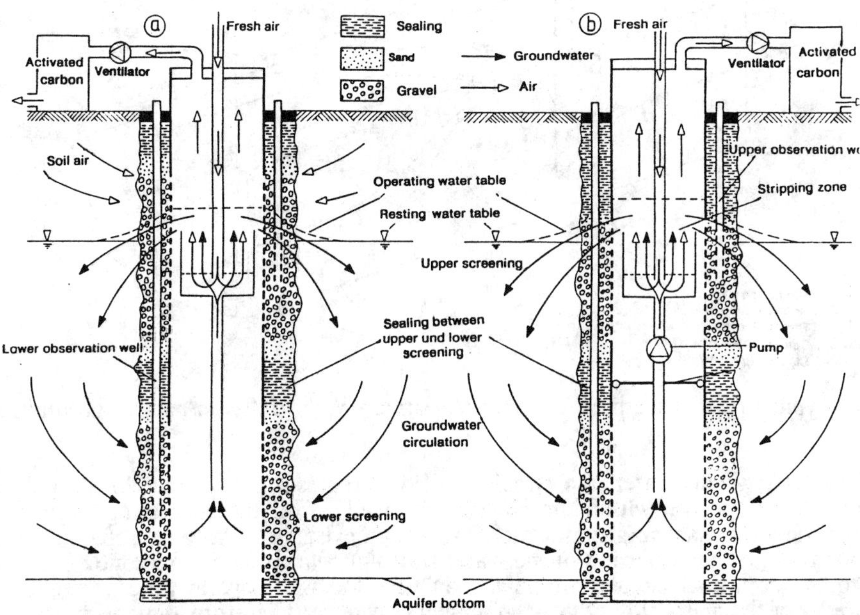

Fig. 1. Typical Vacuum-Vaporizer-Well (UVB) with airlift pump (a) and additional water pump (b)

F. Arendt, G.J. Annokkée, R. Bosman and W.J. van den Brink (eds.), Contaminated Soil '93, 1257–1264.
© 1993 *Kluwer Academic Publishers. Printed in the Netherlands.*

Some disadvantages of groundwater remediation applying current pumping methods (groundwater lowering, limited yield, insufficient remediation) may be avoided if pumping and recharge take place in the same well. The UVB is an application of this circulation well [3,4,5,6].

UVB TECHNOLOGY

The UVB produces a circulation flow within the surrounding groundwater, directed from the upper to the lower screening. Figure 1 shows two of various UVB designs. The water is sucked through the lower screening into the well, transported upwards inside the UVB by the air lift pump (figure 1a) which may be supported by an additional water pump (figure 1b). The water is cleaned by fresh air in the stripping zone under below atmospheric pressure, and it is given back through the upper screening without the water leaving the aquifer. Soil air from the unsaturated area of the aquifer may be sucked into the UVB through the upper screening and thus may also be cleaned. The contaminants are adsorbed by activated carbon. If necessary, the groundwater is cleaned on site and led back to the well. To avoid precipitation, the stripping air loop is closed. Thus contaminants which are not adsorbed can be kept from escaping into the atmosphere.

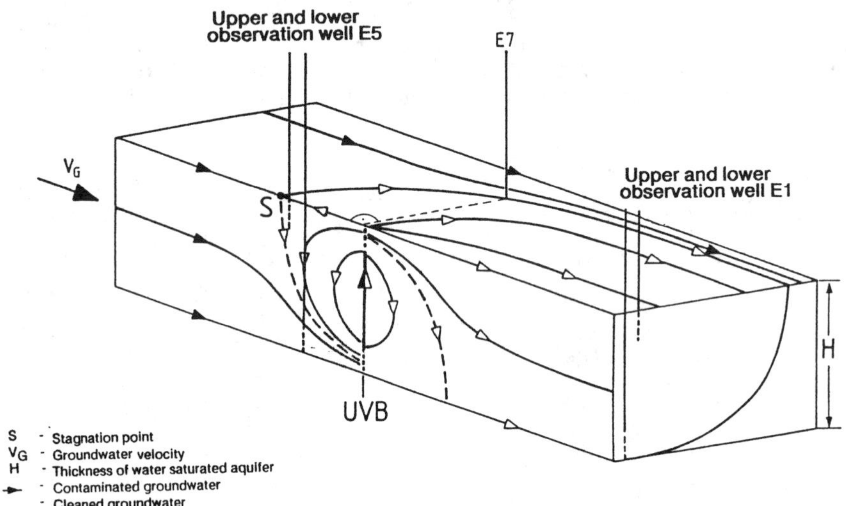

Fig. 2. Typical flow pattern of Vacuum-Vaporizer-Well (UVB) in natural groundwater flow

In resting groundwater, the circulation flow creates a permanent flow and consequently cleans the soil within the zone of the well, as all the circulating water flows through the well. Natural groundwater flow which exists in most cases deforms the circulation flow so that a portion of the water flowing towards the intake zone of the well, due to the continual circulation flow, may pass the well several times, whereas the remainder of the water flows through the well only once. Therefore, dimensioning of the cleaning equipment of the UVB must be made so that one flow through the well is sufficient to ensure decontamination of the water.

GROUNDWATER FLOW AT THE UVB

The circulation flow in moving groundwater shows two separating streamlines, at the bottom and at the top of the aquifer, similar to the perfect well (figure 2). In a well with upward flow, the lower separating streamline corresponds to the withdrawal well and the upper one to the infiltration well. Between these two separating streamlines at the lower and upper boundaries of the aquifer lies the separating stream surface of the flow around the well in the natural groundwater. This surface consists of spacial streamlines and separates between the zone of contaminated and cleaned groundwater in the aquifer.

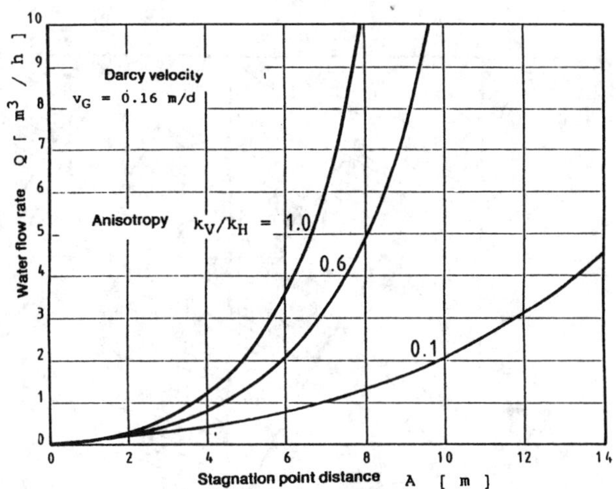

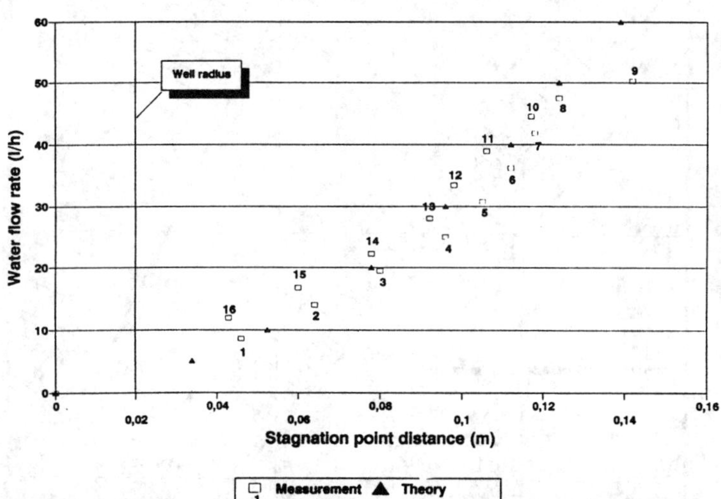

Fig. 3. Water flow rate versus stagnation point distance of Vacuum-Vaporizer-Well (UVB) in natural groundwater flow: influence of anisotropy (a), measurement at physical aquifer model (b)

1260

At the stagnation point S of the UVB (figure 2) the artificial groundwater velocity of the well meets the natural groundwater flow velocity. The dimension of the separating stream surface is characterized by the distance of the stagnation point from the well. As an example for a special case figure 3 shows the water flow rate through the UVB versus the stagnation point distance. In the case of low vertical conductivity (high anisotropy) the stagnation point distance and thus the sphere of influence of the well increase remarkable. The water flow rate through the UVB rises approximately parabolic with the stagnation point distance. Therefore, instead of one well with high water flow rate several small ones with low discharge should be applied.

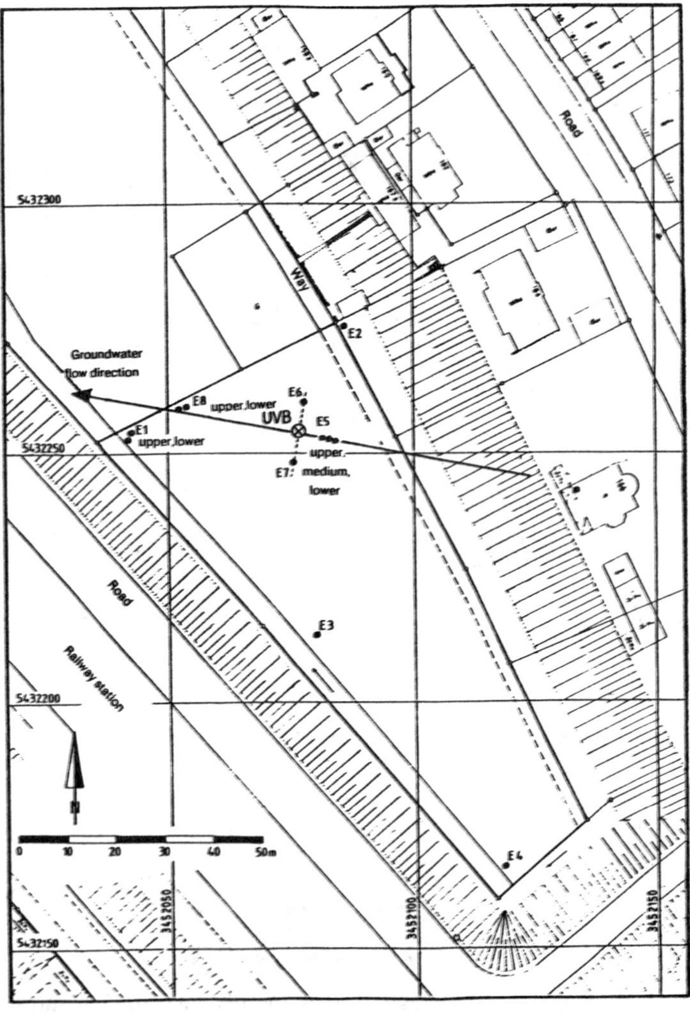

Fig. 4. Karlsruhe Testfield: location of UVB and observation wells

PHYSICAL GROUNDWATER FLOW DEMONSTRATION MODEL

Physical desktop aquifer models show the basic flow phenomena of the UVB remediation technology with respect to the natural groundwater flow. These demonstrations include the remediation of an artificial plume containing an impermeable, horizontal lens.

Demonstrations may be accompanied by a real time computer simulation of the fluid flow shown in the model. These computations result from theoretical investigations and base on the mathematical description of the groundwater flow at the UVB using analytical solutions of the governing Laplace partial differential equation.

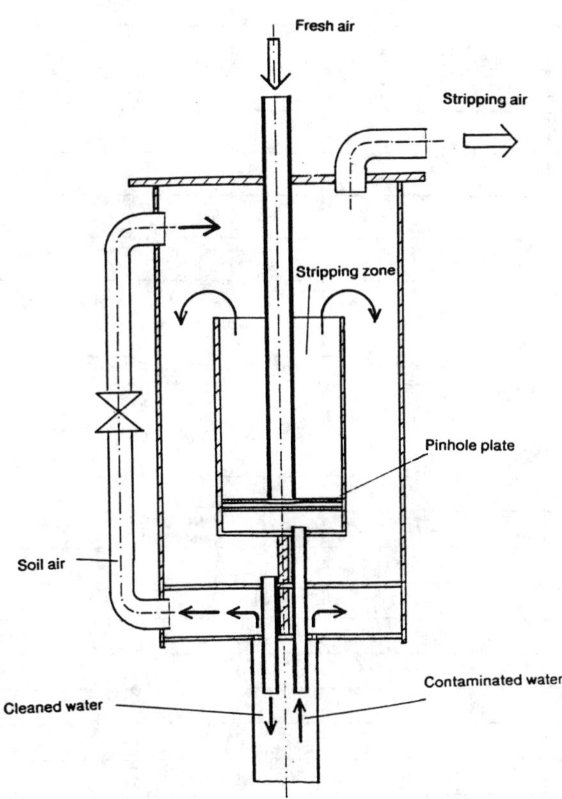

Fig. 5. Karlsruhe Testfield UVB

KARLSRUHE TESTFIELD

All theoretical investigations are needed for the reliable dimensioning of the UVB. To verify the theory measurements are performed at the UVB in the Karlsruhe Testfield. Figure 4 shows some observation wells close to the UVB some of which appear in Figure 2. The groundwater is contaminated by chlorinated hydrocarbons (CHC). The properties of the aquifer (geological profiles, pumping tests, groundwater and soil properties, etc.) have been described in some detail [7,9,8]. With respect to easy operation and low costs the Karlsruhe Testfield UVB in figure 4 is designed different from the UVB of figure 1.

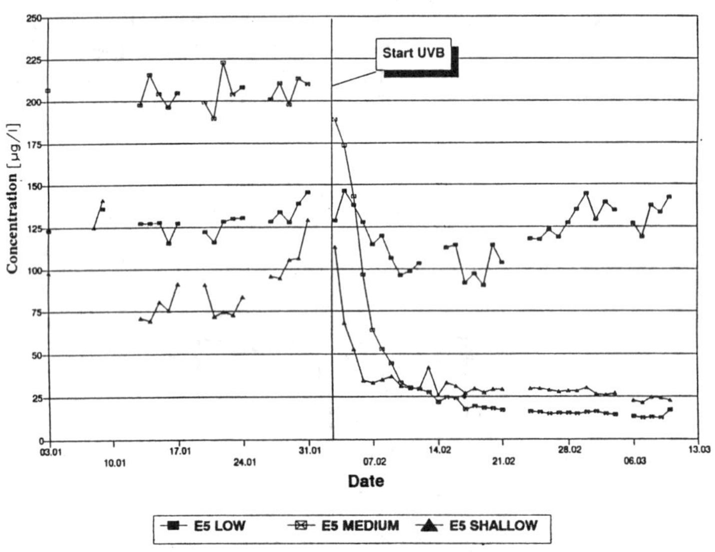

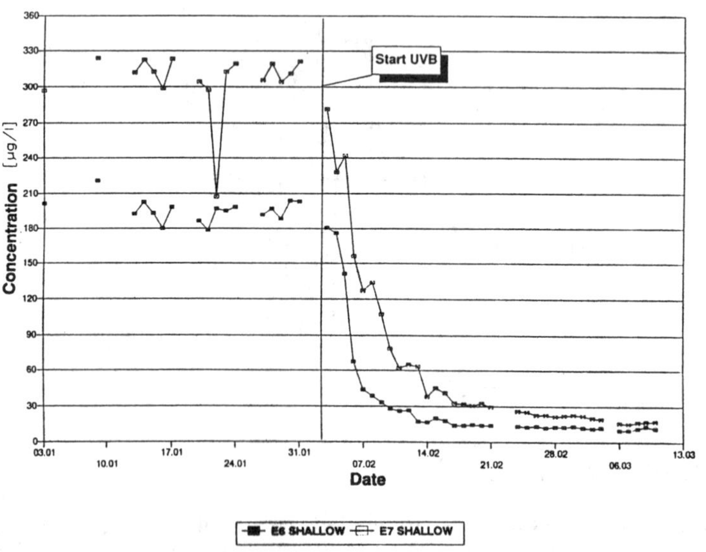

Fig. 6. Concentration curve of chlorinated carbons (CHC) at three upstream (a) and two UVB cross-stream (b) Karlsruhe Testfield observation wells

From the manifold of various groundwater investigations (electric conductivity, oxygen content, pH, temperature, etc.) a typical result as an example is chosen. Figure 6a shows the groundwater decontamination at the upstream observation wells E5, and Figure 6b demonstrates the decontamination at the observation wells E6 and E7 besides the UVB. The diagram is based on numerous chemical groundwater analysis by gaschromatograph, and additionally, using a CHC-Indicator which is using ultraviolet light to decay the CHC [7,8,9].

CONCLUSIONS

At this time IEG is preparing a number of new Remediation Techniques some of which may be shortly sketched here.

One of these new techniques under concideration is using the influence of vibrations on groundwater flow and contaminat transport. Another new technique makes use of the natural groundwater flow gradient to collect, clean and reinfiltrate the contaminated groundwater to save pumping energy. Last not least a new capillary zone flooding method is developed to better attack and remediate the vadoze zone.

ACKNOWLEDGEMENTS

The author gratefully acknowledges IEG mbH, D-7410 Reutlingen, for funding the investigations, and, in particular, B. Bernhardt, IEG mbH, and many others for their work and their numerous helpful discussions and contributions concerning the UVB technology.

REFERENCES

[1] Bürmann, W., Bott, G., Krug, R.: Groundwater Remediation Using the Vacuum-Vaporizer-Well: Operation of the Well and Biological Remediation of a Groundwater Contamination by Triazine, Proc. 3rd Conference ENVIROTECH VIENNA 1992 (ISEP), April 22-24, 1992, Wien, W. Pillman (Ed.), 1992, 723-732

[2] Bürmann, W., Wagner, H., Bott-Breuning, G., Rehner, G.: Bodensanierung in situ durch Grundwasserzirkulation mit der Unterdruck-Verdampfer-Brunnen (UVB) Technologie: Prinzip und Sanierungspraxis, M.U.T.-Kongress, October 6-9, 1992, Basel, in press

[3] Bürmann, W.: Untersuchung der Zirkulationsströmung um den kombinierten Entnahme- und Einleitungsbrunnen zur Grundwassersanierung am Beispiel des Unterdruck-Verdampfer-Brunnens (UVB), Altlastensanierung '90, Dritter Int. KfK/TNO-Kongreß über Altlastensanierung, Karlsruhe, 10.-14. Dezember 1990, in F. Arendt et a. (Ed.), Kluwer Academic Publ., Dordrecht, Boston, London, 1990, 1165-1172

[4] Buermann, W.: Investigation of the circulation flow around a combined withdrawal and infiltration well for groundwater remediation demonstrated for the Underpressure-Vaporizer-Well (UVB), Contaminated Soil '90, Third Int. KfK/TNO Conference on Contaminated Soil, Karlsruhe, December 10-14, 1990, F. Arendt et a. (Ed.), Kluwer Academic Publ., Dordrecht, Boston, London, 1990, 1045-1052

[5] Bürmann, W.: Bodensanierung durch Grundwasserzirkulation mit dem Unterdruck-Verdampfer-Brunnen (UVB), Wasserbau-Mitteilungen 36, Institut für Wasserbau, Technische Hochschule Darmstadt, 1991, 93-102

[6] Herrling, B., Buermann, W., Stamm, J.: In-situ remediation of volatile contaminants in groundwater by a new system of "Vacuum-Vaporizer-Wells (UVB), Subsurface

Contamination by Immiscible Fluids, K. U. Weyer (Ed.), A.A. Balkema Publ., Rotterdam, 1991

[7] Bürmann, W.: Kurzbericht über die Untersuchungen am Unterdruck-Verdampfer-Brunnen (UVB) im Karlsruher Versuchsfeld, Bericht Nr. 680, Institut für Hydromechanik, Universität Karlsruhe, 1992

[8] Bürmann, W.: Zusammenstellung der Meßergebnisse der Untersuchungen am Unterdruck-Verdampfer-Brunnen (UVB) im Karlsruher Versuchsfeld, Bericht Nr. 698, Institut für Hydromechanik, Universität Karlsruhe, 1992

[9] Wagner, H.: Hydraulische, hydrogeologische und geologische Untersuchungen im Rahmen des UVB-Forschungsvorhabens, Diplomarbeit, Teil I, Institut für Hydromechank, gemeinsam mit dem Lehrstuhl für Angewandte Geologie, Universität Karlsruhe, 1992

Application of specialized microorganisms for the bioremediation of soil and groundwater

S. Keuning and D. Jager
Bioclear Environmental Biotechnology
Zernikepark 2, 9747 AN Groningen, The Netherlands

ABSTRACT

In some cases biological soil decontamination processes can be enhanced by using specialized microbial strains that may be obtained from the polluted site or from other sources. In a laboratory research we investigated the feasibility of an in situ biorestauration of an industrial site that was polluted with toluene, p-isopropyltoluene and 1,2-dichloroethane. Degradation of these contaminants by the indigenous micro flora as well as by specialized inocula from other sources was tested and followed under different circumstances. Toluene degradation in contaminated soil slurries could be stimulated by adding inorganic nutrients. Degradation of 1,2-dichloroethane could only be achieved after inoculating the soil with the 1,2-dichloroethane degrading bacterial strains.

1. INTRODUCTION

Biotechnological treatment techniques utilize the capacity of microorganisms to convert both natural and xenobiotic compounds into harmless substances such as water, carbon dioxide and inorganic salts, a breakdown process known as biodegradation.

Numerous polluting compounds can be mineralized by microorganisms which use these compounds as a growth substrate. Many of these organisms have been cultivated in the laboratory in pure culture and have been studied extensively. Examples are components of mineral oil, benzene, toluene, xylenes, phenoles, cresoles, acrylic compounds, alcohols, ketones and aldehydes and several chlorinated hydrocarbons like methylene chloride, 1,2-dichloroethane, chlorobenzenes, chlorophenols etc.

Before biodegradation starts often a lag-time is observed which is referred to as adaptation. This lag-time can be caused by necessary induction of the enzymes that are involved in the degradation processes, by selective growth of the organisms capable of degradation or by genetic changes (mutations) that produce organisms with the appropiate degradation mechanisms.

Sometimes long adaptation times (for instance with halogenated hydrocarbons) can be

F. Arendt, G.J. Annokkée, R. Bosman and W.J. van den Brink (eds.), Contaminated Soil '93, 1265–1268.

succesfully overcome by introducing suitable cultured organisms with specific degradation capacities. The application of organisms for biological purification processes is of course only useful when the presence of (sufficient) organisms is a limiting factor for degradation.

Common contradictions against application of specialized microorganisms are that 'the introduced organisms are not able to maintain themselves' and that 'naturally everything (all microorganisms) is everywhere'. In several other fields of biotechnology inoculation, however, is already applied successfully for a long time: in the production of bread (application of yeast), in dairy industry (application of starter cultures of lactic acid bacteria), in the production of alcoholics (wine and beer yeasts), for the optimal modulation of plants for nitrogen fixation (symbiotic nitrogen fixations) in cattle-feeding (inoculation of the stomach of young calves), in the production of silage to accelerate the start of the lactic acid fermentation and for waste water purification. Inoculation with granulate sludge shortens the adaptation procedure of UASB reactors with at least a few months. Shortening the adaptation periods or specific affecting of the microflora by application of cultivated cultures appears to be possible, although sterile conditions are absent.

At present, however, well documented experiences with introduction of cultured xenobiotic degrading microorganisms in the field under well defined conditions are scarce.
In laboratory scale experiments with polluted soils from an industrial site it has been shown that addition of cultured strains with specific degradation capacities can be an efficient tool to stimulate the in situ (or on site) biodegradation of rather persistent chemicals like chlorinated solvents. In a laboratory research using soil slurries we investigated the feasibility of an in situ biorestauration of an industrial site that was polluted with toluene, p-isopropyltoluene and 1,2-dichloroethane.

2. BIODEGRADATION RESEARCH

2.1 Slurry experiments

Microbial degradation of chemical compounds was followed in soil slurries from polluted subsoils. A sandy soil from an industrial site in the Netherlands was mainly polluted with toluene, p-cymene and 1,2-dichloroethane. The soil contained 10-200 mg/kg (d.w.) p-cymene. Toluene and 1,2-dichloroethane concentrations were maximal 45 mg/kg d.w. Soil slurries were made by mixing polluted soil with groundwater from the site to a homogenous suspension (45% dry mass). Slurries were aerobically incubated at room temperature in serum flasks on a vertical rotor. Degradation of these compounds by the autochtonous microflora as well as by specialised inocula isolated from other sources was followed under different circumstances and measured by gaschromatographic analysis.

2.2 Isolation of autochtonous microorganisms

The presence of microorganisms capable of degrading toluene, p-cymene and 1,2-dichloroethane in the subsurface of the contaminated site was investigated. Selective enrichment techniques were used to isolate indigenous bacterial strains from soil samples that were able to use these compounds as substrate for growth. This could be rapidly achieved for toluene and p-cymene. After three weeks of enrichment and repeated subculturing at least three different bacterial strains that utilized p-cymene as sole carbon and energy source could be isolated from the soil samples (growth rates ca. 0.5 h^{-1} at 30°C). Also several toluene degrading organisms that used toluene as primary substrate for aerobic growth were obtained in pure culture (growth rates ca. 0.3 h^{-1} at 30°C). Bacterial counts with the most probable number method revealed that p-cymene and toluene utilizing bacteria were present in the soil samples at levels of 10^4-10^5 organisms per g dry soil.

However organisms able to degrade 1,2-dichloroethane could not be isolated from the soil. Even after prolonged incubation periods (over five months) there were no indications, that such organisms were present in the soil material.

2.3 Effect of inorganic nutrients

The effect of extra inorganic nutrients on the degradation rates of toluene and p-cymene by the indigenous microflora was determined. The soil slurries were incubated with different amounts of nitrate and phosphates (0.5-2 g/l $NaNO_3$, 2-8 g/l Na_2HPO_4, 0.5-2 g/l KH_2PO_4). These additions resulted in a substantial increase of the microbial degradation rate of toluene and p-cymene. During the incubation period the number of toluene and p-cymene degrading organisms in the soil slurries increased over 1000-fold to a level of 10^8 organisms per g dry soil as determined with the MPN method.

Degradation of 1,2-dichloroethane was negligible and did not improve under these conditions.

2.4 Inoculation with specialized strains

In order to establish degradation of 1,2-dichloroethane the slurries were inoculated with a mixture of two 1,2-dichloroethane degrading bacterial strains (*Xanthobacter autotrophicus* GJ10 and *Ancylobacter aquaticus* AD20, 5 mg c.d.w./l) (3,4). After inoculation, rapid and complete mineralization of 1,2-dichloroethane took place (figure 1). In a control slurry that was not inoculated no significant degradation of 1,2-dichloroethane occurred.

The soil slurries were also inoculated with autochtonous and seperately cultured toluene and p-cymene degrading strains. This did not further enhance the degradation rate of toluene and p-cymene in the soil slurries compared to addition of only inorganic nutrients.

EFFECT OF INOCULATION

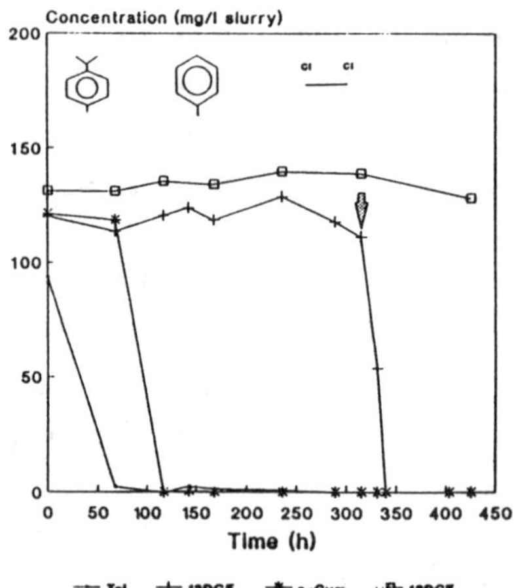

FIGURE 1. Degradation of 1,2-dichloroethane, toluene and p-cymene in soil slurries during aerobic incubation at 20°C with extra inorganic nutrients and after inoculation (arrow) with 1,2-dichloroethane degrading bacteria (strain GJ10 and AD20). A control slurrie (□) was not inoculated.

3. CONCLUSIONS

Soil inoculation with specific bacterial strains may be useful in some cases. In contaminated soil from an industrial site 1,2-dichloroethane was only degraded after inoculation with specialized bacterial strains capable of utilizing 1,2-dichloroethane as substrate for growth. This inoculation resulted in rapid and complete breakdown of 1,2-dichloroethane. Aerobic degradation of p-cymene and toluene in soil slurries could be enhanced satisfactorily by stimulating the growth of indigenous microorganisms through addition of inorganic nutrients. Further research should point out if inoculation of the contaminated subsurface or above-ground bioreactors with 1,2-dichloroethane degrading bacterial cultures is a feasible option for field scale application.

References

1. Keuning, S. and D.B. Janssen. 1987. Microbiologische afbraak van zwarte en prioritaire stoffen voor het milieubeleid. VROM rapport 80007/1-88, 469 pag.

2. Madhyastha, K.M., and P.K. Bhattacharyya. 1968. Microbiological transformation of terpenes X111. Pathway for degradation of p-cymene in a soil Pseudomonad (PL strain). Ind. J. Biochem. 5:161-166.

3. Janssen, D.B., A. Scheper, and B. Witholt. 1985. Degradation of halogenated aliphatic compounds by *Xanthobacter autotrophicus* GJ10. Appl. Environ. Microbiol. 49:673-677.

4. Van den Wijngaard, A.J., R.U. Groningen. In preparation.

PILOT EXPERIMENT ON IN-SITU OXIDATION OF PROPANTHIOL IN THE SOIL

M. Mackeprang, Steinfeld + Partner Umwelttechnik, Hamburg
I. Schönwald, Steinfeld + Partner Umwelttechnik, Hamburg
I. Wagner, Gesellschaft für Bioanalytik, Hamburg
M. Zarth, Umweltbehörde Hamburg

Contents

1. Introduction

During investigations of the subsoil prior to planning a tunnel for the road layout in Hamburg, contamination has been discovered with an intense stench similar to garlic or onions which is an organic sulphur compound.

In contrast to the low concentration measured in the air, the smell emitted by contaminated water and soil can be perceived intensely at a great distance. Everyone exposed to the smell without any kind of protective mask for longer periods of time complained of irritation to the upper respiratory tracts, coughing and in some cases sickness.

A sanitation concept was compiled according to guidelines for decision-making determined in the Amt für Altlastensanierung (authority dealing with contamination deriving from former landfills) in the Free and Hanseatic City of Hamburg. Priorities here were the comparative evaluation of alternative sanitation processes and the selection of the most favourable operational sequence.

2. The pollutant propanthiol

Chemical examinations of soil and water samples proved evidence of organic sulphur compounds. Analytically, propanthiol was proved to be the substance emitting the stench. An anaerobic environment is essential for the formation of thiols. Examination of the groundwater confirmed an extreme deficiency of oxygen. The groundwater also showed a high content of ammonium and sulphate.

Propanthiol - $CH_3-(CH_2)_2-SH$ - belongs to the group of

F. Arendt, G.J. Annokkée, R. Bosman and W.J. van den Brink (eds.), Contaminated Soil '93, 1269–1276.

aliphatic thiols. Its most striking and most unpleasant property is its intense garlic or onion-like smell which can be perceived in even low concentrations. Propanthiol is considerably more volatile and inflammable than the corresponding alcohol, but in comparison to alcohol, is minimally soluble in water. The thiols are similar to the corresponding alcohols in their toxicity and effect on the organism.

A characteristic of thiols, amongst other properties, is that they oxidize easily using oxygen in the air into disulphides. In the presence of strong oxidation agents, like hydrogen peroxide, potassium permanganate or peroxosulphates the thiols are converted to the relevant sulphonic acids. The source of the propanthiol has been determined as an absorbing well belonging to a group of allotments which no longer exists. The formation of thiols in the soil sould be explained by microbiological conversion processes of excrement containing sulphur (e. g. proteins) (1).

3. Horizontal and vertical extent of the contamination

An investigation below ground to define the area of contamination with propanthiol showed an area of around 20.000 m². The propanthiol can be detected by sensors not only in the groundwater but also in soil samples.

<u>Fig. 1:</u> Plan of underground characteristics

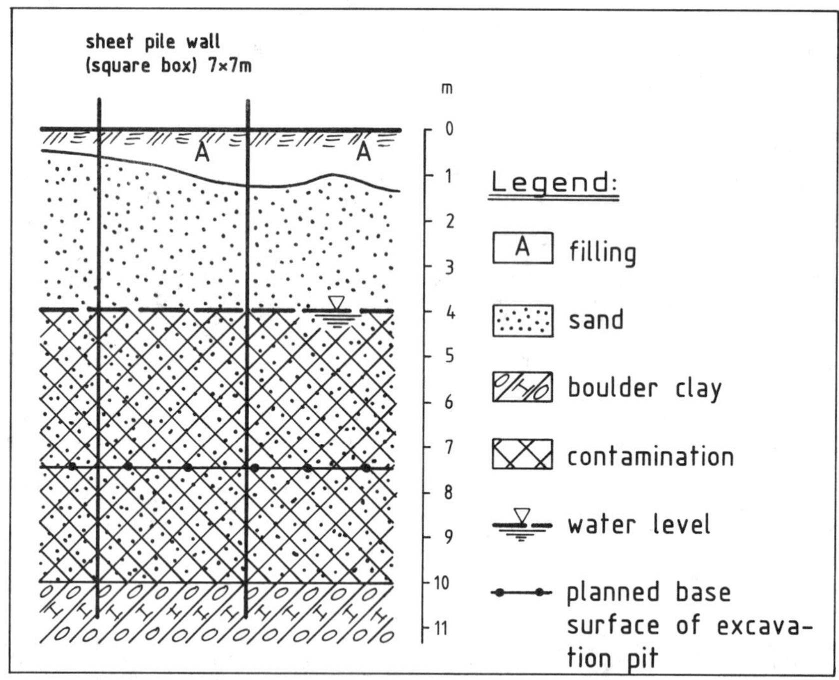

Due to building and the relevant falls in the groundwater level to the north of the contaminated area a reversal of the direction of groundwater occurred leading to an extension of the contamination in a northern direction. The proanthiol is not only evident in water-saturated but also partly in water-unsaturated earth. Fig. 1 illustrates the structure and the extent of contamination underground.

The filling consists mainly of medium and fine sands with fluctuating proportions of silt and gravel. In some areas boulder clay could have worked into the filling. The filling is not considered to be contaminated. Below the filling there are medium and fine sands. In general it can be assumed that the grain size increases with depth. In the sands where the actual propanthiol contamination occurs there are isolated traces of peat. The groundwater level runs within the sands at around 4 to 5 m below the surface. The layers below the sands consist of cohesive structures at around 10 m below the surface, but this boulder marl is not uniformly formed. Boulder clay and locally occurring layers of silt seal the sand at the bottom.

4.   Preliminary investigations within the 1st pilot experiment

Since the influence of the propanthiol contamination on the planned building of the tunnel was unclear, preliminary investigations aimed at clarifying the overall procedure of sanitation were carried out. The central emphasis of this procedure was also to allow a smooth phase of building later. Initial work in determining possible methods of sanitation was carried out as preliminary tests to find out the degree of application for the following techniques:

-   treatment of the contaminated soil in a soil washing plant,
-   micro-biological degradation of the proanthiol in the soil and in the water,
-   oxidation of the groundwater by means of ozone.

Treatment of the soil in a washing plant showed that during the washing process, the contaminants and also the stench were removed almost completely without the use of any chemical additives. Oxidation of the propanthiol was quite possible by means of a damp-mechanical process and atmospheric air as the oxidation agent, but the excavation work and transport of the soil involve adverse stench which could not be accepted.

Micro-biological degradation experiments with propanthiol in the soil and in the groundwater on the other hand, did not provide positive results. The low density of bacteria evident in the soil indicated hostile living conditions for these.

The degradation of pollutants by means of on-site water purification and an ozone generator turned out to be very successful. The propanthiol in the water extracted was oxidized by adding ozone, and sensors did not react to the water subsequent to treatment.

On the whole, the preliminary experiments to define possible sanitation processes showed that oxidation can lead to good success. A central disadvantage of the a. m. process lies in the extraction of contaminated soil and water which leads to propanthiol being evaporated and thus to an intense development of the stench.

To summarise, the 1st pilot experiment showed that a prior in-situ sanitation process is essential for the construction of the tunnel to avoid health risks to construction staff and inhabitants of nearby housing.

The target to aim for in sanitation processes is to carry out excavation work for the building measures without any negative influence of smell or danger to building staff or inhabitants of nearby housing areas from propanthiol.

5. Peripheral conditions for in-situ-sanitation

Affected by contamination are the northern ramp and the northern part of the tunnel. As there is no need for sanitation due to a hazard to water, sanitation is to take place only in the area affected by the building of the tunnel.

Sanitation as in-situ oxidation is to be carried out within a rectangle of approx. 160 x 30 m with a depth of approx. 8 m below the surface. The soil in this area is mostly medium-grained sand and has a permeability of approx. $k_f = 5 . 10^{-4}$ m/s. The groundwater occurring at approx. 4 m below the surface shows a high content of sulphate and iron in addition to propanthiol.

5.1 Suitable oxidation agents

Since oxygen reacts with propanthiol only to dipropylsulphide which also emits an intense smell and easily reacts back to propanthiol, an oxidation agent is to be selected which will react with propanthiol to become propylsulfonic acid.

Many of the oxidation agents for propanthiol cannot be used due to their toxicity or their toxic by-products or other by-products in connection with soil sanitation.

Within the scope of laboratory experiments, contaminated water and soil samples were investigated for the oxidation effect of elementary oxygen, hydrogen peroxide/Caroat[R] (a triple salt of K_2SO_5, $KHSO_4$ und K_2SO_4) with propanthiol. With elementary oxygen, no adequate oxidation effect was achieved, whilst oxidation with H_2O_2 in the presence of catalytic Fe^{2+} ions, i. e. with Caroat, produced good results. A problem arising in soil sanitation is the rapid dissolution of H_2O_2 in the soil (in only a few hours) and the salinization and pH alteration in the soil caused when using Caroat.

5.2 Suitable in-situ sanitation processes

The sanitation processes suitable for this particular case of contamination have been investigated within a sanitation concept with a view to varying criteria (e. g. effectiveness, costs, time requirements, hazards) and evaluated. Table 1 shows the main advantages and disadvantages of the processes under investigation.

Table 1:   Advantages and disadvantages of the sanitation processes selected for treating the propanthiol contamination:

Process	Advantages	Disadvantages
hydraulic sanitation	- relatively low time required -	- only possible with Caroat (extent of eff.) risk of iron hydroxide deposit - treating of large quantities of water
injection	- high effctiveness - possible with H_2O_2	- treating of large quantities of water
sanitation during excavation	- treating of low quantities of water - integrated in building measures	- strong smell - only possible with caroat - low penetration depth
high-speed erosion process	- very high effectiveness - possible with H_2O_2	- high time requirements - promotion of sand/water suspension
soil mix	- high effectiveness - treating of low quantities of water - no production of surplus material - possible with H_2O_2	- high time requirements - process still in test phase
dynamic soil liquefying	- effective mixing of the soil with oxidation agent - possible with H_2O_2 - no production of surplus material	- hazard to neighbouring land from vibration - process not yet tested for sanitation purposes

Based on the concept for sanitation compiled, two processes were tested locally in a 2nd pilot experiment in coordination with the relevant authorities. These were the injection process and the dynamic soil liquefying process. The 2nd pilot experiment served to optimise the technical procedure and to test the data for the oxidation agents (effectiveness, extent of effect, degrees of concentration necessary etc.) which had been determined in the laboratories.

6. Execution of the 2nd pilot experiment

The subsequent condition of building were simulated accordingly in the 2nd pilot experiment (2). Four large test areas of $7 \times 7 \ m^2$ were set up of which two had been sealed by means of a sheet pile casing and in which the level of groundwater was reduced. The injection process and the dynamic soil liquefying process were applied in one of each of the pairs of test areas, i. e. in one water-saturated and one water-unsaturated test area, and tested for process optimisation (extent of effect, time required, injection pressures for the oxidation agent solutions). Fig. 2 shows the structure of the test areas.

Fig. 2: Structure of the test area for the 2nd pilot experiment

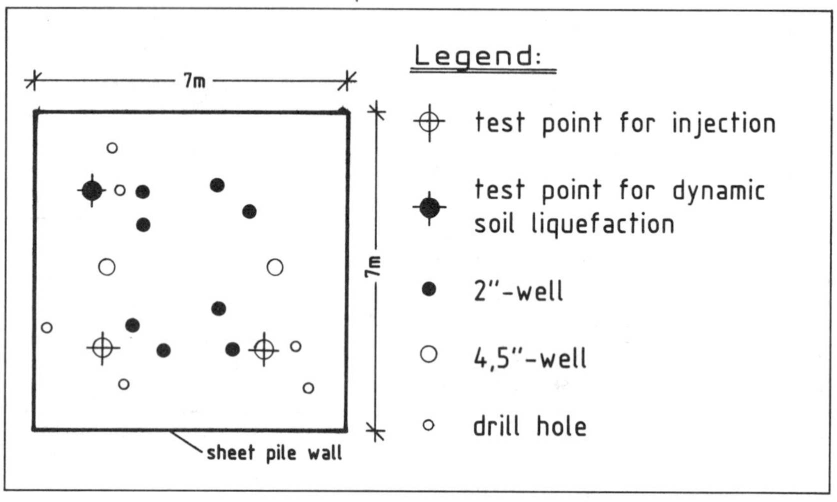

6.1 Injection of oxidation solution into the soil

Water containing oxidation agent is injected into the ground at a pressure of approx. 5-20 bar and through injection lances located in a row at a low distance apart. The lances are then removed and moved in the direction of a gallery of suction lines.

To ensure the hydraulic head is guaranteed during standstill (e. g. at night) and to prevent flow of the contaminated water back into the areas already decontaminated, water is continually removed via the suction lines, treated on site and re-injected after the oxidation agent is added.

6.2 Penetration of oxidation solution in the dynamic soil liquefying process

With this method, the oxidation solution is applied below ground by using a deep-bed vibrator. By vibrations at the foot of the vibrator and by its own weight, this is sunk below ground. The vibrations bring the soil within the application area of the vibrator into a brief period of suspension which ensures the oxidation agent is mixed with the soil.

6.3 Description of test processing

Where the actual building of the tunnel is concerned, there are varying situations existing below ground. The tunnel itself will be built within a diaphragm wall which is integrated into the boulder marl approx. 15 m below ground level and which is closed on all sides.

The sanitation process for the construction of the diaphragm wall and the drainage line running through the contaminated area prior to excavation work, will be carried out in the water-saturated zone. Before commencing sanitation procedures for the construction of the tunnel, the groundwater in the area closed in by the diaphragm wall will be lowered and the contaminated water which is drained off will be re-processed on site. The sanitation process itself will be carried out in the non-saturated zone.

6.4 Results of the 2nd pilot experiment

The success in sanitation was tested by means of sensors (discolouring, smell) in small drillings at distances of 0,5 to 2,0 m and by means of water samples (discolouring, smell, H_2O_2 content) from the relevant water lines at distances of 0,6 and 1,2 m.

The injection tests provided, in addition to the parameters for technical procedures, the necassary H_2O_2 and catalyst concentrations. It was determined, that if the Fe(II) catalyst was absent, a part sanitation of the soil was evident, but there was not sufficient decontamination of the groundwater. For this reason, a catalyst cannot be omitted if success is to be achieved.

The injection tests in the water-saturated zone showed, apart from good sanitation success otherwise, some narrow strips which were contaminated (a few cm) above or in the area of silt and/or boulder marl stratum. In contrast, the water-unsaturated zone showed an even sanitation over the total depth area of the injection processes.

Under the existing conditions below ground, the dynamic soil liquefying process did not attain a homogenous distribution of the oxidation solution underground so that even at relatively close distances from test points ($< 0,5$ m), the soil was not decontaminated. The dynamic liquefying process is thus not a suitable method of sanitation for the case in question.

Selected tests and process data can be seen from Table 2.

Table 2:   Results of the 2nd pilot experiment

Process	test area	concentration H_2O_2/catalyst.	removal success in ground	removal success in soil
dyn. soil liquefying	- within closed area - water-saturated	20 g/l H_2O_2 30 mg/l Fe^{2+}	low	low
dyn. soil liquefying	- outside closed area - water-saturated	20 g/l H_2O_2 30 mg/l Fe^{2+}	low	low
injection	- outside closed area - water-saturated	10 g/l H_2O_2 50 mg/l Fe^{2+}	present	present
injection	- outside closed area water-saturated	10 g/l H_2O_2 25 mg/l Fe^{2+}	partly present	low
injection	- within closed area - non-water-saturated	10 g/l H_2O_2 50 mg/l Fe^{2+}	good	good

7. Summary

During the course of the construction of a tunnel planned in Hamburg, a contamination with propanthiol in the groundwater and in the soil has to be removed to protect construction personnel and inhabitants of nearby housing.

Propanthiol, a sulphur compound with an intense stench has occurred due to excrement from former allotments in the anaerobic environment underground.

To combat this contamination, an in-situ sanitation process of oxidation of the propanthiol is essential below ground. During the course of a sanitation concept, various possibilities of bringing an oxidation agent into contact with the contaminated soil have been investigated. Whilst simulating future construction conditions, pilot tests have shown that an injection of hydrogen peroxide into the water-saturated as well as the water-unsaturated soil is sufficiently positive for oxidation of the contaminant. It has also been determined that injection agent solutions of 10 g/1 H_2O_2 and 50 mg/1 catalyst concentration are sufficient. A radial decontamination around the injection lances of around 1 m are possible.

8. Bibliography

1. Brügmann, A.; Schönwald,I; Zarth, M. (1992):
    Erkundung eines Propanthiolschadens
2. Schönwald, I; Wagner, I; Zarth, M. (1992):
    Sanierungsmöglichkeiten eines Propanthiolschadens durch In-situ-Oxidation. Handbuch der Altlastensanierung; 12. Lieferung, 6/92
3. Dr. Zarth, M. (1992):
    Paper presented on the 4th KfK/TNO Conference Berlin. "System zur Bewertung alternativer Sanierungsverfahren bei der Erstellung eines Sanierungskonzeptes (unpublished).
4. Hedicke, H.; Wagner, I.; Schönwald, I.; Otzdorff, J. (1991):
    Sanierungskonzept Propanthiolschaden. Expertise by Steinfeld + Partner Umwelttechnik GmbH (unpublished).

DIFFERENT LIMITATIONS AFFECTING BIODEGRADATION OF POORLY WATER-SOLUBLE SUBSTRATES

CHRISTIAN PLAS, PETER HOLUBAR, EDITH BAUER, CHRISTINE PENNERSTORFER AND RUDOLF BRAUN

INSTITUTE OF APPLIED MICROBIOLOGY, UNIVERSITY OF AGRICULTURE VIENNA/AUSTRIA

1. ABSTRACT

Biological techniques for waste water and waste air treatment as well as soil remediation have been employed successfully in technical scale. Nevertheless engineers face several problems applying biotechnological methods to environmental affairs. A short overview will be given on the common limitations which decrease the efficiency of pollutant purification and investigations are described which enable to observe the crucial effects seperately. Experiments deal with biological (biochemical stability of the contaminant, toxicity, adaption and growth), chemical (solubility and vapour pressure) and environmental constraints (mass transfer, adsorption and nutrient availability).

2. INTRODUCTION

The occurrence of contaminants in all areas of environmental concern has encouraged a whole industrial branch to develop biotechniques for overcoming pollution problems. Even though high efforts have been spent on increasing knowledge in related sciences (e.g. microbiology, chemical engineering,...) and to optimize remediation equipments, a lot of biological applications have failed or at least residues remained within the polluted medium. The presented paper deals with the possible reasons that have been identified to limit biotechnological ameliorations by numerous authors. Assays are presented to consider limiting effects more or less isolated and help to understand and estimate their influence. All treated examples are subject to biodegradation sensu stricto, i.e. either degradation by microorganisms delivering energy or cometabolic degradation (the considered pollutant serves as side substrate). This means that no chemical or photochemical induction processes are dealt with.

3. INTRODUCTION: WHAT ARE THE PROBLEMS?

3.1. Chemical Constraints

One major problem is poor solubility (Tab. 1.) per se, which is usually superimposed with mass transfer problems. Due to uptake mechanism cells are dependent on receiving their substrates in a water soluble form. However, organisms frequently develop ways to reach insoluble compounds by solubilizing them. When bacteria produce surfactants the organic molecules are captured in micelles and thus can pass the membranes (either directly via pinocytosis or by direct contact between cell wall and organic phase (capillary uptake) (Gerson and Zajic, 1979)).

Tab. 1. Water solubility of investigated organic substances

compound	solubility (mg·l^{-1})
n-pentane	360
n-decane	0,05
cyclohexane	0,052
toluene	570
naphthalin	30
carbondisulfide	1200

F. Arendt, G.J. Annokkée, R. Bosman and W.J. van den Brink (eds.), Contaminated Soil '93, 1277–1284.
© 1993 *Kluwer Academic Publishers. Printed in the Netherlands.*

A second problem faced during decontamination is the high vapour pressure exhibited by many organic solvents. Since decontamination procedures often need extensive aeration, a great extent of the volatile pollutant fraction may be stripped.

3.2. Biological Constraints

Poor degradability of substances is caused by biochemical recalcitrance. It has distinctly been shown by scientists that different structural conformations of the same substance decide upon its degradability (Zehnder, 1990). Biochemical problems usually mean lack of enzymes or inducers, which are found in case of e.g. cycloalkanes. Degradation of such compounds is therefor dependent on the presence of additional substrates or inducer molecules.

Another mechanisms which directly affects substance-microorganism-interaction is toxicity. In most cases due to their low solubility organic molecules exhibit no toxic effects on the biological consortium. Toxicity phenomena have been described for e.g. benzene (Baumann, 1992) and organic sulfur compounds (Plas et al., 1992). Situations showing metabolites more toxic than the original contaminant occur rarely.

There are some environmental remediation techniques employing multi-phase systems to solubilize organic compounds. Two problems may arise with such methods: on one hand the organic solvent might be toxic itself, on the other hand the contaminant may become toxic because of its increased concentration. Except for a certain time for adaption and/or enrichment it may usually be expected that the autochthonic flora is capable of mineralizing most of the common organic contaminations. In dependence on environmental conditions the adaption phase can be accelerated by engineers.

Considering bacteria in waste water treatment plants it is evident that with decreasing hydraulic residence time fast growing organisms will be favoured. Therefor the application of specific microorganisms in environmental technology is frequently limited by their slow growth. Table 2 gives some examples of bacteria which are of importance in waste air or waste water treatment processes.

Tab. 2. Generation times of bacteria important for environmental technology.

organism	application	generation time (h)	reference
Nitrobacter sp.	nitrite oxidation	> 8	Bock, 1988
Nitrosomonas sp.	ammonia oxidation	120	Bock, 1988
Methanobacterium soehngenii	anaerobic wastewater treatment	215	Zehnder et al., 1980
Thiobacillus versutus	sulfide elimination	14	Karagouni et al., 1989
Thiobacillus K4	CS2 degradation	70	Plas et al., 1992
C1-organism (not identified)	groundwater decontamination	60	McCarty et al., 1990
Alcaligenes denitrificans	PAH degradation	35	Weissenfels et al., 1990

As a consequence of Monod-type substrate kinetics conversion rates are sometimes found to be distinctly lower at low concentrated contaminations (see below).

Mixed culture phenomena like substrate competition are of general importance in environmental technology. Inequilibrated nutritional conditions cause shifts in population composition and may supress microorganisms with important metabolic capacities. As an effect of auxotrophy easily degradable substrates are metabolized prior to residual contaminations which thus cannot be removed in continuous systems.

3.3. Environmental Constraints

The way of an organic molecule to the metabolizing bacterium is mainly described by diffusion. Diffusion, however, is influenced by both medium density and distance (besides the concentration gradient), and both parameters have important features concerning remediation techniques.

As an example for density change mass transfer may be impeded by formation of biofilms covered by extracellular material (e.g. polysaccharides). On the other hand mass transfer problems as a result of increased diffusion distance are a phenomenon common to everybody working on waste water and waste gas treatment. The generation of flocs distinctly changes substrate kinetics within a CSTR, the same can be said for biofilms on any support material. Nutrient conditions are totally different for inner layers as shown for various cases (de Beer et al., 1991, Diks and Ottengraf, 1991, Revsbach et al., 1989).

An extremely disagreable situation is faced during soil treatment, since diffusion may represent the main mechanism for substrate transport. Neither mixing like in waste water treatment nor turbulences like in waste gas purification can ameliorate mass transfer rates.

Essential for every organism to maintain its activity is the presence of a certain amount of nutrients. Mostly not posing difficulties in waste waters the lack of nitrogen and phosphorus in heavily contaminated soils may result in poor degradation success. Investigations have shown that an overload of carbon (from hydrocarbons) may be compensated with exact fertilization (Pritchard, 1990).

Dependent on the specific situation oxygen as a most important parameter must be observed. Numerous treatment techniques working under anaerobic conditions have been successfully applied (waste water, groundwater), the main field of applications, however, is covered by aerobic processes, for which it is necessary to guarantee a satisfying pO_2. The oxygen regime is mainly influenced by the water content of the treated material. Both in soils and in biofilters engineers try very hard to regulate moisture, since it is essential for microorganisms to face a convenient water activity (Schlegel, 1985).

Limitation under physiologically optimum conditions can furthermore be mediated by adsorption phenomena. It has been described that clay content and humus strongly influence availability of organic compounds in soils (Bauer et al., 1992). As well polyaromatic hydrocarbons as alkanes are sensitive to these soil components and are therefor difficult to remove from loamy soil.

4. INVESTIGATED EXAMPLES
4.1. Chemical Constraints
4.1.1. Solubility. Limitations due to insolubility were demonstrated with a *Pseudomonas cepacia* isolate growing on n-decane as carbon source. Growth appears very slowly as long as the surface tension remains constant. Only after an adaption period of about 30 hours surfactants are excreted and decrease surface tension to 40 mN·m^{-1} , which facilitates augmented solubilization of decane (up to 200 mg·l^{-1} which is ca. the 400-fold value of its regular water solubility) and subsequently promotes growth (Fig. 1).

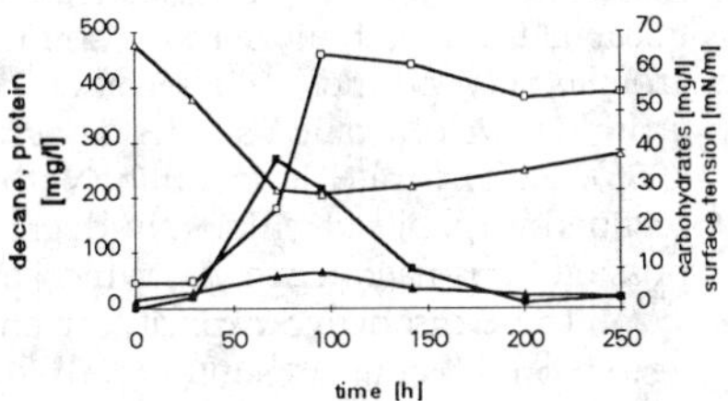

Fig. 1. Growth of *Pseudomonas cepacia* on n-decane. Course of surface tension [mN·m^{-1}] (Δ), carbohydrates [mg·l^{-1}] (▲), protein [mg·l^{-1}] (□) and n-decane concentration [mg·l^{-1}] (■).

A very distinct correlation has been found between growth rate and surface tension for *Pseudomonas paucimobilis* growing on naphthalin. Working with different surfactants the relationship appeared almost linear (Fig. 2).

Similar effects have been described by Breure et al. (1991) who observed growth on sieved naphthalin in dependence on the sieve fraction. The phenomena are comparable: the bigger the substrate surface, the faster the growth.

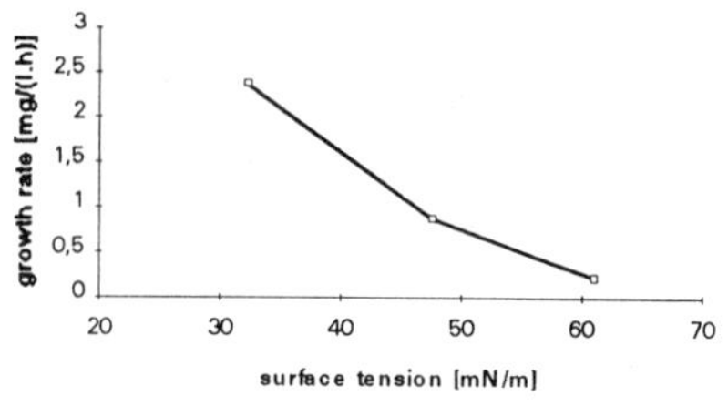

Fig. 2. Growth of *Pseudomonas paucimobilis* isolate B2 on naphthalin dependent on surface tension.

4.1.2. Vapour Pressure. Reports about toxic effects of short chain alkanes due to their denaturing properties are well known (Geller et al., 1991). Given the right concentration, however, makes their removal from e.g. waste gas feasible. Biofilters were fed with an equimolar mixture of C_5-, C_6-, C_7- and C_8-alkanes and the respective gas concentrations were measured by gaschromatography. The results appeared surprisingly clear: the substances were degraded in accordance to their vapour pressure (Fig. 3).

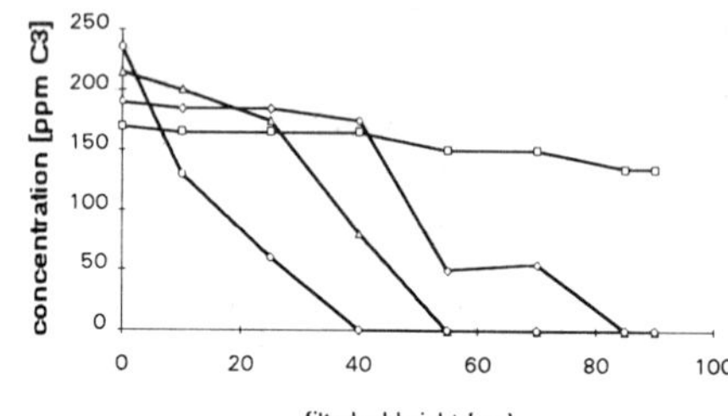

Fig. 3. Degradation of alkanes in a biofilter in dependence on filter height. n-pentane (□), n-hexane (◊), n-heptane (Δ), n-octane (O).

Thus it turned out that the degradation of volatile compounds is possible even under stripping conditions like in biofilters provided that parameters which influence mass transfer and biological activity are suitable.

4.2. Biological Constraints

4.2.1. Poor Degradability. Cycloalkanes very frequently remain in treated soils and waste gas. It is well known that the initial enzymatic attack appears difficult to most microorganisms, because the metabolic step of ring-fission is not yielding energy in this case (in contrary to aromatic compounds) (Geller et al., 1991). Only two bacteria have been described so far to grow on cyclohexane as sole carbon- and energy-source (Stirling et al., 1977, Trower et al., 1985). At any rate it is unlikely that under non-sterile conditions a consortium can be retained, capable of efficiently degrading this certain group of contaminants. However, it is possible to isolate organisms which mineralize cyclo-alkanes by cometabolism.

The degradative capacity of an isolate (T 35) identified as *Pseudomonas cepacia* was tested on several substrates with cyclohexane as model-cosubstrate. Since the first step during degradation of cyclohexane is supposed to be catalyzed by a mono-oxygenase it was assumed that substrates for the induction of this enzyme would work more efficiently than others. Figure 4 shows the respective results using glucose, n-decane, toluene and naphthalin as main substrates to induce cometabolic elimination of cyclohexane.

Apparently naphthalin and toluene did not induce cometabolic elimination of the alicyclic compound, whereas glucose and n-decane did. As a matter of fact in theory the metabolic pathway of cyclohexane degradation starts with the same enzymatic step as with aliphatic substances (like n-decane) which can explain the described activity increase. For sugar the explanation remains unclear: perhaps its role as a general substrate for a whole series of reactions can facilitate its role in cometabolic elimination of alicyclic compounds.

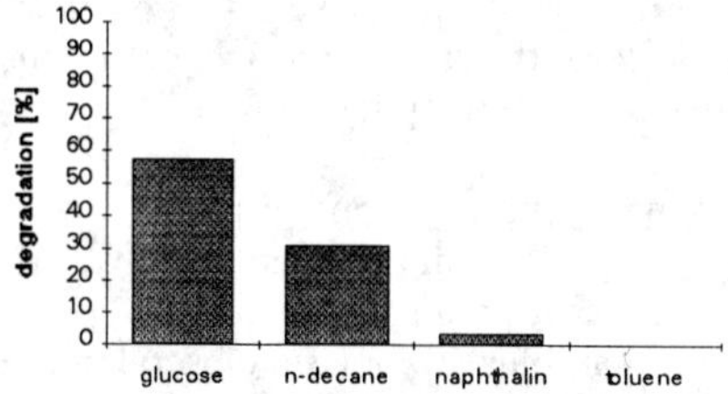

Fig. 4. Influence of different substrates on the cometabolic degradation of cyclohexane by *Pseudomonas cepacia* isolate T 35 (5 days assay).

4.2.2. Toxicity. There are certain examples of poorly water soluble chemicals which nevertheless can reach toxic concentrations. Carbondisulfide may serve as an example for this phenomenon. Investigations on CS_2-degradation led to the isolation of a *Thiobacillus* sp. (isolate K4) using CS_2 as sole energy-source (Plas et al., 1991). Increasing the substrate concentration, however, delivers Monod-type degradation kinetics with a toxic level concentration of about 150 mg·l^{-1} (Plas et al., 1992)(Fig. 5). All values for conversion rates depicted in figure 5 refer to $mgCS_2 \cdot g_{protein}^{-1} \cdot min^{-1}$.

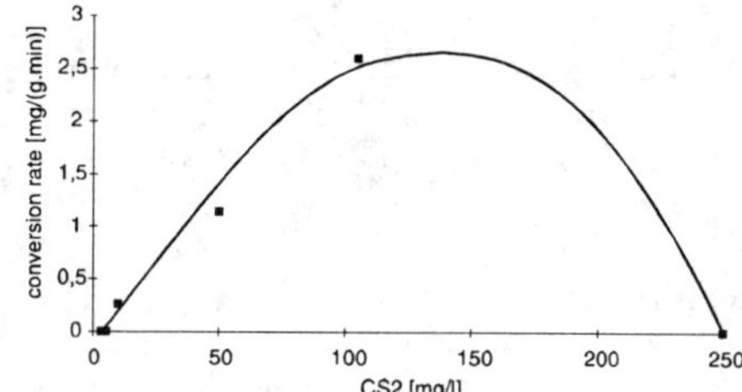

Fig. 5. Degradation kinetics of *Thiobacillus* sp. isolate K4 for carbondisulfide as sole energy source.

Since emissions of CS_2 have - due to its well known toxicity - to be carefully monitored, its concentration in technical plants for waste water or waste gas treatment will hardly approach this toxic barrier.

4.2.3. Adaption can distinctly be recognized observing the course of the metabolic quotient (this is the respiration divided by the glucose induced respiration) expressing specific activity. As shown in soil decontamination experiments with naphthalin as contaminant different soil were investigated to emphasize the influence of other parameters (like pH and content of organic matter) on purification effectiveness. The following graph indicates the time necessary for adaption during a certain remediation procedures (Fig. 6).

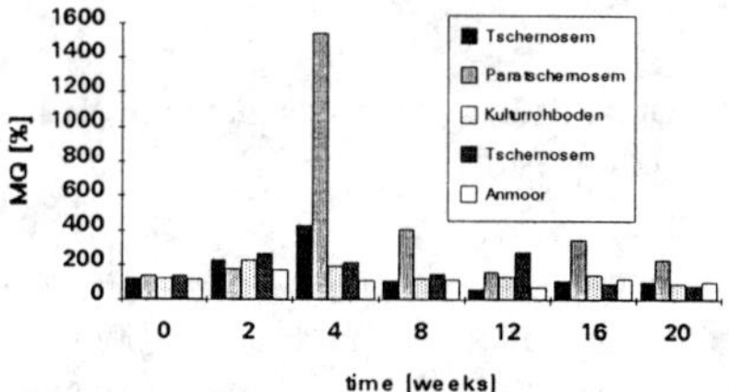

Fig. 6. Course of metabolic quotient MQ as a measure for adaption after contamination of soil with naphthalin.

Interesting to notice is that all tested soils, which are really different indeed, needed approximately the same time to reach culmination of activation.

4.2.4. Slow Growth. Referring to table 2 generation times in the order of several hours or even longer have to be dealt with. Under high selection pressure even slowly growing microorganisms may compete in mixed cultures, if there is no washout problem. In soils and other immobilized systems adherent bacteria are capable of forming films and thus decouple growth rate from hydraulic retention time. The *Thiobacillus* isolate K4 described in table 2 could retain its activity due to continuous selection pressure (waste gas loaded with CS_2) also in a non sterile pilot plant.

4.3. Environmental Constraints

4.3.1. Mass Transfer Limitations.
A series of environmental applications has employed trickling filters recently. One of the special merits of this fermenter-type is process stability due to immobilizedbiomass. Observing the formed biofilm, however, sometimes reveals that it is covered by a homogenuous layer of extracellular material of apparently rather high density (Fig. 7). Microscopic evaluation of the immobilized bacteria confirms this assumption since the present organisms seem to be embedded in a gelatinous phase which probably has the advantage of stabilizing the microenvironment, but on the other hand constitutes a distinct barrier for substrate transfer.

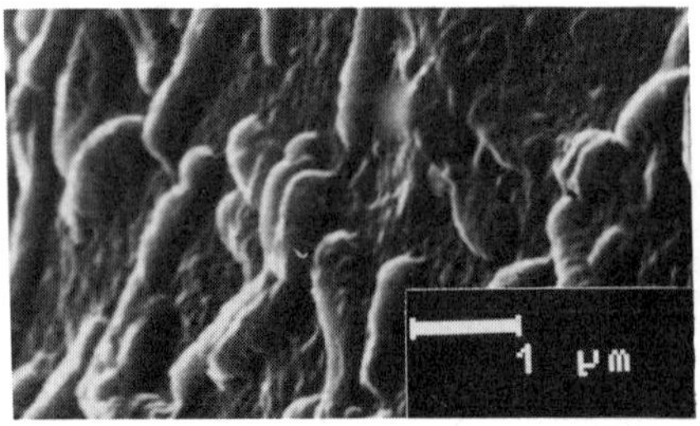

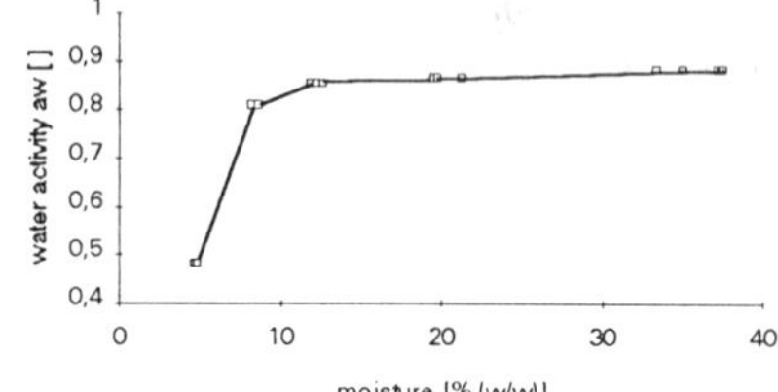

Fig. 7. Scanning electron micrograph of a biofilm grown on sintered glass. Colourless sulfur bacteria were fed with CS_2-loaded waste air.

According to calculations of Ottengraf (1977) biofilm thickness is the second parameter of eminent importance with respect to mass transfer. Dependent on the diffusion distance (in combination with the concentration gradient) nutrient supply decreases for inner biofilm layers

4.3.2. Environmental Factors.
For biofiltration of hydrocarbons the importance of moisture could be outlined. Since water activity is one of the main parameters determining microbial activity, it is evident that certain limits of water content within the filter material have to be respected (Fig. 8).

Fig. 8. Dependence of water activity on moisture in compost.

After checking the influence of water content, the influence of nutrient addition on the efficiency of soil remediation has been tested. Nitrogen and phosphorus fertilizers were added for degrading model hydrocarbons in soils. It could clearly be seen that if an equilibrated balance between carbon, nitrogen and phosphorus is adjusted microorganisms adapt much quicker to contaminations and the remediation efficiency can be enhanced by far (Bauer et al., 1992).

4.3.3. Adsorption.
A frequent problem is that the results of soil remediation processes (expressed in terms of residual contamination) especially in case of polyaromatic hydrocarbons do not match predictions from laboratory experiments (Volkering et al., 1992). It has been suggested that adsorption to soil particles and organic matter has a major

responsibility for bioavailability restrictions and thus degradation effectiveness (Martin et al., 1978, Ogram et al., 1985).

For this reason adsorption isotherms for naphthalin have been measured for different soil types, K_f-values were calculated and activation of organisms was estimated by dehydrogenase activity and the metabolic quotient MQ (respiration divided by glucose induced respiration) (Fig. 9).

Fig. 9. Activity increase measured as dehydrogenase activity (TTC) (□) and MQ (■) (not contaminated reference soils are set 100 %) after 8 weeks of naphthalin-decontamination in 5 different soil types in dependence on the content of organic matter in the soils.

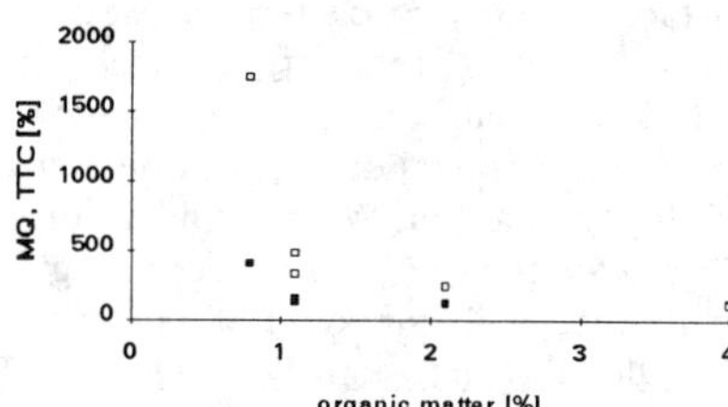

5. SOLUTIONS

5.1. Basic Science

The first step for retaining reliable results is checking analytical methods. Especially handling poorly soluble compounds is critical and a lot of experimantal data might be falsified by application of the wrong method or by non-representative sampling. One has to bear in mind the very problematic situatins of condensing hydrocarbons in waste air treatment or of taking soil samples from a pile of 1000 m^3 (which is in fact a rather small example).

An important feature very often neglected in laboratories is to adapt cultivation techniques to practical parameters and needs (gradient systems, autoselective consortia, volatile substrates, substrat concentration,...). There is usually little need for predictions concerning technical scale equipments from lab-scale experiments under conditions which do not meet the practical situation. So the scientist has to develop enrichment and cultivation techniques which are able to model environments envisaged after up-scaling.

5.2. Different Scales

By skipping experimental scales incomparabilities of optimized lab systems and technical scale are easily overseen. Scaling factors derived from chemical engineering should be considered with respect to the apparently critical parameters of the purification procedure.

5.3. Technical Plants

One of the most neglected points running technical plants is monitoring the relevant parameters. In order to cut down expenses absolutely necessary equipment and personnel for controlling important and limiting parameters are not financed. This behaviour leads to uncontrolled operation, loss of efficiency and as a consequence to an immense loss of money.

A second item must be to employ techniques which guarantee process stability under operational conditions. Recently a strong tendency to the application of immobilized systems in environmental applications could be perceived which is certainly a positive development. These systems are distinguished by high flexibility concerning input fluctuations and toxicity. As another positive characteristic it should be mentioned that process stability can usually be ensured by relatively low maintainment.

6. REFERENCES

Bauer E, Pennerstorfer Ch, Braun R (1992) Dependence of organic contaminant mineralization on soil type. submitted for publication

Baumann H (1992) ECOINFORMA ´92, 2nd Int.Congress and Exhibition on Environmental Information & Communication, Bayreuth/FRG.

de Beer D, Huisman JW, den Heuvel JCvan, Ottengraf SP (1991) Substrate transport and structure of nitrifying bacterial aggregates. In: Engineers Technological Institute - Royal Flemish Society of (ed) International Symposium Environmental Biotechnology, vol 1, Koninklijke Vlaamse Ingenieursvereniging, Oostend/B, pp 337-340

Bock E (1988) Nitrifikation. In: Reinheimer G, Hegemann W, Raff J, Sekoulov I (eds) Stickstoffkreislauf im Wasser, Oldenburg Verlag, Munich/FRG

Disk RM, Ottengraf SP (1991) Process engineering aspects of biological waste gas purification. In: Engineers Technological Institute - Royal Flemish Society of (ed) International Symposium Environmental Biotechnology, vol 1, Koninklijke Vlaamse Ingenieursvereniging, Oostend/B, pp 353-368

Gerson DF, Zajic JE (1979) Microbial Biosurfactants. Process Biochemistry (July):20-22

Karagouni AD, Kelly DP (1989) Carbon dioxide fixation by Thiobacillus versutus: apparent absence of a CO2-concentrating mechanism in organisms grown under carbon-limitation in the chemostat. FEMS MicrobiolLetters 58:1797-182

Martin JP, Parsa AA, Haider K (1978) Influence of intimate association with humic polymer on bidegradation of 14C-labelled organic substances in soil. Soil Biochem 10:483-486

McCarty PL, Semprini L, Hopkins G, Roberts PV (1990) Field evaluations of in-situ biotransformations of chlorinated aliphatic compounds. In: EERO-GBF Symposium on Environmental Biotechnology, GBF Braunschweig/FRG

Ogram AV, Jessup RE, Ou LT, Rao PSC (1985) Effects of sorptin on biological degradation rates of (2,4-dichlorophenoxy)acetic acid in soils. ApplEnironMicrobiol 49:582-587

Plas Ch, Wimmer K, Danner H, Harant H, Holubar P, Braun R (1991) Biologische Oxidation von Schwefelkohlenstoff durch ein Thiobacillus-Isolat. gwf Wasser-Abwasser 132(7):419-421

Plas Ch, Wimmer K, Holubar P, Danner H, Mattanovich D, Jelinek E, Harant H, Braun R Degradation of carbondisulfide by a Thiobacillus isolate. ApplMicrobiolBiotechnol accepted for publication

Pritchard PH (1990) Environmental Catastrophes: Prevention and Development of Rapid, Coordinated Responses to Effectively Contain and Remedy. In: EERO-GBF Symposium on Environmental Biotechnology, GBF Braunschweig/FRG

Revsbach NP, Christensen PB, Nielsen LP, Sorensen J (1989) Denitrification in a trickling filter biofilm studied by a microsensor for oxygen and nitrous oxide. WatRes 23:867-872

Schlegel HG (1985) Allgemeine Mikrobiologie, 6th ed, Thieme, Stuttgart

Volkering F, Breure AM, Sterkenburg A, van Andel JG (1992) Microbial degradation of polycyclic aromatic hydrocarbons: effect of substrate availability on bacterial growth kinetics. ApplMicrobiolBiotechnol 36:548-552

Weissenfels WD, Beyer M, Klein J (1990) Degradation of Phenanthrene, Fluorene and Fluoranthene by Pure Bacterial Cultures. ApplMicrobiolBiotechnol 32:479-484

Zehnder AJB (1990) Factors affecting biodegradation of chlorinated benzenes and hexachlorocyclohexane. In: EERO-GBF Symposium on Environmental Biotechnology, GBF, Braunschweig/FRG

Zehnder AJB, Huser BA, Brock TD, Wuhrmann K (1980) Characterization of an acetate-decarboxylating, non-hydrogen-oxidizing methane bacterium. ArchMicrobiol 124:1-11

USE OF OZONE FOR SOIL REMEDIATION

J.P. SEIDEL, B. ROTHWEILER, E. GILBERT, S.H. EBERLE

Kernforschungszentrum Karlsruhe, Institut für Radiochemie,
Abteilung Wassertechnologie, Postfach 3640, 7500 Karlsruhe

Besides the common methods of soil remediation like soil-washing or
-venting, incineration and biological treatment the use of oxidizing
agents are discussed at the moment. In-situ or on-/off-site chemical
oxidations with ozone containing gas should be a promising treatment-
method of soils, contaminated with toxic or persisitent organic wastes.

On account of the results of investigations on model and natural soils,
the economic and useful applicability of ozone for soil-treatment is
discussed. To examine the effectiveness of ozone, the catalytic
ozone destruction on solid surfaces should be known. In a fixed bed
reactor (flow through and batch) the influence of different surfaces
of minerals (quartz, felspars, dolomite ...) on the ozone decomposition
was tested. It is shown that the rate of ozone destruction is dependent
on the water content of the minerals and is reduced by factor 17 on
quartz with 6 % water content compared with dry quartz.

In the second part of the presentation the reaction of ozone with
organic compounds on solid surfaces were investigated. As models,
polycyclic aromatic compounds were choosen. For example, the reaction
of ozone with anthracene leads to the formation of anthraquinone,
anthrone, phthalic acid, 2-carboxybenzaldehyde and benzene-1,2,4,5-tetra-
carboxylic acid, which are partly mineralized under additional ozone-
consumption to oxalic acid and carbondioxide.
It was found that the oxidation products on solids are formed the same
as in aqueous solution. Also the biodegradability of the oxidation
products are better than the initial compounds.

The following experiments with contaminated soil of a gas work shows
that in afixed bed or in a rotating reactor, the polycyclic
aromatic compounds would be reduced to 99 % with a ozone consumption
of 45 g/kg. The oxidation products, partly extractable with water, are
biodegradable.

A further example showes that the ozone-treated soil is still biological
active (growing of plants, seeds and fungi).

F. Arendt, G.J. Annokkée, R. Bosman and W.J. van den Brink (eds.), Contaminated Soil '93, 1285.

CLEANING OF CONTAMINATED SOILS
— A TREATMENT CONCEPT —

Simone Wömmel, Wolfgang Calmano and Kai Heining

Technische Universität Hamburg-Harburg, Arbeitsbereich Umweltschutztechnik,
Eißendorfer Str. 40, W-2100 Hamburg 90, BR Deutschland

ABSTRACT

A new treatment process for heavy metal contaminated soil is suggested. It consists of several process steps. First, an extraction of heavy metals with weak organic acids like citric acid or acetic acid from the soil is carried out. In soil samples containing up to 149 g lead/kg and up to 18 g antimony/kg 83 % of the lead and 85 % of the antimony could be extracted with acetic and citric acid.

The extraction step is followed by a simple plate electrolysis of the highly concentrated heavy metal eluates. Thus, lead is refined with an electric efficiency of 777 kWh/t lead.

It is planned to use foam fractionation for further reduction of heavy metal concentration in the leachates. First tests with model wastewaters containing heavy metals yielded for lead in enrichment factors of 13.8 for solutions of mineral acids and of 8 for solutions of organic acids, respectively. In this way the concentration of the wastewaters could be reduced to 0.01 mg/l and 0.3 mg/l lead, respectively.

1. INTRODUCTION

The development and application of methods to treat contaminated soil has become a significant task for future because of rapidly increasing numbers of superfund sites. According to data of the German Environmental Protection Agency already 150,000 contaminated sites are detected, among them about 52,000 in Eastern Germany [v. Lersner, 1990]. It can be expected that these numbers will still increase with increasing degree of detection of abandoned industrial areas or warfare-related contaminated sites.

Besides organic pollutants heavy metals represent a substantial potential risk. Developing new remedial techniques ecologically consistent treatment should have high priority. In order to take care of the treated soils the chemical agents used should not be toxic and biologically degradable. Moreover, these techniques should need only small amounts of chemical agents which together with the heavy metals should be recovered as valuable substances. Thus, the waste materials to be deposited will be minimized taking into account both the increasing shortage of resources and the lack of disposal sites. The authors want to regard the present paper as contribution according to this aim.

2. TREATMENT CONCEPT

The suggested treatment method for heavy metal contaminated soils consists of the following process steps. First, an extraction of the heavy metals from contaminated soil is carried out by weak organic acids like acetic acid or citric acid. In this way,

1287

F. Arendt, G.J. Annokkée, R. Bosman and W.J. van den Brink (eds.), Contaminated Soil '93, 1287–1294.
© 1993 *Kluwer Academic Publishers. Printed in the Netherlands.*

neither in the water nor in the soil remain any toxic extraction agents except biologically easily degradable substances. An extensive increase in salinity of the wastewater as well as a severe denaturation of the soil components, both arising by the use of mineral acids, will be reduced. Additional to the occurrance of dissolution processes of carbonates and hydroxides and to the competition of H^+-ions with fixed metals complexing reactions of the organic acids may contribute to the dissolution of metals from the soil constituents.

Especially the use of organic acids offers good conditions for the second electrolysis step because no development of aggressive, persistent or toxic electrolysis products like chlorous gases (from HCl) or nitrosous gases (from HNO_3) may take place. By choosing the refining voltage it is possible to deposit the heavy metals selectively and to obtain a commercializing product. The emission of high salt loads into the wastewater occurring with the application of precipitating or ion exchanging techniques is avoided. This allows an easy wastewater treatment.

Foam fractionation as third possible stage is to complete electrolysis by further lowering the heavy metal concentration of the wastewater from electrolysis. The foam fractionation technique is specially suited to eliminate small amounts of metal concentrations in the range of 1-20 mg/l. Foam fractionation experiments with artificial wastewaters had been carried out in order to examine the applicability of this technique on acetic acid and citric aqueous solutions. The partition effect of this technique is based on the one hand on complexing the dissolved metal species and on the other hand on hydrophobizing the complexed metals. These complexes can adsorb at the surface of the gas bubbles, raise to the liquid surface and be separated. Therefore it is necessary to lower the surface tension of the aqueous solution. It is intended to combine the three process steps as shown in figure 1.

The concentrate from the foam fractionation unit will be returned to the electrolysis unit. Perhaps it will be necessary to have an additional process or a second electrolytical unit as intermediate stage, because the minimum concentration leaving the electrolysis is at the present experimental state higher than the maximum inintial concentration practicable in foam fractionation. The purified acid can be used again for extraction.

3. METHODS AND MATERIALS

The extraction of heavy metal contaminated soil samples by acetic acid and citric acid has been basically examined by Heining [1990]. Samples were collected from three different points of a lead contaminated site. The samples A, B1 and B2 are a mixture of a sandy and a loamy soil whereas sample D is a relatively sandy soil. After drying and crushing of agglomerates a fraction of a grain size less than 1 mm was separated by sieving. Samples were digested with HNO_3 and an extraction method (Sb, As) [Calmano et al., 1992] and total metal contents determined by AAS. Antimony and arsenic were analysed by graphite furnace AAS, the other metals by flame AAS and a part of the foam fractionation samples by ICP-AES.

The extraction experiments with soil sample A were carried out in three steps (24 h for each extraction step) followed by centrifugation of the soil and addition of fresh solvent. The soil residue of the last exctraction step was washed and for the

Fig. 1: Cleaning of heavy metal contaminated soils:

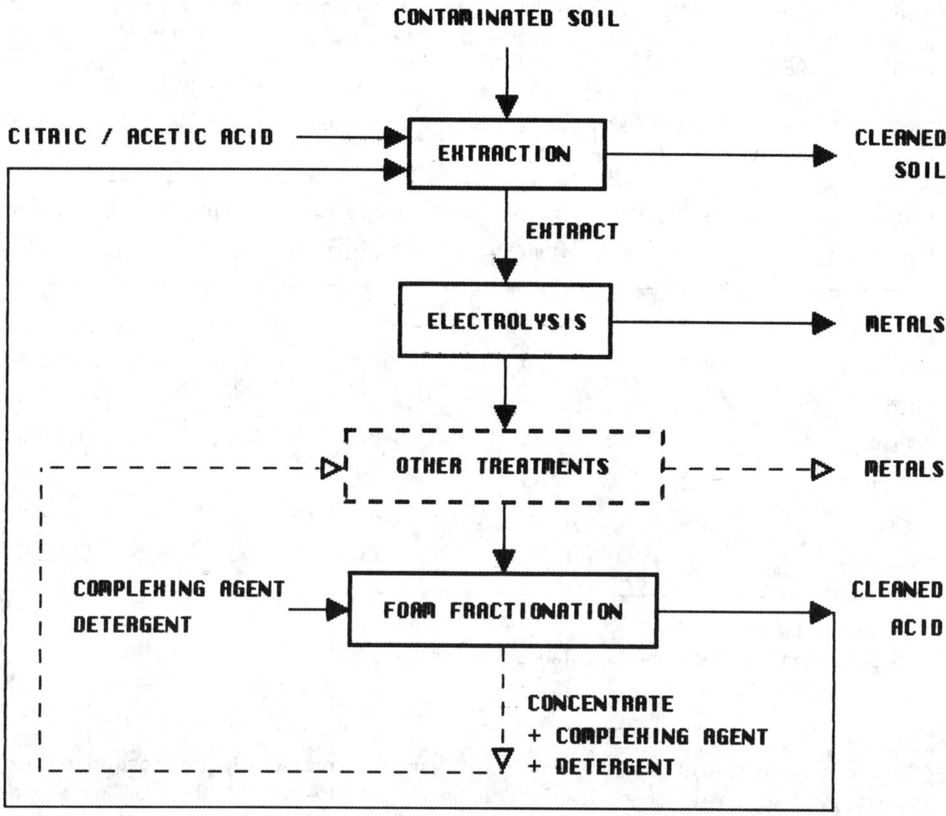

determination of the residual contents digested with concentrated HNO_3. For comparison a leaching experiment with HCl had been realized, too. The concentrations of the acids were 2 mol/l for citric acid, 8.7 mol/l for acetic acid and 10.4 mol/l for hydrochloric acid.

The electrolytic unit was equipped with a graphite anode (345 cm^2), a titanic cathode (200 cm^2) and a power supply which produced 0-40 A and 1-6 V. Formation of dendrites on the cathode could be avoided by addition of gelatine to the electrolytical bath. The electrolytic experiments had been realized with eluates of soil sample B1.

For the foam fractionation experiments artificial aqueous solutions of the metals lead, cadmium, copper and antimony and the metalloid arsenic had been used. Hexamethylenedithiocarbamate (HMDC) and Sodiumdodecylsulfate (SDS) served as complexing agent and detergent, respectively. The application of the technique on acidic solutions of about pH 2 is of special interest because of the possibility of using the eluates of the electrolytical stage directly without any neutralization. The experiments with HCl only had been carried out with lead. As flotation vessel served a filtration column of 110 mm height and of 60 mm in diameter with a frit of the porosity of 4. The flotation gas was nitrogen.

4. RESULTS

Table 1 shows the total contents of lead, cadmium, chromium, nickel, zinc, antimony and arsenic in the soil. They had been determined by Heining [1990]. For comparison the background values of uncontaminated soils are given [Eberius, 1989 and Anonymus, 1988].

Tab.1: Total contents of the samples and background levels of metals in uncontaminated soils [Eberius, 1989 and Anonymus, 1988]

Metal/metalloid Contents	sample A [mg/kg]	sample B1 [mg/kg]	sample B2 [mg/kg]	sample D [mg/kg]	background [mg/kg]
Antimony	9411	2093	18245	25.43	1-3
Arsenic	123	29	94.5	3	15+0.4(L+H)
Lead	51270	80770	148875	75	50+L+H
Cadmium	69	74	288	3.9	0.4+0.07(L+3H)
Chrom	<1	<1	<1	<1	50+2L
Nickel	64	91	320	<1	10+L
Zinc	314	238	313	23.7	50+1.55(2L+H)

L = percentage clay ($<2\mu m$)
H = percentage of anorganic substances in soil

It can be seen that the samples A, B1 and B2 considerably exceeded the background levels in particular of lead and antimony. The results of the leaching experiments of sample A are summarized in table 2 [Heining, 1990]:

Tab. 2: Extraction experiments of sample A [Heining, 1990]

Acid/metal Elution	1st extr. [%]	2nd extr. [%]	3rd extr. [%]	sum L [%]	sum S [%]
Acetic acid/Pb	54.5	18.9	5.5	65.1	83.8
Citric acid/Pb	63.2	36.2	17.3	80.6	63.0
HCl/Pb	about 100	about 100	about 100	about 100	about 100
Acetic acid/Sb	0.9	0.1	0.2	1.2	2.7
Citric acid/Sb	43.8	43.0	14.0	72.5	85
HCl/Sb	about 100	about 100	about 100	about 100	about 100

The sum L was calculated from the metal concentrations of the eluates whereas the sum S results from the leaching of the solid before and after the extraction. Antimony has been determined divergent to all other experiments by X-ray fluorescence. The elution is defined as follows:

$$L = (C_{m,l} * V_{extr}) / (C_{m,b} * m_b) * 100 \quad [\%]$$

$C_{m,l}$ — metal concentration of the solution [mg/l]
V_{extr} — volume of extract [l]
$C_{m,b}$ — metal concentration of the soil [mg/l]
m_b — mass of the soil [kg]

Table 3 shows a collection of the results of the electrolysis experiments [Heining, 1990]:

Tab. 3: Parameter and results of the electrolysis experiments [Heining, 1990]

Experiment	1	2	3
Stirring velocity	—	—	800 min^{-1}
Time	5 h 45 min	7 h	25 h 30 min
Conc of acid	488 g/l	490 g/l	101 g/l
Electrolyte	acetic acid	acetic acid	citric acid
Initial conc. of Pb	12500 mg/l	11260 mg/l	3220 mg/l
Final conc. of Pb	6700 mg/l	5600 mg/l	930 mg/l
Removal of Pb	46.4 %	50.3 %	71.2 %
Initial conc. of Sb	4.55 mg/l	4.51 mg/l	8000 mg/l
Final conc. of Sb	4.11 mg/l	4.23 mg/l	—
Removal of Sb	9.5 %	6.3 %	—

The pH-value did not show any significant change during the electrolysis experiments.

A survey over the foam fractionation results is given in the figures 2 and 3. Figure 2 shows the enrichment factors as a function of the initial lead concentration and the pH-value [Wömmel, 1992]. The enrichment factor is defined as:

$$F_{an} = C_{m,k} / C_{m,a}$$

$C_{m,k}$ = concentration of head products [mg/l]
$C_{m,a}$ = initial bulk concentration [mg/l]

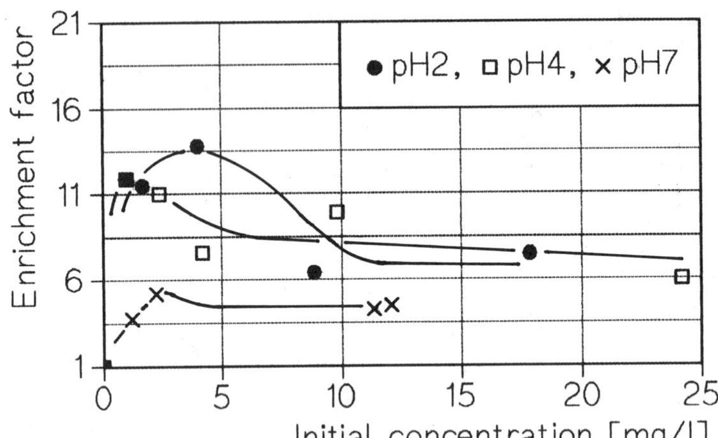

Fig. 2: Enrichment factor as a function of the initial bulk concentration of Pb in HCl

The highest enrichment factors are found if the inintial concentration is between 1 and 5 mg/l. The values show enrichment factors up to 13.8 when the pH-value is 2 and lowest enrichment factors when the pH-value is about 7.
Figure 3 shows the enrichment factor as a function of the initial concentration of lead and organic acid.

The results of cadmium and copper are not presented in the figure but they show a similar course. Arsenic and antimony could not be enriched by foam fractionation neither in acetic acid nor in citric acid.

Fig. 3: Enrichment factor as a function of the initial bulk concentration of Pb in citric acid and acetic acid

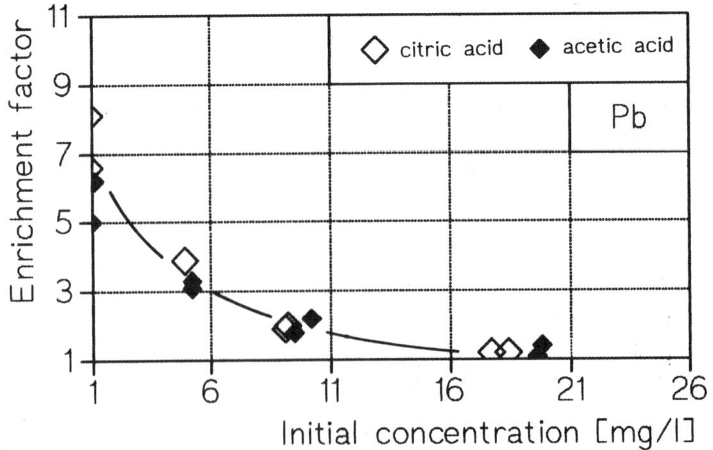

The tests in citric acid show little higher enrichment factors for lead, cadmium and copper (not shown here). As it can be seen in figure 2 the highest enrichment factors correlate with lower initial metal concentrations.

5. DISCUSSION

The soil extraction experiments with leachable fractions of about 70% for lead in acetic acid and about 80% for antimony in citric acid demonstrate the applicability of the method for these metals. With regard to a technical application a process unit consisting of two stages is conceivable. In the first stage the neutralizing capacity of the soil as well as the largest fraction the lead could be degraded and leached with acetic acid. By use of citric acid the residual lead and 80% of the antimony could be leached in a second stage. During the extraction tests very good sedimentation properties of the soil particles could be observed. This is advantageous for the separation of the sludge [Heining, 1990].

A recovery of the dissolved metals by electrolysis has been shown to be realizable. Even with simple plate electrodes without any increase in temperature or stirring of the electrolyte a refining of about 45% of the lead and an electrical efficiency of 777 kWh/t lead could be attained. The electrical efficiency is defined as follows:

$$\eta_{el} = (Q * U) \; / \; (\Delta m_{el} * 3.6)$$

Q   = transferred charge   [C] = [A * s]
U   = voltage between the electrodes   [V]
Δm_{el} = refined amount of lead   [g]

In comparison with the electrical efficiency of 195 kWh/t lead of a lead refining electrolysis given by Melin [1974] the obtained electrical efficiency was very good, taking into account the simplicity of the experimental electrolysis unit.

It could be shown that foam fractionation with the complexing agent Hexamethylenedithiocarbamate and the detergent SDS can be realized in principle with solutions of pH 2. The highest enrichment factors of up to 13.8 had been measured with lead in hydrochloric acid at pH 2 whereas the samples containing Cd, Pb, Cu, As and Sb in citric acid and acetic acid showed maximum enrichment factors of 5 til 6 for copper, lead and cadmium. Arsenic and antimony could neither be enriched nor removed. It should still be clarified whether arsenic and antimony exist as cations under given conditions or whether different, non ionic flotation agents must be applied. The worse flotability of samples with higher initial concentrations had been proved by other authors, too [Carleson et al., 1988]. This phenomenon might be explained if foam fractionation is a surface controlled process. This means that in comparison to the initial concentration of metals in the bulk solution too little sorption surface is available so that the concentrations of the foam and the liquid are of approximately dimensions which would result in an enrichment factor of about 1. The question remains why the flotability of the experiments with hydrochloric acid showed better results than those with organic acid.

1294

6. LITERATURE

Anonymous (1988): Leidraad Bodensanering. Aflevering 4th Nov. 1988, Environmental Protection Agency of the Netherlands, Staatasuitgeverij, s`Gravenhage.

Calmano, W. Hong, J. and Förstner, U. (1992): Einfluß von pH-Wert und Redoxpotential auf die Bindung und Mobilisierung von Schwermetallen in kontaminierten Sedimenten. Vom Wasser. 78, 245-57.

Carleson, T.E. and Moussavi, M. (1988): Chelation and Foam Separation of metal Ions from Solutions. Separation Science and Technologie. 23, (10 & 11), 1093-1104

Eberius, E. (1989): Verfahren zur Dekontamination schlammartiger Sedimente. Eur. Patent: 0 332 985 A1 (20.09.89).

Heining, K. (1990): Reinigung schwermetallhaltiger Böden — Eine Verfahrensstudie am Beispiel Blei. Diplomarbeit. Arbeitsbereich Umweltschutztechnik, Technische Universität Hamburg-Harburg.

v. Lersner, H. (1992): Wir brauchen bundeseinheitliche Grenzwerte. Altlasten. 1, 8-10.

Melin, A. (1974): Blei. from: Ullmanns Enzyklopädie der technischen Chemie. Hg. Bartholome, E.. Verlag Chemie, Weinheim. 4. Aufl., Bd. 8., 542-83.

Wömmel, S. (1992): Untersuchungen zur Abtrennung von Schwermetallen aus Bodenextrakten. Diplomarbeit. Arbeitsbereich Umweltschutztechnik, Technische Universität Hamburg-Harburg.

SEPARATION PROCESSES FOR THE TREATMENT OF CONTAMINATED SOIL

M. Pearl, P. Wood

Warren Spring Laboratory, Stevenage, UK

1. ABSTRACT

A possible means of extending the range of contamination problems treatable by soil by
remediation process based on extraction, destruction and chemical stabilisation is to
separate complex contamination problems into simpler subunits, so enabling their
treatment by existing unit processes. Physical processing technology developed for the
mineral processing industry offers a route to achieve this separation.

This paper reports some interim results for a substantial two year programme investigating
this application of mineral processing being carried out for the Department of the
Environment and coordinated with a similar project being carried out under the United
States Environmental Protection Agency Emerging Technology Programme.

A number of physical processing characterisation tests have been developed for assessing
the potential for separation of particulate contaminants from contaminated soils.
Laboratory testwork so far indicates that various contaminants can be concentrated in
excavated soils using physical processes, in much the same way as mineral ores can be
upgraded. However, these concentration effects may not always be exploitable, for
example, if some contaminants are concentrated and others are not.

2. INTRODUCTION

Ex situ physical processing techniques are being applied to the treatment of
contaminated land with the aim of separating, isolating or concentrating the contaminants,
leaving cleaned non- or less contaminated material for reuse or simple disposal. Such
processes can therefore be considered as one of the treatment techniques used in a multi-
stage, or integrated, process route, where concentrates can go on to further treatment by
chemical, biological or thermal destruction processes or be stabilised. Obviously to be cost
effective the combined cost of all of the individual process steps must be less than a
direct, one step destruction or stabilisation process.

The basis of physical process techniques is that the contaminants occur in, or on,
particles and that these particles can be separated from contaminant-free particles by
exploiting differences in: grain size, settling velocity, particle density (specific gravity),
surface chemical properties (exploited mainly by froth flotation) and other properties such
as magnetic susceptibility.

Each of these processes and the methods by which they can be exploited are
addressed in turn:

F. Arendt, G.J. Annokkée, R. Bosman and W.J. van den Brink (eds.), Contaminated Soil '93, 1295–1304.
© 1993 *Kluwer Academic Publishers. Printed in the Netherlands.*

1 - **Grain Size**: Contaminated soils contain particles which can range in size from very coarse stones to very fine clays. The contaminants themselves do not generally evenly distribute themselves over each grain size. In some cases some of the size bands may be depleted in contaminants to such an extent that they can be considered "clean". In other cases, where there are mixed contaminants, the partitioning by grain size may be such that one type of contaminant concentrates in one size band whilst another type concentrates in a different size band.

2 - **Specific Gravity**: Some site such as those from old mine dumps, metal smelting and fabrication areas may contain contaminants in a dense, heavy phase. Other sites may contain contaminants preferentially adsorbed onto light carbonaceous phases. In both cases concentration of the contaminants can be achieved by separating the higher or lower specific gravity material.

3 - **Surface Chemical Properties**: Froth flotation can be particularly useful for particles that are coated by contaminants as the technique is based on differences in a surface phenomenon. During froth flotation contaminated particles are conditioned such that they become selectively hydrophobic and non-contaminated particles hydrophillic. Adding a frothing agent and passing air through a slurry of the material causes the hydrophobic particles to attach themselves to the air bubbles and concentrate in a discharging surface froth. Hydrophillic particles discharge through a subsurface displacement system. Contaminants may be naturally hydrophobic or can be conditioned so by the addition of very small quantities of chemical reagents.

4 - **Magnetic Properties**: In some contaminated soils it may be possible to remove some of the contaminants by magnetic separation by exploiting differences in the magnetic susceptibility of contaminated and non-contaminated particles. An obvious example could be the removal of heavy metal contaminated magnetite and ferrites from slags.

Physical processing technologies that can exploit differences in these properties have been reviewed in detail previously[1-4] and are commonly used in the mineral processing industry for upgrading and concentrating minerals (or coal) from relatively low grade parent ore. Typical commercially available separation equipment is presented in Table 1[1,2,3].

2.1. The Warren Spring Laboratory Research Programme

Warren Spring Laboratory (WSL) has been investigating the potential for such processes in soil treatment under programmes for the Department of the Environment and the US Environmental Protection Agency.

A phased approach was adopted which is:

Phase 1 - Obtain samples from identified sites and undertake laboratory testwork to ascertain any preferential distribution of contaminants according to differences in physical properties. Soils from some sites were resampled on a larger scale for use in phase 2. Other soils were so called "library" or "database" samples which will be used to demonstrate potential separation using physical differences.

Phase 2 - Design and configure the WSL physical processing pilot plant using equipment most appropriate to the types of separation processes identified in the laboratory tests undertaken in phase 1. Large samples of 20-50 tonne of contaminated soil will then be processed following characterisation in phase 1.

This paper presents examples of recent tests conducted at Phase 1 and carried out on "soils" from several contaminated sites.

TABLE 1. Commercially Available Particle Separation Equipment

Exploitable Feature	Process Equipment
Size	Screens (sieves)
Hydraulic Size	Classifiers
(Settling Velocity)	Hydrosizers
	Hydrocyclones
Specific Gravity	Jigs
	Sluices
	Dense Media Separators
	Spirals
	Shaking Tables
	Tilting Frames
	Vanners
	Duplex Separator
	Multi Gravity Separator
Surface Chemistry	Froth Flotation Systems
Magnetic Susceptibility	Low Intensity Magnetic Drums
	Induced Magnetic Separators
	High Intensity Magnetic Separators

3. MATERIALS AND METHODS

3.1 Contaminated Sites

Only restricted information about the contaminated sites sampled can be given owing to confidentiality arrangements with site owners.

Site 1. Former Chemical Works. This is a large site in the UK contaminated with a wide variety of processes connected with the production of inorganic chemicals. Samples were collected from trial pits from different areas of the site being dug as part of a site survey.

Site 2. Canal Dredgings. This was from dredged material sampled from a site known to have metal contamination. A 5 kg sample was supplied for characterisation.

Site 3. Former Pickling Works. This sample was from a metal cleaning site where acids were used to clean metal surfaces. A 10 kg sample was collected for characterisation.

Site 4. Former Coke Works. As supplied from an equipment manufacturer as typical material of this type. A limited amount of sample was supplied (2 - 3 kg) for assessment.

Site 5. Former Gas Works. Extensively sampled and then composited from a former UK "towngas" production site.

3.2 Laboratory Scale Characterisation Tests

WSL has adapted a number of associated laboratory techniques to characterise the physical potential for separation of contaminants from contaminated materials. These are as follows:

(1)   size separation by wet screening down to 45 microns, with intermediate attrition scrubbing, and hydrocycloning the material finer than 45 microns;

(2)   size and specific gravity separation using "heavy" liquids at several densities on various size bands;

(3)   separation by froth flotation using a variety of reagents and conditions;

(4)   magnetic susceptibility assessment on specific size bands using the Franz Isodynamic Separator over a number of magnetic field strengths.

3.2.1 <u>Size Separation.</u> The method adopted for grain size determination involves screening, or sieving, with water at 10, 2, 0.5, 0.075, and 0.045 mm. Finer screening is difficult and very time consuming. A 1 inch hydrocyclone is therefore used to give the finer cut at between 10 and 15 microns. Intermediate attrition scrubbing for a standard period is applied on deslimed <2 mm material in order to remove surface coatings and to break down "fine" clay balls. Each screened fraction and the two hydrocyclone fractions are then weighed, prepared and analyzed for the contaminants of interest.

3.2.2 <u>Size and Specific Gravity.</u> To ascertain whether there is any preferential partition to specific gravity fractions a series of so called "sink-float" tests are carried using heavy liquids on a number of size fractions. Contaminated material is suspended in these heavy liquids. Particles less than the density of the liquid float whilst those heavier sink. The densities of the liquids can be altered by addition of a diluent. Tests are carried out at a number of densities to ascertain preferential partitioning.

The heavy liquids used for separation are organic liquids such as bromoform (density 2.89 g ml^{-1}), tetrabromoethane (density 2.964 g ml^{-1}) and di-iodomethane (density 3.325 g ml^{-1}). With organic liquids triethlylorthophosphate can be used as a diluent. Inorganic liquids such as zinc chloride solution, or sodium polytungstate (density 3.1 g ml^{-1}) have been used with water as the diluent. The type of liquid used will, to a large extent, depend on the type of contaminant under examination ie some organic compounds will dissolve in the organic liquids and thus an inorganic liquid would need to be used. It is worth noting that heavy liquids are generally quite expensive and washing and recovery systems are commonly used to reconstitute washings to the original liquid density after a test.

3.2.3 <u>Froth Flotation.</u> Laboratory flotation tests are carried out batchwise in various flotation machines that can treat 0.2 - 2 kg batches of material.

During testing, material is slurried for a given period with a variety of reagents, using acid and alkali to control and maintain the pH. Initially the naturally hydrophobic materials are floated by adding only a frothing agent. Passage of air through the slurry then produces a froth which is scraped off into a collecting vessel. After the froth becomes barren the air is switched off and conditions modified such that a quantity of a collector can be added. Air is then passed through the slurry and a further froth concentrate produced. All froth concentrates and the final non-floating material in the flotation cell are separately filtered and each product chemically analyzed to ascertain partitioning. Further tests are carried out at different reagent additions, conditions or reagent types. It is also common practice to ascertain the effects of reducing the quantity of very fine particles in the feed material by desliming using a fine-cutting hydrocyclone. These fine particles (<10 microns) can substantially increase reagent consumption, and interfere with the collector attachment to coarser particles.

3.2.4 <u>Magnetic Separation.</u> The laboratory separator used for the magnetic characterisation is the Franz Isodynamic Magnetic Separator. It consists of an electromagnet with two pole pieces which have a long narrow gap. A vibrating chute fits into the gap. The supports for the magnet allow it to be inclined such that the chute has a forward and sideways tilt. Dry material is fed from a hopper located at the uppermost part of the chute whilst two containers collect the "magnetic" and "non magnetic" products from the lowermost part of the chute. The chute has a plastic removable cover which prevents material sticking directly onto the magnet. The magnetic field is applied such that the "magnetic" particles move down and across the chute to the upward side of the sideways inclination. The "non-magnetic" particles gravitate down the chute to the downward side of the sideways inclination. Tests are carried out over a number of field strengths which are varied by altering the DC current around the magnets. Thus all material is run through the apparatus at low current and the products collected. The non-magnetic product is then repassed at a higher current and so on. In this way a number of products are obtained. It is worth noting that the first run is carried out with no current ie residual magnetism. The products from this first run are the ferromagnetics. Particles separating at higher currents are either small amounts of ferromagnetics in a larger non-magnetic particle, or paramagnetic particles.

Tests are carried out for a number of particle sizes. At sizes less than 45 microns separation becomes very difficult as the particles tend to aggregate. An adaptor can be fitted between the air gap such that separation can be carried out in a "wet" mode - the particles thus being in a dilute suspension[4].

4. RESULTS AND DISCUSSION

4.1 Size Separation

Laboratory tests have shown that for some sites there is enrichment of the contaminants to particular size fractions and consequently depletion in the remainder.

For example, in a sample of inorganically contaminated canal dredgings over 50% of the content of cadmium, copper, lead, nickel and zinc were contained in the <0.01 mm grain size fraction, which accounted for 31% of the sample by weight (see Table 2). The removal of this ultrafine material would reduce the heavy metal contents of the remaining material from 124 mg.kg^{-1} nickel to 86 mg.kg^{-1}, 333 mg.kg^{-1} copper to 233 mg.kg^{-1}, 832 mg.kg^{-1} zinc to 511 mg.kg^{-1}, 502 mg.kg^{-1} lead to 318 mg.kg^{-1}, and 10 mg.kg^{-1} cadmium to 7 mg.kg^{-1}.

On some sites size separation may produce inconsistent results between different areas of the site. These variations may reflect the variety of processes that formerly took place on the site and/or a variety of "chemical wastes" and "spills" that might have taken place during former use. Such sites are a complex number of individual problems and homogenising material over the whole may further complicate the ability to separate different types of concentrate.

For example, soil samples from a former inorganic chemical works taken from an area contaminated with solid spent catalysts contained 99% of their lead in the 2-10 mm fraction (see Table 3). However, arsenic, copper and zinc were relatively depleted in this fraction and were enriched in the fractions finer than 2 mm. Other areas of the site however showed different contaminant distributions with size. For example, 90% of the nickel in samples from a second area was concentrated in the 2-10 mm fraction (at 10,000 mg.kg^{-1}) along with 76% of the arsenic (at 247 mg.kg^{-1}). In samples from a third area of the same site arsenic was the major contaminant and widely was distributed over all grain sizes with slight enrichment in the finest (<0.01mm) fraction, which contained 31% of the arsenic and accounted for 17% of the mass of the sample.

4.2 Separation by Specific Gravity

Laboratory tests carried out on samples from a former metal works site contaminated with copper, nickel and zinc indicated that a combination of size and specific gravity separation could be used to concentrate these metal contaminants (see Table 4). The heavy fraction (>2.8 g.ml^{-1}) of the size range 0.01-2.00 mm accounted for only 6.3% of the sample mass yet contained 22.4% of the copper, 18.4 % of the zinc and 20.7% of the nickel. The light fraction (<2.0 g.ml^{-1}) which contains carbonaceous matter, is 5.8% of the sample mass and contains 13.9% of the copper, 9.0% of the zinc and 7.0% of the nickel. In addition the fraction <0.01 mm accounted for 10.9% by mass and 25.1% of the copper, 27.3% of the zinc and 16.8% of the nickel. The removal of these fractions would leave a cleaner residue which consists of 77% of the original sample by weight. Compared to the original material this residue would show reductions of copper from 333 to 167 mg.kg^{-1}, of zinc from 398 to 234 mg.kg^{-1} and of nickel from 70 to 50 mg.kg^{-1}.

4.3 Separation by Froth Flotation

As an example laboratory flotation testwork was undertaken on soils from a former coke works contaminated with heavy metals, mineral oils and polynuclear aromatic hydrocarbons (PAHs) - fluoranthene, benzo(k)fluoranthene, benzo(a)pyrene, indeno(1,2,3,cd)perylene and benzo(ghi)perylene. It was found that 96% of the mineral oils and 80% of the PAHs could be removed in a froth concentrate which constituted 32% by weight of the original material (see Table 5). The recovery of heavy metals reporting to the froth concentrate varied from 50 to 70%. The non-floating product, which effectively constitutes the cleaned product, had a final value for mineral oils of 0.14% from an original 2.31%, and for PAHs of 47 mg.kg^{-1} from 162 mg.kg^{-1}. The most significant metal value reductions were zinc from 907 mg.kg^{-1} to 473 mg.kg^{-1}, copper from 256 to 129 mg.kg^{-1} and lead from 388 mg.kg^{-1} to 168 mg.kg^{-1}.

4.4 Separation Using Differences in Magnetic Susceptibility

Results on assessing the potential for heavy metal reduction from a former gasworks site have shown that lead, copper, zinc and nickel are preferentially concentrated to magnetic products (Table 6). The non-magnetic product, which represented almost 70% by weight of the original material has approximately 60% less zinc (76 mg.kg^{-1} from 180 mg.kg^{-1}) and nickel (11 mg.kg^{-1} from 29 mg.kg^{-1}) and, nearly 40% less lead (350 mg.kg^{-1} from 545 mg.kg^{-1}) and copper (66 mg.kg^{-1} from 108 mg.kg^{-1}) than the original material.

5. CONCLUSIONS

Mineral processing equipment has potential for the treatment of contaminated soils in order to separate grossly contaminated from less or uncontaminated particles. Suitable equipment with a proven track record in the mineral industry is commercially available. Not all sites will be amenable to physical fractionation either because (i) the contaminants do not preferentially partition to a high enough degree to specific fractions such that "cleaned" or "simple" fractions are produced or (ii) the site may have had such a mixed history and mixture of contaminants on it that treatment of localised hot-spots is uneconomic.

This paper has described four laboratory techniques which can be used to assess the potential of physical separation techniques for the treatment of contaminated land. The aim of physical separation is to produce either "clean" fractions that require no further treatment or, a number of concentrates of different contaminants so that further treatment

TABLE 2. Distribution of Inorganic Contaminants after Size Fractionation in Canal Dredgings.

Grain size range (mm)	Weight %	Contaminant									
		Ni		Cu		Zn		Pb		Cd	
		Conc mg.kg^{-1}	Dist^n %	Conc mg.kg^{-1}	Dist^n %	Conc mg.kg^{-1}	Dist^n %	Conc mg.kg^{-1}	Dist^n %	Conc mg.kg^{-1}	Dist^n %
3.0-10.0	13.6	40	4.4	69	2.8	230	3.8	74	2.0	2	3.4
0.5-3.0	10.3	130	10.8	380	11.8	870	10.8	580	11.9	14	14.7
0.075-0.5	29.3	67	15.8	216	19.0	417	14.7	295	17.2	6	19.0
0.01-0.075	16.2	134	17.4	311	15.0	688	13.4	398	12.8	6	10.3
<0.01	30.6	210	51.6	560	51.4	1560	57.3	920	56.1	17	52.6
Original soil	100.0	124	100.0	333	100.0	832	100.0	502	100.0	10	100.0

TABLE 3. Contaminant Distribution by Grain Size for a Sample of "Soil" from a Former Chemical Works.

Size mm	Weight %	As		Pb		Cu		Zn	
		Assay mg.kg^{-1}	Distrb. %	Assay mg.kg^{-1}	Distrb %	Assay mg.kg^{-1}	Distrb. %	Assay mg.kg^{-1}	Distrb. %
2.0-10.0	48.9	3	2.9	68950	99.7	208	16.3	48	0.5
0.5-2.0	12.9	62	15.9	256	0.1	1122	23.2	18560	59.3
0.075-0.5	11.5	96	21.7	208	0.1	1302	24.0	6550	18.6
0.045-0.075	4.7	88	8.2	128	<0.1	822	6.3	2885	3.4
0.010-0.045	7.8	116	17.9	117	<0.1	408	5.1	2313	4.8
<0.010	14.1	120	33.4	180	0.1	1110	25.1	3930	13.7

TABLE 4. Summary of Heavy Metal Distribution by Size and Specific Gravity In Contaminated Material from a Former Metal Pickling Works

Size-Specific Gravity Fraction	Weight %	Cu		Zn		Ni	
		Assay mg.kg^{-1}	Distrb. %	Assay mg.kg^{-1}	Distrb. %	Assay mg.kg^{-1}	Distrb. %
2.0-10.0 mm	52.06	167	26.37	234	23.80	50	39.78
0.010-2.0 mm <2.8 g ml^{-1}	24.99		12.03		21.50		15.66
0.010-2.0 mm <2.0 g ml^{-1}	5.75	805	13.90	620	8.96	86	7.04
>2.8 g ml^{-1}	6.30	1186	22.43	1164	18.43	230	20.72
<0.010 mm	10.90	766	25.06	996	27.31	108	16.80

Original soil concentrations were: 333 mg.kg^{-1} Cu, 398 mg.kg^{-1} Zn and 70 mg.kg^{-1} Ni

TABLE 5. Froth Flotation Results for Sample from a Former Coke Works

	Wt %	Cu		Pb		Zn		Mineral Oils		PAH	
		Assay $mg.kg^{-1}$	$Dist^n$ %	Assay $mg.kg^{-1}$	$Dist^n$ %	Assay $mg.kg^{-1}$	$Dist^n$ %	Assay %	$Dist^n$ %	Assay $mg.kg^{-1}$	$Dist^n$ %
Conc 1	28.2	514	56.6	841	61.2	1832	56.9	7.5	91.4	419.2	73.1
Conc 2	3.7	627	9.1	975	9.3	1853	7.6	2.8	4.5	313.4	7.2
Non-Floats	68.1	129	34.3	168	29.5	473	35.5	0.14	4.1	46.7	19.7
	100.0	256	100.0	388	100.0	907	100.0	2.31	100.0	161.6	100.0

TABLE 6. Summary Distributions to Illustrate Laboratory Magnetic Separation. The Material used was 0.01-0.5 mm Fraction of a Sample from a Former Chemical Plant

Product	Wt. %	Fe		Pb		Cu		Zn		Ni	
		Assay %	$Dist^n$ %	Assay $mg.kg^{-1}$	$Dist^n$ %	Assay $mg.kg^{-1}$	$Dist^n$ %	Assay $mg.kg^{-1}$	$Dist^n$ %	Assay $mg.kg^{-1}$	$Dist^n$ %
Ferromags.	2.3	20.4	15.4	948	4.0	68	11.4	603	7.7	229	18.2
0-2.0 amps	28.5	7.6	71.5	987	51.6	174	46.0	397	62.9	57	55.7
Non-mags	69.2	0.6	13.1	350	44.4	66	42.6	76	29.4	11	26.1
TOTAL	100.0	3.0	100.0	545	100.0	108	100.0	180	100.0	29	100.0

is relatively simple. The laboratory tests have shown that for some contaminated materials there is a potential for producing these "clean" or "simple" products. However, not all materials respond in the same manner and it is for this reason that laboratory characterisation tests are an important precursor to determine whether soils can be treated in this way.

Having demonstrated the potential for successful separation of a particular soil the next stage in determining a clean-up route would be to undertake pilot scale tests. The choice of separation equipment and the circuit configuration for any pilot scale test would be determined from the laboratory characterisation results. The next phase of the current investigation is to undertake pilot scale operations at up to 2 t hr^{-1} on soils from a number of contaminated sites.

6. ACKNOWLEDGEMENTS

This paper is based on work carried out for the U.K. Department of the Environment and the U.S. Environmental Protection Agency. The views expressed are those of the author and do not necessarily reflect those of the sponsors.

7. REFERENCES

1. Wills B A (1979), Mineral Processing Technology. International Series on Materials Sciences and Technology, Vol 29. Pergamon International Library. 418p

2. Weiss N L (1985), Mineral Processing Handbook, Society of Mining Engineers, American Institute of Mining, Metallurgical and Petroleum Engineers Inc., USA. 2019p

3. Armishaw R, Bardos R P, Dunn R M, Hill J M, Pearl M, Rampling T, and Wood P (1992), Review of Innovative Contaminated Soil Clean-up Processes. LR 819 (MR), Warren Spring Laboratory, Stevenage, UK. ISBN 085624 7294

4. Greene, G M and Cornitius, L E; A technique for magnetically separating minerals in a liquid mode. Journal of Sedimentary Petrology, 1971, v. 41, p.310-312

CLEANING OF MERCURY CONTAMINATED SOIL BY USING A COMBINED
WASHING AND DISTILLATION PROCESS

Mr. Dr. Roland Hennig

Harbauer GmbH & Co. KG, Flughafenstr. 21, D-1000 Berlin 44, Tel.: +(030) 61 37 30-0

1.   Introduction

Various kinds of large-scale technical equipment for cleaning contaminated soil now exist
and these have proved themselves in practice within their respective application limits.
Further development of such processes is aimed at expanding the range of applications,
improving efficiency and increasing the acceptance of such equipment in the political
environment.

A soil distillation process that can be used for a wide range of pollutants and soils, either as a
discrete unit or in combination with a washing plant, has now been developed by Harbauer,
an environmental technologycompany. Fine-grain and highly contaminated soils, which so
far have been particularly problematic, can be treated with the help of this process.

A large-scale washing and distillation plant is currently being built by Harbauer for the
cleaning of mercury-contaminated soils. This equipment will enter service in June 1993. The
washing technology deployed was developed through years of experience that Harbauer
acquired with their washing plants in Berlin and Vienna. **Figure 1** conveys an impression of
the Harbauer soil cleaning plant in Vienna. Using this equipment, more than 50,000 tons of
soil contaminated with cyanide have been cleaned.

Figure 1: Harbauer soil washing plant in Vienna ready for work

F. Arendt, G.J. Annokkée, R. Bosman and W.J. van den Brink (eds.), Contaminated Soil '93, 1305–1314.
© 1993 *Kluwer Academic Publishers. Printed in the Netherlands.*

The old stationary soil decontamination plant in Berlin is being dismantled at present. An ultra-modern soil cleaning centre will be erected on the same property, with commencement of operations scheduled for May 1993. **Figure 2** shows a side-view of the new soil cleaning plant in Berlin.

Figure 2: View of the stationary soil washing plant in Berlin, starting March 1993

Extensive laboratory and pilot tests were carried out for the cleaning of mercury-contaminated soil using a vacuum distillation technique. The results and the large-scale technical implementation of this process are described below.

2.   Basic principles

The development of vacuum distillation technology for cleaning contaminated soil arose from the necessity to clean highly contaminated fine-grain fractions. Such fractions are either found in soil or are obtained from soil washing as so-called fine-grain residues.

Biological and thermal processes already exist for the treatment of such fractions. Biological processes can be applied to only a limited extent on account of the low degradability of organic residues that are obtained. To this is added the bad transportation of substances in wet fractions of fine-grain soils, which excludes any meaningful application of biological processes for highly contaminated fine-grain fractions in the vast majority of cases.

Metal compounds cannot be removed from contaminated soils using conventional thermal processes. Thermal processes (incineration) are well suited to the cleaning of soil contaminated by purely organic pollutants. However, thermal treatment of very fine-grain particles involves costly and complex technological processes.

The substantial volumes of flue gases produced by conventional thermal processes for soil decontamination are a significant drawback. This leads to high investment costs for sophisticated waste gas purification plant and to considerable delays in official clearance procedures.

Soil cleaning using vacuum distillation excludes the drawbacks mentioned above. The range of applications for this process is limited, however, to pollutant compounds with defined boiling points. These include a large number of organic compounds, among them crude oil fractions, halogenated hydrocarbons, PAHs, cyanides and metal compounds that can vapourize. **Table 1** shows a list of pollutants that can be successfully removed from soils using the vacuum distillation process.

Element / Group of contaminants	Boiling point [°C] at 1013 hPa
PAH (EPA)	218 - 505
TPH	25 - 500
HALOGEN. VOL. ORGANICS	25 - 150
BTEX	80 - 140
PHENOLE (EPA)	182 - 230
CN	decompose at > 450 °C
Hg	356

Table 1: Survey of contaminants with their boiling points suitable for vacuum distillation

The vacuum distillation process involves heating the soil under a vacuum to a pre-defined temperature. The pollutants evaporate and are then transported to a condenser using inert gas. In the condenser, the contaminants are condensed and then extracted. The condensates are either treated for possible re-use or are sent to a sophisticated stationary incinerator for disposal. Non-combustible pollutants must be disposed of properly. Surplus inert gas and air mixture is either fed back into the cycle or is released into the atmosphere via a small-scale waste gas purification plant.

The optimal temperature is determined in preparatory experiments. The desired temperature should be as low as possible in order to prevent the pollutant compounds from decomposing. The boiling points of the various pollutant compounds can be influenced by varying the degree of vacuum, as shown in **Figure 3**.

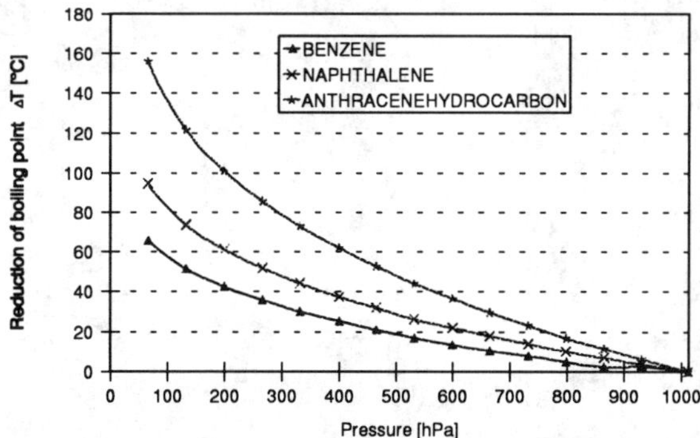

Figure 3: Boiling point reduction of organic compound versus pressure

The principal benefits of the vacuum distillation process are as follows:

- low volumes of process gas

- low energy costs

- oxygen concentration of process is low

- leaks in the plant do not lead to emission of process gases

- inert carrier gases increase the effectiveness of the process and prevent undesired chemical reactions

3.   Description of the combined washing and distillation process

The process described below is suitable for virtual decontamination of mercury-polluted soils. A large-scale technical installation is currently being constructed and will commence operation in mid-1993.

The cleaning process involves two separate stages: the washing stage and the distillation stage.

The soil is pre-treated in a crushing and sorting plant. When the soil is washed, mercury compounds are transported into soil fractions that are then capable of being separated. The soil fractions enriched with mercury are then fed into the distillation plant and cleaned. The condensed mercury is either re-used or sent to a special waste storage unit.

3.1   Washing stage

Cleaning the soil using a wet mechanical process essentially involves a six-stage grading and sorting process, the most important material flows of which are shown in simplified form in **Figure 4** below.

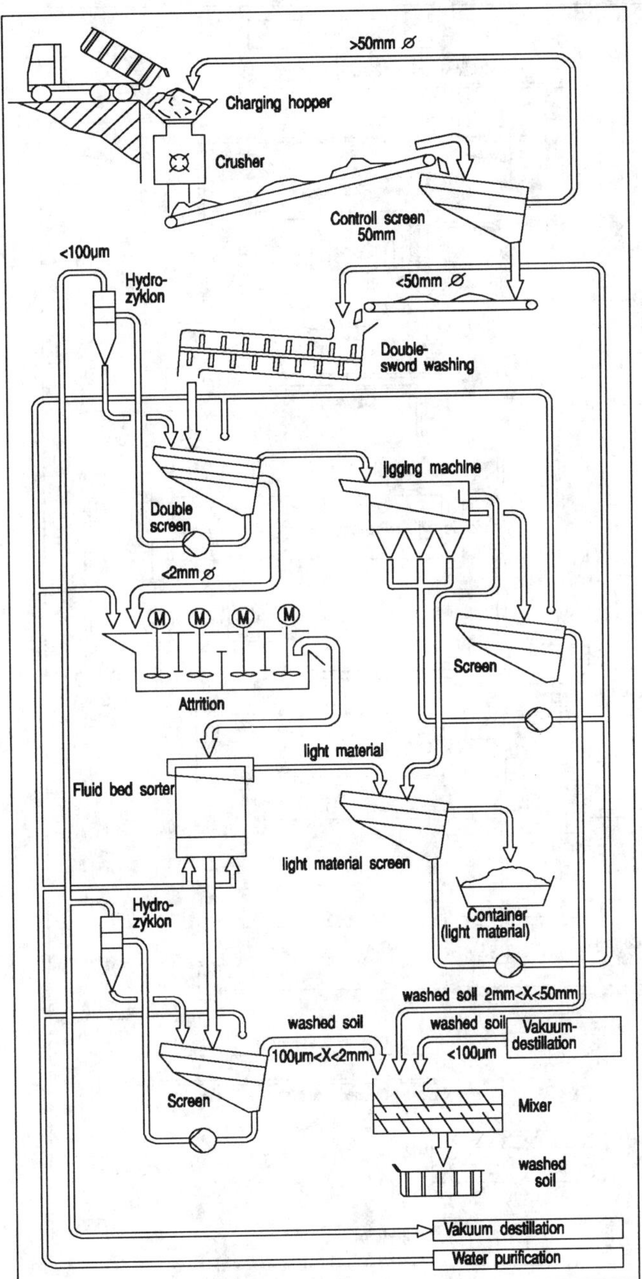

Figure 4: Flow chart for soil washing process

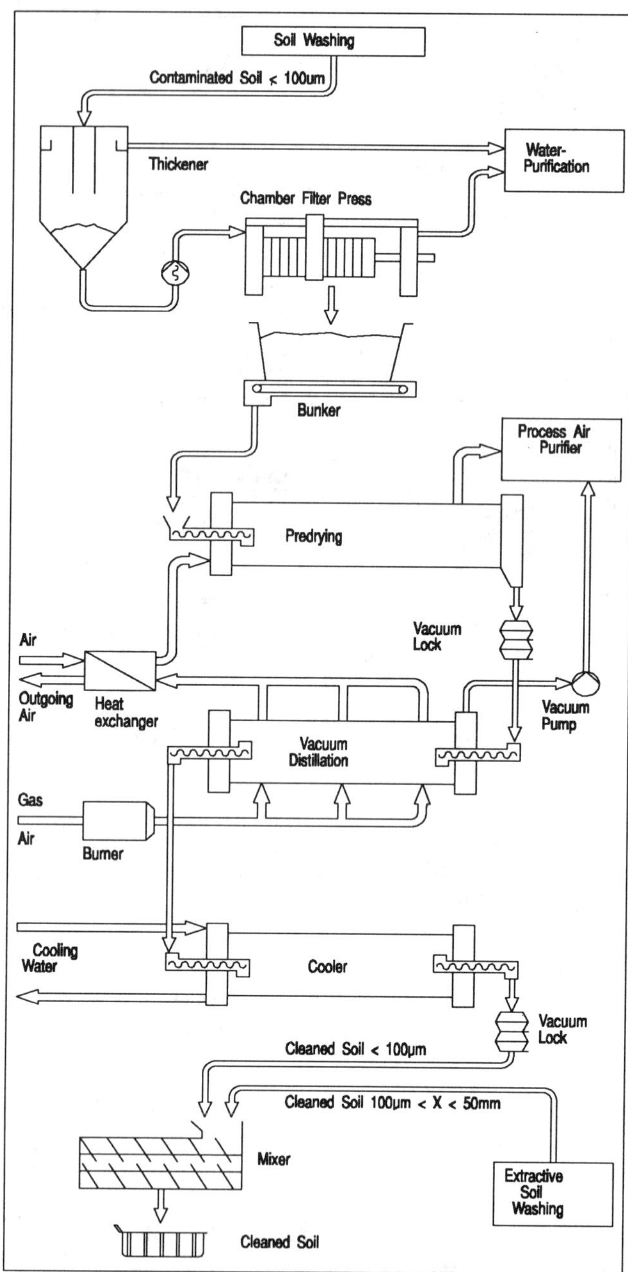

Figure 5: Flow Chart Vacuum Distillation Plant

3.2   *Vacuum distillation*

Figure 5 shows the flow sheet for the vacuum distillation plant.

The fine-grain suspension (< 100 μm) extracted from the soil washing stage is then pre-thickened with the help of flocculation agents in a sedimentation stage consisting of six thickener units, to form a compressible sludge. The pre-thickened sludge from the sedimentation plant is dehydrated in a chamber filter press in order to minimize the amount of energy required in the pre-drying phase.

After pre-drying, the soil to be decontaminated is conveyed through vacuum chambers into the vacuum distillation plant, where the soil is treated at a pressure of 100 hPa and a mean temperature of 310°C. The exhaust vapour is condensed, the recovered mercury is reutilized and the aqueous condensate is treated in the ancillary water purification plant. All process air flows are amalgamated and fed through a sophisticated process air purification plant. After the fine-grain fraction has been cleaned, the resultant product is mixed with the material obtained from the soil cleaning stage, so that the soil is available for reutilization in its original composition.

3.3   *Large-scale technical implementation*

A large plant is currently being constructed in Bavaria and is due to commence operation in June 1993. **Figure 6** provides an impression of this plant.

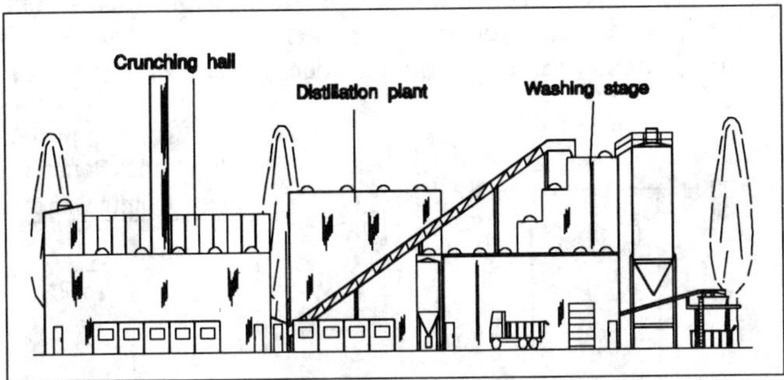

Figure 6: View of the washing and distillation plant

4.   Results

4.1   Results from decontamination of mercury-polluted soil

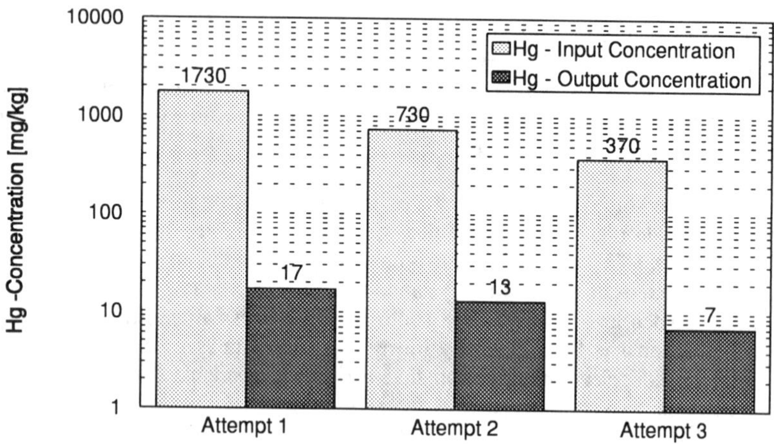

Figure 7: Soil washing results for mercurial soils
Grainsize >100 µm

Figure 7 shows the results obtained from cleaning contaminated soil by means of the washing process at grain sizes < 100 µm. The fine-grain residue highly contaminated with mercury that is produced by the process is then fed into a distillation plant. The results obtained through distillation of these fine-grain residues are shown in **Figure 8**.

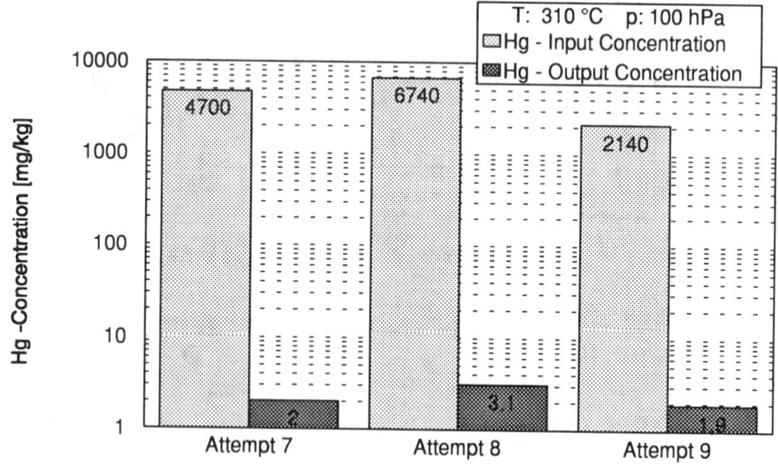

Figure 8: Distillation Result for Hg-loaded Fines Residues

Treatment of the fine-grain residues in the mercury distillation plant can effect with certainty a reduction in pollutant concentration from a mean input value of 3,000 mg/kg to an output value under 2 mg/kg.

4.2 Laboratory trials for treatment of organic pollutants

The suitability of the process for the removal of organic compounds was also tested in various experiments. **Figure 9** shows a sample of the results obtained in the treatment of gas works soil that was highly contaminated with PAH.

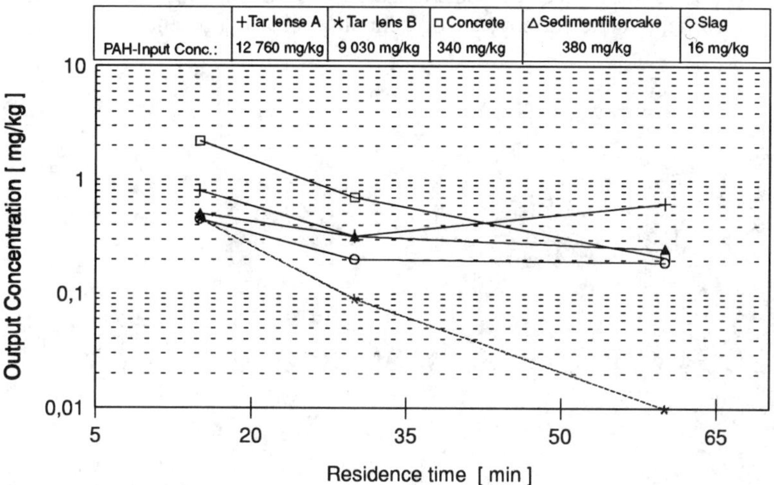

Figure 9: Concentration of PAH's versus decontamination period for a gas soil

The cleaning effect is dependent on temperature and pressure, but also on the decontamination period. In the case presented here, a 99% degree of purification could be obtained at a decontamination period of 25 minutes. Analyses of the cleaned material showed that contamination levels were below the detection limit.

5. Summary and prospects

Following years of experience acquired with large-scale technical applications for cleaning contaminated soils, Harbauer has now developed a combined soil washing and distillation plant. A full-scale version of such a system is currently being implemented. This technology can be applied to a wide range of different substances, including numerous organic, inorganic and metallic compounds. Particularly worthy of note is the outstanding suitability of the process for fine-grain, highly-contaminated soils.

Extracts from the results that can be expected with the plant in the treatment of mercury-contaminated soils are presented. These show that such soils can be cleaned with certainty to levels below the statutory limits.

Soil distillation is an economic solution for the cleaning of fine-grain, highly-contaminated soils or of fine-grain residues obtained from soil cleaning plants. The significant benefit of this process is the low volume of process gas required. This, in turn, means low investment costs and relatively quick official approval.

Contaminants are extracted by this process in the lowest possible volumes and in highly concentrated form, and can either be treated for reutilisation or must be disposed of in the correct manner. Fixed incineration plants that have a high level of technical sophistication and the full range of waste gas purification equipment are suitable for disposing of organic pollutants. This guarantees that the contaminants removed from the soil are destroyed in plants that correspond to the current state of technological development.

Soil remediation using microemulsions - concept and first results

W.D. Clemens, F.H. Haegel, M.J. Schwuger, K. Stickdorn, G. Subklew, L. Webb[*], Institute
of Applied Physical Chemistry and Institute of Biotechnology 3[*], Research Centre Jülich,
P.O. Box 1913, D-W-5170 Jülich

Introduction

Soil washing processes have been sucessfully applied to remove pollutants from
soil on a technical scale [1], [2]. However, their efficiency is often limited, if soils with a
high content of clay and/or fine particles must be cleaned. For this reason alternative
methods of soil extraction with microemulsions for the remediation of these highly
contaminated soil fractions have been considered. Due to their great wetting and
solubilization property, microemulsions seem to be very suitable for the removal of harmful
organic compounds from this soil fraction [3], [4].

It is inevitable that after the washing process residues of the microemulsion remain
in the soil. To avoid long-term toxic consenqueces, the microemulsions should be
composed of biodegradable compounds (surfactant and oil in water).

In the following, the fundamantal behaviour of microemulsions is described,
followed by the first results of extraction experiments with synthetic and real soil samples
contaminated with PAHs. After describing the examination of biodegardability of vegetable
oils and surfactants, a conception for soil remediation with microemulsion composed of
biodegradable components is presented.

Properties of microemulsions

Microemulsions are thermodynamically stable, macroscopically homogeneous and
optically transparent mixtures of water, oil and surfactant [5]. Mostly they contain also a
short-chain aliphatic alcohol and an electrolyte. Compared with other solvents
microemulsions exhibit some special properties, which make them appear suitable for
application in soil remediation.

Based on their composition they have good solubilization properties for polar or
ionic molecules as well as for nonpolar organic substances. The solubilization capacity of

F. Arendt, G.J. Annokkée, R. Bosman and W.J. van den Brink (eds.), Contaminated Soil '93, 1315–1323.
© 1993 *Kluwer Academic Publishers. Printed in the Netherlands.*

such solutions is higher than the that of aqueous micellar systems , i.e. systems containing only the surfactant.

In contrast to emulsions, microemulsions are thermodynamically stable and can therefore be handled easily and be prepared by gentle shaking. The phase behaviour of such ternary mixtures at constant temperature is described in a Gibbs triangle. The three sides of the equilateral triangle are formed by the molar ratios of the compounds. Such a phase triangle is presented schematically in figure 1. Depending on the composition of the mixtures, one phase, two phase and three phase regions are formed. In a three phase region an oil rich upper phase, a surfactant rich middle phase and a water rich lower phase are observed.

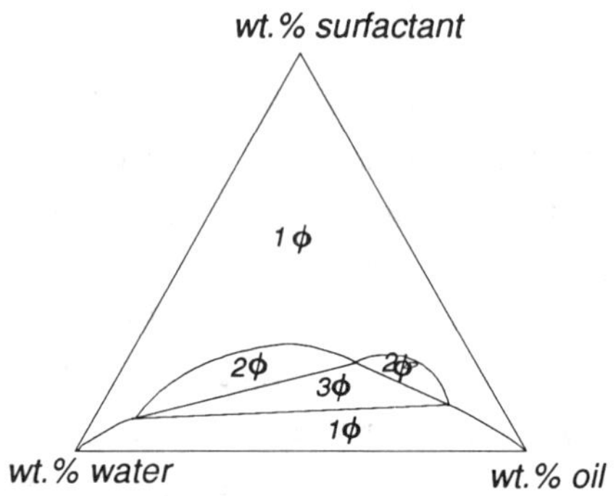

Figure 1: schematic phase diagram of a ternary system at constant temperature

Even though microemulsions are macroscopically transparent, they exhibit a submicroscopical structure. Water-rich and oil rich domains are formed which are separated by a surfactant monolayer [6]. Depending on the composition of the mixture different structures are formed: with increasing oil content first oil droplets in an aqueous matrix are formed, then a bicontinuous structure and at high oil contents water droplets in the oil-continuous phase are formed. The droplet size ranges between 5 and 500 nm. These structures cause a very large internal interface which gives rise to an increase in extraction rates of a factor of 10 to 100 compared to a conventional stirred two-phase system [7].

Another important property for soil remediation is the ultralow interfacial and surface tension which can be achieved in microemulsions [8]. The interfacial tension can be two to three orders of magnitude lower than that of pure water and oil. This enables a good wetting and therefore a good mobilization of organic substances entrapped in soils or porous rocks.

Based on these above mentioned properties microemulsions are already used to a small extent for tertiary oil recovery [9]. By means of the enhanced recovery, approximately 20% more of the content of an oil reservoir can be expoited.

Starting with this concept we plan to realize soil remediation with microemulsions. The contaminats, e.g. polycyclic aromatic hydrocarbons (PAHs), are solubilized in the organic compartments of the microemulsions and washed out of the soil. This is an important difference to other remediation concepts which are based on mechanical attrition and flushing, since the pollutant is dissolved in the microemulsion. Therefore, this technique is especially applicable to soils containing high amounts of clay. This fraction, which is highly contaminated is removed by the conventional washing process and can usually only be separately deposited as a toxic waste. It could, however, be cleaned using the microemulsion concept. Furthermore the microbiological degradation of the contaminants is more effective, since they are dissolved in the organic phase and are therefore in later steps more available for biodegradation. Another advantage of the microemulsion concept is the simple enrichment of the pollutant in the washing liquid. The microemulsion can be separated specifically due to the thermodynamic properties by changing the temperature or pressure. The separation can be carried out in such way that a water-rich phase containing the surfactant and an oil-rich phase containing the contaminant are formed. This yields on the one hand the above mentioned enrichment of the pollutant in the organic phase and on the other hand it is possible to recycle the surfactant with the aqueous phase.

Microemulsions which will be used for soil remediation are composed of biodegradable and toxicologically harmless components, because their introduction into the soil shall not pose additional problems. Especially mineral oils cause contamination of soils with PAHs, therefore we intend to work with vegetable oils like rape-oil as nonpolar component. The surfactants have also to be biodegradable since they are able to change the soil structure and therefore cause changes in the transport properties of the soil. This effect is used during the washing process, but a permanent and therefore uncontrollable contamination has to be avoided.

Extraction experiments with PAH-contaminated soil

Experiments were started with by basic investigations on solubilisation and adsorption properties of polycyclic aromatic hydrocarbons in mixtures of water, oils, surfactants and clay minerals in order to get information about their transport behaviour in soil. Therefore adsorption isotherms of pyrene onto sodium montmorillonite from the following solutions ware determined:

1 water

2 32 µM $C_{12}E_4$ (0.8 times critical micellar concentration)

3 320 µM $C_{12}E_4$ (8 times critical micellar concentration)

4 320 µM $C_{12}E_4$ + 32 µM isooctane

Measurements were performed with a new fluorescence spectroscopical technique [10].

Figure 2 shows the influence of the nonionic surfactant tetraethyleneglycol mono-dodecylether ($C_{12}E_4$) and $C_{12}E_4$/isooctane on the adsorption of pyrene. As the adsorbed amount is very small in any case, all data were fitted by linear isotherms. Particularly for small concentrations deviations may be considerable due to the complicated measzrement method. With increasing surfactant concentration a decrease of adsorption is found which becomes even more profound by addition of oil. With increasing volume fraction of organic material, solubilisation overcomes adsorption and the phase equilibrium is shifted towards the liquid phase.

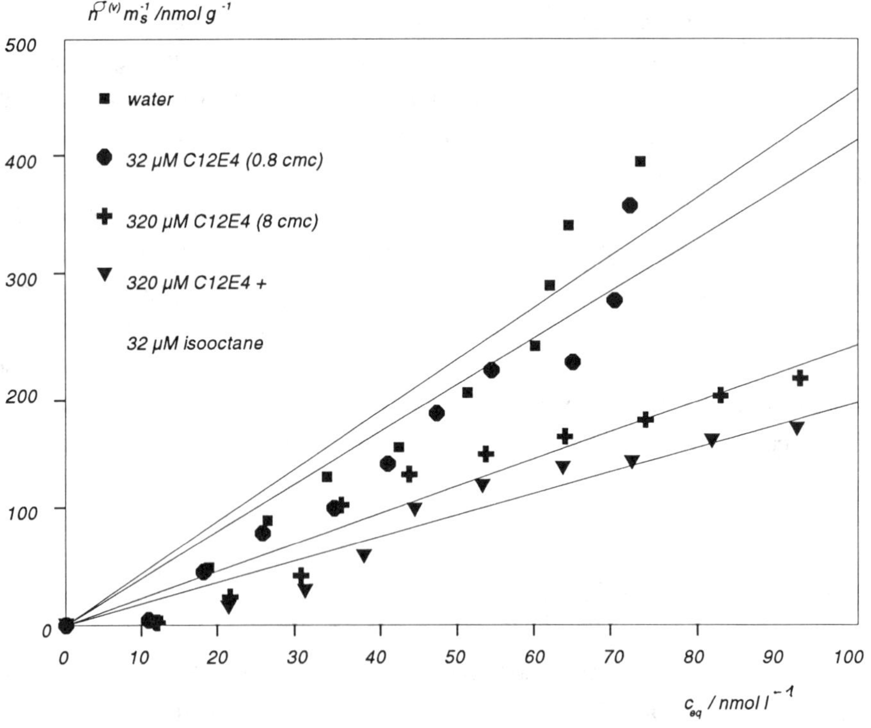

Figure 2: adsorption isotherms of pyrene on Na-montmorillonite with different surfactants

The decrease of adsorbed pyrene after changing from an aqueous solution to a 0.0012 % solution of $C_{12}E_4$ (320µM) is about 50%. Addition of $3.6 \cdot 10^{-4}$ weight percent (w/w) of isooctane results in a further decrease of adsorption of about 17 % with respect to the surfactant solution. This decrease is overproportional and can certainly be clearly enhanced by choice and optimisation of suitable systems.

The adsorption equilibrium however does not represent the only criterium for the application in extraction of PAH-contaminated soil. It must be tested in a real system. Therefore PAH-contaminated material from the outlet of a hydrocyclone in a soil extraction plant was extracted with water and the above mentioned $C_{12}E_4$-isooctane solution. 100 mg of soil were shaken with 60 ml of the extraction medium at $25\,^{\circ}C$ for 48 hours. After centrifugation at 20,000 g the concentration of extracted pyrene could be estimated by second derivative UV spectroscopy (figure 3). The contamination with pyrene was given to be 340 mg per kg soil by the operator of the soil extraction plant. The amount extracted was in the same order of magnitude. Compared to water the oil containing system shows a three to four times higher extraction capacity.

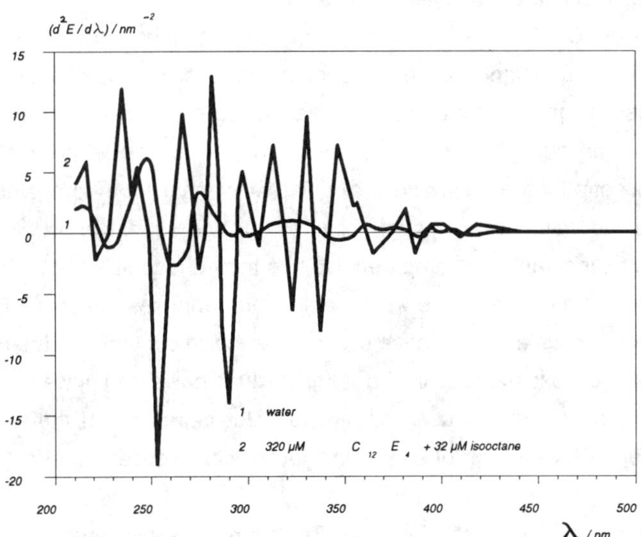

Figure 3: extraction of PAH-contaminated material, UV-spectra (second derivative) of extraction solution.

On the basis of these results, a concept for soil remediation with microemulsions was determined. As the substances used must be biologically biodregadable, alkylpolyglucosides (APG) (Hüls AG, Marl) were used as surfactants and rape oil or rape methyl ether (GET GmbH., Aldenhoven) as oil component. Up to now only little is known about microemulsions with plant oils. For this reason basic investigations on the phase behaviour of surfactant water oil mixtures must be carried out first.

The binary mixtures of water and surfactant were investigated in the temperature range from 20 to 35°C for two mixtures of alkylpolyglucosides with chain lengths of C_{10} - C_{12} and C_{12} - C_{16} respectively. The phase boundaries of both surfactants are only slightly dependent on temperature [11]. For $C_{12}C_{16}$-APG the cloud point is already exceeded at 20°C and a two-phase-region is found for surfactant contents of about 1% (weight per weight) which is extended up to 22%. Solutions in this range split into a surfactant poor phase (1 %) and a surfactant rich phase (22 %). In contrast to other nonionic surfactants like alkyl polyglycol ethers the surfactant rich phase separates belowthe aqueous phase due to its high density.

In ternary systems of $C_{12}C_{16}$-APG, rape oil and water three-phase- regions were found, the water phase being situated in the middle and separating the surfactant rich and the oil rich phase. This behaviour may have consequences for separation and recycling of surfactant. The two phase region of the binary mixtures is followed by a single phase. Above 40 % liquid crystalline phases are present.

The mixture $C_{10}C_{12}$-APG behaves completely differently. No demixing is found up to 35°C and 70 % surfactant. Above this content, liquid crystalline phases are observed. In ternary phase diagrams no three-phase regions were observed. As the solubilisation capacities for oil in surfactant solutions are best near the clouding point, the highest surfactant concentration possible and a clouding point near room temperature would be ideal. In order to fulfil this criterium mixtures of both surfactants with 11 % total concentration of surfactant were prepared. Up to a mixing ratio of 70 % $C_{12}C_{16}$-APG and 30 % $C_{10}C_{12}$-APG demixing in the whole temperature range is observed. From a mixing ratio of 60/40 and above a single phase occurs over the whole temperature range.

Solubilisation experiments in the latter mixture however showed a solubilisation capacity of only about one percent for rape oil. But it was found that the viscosity is drastically reduced by addition of oil, so that even more concentrated solutions can be used.

The above mentioned solution with 1% rape oil was investigated for its extraction capacity. At the same time a parallel experiment was performed with a microemulsion of Igepal CA 520 (Aldrich), heptane and water. Both sytems were able to extract about the same amounts of pyrene within two hours. The values of about 1000 mg/kg were clearly higher than the analytical values of 340 mg/kg. But higher contents could be confirmed by extraction with toluene. Values of about 1200 mg per kg soil were found.

Microbiological degradation

It is probable that treatment of soils with microemulsions causes a considerable reduction in the indigenous microflora due to the high surfactant concentration. The microflora is, however, important for removing the residual amounts of surfactant and rape oil from the soil after the extraction process. As these substances are principally biodegradable, it appears purposeful to inoculate the soil subsequently with microorganisms which not only degrade the remaining amounts of microemulsion, but also mineralize the residuals PAHs.

In order to select a microorganism population which simultaneously degrades rape oil and surfactant, a 2 liter bioreactor with biomass feedback and continual substrate feed was constructed.

Experimental data:

Total reactor volume:	2.5 l
Substrate feed rate: rape oil	3.5 ml/d
surfactant	2.2 ml/d
Mineral medium feed rate	1640 ml/d
Hydraulic retention time	1.5 days
Mineral medium: $(NH_4)_2HPO_4$	0.5 g/l
KH_2PO_4	0.25 g/l
$MgSO_4 \cdot 7H_2O$	0.05 g/l
Trace elements	5 ml/l
Biomass concentration	8-10 g/l

After starting the reactor, the contents foamed due to the presence of the surfactant. This was controlled by reducing the aeration to a minimum. After the population had developed, no more foam was produced, which indicated that the surfactant was being degraded.

First attempts to isolate the microorganisms responsible for degradation of the rape oil and surfactant using normal microbiological media were unsuccessful, which could indicate a high specificity of the microorganisms for these substrates. At the moment other media are being tested.

Microscopical examination of the population showed that bacteria are the main component. However, a fungal culture has been isolated from the population which is capable of growing on both rape oil and surfactant.

Process of soil decontamination with microemulsions

The basic ideas of the concept for remediation of highly contaminated fine-particle fractions of a soil are as follows:
- recovery of a biologically active, clean soil.
- re-use of the surfactant from the microemulsion.
- efficient biodegradation of the oil used for the microemulsion.
A flow sheet of the process is shown in figure 4.

The soil contaminated with PAH is upgraded into a low polluted fraction (large grain size) and a highly polluted fraction (small grain size); this can be the clay compound of the soil. The material to be cleaned and the microemulsion are intensively mixed. During this extraction process, the contaminants pass from the soil particles into the microemulsion. After the extraction step, the contaminant-rich microemulsion and the solid soil particles are separated. Washing with water and inoculation with microorganisms lead again to biologicallly active soil.

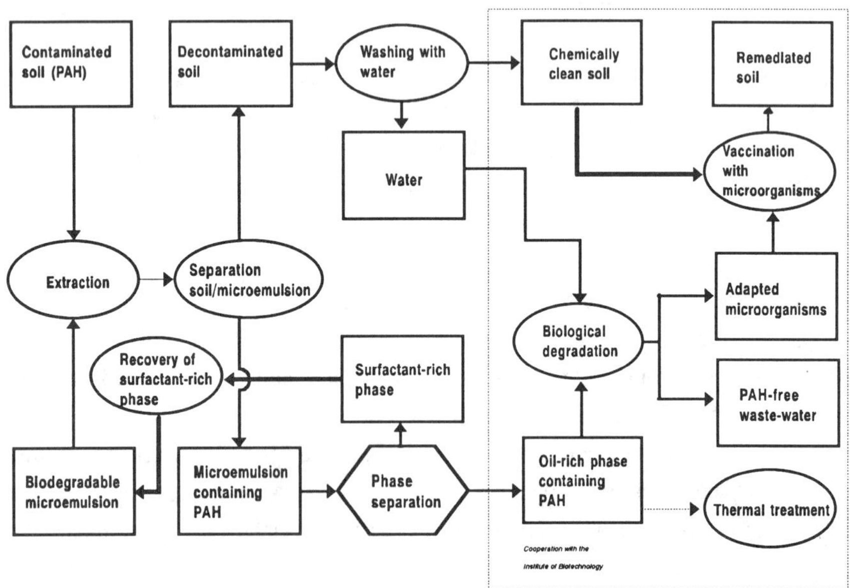

Figure 4: flow sheet of the remediation process with microemulsions

The microemulsion carrying the contaminant can be separated into a surfactant-rich phase (low content of contaminants) and an oil-rich phase (high content of contaminants). The surfactant-rich phase is recycled to form a new microemulsion. The oil-rich phase and the water from the soil washing step are treated either separately or together with selected microorganisms to degarde the organic compounds (oil, contaminat, surfactant). Alternatively, a thermal treatment of these compounds may happen.

As result of these biodegradation-processes a PAH-free water is produced and a surplus of adapted microorganisms.

Literature

[1] V. Franzius, Chem.-Ing.-Tech. 63 (1991),348

[2] H.-D. Sonnen, Chem.-Ing.-Tech. MS 1957/91

[3] M.J. Schwuger, G. Subklew, Workshop für Interessenten am Forschungsprogramm, Kernforschungszentrum Karlsruhe , Oktober 1992

[4] W. Clemens, F.-H. Haegel, M.J. Schwuger, C. Soeder, K. Stickdorn, Patent, applied

[5] J. H. Schulman, W. Stoeckenius, L.M. Prince, J. Phys. Chem. *63*, 1959,1677

[6] W. Jahn, R. Strey, J. Phys. Chem. *92*, 1988, 2294

[7] D. Bauer et al. in "Separation Sci. and Technol.", 2nd Conference, Vol. 2, 425

[8] D. Langevin in "Reverse Micelles, Proceedings of the European Science Foundation Workshop", eds. P.L. Luisi, B.E. Straub, Plenum Press, New York 1984, 287

[9] M.K. Sharma, D.O. Shah in "Macro- and microemulsions in enhanced oil recovery", ACS Symposium Series 272, 1985, 149

[10] W.D. Clemens, F.-H. Haegel, M.J. Schwuger, Langmuir, in print

[11] D. Balzer, Tenside Surf. Det. 28(6), 419-427 (1991)

Cleaning Organically Contaminated Soil by Steam Extraction
- Results of Semi-Industrial Tests -

K. Hudel, F. Forge, R. Klein[1] , H. Fr. Schröder, J. Tränkler, M. Dohmann
Institut für Siedlungswasserwirtschaft der RWTH Aachen,
Templergraben 55, D - 5100 Aachen

1. Introduction

The findings presented here were compiled by the Institute of Aachen Technical University for Water Management in Residential Areas (ISA) in collaboration with Bonnenberg u. Drescher Ingenieurgesellschaft mbH, Aldenhoven in the context of a research project sponsored by the Federal Ministry for Research and Technology (BMFT).

The elaborated remedial action technique can be classified as a thermophysical process in the low-temperature range (T = 100 to 160°C). Unlike biological methods, the success of the remedy implemented with thermal treatment can be achieved quickly and it is easier to control. This technique is highly efficient. Compared with soil washing installations employing chemical and physical principles, it offers the advantage of also registering the contaminants that are adsorbed on the fine and finest grains. Moreover, the high temperatures required by pyrolytic and incineration methods (T = 350 to 1200°C) disintegrate not only the anthropogenic contamination, but also the natural organic soil constituents (humic matter). Cost-intensive measures for dust recovery, chemical washing, nitrogen removal and the adsorption of heavy metals and ultra-toxic substances from the waste gas lead to treatment costs of up to and above DM 500/metric ton of contaminated soil. In addition, this method of treatment is rejected by much of the population because of the feared emission of ultra-toxic substances in the waste gas. Steam treatment is not encumbered with these disadvantages as only an extremely small amount of the concentrated contaminants extracted from the soil have to be disposed of in existing hazardous waste incineration plants.

2. Process description

Mirroring conventional thermal methods, the extraction of organically contaminated soils and residues by steam initially involves the contaminants changing from the liquid or adsorbed state to the gas phase. The steam technique does without gas scrubbing as the contaminated vapours can easily be condensed in a _single_ subsequent process stage.

Soil cleaning by steam is carried out by steam distillation. This method is the most important special application of carrier distillation. It enables high-boiling substances that are immiscible or difficult to mix with water to be distilled often at the relatively low temperature of 100°C. From a thermodynamic viewpoint the reduction of the boiling point of not easily volatile substances can be explained by the fact that they form a heterogeneous aceotrope in the water mixture /1, 2/. The boiling point of this aceotrope is similar to that of the pure water at the relevant pressure (e.g. p = 1 bar, i.e. T = 100°C). As a precondition, the inner surface of the solid being treated must be completely wetted with water. When the material is subsequently heated in the mixing apparatus (steaming and jacket heater), "microboiling" takes place inside the pores of the adsorbate, whereby a lot of water and a little of the high-boiling

---

[1]   Bonnenberg u. Drescher Ingenieurgesellschaft mbH, Aldenhoven

F. Arendt, G.J. Annokkée, R. Bosman and W.J. van den Brink (eds.), Contaminated Soil '93, 1325–1335.
© 1993 _Kluwer Academic Publishers. Printed in the Netherlands._

substance distils over. Described in graphic terms, the molecules of the substance being distilled are enveloped in water molecules; that is to say, solvation takes place. The chemically different character of the molecules is masked, so that these "quasi water atoms" can distil off as well. The second mechanism that is effective when treating soils with steam is the desorption of the contamination from the solid matrix. We have learnt from regenerating exhausted activated carbons that "dry" treatment with superheated steam obtains the best results /3/. This is because pores, especially bottleneck pores, are closed by condensation water if saturated steam is used. This prevents the steam from reaching the contaminants adsorbed inside the pores and thus from bringing about desorption. Experiments currently being conducted are expected to determine which operating mode achieves the best results depending on the contaminant concerned.

In chemical engineering, steam distillation takes place either in a water mixture or by injecting steam into the bulb bottom. Industrial applications include, for example, fatty acid refining, mineral oil processing and purifying chemical effluent /4, 5, 6, 7/. In industrial processes the brief steam-organic phase contact time often causes the actual steam consumption to exceed the theoretical requirement: the vapours are not charged with the maximum possible organic content (thermodynamic equilibrium) /8/. With the steam extraction of granular solids, it would appear feasible to reduce the steam consumption by recirculating the vapours.

Following condensation, the steam carrier component can be fractionated from the highly concentrated organic contaminant phase by gravity separation. The process water must be purified afterwards. Conventional methods of industrial effluent treatment can be used for this purpose, such as membrane processes, contaminant adsorption on activated carbon, and extraction of the water phase with organic solvents followed by recovery by rectification. ISA is also investigating the scope for thermal and catalytic cracking in a steam current as a means of decomposing the contaminants into non-toxic components.

3.   Semi-industrial tests

3.1   Description of the pilot plant

An intermittent mixer in which a mechanically generated fluidized bed with excellent heat and mass transfer behaviour forms was used for decontaminating soil with steam /9/. This setup achieved thorough contact between the steam, supplied by a steam generator and injected into the mixer by a steam lance, and the contaminated soil. The contaminated vapours were condensed in a heat exchanger downstream of the reactor, whereby most of the contaminants were immediately separated from the condensate as a floating, water-insoluble phase. This concentrated contaminant was disposed of in a hazardous waste incineration plant. In the transitional period the same procedure was used for the aqueous phase; introduction into a communal sewage treatment plant following adsorptive pretreatment is, however, envisaged.

Figure 1 is a flowchart depicting the pilot plant set up at the Institute of Aachen Technical University for Water Management in Residential Areas. The mechanical fluidized bed reactor (ploughshare mixer) comprises a cylindrical drum with axial whirler (n = 160 rpm). Solid agglomerations are disintegrated by a rotating knife head. The form taken by the mechanical fluidized bed in a ploughshare mixer is illustrated on the left of Figure 2. The mixing drum with whirler and radial knife head is shown in the right of the Figure (top view). The reactor (type DVT 30 manufactured by Lödige, Paderborn) has a gross capacity of 130 l; the manufacturer recommends a filling of 50 to 70 % by volume. The steam generator delivers saturated steam (m $\leq$ 100 kg/h) up to p = 6 bar / T = 164°C, most of which is injected directly into the mechanical fluidized bed of the mixer by the steam lance. A small partial flow is

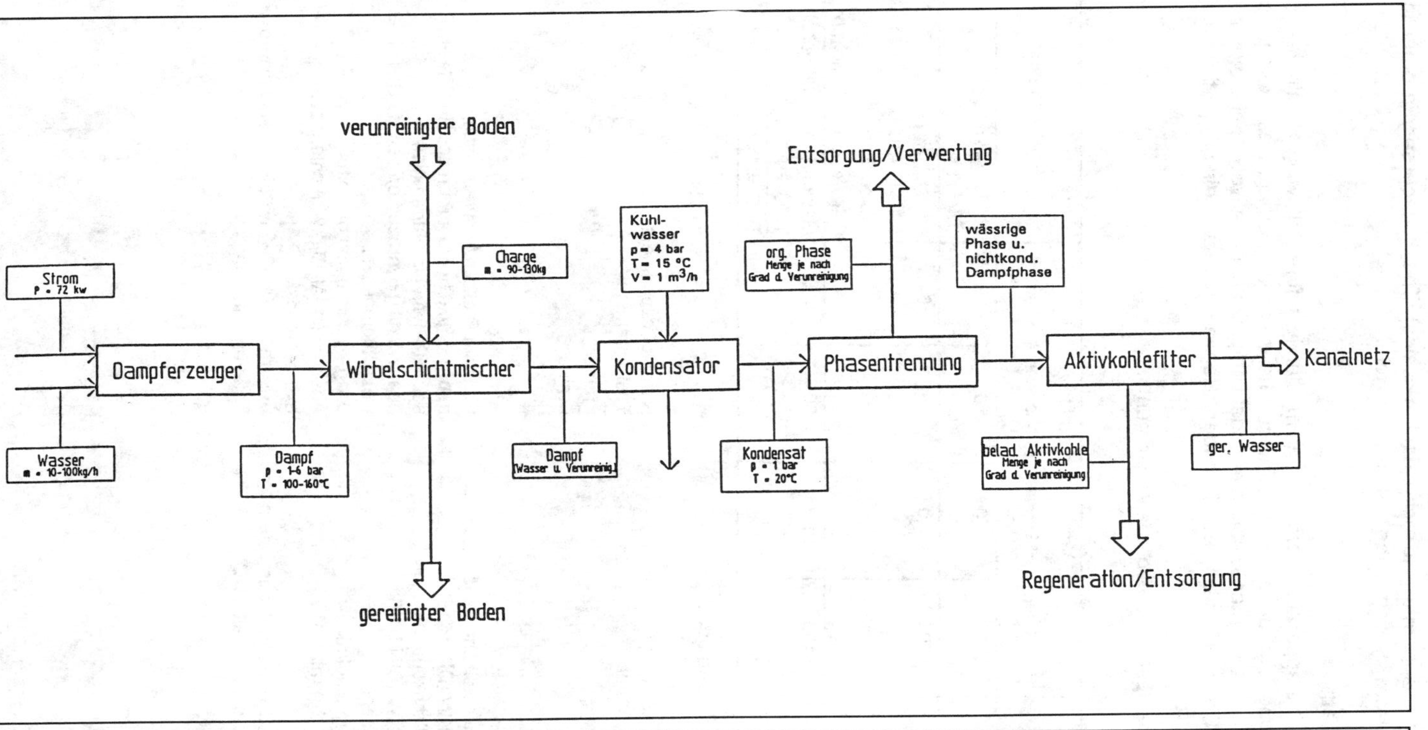

verunreinigter Boden	Contaminated soil	Entsorgung	Disposal	Phasentrennung	Phase separation
Kühlwasser	Cooling water	Verwertung	Reuse	Aktivkohlefilter	Activated carbon filter
Strom	Power	org. Phase	Organic phase	gereinigtes Wasser	Purified water
Charge	Batch	Kondensat	Condensate	nichtkond. Dampfph.	Non-cond. steamphase
Wasser	Water	Kondensator	Condenser	beladene Aktivkohle	Exhaused act. carbon
Dampf	Steam	Kanalnetz	Sewerage	Gereinigter Boden	Decontaminated soil
Verunreinigung	Impurities	Dampferzeuger	Steam generator	Wirbelschichtmischer	Fluidized bed mixer

Figure 1:   Flowchart of the pilot plant for steam extraction

branched off to heat the apparatus indirectly (double casing). This is necessary in order to prevent the moist soil from sticking to the wall of the mixer.

3.2 Experimental method

Samples of the soil in the apparatus were taken immediately before (following 10 minutes of thorough material mixing) and after the experiment. The injection of steam was interrupted occasionally during the test in order to obtain material samples after relatively brief exposure to treatment as well. Composite samples of condensate were collected directly at the condenser outlet.

Table 1 lists the process parameters of the experiments.

Versuchsbezeichnung	Versuchsparameter				Bemerkungen
	$T_{Mischer}$ [°C]	$P_{Mischer}$ [bar]	$n_{Mischer}$ [U/min]	m_{Dampf} [kg/h]	
Boden mit Dieselöl-kontamination	103	1,0	80	29	ca. 1,6 g/kg KW (DIN H 18)
Boden mit Misch-kontamination	104	1,0	80	64	ca. 8,5 g/kg KW (DIN H 18), Hexachlorbenzol, 0,4g/kg Chlorphenolmix
Tonerde (Al_2O_3) mit CKW	103	1,0	80	105	Gesamtkohlenstoffgehalt TOC = 2,8 %
mit Fettsäureestern					TOC = 14,6 %

Table 1: Experimental conditions

Legend:

Versuchsbezeichnung	Name of experiment
Versuchsparameter	Test parameters
$T_{Mischer}$	T_{mixer}
$P_{Mischer}$	P_{mixer}
$n_{Mischer}$	n_{mixer}
m_{Dampf}	m_{steam}
Bemerkungen	Remarks
Boden mit Dieselölkontamination	Soil with Diesel oil contamination
ca. 1,6 g/kg KW (DIN H 18)	app. 1.6 g/kg hydrocarbon acc. to DIN H 18
Boden mit Mischkontamination	soil with hybrid contamination
ca. 8,5 g/kg KW (DIN H 18)	app. 8.5 g/kg hydrocarbon acc. to DIN H 18
Hexachlorbenzol	Hexachlorobenzene
0,4g/kg Chlorphenolmix	0.4 g/kg chlorophenol mixture
Tonerde (Al_2O_3) mit CKW	Adsorbents (Al_2O_3) with chlorinated hydrocarbons
Tonerde (Al_2O_3) mit Fettsäureestern	Adsorbents (Al_2O_3) with Fatty acid esters
Gesamtkohlenstoffgehalt TOC	Total organic compound

Figure 2: Left: Ploughshare mixer showing the mechanical fluidized bed
           Right: Mixer with whirler and knife head, Lödige, Paderborn

3.3   Results

3.3.1   Soil contaminated with diesel oil

The experimental material was soil removed from a former dump. It primarily comprised fine sand with auxiliary components of silt and grit (soil type according to German classification fS,u,g'). This soil contained almost exclusively n and iso-alkanes (mineral oil constituents). The hydrocarbon content was found to be 1600 mg/kg (0.16 %) according to DIN H 18. Figure 3 shows ion current chromatograms[2] of the toluene extracts obtained in the laboratory from the original soil as well as those obtained from the soil after one and two hours of steam treatment. After t = 1 h the alkanes up to a chain length of C16 are depleted by only 47 to 71 %. Higher alkanes are not depleted. In contrast, the gas chromatogram of the soil after t = 2 h indicates an almost complete depletion of the alkanes. Note that the peak heights at the bottom of Figure 3 (after two hours of treatment) have to be mentally divided by 3 when comparing with the other two chromatograms (higher selected detector sensitivity). These results were obtained with a low specific steam quantity, of 0.53 kg steam/kg soil.

---

2 The abscissa represents the retention time (time steps) of the sample constituents in each case. The intensity value marked on the ordinate (right) is provided for comparing different chromatograms. Identical or similar values allow a direct comparison of peak heights and thus, as an initial approximation, of concentrations.
The ratio of the different intensity values of two chromatograms is also valid in inverse proportion for the relevant peak heights; with twice the intensity value, for example, the peak heights have to be halved.

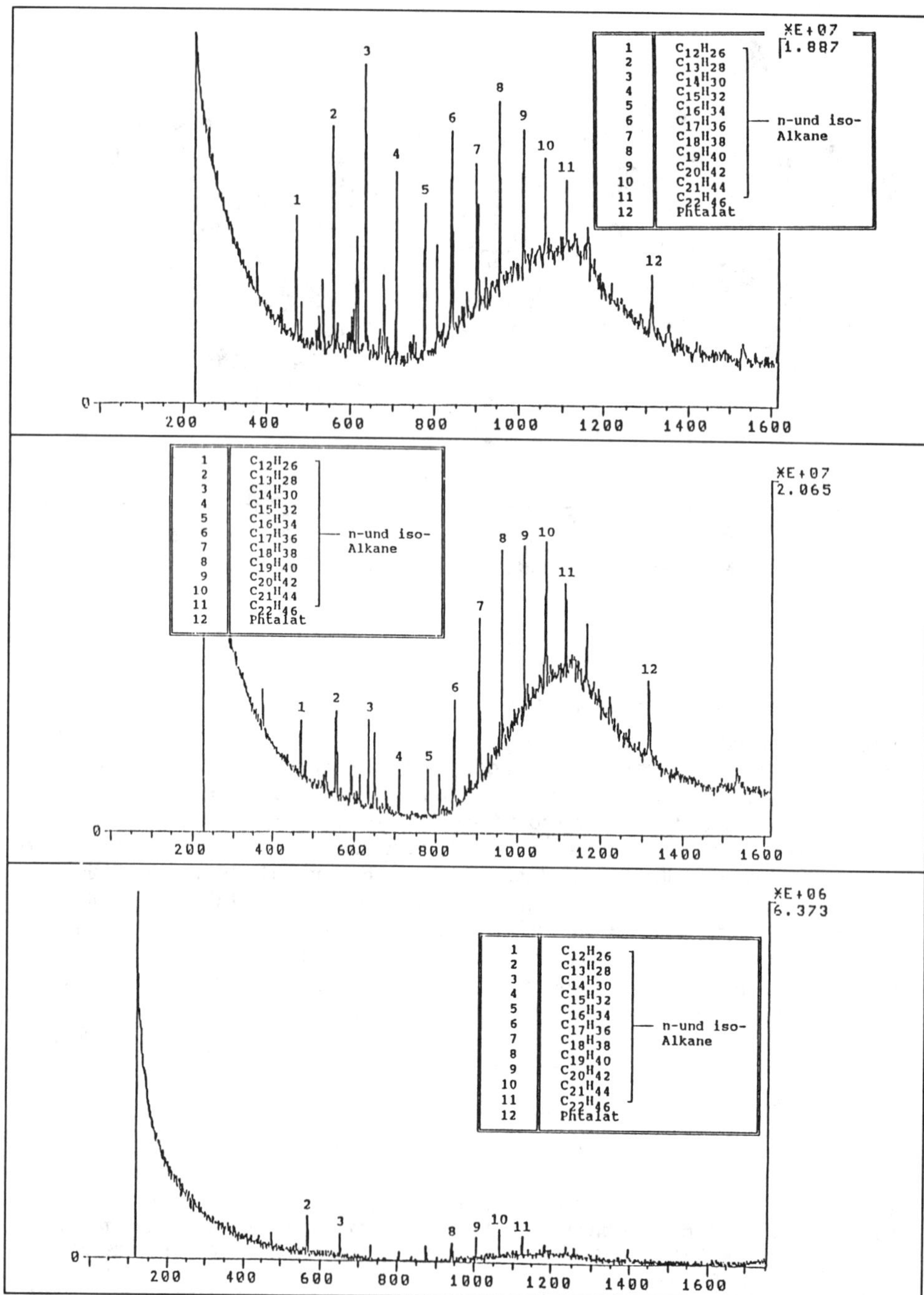

Figure 3: Ion current chromatograms;
Soil before treatment (top)
Soil after treatment, t = 1 h (centre)
Soil after treatment, t = 2 h (bottom)

3.3.2 Soil with hybrid organic contamination

Here again, the experimental material was contaminated soil from a former dump. Before treatment the soil was dark brown to black, becoming light brown after treatment. Before the material was treated it was mechanically screened with d = 5 mm (full industrial scale: d = 15 to 20 mm). Figure 4 shows the grain size distribution of the material before and after treatment in the fluidized bed mixer.

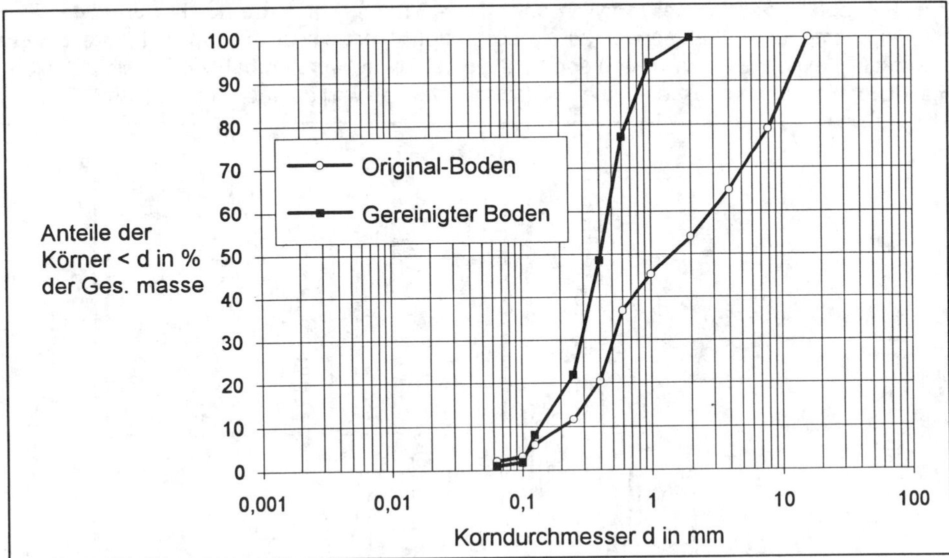

Figure 4: Screening curve of the experimental soil
Legend: Anteile der Körner < d in % der Ges. masse   Share of grains < d in % of total
         Original-Boden, Gereinigter Boden   Original soil, treated soil
         Korndurchmesser d in mm   Grain diameter d in mm

The following table contains the analysis results for the original soil:

Probe	TOC	CSB	Phenol-index	EOX	KW	TR
	[mg/kg]	[mg/kg]	[mg/kg]	[µg/kg]	[mg/kg]	[%]
E6	220	1020	6,8	320	8500	91,3

Table 2: Chemical parameters of the experimental original soil /10[3]
Legend: Probe   Sample   CSB   COD
         Kohlenwasserstoffe   Hydrocarbons   Trockenrückstand   Dry residue

A phenol mixture with an overall concentration of 0.4 g/kg soil (0.04 %) was artificially added to the soil. The mixture comprised 2-chlorophenol, 2,4-dichlorophenol, m-cresol and 2,4,6-trichlorophenol of equivalent percentage by mass with boiling points between 180 and 230°C.

Figure 5 contains ion current chromatograms of the toluene extracts from the soil obtained in the laboratory before and after treatment. Before the steam treatment the soil contained

---

[3] Before additional contamination with chlorinated phenols

various alkanes and hexachlorobenzene (original contamination), the added chlorophenols/cresol and tributyl phosphate (plasticizer from the dumped material); cf. top of Figure 5.

Following two nours of steam stripping, the chlorophenols/cresol were depleted to below the detection limit (bottom of Figure 5). With the exception of minute traces of C_{14} and C_{15}, no alkanes were present any more. Hexachlorobenzene was found in the condensate from the first half hour (not illustrated) (boiling point $T_S = 322°C$). The large number of unidentified organic compounds that appear as a background spectrum at the top of Figure 5 were completely expelled at between 0 and 30 minutes. These were probably products of biological decomposition processes at the site (metabolites) as well as trace contaminants.

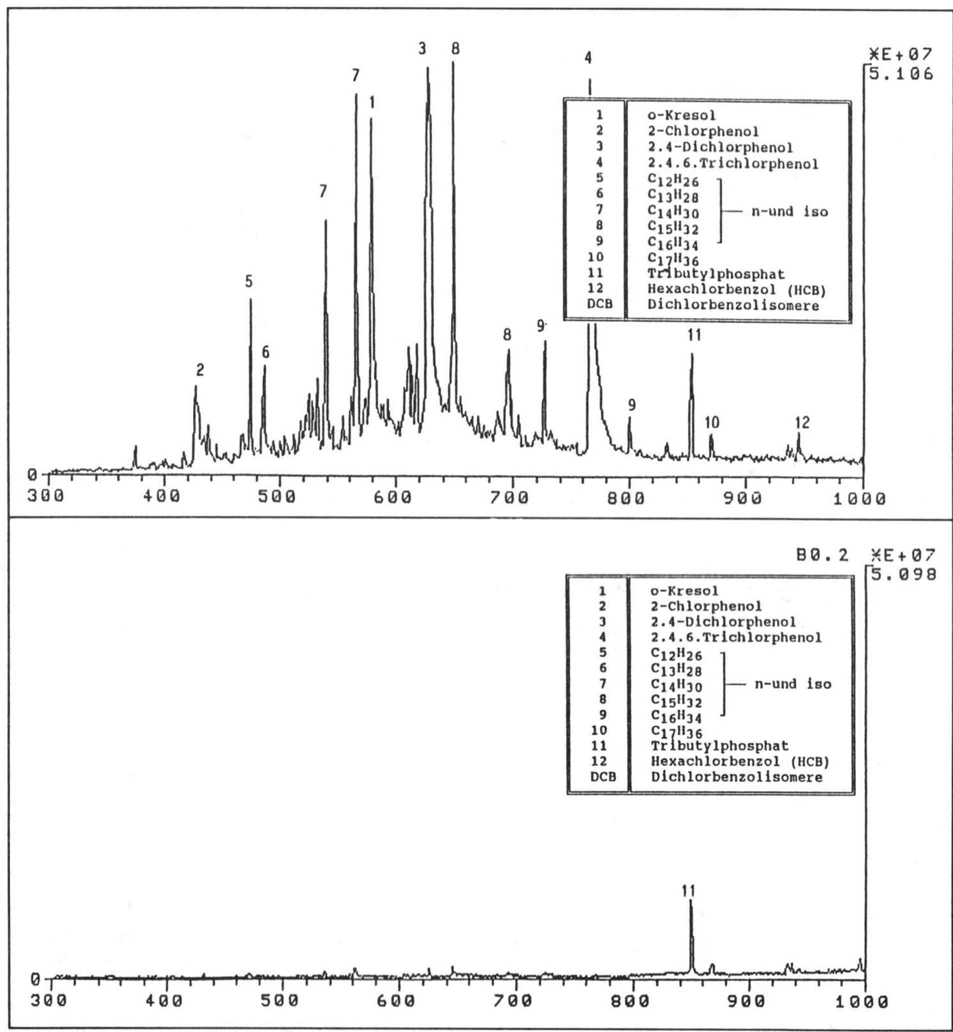

Figure 5: Ion current chromatograms   Soil before treatment   (top)

                                  Soil after treatment, t = 2 h   (bottom)

The composite samples of condensate from the experiment periods 30 to 60 minutes and 60 to 120 minutes (not illustrated) contained the remaining identified contaminants that were present during the first half hour in clearly decreasing concentrations. No more aliphatic compounds were detected in the condensate collected during the second half of the experiment (t = 60 to 120'). The mineral oil contaminants contained in the soil had thus already been completely expelled after one hour of steam treatment.

3.3.3 Organically contaminated production residues

Experiments were conducted to investigate the steam treatment of aluminium oxides (Al_2O_3) contaminated during waste air and process water purification for two different chemical production processes, namely chlorinated hydrocarbon manufacture and fatty acid ester production.

The first of these materials comprised a relatively coarse core with narrow grain size distribution ($d_{10\%}$ = 1.1 mm, $d_{50\%}$ = 1.5 mm, $d_{90\%}$ = 2.4 mm) and was contaminated exclusively with chlorinated hydrocarbons. The batch from the fatty acid ester production facility had a much finer grain size and a broader grain curve ($d_{10\%}$ = 0.043 mm, $d_{50\%}$ = 0.072 mm, $d_{90\%}$ = 0.101 mm). This adsorbent was qualitatively examined by gas chromatography/mass spectroscopy (GC/MS) in order to determine its constituents:

- Methyl and ethyl esters of higher fatty acids
  (chain length C 16 to C 18, boiling point T_B = 420 - 440 °C)
- Butane
- n and iso-hexanes
- Ethanol
- Phthalate

Figure 6 shows the attainable contaminant depletion for the two adsorbents. The total organic carbon (TOC), a standard widely applied in waste analysis, is used as the measure of contamination.

The initial concentration of 2.8 % (TOC) of adsorbents contaminated with chlorinated hydrocarbons during solvent production was reduced to 0.6 % after one hour of treatment. As expected, the content of elutable organic halogen compounds (EOX = 41 g/kg) was reduced by 99.8 %. The experiment had to be terminated after one hour of treatment owing to technical difficulties; further contaminant depletion is to be anticipated if the exposure to treatment is increased to t = 2 h. Initial contamination of TOC = 14.6 % of adsorbents from fatty acid production was reduced to a residual content of between 1.5 % and 2.0 %. It is noteworthy that this depletion was already achieved after one hour of treatment.

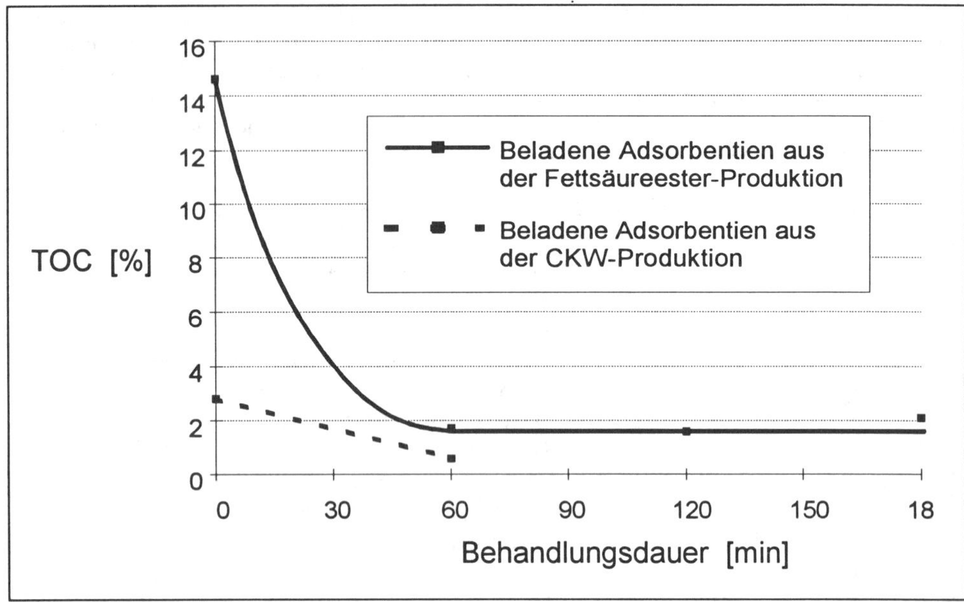

Figure 6: Reduction of the total organic carbon (TOC) of contaminated aluminium oxides by steam treatment

Legend: beladene Adsorbentien aus der   Exhausted adsorbents from
Fettsäureesterproduktion   fatty acid ester production
beladene Adsorbentien aus   Exhausted adsorbents from
der CKW-Produktion   chorinated hydrocarbon production
Behandlungsdauer [min]   Duration of treatment [min]

4.   Process application area

In principle, the process can be used with volatile contaminants (chlorinated hydrocarbons, BTX), but especially with contaminants with a higher boiling range that are volatile in steam. The following substance classes of contaminations are considered volatile in steam /1, 5, 6, 7/:

- Hydrocarbons and mineral oils (boiling range T_B = 200 to 350 °C)
- Chlorinated aromatic compounds and phenols (T_B = 180 to 350 °C)
- Polycyclic aromatic hydrocarbons (T_B = 270 to 400 °C)
- Polychlorinated biphenyls (T_B = 300 to 410 °C)
- Polychlorinated dibenzodioxins

Steam extraction reaches its application limits in the presence of extremely high boiling PAHs, such as occur in the form of the tar pitch residue from coal tar processing. While production residues with an volatile contamination spectrum can be advantageously treated, contaminated sites with exclusively volatile contaminations can be cleaned less expensively by the in-situ method of soil air extraction.

5.   Summary

Semi-industrial steam extraction tests were conducted with organically contaminated soils and residues in a mechanical fluidized bed reactor. In three series of experiments contaminated batches were exposed to saturated steam at T = 100°C / p = 1 bar over a period of two to three hours.

In particular, soil from a former dump that was contaminated with mineral oil (hydrocarbon content according to DIN H 18: 0.16 %) was completely decontaminated by two hours of steam extraction. Furthermore, a sample of original soil was additionally contaminated with a chlorophenol mixture (total concentration 0.04 %, boiling range 180 to 230°C). The alkanes, chlorophenols, o-cresol and hexachlorobenzene contained in the soil before the steam treatment were depleted, in some cases to below the detection limit, following steam stripping at normal pressure and T = 100°C. Especially notable in this connection was the hexachlorobenzene contained in the original soil, which is not easily volatile and has a boiling point of approximately 322°C. Finally, two aluminium oxides that had been used as adsorbents in chemical production processes and thereby exhausted, were treated by steam extraction. Both the charge that was contaminated with chlorinated hydrocarbons and that containing fatty acid esters were treated successfully. The organic carbon content of the latter material, for example, was depleted from 14.6 % to values between 1.5 % and 2 %. The adsorbents can therefore be harmlessly dumped or systematically reused.

In the context of site decontamination this method is favourably used in the presence of contaminants that are not easily volatile and can be distilled by steam. These include aliphatic and aromatic hydrocarbons (mineral oils), chlorinated aromatic compounds, PCBs and many PAHs. Full-scale implementation of the process so far examined by semi-industrial testing has commenced. Specific decontamination costs of approximately 150 to 200 DM/metric ton of soil are anticipated.

6. Bibliography

/1/ Sattler, K. — Thermische Trennverfahren, VCH Verlagsgesellschaft mbH, Weinheim, 1988

/2/ Grassmann, P. — Physikalische Grundlagen der Verfahrenstechnik; Salle+Sauerländer, Frankfurt, 1983

/3/ Figuera, Maria E. — Über die Wasserdampfregeneration von Aktivkohle -Beitrag zum Entfernen von Schadstoffen bis in den umweltrelevanten Konzentrationsbereich, Dissertation Universität Kaiserslautern, 1987

/4/ Stage, H. — Kombinationsschaltungen mit Aufbauelementen zur Fettsäure-Destillation und -Fraktionierung sowie zur Speiseöl-Entsäuerung und -Desodorierung; Fette - Seifen - Anstrichmittel; 72. Jahrgang; Nr. 4; 1970

/5/ N.N. — Kraftstoff - die treibende Kraft, Firmenbroschüre der BP Tankstellen GmbH; Hamburg, 2.Auflage 1990

/6/ Victorelli, J.C.; de Andrade, P.S.; Elkaim, J.-C — Chlorinated Hydrocarbons Liquid Wastes: Steam Extraction In Place Of Incineration; Water Science and Technology; Volume 24; Nr.12; 1991

/7/ Veith, G.D.; Kivus, L.M. — An Exhaustive Steam-Destillation and Solvent Extraction Unit for Pesticides and Industrial Chemicals; Bulletin of Environmental Contamination and Toxicology; Bd. 17; 1977;

/8/ N.N. — Ullmanns Enzyklopädie der technischen Chemie, Verlag Chemie Weinheim, Band 2, 4. Auflage 1977

/9/ Lücke, R. — Örtliche Wärmeübergangskoeffizienten in einem Pflugscharschaufeltrockner; Verfahrenstechnik, 10.Jahrgang; Nr.12; Dez. 1976

/10/ Pöppinghaus, K. Weißenfeld, S. — Sanierungsbegleituntersuchung Widdig; Gutachterliche Stellungnahme, Forschungsinstitut für Wassertechnologie an der RWTH Aachen, August 1990

Ecotoxicological Aspects of Chemical Pre-Oxidation Combined with Subsequent Microbial Degradation of Polycyclic Aromatic Hydrocarbons

Frank HAESELER, Michael STIEBER, Peter WERNER, Fritz Hartmann FRIMMEL

DVGW-Forschungsstelle am Engler-Bunte-Institut der Universität Karlsruhe, Richard-Willstätter-Allee 5, W-7500 Karlsruhe

1   SUMMARY

This paper deals with basic investigations carried out in batch reactors on ozonation of the three polycyclic aromatic hydrocarbons (PAH) naphthalene, phenanthrene, and pyrene. The resulting oxidation products could be measured as dissolved organic carbon (DOC) and their ecotoxicological effects could be characterized with the bioluminescence test. After treatment with ozone the samples were inoculated with a bacterial mixed culture adapted to PAH; biological degradation was followed in a Sapromat by registering biological oxygen consumption. The mixed culture could metabolize the PAH as well as part of the oxidation products, although ozonation caused the formation of oxidation products with high toxicity, which was shown in the bioluminescence test.

2   INTRODUCTION

Polycyclic aromatic hydrocarbons (PAH) are built during industrial pyrolysis processes and cause enormous contaminations of soil and groundwater especially during the operation of gasworks. Since many of these substances are said to be cancerogenic and mutagenic, they represent a great danger to the environment. The EPA (Environmental Protection Agency / USA) has provided a list of the sixteen most important and representative substances of this group.

The remediation of sites contaminated with PAH can be performed by means of washing procedures and thermo-treatment, but also with microbial methods, which lead to a natural mineralization of the contaminants. The bacterial degradation of these contaminants is limited by certain factors. It is known that the degradation of PAH is reduced with an increasing number of rings and is only co-metabolic from 4 rings onward. However, Weißenfels et al. (1990) and Walter et al. (1991) reported that PAH with 4 nuclei can be used as sole energy and carbon source. According to general belief, bacteria can take up only dissolved PAH by diffusion into the cell, which explains why bad solubility and low dissolution rate are to be regarded as limiting factors (Breure et al. (1992)). Furthermore, bioavailability of the contaminants is of main importance for the success of biological remediation measures (Stieber et al. (1990)). The goal of the investigations presented in this paper was to test whether a combination of chemical pre-oxidation with subsequent microbial degradation would theoretically lead to an increase in PAH elimination rates.

To this end, the model substances naphthalene, phenanthrene, and pyrene, which can be regarded as representative for PAH with 2, 3, or 4 nuclei, were mixed with different ozone doses. After the ozone had reacted completely with the aromatic hydrocarbons, a bacterial mixed culture adapted to PAH was added. The biodegradation of the resulting oxidation products and of the remaining PAH was then followed in a Sapromat.

This study was initiated by the conclusion that in present studies ozone was applied on a practical scale without the existance of sufficient basic laboratory investigations on model substances (Battermann & Werner (1987), Lund et al. (1991)).

F. Arendt, G.J. Annokkée, R. Bosman and W.J. van den Brink (eds.), Contaminated Soil '93, 1337–1344.
© 1993 *Kluwer Academic Publishers. Printed in the Netherlands.*

1338

The studies by Legube et al. (1986), Doré (1989) and Sturrock et al. (1963) described the chemical oxidation path of the examined PAH with ozone. The basic reaction mechanism of the 1,3 dipolar cycloaddition of ozone with the most reactive double binding was predominant under the chosen experiment conditions. This mechanism had already been demonstrated in 1960 by Criegée (Allinger et al. (1980)).

3   MATERIAL AND METHODS

The ozone was produced from technical oxygen with an Anseros Ozomat COM / R. The ozone concentration in the gaseous phase (incoming/outgoing air) was measured spectro-photometrically with an Anseros Ozomat GM. Figure 1 shows the unit for ozone treatment. The oxygen-ozone-mixture was conducted into the reactor with a flow of 40 L/h at an ozone concentration of 45 mg/L.

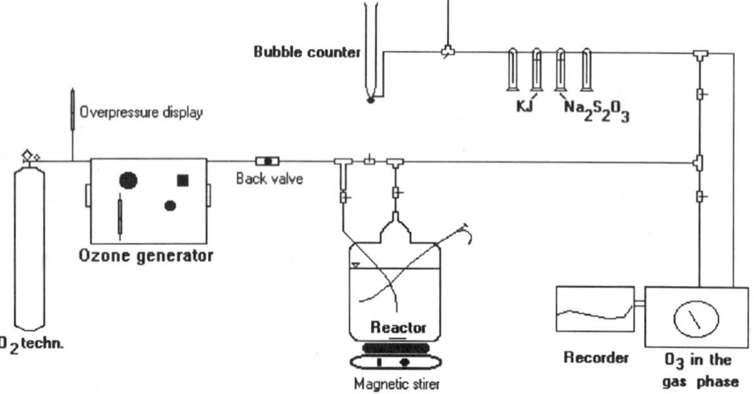

Figure 1:   Diagram of the Ozonation Unit

2 g of pestled PAH were distributed regularly by strong stirring in a 5-L-reactor containing 4 L of de-ionized water at 20 °C (figure 1) and were mixed with ozone. Samples of 450 mL each were taken at different intervals (30, 60, 120, and 240 minutes) without interrupting the experiment. 400 mL were filled into the batch reactor vessels (total volume 1.25 L), 50 mL were used for analytical purposes. Each ozonation was carried out twice in order to obtain 800 mL for investigation of the sample.

The water samples were extracted with cyclohexane; their PAH content was measured with a Hewlett-Packard gas chromatograph with flame ionization detector. The content of dissolved organic carbon (DOC) was measured in a Dohrman DOC analyser after membrane filtration through a polycarbonate filter of 0.45 μm pore size and acidification with concentrated phosphoric acid.

Two days after ozonation the samples in the batch reactors were mixed with ten times concentrated mineral salt medium according to Lockhead and Chase. These solutions were inoculated with 1 mL of the bacterial population DVGWMS1 (selectively enriched on PAH). The reactor vessels were then incubated at 20 °C in a Sapromat (Voith), while the biological oxygen consumption was evaluated.

Furthermore the following biological parameters were determined: total cell number and PAH-degrading bacteria were evaluated on microtiter plates according to the most probable number procedure (MPN) in five parallel series for each sample (Stieber et al. (submitted 1992)). The samples of both experiments were incubated at 20 °C during 7 days for the determination of total cell numbers, naphthalene and phenanthrene degraders, and during 30 days for pyrene degraders.

The toxic effect of the water compounds was evaluated with the bioluminescence test. The samples that had been filtered through polycarbonate membrane filters of 0.45 μm pore size were examined according to DIN 38 412, part 34.

4   RESULTS

Figure 2 presents the naphthalene concentration in comparison with ozonation time and shows that ozone reacted with naphthalene. The oxidation reaction caused an introduction of oxygen atoms into the PAH molecules. This resulted in more hydrophilic oxidation products which were determined by means of the parameter DOC (dissolved organic carbon)

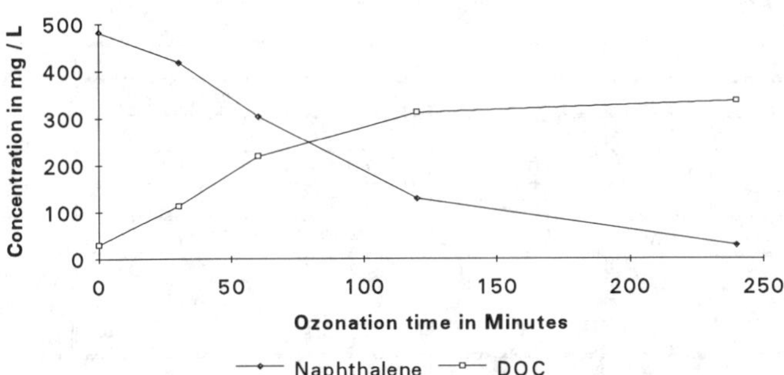

Figure 2:   Formation of Water Soluble Oxidation Products During Ozonation of a Naphthalene Suspension

Figure 3 demonstrates the kinetics of the oxygen consumption in the course of biodegradation in the batch reactors. The sample indicated with a.d. served as control. For that purpose mineral salt medium and inoculum were added to de-ionized water. The reference sample with unozonated naphthalene is indicated as O_3 t0. The ozonated naphthalene samples were marked according to their ozonation time as O_3 t30 for 30 minutes up to O_3 t240 for 240 minutes.

As can be seen from figure 3, microbial degradation activities started after 1.5 days in the reference sample O_3 t0. In the ozonated samples biological oxygen consumption started later, which could be attributed to the presence of toxic oxidation products. These compounds had a bactericide effect on the added bacterial mixed culture. Due to repeated inoculation after two days growth of the microorganisms finally started. The number of naphthalene degrading bacteria at the beginning and in the end of the experiment (table 1) demonstrates that the biological naphthalene degradation was combined with a marked multiplication of the mixed population.

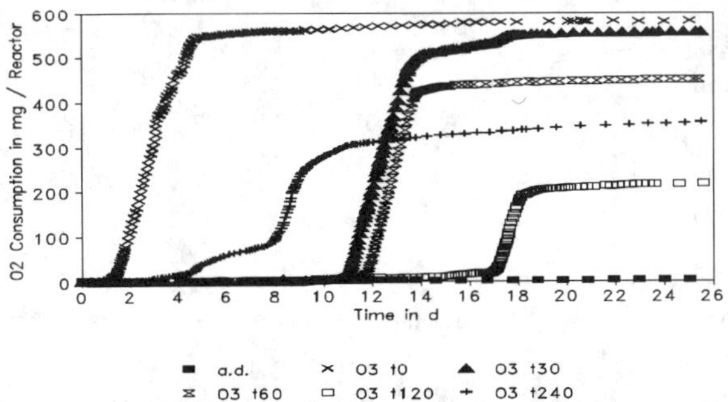

Figure 3:   Influence of Ozonation of Naphthalene on Biological Oxygen Consumption

After the experiment had been run, none of the samples contained any naphthalene. The DOC values of the reference sample t0 demonstrated that during microbial degradation of naphthalene hardly any dissolved metabolites were produced in this system (figure 4). In the samples with 30, 60 or 120 minutes of ozone treatment the DOC could not be degraded by the available microorganisms (figure 4). The DOC of the sample with 240 minutes ozonation time, however, was biologically degraded by 76 %. The results of the bioluminescence test showed that toxicity of the solutions was reduced to a great extent when naphthalene or its oxidation products were degraded biologically (t0 and t240). Otherwise, toxicity remained the same (t20) or was only reduced slightly (t30 and t60).

Table 1: Microbiological Parameters Before and After Biodegradation

	Bacteria / mL						
Assay Start	**Contr.**	**t 0**	**t 30**	**t 60**	**t 120**	**t 240**	
Naphthalene assay	$2,6\ 10^5$	$1,6\ 10^5$	0	0	0	0	Naphthalene-Degraders
Phenanthrene assay	$1,6\ 10^3$	$9,0\ 10^3$	$2,3\ 10^3$	$2,6\ 10^3$	$7,7\ 10^2$	$4,2\ 10^3$	Phenanthrene-Degraders
Pyrene assay	$1,3\ 10^4$	$4,3\ 10^3$	$4,3\ 10^3$	$2,3\ 10^3$	$7,7\ 10^3$	$3,6\ 10^3$	Pyrene-Degraders
Assay End	**Kontr.**	**t 0**	**t 30**	**t 60**	**t 120**	**t 240**	
Naphthalene assay	$4,2\ 10^4$	$5,7\ 10^7$	$4,6\ 10^6$	$1,1\ 10^7$	$2,6\ 10^5$	$2,6\ 10^5$	Naphthalene-Degraders
Phenanthrene assay	$1,6\ 10^3$	$7,9\ 10^7$	$1,6\ 10^7$	$1,6\ 10^6$	$1,1\ 10^6$	$1,5\ 10^5$	Phenanthrene-Degraders
Pyrene assay	$5,2\ 10^3$	$1,5\ 10^4$	$5,6\ 10^4$	$3,6\ 10^4$	$2,3\ 10^4$	$5,2\ 10^4$	Pyrene-Degraders

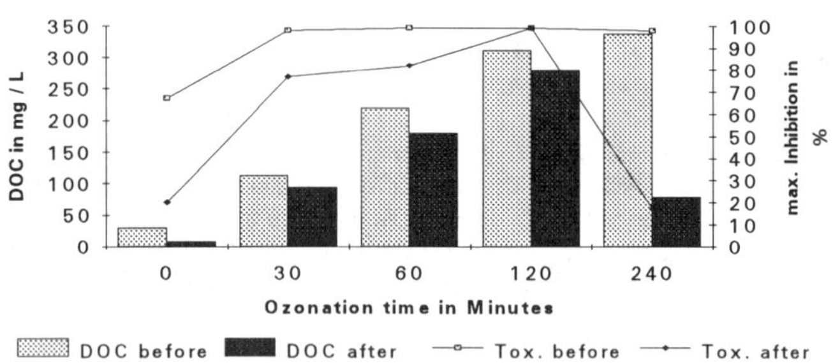

Figure 4: DOC-Content and Maximum Inhibition of Light Emission in the Bioluminescence Test, Before and After Biodegradation of Naphthalene

Figure 5 compares phenanthrene concentration with ozonation time. The decreasing phenanthrene concentration shows that ozone also reacted with this PAH and caused water soluble oxidation products.

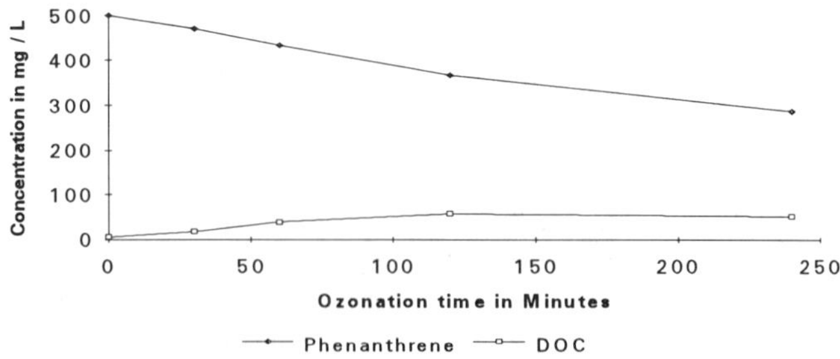

Figure 5: Formation of Water Soluble Oxidation Products During Ozonation of a Phenanthrene Suspension

Figure 6 shows the kinetics of oxygen consumption in the Sapromat samples with phenanthrene.

The samples were marked in the same manner as in the above described naphthalene experiment. The figure illustrates that the lag phases of the mixed population were prolonged along with the increasing ozonation of phenanthrene. Unlike in the naphthalene experiment, however, there was no bactericide effect of the oxidation products (table 1). The number of phenanthrene degrading bacteria (table 1) increased by 10^2 to 10^3 in the course of biological degradation.

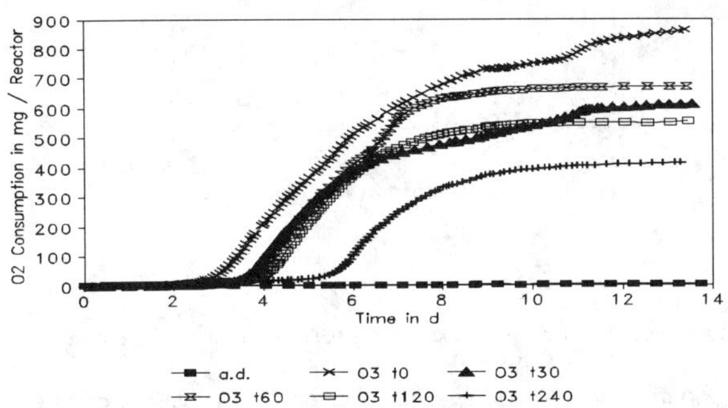

Figure 6:   Influence of Phenanthrene Ozonation on Biological Oxygen Consumption

At the end of the experiment, none of the samples contained any phenanthrene. The DOC values of the reference sample indicated that microbial phenanthrene degradation caused the formation of water soluble metabolites. In the end, dissolved compounds also remained in the ozonated samples. Their concentration raised with increasing ozonation time and they obviously could not be degraded by the existing microorganisms. As can be seen from figure 7, the total DOC value in sample t240 was even increased distinctly due to the activity of the added microorganisms. The investigations carried out so far are not sufficient to estimate whether the produced compounds resist permanently to microbial degradation, or whether the microorganisms simply require a longer adaption phase. Further investigations would first have to clarify the composition and environmental significance of the produced DOC before a statement on its degradation could be made.

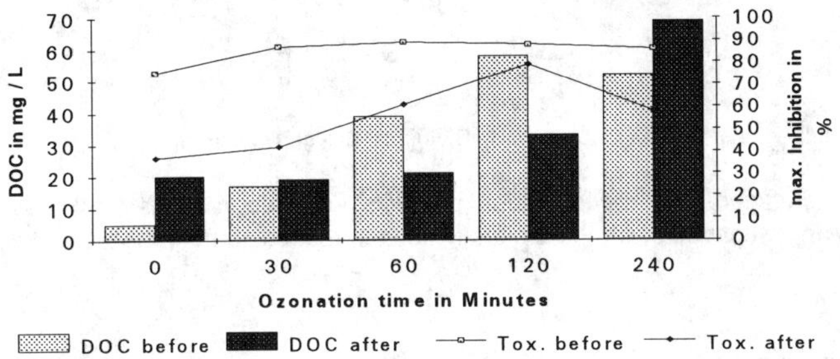

Figure 7:   DOC-Content and Maximum Inhibition of Light Emission in the Bioluminescence Test, Before and After Biodegradation of Phenanthrene

Figure 8 presents the ozonation of pyrene which also caused an oxidation and the formation of water soluble oxidation products.

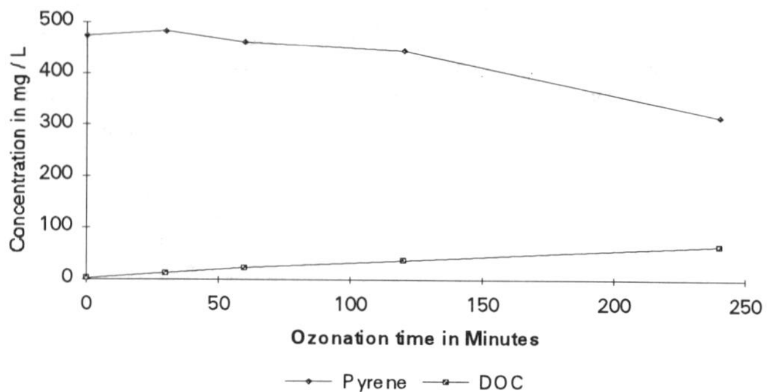

Figure 8:   Formation of Water Soluble Oxidation Products During Oxidation of a Pyrene Suspension

Figure 9 shows the kinetics of the oxygen consumption in the course of biodegradation in pyrene samples. On the one hand this diagram illustrates that the lag phase of the bacterial mixed culture was reduced with increasing ozonation. On the other hand ozonation appeared to have also a clearly positive influence on the total microbial degradation.

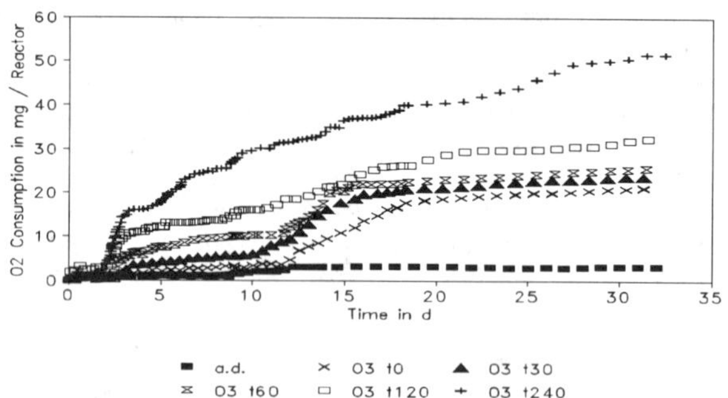

Figure 9:   Influence of Pyrene Ozonation on Biological Oxygen Consumption

As can be seen in figure 10, the oxidation products were degraded almost completely by the microorganisms. Investigations with the bioluminescence test could not prove a toxic effect in any of the cases.

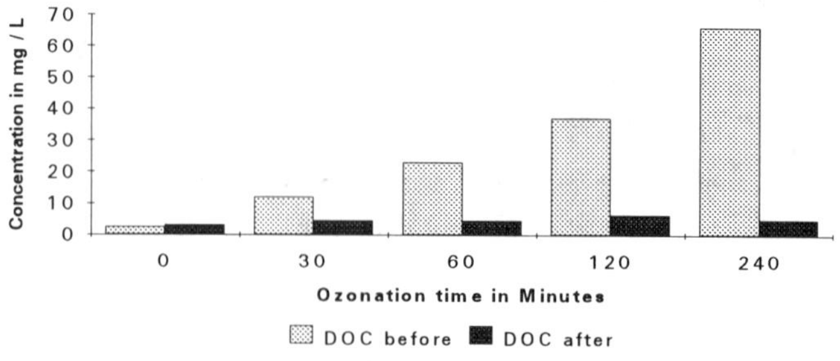

Figure 10:   DOC Content Before and After Biolgical Degradation in the Pyrene Experiment

5   CONCLUSIONS

The reaction of ozone with the various PAH manifested itself in different kinetics and effectivity rates. After 240 minutes of treatment under the described conditions 94 % of naphthalene, 44 % of phenanthrene, and 34 % of pyrene were oxidized. The declining oxidation rate correlated very well with the decreasing solubility of the three PAH (30 mg/L for naphthalene, 1.9 mg/L for phenanthrene, and 0.14 mg/L for pyrene).

In contrast to the oxidation rate, the ozone consumption was in the same range for all three PAH. For that reason the different concentrations of oxidation products (measured as DOC) can only be explained with different mineralization degrees. The calculated total carbon (PAH + DOC) decreased with continued ozonation, which indicates a mineralization up to CO_2.

The degradation experiments with naphthalene, phenanthrene, and pyrene lead to highly inconsistent results regarding biological degradation as well as their ecotoxicological effects. The oxidation products built during ozonation of naphthalene were only degradable biologically after the longest ozonation time (240 minutes). Besides this, the lower ozone doses resulted in products with bactericide effects on the PAH-adapted bacterial population and with a high toxicity shown in the bioluminescence test. The oxidation products of phenanthrene were partly biodegradable. They prolonged the lag phase of bacterial growth and exhibited a high toxicity in the bioluminescence test, which was, however, reduced as a consequence of the biodegradation process. The oxidation products of pyrene were completely biodegradable, they reduced the lag phase of the microorganisms and did not inhibit the light emission of the luminescent bacteria.

These conclusions are summarized in table 2.

Table 2: Characteristics of the Oxidation Products With Regard to Biodegradation

	Toxic Effect	Biodegradability	lag phase
Naphthalene	relevant and persistent	very low	baktericide
Phenanthrene	high, partly eliminated	in part	prolonged
Pyrene	not detectable	good	reduced

These facts have to be taken into account when bioremediation measures are applied in practice, i. e. pre-oxidation does not necessarily lead to an improvement of degradation qualities. The investigations presented in this paper, which so far have only been carried out with PAH suspensions, can certainly not lead to conclusions for remediation of polluted soils, in which the contaminants are mainly adsorbed to the soil matrix. Model investigations with PAH that are adsorbed to the surface first have to show whether similar results can be obtained in such a system.

A prerequisite for the evaluation of remediation measures is the measurement of the ecotoxicological effect of oxidation products before and during remediation procedures. A contamination of the aquifer by formation of water soluble and toxic intermediate products has to be avoided.

6   REFERENCES

Allinger N.L., Cova M.P., de Jongh D.C., Johnson C.R., Lebel N.A., Steevens C.L.; 1980; Organische Chemie; Koßmehl Gerhard Hrsg.; Walter de Gruyter Berlin New-York.

Anonymus DIN 38 412 Teil 34 Deutsche Einheitsverfahren zur Wasser, Abwasser- und Schlammuntersuchung; Testverfahren mit Wasserorganismen (Gruppe L). Bestimmung der Hemmwirkung von Abwasser auf die Lichtemission von Photobacterium phosphoreum.

Battermann G., Werner P.W.; 1987; Feldexperimente zur mikrobiologischen Dekontamination;

Abfallwirtschafft in Forschung und Praxis; Franzius V. Hrsg.; 22; 367-373.

Breure A.M., Sterkenburg A., Volkering F., van Alden J.G.; 1992; Bioavailability as a Rate Controlling Step in Soil Decontamination Processes; Vortrag Dechema International Symposium; Karlsruhe; FRG.

Doré M.; 1989; Chimie des oxydants et traitement des eaux; éd. Lavoisier Tec. et Doc.; 11, rue Lavoisier; F - 75348 Paris Cedex 08.

Legube B., Guyon S., Sugimitsu H., Doré M.; 1986; Ozonation of Naphthalene in aqueous Solution - I & II; Wat. Res.; 20; 197-214.

Lund N.C., Swinianski J., Gudehus G., Maier D.; 1991; Laboratory and Field Tests for a Biological In-situ Remediation of Coke Oven Plant; In Situ Bioreclamation; Butterworth - Heinemann (editors); 396-412.

Stieber M., Böckle K., Werner P.W., Frimmel F.H.; 1990; Biodegradation of Polycyclic Aromatic Hydrocarbons (PAH) in the Subsurface; Third International KfK / TNO Conference on Contaminated Soil; Arendt F., Hinsenveld M., van den Brink W.J. (editors); 1; 473-479.

Stieber M., Haeseler F., Werner P.W., Frimmel F.H.; A Rapid Screening Method for Microorganisms Degrading Polyaromatic Hydrocarbons in Microtiter Plates; submitted Appl. Microbiol. Biotechnol.

Sturrock M.G., Cline E.L., Robinson K.R.; 1963; The Ozonation of Phenanthrene with Water as Participating Solvent; J. Org. Chem.; 28; 2340-2343.

Walter U., Beyer M., Klein J., Rehm H.-J.; 1991; Degradation of Pyrene by Rhodococcus sp. UW1; Appl. Microbiol. Biotechnol.; 34; 671-676.

Weißenfels W.D., Beyer M., Klein J.; 1990; Degradation of Phenanthrene, Fluorene and Fluoranthene by Pure Bacterial Cultures ; Appl. Microbiol. Biotechnol.; 32; 479-484.

BIODEGRADATION OF TNT (2,4,6-TRINITROTOLUENE) IN CONTAMINATED SOIL SAMPLES BY WHITE ROT FUNGI.

Andrzej Majcherczyk, Andreas Zeddel, Aloys Hüttermann

Institut für Forstbotanik der Universität Göttingen, Abteilung für technische Mykologie, 3400 Göttingen, Büsgenweg 2, FRG

ABSTRACT

Degradation of 2,4,6-trinitrotoluene (TNT) in liquid cultures of *Pleurotus ostreatus* and *Phanerochaete chrysosporium* was investigated. Both fungi were able to degrade this compound in only 45 and 120 hours respectively. No polymerisation products were detected. A special soil preparation technique allowed white rot fungi to penetrate the soil easily. *Pleurotus ostreatus* and *Trametes versicolor* degrade TNT in contaminated soil samples in about 4-6 weeks.

INTRODUCTION

In many countries, and especially in Germany, the destruction of TNT producing plants during and after the Second World War left hundreds of highly contaminated sites and, consequently, to highly contaminated ground water reservoirs, sea and river sediments. Until now, TNT (2,4,6-trinitrotoluene) is one of the most important conventional explosive used by military forces and for mining technologies. In addition to the above mentioned source of contamination, the careless disposal practices of manufacturing and loading facilities over the past 50 years has led to various grades of contamination of industrial areas and ground water (Pennington et al. 1990). The potential hazard of TNT and other nitroaromatics to human health as well as their toxicity for higher and lower animals and plants has been demonstrated in recent years (Kozuka et al. 1978). In addition, a mutagenic activity of this compounds was shown (Rickert et al. 1984) and a bioaccumulation of other nitro-explosives in plants was demonstrated (Harvey et al. 1991).

The nitroaromatic compounds (e. g., TNT, dinitrotoluene, *p*-nitrophenol, hexahydro-1,3,5-trinitro-1,3,5-triazine) are recalcitrant to complete degradation in the environment. The abilities of biological systems to degrade nitroaromatic compounds and other possible means of diminishing their concentration, e.g., photodegradation, were the subject of many studies (Lipczynska-Kochany 1991, Parrish 1977, Kaake et al. 1992, Spain et al. 1979, Spain & Gibson 1991). The usual pathway of degradation of these compounds by biological systems is the reduction of one or two nitrogroups to amino- (Naumova et al. 1982) and hydroxyamino- residues producing however, compounds with at least the same toxicity as the original by present ones. The biological interactions influence the further transformation of these products yielding azoxy compounds and probably azoxy polymers, developing a source for long term release of toxic compounds in the environment (McCormick et al. 1978). Already at low concentrations, TNT was reported to be inhibitory to many bacteria, actinomycetes, yeast and fungi, and until the last decade only a few organisms were found to be able to mineralise this compound.

Positive results of degradation of TNT were reported on the application of enriched microbial cultures from sludge and the use of *Pseudomonas* species. The mineralisation of TNT was detected at

F. Arendt, G.J. Annokkée, R. Bosman and W.J. van den Brink (eds.), Contaminated Soil '93, 1345–1351.

only very low levels and transformation of TNT to macromolecular structures of the polyamide type, formed by reaction of biotransformation products with lipids and protein constituents of the microflora, were published (Carpenter et al. 1978).

Studies during the recent years have shown a high degree of mineralisation of TNT and dinitrotoluenes by white rot fungus *Phanerochaete chrysosporium* (Valli et al. 1992, Fernando et al. 1990). The present communication reports on studies of degradation of TNT by other white rot fungi, *Pleurotus ostreatus* and *Trametes versicolor*. The big advantage of these organisms is their ability to hydroxylate and cleave aromatic ring systems, leading to water soluble and easy degradable aliphatic compounds.

MATERIALS AND METHODS

For a preliminary study in liquid cultures twelve days old cultures of *Pleurotus ostreatus,* grown on 50 ml BSM medium under normal air conditions and at 25° C, were used for the experiments. *Phanerochaete chrysosporium* was cultivated at a constant at 35° C and atmosphere of 70 % oxygen in 50 ml of nutrient nitrogen-limited medium. Six days old cultures were incubated with TNT. Trinitrotoluene was added to all cultures to give a final concentration of 100 mg/L. Cultures, in triplicates, were harvested after 0, 4, 16, 24, 45 and 120 hours and the mycelium was separated from the culture medium. Both, mycelia and medium filtrates, were extracted three times with 20 ml of dichloromethane. Internal standards (dinitrobenzene and nitroaniline) were added before the extraction step. Samples were analysed with GC-MSD, and the TNT concentration was calculated according to the recovery of an internal standard. In parallel experiments, under analogue culture conditions, an extract from a TNT-contaminated soil in a concentration corresponding to 50 mg/L of pure TNT was added. Samples were processed as described above.

A part of dichloromethane extract was evaporated, dissolved in tetrahydrofuran and analysed with GPC using TSK 1000H column.

In the next step, the studies were extended to soil samples contaminated with TNT, collected from a manufacturing facility destroyed in 1945. Samples of soil containing 800 mg/kg TNT were mixed with water to obtain a slurry, supplemented with potato pulp and finally solidified with wood chips to an absolutely homogenous, crumbly structure. The material was inoculated with 3 % millet culture of fungi and incubated at 25° C. Soil samples prepared in this way, were fully penetrated by fungi already after 2 weeks. Samples of soil were extracted and analysed after 2, 4 and 6 weeks.

RESULTS AND DISCUSSION

The degradation of TNT was measured by the disappearance of TNT. Results of the degradation in liquid cultures are presented in Fig.1a, 1b, 2a and 2b. *Pleurotus* and *Phanerochaete* were able degrade TNT in only 120 hours, but differed significantly in the kinetics of this process. *Pleurotus ostreatus* degraded TNT almost completely after only 48 hours, both in the case of the synthetic compound and soil extracts. *Phanerochaete chrysosporium* degraded in this time about 93 % and 70 % of TNT respectively. Mineralisation of TNT was not studied here. A full detoxification of soil samples has to be proved using a 14C-labelled TNT and especially through toxicity tests.

The degradation pathways of these two fungi studied here in liquid cultures probably differ. This was already shown for the degradation of other compounds and could be expected on the basis of different enzymes systems.

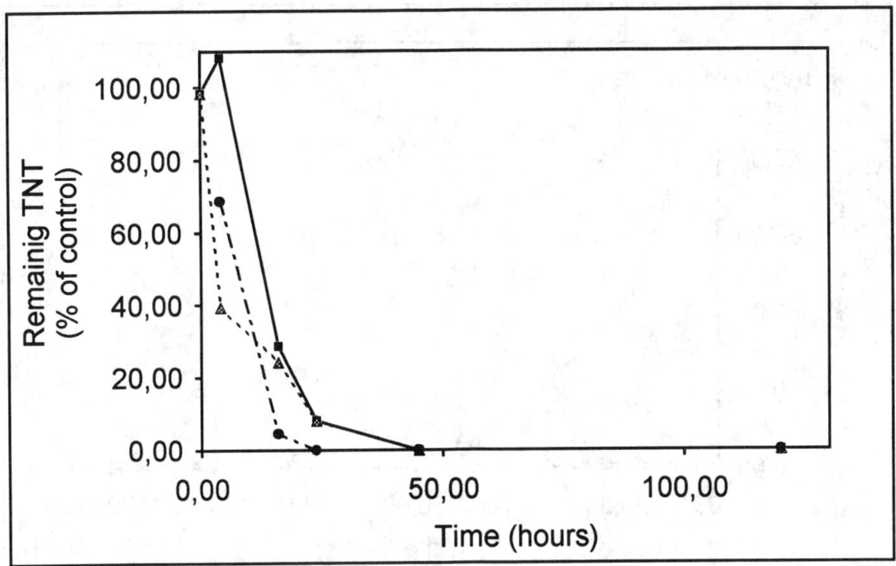

Fig. 1a. Degradation of synthetic TNT in stationary liquid cultures of *Pleurotus ostreatus*. Starting concentration of TNT was 100 mg/L. Determined in triplicate as a disappearance of 2,4,6-trinitrotoluene in culture media and as bound to mycel, in comparison to control flasks without fungi. ----■----: total TNT; - - ▲ - - : TNT in culture medium; --- ● - --: TNT found in filtered and washed mycel.

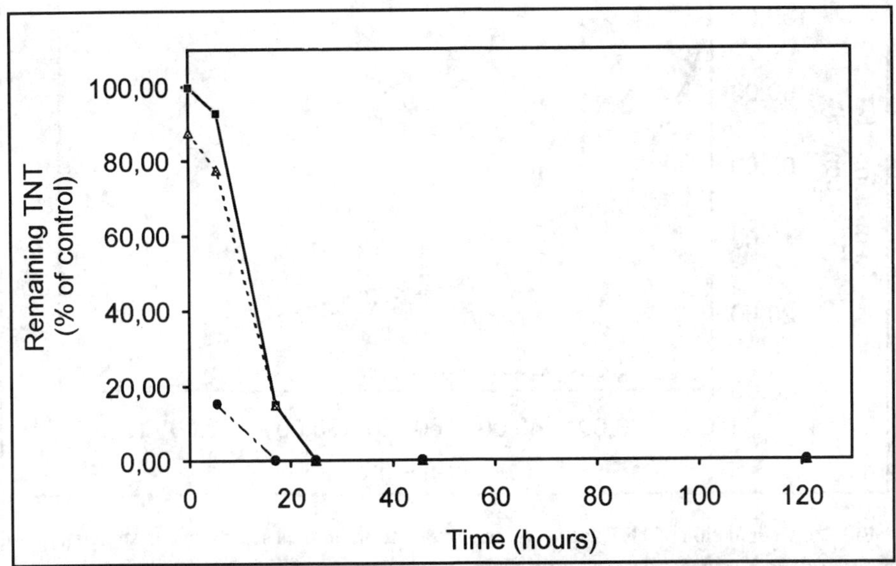

Fig. 1b. Degradation of TNT added as an extract from soil samples in stationary liquid cultures of *Pleurotus ostreatus*. Starting concentration of TNT was 50 mg/L, other parameters as in Fig. 1a.

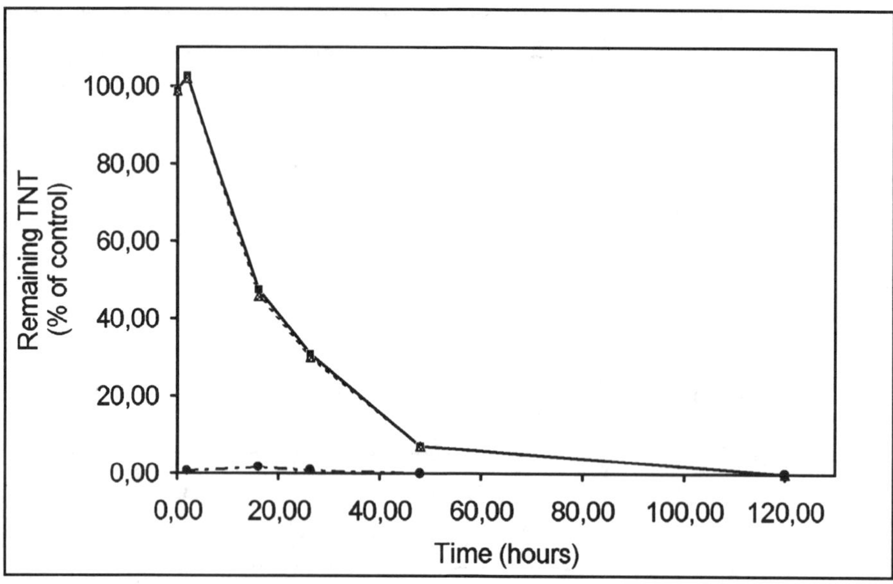

Fig. 2a. Degradation of synthetic TNT in stationary liquid cultures of *Phanerochaete chrysosporium*. Parameters see Fig. 1a.

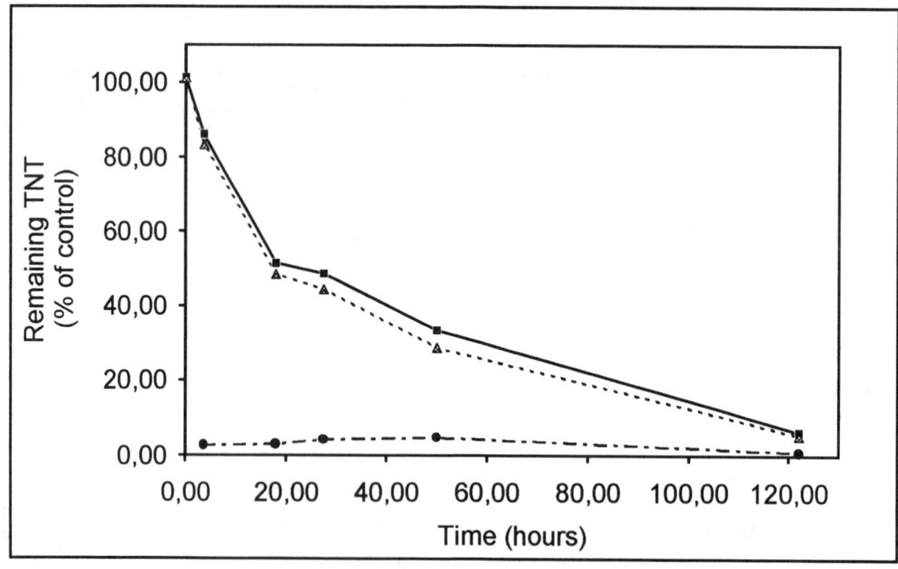

Fig. 2b. Degradation of TNT added with an extract from original soil sample in stationary liquid cultures of *Phanerochaete chrysosporium*. Starting concentration of TNT was 50 mg/L, other parameters as in Fig. 1a.

A large part of TNT was found to be adsorbed to the mycelium of *Pleurotus* only after 4 h. During the whole degradation period, a part of the TNT was found adsorbed to the mycel. The mycelium of *Phanerochaete* did not adsorb TNT significantly. The degradation of TNT in the original soil extract was slower than in case of the synthetic compound.

It was of great interest to find whether any part of TNT was polymerised by fungal enzymes. Therefore, a part of the extracts were analysed using gel permeation chromatography (GPC). In the case of *Pleurotus ostreatus,* some products possessing higher molecular weight than the original substrate were found within 120 hours of incubation. The products revealed a molecular weight of ca. 480, 400, 350 and 280 Daltons. No polymerisation products of polyamide or azoxy type (at least two TNT units) could be detected. Two of the products reached their maxima at about 48 h. One declined to almost zero at the end of the experiment, the other one remained at a constant concentration. No compounds over 1000 Daltons resulting from the degradation were detected (Fig. 3), and in case of soil extract, small amount of such substances declined during the cultivation period. Corresponding analysis for *Phanerochaete chrysosporium* are under investigation.

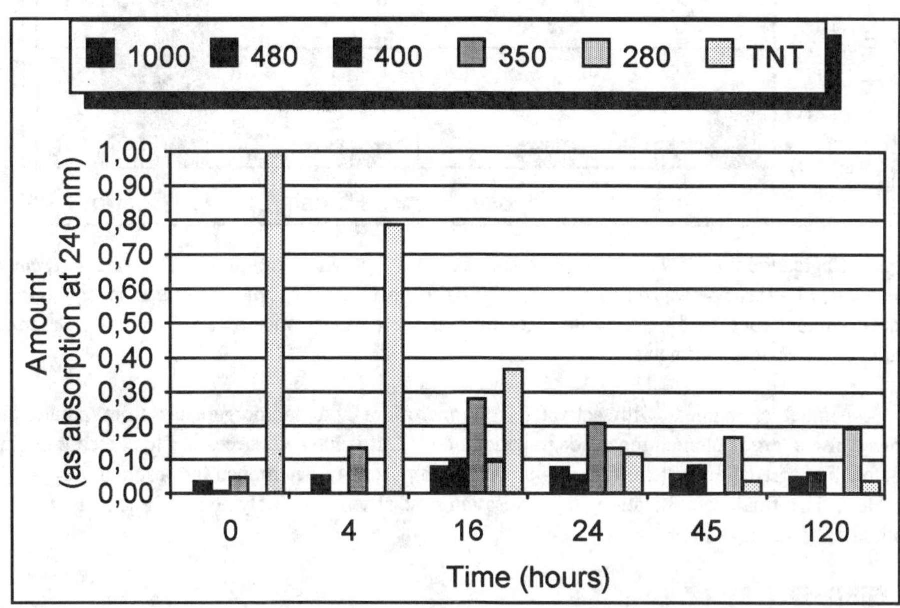

Fig. 3. GPC analysis of TNT and its metabolites by degradation with *Pleurotus ostreatus* in liquid cultures. The legend shows the approximate molecular weight of the compounds.

The existing potential of *Phanerochaete* to degrade trinitrotoluene in a solid state system (Fernando et al. 1990) was tested for *Pleurotus ostreatus* (two strains) and *Trametes versicolor* in a separate series of experiments. Results are presented in Fig. 4. Primary degradation of over 90% of the initial TNT concentration was reached already after 4 weeks by all fungi. *Trametes versicolor* removed TNT completely after 6 weeks of cultivation.

The extraction of soil samples showed together with a high TNT concentration (500-8000 mg/kg) a large amount of brown-red, probably already polymerised degradation or even biodegradation products. The ability to degrade this compounds, kinetic studies of TNT degradation as well as identification of some intermediate degradation products and a resulting toxicity are under study. A comparison of degradation ability and the degradation pathway of white rot fungi are under investigation.

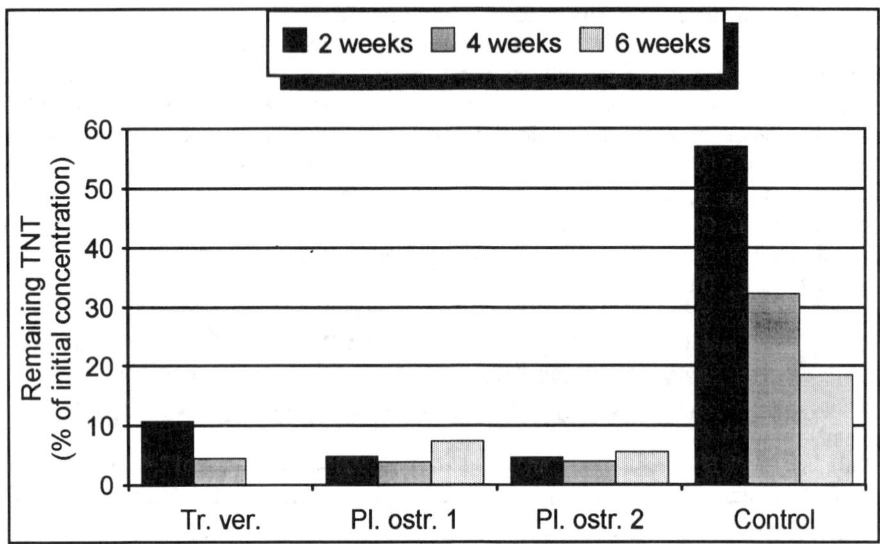

Fig. 4. Degradation of TNT in original soil samples by white rot fungi: **Tr. ver.** - *Trametes versicolor*, **Pl. ostr. 1** - *Pleurotus ostreatus* Florida, **Pl. ostr. 2** - *Pleurotus ostreatus* spec., **Control** - control without fungi and potato pulp. Measured as the disappearance of TNT in the total sample, results of triplicate experiments.

A parallel experiment with activated soil microflora and microorganisms from potato pulp showed also a large potential for the degradation of TNT. It is however necessary to compare this two possible biodegradation systems and it is especially important to determined the resulting degradation products. The final detoxification, mineralisation and absence of polymers will lead to the best method of decontamination.

REFERENCES

Carpenter, D. F., McCormick N. G., Cornell J. H., and Kaplan A. M. (1978). Microbial Transformation of 14C-Labelled 2,4,6-Trinitrotoluene in an Activated-Sludge Systems. Appl. Environ. Microbiol. 35(5): 949-954

Fernando, T.,. Bumpus J. A., and Aust S. D. (1990). Biodegradation of TNT (2,4,6-Trinitrotoluene) by *Phanerochaete chrysosporium*. Appl. Environ. Microbiol. 56(6): 1666-1671

Harvey, S. D., Fellows R. J., Cataldo D. A., and Bean R. M. (1991). Fate of the explosive Hexahydro-1,3,5-trinitro-1,3,5-triazine (RDX) in Soil and Bioaccumulation in Bush Bean Hydroponic Plants. Environ. Toxicol. Chem. 10: 845-855

Kaake, R. H., Roberts D. J., Stevens T. O., Crawford R. L., and Crawford D. L. (1992). Bioremediation of Soils Contaminated with the Herbicide 2-*sec*-Butyl-4,6-Dinitrophenol (Dinoseb). Appl. Environ. Microbiol. 58(5): 1683-1689

Kozuka, H., Mori M.-A., Katayama K., Matsuhashi T., Miyahara T., Mori Y., and Nagahara S. (1978). Studies on the Metabolism and Toxicity of Dinitrotoluenes - Metabolism of Dinitrotoluenes by *Rhodotorula glutinis* and Rat Liver Homogenate. J. Hygienic Chem. 24: 252-259

Lipczynska-Kochany, E. (1991). Degradation of Aqueous Nitrophenols and Nitrobenzene by Means of the Fenton Reaction. Chemosphere 22(5-6): 529-536

McCormick, N. G., Cornell J. H., and Kaplan A. M. (1978). Identification of Biotransormation Products from 2,4-Dinitrotoluene. Appl. Environ. Microbiol. 35(5): 945-948

Naumova, R. P., Belousova T. O. and Gilyazova R. M. (1982). Microbial Transformation of 2,4,6-Trinitrotoluene. Appl. Biochem. Microbiol. 18(1): 73-77

Parrish, F. W. (1977). Fungal Transformation of 2,4-Dinitrotoluene and 2,4,6-Trinitrotoluene. Appl. Environ. Microbiol. 34(2): 232-233

Pennington, J. C., and Patrick Jr. W. H. (1990). Adsorption and Desorption of 2,4,6-Trinitrotoluene by Soils. J. Environ. Quality 19(3): 559-567

Rickert, D. E., Butterworth B. E., and Popp J. A. (1984). Dinitrotoluene: Acute Toxicity, Oncogenicity, Genotoxicity, and Metabolism. Crit. Rev. Toxicol. 13(3): 217-234

Spain, J. C., and Gibson D. T. (1991). Pathway for Biodegradation of *p*-Nitrophenol in a *Moraxella* sp. Appl. Environ. Microbiol. 57(3): 812-819

Spain, J. C., Wyss O., and Gibson D. T. (1979). Enzymatic Oxidation of *p*-Nitrophenol. Res. Commun. 88(2): 634-641

Valli, K., Brock B. J., Joshi D. K., and Gold M. H. (1992). "Degradation of 2,4-Dinitrotoluene by the Lignin-Degrading Fungus *Phanerochaete chrysosporium*." Appl. Environ. Microbiol. 58(1): 221-228

PROCESS FOR MICROBIOLOGICAL SOIL-REMEDATION USING
GAS/SOLID-FLUIDIZED BED

W. Grau, W. Behns, B. Ebenau, L. Metzler, M. Müller
FZB Biotechnik GmbH Berlin
H. Haida, K. Friedrich, R. Lakowitz
TU "Otto von Guericke" Magdeburg
H. J. Künne, Magdeburger Energie- und Umwelttechnik GmbH

1. INTRODUCTION

The gas/solid-fluidized bed turns out to be suitable for
the performance of the solid-state-fermentation because of
its special features. A procedure can be established permitting an advantageous utilization for microbiological soil
remedation.

In order to prove the ability of this contacting technique
following fields of problems have to be investigated with regard to microbial metabolism and biodegradation of harmful
compounds:

- selection of the material for testing (soil/harmful com-
  pound/microorganism)

- microbiological pre-investigation

- process control of fluidized bed reactor.

A pilot plant (figure 1) is used for experimental investigation.

A special test system was available:

- heath soil
    * dry matter: about 93 man-p.c.
    * humus content: about 3   "
    * germ count: "   10 E + 6 cfu/g dry matter soil
    * nitrogen content: "   5 mg/100 g soil
    * phosphate content:"   3 mg/100 g soil

- technical dibutylphthalate (DBP)

- mixed culture.

2. RESULTS

A DBP-degrading mixed culture was prepared for the investigation of the fluidized bed solid-state-fermentation, whose degradative ability was proved in submerged cultivation.

F. Arendt, G.J. Annokkée, R. Bosman and W.J. van den Brink (eds.), Contaminated Soil '93, 1353–1356.
© *1993 Kluwer Academic Publishers. Printed in the Netherlands.*

The culture was isolated from samples of soil from industrial areas. The isolation was performed under selective
conditions using dibutylphthalate as C-source.
Figure 2 shows the different growth behaviour of the isolated
lated mixed culture in submerged fermentation that especially developes in DBP-containing soil. It was shown that
the indigenous mixed culture metabolizes DBP only in co-operation. Addition of P and N to the soil results in a shorter
lag phase and further DBP degradation.
Latest investigations of fluidized bed fermentation pointed
out that the cultivation of the mixed culture is possible both
at sterile and nonsterile soil under fluidization conditions. Adaption periods could be considerably shortened by
nutrient supply (figure 3). The pH-value is typical. The increase of pH-value indicates the growth of mixed culture.

The soil fluidization was maintained approximately constant
at 10 man-p. c. during the whole test period. The decrease of
DBP concentration in the soil during the fluidization period
is outlined in figure 4.

An energy-optimized process is examined concerning required
energy for solid fluidization alternating with non-fluidi-
zation-periods simulation of piling).

3. CONCLUSIONS

A water scarced cultivation is possible under conditions
of gas/solid-fluidization. A DBP biodegradation was observed simultaneously.

The fluidized bed is especially suitable for the treatment
of soil with high silt content.

The process utilization is advantageous in combination with
other remedial action techniques, such as

- washing process: Treatment of high-contaminated fine particals

- piling process: Using the gas/solid-fluidized bed for
  intensive fermentation, mixing and homogenization

The utilization concept of the fluidized bed process in
large scale soil remedation involves the modular construction
of a movable plant.

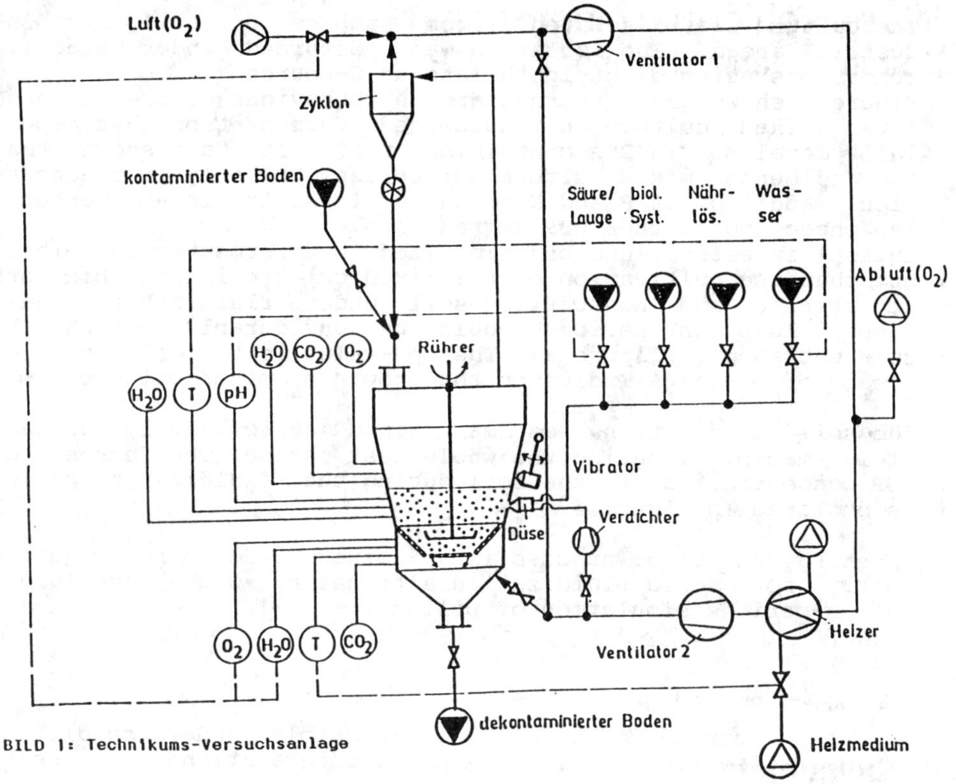

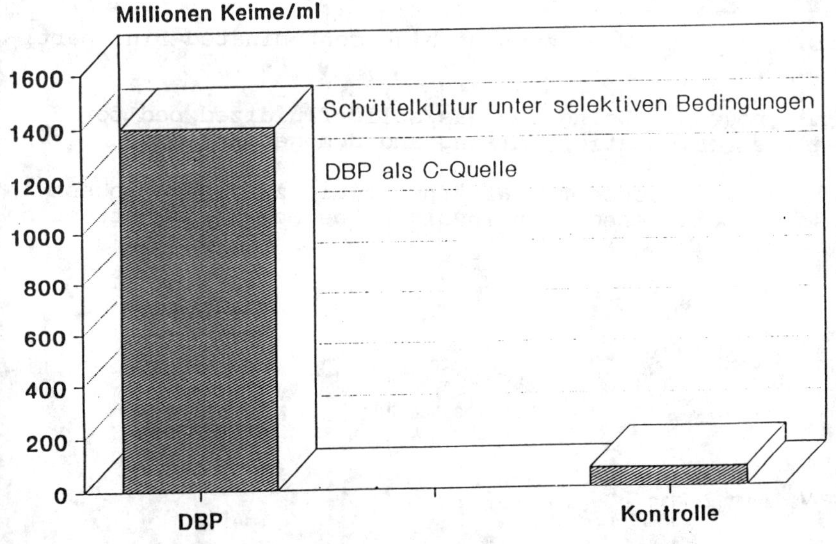

BILD 1: Technikums-Versuchsanlage

Nachweis der Abbauaktivität
Bodenmischprobe

BILD 2: DBP-Abbau im Boden

Wirbelschichtfermentation
Boden + Nährsalze mit 1% DBP

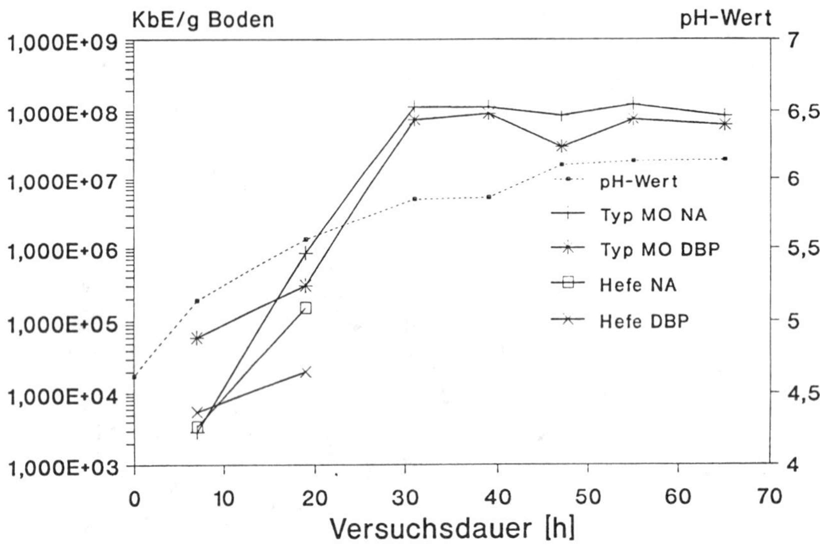

BILD 3: Kultivierungsverlauf

Wirbelschicht
DBP-Konzentration

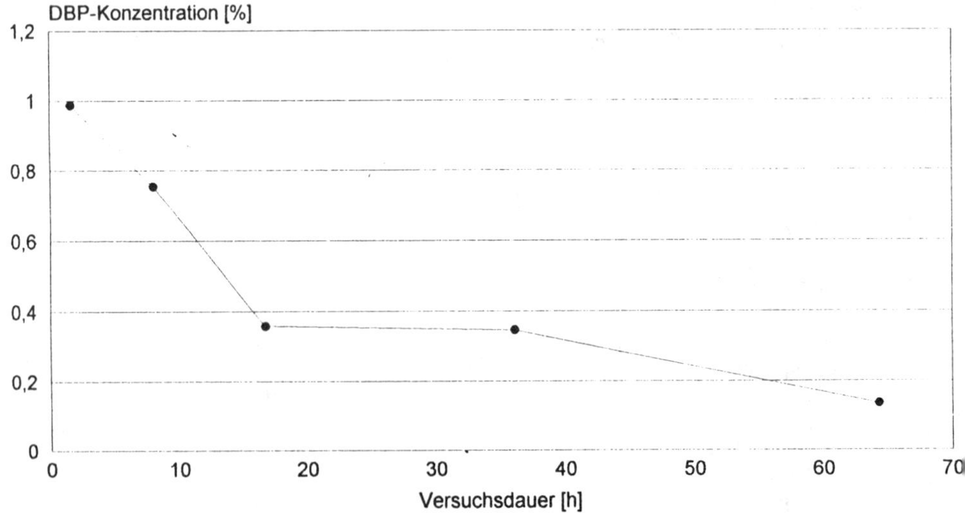

BILD 4: DBP-Verlauf im Boden

Results of a Feasibility Study into the Waste Disposal at the Johannes Pit using the VerTech Wet Oxidation Process

Dr. M. Daun , W. Bitter J. Pönisch

Mannesmann Anlagenbau AG, Abt. DET, Theodorstrasse 90, 4000
Düsseldorf 30. Germany.

1.   Introduction

The Johannes pit, a worked-out open-cast lignite mine in the industrial region of Bitterfeld/Wolfen, was used for decades as a dump and settling tank for the waste water from the cellulose production at the film factory in Wolfen. Today the pit covers an area of approx. 30 ha, measuring 1400 m North/South and 450 m East/West. With an average depth of 10 m, it is filled right up to the water surface with $2.7 \cdot 10^6$ m³ of sediments, of which $2.1 \cdot 10^6$ m³ lignin sludge. Whilst the other sediments contain a very high percentage of solids, the lignin sludge can be described as a deposit containing between 45% and 95% water with a gel-like to pulpy consistency.

As a result of the discharge of untreated sewage from the film factory, ash from the works' own power station and various sludges, the "lignin sludge lake" became heavily contaminated during the course of the years. The investigations conducted to date have revealed significant contents of harmful substances in the pit sludge on the one hand, and considerable contamination of the groundwater in the area surrounding the pit on the other. A further major hazard potential is posed by the gaseous emissions from the "lake" produced primarily by the anaerobic fermentation of the organic constituents of the sludge and volatile contaminants in the sludge.

As part of the study contracted by the Bitterfeld Qualification and Project Planning Company to clarify the real extent of the environmental pollution caused by the pit, the possible related danger to property requiring protection and the necessary rehabilitation requirements, methods of decontamination and/or safe disposal of the pit sludge are also being investigated.

One possible method of treating the lignin sludge contaminated with the harmful substances is to convert the organic constituents, and partially also the inorganic matter, by wet oxidation using oxygen in the deep shaft reactor at temperatures of approx. 290° C and pressures of approx. 100 bar. Experience with the treatment of sewage sludges in this type of reaction system shows that a more or less inorganic solid, non-polluted waste gases and waste water which can be subjected to easy biological secondary treatment methods are produced.

F. Arendt, G.J. Annokkée, R. Bosman and W.J. van den Brink (eds.), Contaminated Soil '93, 1357–1368.
© 1993 *Kluwer Academic Publishers. Printed in the Netherlands.*

2. Contents of the Lake and Resulting Rehabilitation Activities

The studies contracted by the Bitterfeld Qualification and Project Planning Company were to allow conclusions to be drawn as to the hazard potential of the deposits in "Pit Johannes" as well as to the resulting emission and exposure risks in order to be able to develop the necessary safety and rehabilitation measures.

- Results of the investigation - lignin sludge

The gel-like lignin sludge is enriched with a wide range of substances, many of which toxic, resulting from the discharged waste and the conversion processes taking place in the sludge. The lake must be regarded as distinctly to heavily contaminated which rules out the possibility of an uncontrolled drainage, use or deposit. Of particular relevance for the hazard potential are the following substances and material classes:

— Heavy metals: Lead and zinc (the C value of the Holland list is considerably exceeded);
— Inorganic and organic sulphur compounds: Sulphate, sulphide (all samples exceed the C value of the Holland list), carbon disulphide, hydrocarbons, methyl mercaptan, dimethyl sulphide;
— Solvents: Methanol and toluene;
— Phenols: Phenol, methyl phenols, chlorophenols (several samples exceed the B value of the Holland list) and
— organically bound halogen compounds measured as EOX (the majority of the samples exceed the B value of the Holland list).

Dibenzodioxines and dibenzofuranes are not relevant for the lignin sediment. Despite its toxic constituents the lignin sludge is biologically active, indicated clearly by the visible emissions from the pit.

– Results of the investigation - air

As far as the air is concerned it is to be observed that the typical odour which used to surround the vicinity of the pit today plays only a subordinate role, insofar as it was caused directly by the waste water from the cellulose production which was stopped in the spring of 1991. The sources of both the current and also future gas releases are thus primarily anaerobic fermentation processes and the escape of volatile contaminants from the sludge. The investigations of the gas formations in the dump revealed not only the main constituents methane, nitrogen, oxygen and carbon dioxide but also traces of the gases hydrogen sulphide, toluene, dichloromethane and tetrachloromethane.

– Results of the investigation - groundwater

The investigations of the groundwater show that high concentrations of AOX and steam-volatile fatty acids are to be found in the groundwater downstream of the "Pit Johannes". Other characteristic and hazard-relevant constituents of the lignin sediment and other indicative parameters were not discovered here.

As the results of the investigations into the constituents of the "lake" and the emission paths described above indicate, there is an urgent need of rehabilitation and safety measures in order to protect the air and the groundwater from further immissions. Possible safety measures which were also subjected to critical examination during the course of the project as a whole are not discussed further here, as they do not offer a long-term solution to the rehabilitation problem. In the medium to long term, only a conversion of the solid contents of the "lake" can help to achieve the objective of the rehabilitation measures. Wet oxidation is one possible treatment method here.

3.   Wet Oxidation in the Deep Shaft Reactor

Wet oxidation is a treatment process which has been known since the beginning of this century in which organic substances, in some cases even inorganic substances, are oxidised in the liquid phase using oxygen at temperatures of 150° C to 330° C and pressures of 10 to 220 bar. Some methods operate even in the supercritical range with temperatures of up to approx. 600° C and pressures around 250 bar. In other methods, the effectiveness of the process is increased by the use of catalysts. The products of the process are primarily carbon dioxide and water. In the aqueous phase, depending on starting product and reaction conditions, relatively low concentrations of low-molecular organic acids are left.

The effectiveness (i.e. degree of destruction) of the wet oxidation process relative to frequently occurring organic compounds is better than 98 %. Only low-molecular organic acids, which themselves are highly biologically degradable, are not attacked in the wet oxidation reactor and are actually formed in small quantities. The compounds are thus primarily responsible for the residual chemical oxygen demand (COD) of the medium leaving the reactor.

If the organic compounds to be converted contain sulphur, chlorine or phosphorus, then sulphuric, hydrochloric and phosphoric acid are formed. Organic nitrogen is primarily converted into ammonia.

In the wet oxidation processes employed to date, high pressure vessels were used. In the wet oxidation method using a deep shaft reactor such as that developed for the VerTech process, the surface installations operate at low pressure. The configuration of the VerTech reactor is shown in Figure 1.

The sludge to be treated is admitted to the reactor system by being pumped into the inner pipe, the inlet pipe. Oxygen is injected into the inlet pipe at various depths.

In order to start the reaction, the sludge suspension at the bottom of the reactor is heated to the requisite reaction temperature via a heat exchanger system. An external gas-fired burner is used for this initial heating up process.
The oxidation process begins at a temperature of around 175° C, whereby the temperature increases sharply with increasing depth due to the exothermic reaction. The oxidation process is achieved by the addition of oxygen. The heat generated by the oxidation is now sufficient to maintain the desired reactor temperature. The heat exchanger system is now used to cool the column in which the oxidised material rises in the return pipe. The liquid column exerts a hydrostatic pressure of approx. 85 to 110 bar at the bottom of the reactor. This high pressure is achieved without any high pump power.

The pump which delivers the sludge into the reactor serves only to compensate the friction losses and to overcome the slight difference in height between the inlet and return pipes. The temperature at the bottom of the reactor is kept at approx. 275° C by the cooling water, whereby the high pressure prevents the sewage sludge from boiling. The heat exchange between the liquids in the inlet and return pipe and the masstransfer from the liquid into the gaseous phase is improved by the intense mixing of the gas bubbles present in the liquid.

The diameter and length of the pipe system are designed in such a way that an adequate residence time and sufficiently high temperatures are assured to achieve the desired degree of oxidation. The reactor behaves almost like an ideal plug flow, thus ensuring that all the sludge particles are completely incorporated into the reaction processes. The large length/diameter ratio of the reactor is a characteristic for a favourable counterflow heat exchange between inlet and outlet.

The oxidised liquid flows through the return pipe of the reactor system back to the surface and is cooled both by the heat exchanger circuit operating in counterflow and by the sludge also flowing in the opposite direction in the inlet pipe.

The VerTech system was tested on a laboratory scale for the first time in 1973. Since 1982, experience has been gathered using a reactor of approx. 1500 metre depth under realistic operating conditions in Longmont/Colorado (USA). The first commercially used industrial-size plant is currently being commissioned in Apeldoorn (Netherlands). From 1993, 9% of the total sewage sludge production in the Netherlands will be disposed of in this plant.

4. Performance of the Wet Oxidation Investigations

In order to test the convertibility of lignin sludge in the deep shaft reactor, various wet oxidation tests were performed with lignin sludges from various sources during the course of the feasibility study.

During the choice of specimens, attention was paid to an adequate distribution over the area of the pit and to the depth of the sampling point in order to be able to carry out wet oxidation tests on representative quantities and qualities of sludge. During the sludge sampling and analysis it was discovered that with increasing sampling depth, the loss on ignition and the lignin content of the sludge samples decreased significantly, whilst the dry substance content of the sludge increases sharply. The whole lignin sludge lake can thus be described as highly heterogeneous with respect to the sludge composition. Particularly noticeable were the widely fluctuating local heavy metal contents.

In a further phase of the investigation, not only pure lignin sludge but also mixtures of sewage sludge and lignin sludge were subjected to wet oxidation. These tests were intended to give an indication of how the sludge disposal from the Pit Johannes could possibly be combined with the sludge disposal from a sewage plant with a view to achieving a longer period of operation of the VerTech system.

The wet oxidation tests were performed in two different reaction systems. The general examination of the sludge specimens with respect to their suitability for wet oxidation was performed in an electrically heated autoclave (specimen volume 1,000 ml) in which the conversion of the lignin sludge is investigated in relation to a temperature profile which the specimens have to pass through within a given period of time and which corresponds fundamentally to the temperature curve in the industrial scale deep shaft reactor.

The maximum temperatures achieved in the tests lay between 280 and 298° C. The residence time of the sludge in the reaction vessel was approx. one hour by analogy with the residence time in the industrial scale reactor. The oxygen concentration was constantly set to a stoichiometric ratio to the chemical oxygen demand (COD) of the substances in the reactor.

Some of the lignin sludges were wet oxidised in a laboratory plant (pipe reactor). Here again, residence times of one hour and maximum temperatures of 270 - 290° C were set. Larger volumes of sludge (approx. 1250 litres per oxidation process) were treated in this plant so that adequate volumes of degradation products (waste gas, process water, solids) were also available for the further investigations.

1362

5.   Results of the Investigations

Figure 2 shows schematically the material flows occurring during the VerTech process.   The result of the conversion of a sludge specimen ("influent") with oxygen in the wet oxidation reactor is a raw gas and a reactor "effluent".   The separation of the effluent produces process water which may have to be further treated, depending on the demanded water quality, and a filter cake which is either further processed or dumped.

5.1   Evaluation of the Wet Oxidation Experiments

The wet oxidation results achieved in the laboratory autoclaves can be evaluated as good to very good with respect to the reduction of the organic mass and the dry residue.   This is documented exemplarily in Figures 3 and 4 with the results of the sludge from sampling point RKB 26/92.

The reduction of > 90% in COD in the not process-optimised LBR tests indicates that with an optimised design of a VerTech reactor, even higher reduction values can be achieved.   It should also be noted here that this residual COD value is attributable primarily to the low-molecular acids produced in the process.   The components do not, however, represent a hazard potential.   They are degraded quite simply in a downstream biological process water treatment plant.   The considerable improvement in the biological degradability of the sludge constituents can be illustrated by the COD/BOD_5 ratio.

The COD/BOD_5 ratio of 11.3 in the influent indicates that it is very difficult to treat the lignin sludge by biological methods.   Following wet oxidation, a COD/BOD_5 ratio in the filtrate of approx. 1.6 is achieved, whereby the remaining organic substances can be easily biologically degraded.

From the autoclave experiments it is also possible to derive the volume of the dry residues to be dumped.   At sampling point RKB 26, for example, approx. 40 kg of dry residues are produced per cubic metre of sludge treated.

Similarly good results as those achieved in the conversion of the lignin sludge were also achieved with the conversion of sewage sludge/lignin sludge mixtures, so that an admixture of sewage sludge to prolong the operational period of the reactor is generally possible.

5.2   Raw Gas Composition

The conversion of the sludge in the liquid phase (maximum temperature 60° C) means that the raw gas is dust-free.   The main constituents of the gas phase are carbon dioxide, non-converted oxygen, inert nitrogen, small quantities of volatile hydrocarbon compounds and carbon monoxide as a product of the incomplete oxidation reaction (see Table 1). The raw gas contains only minor traces of any other compounds. Although the carbon monoxide content in the raw gas exceeds the limit on hazardous substances demanded by the 17th Federal Immissions Control Act and its associated directives, its concentration is reduced in a catalytic raw gas post-treatment to well below the demanded limits. For all the other raw gas constituents subject to limit value control, the values are (in some cases far) below the demanded limits without additional technical measures.

Component	Concentration (Vol.%)	
	at reactor outlet	after catalytic raw gas treatment
CO_2	64,8	67,6
CO	3,7	0
O_2	3,8	2,4
N_2	2,6	2,2
C_xH_y	0,4	0
H_2O	24,7	27,8

Table 1: Mean gas phase composition at the reactor outlet and after catalytic raw gas treatment

The VerTech process thus has particular advantages on the waste gas side in comparison with thermal material conversion processes (e.g. combustion). The waste gas volume is only 15% of the flue gas volume released during conventional sludge combustion, and the waste gas contains no notable volumes of NO_x or SO_2. It should also be pointed out that no dioxins or furanes can be detected in the waste gas. A reformation of dioxin in the waste gas by "Denovo synthesis" cannot take place under the prevailing process conditions.

5.3 Assessment of the Resulting Process Water

The analyses performed show that a considerable part of the heavy metals originally in the lignin sludge is now found in the process water (filtrate). Before biological process water treatment can take place, the high zinc and nickel concentrations, in particular, must therefore be reduced by precipitation with milk of lime $Ca(OH)_2$ which is added directly to the effluent.

The heavy metal concentration can be drastically reduced by precipitation at pH 9. The precipitated compounds are then removed during the separation of the remaining suspended solids which is performed anyway in a chamber filter press.

The filtrate left after separation of the solids from the effluent still contains a certain concentration of soluble organic compounds. These are primarily low-molecular organic acids which can be easily degraded by biological means.

Compared with the average composition of communal sewage, the filtrate from the lignin sludge effluents contains no increased concentrations of dissolved nitrogen compounds in the form of ammonia and ammonium or phosphorus compounds.

Only after admixture of sewage sludge to the lignin sludge do ammonium concentrations occur in the waste water which are distinctly higher than the average values in communal sewage. Nevertheless the discharged process water can be post-treated without problem in an existing sewage plant or in a specially erected biological treatment stage.

It should finally be pointed out that with the pure wet oxidation of sewage sludge, a precipitation of the heavy metals is not necessary. In view of the considerably lower starting concentrations and the improving pH values in the effluent (pH = 7 - 8), primary heavy metal oxides are precipitated during the wet oxidation and are deposited on the suspended inorganic residue particles (SiO_2, Al_2O_3) with which they are then separated.

5.4 Residues

The residues which are separated during the wet oxidation of lignin sludge exhibit high heavy metal contents. Furthermore they contain a small amount of organic compounds. Therefore, the residues should be taken to a possible industrial reuse or should be delivered to post-treatment. The unfavourable composition of the residues in this case, however, is not a disadvantage of the VerTech process but due to the problem of the very high heavy metal contents in the original lignin sludge which cannot be reduced by other methods either.

The situation is considerably more favourable in the case of the wet oxidation of purely sewage sludge. A typical composition of the residues of wet oxidised sewage sludge is shown in Table 2.

Metal	measured
Cadmium	4,5
Chrome (III)	200
Copper	520
Mercury	3,0
Lead	425
Nickel	75
Zinc	2200

Table 2: Heavy metals in the residues (mg/kg dry)

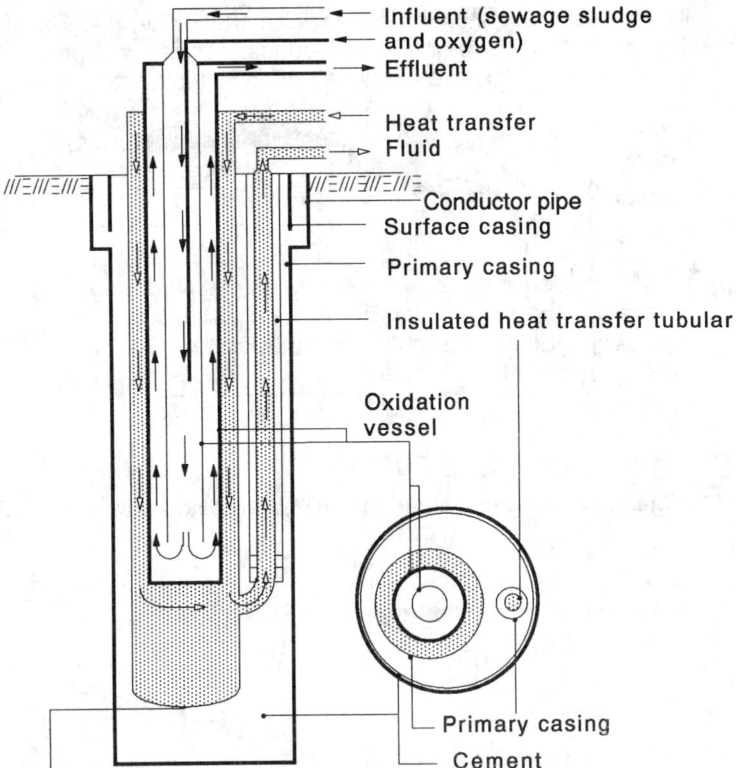

figure 1: View of the VerTech Oxidation Vessel

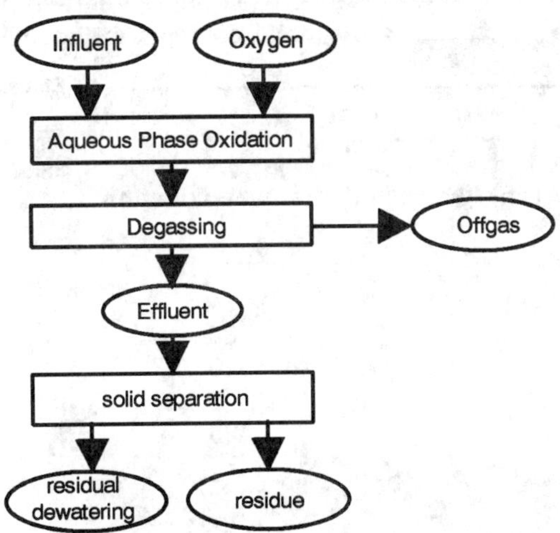

figure 2: Material flow of the VerTech - Process

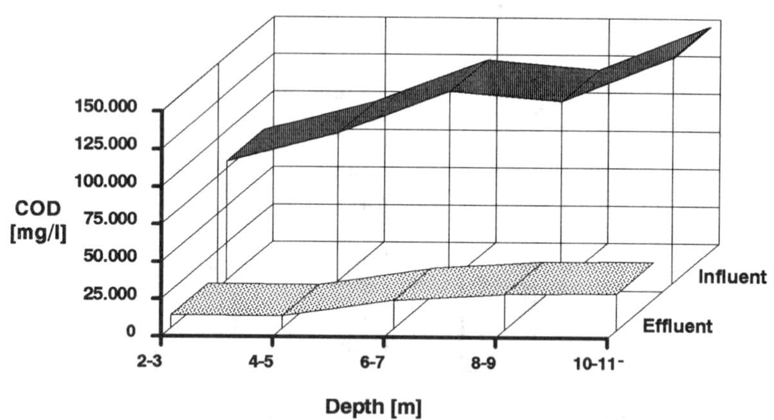

figure 3: Chemical Oxygen Demand (COD) before and after wet oxidation

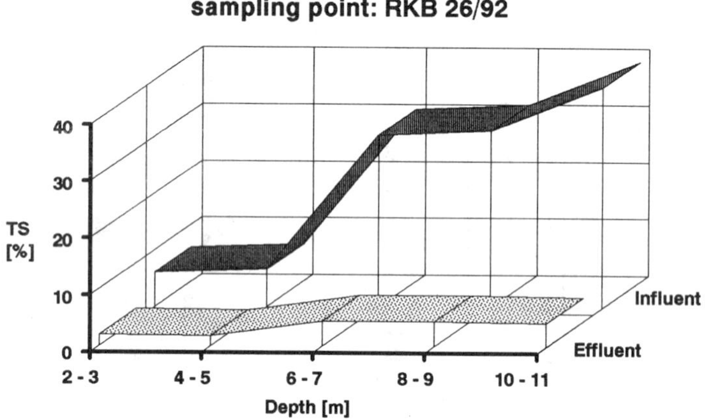

figure 4: Total solids (TS) before and after wet oxidation

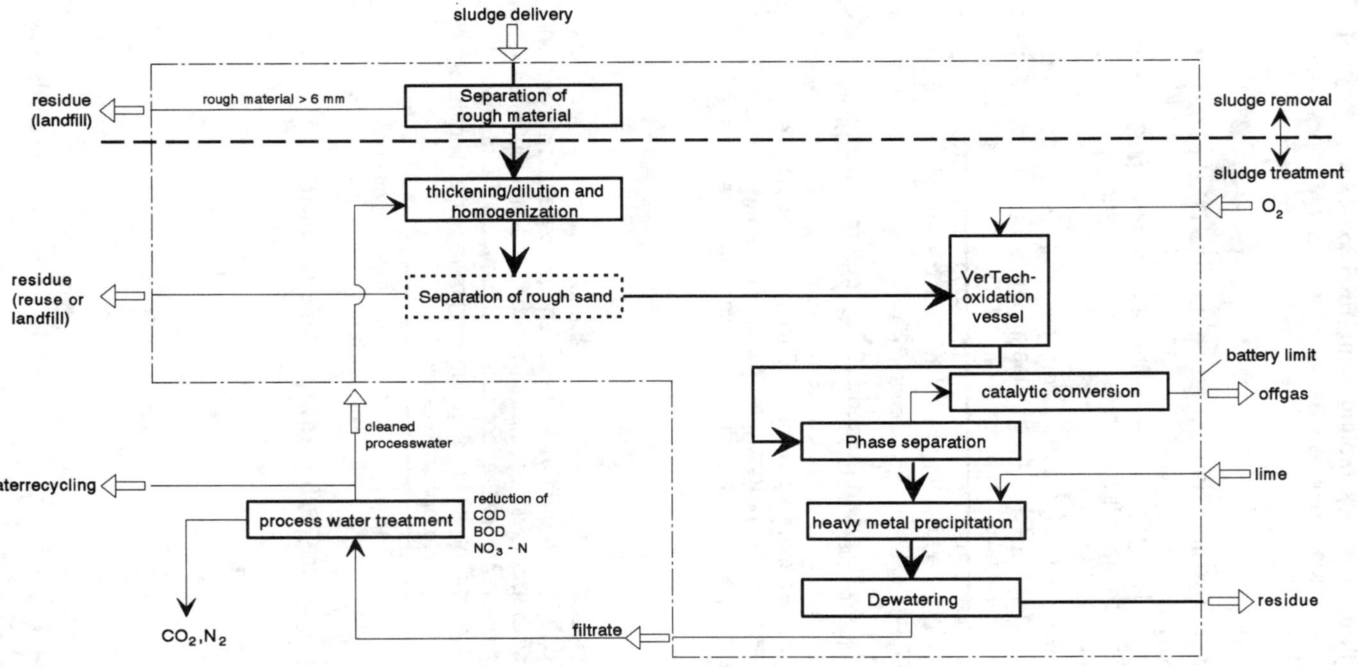

figure 5: Treatment of lignin-containing sludge with VerTech - Proces (process flow sheet)

Elution tests with the residues produced outstanding results. The heavy metals remain unleachably bound. These residues can be safely deposited on a Class II refuse dump.

6.   Concept of a VerTech Plant for Lignin Sludge Treatment

The plant flow chart for lignin sludge treatment shown in Figure 5 was developed on the basis of the results of the wet oxidation tests performed.

In view of the very heterogeneous composition of the sludge specimens, the sludge is set to a mean COD of 40 g/l (design value for the VerTech reactor) in a dilution and homogenization stage. Treated process water can be used for dilution. If the sludge contains more than 6% solids, the coarse sand is removed in a centrifuge. Only then does the sludge enter the actual reactor system which has already been described in detail in section 3.

The gases from the oxidised liquid leaving the reactor are separated in a phase separator before passing to a catalytic post-treatment. The heavy metals in the process water are precipitated by the addition of $Ca(OH)_2$ and removed in a separator together with the inorganic residues. The solids-free, easily degradable filtrate subsequently passes to a biological sewage treatment stage.

7.   Summary

The VerTech wet oxidation process is well suited to the treatment of the contaminated lignin sludge from the Pit Johannes. The process is environmentally safe, as only carbon dioxide and nitrogen are emitted to the atmosphere. A favourable utilisation period for the VerTech plant can be achieved if the plant continues to be used for the disposal of sewage sludge following the sanitation of the Pit Johannes (period required approx. 10 years).

A COMBINED REMEDIATION TECHNIQUE FOR SOIL CONTAINING ORGANIC CONTAMINANTS: COMBINED HYDROCYCLONE SEPARATION, PHOTOCHEMICAL TREATMENT AND BIO-REMEDIATION.

B. AKSAY, E. TEN BRUMMELER and J. BOVENDEUR

Heidemij Realisatie BV, P.O.Box 660, 5140 AR Waalwijk, The Netherlands.

1. ABSTRACT.

In a pilot plant study the potentials of a novel technique for the complete elimination of organic contaminants in soils have been investigated. Several different processes are combined for this purpose: a separation process by means of hydrocyclones, and an aerobic biodegradation process for the fine fraction of the soil, obtained by hydrocycloning. Photochemical oxidation has been used as an additional techniques to break down the contaminants to a certain level, resulting in easy biodegradable fragments. A pilot installation was designed, built and operated under several conditons with several types of soil contaminated with hydrocarbons and PAH. Four different soils with several contamination degrees of mineral oil have been treated.

2. PROCESS DESCRIPTION

The concept described above was investigated by Heidemij at pilot plant scale. The pilot plant had a maximum capacity of 5 tons per charge.

In the case of soil contaminated with oil components (alifatic hydrocarbons, PAH) the main aim of the separation technique is to produce a clean fraction (the sand fraction) out of the oil contaminated soil which can be reused. On the other hand the separation process produces a heavily polluted sludge fraction that has to be treated additionally in the next unit operations. The efficiency of the hydrocyclone separation step is optimized by a multi stage configuration of hydrocyclones.

The ultra-violet (UV) radiation/oxidation treatment technology that is developed by Heidemij applies a combined process of UV radiation and hydrogen-peroxide to oxidize organic components in water. Various operating parameters are adjusted in the pilot plant to enhance the partial oxidation of organic contaminants in the soil. The oxidation process was applied on the already separated fines contaminated with oil and PAH.

The fine fraction that already has been pretreated by photochemical oxidation needs a final biotechnological treatment step, where the pollutants (fragments) are eliminated by biodegradation. The biodegradation was carried out in a circular tank reactor with a volume of 25 m^3. The total solids concentration of the slurry amounted to 3%. The bioreactor is aerated by a compressor and perforated aeration plates placed at the bottom of the reactor.

3. RESULTS

Several charges with oil contaminated soils were investigated in the pilot plant, with initial mineral oil concentrations (assessed by gaschromatography) of 400-5,000 mgkg^{-1}. A charge is considered to be completed as the threshold level of 100 mgkg^{-1} is achieved. The period needed to reach this level appears to be in the range of 3-8 days for the pilot plant.

F. Arendt, G.J. Annokkée, R. Bosman and W.J. van den Brink (eds.), Contaminated Soil '93, 1369–1370.
© *1993 Kluwer Academic Publishers. Printed in the Netherlands.*

During the second experiment with completely supported biodegradation a significant removal of PAH was obtained. Within a period of 15 days a reduction from 30 mgkg^{-1} PAH (10) to 5-10 mgkg^{-1} is achieved (Fig. 1). As the sand fraction did not contain a significant amount of PAH, the final concentration of the PAH in the recombined soil mass (sand and fine fraction) amounts to 2-4 mgkg^{-1}. This figure is below B level and is close to A level for PAH.

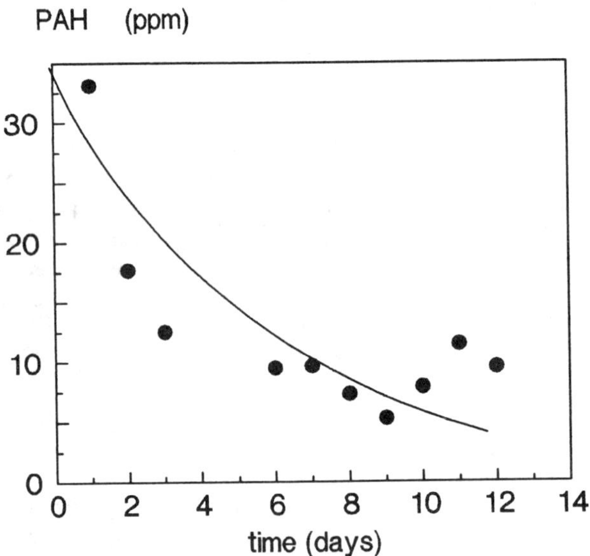

FIGURE 1. Elimination of PAH by bioremediation supported by photochemical oxidation.

5. CONCLUSIONS

The combination of soil washing techniques (hydrocyclone separation) with bioremediation (bioreactor), supported by photochemical oxidation has shown promising results for the elimination of mineral oil and PAH.

Based on these results the combined technique will be developed further for treatment of recalcitrant organic xenobiotics in soil.

In general, combined techniques enable tailor-made solutions to soil pollution problems. As a result combined techniques require detailed information on the characterization of contaminants and geophysical components.

6. ACKNOWLEDGEMENT

This study on the possibilities of a combined technique has been financially supported by NOVEM.

SOIL WASHING INCLUDING BIOLOGICAL TREATMENT OF THE CONTAMINATED SILT/CLAY-FRACTION

GÜNTER BEUDERT, PETER KÜBLER, KARL SCHMID UND HERMANN H. HAHN

INSTITUT FÜR SIEDLUNGSWASSERWIRTSCHAFT, UNIVERSITÄT KARLSRUHE, AM FASANENGARTEN, D-7500 KARLSRUHE

1. INTRODUCTION

Soil washing has been developed to a widely used technique for the cleanup of hazardous sites. The mechanism of soil washing is the separartion of the highly contaminated silt/clay-fraction of the soil which has to be disposed of or burned. Organic soil pollutants like mineral oil are principally susceptible to biological degradation. If such a contamination in the silt/clay-fraction could be effectively reduced by biological treatment, the way to get rid of that fraction could be the remixing with the cleaned course soil fractions for redeposit or the disposal to a landfill with lower safety demands compared to a landfill necessary for the untreated fine fraction.

2. MATERIALS AND METHODS

The contaminated silt/clay-fraction was separated using a small scale washing unit consisting of a mixer, screens (2 mm, 1 mm, 0,2 mm) and a hydrocyclone (20 μm). The resulting suspension of fine particles < 20 μm was continously fed to an aerated stirred reactor for biological treatment. Treatment efficiency was determined by analysis of the hydrocarbon content (IR spectroscopy of the freone extract according to DIN 38 409, Teil 18) of the reactor influent and effluent. The extracted hydrocarbons were then characterized by GC-FID.

The experiments were carried out with silty soil (Loess) artificially contaminated with diesel and with an oil contaminated sandy soil from a hazardous site. The silty soil consisted of 70 % coarse silt and 12 % particles smaller than 20 μm. The sandy soil contained 8 % of the soil mass in the fraction < 20 μm.

3. RESULTS AND DISKUSSION

3.1. Fractionation by soil washing

Due to the high silt content the Loess soil is not a typical soil for application of the washing technique. Nevertheless the separation of the silty soil (hydrocarbon content 6 700 mg/kg dry soil) resulted (Fig.1) in a fine particle fraction (< 20 μm) with high hydrocarbon concentration (35 000 mg/kg) and a coarse particle fraction of low hydrocarbon concentration (2 400 mg/kg). About 67 % of the hydrocarbon mass were adsorbed to the fine fraction which contains only 10 % of the soil mass.

The sandy soil from the hazardous site (hydrocarbon content 3 200 mg/kg) was separated into the fine particle fraction (hydrocarbon content 27 000 mg/kg) and four coarse fractions with hydrocarbon contents of 600 - 4000 mg/kg (Fig. 2). The high residual contamination of the coarse fractions was due to the presence of porous coal particles with high hydrocarbon content which couldn't be separated of by the applied procedure. The fine particle fraction contained 68 % of the hydrocarbon mass accumulated in 8 % of the soil mass.

3.2. Biological Treatment

The biological treatment efficiently reduced the particle bound hydrocarbon content to 86 % of the original contamination at a dilution rate of 0.17 day^{-1}. The GC-FID analysis indicated a preferential degradation of the n-alkanes. The elimination rate decreased from 100 % for the C_{11}-alkane to 87 % for the C_{25}-alkane, whereas that of the unresolved area of the chromatogramm decreased by 84 %. During a consecuting periode of 14 days in batch culture the treatment efficiency with regard to the total hydrocarbon content raised to 95 %.

For the fine particle suspension of the hazardous site only a reduction of 25 % of the original hydrocarbon contamination could be achieved by biological degradation at a dilution rate of 0.17 day^{-1}. According to the GC-FID analysis mainly n-alkanes were degraded under these conditions (Fig. 3). After the consecuting batch period of 18 days no resolved peak could be detected and the treatment efficiency raised to 85 %.

F. Arendt, G.J. Annokkée, R. Bosman and W.J. van den Brink (eds.), Contaminated Soil '93, 1371–1372.
© 1993 Kluwer Academic Publishers. Printed in the Netherlands.

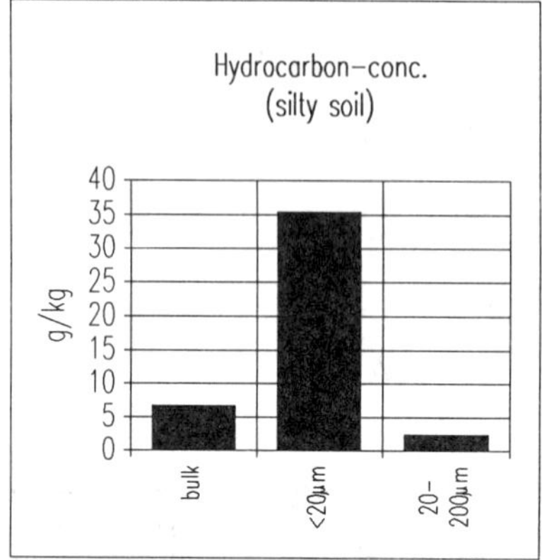

FIGURE 1. Hydrocarbon concentration in the silty soil and the fractions of the soil washing procedure.

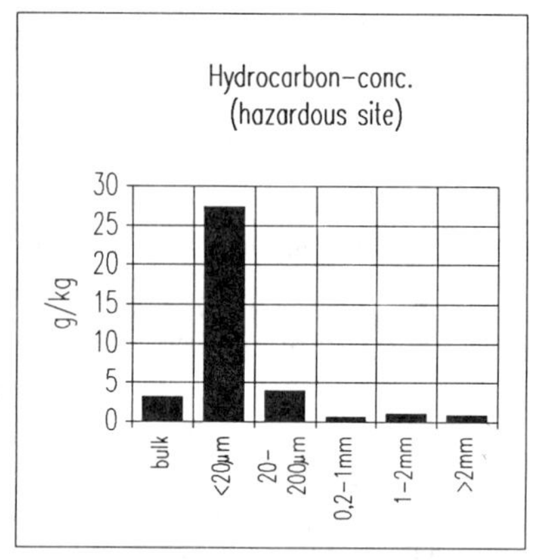

FIGURE 2: Hydrocarbon concentration in the soil from the hazardous site and the fractions of the soil washing procedure.

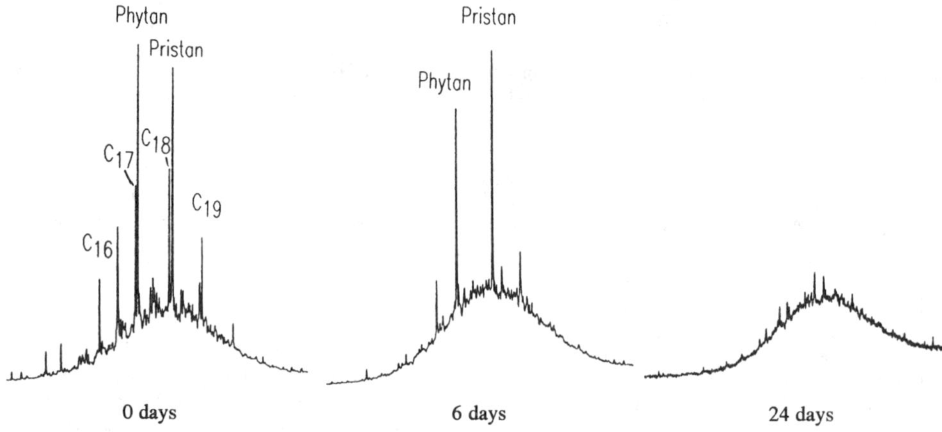

FIGURE 3: Chromatogramme (GC-FID) of the hydrocarbons in the influent of the reactor and after 6 and 24 days of biological treatment.

4. CONCLUSIONS

The direct biological treatment of the fine particle suspension from soil washing could be successfully applied to the artificially oil-contaminated silty soil. Treatment efficiency was low for the fine particle fraction of the soil from a hazardous site due to the low bioavailability of the hydrocarbons (inclusion in porous soil constituents) or due to the low biodegradability of the hydrocarbons at the hazardous site.

5. REFERENCES

DIN 38 409, Teil 18 (1981): Bestimmung von Kohlenwasserstoffen. <u>Deutsche Einheitsverfahren zur Wasser-, Abwasser- und Schlammuntersuchung</u>. 9. Lieferung. Verlag Chemie Weinheim.

Schmid, K., Abbas,F., Beudert, G. and Hahn, H.H. (1993): Investigations on the Bilogical Remediation of the Fine Particle Fraction of Hydrocarbon-containing Hazardous Sites in a Continously Fed Bioreactor. In: <u>Contaminated Soil '93</u>. Fourth International Conference on Contaminated Soil. Berlin, 1993. Kluwer Academic Publishers, Dordrecht, Boston, London.

Bacterial Degradation of PAH in Mixed Phase Systems in Airlift Bioreactors

Dieter Bryniok, Wolfgang Brunner, and Hans-Joachim Knackmuß

Fraunhofer-Institut für Grenzflächen- und Bioverfahrenstechnik,

Nobel Str. 12, W-7000 Stuttgart 80

Polycyclic aromatic hydrocarbons (PAH) are formed during combustion of organic matter and, therefore, are ubiquitously distributed in low concentration as environmental contaminants (1, 2). They are predominatly adsorbed to humic substances and may be subject to microbial degradation or humification processes (3). Besides, PAH are found in considerable higher levels in the ground of PAH-releasing industrial plants like gas works, coking plants and wood preservating facilities. These sites are contaminated by PAH-containing coal gasification products like crude benzene or coal tar. These are complex mixtures of phenols, aromatic compounds, substituted (poly)aromatics, and PAH. A major fraction of coal tar remains unidentified. Phenols and aromatics act as solvents and transported PAH even in great depth. The content of volatile coal tar components decreased over decades leading to increased viscosity and reduced mobility of coal tar. Consequently, PAH are present in high concentrations even in great depth not only adsorbed to the organic content of soil like humic acids, but also in the pores of soil and adhesively fixed to soil minerals.

Pyka et al. (4) used industrial coal tar for bench scale column experiments to investigate release rates and transport of some representative PAH in various aquifer materials. They observed very low PAH releasing rates leading to a assumed minimum time of 3, 300, and 12000 years for the complete removal of naphthalene, fluoranthene, and benzo(a)pyrene by a constant water flow. Consequently poor perspectives for in situ soil remediation by conventional pump and treat techniques have to be expected.

Soil washing is one of the techniques representing the actual state of the art for remediation of PAH contaminated soil. PAH are desorbed from gravel or sand by application of sufactants and/or high kinetic energy (jet grouting). These procedures do not destroy PAH, but concentrate them in a residual sludge containing small soil particles, usually < 50 μm. Because of the high specific surface of these particles and their content of organic carbon compounds, the sludge is highly contaminated with PAH and is actually managed as hazardous waste by deposition in appropriate landfill sites or by thermal processes. Therefore soil washing procedures cause inadequately high costs and can not be applied, if the clay and fine silt content exceeds 25 %. The microbial treatment of this sludge by conventional on-site/off-site techniques like landfarming or composting processes is limited by the poor water transport rates.

Biodegradation of PAH in soil/water systems is limited by extremely low solubility and solubilization rates, which are dependent on their very high sorption constants. Biodegradation of PAH in slurry bioreactors requires residence times of several weeks and is not practicable under economic aspects.

In preliminary laboratory experiments we demonstrated that PAH biodegradation can be dramatically accelerated by the use of water-immiscible solvents as lipophilic mediators for facilitated mass transfer of the PAH (5). A 15-l-laboratory scale plant containing two airlift

1373

F. Arendt, G.J. Annokkée, R. Bosman and W.J. van den Brink (eds.), Contaminated Soil '93, 1373–1374.
© 1993 *Kluwer Academic Publishers. Printed in the Netherlands.*

bioreactors was planned and constructed to prove the applicability of these organic solvents for soil decontamination.

In the first batch experiments using high contaminated soil suspensions (maximum particle size 100 μm) as substrate, degradation of low molecular weight PAH up to 3 rings was observed within weeks. PAH containing 4 or more rings persisted in these periods of time.

When an organic solvent was added to the reactor fluid PAH degradation was significantly accelerated and the times required were in the same order of magnitude as in preliminary experiments in 500 ml scale. Also in these experiments higher condensed PAH persisted or were degraded with significantly lower rates.

Highest degradation rates were obtained when the PAH were dissolved in an appropriate water insoluble solvent, supplied as substrates in the absence of soil particles, and the biological degradation process in the airlift-reactors was carried out in an organic/aqueous mixed phase. It was evident, that desorption of PAH from soil particles was the rate limiting factor for PAH degradation. Consequently desorption of PAH from contaminated soil have to be performed as a separate unit operations.

Desorption of PAH from fine particles may be carried out by washing with surfactants in high concentrations and extraction with steam, solvents, or supercritical fluids. In laboratory studies basic investigations were performed for an additional alternative, the fluid-fluid extraction of soil washing process water or other soil suspensions. Suspensions of PAH contaminated soil from a former coking plant had been supplied with small amounts of a biodegradable microbial biosurfactant and served as a model process water of soil washing processes. Various solvents were tested. Temperature, stirring time, number of extraction steps, and solvent-suspension ratio were varied. The degree of PAH desorption from soil particles and PAH transition into the solvent depended on the particular solvent and was significantly improved with increasing number of subsequent extraction steps and increasing temperature. It was confirmed by GC analyses, that the PAH were desorbed from soil particles without replacing solvent molecules for PAH molecules.

A soil remediation concept basing on soil washing, liquid-liquid extraction and PAH biodegradation in technical scale will also be rather expensive. Low expenditure bioremediation techniques are generally applicable only if target values for soil clean up are less stringent than usual in Germany at present and are restricted to those compounds, which are mobilized in concentrations high enough to represent a substantial risk for groundwater, environment, and human health.

References:
(1) Andelman, J. B. and Snodgrass, J. E., 1974, "Incidence and Significance of Polynuclear Aromatic Hydrocarbons in the Water Environment" CRC Crit. Rev. Environ. Control 5, 69-81
(2) Blumer, M., "Polycyclic Aromatic Compounds in Nature" Sci. Am. 234, 1976, 34-44
(3) Sims, R. C. and Overcash, M. R., 1983, "Fate of Polynuclear Aromatic Compounds (PNAs) in Soil-Plant Systems", Residue Rev. 88, 1-68
(4) Pyka, W., Schüth, C., Wilhelm, T., Grathwohl, P., 1992, "Dissolution and Transport of Coal-Tar Constituents and their Impact on Groundwater Quality" Int. Symp. Environ. Contam. Budapest, October 12-16
(5) Köhler, A., Schüttoff, M., Bryniok, D., Knackmuß, H.-J., "Different Approaches to Accelerate Biodegradation of Phenanthrene by *Pseudomonas aeruginosa* strain AK1", submitted for publication

Microemulsions for soil remediation

W.D. Clemens, F.H. Haegel, P. Nolte, K. Stickdorn, L. Webb[*], Institut für Angewandte Physikalische Chemie und Institut für Biotechnologie 3[*], Forschungszentrum Jülich, Postfach 1913, D-W-5170 Jülich

Microemulsions are macroscopically homogeneous, thermodynamically stable mixtures of water, oil and surfactant. They exhibit very large internal interfaces, extremely low interfacial tensions, good solubilization properties and an easy preparation and separation. For these reasons microemulsions seem to be especially appropriate for soil washing. Microemulsions composed of biodegradable raw material are used for the extraction process.

The phase behaviour of mixtures with surfactants and oils on vegetable basis is different from that of the mostly investigated mixtures of nonionic surfactants and mineral oils. Three-phase regions are observed for a ternary mixture of alkyl polyglucosides with a chain-length of C_{12} to C_{16}, rape-oil and water. Here the surfactant-rich phase forms the lower phase of the system and the water-rich the middle-phase. This is different from most of investigated systems with mineral oils and synthetic nonionic surfactants and may be important for a further step of separation and recycling of the surfactant.

Extraction experiments with a PAH-contaminated material from the outlet of a hydro-cyclone in a soil extraction plant were performed with a homogeneous mixture of APG/water/ rape-oil, a system of Igepal CA 520 ($C_9H_{19}C_6H_4$-(O-CH_2-CH_2)$_5$/ water/ heptane and for a comparison with toluene. Both microemulsion systems show an extraction yield of approximately 1000 mg/kg, during the control experiment with toluene 1200 mg/kg were extracted. These primary results already show that extraction rates of 85% are possible, even though the extraction liquid has to be optimized.

Investigations concerning the microbiological degradation of surfactant and rape oil have been performed. In order to select a microorganism population which simultaneously degrades rape oil and surfactant, a 2 liter bioreactor with biomass feedback and continual substrate feed was constructed. After starting the reactor, the contents foamed due to the presence of the surfactant. This was controlled by reducing the aeration to a minimum. After the population had developed, no more foam was produced, which indicated that the surfactant was being degraded. First attempts to isolate the microorganisms responsible

F. Arendt, G.J. Annokkée, R. Bosman and W.J. van den Brink (eds.), Contaminated Soil '93, 1375–1376.

for degradation of the rape oil and surfactant using normal microbiological media were unsuccessful, which could indicate a high specificity of the microorganisms for these substrates. At the moment other media are being tested. Microscopical examination of the population showed that bacteria are the main component. However, a fungal culture has been isolated from the population which is capable of growing on both rape oil and surfactant.

COMBINED BIOLOGICAL AND CHEMICAL OXIDATION OF THE LEACHATE OF AN INDUSTRIAL SITE IN AN EXPERIMENTAL PLANT

O. Debus*, H. Krebs** and L. Hemmerling*

*DFG-Graduate College Biotechnology, AB Gewässerreinigungstechnik (Dep. for Wastewater Treatment), Technical University Hamburg-Harburg, Eißendorfer Str. 42, D-2100 Hamburg 90
**Holsteiner Gas-Gesellschaft mbH, Kaiser-Wilhelm-Str. 115, D-2000 Hamburg 36

SUMMARY

In a research and development project (r&d) at the Technical University Hamburg-Harburg (Department For Wastewater Treatment) on a treatment system for leachates of a highly contaminated site several process variants have been tested in an experimental plant. A multiple stage process with two biological stages for biological degradation of organic substances and nitrification of ammonia was used. It was found that 50% of the chlorinated organics were susceptible to biological degradation. Additional processes must therefore be used to meet the limit concentrations of the sewer system for these chlorinated compounds. First, an activated carbon column was used as final stage and them a chemical oxidation was installed. For this aim an ozone/UV-oxidation was operated between the two biological stages to minimize the energy costs. With this new combination of processes the limit values for the sewer system could be met. Additionally the largest amount of the volatile organics could be converted biologically by the use of a membrane aeration system for the oxygen supply in the first biological step.

1. INTRODUCTION

For the treatment of the leachate of an industrial site an extensive r&d project has been carried out. The goal of the project was to develop a leachate treatment plant, which can eliminate a large fraction of the organic matter, in particular the chlorinated organic substances, and the ammonia from the leachate. In the process the amount of waste and energy should be minimized. The limit values for discharging into the sewer system are 200 mg COD/l, 250 μg AOX/l, 50 mg NH_4^+-N/l. Denitrification was not necessary, because this is done in the sewage treatment plant. A further goal of the plant developing was the destruction of the volatile chlorinated and non-chlorinated hydrocarbons without the elimination by stripping them out of the leachate [Debus et al. 1990]. To meet these requirements oxidative destruction processes were better suited than concentration processes. An experimental plant was constructed, which was successfully operated based on the success and experience of preinvestigations. The results can be summerized as follows:

* The installed fixed bed technology led to stable conversions and clear discharges by the establishment of a stable biofilm culture;
* the membrane aeration retained the largest fraction of the volatile substances in the system and supported the growth of microorganisms in combination with an other carrier material (a varied culture of nitrifiers and degrader of substance, among them volatile chlorinated and nonchlorinated hydrocarbons);
* the Sequencing Batch Reaction (SBR) mode allowed high rates for the nitrification and for the degradation of hardly degradable compounds;
* the chemical oxidation converted a large fraction of hardly degradable large molecules into smaller, more easily degradable substances, which could be degraded in the second biological stage.

F. Arendt, G.J. Annokkée, R. Bosman and W.J. van den Brink (eds.), Contaminated Soil '93, 1377–1380.
© 1993 Kluwer Academic Publishers. Printed in the Netherlands.

2. MATERIALS AND METHODS

Fig. 1 shows the process scheme of the experimental plant based on the above mentioned experiences. After the pretreatment of the leachate in a technical scale flotation plant (240m^3/d), where heavy metals and emulsified oils are separated, the pretreated leachates flow into a container. Out of this container the first biological stage is filled in batch mode every 12 h up to 2.4 m^3 d^{-1}. The exchange rate of the batch process is 50% per cycle. The biology is constructed as a fixed bed. Polypropylene bodies with a specific area of 250 m^2/m^3 were used as carriers for the microorganisms. During the complete cycle the leachate is enriched with oxygen in the external membrane aeration unit. Within the membrane aeration system, silicone tubings are parallely fixed in which the leachate flows. Pure oxygen is flowing around the outside of the silicon tubings under pressure (f.e. 1.7 bar). The oxygen diffuses through the membranes and is enriched without the foaming in the liquid phase (f.e. from 3 mg O$_2$/l to 8 mg O$_2$/l). A part of the formed carbon dioxide is discharged via the membrane into the gas phase. This contributes to the stabilization of the pH-value in the liquid phase. The first biological stage is followed by the chemical oxidation. The ozone-reactor (2 kW) and the following UV-reactor are continuously charged from an intermediate container. The energy charge is set in that way, that only incomplete oxidation is obtained.

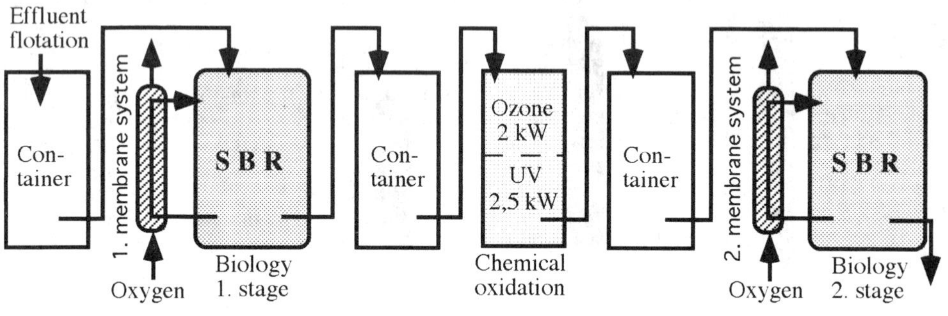

Aeration: 1. membrane system 12m^2 2. membrane system 10 m^2
Reactors: 1. bio. stage 2,5 m^3 OZONE/UV. 0,3 m^3 2. biol. stage 1,25 m^3

FIG. 1: Process scheme of the experimental plant for the leachate treatment.

The second biological stage is half as big as the first biological stage and the membrane aeration had 20% less membrane area (Fig. 1). The complete plant is constructed out of stainless steel and is operated by a stored program control.

3. RESULTS

The effluent of the flotation plant had the following mean composition:
The concentrations of the organic compounds were generally low, as it is often found at closed older sites. TOC was 160-200 mg/l C and the COD/TOC-ratio about 3 points to the fact that parts of the organic compounds were little oxidized. Mean of COD/BOD$_5$-ratio was about 10, which stands for a low part of well degradable organic compounds. The fraction of the chlorinated organic compounds was above 1 mg/l Cl$^-$ - measured as AOX. As a total (nonvolatile and volatile compounds) values of 1.7-1.8 mg/l Cl$^-$ were measured. In various leachate samples a phenol index between 1 and 3 g/l was estimated. The concentration of nonpolar hydrocarbons was about 1 mg/l. The high molecular fraction - presumable humic substances (fulvic acids) - was up to 20 % of the total organic compounds. The fraction of the volatile organic compounds - measured as POC - was several percent of the TOC. The subgroup of the volatile chlorinated substances (POX) was 30% (0.5 mg/l) of the total AOX. The ammonia concentration varied from 100 to 300 mg-N/l, a multiple concentration of communal sewage; nitrite and nitrate were not detectable. Phosphate was normally concentrations (ca 1 mg/l) too low to enable metabolic activity.
Table 1 shows the degree of purification in single stages in the last experimental period.

TABLE 1: Effluent values of the purification processes, * detection limit.

Effluent of the	COD, mg/l	AOX, μg Cl-/l	Phenols, mg/l	Ammonia-N, mg/l
Flotation	620	2100	2.0	210
1. biol. stage	360	1050	< 0.1*	205
Ozone/UV	50	150	< 0.1*	100
2. biol. stage	40	65	< 0.1*	20

In the Fig. 2 the related purification values (flotation effluent 100 %) of the different stages are presented.

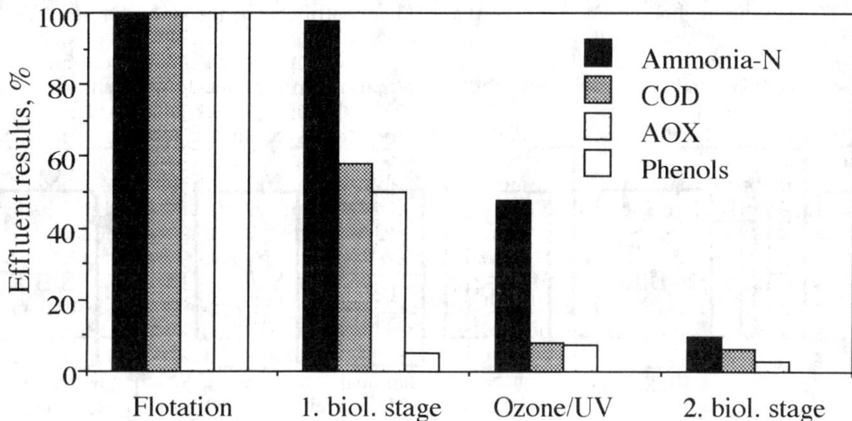

FIG. 2: Purification of the different stages of the experimental plant, related to the effluent of the flotation plant.

In the first biological stage 40% of the COD and 50 % of the AOX were degraded (mean values). The elimination of ammonia in the first biological stage was low. Here the nitrification was inhibited by organic substances. In the chemical stage 40% of the COD and 50% of the AOX were eliminated. The second biology degraded about 5% of the COD and of the AOX and exhibited a nearly completely nitrification. In the chemical oxidation a partial oxidation of the ammonia, parallel to the oxidation of the COD, took place. Other single and summarized compounds like phenols, were nearly completely converted already in the first biological stage. In table 2 the compounds concentrations of single volatile in the effluent of the treatment stages are listed. According to this, in the first biological stage most of the volatile BTX-aromatics was eliminated, while the chlorinated aliphatics were left in the liquid phase to a higher degree and only the chemical oxidation reduced the concentrations below the detection limit.

TABLE 2: Volatile compounds in the effluents of the single process stages.

Effluent	Benzene μg/l	Toluene μg/l	Xylene μg/l	Trichlororethene μg/l	Tetrachloroethene μg/l
Flotation	190	705	540	11	3.5
1. biol. stage	3.8	2.4	10.1	5.3	0.7
Ozone/UV	< 0.2	< 1.0	< 0.5	< 1.0	< 0.2
2. biol. stage	< 0.2	< 1.0	< 0.5	< 0.2	< 0.2

The way in which the substances are removed from the liquid phase is clarified by table 3. Listed are the pollutant discharge via the liquid effluent and the gas phase of the biological stages, which had a joint exhaust gas system.

TABLE 3: Balance single volatile compounds over the gas and liquid phase - mean liquid phase flow 1.2 m^3/d, gas phase flow 2.4 m^3/d -.

Balance of the biological stages	Benzene	Toluene	Xylene	Trichloroethene	Tetrachloroethene
Influent liquid phase conc., μg/h	9500	35250	27000	550	175
Concentration in the off gas, μg/l	0.83	1.3	1.5	<0.0005	0.42
Mass flow off gas, μg/h	83	130	150	< 0.5	42
Effluent liquid phase 1.biol. st., μg/h	190	120	505	265	35
Discharge gas phase, %	0.87	0.37	0.56	< 0.09	24
Discharge liquid phase. 1. biol. st., %	2.00	0.34	1.87	48	20
Discharge liquid phase. 2. biol. st., %	< 0.11	< 0.14	< 0.09	< 1.82	< 5.71

Only a few percent of the BTX-aromatics left the plant via gaseous and liquid effluent streams, while 50% of the highly chlorinated aliphatics were detectable in the effluent and exhaust gas. It is shown, that the aerobically well degradable BTX-aromatics were not stripped but were mainly biologically degraded.

Extensive experiments have been performed in a laboratory membrane reactor. These showed that a large fraction of BTX compounds could be degraded by membrane bound biofilm, which had been formed on the water side of the membrane surface. In such reactors the biofilm degrades these compounds and the membrane bound biofilm acts as a barrier to the transfer of volatile substances to the gas phase [Debus et al. 1992].

By examination of the membrane aeration modules at the end of the 7-month period of operation, it was found a stable and equally spread biofilm has been formed on the membrane surface of the silicone tubings. A blockage of single tubings by biomass could not be detected. In a technical plant the specific gas flow could be further reduced to retain also the nondegradable volatile substances to a higher degree in the liquid phase. By this, higher concentrations can be converted by the chemical oxidation.

4. CONCLUSIONS

The presented process combination SBR-operated fixed film biology - chemical oxidation with an ozone/UV combination - SBR-operated fixed film biology was successfully operated continuously for months. The limit values for the sewer system could be kept very well. The membrane aeration system exhibited high process stability concerning both, technical and biological aspects.

6. REFERENCES

Debus, O. and Wanner, O. (1992). Degradation of xylene by a biofilm growing on a gas-permeable membrane. Water, Science & Technology, Vol. 26 Numbers 3-4 (1992), pp. 607-616, Pergamon Press Ltd., Oxford, ISBN 0080420443

Debus, O;. Wanner, O. and I. Sekoulov (1992). Aerobic degradation of volatile aromatics in a membrane-biofilm reactor - comparison of measured and modelled results -. Dechema Biotechnology Conferences Vol. 5. Pp. 971-976, VCH Weinheim, New York.

Debus, O.; Krebs, H.; Rubio, M. and Wilderer, P. (1990). Development of a membrane oxygenation system for the aerobic biodegradation of volatile compounds in a leachate treatment facility. 3. International KfK/TNO Conference on Contaminated Soil, 10-14. 12. 90, Karlsruhe. In: Arendt, F.; Hinsenveld, M.; v.d. Brink, W.J. (eds.), Contaminated Soil´90, Kluw. A. P..

Biological treatment of silt taken from soil contaminated by organic substances in a 4-stage reactor cascade

Dipl.-Ing. F. Elias, Prof. Dr. Ing. U. Wiesmann, TU-Berlin

Technische Universität Berlin, Institut für Verfahrenstechnik,
Sekr. MA 5-7, Straße des 17. Juni 135, 1000 Berlin 12,

For some time the technique of soil washing and grading has been used to clean soil contaminated by organic substances. Reported cleaning efficiencies refer only to that part of the soil which can be characterised as sand with a particle size $d_p \geq 63$ µm. Highly contaminated silt ($d_p < 63$ µm) still remains which has to be landfilled at high costs. This situation calls for the development of adequate and affordable procedures. For this reason the biological treatment of silt was examined at the Institute for Chemical Engineering of the Technical University of Berlin.

The existing laboratory scale plant consists of a 4-stage stirred tank cascade with a volume of $V_R = 10$ l (Fig. 1). The silt is kept suspended in a stirred storage tank with a solids content of TS = 10-15 %. In a free hydraulic gradient the suspension flows through the cascade. There the suspension is stirred with a standard paddle mixer, gassed with compressed air and then separated in a thickener.

The silt used is from a former waste oil processing plant named Pintsch AG Berlin. It has been treated and separated as highly contaminated waste in a soil washing plant on these grounds.

Our biological treatment

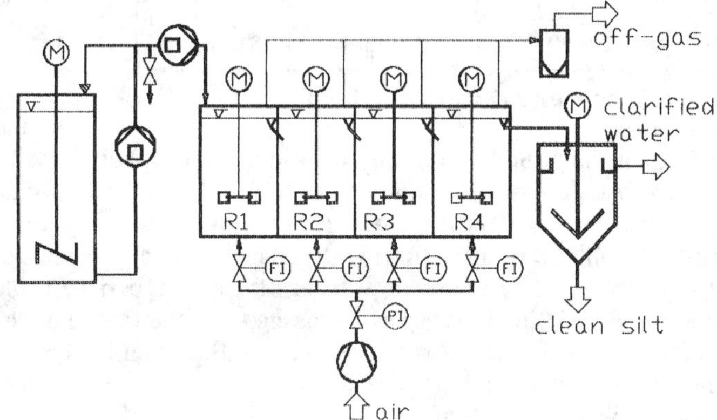

Fig. 1: plant engineering

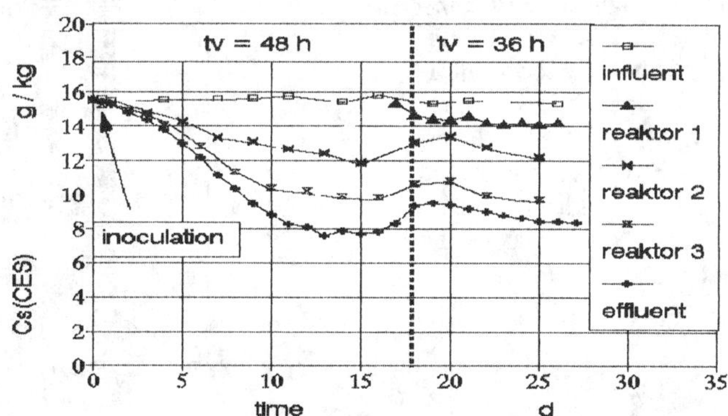

Fig. 2: concentration of the substrate on the silt at the time of investigation

1381

F. Arendt, G.J. Annokkée, R. Bosman and W.J. van den Brink (eds.), Contaminated Soil '93, 1381–1382.
© 1993 Kluwer Academic Publishers. Printed in the Netherlands.

was controlled by measuring the parameters of "in cyclohexan extractable substances" (CES), its chemical oxygen demand (COD), and also some PAH.

PAH	c_{s0}	c_{s1}	c_{s2}	c_{s3}	c_{s4}	α
	in g/kg					
acenaphthylene	8,23	6,94	5,73	4,66	2,97	0,64
acenapthene	13,00	10,72	8,26	5,49	3,81	0,71
fluorene	14,90	11,71	9,59	8,32	3,21	0,78
phenanthrene	153,97	138,41	103,68	39,32	5,87	0,96
anthracene	14,61	13,07	11,70	7,79	5,93	0,59
fluoranthene	162,31	161,77	117,97	84,53	69,98	0,57
pyrene	75,43	63,94	54,70	37,96	25,61	0,66
chrysene	33,37	29,00	25,39	18,00	11,32	0,66

<u>Table 1</u>: concentration of PAH at the silt

At first the retention time was held at $t_V = 48$ h; later it could be cut down to $t_V = 36$ h. The concentration of the substrate on the silt could be reduced by 50 % - comparable to former test results (Fig. 2).

The results of the analysis prove clearly that the concentration of PAH is reduced (Table 1). A closer examination of the PAH concentration in the individual reactor stages shows, that the highest rate of reduction is found in the 3rd stage of the cascade. In the first two stages, the microorganisms remove the easily degradable substrates. Based on the general parameter CES, substrate removal amounts to 50 % . When the easily degradable substances run out in the 3rd and 4th stage, the microorganisms have to reorganise their enzymatic system to be able to reduce PAH. Only adaptable microorganisms can be successful in the last two stages of the cascade whereas the others perish. Table 1 shows the concentration of the PAH in all stages of the cascade on the first and the 26th day of our test. The removal efficiency is between $\alpha = 0,57$ - 0,96. Even for the four-ring PAH (pyrene and chrysene) it is relatively high with $\alpha = 0,66$. Phenanthrene has the removal highest efficiency with $\alpha = 0,96$. Fig. 3 illustrates how the concentration of the six most concentrated PAH is reduced in each of the four stages of the cascade as measured on the 26th day of the test.

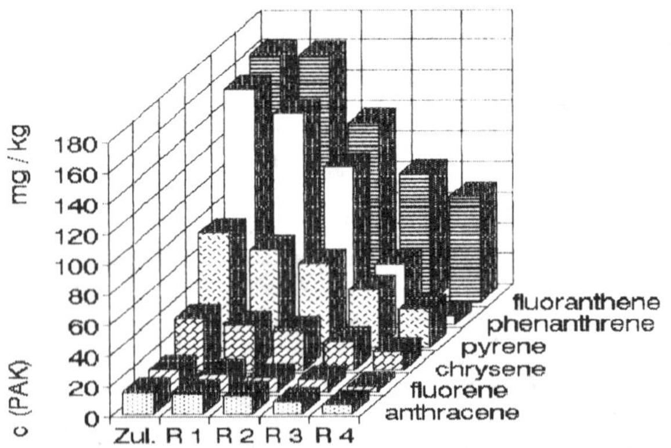

<u>Fig. 3</u>: amounts of PAH on the 26th day of investigation

THE CONVERSION OF GASWORKS CONTAMINATED SOIL INTO A CARBONACEOUS ADSORBENT USING A NOVEL APPLICATION OF THERMAL TREATMENT TECHNOLOGY

G.D. Fowler, S.K. Ouki, C.J. Sollars, and R. Perry

Imperial College Centre for Toxic Waste Management, Imperial College of Science, Technology, and Medicine, London SW7 2BU, United Kingdom

SUMMARY

Research work within the Centre for Toxic Waste Management has been investigating the reuse of gasworks contaminated soils by thermal conversion of the organic components into a porous carbonaceous material which possesses adsorption characteristics.

The processing of the contaminated soils consists of impregnating the samples with a 25% saturated zinc chloride solution prior to carbonization at temperatures below 600 °C in an inert nitrogen environment. The carbonaceous product is then treated with a hydrochloric acid solution at 60-80 °C to remove any residual zinc chloride, sulphur, and soluble ash remained after processing. The final product is then dried overnight at 105 °C and ground to pass a 150 μm sieve.

The initial characterization of the soils showed that the samples were heavily contaminated with cyanide, sulphur, sulphate, and other metals such as Cr, Co, Cu, Ni, Zn, and Pb, and had a variable initial carbon content ranging from 14% to 50%. Characterization of the final carbonaceous product demonstrated that the thermal treatment significantly decreased the toxicity of the contaminated soils. A 95% reduction of original cyanide content was achieved and the organic contamination of the soils was converted into a carbonaceous lattice removing the potential hazard associated with coal tars. In addition, the results also showed that the non-volatile metallic contamination was entrapped within the carbon matrix.

Aqueous adsorption characterisation using phenol and 4-nitrophenol revealed that the carbonaceous materials developed adsorption capacities ranging variably from 10% to 94%, with the major controlling factor being the initial carbon content and the amount of $ZnCl_2$ used during the activation process. Gas adsorption studies showed that the adsorbents exhibited BET surface areas ranging from 110 m^2/g to 570 m^2/g. These values compare fairly well with the ones obtained for commercial carbons such as Norit SA4 and BDH charcoal when taking into account the heterogeneous nature of the original material, contaminated soils.

F. Arendt, G.J. Annokkée, R. Bosman and W.J. van den Brink (eds.), Contaminated Soil '93, 1383–1384.
© 1993 *Kluwer Academic Publishers. Printed in the Netherlands.*

This study has demonstrated that low temperature thermal treatment can convert contaminated soils into carbonaceous adsorbents with potential application for reuse. It also showed that the process substantially reduces the toxicity and volume of the contaminated soil material. The final products are very variable and therefore their ultimate end-use would be in applications where bulk, low grade adsorbents with a once-only use are required. To this end, the Centre for Toxic Waste Management is currently investigating the possibility of using these carbonaceous materials in landfill management where daily cover for freshly deposited waste is needed. This novel application has great potential in retarding pollutant transport from leachate into groundwater by adsorption and entrapment within the carbonaceous matrix.

**Summary Table of Relevant Adsorbent Characteristics
of the Final Carbonaceous Products**

Label	BET Area (m^2/g)	Gas Volume Ads. (ml/g)	Carbon (%)	Phenol Ads. (%)	4-Nitrophenol Ads. (%)
STA01C	110	25.31	21.71	16.63	29.42
STA02C	196	45.15	57.54	25.05	39.52
STA03C	219	50.38	40.09	31.55	49.42
STA04C	570	130.94	57.13	69.12	93.22
STA05C	129	29.74	17.96	18.49	30.98

STA01C to STA05C: Activated and carbonized contaminated soil samples
Ads.   : Adsorption

KEYWORDS

Thermal treatment, activated carbon, cyanides, remedial action techniques, waste reuse

ACKNOWLEDGMENT

G.D. Fowler acknowledges the provision of a studentship by UK SERC, and additional research funding by Balfour Beatty Environmental Management, UK.

Microbial soil decontamination in a slurry reactor

M. Geerdink, E. Hardevelt, I. Schouten, M. v. Loosdrecht, K.Ch.A.M. Luyben
Department of Biochemical Engineering, Kluyverlaboratory for Biotechnology
Delft University of Technology, Julianalaan 67, 2628 BC Delft, The Netherlands

Introduction

In biological soil decontamination a notorious problem is the high rest
concentration of some or all of the contaminants after treatment. This is the
main obstacle for the application of this technology.
Generally three possible limitations are considered, i.e. degradability (the
contaminants cannot be degraded under the existing conditions, are degraded
very slowly by the microorganisms present, or inhibiting metabolites are formed);
diffusivity (the diffusion rate of the components out of the soil aggregates is very
low); desorptivity (the desorption rate of the polluting compounds from the
surface of the soil particles is very low). In order to develop methods for
reducing the rest concentration after the biological treatment effectively, it is
necessary to determine which of the three mechanisms is the rate limiting step in
the decontamination process. In the experiments described in this contribution
diesel oil and hexadecane were used to distinguish between the three
mechanisms.

Results

Degradability
Degradability was first determined by comparing biodegradation of hexadecane
and of diesel oil in dispersion in a batch reactor, using an initial concentration of
the substrate of 4 g/l. Both substrates showed similar oxygen uptake and carbon
dioxide production behaviour. Two days after inoculation no substrate could be
detected in either case. No metabolites could be detected in the water phase.
From these observations it follows that all components in the diesel oil have
been degraded. No inhibitory metabolites have accumulated to toxic levels [1].
This indicates that biodegradability as such is not rate limiting.
A maximum oxygen uptake rate (OUR) could be calculated for both hexadecane
and diesel oil degradation in dispersion. Assuming that in the exponential phase
this OUR is proportional to the biomass growth rate, a maximum growth rate for

1385

F. Arendt, G.J. Annokkée, R. Bosman and W.J. van den Brink (eds.), Contaminated Soil '93, 1385–1386.
© 1993 *Kluwer Academic Publishers. Printed in the Netherlands.*

Next, degradability in a chemostat was looked at, using dilution rates between 0.5 and 6 day^{-1}. In all these experiments the diesel oil concentration in the effluent remained about 4% of the influent concentration. However for the highest dilution rate, $\tau=4$ h, a much higher oil concentration was found (about 50% of the influent concentration). In this case no partially oxidized components were detected either.

Diffusion or desorption

From the results above it was concluded that the limiting mechanism in a soil treatment system is either diffusion or desorption. To distinguish between these two the following experiments were performed.

In shake flasks, inoculated with the same consortium as before, two types of clay were used as adsorbents using several initial loadings of hexadecane. O_2 consumption and CO_2 production were monitored. After an initial break down period, till the oxygen uptake was back to base level, duplicates were milled in a ball mill for several hours. All flasks were reinoculated and O_2 consumption and CO_2 production were monitored again.

Results were that an initial increase in the degradation in the milled clay occurred. A linear relationship between initial loading and total oxygen uptake was found. These results suggest a diffusion limited process.

Conclusions

The observed rest concentration of oil on soil after biological treatment [2] is not caused by the inability of the micro-organisms present in the soil to degrade the oil, or by any toxic metabolites formed during the degradation. From degradation of diesel oil in dispersion it follows that all the oil can be mineralized.

The degradability of diesel oil in continuous mode was determined. Conclusion is that up to a dilution rate of 2 day^{-1} the oil can be degraded.

Milling contaminated clay, after initial biodegradation, resulted at first in a faster degradation rate compared to unmilled clay. In these experiments the total oxygen uptake was proportional to the initial loading. This all suggests that diffusion out of the clay aggregates is the limiting mechanism for the ultimate oil degradation.

Literature

1. Geerdink, M.; Hardevelt, E.; Loosdrecht, M. v.; Luyben, K.Ch.A.M. (1992): Biological soil decontamination in a slurry reactor. In <u>Soil Decontamination Using Biological Processes</u>, Proceedings DECHEMA International symposium, Karlsruhe.

2. Kleijntjens, R.H. (1991): <u>Biotechnological slurry process for the decontamination of excavated polluted soils</u>. PhD-thesis, University of Technology, Delft.

Remediation and Re-Urbanisation of Derelict Industrial Sites — An Alternative Concept

Dieter D. GENSKE, Herbert KLAPPERICH, Christoph OLK
Deutsche Montan Technologie DMT-IWB, Franz-Fischer-Weg, 4300 Essen 13, Germany

Peter NOLL
Montan Grundstücksentwicklungsgesellschaft, MGE, Zur Pannhütt 64, 4355 Waltrop, Germany

One of the major industrial regions in Germany is the Ruhr-District, an area of about 5000 km^2 and a population of 5 million. The development of this industrial belt began in the middle of the last century when the commercial exploitation of the vast coal resources started. The mining activities also initiated a great variety of secondary industries, such as coal refinement plants, steel industries, chemical plants, etc. Today, due to the decline of the mining industry, a large number of factories are abandoned and have turned into industrial wasteland.

The contamination of the soil beneath the former industrial sites, however, has remained, jeopardizing the groundwater resources. During the last decades pollutants such as hydrocarbons and heavy metals have migrated through the cenozoic sediments into the jointed mesozoic marls, causing massive contamination problems. The extraction of these pollutants has to be considered quite a complicated and expensive affair. On the other hand, most of the former industrial sites are now located close to the city centers, some even used to be the nucleus from which the city started to spread. Prospective investors are, however, discouraged from purchasing these sites due to the economical risks connected with the remediation work.

In this poster presentation a cost effective remediation strategy is introduced. It is a combination of a surface isolation technique and hydraulic measures. The poster mainly deals with the first point. The cover system described has to meet three tasks:
  (i)   it has to be waterproof to prevent the precipitation to penetrate into the contaminated ground;
  (ii) it has to be gasproof to stop the migration of toxic gas to the surface;
  (iii) it has to be stiff enough to allow streets and structures to be built on.

The third aspect accounts for the fairly inhomogeneous ground conditions. Typically, massive fragments of the former foundations of the dismantled buildings have remained in the subground next to loose fillings, thus causing severe structural problems as to the possible differential settlement of future structures.

Figure 1 depicts a reinforced geotextile sandwich system which was designed to satisfy all the objectives mentioned above. Basically it is composed of three elements: a lower reinforced support layer, a drain and seal system, and an upper reinforced element to account for the vehicular or structural loads.

The cover system is applied to an abandoned mining site close to the city of Essen, the remediation of which is funded by the European Community with a budget of about 10 million ecu. In order to monitor the stress-strain behavior of the sandwich system and the performance of the sealing geomembrane, a 500 m^2 test field was set up on the site (fig. 2). GENSKE et al. 1993 report on the results obtained from this test field and discuss the risks and benefits of the cover technique proposed.

Genske, D. D., H. Klapperich & P. Noll 1993: Surface Confinement Techniques of Derelict Industrial Sites. — GEOCONFINE 93 International Symposium on the Geology and Confinement of Toxic Wastes, Montpellier, France, 8-11 June 1993.

F. Arendt, G.J. Annokkée, R. Bosman and W.J. van den Brink (eds.), Contaminated Soil '93, 1387–1388.

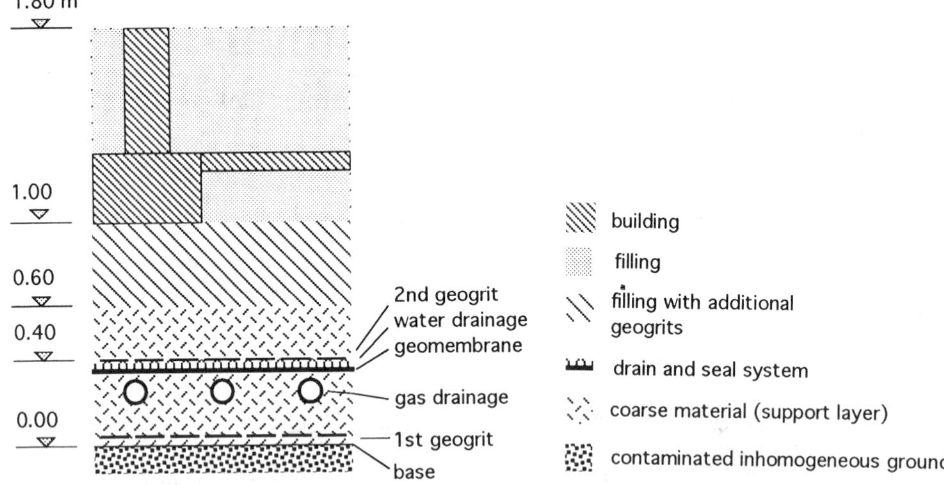

Figure 1: A reinforced geotextile sandwich system. Besides the reinforcing geogrits a geomembrane is included to stop the migration of toxic gas to the surface.

Figure 2: A 500 m^2 test field to monitor the stress-strain behavior of the sandwich system. The test field is located at an area within the derelict mining site were the ground conditions are extremely inhomogeneous. The arrangement of geogrits and the geomembrane is the same as given in figure 1. After completion of the test field a load test was carried out during which the earth pressure, the settlement and the deformation of the geogrits and the geomembrane were measured.

HEAVY METAL REMOVAL FROM MINERAL SOILS WITH VARIOUS EXTRACTANTS

Greinert H., Poprawska B.

Higher College of Engineering, Department of Environmental Restoration, 50 Podgórna Street, 65-246 Zielona Góra, Poland

INTRODUCTION

About 5000 ha of arable soils around the copper smelter in Głogów/Oder are contaminated heavily with Cu and Pb and in lesser degree with other heavy metals too. Contemporaneously with heavy metals, the smelter emited large amounts of SO_2, what causes acidification of the soil and may increase the mobility of heavy metals.

The work presents the results of laboratory experiments on the effectivity of Cu, Pb, Zn and Cd extraction from the polluted soils with various extractants.

MATERIALS AND METHODS

The contaminated soils, surrounding the copper smelter, belongs to brown soil type and black earths, developed from the alluvial river sediments and in the higher locations, from loess-like material. The extraction experiments were provided on 5 soil: 1a/,A_p horizon of sandy soil, with 1,44 % organic matter and pH 6; 1b/, C horizon (sand) from the same profile; 2/, A_p horizon of black earth, silty clay loam with 2,48% organic matter and pH 5,8; 3/, sandy A_p horizon of brown soil, containing 0,85% organic matter and ph 5,3; 4/, A_p horizon of black alluvial soil-silty clay with 3,08% organic matter and pH 4,8. The extraction were conducted 1 hour, the proportion soil: extractant was 1:5. As extractants were used: H_2O, H_2O + CO_2, 1n KCl, 2,5% CH_3COOH, 0,1 m Na_2EDTA and 2n HNO_3 (hot digestion).
The heavy metal determination have been done with atomic absorption spectrophotometer.

RESULTS AND DISCUSSION

There were not stated essential differences between results of heavy metal solubility in H_2O and H_2O + CO_2. The water solubility, expressed as the percent of total amount, was for Cu 1,23%, Pb 0,54%, Zn 1,03% and Cd 1,00%; in solution of 1 n KCl respectively 7,73%, 9,06%, 13,89% and 29,53%; in 2,5% CH_3COOH - 21,55%, 3,67%, 34,82% and 13,89%; in 0,1 m Na_2EDTA - 68,92%, 56,13%, 54,70% and 67,74% and in

F. Arendt, G.J. Annokkée, R. Bosman and W.J. van den Brink (eds.), Contaminated Soil '93, 1389–1391.

2n HNO_3 - 97,40%, 82,00%, 94,21% and 97,96% respectively. The heavy metal solubility in water is still low, what indicates, that the "acid rain" is neutralized by soil buffer capacity. The pot experiment with this polluted soils showed, that on good buffered, heavy soils the plants are taking up normal amounts of heavy metals, but this is not the case on sandy soils, where the heavy metal concentration in the plants increased several times.

Tabele 1. The total amount and solubility of Cu, Pb, Zn and Cd in the soils.

SOIL	FORMS AND UNITS		Cu	Pb	Zn	Cd
1a	total	ppm	908,00	248,90	58,90	1,70
	sol. in H_2O	%	0,78	0,20	0,00	0,00
	sol. in 1n KCl	%	0,85	2,49	8,11	26,76
	sol. in 2,5% CH_3COOH	%	14,39	4,51	14,26	24,70
	sol. in 1m Na_2EDTA	%	74,89	80,13	51,87	70,50
	sol. in 2n HNO_3	%	98,35	95,56	96,77	97,10
1b	total	ppm	21,50	28,30	67,00	0,97
	sol. in H_2O	%	2,81	1,63	2,72	0,77
	sol. in 1n KCl	%	3,70	12,36	22,40	33,51
	sol. in 2,5% CH_3COOH	%	20,91	1,57	89,55	2,58
	sol. in 1m Na_2EDTA	%	62,00	22,40	72,46	72,16
	sol. in 2n HNO_3	%	100,00	75,85	97,80	97,94
2	total	ppm	1057,00	369,60	104,90	1,50
	sol. in H_2O	%	0,19	0,12	0,00	o,81
	sol. in 1n KCl	%	0,19	1,01	0,00	24,00
	sol. in 2,5% CH_3COOH	%	11,64	2,87	40,20	30,33
	sol. in 1m Na_2EDTA	%	57,70	45,34	41,44	45,00
	sol. in 2n HNO_3	%	88,90	72,12	83,30	100,00
3	total	ppm	262,00	180,20	25,30	1,10
	sol. in H_2O	%	1,94	0,58	2,41	0,22
	sol. in 1n KCl	%	10,53	14,44	26,00	33,64
	sol. in 2,5% CH_3COOH	%	41,10	7,77	26,91	3,18
	sol. in 1m Na_2EDTA	%	71,72	55,37	79,94	77,27
	sol. in 2n HNO_3	%	99,77	66,50	100,00	100,00
4	total	ppm	578,00	218,70	94,50	1,90
	sol. in H_2O	%	0,42	0,21	0,00	0,26
	sol. in 1n KCl	%	23,40	15,00	12,75	29,74
	sol. in 2,5% CH_3COOH	%	19,71	1,64	3,17	8,68
	sol. in 1m Na_2EDTA	%	78,30	77,43	27,83	73,68
	sol. in 2n HNO_3	%	100,00	100,00	93,17	94,74

The amounts of heavy metals, dissolved by 1n KCl are much higher in the soils, located in the eastern part of the investigated area. This soils are more acid and it may be the result of rainy weather, which comes with the winds from the west ("acid rain").

The exchangeable forms of heavy metals could be evaluated

as potential risk for the environment. 2n HNO3, used widely
for the pollutoin studies by the Scandinavian scientists,
extracted in most cases almost 100% of the heavy metals.
Exception was lead, which concentration in this extractants
were lower.

   Looking on the results from the remediation point of view,
it has to be stated, that the one reasonable methods in this
case are the methods of lowering the solubility of heavy
metals (liming, organic matter application, phosphate
fertilization).

<u>SOIL CONTAMINATION IN THE PYRENEES, A CASE STUDY</u>

A.J. van de Haar and G. van Roekel, Fugro B.V.; A. Traspaderne, T.P.A.

Introduction

In the upper rio Gállego river basin, an area of over 3000 km^2 in the Spanish Pyrenees, an extensive soil and groundwater remediation study is in execution. The study, which has been ordered by the Diputacion General de Aragon, the D.G.A., started in 1991 and has now reached its final stage. The study has been appointed to the Spanish company T.P.A. in cooperation with the Dutch consultancy Fugro B.V.

The study is actuated by the presence of Inquinosa, a former chemical plant in the industrial town of Sabiñanigo. In 1975, Inquinosa started the production of the insecticide Lindane, the gamma-isomer of hexachloro-cyclohexane, generally abbreviated as HCH. From 1975 onwards Inquinosa disposed about 60.000 ton of bulk solid HCH waste and over 150.000 ton of bulk liquid HCH waste in the surroundings of Sabiñanigo. In this period, large fish mortality was observed in the rio Gállego, which was related to the HCH dumping activities. After fierce protests from environmental movements, the D.G.A. decided to take back Inquinosa's Lindane production license in 1988.

The primary objective of the study is the elaboration of an integral remedial plan for those sites, affected by Lindane and other industrial production and disposal activities. Because a limited amount of information was available concerning industrial production processes and disposal activities, the study started with an inventory, including inspection of aerial photographs, inquiries with local industries and authorities and an extensive field reconnaissance. Based on the results of the inventory, about ten sites were selected for preliminary investigations. From the preliminary investigations, four sites emerged as seriously contaminated, requiring closer investigations. Two of them concern the uncontrolled Sardas and Bailin landfills.

F. Arendt, G.J. Annokkée, R. Bosman and W.J. van den Brink (eds.), Contaminated Soil '93, 1393–1397.
© 1993 *Kluwer Academic Publishers. Printed in the Netherlands.*

The third one, the area around the Lindane plant, has been contaminated due to aerial waste dispersion. Finally, the sediments of an artificial lake, the embalse de Sabiñanigo, have been contaminated with HCH and mercury, due to discharge of liquid industrial wastes, containing HCH and mercury.

Bailin landfill area

This presentation will focus on one of the four contaminated sites, the Bailin landfill, which shows some interesting geohydrological features.

The Bailin area is situated in a zone of heavily folded Tertiary sediment bedrock, partly covered with a thin erosion layer. The orientation of the sediment layers is quite continuous with a more or less vertical dip. The sediments are of a continental origin and consist of an intercalation of claystone, sandstone and conglomerate. The sandstone and the conglomerate sandstone matrix are both cemented with carbonate, resulting in a very low layer permeability. However, due to the folding process, a local fault system has formed inside the incompetent sandstone and conglomerate layers. Due to the fracturing, the permeability of these layers has locally increased to above 1 m/day.

The Bailin landfill is emplaced on a steep mountain slope above a small brook, the arroyo de Paco, which is drained by the rio Gállego, about one km to the west. The landfill consists of two separate parts, the upper one containing HCH wastes and the lower one containing domestic wastes. Dumping of HCH wastes, first bulk and later in large plastic bags, took place from 1984 until 1988. Over 20.000 ton solid HCH waste were disposed at Bailin. Disposal of domestic wastes started in 1987 and still continues.

Due to the absence of effective top and basement isolations and due to the position of the HCH landfill in a valley-like depression, large amounts of surface runoff and rainwater infiltrate in the HCH landfill body, dissolving and mobilizing the stored HCH waste. With help of a provisional drainage system, part of the HCH percolate is diverted towards a concrete percolate collection basin.

However, due to infiltration into the erosion layer and the bedrock underneath the landfill and due to leakage of the basin, a significant part of the HCH percolate finally escapes.

The ineffective percolate drainage measures have led to serious contamination of soil and groundwater in the Bailin landfill area. In percolate surface runoff and in percolate sources, drained by the arroyo de Paco, extremely high HCH concentrations, unto the HCH solubility limit, have been observed. The escaped HCH finally ends up in the rio Gállego, threatening its drinkingwater production and irrigation functions.

Remedial option

Due to the hydrological and geohydrological situation of the Bailin landfill area, removal of the HCH landfill, the domestic landfill, the surrounding contaminated erosion layer, and isolation of the contaminated bedrock groundwater body appeared as the most favourable option.

However, given the very high costs attended with this option, the D.G.A. requested to elaborate an alternative option, consisting of in-situ isolation of the contaminated units. An important boundary condition with the elaboration of the remedial option was the requirement of further operation of the present domestic landfill. Based on the request of the D.G.A., Fugro and T.P.A. elaborated an in-situ isolation plan.

Remedial plan

The remediation will start with the installation of a hydrological barrier and groundwater withdrawal system in the permeable bedrock layers downstream from the HCH landfill. The system will aim at the control of the groundwater contamination in this area.
The hydrological barrier will be emplaced at the lower end of the contaminated groundwater area, to avoid further dispersion of contaminated groundwater. It will be positioned on top of a small crest, running towards the arroyo de Paco.

The barrier will consist of an infiltration trench, cut into the bedrock, more or less perpendicular to the bedrock layering. Above the permeable layers, small infiltration basins will be constructed to maximize infiltration. The infiltration system will be fed from an artificial lake, which will be constructed in the arroyo de Paco, below the infiltration trench.

Apart from the barrier, a groundwater withdrawal will be constructed inside the contaminated groundwater area. It will consist of two trenches. The trenches will be emplaced in small valleys, running towards the arroyo de Paco. In these valleys, percolate sources have been observed.

Based on model calculations, the combination of the blockade and withdrawal elements will induce an upward movement of contaminated groundwater, which will lead to a geohydrological isolation and gradual reduction of the bedrock groundwater HCH contamination.

After the emplacement of the hydrological barrier and groundwater withdrawal system, a second withdrawal system will be constructed in front of and underneath the HCH landfill. The withdrawal in front of the landfll will consist of a trench, cut into the bedrock, to catch shallow and surfacial HCH percolate flow. Subsequently semi–horizontal borings will be performed underneath the HCH landfill. The borings will aim at the drainage of deeper percolate flow and at the lowering of the watertable to a level below the bottom of the landfill.

A third withdrawal system, consisting of a trench comparable with that below the HCH landfill, will be emplaced along the foot of the domestic landfill.

Subsequent to the emplacement of the withdrawal systems, a large surface runoff interception ditch will be constructed around the landfills. The ditch will aim at the reduction of inflow of shallow groundwater and surface runoff into the HCH landfill and the domestic landfill. Given the topography and surface of the drained area and the peak precipitation, the interception ditch must be suited for peak discharge volumes of around 20 m^3/s. The upper ditch wall coating will consist of permeable material, to enable inflow of shallow groundwater.

In the final phase of the remediation, the HCH landfill is planned to be provided with a top isolation, to drain incoming rainwater and surface runoff. Previous to its emplacement, the present irregular landfill top and instable landfill front will be smoothed and flattened with contaminated erosion layer material, to ensure a sufficient top isolation rainwater drainage capacity and front talus stability. The top isolation will include a HDPE liner and two mineral isolation layers to ensure sufficient future isolation capacity.

With the design of the described remedial elements, a simple and robust design of the remedial elements has had special attention. For example, drainage and withdrawal constructions will be gravity-driven as much as possible and will be easy to access and maintain. Therefore, it is expected that the elaborated remedial plan will fulfil the aim of the remediation at minimal construction and maintenace costs.

Combined Process for Treatment of
Highly Charged Dump Leachate

Dr. Klaus Hagen (Lecturer)
Dr.-Ing. Karl Scharff

WABAG Wassertechnische Anlagen
GmbH & Co. KG
P.O. Box 20 49, W-8650 Kulmbach

1.   ABSTRACT

With the help of a multi-step treatment process, organically and inorganically highly loaded dump leachate can be
treated in a way to allow discharge into a clarification
plant.

2.   PROBLEM

The sewage water of a special waste dump, closed since
1984, contains up to 250 mg/l iron as well as big quantities
of low volatile and non volatile hydrocarbons, for example
dichloromethane, hexachlorocyclohexane (HCH), chlorophenol,
aromatic compounds, BTX-compounds and PCB's.
The cumulative parameters CSB and AOX come up to 15.000 mg/l
respectively 300 mg/l on an average.

3   PROCESS DESCRIPTION

The treatment of the sewage water happens in four
steps, two of them being carried out in the Batch-process,
two in a continuous process (see Fig. 1).
After treatment, the water can be discharged into a sewage
treatment plant.
The accumulation and separation of the lipophilic matters
happen in the water storage tanks at the beginning of the
treatment line.
All chemico-physical treatment steps for the treatment of
the sewage water, i.e. pH-value-adjustment, precipitation,
flocculation and sedimentation proceed one after another in
reactors without things built-in.
By means of pH-increase to pH 9 - 11, except the iron also
the existing $Ca(HCO_3)_2$ will be precipitated as $CaCO_3$ in
order to avoid operation breakdowns caused by production of
coatings in the subsequent steps.
The precipitation products are furthermore used for adsorption of lipophilic matters. After sedimentation and thickening, the sludge will be dewatered to $\leq$ 50 % TS by a chamber
filter press.
Subsequently, a rectification (steam stripping), adsorption
and neutralization will be carried out continuously, treat-
ment- and stop-periods relieving one another, depending on
the waste water quantity. During the rectification period,

F. Arendt, G.J. Annokkée, R. Bosman and W.J. van den Brink (eds.), Contaminated Soil '93, 1399–1400.
© 1993 Kluwer Academic Publishers. Printed in the Netherlands.

especially the water-steam- and the low volatile compounds
will be removed and upgraded. This way, only 3 - 5 litres of
organical concentrate per m3 sewage water need to be discharged. By subsequent adsorption of activated carbon, mainly the non volatile, easily absorbable organic matters will
be maintained and, this way, the AOX-producing matters will
be reduced to < 1 mg/l. As final treatment, the pH-value
will be adjusted with the help of HCl or NaOH (see Fig. 1).
Further important parameters:

    BTX-compounds: n.n. (non detectable)
    Phenol-index: n.n.
    Low volatile chlorinated
    hydrocarbons: n.n.

As a matter of principle, the residual products are being
upgraded as much as possible and burnt as special waste for
elimination.
The whole plant is carried out as closed construction including complete air cleaning with aeration and deaeration.

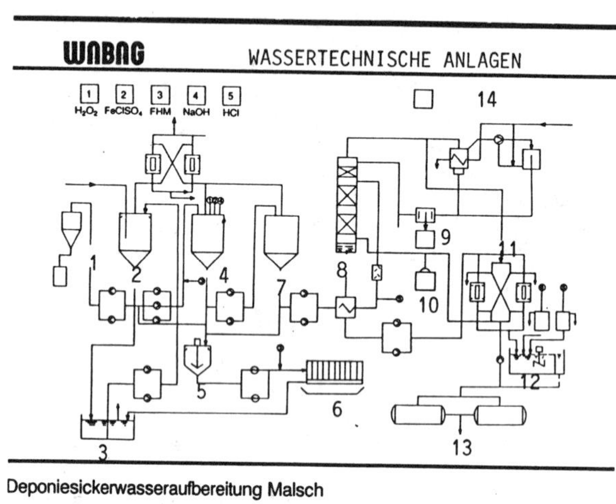

3.   ADVANTAGES

The above-mentioned process allows to profit from the
following advantages:

- Advantageous variation spectrum as far as the
  treatment periods are concerned, considering reaction
  and dosing sequence,
- Good adaptation to differing raw water quality and
  sewage water quantities,
- Easy control of the process steps
- Easy realization of shift operation,
- Reduction of the products to be discharged and
  destruction of the problematic matters by burning
- Starting position for further development of the
  single steps

IN SITU TREATMENT OF GROUNDWATER IN A SOLIDS-FREE ZONE
(PILOT PLANT LEUNA-WERKE AG)

Dr Lutz Haldenwang

WABAG GmbH & Co. KG
Dresden Office
0-8122 Radebeul, Hauptstr. 47

Dr Dieter Eichhorn

Dresdner Grundwasser Consulting GmbH
0-8020 Dresden, Meraner Str. 12

1. TASK

Groundwater in the area of the LEUNA-Werke AG, a traditional industrial area in Central Germany, is extremely high loaded with phenols, paraffins, halogenated hydrocarbons, mineral oil products and others, mostly harmful organic substances. The new in-situ technology of a SOLIDS-FREE ZONE is used for the complex treatment of groundwater and investigated within the project 02 WT 9186 initiated by the Federal Ministry of Research and Technology as part of a complete sanitation plan. It is being tested as a ditch-type pilot plant under real conditions.

2. DESCRIPTION OF PROCEDURE FOR A SOLIDS-FREE ZONE

2.1. Design principle

- One or several ditches, transversally to the groundwater flow direction/groundwater passes the ditch transversally.
- Length of ditch over part of or over the total flow profile
- Built up to or into the impermeable layer or of incomplete consctruction
- Vertical and permeable ditch walls

2.2. Basic technologies

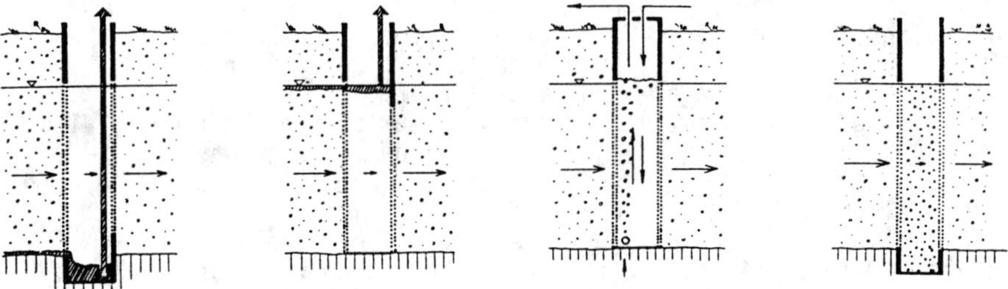

FIG. 1. a) sedimentation/accumulation b) flotation/accumulation
c)stripping/admixture d) floating bed/whirl bed

F. Arendt, G.J. Annokkée, R. Bosman and W.J. van den Brink (eds.), Contaminated Soil '93, 1401–1402.

2.3. Treatment effects

 2.3.1. Separation of phases: Sedimentation of sinkable components,
addition of precipitants, collection on the bottom of the ditch
(e.g. chlorinated hydrocarbons), suction/floating or flotation of
floatable water components on the fluidized surface, storage through
inverted weir, suction (e.g. mineral oil, fuel).
 2.3.2. Biological degradation: Microbiological degradation of water
components (e.g. phenols) with or without benthic-film carrier
material.
 2.3.3. Admixture: Electron donors or electron acceptors
(e.g. nutrients for biological processes or oxygen).
 2.3.4. Separation: Stripping of highly volatile components
(e.g. ether).

3. ADVANTAGES OF THE SOLIDS-FREE ZONE
- Complete registration of the total groundwater flow in the
  designed area enabling intensive processes of material separation,
  material admixture and material transformation (physical, chemical
  and biological).
- Direct treatment in the aquifer without abstraction of water and
  re-infilation.
- The in situ treatment processes take place in a defined space and
  are at any time easily and fully accessible for analytical
  supervision.
- Application of basic technologies of aerobic on-site water
  treatment.
- Long treatment time already with small ditch width
  (1 m about 3 days).

4. PILOT PLANT LEUNA-WERKE AG
- Task   : skimming off mineral oil products,
                      phenol degradation
- Site conditions: pleistocene aquifer, thickness - 5m,
                      groundwater surface 3m below ground surface.
- Design   : Ditch with length - 20m,
                      width - 1.0m,
                      depth - 6.0m.
- Technology   : A-phase - stripping, microbial degradation in the
                                whirl bed, admixture of nutrients
                      B-phase - sedimentation, flotation
- Controlling and sample-taking technique: SGM systems in flowing along
                                            and flowing away

5. APPLICATION OF THE SOLIDS-FREE ZONE
   In principle, all groundwater layers have the prerequisites
required for this, if they have been contaminated by failures,
industrial enterprises and military objects; if the impermeable
layers are not deeper than 10m below the surface of the ground and
if they are not obstructed by other use above the aquifer.

Remediation Wells Equipped With Porous Polyethylen Filters

Heinz Hötzl & Ingo Sass[*]

Introduction

While producing and running soil gas extraction and groundwater wells especially in fine grained soils different problems occur. The presented filter technique offers some solutions fitting for remediation tasks in the unsaturated zone. Large diameter wells in fine grained soils do not influence areas as large as in comparatively coarser grained soils. For this reason several wells with smaller diameters (less than 3½ inch) are more suitable: with the same energy input a larger contaminated area can be extracted.

But it is a problem of the practical well design to realize a fitting filter gravel pack, because small diameter wells have not got enough free space. Enlarging the specific diameter of the borehole will increase the expenses. On the other hand a very thin or too coarse filter gravel pack will cause well damages or a clogging up with mica or silty material.

To achieve a large and steady laminar capture area in spite of a low soil permeability higly porous filter cylinders were developed. This filter tubes substitute conventional slot well screens and they are simulating a kind of fine grained filter sand. They are rigid and stable enough to bear the soil load in a horizontal well. This technology fits especially for remediation projects in stratified soils with backwater bodies or for example for soil gas extraction in built-up areas (Fig. 1.).

Porous Polyethylen Filters

The porous filters are sintered of high density polyethylen granules. They are made of 100% polyethylene (first prototypes of sintered polypropylene and polyvinylidenfluoride are in laboratory tests). The porous structure has a double function: on the one hand it works comparably to the common filter gravel and on the other hand it stabilizes the well like conventional slot PVC-, PE- or other screens. This homogenous and highly porous matrix guarantees a laminar

[*] Department of Applied Geology
University of Karlsruhe
Kaiserstraße 12
D - 7500 Karlsruhe 41

fluid flow and helps to avoid precipitations. A turbulent breakthrough or abrasion of the filter tube with silty material is impossible. The hydrophobic plastic material avoids penetration of soil humidity and soil water into the well. Only under saturated conditions with a certain water pressure level (depends on the air entry value of a certain filter type) water can pass into the filter cylinder. For this reason those filters are suitable for measures with backwater and groundwater.

Material	Sintered-PE
Porosity	45 - 60 %
Pore size diameter	
maximal	90 - 120 μm
average	45 - 80 μm
K-value	$5 \cdot 10^{-2}$ -
(by air measurement)	$2 \cdot 10^{-4}$ m/s
Pressure drop	
(air at 250 m/h)	0,2 - 1,5 hPa
Actual length	0,5 & 1,0 m
Outer diameter	60 - 125 mm[#]
Inner diameter	50 - 100 mm
Wall thickness	2,5 - 12,5 mm

Tab. 1: Some physical data of PE-filter tubes (Range represents several available types; [#]larger diameters are in preparation)

Field experiments

First field experiences are made in places with different soil types (clayey silt, silty sand, sandy fine gravel and refilled anthropogenic soil). Eight vertical vapour extraction wells with different borehole diameters (from 2 to 8½ inch) are installed. They all are situated in residual industrial pollutions with chlorinated VOC's and other minor components. First field experiments started in spring 1992, and a new back water test site will be equiped with horizontal wells until spring 1993. Results of this experiments and of the horizontal vapour extraction well shown in Fig. 1. will be presented. They include comparative tests of several filter materials in backwater and soil gas extractions which proved suitability of the polyethylen well screens.

F. Arendt, G.J. Annokkée, R. Bosman and W.J. van den Brink (eds.), Contaminated Soil '93, 1403–1404.
© 1993 Kluwer Academic Publishers. Printed in the Netherlands.

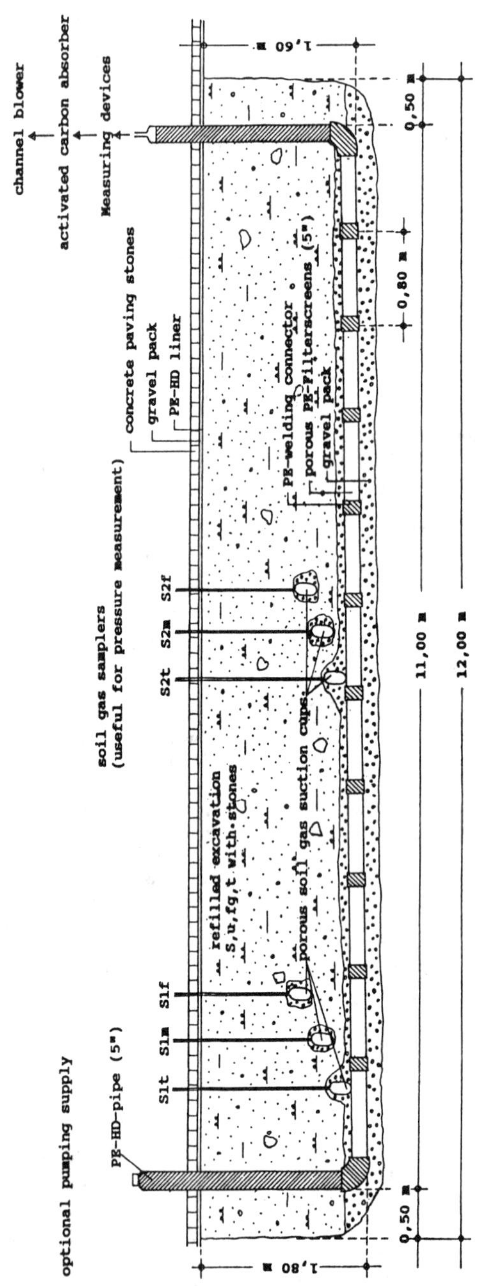

Fig. 1: Well design example for the use of porous polyethylen screens in a horizontal vapour extraction well in a residual pollution with VOC's in Hanue/Germany. The use of is second blower unit is possible.

APPLICATION OF TEST SYSTEMS FOR BALANCING AND OPTIMIZATION OF BIOLOGICAL SOIL TREATMENT

K. HUPE, J. HEERENKLAGE, S. LOTTER, R. STEGMANN

Technische Universität Hamburg-Harburg
Abfallwirtschaft und Stadttechnik
Harburger Schloßstraße 37
D-2100 Hamburg 90

1. ABSTRACT

Various closed test systems in laboratory scale are presented which are applied to evaluate and optimize methods of biological soil treatment: glass vessels, respirometers and reactor systems.

2. MATERIALS AND METHODS

Biological treatment of contaminated soils requires methods to optimize and evaluate the turnover of contaminants. The mere measurement of the contaminants concentration in the soil is not sufficient. For a complete balancing of the contaminants turnover it is necessary to measure all emissions as well as the remaining residues. The application of closed test systems is a necessary precondition for making balances. Parameters for the carbon balance are: biomass, mineralization (degradation), emitted contaminants in the gas phase and residual concentrations in the soil.

For the optimization of the milieu conditions batch setups and respirometers have proven to be successfull. Gas-tightly sealed glass vessels (batch setups) are a particularly simple system to investigate degradation processes. In these vessels the CO_2-evolution is discontinuously determined via absorption in NaOH. In addition the soil samples are analysed. Contaminants emitted in the gaseous phase can also be measured (GC-measuring). Since the test setups are relatively simple it is possible to investigate a large number of tests in parallel.

In the respirometer the biological O_2-consumption is continuously measured. Due to the absorption of the produced CO_2 onto NaOH a vacuum is produced in the closed vessels. By means of a manometer, the electrolytic oxygen production from $CuSO_4$ is initiated. Connecting the measuring device to a personal computer it is possible to continuously record the oxygen consumption (mg O_2). To describe the degradation progress in the respirometer, vessels can be removed during a test run and the soil material can be analysed. Abiotic CO_2-evolution does not influence the results. The system consists only of 12 test vessels.

Reactor systems are used in different scale (volume 3 l, 6 l, or 80 l) in order to simulate the conditions in an aerated windrow or in reactors. They are continuously aerated by compressed air. In these systems emitting volatile components can be measured.

In addition to the static test system also dynamic test systems are in use (batch setup, 3 litre paddle stirrer reactors). Figure 1 shows a test system (static and dynamic) including the continuously operated monitoring system. Especially the treatment of clay-like soils can be improved using dynamic bioreactors. Further investigations are necessary to avoid or reduce pellet formation during dynamic treatment of soils.

F. Arendt, G.J. Annokkée, R. Bosman and W.J. van den Brink (eds.), Contaminated Soil '93, 1405–1406.
© 1993 *Kluwer Academic Publishers. Printed in the Netherlands.*

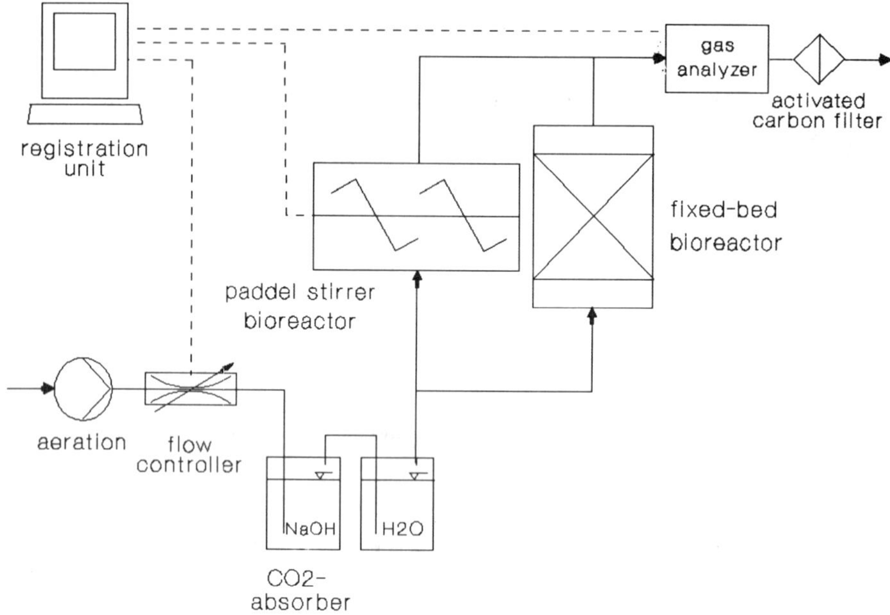

Figure 1.   Principle of the bioreactor test system (static and dynamic)

3.   RESULTS

The investigations carried out until now have shown that these test systems are an important tool to evaluate and optimize the biological degradation of contaminants. The batch setups and respirometers proved to be suitable particularly for pretests to optimize the milieu conditions. For material balancing closed aerated bioreactor systems with continuous analytical recording of the exhaust gas (CO_2, TOC) have been used successfully. The carbon balance of the oil - especially in test reactor systems where a soil/compost mixture was treated - is not satisfactory. The fate of the non-detected carbon is still unknown. The observed balance gap indicates the formation of bound residues in the humic matrix; this phenomenon will be subject of further investigations.

4.   ACKNOWLEDGMENT

This project is funded by the Deutsche Forschungsgemeinschaft. It is a subproject of Research Centre 188, Hamburg, "Treatment of Contaminated Soils". Results from the intensive discussions in the different working groups of the research centre are to a certain extend incorporated in this paper; their input is most appreciated.

5.   REFERENCES

Stegmann, R., Lotter, S., Heerenklage, J. (1991). Biological treatment of oil-contaminated soils in bioreactors. In R.E. Hinchee and R.F. Olfenbuttel (Eds.), On-Site Bioreclamation, First In-Situ and On-Site Bioreclamation Symposium, Butterworth-Heinemann, pp 188-208

Suitability of biological methods for the regeneration of loaded activated carbon using chlorobenzoic acid and thioglycolic acid as model substances.

Mustafa Jaar; Harald Krebs

Holsteiner Gas-Gesellschaft mbH, Kaiser Wilhelm Strasse 115,
2000 Hamburg. *Germany.*

Abstract

Among the methods for the biological regeneration of granular activated carbon (GAC) a distinction can be made between the "simultaneous", the "quasi-simultaneous" and the "consecutive" method. In the first method, adsorption, desorption and biological degradation occur simultaneously. The reactor is charged in a continuous mode. In the second method, loading and regeneration are conducted successively in the same reactor. Adsorption and regeneration processes occur in short time intervals. The reactor is operated in the form of a sequencing batch reactor (SBR-technology). In the third method, the GAC is packed in a special regeneration reactor after loading Jaar et al. [1]. The first two methods are summarized in the following.

The results of the studies showed that in comparison with a continuous flow GAC reactor and a sand Sequencing Batch Reactor (SBR), the GAC-SBR achieved a higher rate of biological degradation. After an operating period of 14 months during which loading and simultaneous regeneration of the GAC took place and during which a substrate amount of 59 mg C/g GAC in total was fed, the regeneration efficiency was 93%.

Introduction

The regeneration method for GAC and the type of regeneration technology go by the type and composition of the adsorbed substances. The method used in most cases is the thermal regeneration which in addition to high energy consumption leads to a general loss of GAC mass.

The use of biological regeneration methods represents a possibility to create an environmental friendly solution for the regeneration of loaded GAC and to probably minimize the costs. Biological regeneration of GAC is achieved through biological activity on the GAC surface in interaction with adsorption, desorption and pore diffusion. When a critical activity of the biofilm - who takes up around the individual grains of the GAC - is reached, a reversion of the concentration gradient in the pores of the GAC takes place. Adsorbed substances desorb, are transported back to the biofilm and degradated.

Material and Methods

For the studies, three full mixed fixed bed reactors were operated in parallel. All three reactors were equiped with a gas permeable membrane system for the oxygen infusion. The working volume of each reactor was 2.6 l. The first and the second reactor (SBR and continous mode) were packed with 2 kg GAC of the type HKW1 each and the third reactor (SBR) was packed with 5 kg of sand.

The bacteria culture *pseudomonas putida* PRS2015/pAC27 was used for the regeneration studies with the model substrate 3-chlorobenzoate (3-CB). That strain is known to be able to
decompose 3-CB completely Daniel et al. [2]. In the studies with the second model substrate thioglycolic acid (TGS), *pseudomonas putida* from the previous studies were inoculated. The specific mineralization products chloride and sulfate were used for the degradation of the model substances.

F. Arendt, G.J. Annokkée, R. Bosman and W.J. van den Brink (eds.), Contaminated Soil '93, 1407–1408.
© 1993 *Kluwer Academic Publishers. Printed in the Netherlands.*

Results

In order to describe the influence of the substrate concentration on the microbial degradation in the three reactors, the model substances were fed in stepwise increased concentration. Fig. 1 and 2 show the concentration of chloride and sulfate in the drain measured during the study period. The figures show that only in reactor 1 (GAC-SBR) the degradation of substrate was almost completed. Compared to the two GAC reactors, the toxic influence of the substrate on the microorganisms was strongest in the sand reactor, due to the lack of adsorption effects. The dominance of adsorption effects and the SBR-technology - which presumably leads to a strong selection among the organisms - turned out to be of great advantage.

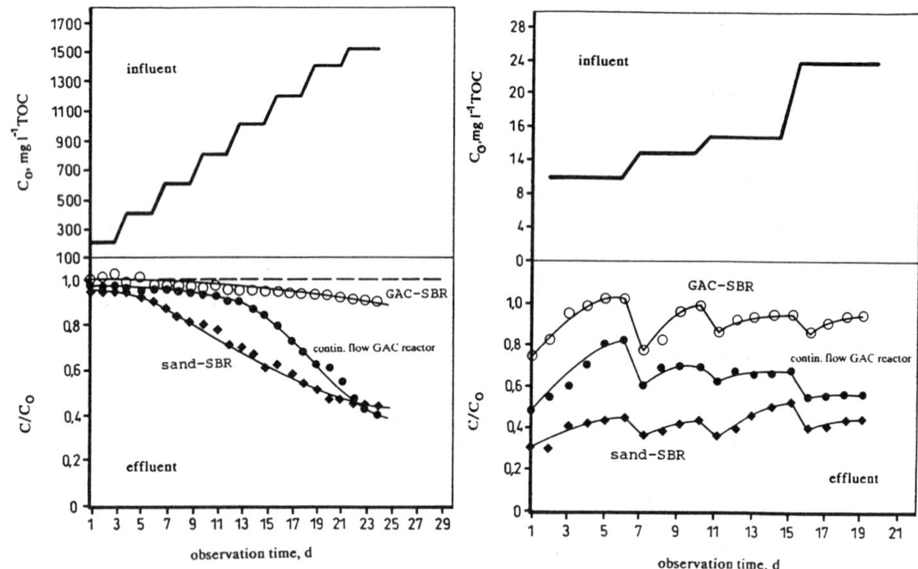

Fig. 1   Effect of a stepwise increase of the 3-CB concentration on the reactor performance

Fig. 2   Effect of a stepwise increase of the TGA concentration on the reactor performance

Summary

The studies confirm the possibility of the biological regeneration of GAC. A prerequisite for this process are the biological degradability of the adsorbed substances and their reversible bond. Influential factors for the microbial regeneration of loaded GAC are the availability of the substances to be degradated and the biological activity of the microorganisms on the GAC. The higher the degradation activity of the microorganisms and the higher the availability of the substances to be degradated, the higher is the driving concentration gradient between the GAC and the surrounding biofilm and the faster runs the regeneration of the carbon.

Literaturverzeichnis

[1] Jaar, M.A.; Krebs, H.; Rubio, M.A.; Wilderer, P.A.: Biologische Regeneration einer mit 3-Chlorbenzoat beladenen Aktivkohle; Z. Wasser-Abwasser-Forschung 22(1989), S. 1-4

[2] Daniel, R. S. und Tiedje, J. M.: Isolation and Partial Characterization of Bacteria in an Anaerobic Consortium that Mineralizes 3-Chlorbenzoat. Applied and Environmental Mikrobiologie 48 (1984), S. 1159-1165.

MICROBIOLOGICAL REMEDIATION OF OIL CONTAMINATED SOIL IN THE bioStack-FLACHBETT-BIOREAKTOR

Janzen, St.; Raphael, Th.; Knackstedt, H.-G.; Bröcking, P.; Sprenger, B.:

HP-biotechnologie GmbH, Witten, Germany

INTRODUCTION
Soils contaminated with hydrocarbons usually can be cleaned by microbiological remediation techniques. The treatment of small quantities and fine grained soil parts however turns out to be problematic. In order to provide suitable methods even for these applications the following presented flat-bed reactor system was developed, tested and put into practice.

FLAT-BED-BIOREACTOR-SYSTEM
FLAT-BED-BIOREACTOR
The basic unit of bioreactors (fig. 1) consists of a standardized flat-bed container having a hook carrier (6,5 m x, 2,5 m x 1,3 m). Its highest carrying capacity amounts to 15 m^3 (23-25 Mg). The bottom of the reactor is pierced with 3,0 cm ø bores (1) and equipped with perforated gas tubes (2) which can be provided with pressed air. In order to mix the soil three stirring shafts (3) with special formed stirring elements are installed and are moved by a separate driving aggregate via clutches (4). To use the reactors in a compact way the units are stackable up to four reactors (bioStack).

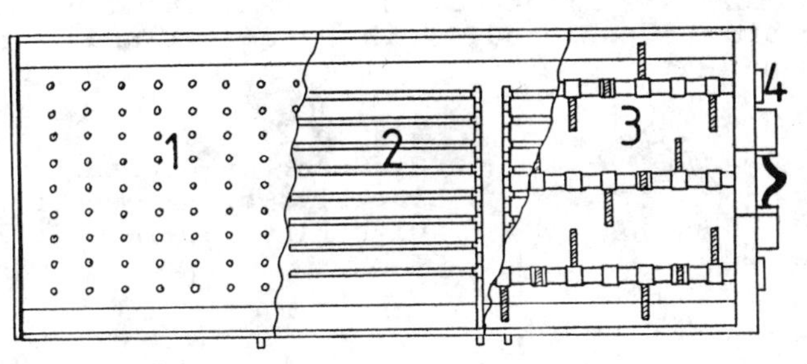

Figure 1:
Flat-Bed-Bio-reactor
(from the top)
1 Perforated Bottom
2 Gasification Tubes
3 Stirrer
4 Drive Clutch

OPERATING SYSTEMS
In addition to the bioreactor for solid treatment a basin, an equipment container and a housing are included in the treatment system as well. The treatment water trickling through the ground is collected by a basin, led via a pump in the equipment container and stored or treated biologically in a tank. The treatment water is spread - depending on demand - over a sprincle plant and led into a circuit.

F. Arendt, G.J. Annokkée, R. Bosman and W.J. van den Brink (eds.), Contaminated Soil '93, 1409–1410.

In order to prevent degasification of harmful substances and dust outlet the reactor and the basin are housed by a tent. Air is continuously led through the tent via a blower and the used air is cleaned by an activated carbon filter. All electric appliences as pumps and blower are stored in the equipment container.

OPERATION CONDITIONS

The contaminated soil is directly filled into the flat-bed bioreactor at the site of origin where it can be treated (on-site) in a mobile basin. The flat-bed reactor which is equipped with natural fibre fleece can be used for transportation to a recycling centre as well (off-site). During transportation the soil is covered by a hood.

Depending on the soil structure different equipped bioreactors are needed. In case of a low contaminated sand the use of a flat-bed with bottom perforation and sprinkling is sufficient, in case of a highly contaminated soil an additional treatment with pressed air, in special cases with oxygen is needed and in order to clean up clayey soils the soil has to be mixed by the stirrer in different time intervals to destroy a channel formation and to get a better exchange of substances.

RESULTS

Up to now several different soils have been treated in the here presented system. It was found out that sand could be cleaned very well. Even fine grained soil could be treated successfully in a very short time. Figure 2 shows for instance the microbiological removal of hydrocarbons from a fine grained soil containing small quantities of organic plant residues like leaves and roots.

Already after nine weeks an oil-reduction of 75 % could be noticed in a soil being airated continuously and mixed only once or twice a week for five minutes. After further nine weeks the aimed concentration was reached. A degasification of harmful substances could be excluded by the chemical analysis of the activated carbon filter.

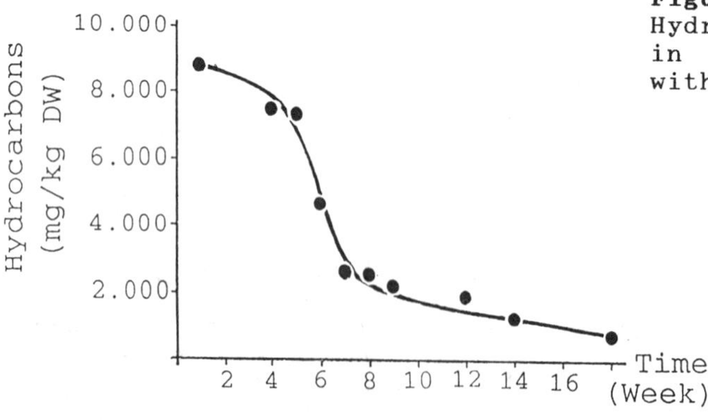

Figure 2:
Hydrocarbon Degradation in Flat-bed-Bioreactor with Fine Grained Soil

ELECTROCHEMICAL SOIL TREATMENT

H.JEHRING [1], R.JEHRING [1], R.ROHDE [2]

[1] ING.-BÜRO DR.H.JEHRING, ELEKTROCHEMISCHES LABOR,
   Tharauer Allee 3, W-1000 BERLIN 19
[2] BERATENDE GESELLSCHAFT FÜR GEOTECHNISCHE INGENIEURLEISTUNGEN,
   Lehrter Straße 16-19, W-1000 BERLIN 21

The soil/water sphere represents a natural electrochemical
system. It may be cleaned advantageously by electrochemical
methods. The driving forces result from electric field effects
onto charged particles, especially at charged interfaces.
One of the most important advantages of electrochemical
treatments over hydraulic methods is the higher flow rate
through small capillaries (low hydraulic conductivity), caused
by electric forces at the solid/liquid boundary.
Electrochemical soil treatment one could combine with other
methods, for example hydraulics.
Different electrochemical processes may be connected together:
Electrochemical metal deposition (electrolysis) at electrodes
and electrokinetic flow (electrophoresis, electroosmosis).
The range of removable pollutants by electrochemical soil
treatment extends from heavy metals and arsenic to organics
incl. biological aspects.
However, there is open the question about the damage to
metallic objects by electric fields in the soil (stray
current corrosion) during electrochemical treatment.
For investigation the circumstances of stray current corrosion
on metallic objects near the electrochemical soil treatment as
well as for assessment of chances of actions against stray
current corrosion different experiments on soil boxes were
carried out:
- Electrochemical measurements of potentials within the soil
of an electrochemical system for estimation the electric
fields causing stray current (equipotential lines).
- Measurements of corrosion currents through metallic probes
within the stray current area of the test cell as equivalent
of corrosion rate.
- Electrochemical potential measurements (switch over
potentials) at the metal/soil interface on probes during
electrochemical soil treatment (free corrosion potential,
protection potential).
- Electrochemical stray current protection by stray current
drainage and stray current soutirage controlled by given
specific cathodic protection potential of the metallic probe
in the soil.

F. Arendt, G.J. Annokkée, R. Bosman and W.J. van den Brink (eds.), Contaminated Soil '93, 1411–1412.
© *1993 Kluwer Academic Publishers. Printed in the Netherlands.*

About the endangering of metallic objects (pipelines, cables, tanks, buildings, components) by electrochemical soil treatment und about the corresponding preventive measures at the moment one can say:
- Occurrence of stray current corrosion on metallic objects in the sphere of influence of electrochemical soil treatment, like electroosmosis, is possible and must taken into account.
- The degree of endangering depends on medium (electrical soil/water conductivity), on kind of endangered metallic objects (material, corrosion protection, construction, function, location, age) and of course on the operational procedure of the electrochemical installation for soil treatment (voltage, current density, kind and position of electrodes, soil pollution, rinsing solution).
- During planning and before starting electrochemical soil treatment sufficient enquiries, preliminary examinations as well as electrochemical field measurements about stray current effects on the surroundings must be made as a matter of principle.
- Provided that there is something to protect, technological and local conditions and potentialities of steps against negative consequences of electrochemical site clearance should be included into the site investigation.
- Electrochemical corrosion protection (stray current protection, cathodic protection) combined to electrochemical soil treatment presents as has been shown a recommendable promising technology for effective soil clearance without negative side effects on any kind of metallic buildings.
- As the guide to set und maintain electrochemical protection during electrochemical soil treatment serves the (cathodic) protection potential of the given metal (with or without coating) under prevailing conditiones (soil/water/additives/pollutiones).
The results of our investigations about electrochemical soil treatment like electroosmosis (arsenic and others) prove the advantages of this method.
In one case we could eluate 1200 mg arsenic within the first two weeks by electroosmosis in a pilot plant filled with soil from warefare-related contaminated sites.
A further test showed the superiority of electroosmotic cleaning over hydraulic leaching in soils with high hydrodynamic resistivity.
The electrochemical in-situ treatment in urban und industrial area requires to take sufficiently into account the other (negative) site of soil electrochemistry in form of stray current. However in case of stray current protection is indispensable its feasibility almost certainly is given.
The investigations were promoted by SENATSVERWALTUNG FÜR STADTENTWICKLUNG UND UMWELTSCHUTZ, W-1000 BERLIN.

VEGAS: A NEW FACILITY FOR LARGE-SCALE PHYSICAL MODELLING OF IN-SITU REMEDIATION OF CONTAMINATED GROUNDWATER AND SOILS

H. Kobus, G. Teutsch, H.-P. Koschitzky, O. Cirpka
Institut für Wasserbau, Universität Stuttgart

1. Abstract

The research facility on remediation of contaminated groundwater and soils (German "Versuchseinrichtung zur Grundwasser- und Altlastensanierung") *VEGAS*, presently under construction at the University of Stuttgart, will consist of a laboratory hall with several artificial aquifers. *VEGAS* is funded by the Federal German Ministry of Research and Technology BMFT and the Ministry of Environmental Affairs of the State of Baden-Württemberg and will be available to any German research group. The main objectives of the facility are to develop and optimize subsurface remediation technologies and to improve the scientific understanding of contaminant transport and transformations in the subsurface.

2. Why large-scale physical modelling of subsurface remediation?

At contaminated field sites data about contaminant content and distribution as well as exact data about subsurface characteristics like hydraulic conductivity, organic carbon content, pH or redox potential are scarce. Hence it is hardly possible to calculate mass balances for remediation activities. Proving in-situ biodegradation or success of any other in-situ remediation technique is very difficult. The injection of substances into the subsurface either to obtain well defined initial conditions or to enhance remediation processes is strictly limited by environmental protection laws. Therefore it is very difficult to introduce new in-situ remediation technologies especially if they include the injection of nutrients or other reactive agents into the subsurface.

On the other hand, it is mostly questionable to transfer results from classical laboratory investigations to field conditions. As subsurface structures are heterogeneous and commonly unknown in detail, it is in general not possible to scale up results of microscale batch or column experiments with homogeneous fillings to the field scale.

VEGAS is a contribution to bridge this gap. Large-scale physical modelling in artificial aquifers makes it possible to conduct experiments under controlled conditions, which include considerations of heterogeneity at a scale close to natural field conditions. It is possible to arrange artificial heterogeneities either in a well defined layout or in the form of arbitrary structures. Another advantage of large-scale physical modelling is the possibility to use technical equipment from field site applications in a laboratory, thus providing a direct link to technology applications in the field.

3. Technical details

Figure 1 shows a typical artificial aquifer configuration the largest of which is 18m long, 9m wide and 5m high. The containment is constructed in stainless steel in order to avoid adsorption of organic contaminants. Natural groundwater flow can be simulated by a circulation system. If neccessary a cover can be provided in order to controll contaminant volatilization. A treatment system for contaminated water and air will be

1413

F. Arendt, G.J. Annokkée, R. Bosman and W.J. van den Brink (eds.), Contaminated Soil '93, 1413–1415.
© 1993 Kluwer Academic Publishers. Printed in the Netherlands.

1414

installed. The basin can be divided into several segments. Furthermore, a considerable number of smaller domains will be installed: midrange-scale basins for three-dimensional experiments, two-dimensional models for investigations on sequential processes in groundwater flows, and lysimeters filled with undisturbed natural soils for infiltration experiments.

VEGAS experiments will be carried out by research institutes and industrial users through-out Germany which may include companies providing remediation technologies, consultants or industrial environmental protection departments.

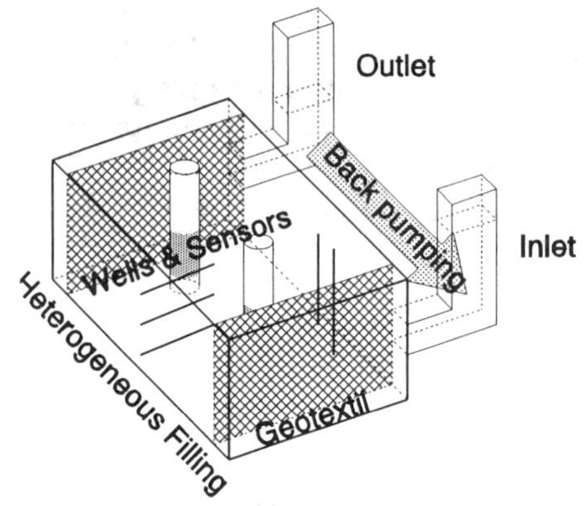

Figure 1: *Artificial aquifer constellation*

4. Some examples of research topics in *VEGAS*

4.1. Fate and behaviour of contaminants in the subsurface

Transport and transformations of contaminants in heterogeneous systems

The effective behaviour of organc contaminants in heterogeneous porous media will be investigated for the case of PAHs. One project investigates mass transfer in well defined heterogeneous configurations, another will deal with the enhancement of extraction by surfactants and cosolvents, and a further project will focus on bioavailability of PAHs.

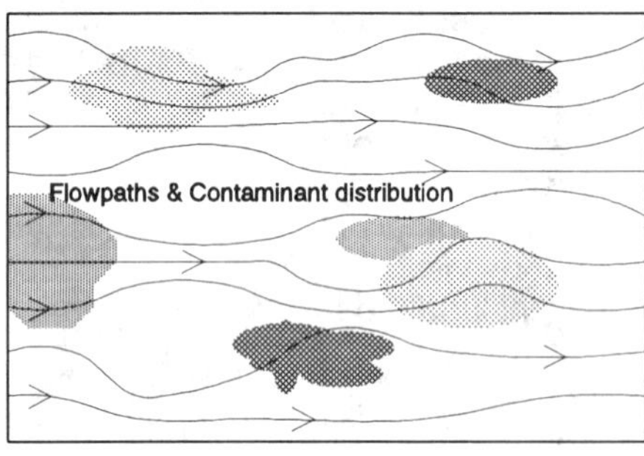

Figure 2: *Flowpaths & contaminant distribution*

Multiphase flow in the subsurface

Measurement techniques for non-aqueous phase liquids (NAPL) will be optimized and experiments on simultaneous infiltration of NAPL and water into an undisturbed natural soil column will be carried out. Numerical modelling codes on multiphase flow and transport will be optimized and validated by the experiments.

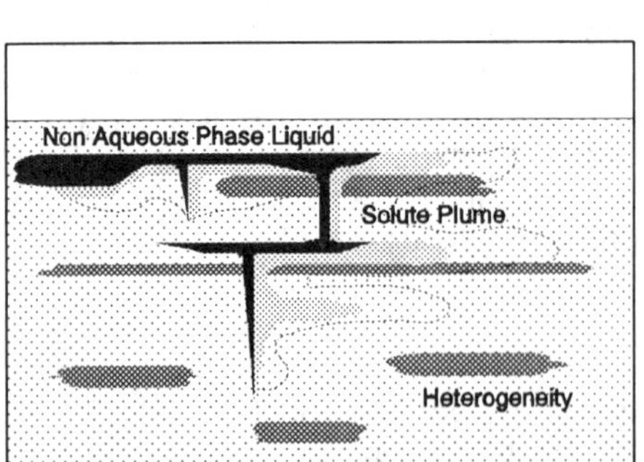

Figure 3: *Infiltration behaviour of dense nonaqueous phase liquids*

4.2. Technology development for *in-situ* groundwater and soil decontamination

Test of different hydraulic configurations
The removal of NAPL by surfactant enhanced extraction and *in-situ* biodegradation will be compared for different hydraulic constellations such as circulation wells, four-well systems, etc.

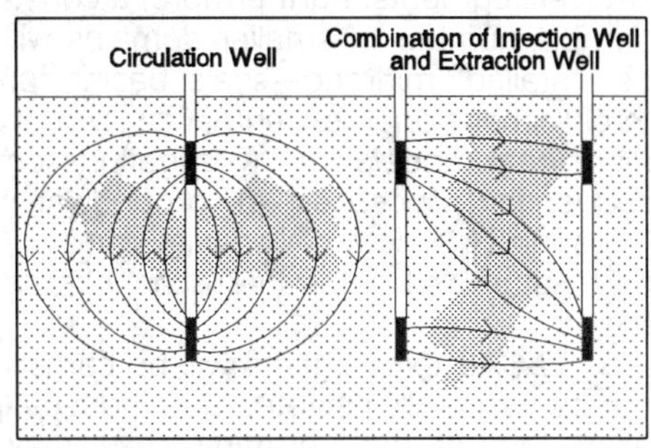

Figure 4: *Possible well configurations*

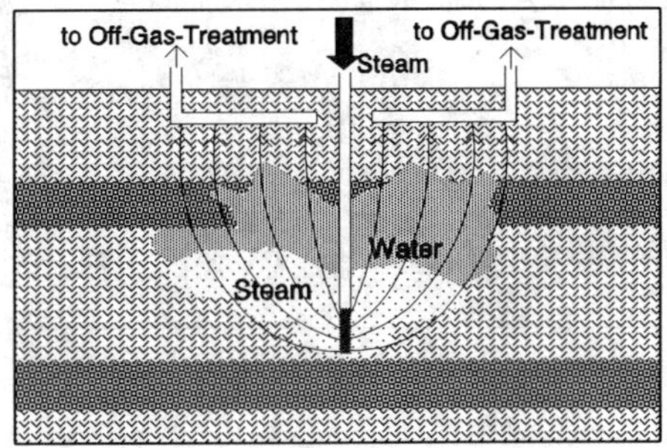

Figure 5: *In-situ steam extraction*

In-situ thermal enhancement of soil vapor exctraction
Experiments on thermal mobilisation of semivolatile contaminants by injection of steam or hot air will be compared to conventional soil vapor extraction. Investigations will focus on pure organic phase mobilisation due to pressure drop at the condensation front and to the influence of inhomogenous structures on the efficiency of the technology.

Optimization of in-situ biodegredation
Effective nutrient and electron acceptor supply in heterogeneous porous media will be investigated at large scale and compared with batch experiments in order to identify scale dependent parameters of biodegradation.

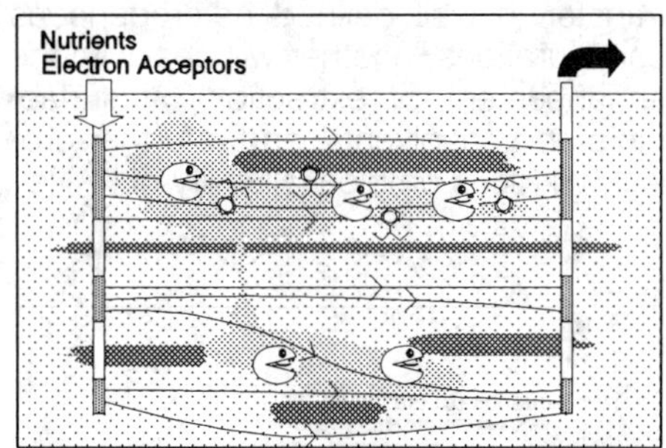

Figure 6: *In-situ bioremediation set up*

EXTRACTION OF HEAVY METALS FROM SOIL MATERIAL, DUST, CINDERS, MUD AND
SEWAGE SLUDGE

KRETZSCHMAR, RAYMUND

DEPARTMENT OF WATER MANAGEMENT AND LANDSCAPE ECOLOGY, CHRISTIAN-ALBRECHTS-
UNIVERSITY, KIEL, FEDERAL REPUBLIC OF GERMANY

1. ABSTRACT

Residual substances have either to be avoid, reduced or utilized. A
chemical treatment for waste and contaminated sites containing heavy metals
is presenented. Many residual materials can be decontaminated by
extraction employing organic acids: heavy metals are reclaimed, after dehydration the extraction agents can be re-used in a cyclic process, and
aceton ist obtained by fractionating distillation.

2. BASICS

If heavy metal-containing residual products are mixed in a reactor with the
fivefold quantity of a monocarbon acid or a mix of monocarbon acids (C_1-C_4)
and incubated for an hour, an equilibrium adjusts comprising e.g. water-
soluble metalloacetates and many organic compounds. The fluid phase is
separated and processed. The success of decontamination and the efficiency
of this method are especially high when relatively dry residual material is
used. Monocarbon acids are simple to produce and easy to dispose of because
they are degradable.

3. PROCESS ENGINEERING

As depicted in Fig. 1, residual material and acid reach the reactor through
a sluice. From time to time aliquots are removed from the reactor and
arrive at a multi-stage filter.

3.1 Filtration

Step 1: The mixture is squeezed off with a maximum pressure of 1.5 MPa
Step 2: The sediment from 1 is washed under pressure with acid
Step 3: The sedimetn from 2 is washed under pressure with water
Step 4: The sediment from 3 is washed under pressure with purified sewage
        "to activate the microbiology"
Step 5: The sediment from 4 is ejected, it is decontaminated
Step 6: The filter is flushed or changed if necessary.
The filtrates of the steps 1 - 3 are separately collected according to
their portion of acid and water.

3.2 Treatment of the filtrate

The reclamation of the acid is achieved by a two-step fractionating disil-
lation (evaporator 1 and 2, Fig. 1). The residue of distillation will be

F. Arendt, G.J. Annokkée, R. Bosman and W.J. van den Brink (eds.), Contaminated Soil '93, 1417–1418.
© 1993 *Kluwer Academic Publishers. Printed in the Netherlands.*

concentrated yielding a mixture of heavy metals, organic contaminations and carbonates. The vaporous acids will be condensed, and re-enter either directly or after dehydration the extraction process. The residual material of the second evaporator has to be conventionally disposed of.

4. LABORATORY RESULTS

Average rates of extraction of sewage sludge and mud from the Elbe-river are given as determined by a Soxhlet:
Aluminium up to 98 %; lead 79-92 %; cadmium 86-94 %; chromium 70-86 %; iron 71-87 %; copper 84-95 %; nickel 57-82 %; mercury 32-34 % and zinc 70-96 %.

5. ESTIMATION OF COSTS

In case of 5 t capacity per day the specific costs are 1300 DM/t.
If the capacity is 5 t per hour the cost will reduce to 270 DM/t.

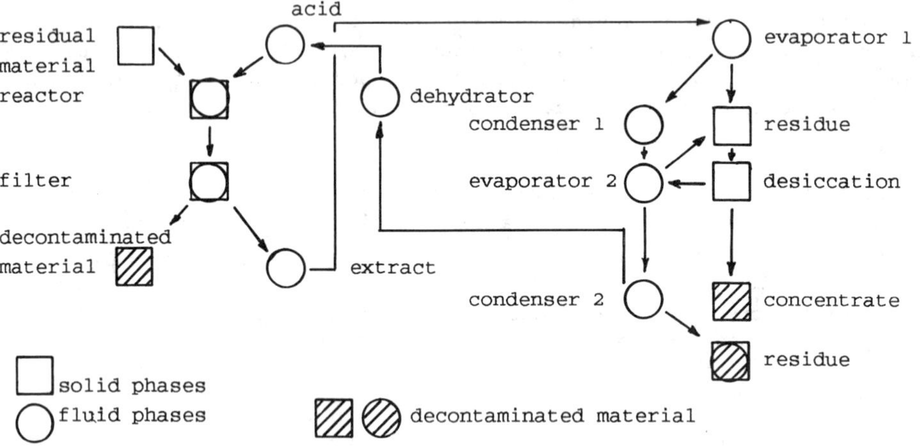

Figure 1: Diagram of the extraction of heavy metals

Physical, Chemical and Biological Methods
of In-Situ Treatment of Soils Contaminated with Mineral Oil

M. Müller, F. Forge, M. Liebeskind, H. Fr. Schröder
Institut für Siedlungswasserwirtschaft der Rheinisch-Westfälischen Technischen
Hochschule Aachen, Templergraben 55, D - 5100 Aachen

For many years soil venting and biological treatment have proven successful in-situ methods of treating soils contaminated with mineral oil. A new type of in-situ soil decontamination involves treating unsaturated soil with gaseous ozone as a means of chemically oxidizing the organic contaminants. The combined use of the stated in-situ techniques was examined on an industrial scale together with Hannover Umwelttechnik GmbH (HUT), Hannover and the chemical laboratory of Dr. E. Wessling, Altenberge by eliminating mineral oil pollution in a section of soil without groundwater saturation in the context of a project sponsored by the Federal Ministry for Research and Technology (BMFT). Specialist technical supervision was provided by the Federal Environmental Agency.

The soil venting technique is based on the propensity of substances with a high vapour pressure to evaporate in soil that is not saturated with groundwater. The contaminated soil air is extracted by wells, cleaned by adsorption or catalytic/thermal treatment and then discharged to atmosphere. Only highly volatile mineral oil components (gasoline fraction) can be treated by soil venting.

Methods of biological soil decontamination promote the natural destructive ability of bacteria and other micro-organisms by improving the soil environment, especially by ensuring an adequate supply of oxygen and a balanced nutrient ratio. Soil air extraction is an especially successful means of supplying the micro-organisms with oxygen in the unsaturated soil zone, whereby the contaminated soil is supplied with an after-current of atmospheric oxygen.

The chemical oxidation of organic contaminants by introducing gaseous ozone into the soil is a new, supplementary technique. The water solubility of the organic contaminants dissociated by the ozone reaction is tendentiously improved and their bio-availability thus enhanced by the inclusion of hydrophilic functional groups (aldehyde, carboxyl and keto groups). It was possible to ascertain the chemical structure of individual intermediates of the PCA decomposition.

Particular attention was paid to the potential environmental hazard posed by products formed by the ozone reaction of the organic contaminants. Soils contaminated with mineral oil and PCA-contaminated soils were treated with ozone and examined for mutagenic and toxic effects by means of the Ames test. The mutation-inducing effect of PCA-contaminated soils was reduced by ozone treatment. The reduction in the mutagenic effect of soil extracts contaminated with benzopyrene is clearly illustrated in Figure 1 (laboratory experiment).

F. Arendt, G.J. Annokkée, R. Bosman and W.J. van den Brink (eds.), Contaminated Soil '93, 1419–1420.
© 1993 *Kluwer Academic Publishers. Printed in the Netherlands.*

1420

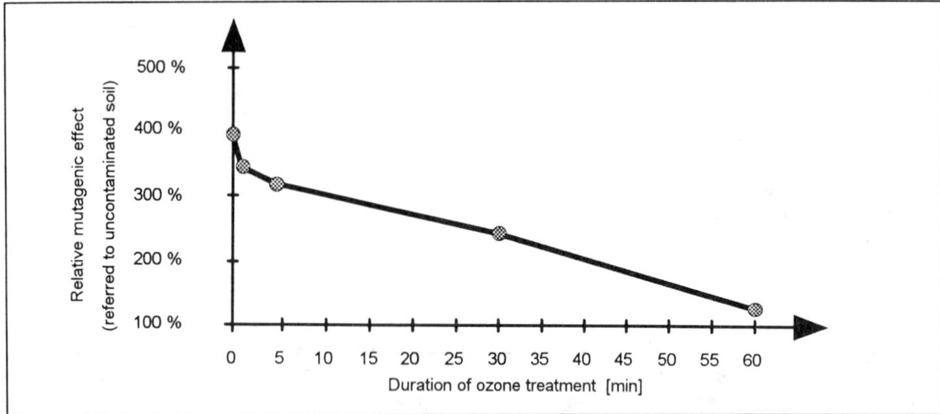

Figure 1:   Mutagenic effect of soil contaminated with benzopyrene

In further laboratory experiments soil contaminated with mineral oil was pretreated with ozone before being cleaned biologically in a percolation apparatus in realistic conditions. With reference to chromatograms, Figure 2 shows that the contaminants were largely eliminated after a very short time.

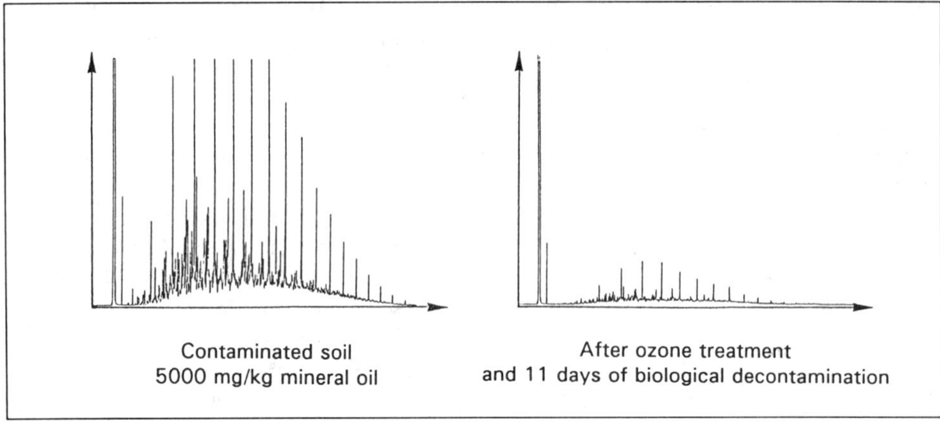

Figure 2:   Cleaning soil contaminated with mineral oil

It has been demonstrated that employment of the ozone technique together with soil venting and biological soil treatment is able to remove mineral oil components that cannot be eliminated by one of the stated methods used in isolation. This combined process can be used in-situ in cases where the pore structure of the contaminated soil is sufficiently permeable to allow mass transfer and provided that the mineral oil constituents of the soil are not present in the form of viscous, gummy contaminant lenses. The combination of the in-situ methods described here, including chemical/oxidative contaminant elimination with ozone, harbours much promise especially for rehabilitating PCA-contaminated sites insofar as the pollutants are difficult to tackle directly by biological treatment.

INVESTIGATIONS ON THE SEALING OF CONTAMINATED AREAS BY GROUTING

H. Müller-Kirchenbauer*, C. Schlötzer*, J. Rogner*, W. Friedrich**

* Universität Hannover, Welfengarten 1, D-3000 Hannover 1
** IGH Ingenieurgesellschaft, Dr.-Ing. K. Weseloh - Prof. Dr.-Ing.
   H. Müller-Kirchenbauer mbH, Volgersweg 58, D-3000 Hannover 1

One of the possibilities for safely disposing of old waste is by their complete encapsulation incorporating a surface cover and lateral vertical sealing elements, themselves integrated into a base geological barrier or an artificially constructed base seal. The construction of artificial base seals can, on the one hand, be achieved using relatively expensive mining techniques, or on the other hand, by grouting techniques - the latter already being in general use in classic foundation construction to seal off areas against the inflow of water. This technique involves injecting material into the intergranular pore spaces of the soil and into fracture systems of the rock. Thus, the cost intensive removal and disposal of large volumes of excavated material is almost completely avoided. Moreover, if angled bore holes are used to form an injection body below the old waste in the form of an inverted roof, then this type of construction, in connection with a surface cover, together creates all-round encapsulation (Fig. 1).

When compared to the so-called classic grouting material for sealing off areas against the entry of non-contaminated water, higher standards are demanded of the injection bodies required for the safe disposal of old waste - higher standards mainly associated with the effectiveness of the seal and its chemical resistance. As part of an experimental programme, laboratory scale prepared suspensions were investigated with respect to the following parameters (MÜLLER-KIRCHENBAUER et al. 1992):
- injectability using compressed air injection in coarse sand as well as in a model fracture system with different fracture widths,
- permeability behaviour, chemical resistance and stress deformation behaviour of the pure grouting material and grouted pore systems.

Various grouting materials were used including conventional sodium and calcium bentonite cement suspensions as well as high solids content calcium bentonite bonding material suspensions, whose mixture formulations are derived from dense suspensions used for one-phase diphragm walls. In addition to the use of a special hydraulic bonding material - especially selected for the purpose - materials with a high solids content were prepared containing an additive for fluidizing the suspension as well as for delaying the bonding process.

Samples of the grouted pore systems were used for the determination of soil mechanical parameters (densitiy and water content). The values for the parameters that were determined were consistent throughout the length of the samples which indicated an homogeneous distribution of solids and thus an almost completely regular filling of the pores.

The permeability behaviour of injected pore systems with respect to water were determined in a tri-axial through-flow test (gradient i = 30). Once the through-flow had begun, the through-flow coefficient k_f returned to an almost time constant value below $k_f = 4 \cdot 10^{-10}$ m/s. There were significant differences in the chemical resistance of the various grouting mixture formulations. These investigations were carried out on pure grouting material as well as on grouted pore systems with three-dimensional toxic material attack. In addition, the pure grouting material was investigated in modified storage experiments, which more closely reflected the in situ conditions of one-dimensional toxic material attack, lateral support of the sample, as well as toxic material attack via the pore spaces only of the exposed rock/soil foundations. The effect of the test fluid was quantified for the grouted pore systems by determining the lateral weight changes. In the case where the coarse sand was grouted with a conventional sodium bentonite cement suspension, the samples, which were 56 days old at the beginning of the experiment,

F. Arendt, G.J. Annokkée, R. Bosman and W.J. van den Brink (eds.), Contaminated Soil '93, 1421–1422.
© 1993 *Kluwer Academic Publishers. Printed in the Netherlands.*

had disintegrated after approx. 30 days of storage (Fig. 2). However, the pore systems grouted with the high solids content calcium bentonite bonding material suspensions were still solid after a period of over 500 days (Fig. 2).

According to MÜLLER-KIRCHENBAUER et al. (1988) it is possible, using the tendency of the stored samples to weaken over time because of the toxic material attack, to calculate - or at least to estimate within an order of magnitude - the lifetime of sealing wall elements (duration of the arithmetical weakening of the total thickness of the seal).

The results of the experiments carried out show, in principle, the possibility of using high solids content suspensions as grouting materials. They indicate that it is possible to form grout sealed areas resistant to toxic materials within porous and/or fractured subsurfaces.

References

Müller-Kirchenbauer, H., Schlötzer, C. and Rogner, J. (1992). Untersuchungen zur Verwendbarkeit von Injektionsmassen als Abdichtungsmaterial für horizontale, vertikale oder geneigte Deponiedichtungen (Investigations of the applicability of grouting materials as sealing material for horizontal, vertical or angled disposal site seals). Research report for a project funded by the State of Lower Saxony, Institut für Grundbau, Bodenmechanik und Energiewasserbau, Universität Hannover.

Müller-Kirchenbauer, H., Friedrich, W., Gremmel, D., Markwardt, W. and Rogner, J. (1988). Neue Ergebnisse und Aspekte auf dem Gebiete der Dichtwandforschung (New results and comments in the field of seal wall research). Zweiter Internationaler TNO/BMFT-Kongreß über Altlastsanierung, Hamburg.

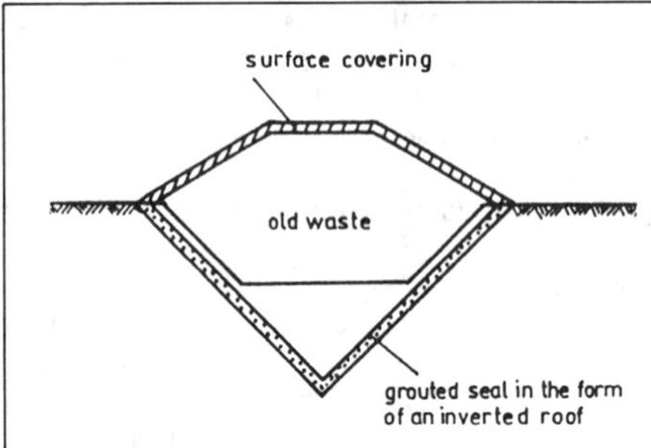

Figure 1.

The encapsulation of old waste with a combination of seal grouting and surface cover

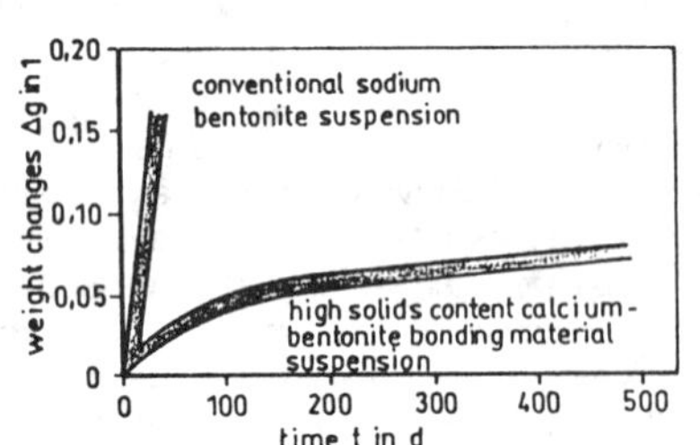

Figure 2.

Weight changes in grouted pore systems in an open storage experiment with three-dimensional toxic material attack

Extraction and Modification of Organic Contaminations with Supercritical Water for the Decontamination of Soils

Nowak, K.; Brunner, G.
Technische Universität Hamburg-Harburg, AB Verfahrenstechnik II
Eißendorfer Straße 38, D-2100 Hamburg 90, Germany

There are still serious problems in the cleaning of soil materials that consist of very fine particles. Polycyclic Aromatic Hydrocarbons (**PAH**'s) and chlorinated hydrocarbons (e.g. **Poly**chlorinated **B**iphenyls, **PCB**'s) cause considerable trouble. In such cases the treatment of contaminated soils with supercritical water ($\vartheta_c = 374\,°C$, $P_c = 22.1\,MPa$) is investigated as an innovative remediation method within the research programm SFB 188 "Contaminated Soils", funded by the Deutsche Forschungsgemeinschaft (DFG), Germany. Water at supercritical conditions acts both as solvent and as reaction medium for hydrocarbons. The objective is not only to clean the soil material , but also to chemically modify the contaminants, either to completely oxidise them, or to pretreat them for a final biodegradation in aqueous solution. Therefore the use of an additional oxidant like oxygen may be essential.

The experiments are carried out with a semi continuously operated extraction plant. The extraction autoclave contains the soil material as a fixed bed ($V = 460\,cm^3$). Water flows continuously through this fixed bed. Downstream to the autoclave a tube reactor ($d_i = 6\,mm$, $L = 12\,m$, $\vartheta_{max} = 450\,°C$) is coupled. The extraction experiments take place at a pressure range of $23 - 25\,MPa$. Due to hardware limitations each extraction experiment is divided into a heating period ($\vartheta \approx 20 \rightarrow 380\,°C$) and a period at constant temperature ($\vartheta \approx 380\,°C$). As additional oxidant compressed synthetic air is added at the tube reactor.

The experiments carried out up to now proved, that supercritical water is an excellent solvent for the decontamination of soil materials. Table 1 shows a summary of results. A typical extraction curve for supercritical water as solvent is illustrated in figure 1. The extract forms stable oil-in-water emulsions with a high content of organic material ($TOC \approx 20\,000\,mg_c/l$). These emulsions contain very small oil droplets ($\approx 90\,vol\%$ with $d_p < 3\,\mu m$) and show a good biodegradability according to a modified Zahn-Wellens test. The biodegradability is clearly increased compared to the untreated contaminations. In some cases biodegradability is obtained only after the treatment with supercritical water .

Toxic contaminations can be oxidised totally in supercritical water if an additional oxidant is used. Supercritical water itself takes part in hydrolytic reactions with hydrocarbons. Within these reactions dehalogenations and ring-opening reactions of PAH's with the addition of hydroxyl groups are of interest with respect to biodegradability. Chemical modifications in supercritical water are under current investigation within this project.

F. Arendt, G.J. Annokkée, R. Bosman and W.J. van den Brink (eds.), Contaminated Soil '93, 1423–1424.
© 1993 *Kluwer Academic Publishers. Printed in the Netherlands.*

Due to the expected high costs soil decontamination with supercritical water will be restricted to soil fractions with a high content of very fine particles or to materials with toxic contaminations. Nevertheless treatmenting soil extracts, effluents from land disposals and toxic wastewaters with supercritical water bears a lot of advantageous aspects.

TABLE 1: Extraction results of real contaminated soil materials

soil material		NMS	CND 36	VSTR
		petrol station	industrial site	old barrel depot
sauter diameter	[μm]	30	7	14
contamination		lubricating oil	lubricating oil	PAH's
		aged over 1 year	aged over 20 years	aged over 45 years
initial concentration				
[mg hydroc./kg TS]		20 200	103 500	11 050
extraction conditions				
final concentration				
[mg hydroc. / kg TS]		< 20	< 100	< 10
extraction	[%]	> 99,9	> 99,9	> 99,9
temperature	[K]	663	655	665
pressure	[MPa]	24	24	25
solvent/soil ratio				
[kg water / kg TS]		6	6	12

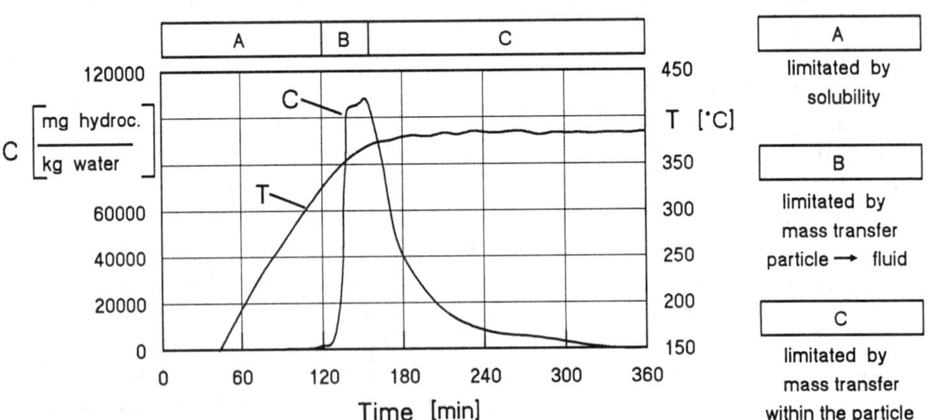

FIGURE 1. Hydrocarbon content in the water effluent **C** and mean fixed bed temperature T versus extraction time (extraction of a real contaminated soil material, CND 36).

Mobilization of Heavy Metals in Contaminated River Sediments by Bacterial Leaching

ONDRUSCHKA, J., SEIDEL, H., and U. STOTTMEISTER

Center of Environmental Research Leipzig-Halle
Permoserstrasse 15, O-7050 Leipzig

Some of the river sediments in the East-German industrial region are highly contaminated by heavy metals and organic pollutants. Dredging sludges are potentially dangerous if dumped in a uncontrolled way. Heavy metals are being mobilized and made available to plants by acid precipitations and natural bacterial leaching. Humification may be delayed; thus, the natural recycling process of the soil is slowed down.

Bacterial leaching is based on the ability of chemolithotrophic bacteria of the genus Thiobacillus to transform little soluble sulphidic metallic compounds into water-soluble metal sulphates via biochemical reaction mechanisms. The objective of our laboratory tests was to explore the influence of the factors involved in bacterial leaching on the mobilization of heavy metals in river sludges and rot substrates prepared from them. This should be relevant for the task of exploring ways of decontaminating river sludges by intensifying bacterial leaching.

The concentration of heavy metals of sediments dredged up from the Weisse Elster river in the Leipzig region in the years 1990, 1991 and 1992 is shown in figure 1. We chose the sediment dredged up in 1991 for our laboratory tests. 4 kg of it were leached in a percolator with 2 liters of circulating water for 30 days. During this period the pH value dropped from 7.2 to 5.6. The leaching rate for the heavy metals amounted to a total of 20%.

F. Arendt, G.J. Annokkée, R. Bosman and W.J. van den Brink (eds.), Contaminated Soil '93, 1425–1427.
© 1993 Kluwer Academic Publishers. Printed in the Netherlands.

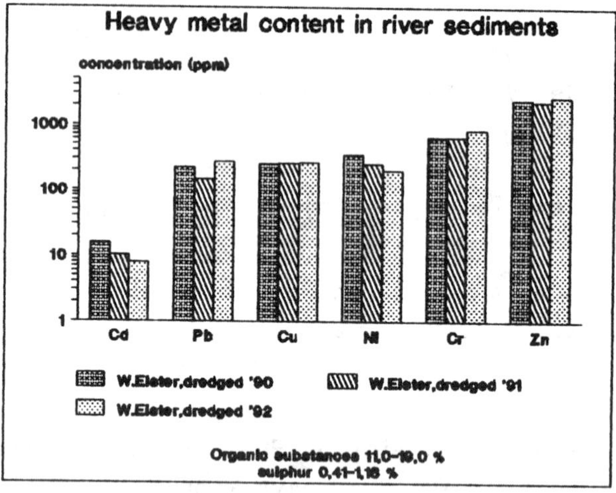

Fig. 1

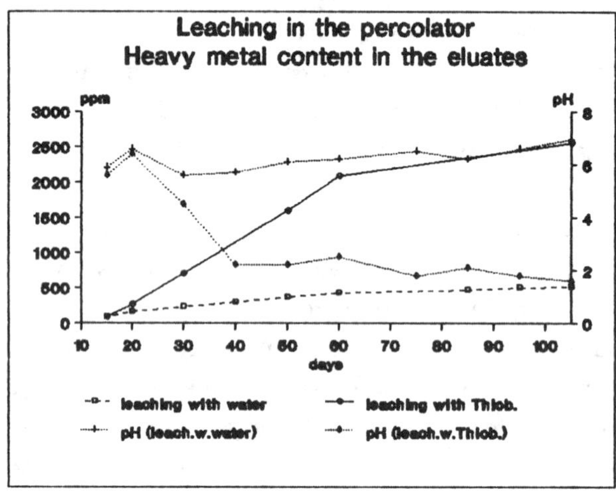

Fig. 2

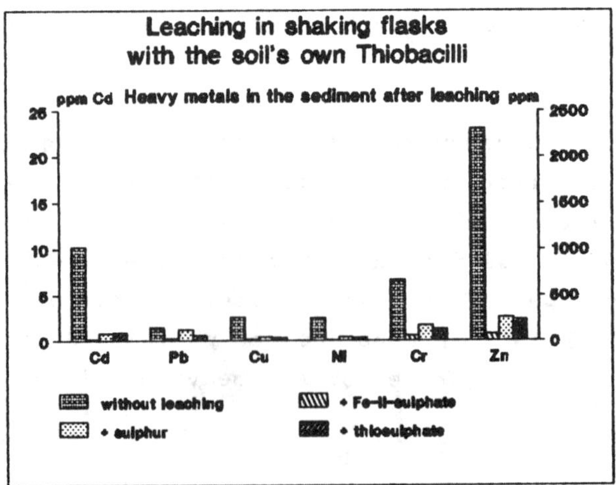

Fig. 3

To simulate natural leaching conditions, perculators were filled with 1 kg of sediment each, on which 20 ml of water or
broth of cultivated Thiobacilli, respectively, were rained
every day. If leaching is performed with water, the pH value
remains constant, after 60 days the metal concentration in the
sediment had dropped by approximately 20 %. But if the
leaching process is assisted microbially by Thiobacilli, the
pH value will drop to values around 2 within a short time, and
the leaching rate will already be about 70 % after 60 days.
The heavy metal concentration in the eluate in its dependence
on the naturally obtained pH value and the duration of
leaching can be seen in figure 2.

If the bacterial leaching process performed by the soil's own
Thiobacilli in the shaking flask was supported by adding energy sources such as iron(II)sulphate, sulphur or thiosulphate
to a suitable nutrient medium, a pH value of 1.8 to 2.2 appeared. After 30 days leaching rates of 80 to 95 % were reached;
the residual metal concentration in the sediment was between
190 and 640 ppm (figure 3).

Furthermore, rot substrates prepared from river sludges and
straw are also included in our current investigations. According to the results obtained so far, no leaching of heavy metals was possible in the percolator after 30 days under normal
conditions.

References

Ondruschka, J., Glombitza, F.:
    Inhibition of natural microbiological leaching processes,
    Environmental Biotechnology, 1991, Part II
    (Internat. Symposium Ostende /B.(1991), 561-563
Glombitza, F., Iske, U., Ondruschka, J.:
    Biotechnology based opportunities for environmental
    protection in the uranium mining industry,
    Acta Biotechnol. 12 (1992) 2, 79-85

A systematic procedure for the selection of solvents to be used in soil decontamination

Andreas Rebhan
BCR-Ingenieurbüro für Verfahrenstechnik und Umweltschutz GmbH
Königsallee 178a, 4630 Bochum 1

Summary
The selection of a solvent is influenced essentially by a economic and ecologicial parameters, and by its physical-chemistry. The experiments that were carried out show a method for the preliminary choice of a solvent. Further more, selected organic solvents were investigated under laboratory conditions, as to their suitability in the removal of various polycyclic hydrocarbons from a sand mixture. The results also illustrate the effectiveness of a solvent or solvent mixture when the influencing parameters such as soil moisture content and the addition of surfactants are varied.

Introduction
Initially, a literature search was carried out to obtain an overview of solvents that were feasible in principle. From the numerous possible solvents or solvents mixtures, those practical for future applications were chosen and subjected to further examination. In the selection of a solvent or solvent mixture, the requirements to consider are those such as price, safety coefficients and ecological parameters on one side, and on the other side the physical-chemistry properties such as vapour pressure, enthaly of vaporisation, selectivity and interaction parameters.

Preliminary choise of solvent
The preliminary selection was carried out by means of a point system as suggested in Ref. [1,2] for 37, mostly organic solvents used for the removal of polycyclic hydrocarbons contamination. In this, the following influencing factors were considered:

-	price	-	flash point
-	boiling point	-	seletivity
-	specific heat capacity	-	enthalpy of vaporisation
-	density	-	Vapour pressure
-	dynamic viscosity	-	surface tension
-	solubility in water	-	interaction parameters
-	water purity class	-	equilibrium curve

For 21 selected solvents, there then followed the determination of cleaning performancs for the removal of acenaphthen, phenanthren, fluoranthen and pyren, under laboratory conditions [3].
In total, 83 series of experiments were conducted to determine the suitability of various solvents or solvent mixtures under conditions of varying solvent/solid ratio, extraction time, soil moisture content and surfactant addition.

Experimental results
The experiments show which solvents were distinguished by favourable properties, particularly when good cleaning was already expected due to the physical and chemical

F. Arendt, G.J. Annokkée, R. Bosman and W.J. van den Brink (eds.), Contaminated Soil '93, 1429–1430.
© 1993 Kluwer Academic Publishers. Printed in the Netherlands.

1430

properties. From the experiments it may be inferred that the material data used in the assessment produces an indication of the cleaning success, so that it is possible to establish a relationship between the cleaning performance and the physical chemisty parameters. This relationship is usually expressed by the distribution coefficients and the capacity.

The cleaning successes as explained in Ref. [4] for pure solvent such as methylethylketone, ethylacetate, acetone und n-hexane, were essentially confirmed by the experiments undertaken. Ref. [4] also shows the influence of the soil moisture. Acetone`s cleaning performance falls sharply with increasing soil moisture content. In comparison, the moisture content has a less significant effect an cleaning with n-hexane. This also applies to the cleaning performance of ethylformiate and dichloromethane.

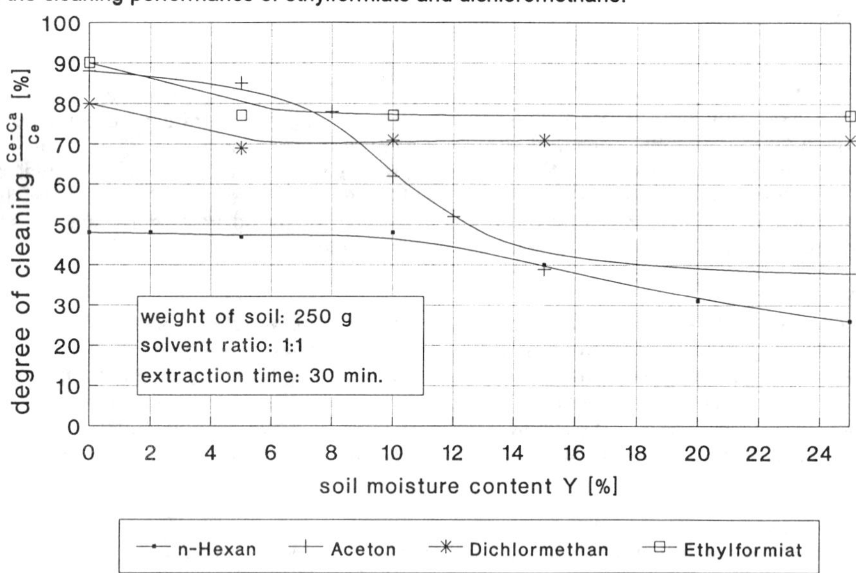

Fig.1: Influence of soil moisture content on the degree of cleaning by various solvents

Literature:

[1]   Hampe, M.J.:   "Flüssig/Flüssig-Extraktion:Einsatzgebiete und Lösemittelauswahl"
                      Chem.-Ing.Tech., 50 (1978), Nr. 9, S. 647-655,
[2]   Hampe, M.J.:   "Lösungsmittelauswahl bei der Flüssig/Flüssig-Extraktion unter
                      physikalisch-chemischen Aspekten"
                      Chem.-Ing.Tech., 57 (1985), Nr. 8, S. 669-681,
[3]   Suttka,R.:   "Experimentelle Untersuchungen zur Extraktion polyzyklischer
                      aromatischer Kohlenwasserstoffe aus Sand"
                      Diplomarbeit 1992, Lehrstuhl für thermische Stofftrennverfahren
                      Ruhr-Universität Bochum, o. Prof. Dr.-Ing. R. Billet
[4]   Haase,B.et.all.:"Reinigung hochkontaminierter Böden mittels Waschverfahren"
                      Chem.-Ing.Tech., 62 (1990), Nr. 8

Investigations on the fixation and release of organic compounds in river sediments from the Leipzig region

---

Matthias Remmler, Frank-Dieter Kopinke, Ulrich Stottmeister

Centre of Environmental Research Leipzig-Halle GmbH, Permoserstr. 15, D-7050(O) Leipzig

Sediments from rivers that had been heavily polluted by the coal converting and petrochemical industries were characterized by thermoanalytical methods. The DTG and DSC curves obtained reveal broad, not very specific ranges of reaction ($\dot{m}_{max}$ at 410°C, $\dot{H}_{max}$ at 480°C), as they have been found for ordinary, unpolluted sediments and soils. More detailed information on the structural units of the organic sediment matrix is provided by the combination of fast pyrolysis, gaschromatographic analysis and mass spectrometric identification of the resulting pyrolysis products (Py-GC/MS). In this way the following hydrocarbon families were detected this way: n-alkanes (C1...C30), 1-olefins (C2...C30), n-alkylbenzenes (C6...C25), phenols (C6...C9), pyrrols (C4...C7), furanes, cyclopentenones (C5...C7), fatty acids (C2...C16), pyridines (C5...C8), naphthalenes (C10...C13) and others. They completely match those structures, that were recently described as basic units of humic acids in soil by Schulten and Schnitzer [1]. Typical marker of mineral oil or coal contamination, such as the isoprenoids norpristane, pristane, phytane and PAHs with more than two condensed rings [2] could be detected as traces only. The sediment does not contain appreciable amounts of loosely bonded, low molecular weight hydrocarbons. Less than 1% of the organic fraction can be evaporated by thermodesorption up to 400°C (Ev-GC/MS). Only the thermal degradation of the matrix at higher temperatures, accompanied by rupture of covalent bonds, renders fragments accessible to GC detection. At 650°C about 10% of the organic carbon in the sediment are converted into volatile organic compounds. Unspecific main products of pyrolysis are always water, carbon oxides and solid coke.
Figure 1 presents recovery rates of volatile compounds, obtained by thermodesorption at 400°C after loading sediment and ash samples with a pentane solution.

Figure 1: Recovery of sorbed compounds by thermodesorption at 400°C

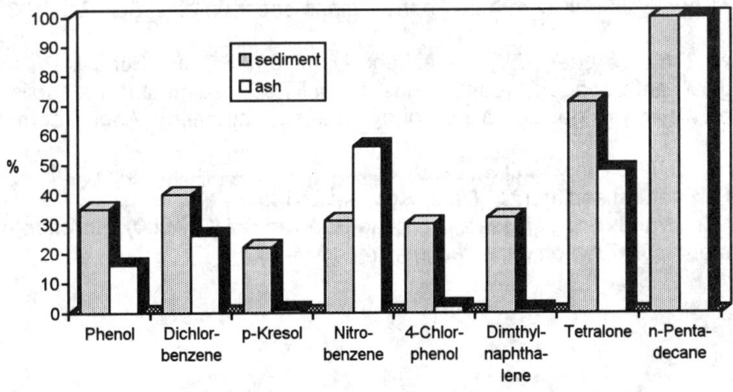

F. Arendt, G.J. Annokkée, R. Bosman and W.J. van den Brink (eds.), Contaminated Soil '93, 1431–1432.
© 1993 *Kluwer Academic Publishers. Printed in the Netherlands.*

All the unsaturated hydrocarbons are fixed either partially or completely during the desorption procedure. In these reactions the ash of the sediment is more reactive than the native sediment . Obviously, the organic as well as the inorganic constituents of the sediment stimulate the chemical conversion of the sorbate. In conclusion, the results confirm the view that Ev/Py-GC/MS is suitable for a qualitative, but not for a quantitative characterization of mixed biopolymers.

For the sorption of naphthalene from an aqueous solution at 22°C on the special sediments under investigation an adsorption coefficient K_{OC} = 1100 $\pm$ 150 was measured. It corresponds very well to values for other organic matrices found in the literature [3,4]. K_{OC} is in each case related to the portion of organic carbon in the sediment fraction. The inorganic matrix (ash) does not contribute to the adsorption potential of the sediment in the aqueous solution. It exhibits, however, a remarkable reactivity in reactions converting PAHs in nonpolar solvents. 9-methylanthracene (MA) is immobilized almost completely in a sediment ash at room temperature (Figure 2). The reaction products are mainly slightly soluble or nonsoluble. This indicates polyaromatic structures.

Figure 2: Sorption and reaction of 9-methylanthracene in hexane for several adsorbents

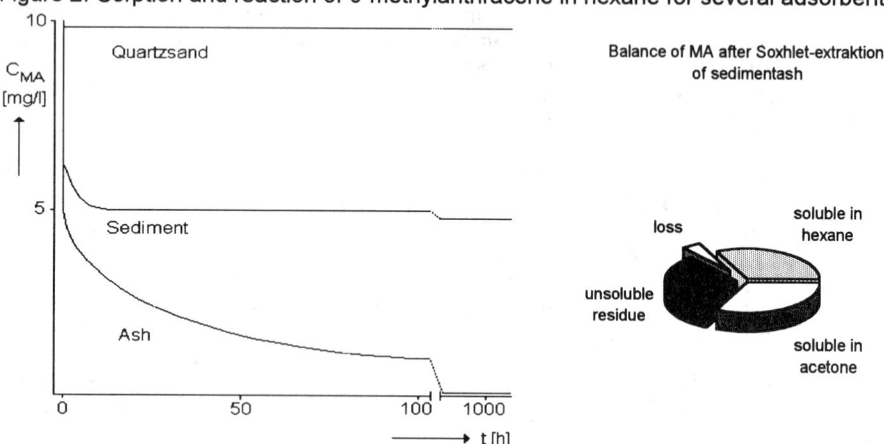

References

[1] Schulten, H.-R., and Schnitzer, M. (1992). Structural studies on soil humic acids by Curie-point pyrolysis-gas chromatography / mass spectrometry, Soil Sci. 153 (3): 205-224.

[2] de Leeuw, J.W., de Leer, E.W., Sinninghe Damste, J.S., and Schuyl, P.J.W. (1986). Screening of anthropogenic compounds in polluted sediments and soils by flash evaporation / pyrolysis gas chromatograhpy - mass spectrometry, Anal. Chem. 58: 1852-1857.

[3] Karickhoff, S.W., Brown, D.S., and Scott, T.A. (1979). Sorption of hydrophobic pollutants on natural sediments, Water Res. 13: 241-248.

[4] Means, J.C., Wood, S.G., Hassett, J.J., and Banwart, W.L. (1980). Sorption of PAH by sediments and soil, Environ. Sci. Technol. 14: 1524-1528.

IN–SITU BIORESTORATION OF SOIL CONTAMINATED BY MINERAL OIL, PAHs AND
PHENOLS IN COMBINATION WITH EXTRACTION USING NON–IONIC SURFACTANTS

GEROLD REUSING

SACO GmbH, Lehrter Str. 16–19, 1000 Berlin 21, Germany

1.   INTRODUCTION

A former asphalt production site was found to be contaminated mainly by
aromatic, polycyclic and phenolic compounds. Contamination extended from
the unsaturated zone (>2.5 m b.g.l.) to the bottom of the topmost aquifer
(6 m b.g.l.). The main aquifer below is protected by a layer of boulder
clay.

Laboratory experiments have shown that the combination of contaminants
found at the site can be degraded microbially. It has also been shown that
the addition of surfactants enhances extraction of contaminants and is
beneficial to the biodegradation.

Field experiments have been initiated in February 1991 in order to verify
laboratory results, in particular the application of surfactants.

2.   GENERAL DESCRIPTION OF THE IN–SITU TREATMENT TECHNIQUE

The restoration scheme used consists of a combination of biodegradation and
extraction. A closed hydraulic circle is induced by a suitable arrangement
of injection and abstraction wells. A biological treatment plant,
consisting of a bioreactor and various filters is inserted between the
wells at the surface.

By adding nutrients and nitrate as oxygene donor to the re–infiltrated
water biodegradation due to microrganisms present in the soil is
stimulated.

Contaminants are extracted using non–ionic surfactants which are also added
to the re–infiltrated water, at which increasing its temperature to about
25 °C was found to enhance the process. The use of surfactants is described
in more detail below.

3.   INCREASE OF THE EFFICIENCY OF BIODEGRADATION BY EXTRACTION OF
         CONTAMINANTS USING NON–IONIC SURFACTANTS

Column experiments have been used to investigate the behaviour of
surfactants before application in actual field tests. The main results are:

–   A considerable rise in solubility of PAHs by a factor of 2100 to 3100
–   Not only low-molecular, but also high–molecular weight PAHs are
        mobilized

F. Arendt, G.J. Annokkée, R. Bosman and W.J. van den Brink (eds.), Contaminated Soil '93, 1433–1434.
© 1993 Kluwer Academic Publishers. Printed in the Netherlands.

\-   No negative effects of biodegradable surfactants on the actual biodegradation process have been observed. The microbial population changed very little during the course of the column experiments.

In the course of the field experiments two tests were performed in August/September 1991 adding the surfactants intermittently. However, these tests did not produce the desired effect. Therefore, surfactants were added continuously to the infiltrated water in the following restoration phase from 15 April 1992 on. In addition, by pre-heating the infiltration water, the solubility of the surfactants and their dispersion in the soil could be enhanced.

Concentration measurements in the circulation water before entering the treatment plant showed significantly enhanced contaminant extraction after application of surfactants. In particular, PAHs showed a rate of extraction increased by a factor of 10 (see Fig. 1).

Surfactant addition not only enhances contaminant extraction, but also improves biodegradation by increasing bioavailability due to increased solubility. The improvement in biodegredation was concluded from increases in nitrate consumption and microbial population.

4.   CONCLUSIONS

Laboratory and field experiments have shown that the efficiency of in-situ treatment of contaminated soil can be significantly improved by addition of non-ionic surfactants to the re-infiltrated water in order to extract contaminants. In addition, there is a beneficial effect to the actual biodegradation by improving the bioavailability of contaminants. Pre-heating the infiltrated water further accelerates the progress of remediation.

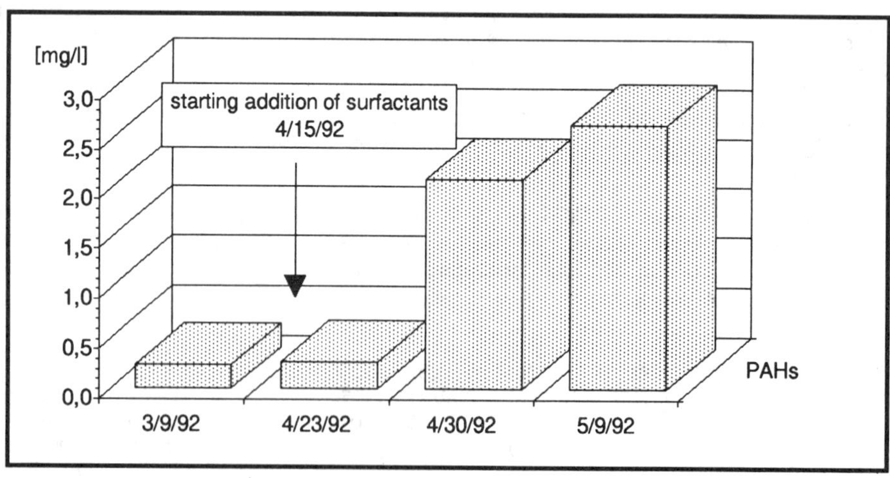

FIGURE 1:   Concentration of PAHs in the circulation water before entering the treatment plant.

Laboratory Examination of Heavy Metal Extraction from Contaminated Soils Using Organic Complexing Agents

H.-J. Roos, F. Forge, H.Fr. Schröder, R. Klein[1] , M. Dohmann
Institut für Siedlungswasserwirtschaft der RWTH Aachen (ISA)

Unlike organic soil contaminants, heavy metals are neither thermally decomposable nor microbiologically degradable. They are subject to chemically and biologically induced conversion processes (redox reactions) that influence their mobility but the problems they pose can be durably resolved only by removing the heavy metals from the contaminated materials. Until now efforts to achieve this have used mineral acid extraction processes, whereby not only the heavy metals but also substantial elements of the solid matrix are dissolved. With a view to subsequently reusing both the decontaminated soil and the residues containing heavy metals, however, endeavours must focus on extracting only the desired substances from the solid material and otherwise intervening in the existing matrix as little as possible. The objective of the process presented here, which is currently being developed, is the selective removal of heavy metals from relevantly contaminated solids. The method employs chelating agents that form specific bonds with certain heavy metals.

A fundamental distinction is made between organically dissolved complexing agents and those that are present in aqueous solutions. In the case of the former, the complexing reagent and the solvent can be identical. Extraction in an aqueous environment may be a precarious method because water represents an excellent solvent for a large number of mineral substances owing to its polar nature. As a consequence, the advantage derived from the specific reaction of the complexing agent with the heavy metal(s) can be thwarted. This problem does not arise with extraction in the organic phase but the necessity of drying the soil first in this case can be disadvantageous. The choice of a decontamination method in either the organic or aqueous phase is thus largely influenced by the properties of the soil being treated.

The use of organic chelating agents for heavy metal extraction in both the aqueous and organic phases was investigated in extensive laboratory experiments. The principal experimental material was filter dust containing heavy metals. The experiments were targeted at determining the influence of the pH value, temperature, water content, bond condition and method on the extraction yield.

[1] Bonnenberg u. Drescher Ingenieurgesellschaft mbH, Aldenhoven

The findings presented here were compiled by the Institut für Siedlungswasserwirtschaft der RWTH Aachen (ISA) in collaboration with Bonnenberg u. Drescher Ingenieurgesellschaft mbH, Aldenhoven in the context of a research project sponsored by the Federal Ministry for Research and Technology (BMFT).

F. Arendt, G.J. Annokkée, R. Bosman and W.J. van den Brink (eds.), Contaminated Soil '93, 1435–1436.

1436

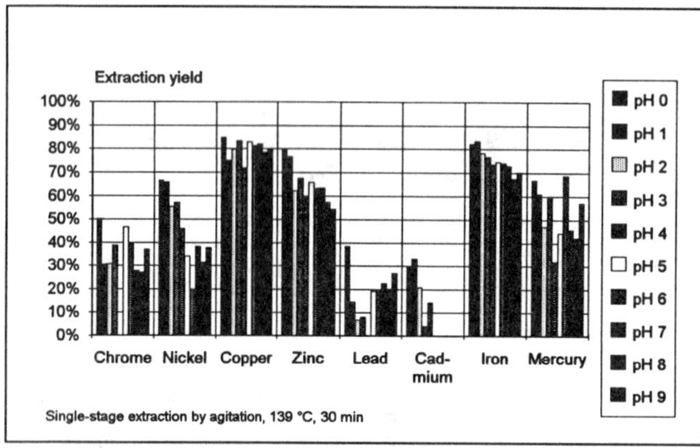

Figure 1: Heavy metal extraction with an organic chelating agent in the organic phase as a function of the pH value

The tests revealed that heavy metal complexes could be formed only if the metals were first converted to the ionic form. In the organic phase, a certain chelating agent achieved very good extraction yields for the elements copper, zinc, nickel, chrome, mercury and iron (Figure 1) with low pH values and relatively high temperatures; the relevant results obtained for lead and cadmium were less satisfactory. The influence of the pH value on the extraction yield reduced as the temperature increased.

Another chelating agent in an aqueous solution provided an outcome that complements the results described above: more than 80% of the lead and cadmium were extracted, while the other elements recorded poor yields (Figure 2). Moreover, an impressive 98% of the lead content was extracted from soil contaminated with this element. The influence of the pH value on the extraction rate was less pronounced with this complexing agent than with the other one; similarly, the extraction was temperature independent in the range from 20°C to 70°C.

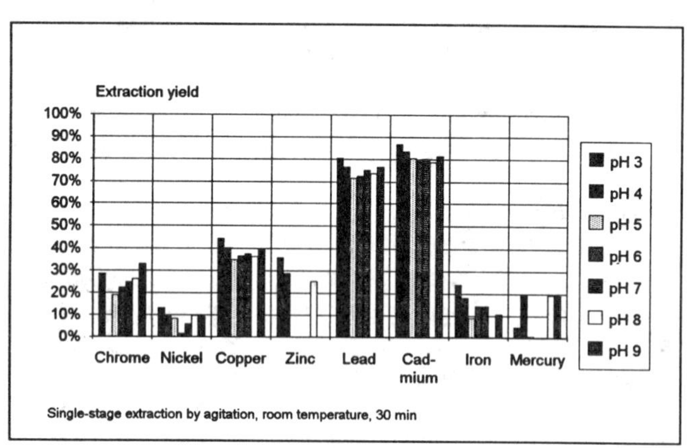

Figure 2: Heavy metal extraction with an organic chelating agent in the aqueous phase as a function of the pH value

Optimization measures such as the cascading of several extraction stages or sequential extraction with different complexing agents are currently being investigated.

Comparison between Porous Polyethylen Filters and Conventional Slot-Screens in Soil Gas Extraction Wells

Ingo Sass[*], Klaus Frank[**] & Franz Bruckner[**]

Introduction

The pollution with trichlorethene at an industrial site in Walldorf near Heidelberg first was recognized in 1987. After four years of withdrawal of groundwater and remedeation of the unsaturated zone with vapour extraction wells a new-built gauge downstream detected a still considerable contamination. A detailed field investigation established a backwater body dammed on the top of an organic clay layer. This kind of perched groundwater works as a reservoir for VOC's.

The soil profile determined by 10 ram core drillings shows 11 to 14 m of quarterny sediments (the upper 8 m are shown in Fig.1) above oligocene clays. The quarterny sequence consists of about 3 m of gravel at the base and a layered, partially very clayey coarse silt above. Within this silt-sequence there is that peaty clay layer. The backwater body on its top sometimes reaches up to 1.0 m thickness.

Contamination with VOC

The remedeation of the soil gas (50 mg/m^3 clorinated VOC's) above the perched water table (VOC: 18 000 mg/m^3) which must be cleaned up, too, was extremely difficult. The ranges of influence of the vapour extraction are small, respectively the soil gas follows preferential flow paths. To realize a sufficient filter gravel or sand pack for soil gas extraction wells very large drilling diameters would be necessary (> 300 mm on the test site), but this is very expensive for the great number of wells. Large wells, in addition, must be

handled with a relatively great vacuum, because the pressure gradient decreases steeply. This causes, that backwater is dragged along in differing but considerable quantities. The efficiency of the soil gas extraction is drastically reduced by these effects. In addition, wells of relative small diameters tend to clog up. For the reason that the backwater at this site must be treated directly, it is (regarding the energy costs) nonsensical to run a soil gas extraction in the way that the perched groundwater is sucked in, too.

The problems of soil gas extraction wells may be solved with special filter tubes. The criteria which guided the filter screen development are:

- chemically inert,
- hydrophobic,
- highly permeable,
- reduction or substituion of grain filter packs,
- stable.

Aim of this development is the reduction of the drilling diameter, to dispense with gravel or sand packs as far as possible and to minimate clogging up (by water, incrustation, mud). In a first field test one test well with a porous polyethylen filter tube and one with a common PVC-slot-screen were built up. The new screen consists of porous, sintered PE-HD and fullfils the demands.

Constructive Characteristics

The center distance of the wells is only 1.80 m. The screens were installed in the zone of the backwater table variation. As shown in Figure 1 both of the wells (BL8 & BL9) are totally equivalent concerning diameter, screen sizes and filter gravel pack. Is was not possible to dispense with a filter gravel pack because of the protection of the PVC-slot-screens. The active surface of the filters are therefore comparable.

*   Department of Applied Geology
     University of Karlsruhe
     Kaiserstraße 12
     D - 7500 Karlsruhe 1
** Hannover Umwelttechnik
     Impexstr. 5
     6909 Walldorf

F. Arendt, G.J. Annokkée, R. Bosman and W.J. van den Brink (eds.), Contaminated Soil '93, 1437–1438.
© 1993 *Kluwer Academic Publishers. Printed in the Netherlands.*

Observations and results

The first running of the channel blower was observed very carefully inside the wells, at the suction and at the pressure side of the surface equipment. The measurements (ΔP, Q, T, C etc.) were made in short intervalls. The wells run never before, even not to test the suction device. The flow regime was never turbulent, because the flow was increased slowly. Caused by the low permeable soil after 48 h the subpressure in the wells stabilized at 110 hPa (BL9) and 113 hPa (BL8) and the surrounding vacuum field became stationary.

The VOC-output is not a good parameter for comparing the wells, because they have different positions relative to the polluted areas. The developement of the gas volume flow and the vacuum can be used to compare the wells. The yield of the conventional well decreased within a few days; the pressure remained comparable to that of the PE-well. The conventional well fills up with water soon, because of the partially lifted backwater table and especially the sucked in soil moisture. The result is a blockade of filtration area. This blockade can not occur at the PE-wells because of their hydrophobic character. Soil moisture enrichment surrounding the well could not be observed. The PE-well does not demand a water separator in front of the activated carbon and no backwater level controlling pump.

In addition BL8 is clogged up with silty mud due to water breakthroughs caused by perched water table fluctuations. It had to be regenerated soon. The PE-well is saver against such fluctuations because of its high water entry value. In some extreme cases backwater heigth reached the water entry value and a water breakthrough occured, but the porous structure filters the suspension. Therefore the water inside the well is free of particles. The well does not show a decrease of yield, so it can be stated that no considerable filter cake exits.

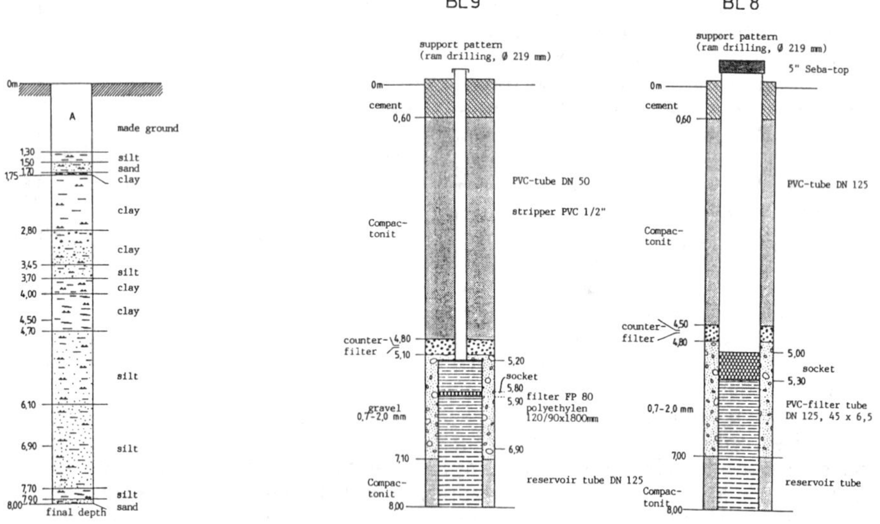

Fig. 1: Soil profile and support patterns of both soil gas extraction tests wells

BIOLOGICAL REMEDIATION OF THE FINE PARTICLE FRACTION OF HYDROCARBON-POLLUTED HAZARDOUS SITES IN CONTINUOUSLY FED BIOREACTORS

K. SCHMID, F. ABBAS, G. BEUDERT, H. H. HAHN

INSTITUT FÜR SIEDLUNGSWASSERWIRTSCHAFT, UNIVERSITÄT KARLRUHE
AM FASANENGARTEN, 7500 KARLRUHE, TEL.: 0721/608-3878

1. INTRODUCTION

The advantages of biological process in soil remediation are the positive ecological balance, the low energysexpense and the possible reuse of the remediated soil. Some restrictions of biological remediation techniques are the long time needed for remediation and the feasibility only for soils with a high water conductivity.

It is difficult to clean up fine textured soils by composting or in-situ due to the limited supply of oxygen and nutrients to the microbes. These problems can be solved in bioreactors with mixing and aeration facilities. Our investigations focus on the efficiency of biological treatment for hydrocarbon contaminated fine textured soils or soil washing fractions in a continous fed slurry reactor.

2. MATERIAL AND METHODS

Investigations were done with the fine particle fraction (<20 μm) of a hydrocarbon contaminated hazardous site (hydrocarbon content 30 g/Kg). For further information about the pretreatment of the soil see Poster No. E 212 (Beudert et al., 1993). In additionm, tests with an artificially contaminated silty soil (diesel or motoroil 5 %) were done.

The investigation can be divided into two steps:
-      **Tests on biodegradibility in batch-cultures**

To describe the time pattern of biomass growth and hydrocarbon degradation the cell count of hydrocarbon degraders and the hydrocarbon concentration were determined in batch experiments. O_2-consumption rates were measured in concomitant Warburg-type experiments.
-      **Tests with continously fed bioreactors**

The degradation rates for hydrocarbons were determined in continously fed bioreactors at different dilution rates.

In addition to the hydrocarbon analysis according to DEV H 18 some samples were analysed by GC-FID to determine the degradation of the different alkanes and of the unresolvable complex underground.

3. RESULTS AND DISCUSSION

The BOD_5-values of the tests in Warburg type flasks show that there are high Oxygen-consumption rates when diesel and motoroil contaminated fine particles are used. Like shown in Tab. 1 diesel is metabolized with the highest, motoroil with a lower and the real hazardous site with the lowest rate.

The time pattern of hydrocarbon degradation in batch cultures is shown in Fig. 1. The residual concentration for the diesel decreases continously with time. The residual concentration of motoroil and the hazardous site asymptotically tend towards a final residual concentration.

The residual hydrocarbon contamination (%) obtained in the continously fed chemostats at different dilution rates are presented in Fig. 2. The residual contamination in the reactor with diesel as only carbon and energy source increases slightly with increasing dilution rate.

F. Arendt, G.J. Annokkée, R. Bosman and W.J. van den Brink (eds.), Contaminated Soil '93, 1439–1440.
© 1993 *Kluwer Academic Publishers. Printed in the Netherlands.*

Table 1: BOD$_5$ of different contaminated fine particle fractions in Warburg-type tests.

Sample	Diesel	Motoroil	Hazard. Site
BSB$_5$	810	390	176
BSB$_5$ inok	1 434	686	326

The terms Diesel, Motoroil, Hazard. Site refer to the kind of contamination of the soil. The samples with diesel and motoroil were artificially contaminated with 5 % (w/w) of hydrocarbons. The fraction of the hazardous site is contaminated with 3 % (w/w) . To the inoculated samples 2 % of logarithmically on the same substrate growing biomass was added.

The residual concentration of motoroil raises continously at a higher level compared to diesel. For the soil from the hazardous site a distinct increase in the residual concentration of hydrocarbons occurs when the dilution rate is adjusted above 0,2 d^{-1}. The hydrocarbon degradation in the chemostat cultures with a dilution rate of 0,2 d^{-1} is more effective than in batch cultures.

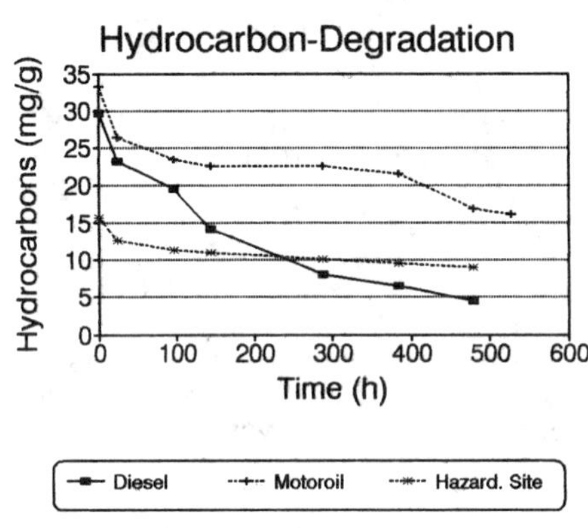

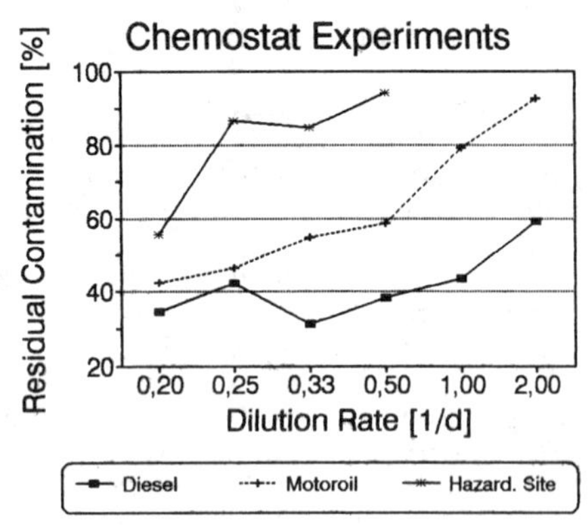

FIGURE 1: Hydrocarbon degradation in batch cultures.

FIGURE 2: Hydrocarbon degradation in chemostat cultures at different dilution rates.

4. OUTLOOK

Further investigations are made to raise the bioavailibility of the hydrocarbons and to recycle the biomass. By use of these techniques we try to establish higher biodegradation rates.

5. REFERENCES

Beudert G., Kübler P., Schmid K., Hahn H. (1993): Soil Washing including biological treatment of the contaminated slit/clay-fraction. In: Contaminated Soil `93. Fourth International Conference on Contaminated Soil. Berlin, 1993. Kluwer Academic Publishers, Dordrecht, Boston, London.

<u>FIXATION OF HEAVY METALS IN THERMALLY TREATET MINERALOGICAL SOIL</u>
<u>CONSTITUENTS</u>

BODO K. SCHÜRGERS

GEOLOGISCHES INSTITUT DER UNIVERSITÄT ZU KÖLN, ABT. QUARTÄR-
GEOLOGIE, ZÜLPICHER STR. 49, W-5000 KÖLN 1.

ABSTRACT: The influence of thermal treatment on the fixation of heavy
metals in mineralogical soil constituents has been examined in a model experiment. In order to gain complex knowledge about the mobility of heavy
metals, a sequential extraction procedure was used, which was complemen-
ted by mineralogical and phase analytical evaluations.

From important soil minerals (kaolinite, illite, orthoclase, quartz, calcite)
$CaCO_3$-free and $CaCO_3$-containing mixtures have been prepared. With easily
soluble heavy metal salts (conc. Pb: 3 g/kg, Zn: 15 g/kg, Cd: 0,4 g/kg)
they were turned into a watery suspension, which was then dried and
heated for an hour at 300°C, 600°C and 1000°C.

The experiment shows that heating at 300°C results in only slight change
of mobility. The heavy metals appear in water soluble, exchangable
(NH_4OAc-sol.) and carbonate (NaOAc-sol.) fraction.

At a 600°C heating, heavy metals in $CaCO_3$-free mixtures are transformed
into aqua-regio soluble compounds, whereas the immobilization in $CaCO_3$-
containing mixtures is considerably smaller at the same temperature.

This divergent behavior is even more evident under a 1000°C-thermal
treatment. In the $CACO_3$-free mixtures (fig. 1) the heavy metals are mostly
fixated in silicate structures ($HF/HClO_4$-sol.). However, in most of the
$CACO_3$-containing samples, a fixation does not occur. Especially lead and
cadmium are easily mobilized, sometimes even in considerable quantities
(fig. 2).

Essentially it can be said, that apart from the heating temperature, it is
the mineral composition which is crucial to the fixation of heavy metals in
stabil compounds.

F. Arendt, G.J. Annokkée, R. Bosman and W.J. van den Brink (eds.), Contaminated Soil '93, 1441–1442.
© 1993 *Kluwer Academic Publishers. Printed in the Netherlands.*

1442

Fig. 1: Solubility of Pb, Zn, Cd - without CaCO3 - thermal treatment at 1000ºC

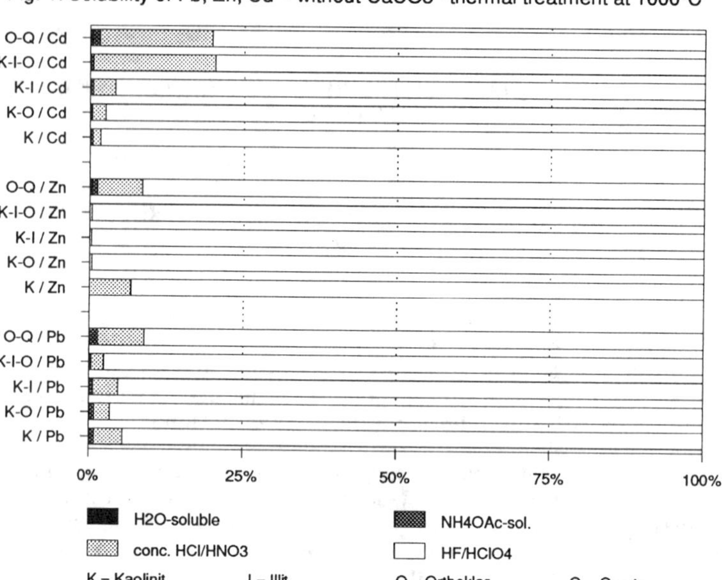

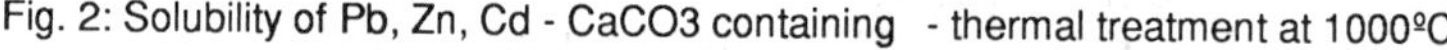

Fig. 2: Solubility of Pb, Zn, Cd - CaCO3 containing - thermal treatment at 1000ºC

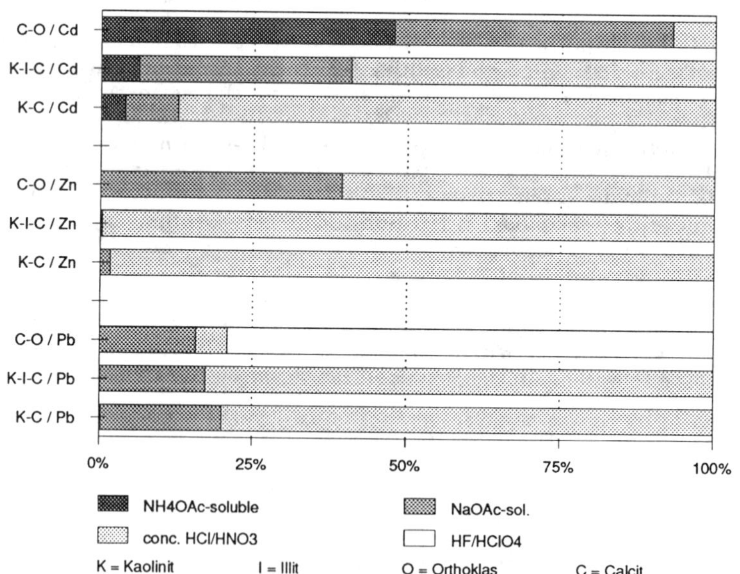

DEVELOPMENT AND APPLICATION OF A COMBINED SOIL CLEANING PLANT FOR SANITATION OF MINERAL OIL CONTAMINATED SOILS

E. SCHUSTER, J. HUTTER, E. HUBER, S. HUBER, M. BLANK-HUBER

Gebr. Huber, Gesellschaft für Versorgungs- und Geotechnik mbH
Tannenwaldstr. 2, 8000 München 70

1. Introduction

Washing as well as biological procedures for soil purification have already been used in practice. Used separately, both treatments show advantages and disadvantages as follows:

Washing is a quick procedure which is effectively used with soils containing a wide range of particle sizes. The fraction ≥ 200 μm can be completely cleaned during the washing process and is ready for reuse immediately afterwards.
The residual fraction of particles < 200 μm is highly contaminated. Because of the high specific binding capacity the technical requirements for washing this fine material - if at all possible - increase extremely. Up till now this fine and highly contaminated fraction of the soil material has been disposed on special waste disposal sites.

Biological treatment of contaminated soil is done either in situ or on/off site. While the soil remains where it is in the first case, on site remediation requires the excavation of the contaminated soil material and subsequent treatment in biobeds. This may include adjustment of moisture, aeration by turning the material from time to time, addition of nutrients and also adapted microorganisms with a high degradation capacity for a specific (group of) contaminant.
The main disadvantages of the technique are the long treatment time on the one hand and the high area requirement on the other. Moreover the control of the whole process is difficult.

2. Pleading for a Combined Method

Having in mind the above mentioned pros and contras the idea was to combine both procedures in a way that the disadvantage of the mechanical treatment is compensated by the advantage of the biological treatment and vice versa. This should be reached as follows:

▸ The high energy demand and the extensive technical requirements (and associated costs) for mechanical washing of the fine fraction are avoided by treating this fraction biologically.

▸ The long treatment duration and the high area requirement of biological sanitation procedures are strongly reduced because the large soil particles (sand, skeleton) are purified by washing.

▸ The control of biological degradation processes is optimized by using a bioreactor (instead of biobeds). The reactor allows a continuous record and regulation of essential parameters like oxygen content, temperature, pH, nutrients etc. Since we talk about a completely closed system the risk of a secondary environmental pollution (e.g. pollution of groundwater by in situ bioremediation) is extremely reduced.

The realization of the idea in a technical scale is described in the following:

F. Arendt, G.J. Annokkée, R. Bosman and W.J. van den Brink (eds.), Contaminated Soil '93, 1443–1444.
© 1993 *Kluwer Academic Publishers. Printed in the Netherlands.*

3. Gebrüder Huber's Combined Soil Cleaning Plant - The Combination of Washing and Biological Treatment of Contaminated Soil

- The contaminated soil material is **pre-sieved** to sort out lumps bigger than 100 mm. Due to the high ratio of weight : surface area the contamination of these big lumps (stones, concrete blocks etc.) is so minute that they usually can be regarded as clean without any treatment. If necessary they can be washed manually by high-pressure-high-temperature-spraying.

- Soil material smaller than 100 mm is **pre-washed** in a rotating washing drum and then transported to the washing unit by an elevator.

- After separating material bigger than 32 mm by a circular motion screen the contaminated material is washed in a **log washer**. The washing effect is mainly due to physical forces (shearing, friction, attrition) but depending on the kind of contaminant also due to dissolution/desorption processes (almost negligible when dealing with mineral oil contaminated soils). There is no addition of chemicals like acids, oxidants or detergents.

- Soil material bigger than 4 mm is clean after this washing procedure and ready for reuse. It is **classified** into three grain size classes (4-8, 8-16, 16-32 mm) to optimize further utilization, for example as a building material.

- The next step is separation of the finest and most contaminated fraction (0-60 μm) by a **hydrocyclone** (density separation) and by flotation. By adding a flocculating agent the subsequent sedimentation of the fine particles in a cross flow lamella clarifier is enhanced.

- After further thickening the slurry containing particles < 60 μm is transfered into an **air lift bioreactor**. Under optimized conditions (aeration, pH, nutrient supply, temperature) the contaminants are microbiologically degraded by the autochthonous microflora.

- Particles sized 60 μm - 4 mm (coming out of the hydrocyclone) are washed a third time in an **inverse stream washer**. Grain sizes between 200 μm and 4 mm are then clean.

- Particle sizes between 60 and 200 μm are thickened to a water content of about 80 %. The contaminants then is microbiologically degraded using a **constant flow bioreactor**. The fractions 0-60 μm and 60-200 μm are treated separately because of energetical reasons.

- The slurry treated in the bioreactors finally is thickened by a **belt filter dehydrator** and is then ready for refill.

- The **process water** is run in a cycle, thus avoiding an input into the waste water canalisation. In the case of mineral oil contaminated soils further water conditioning is not necessary since the contaminant completely adsorbs to the solid surfaces. Even a coalescence separator turned out to be unnecessary. Water leaving the process circle (moisture content of the fine material) is supplemented by adding fresh water.

The described soil purification plant prooved to be suitable for decontaminating mineral oil contaminated soils. Meanwhile this has been tested in many cases, the contaminated soils originating mostly from petrol stations on the one side and from an hydraulic oil contaminated industrial site on the other.

Main limitations at this time exist in microbiological degradation of the contaminants using the bioreactor. These problems are dealt with in more detail in the poster contribution of BLANK-HUBER et al. (section E2).

Warfare-related contaminated Site Stadtallendorf/Hessen
Actual Stage of a Research & Development Project for the Soil Remediation

Dipl.-Geol. Manfred Serwe
Regierungspräsidium Gießen
Landgraf-Philipp-Platz 3-7
W-6300 Gießen, Germany

Dipl.-Geol. Wolfgang Koch
Ingenieurbüro für Verfahrenstechnik
Dr. Born - Dr. Ermel GmbH
Finienweg 7, W-2807 Achim-Baden, Germany

1. Introduction

The contaminated site of Stadtallendorf, located approximately 20 km east of Marburg, is bearing two former explosives and ammunition factories dating from the Third Reich. During the operation and dismantling of the plants pollutants like TNT and related chemicals have been widely distributed leading to the contamination of soil and groundwater. Even nowadays the endangerment for human health and environment is still present.

Since the hazard of warfare-related contaminated sites has been recognized only a few years ago, there are a variety of problems in Stadtallendorf to be worked on for the first time. Efforts are being made on the development of a technology for the clean-up of TNT-contaminated soil, that will be followed by a 2-years lastening large scale test period. These works which are subject of the following text are supported by the German Ministry of Research and Technology. Beginning 1993 the management of the research and development project Stadtallendorf was transferred from the RP Gießen to the Hessische Industriemüll GmbH, Bereich Altlastensanierung, Wiesbaden.

2. Development of a Soil Clean-up Technology - Present Stage

For the treatment of TNT-contaminated soil a combined process consisting of a washing plant and a thermal destruction is planned.

The washing plant consists of two washing drums, a flotation and a final hot vapor extraction. The contaminated residues mainly derive from the higly contaminated fine material, the organic matter of the soil and the process water. The decontamination of this suspension is achieved by a thermal treatment. Afterwards, the different mass streams of clean material are blended in a conditioning plant, revitalized by adding bentonite and refilled into the excavation pits.

The basic flowsheet has been succesfully tested by an extensive lab- and pilot scale program. Particularly it could be demonstrated that the clean-up citeria of 1 mg Σ-nitroaromates per kg soil (dry) can be reached while the recovery of clean soil in the washing plant amounts to more than 70 % of the input material.

3. Construction of the Large Scale Plant

Based on the results of the test program the engineering for the large scale plant (throughput 20 t/h) was conducted. During the planned two years of operation this plant will treat 130.000 t of

F. Arendt, G.J. Annokkée, R. Bosman and W.J. van den Brink (eds.), Contaminated Soil '93, 1445–1446.
© 1993 *Kluwer Academic Publishers. Printed in the Netherlands.*

contaminated soil (see figure). Lurgi-Umwelt-Beteiligungsgesellschaft mbH, Frankfurt/M., is responsible for the engineering, construction and operation of the plant.

Currently the permitting procedure is running according to the German Waste Law (AbfG § 7 Abs. 1). The permit for the plant is expected for the second quarter 1993. Accordingly, the remediation action will be started in 1995.

Figure: Model of the large scale soil cleaning plant

4. Scientific and Neutral Attendance

The scientific support of the project is conducted by the universities of Marburg and Mainz, branches chemistry and geography. The scientific support group is responsible for quality control and ecological evaluation of the remediation. Futhermore, they are working on different scientific problems related to analytical chemistry and pedology. For example, an analytical method suitable for the control of the clean-up criteria in the soil cleaning plant has to be developped as well as a proposal concerning the revitalisation of the soil.

The engineering company Dr. Born - Dr. Ermel GmbH has been charged with the neutral attendance of the project in order to organize and coordinate the various involved parties.

5. Outlook

The results of the research and development project in Stadtallendorf are enabling a soil remediation not only for this site but also for a huge number of similar areas in Germany. Apart from the advances in process engineering particularly the reseach of the scientific support group will lead to a better understanding of the warfare-related contaminations.

COMPARATIVE EVALUATION OF IN SITU TREATMENT OF CONTAMINATED CLAY SOILS BY VAPOR STRIPPING, CHEMICAL OXIDATION, AND SOLIDIFICATION

R.L. Siegrist, M.I. Morris, O.R. West, D.D. Gates, D.A. Pickering, R.A. Jenkins, T.J. Mitchell

Oak Ridge National Laboratory[1], Oak Ridge, Tennessee, USA

1. INTRODUCTION

Fine-textured soils and sediments contaminated by trichloroethylene (TCE) and other chlorinated organics present a serious environmental restoration challenge. While in situ processes such as bioremediation and soil vapor extraction can function at sites with permeable soils (e.g., K $>10^{-3}$ cm/s), they are normally ineffective in wet, clay soils and sediments. Environmental restoration of these fine-textured soils has normally consisted of either (1) excavation followed by onsite storage, offsite land filling or thermal treatment, or (2) inplace containment by capping and slurry wall emplacement.

In November 1990 a research and demonstration project was initiated at Oak Ridge National Laboratory (ORNL) by the U.S. Department of Energy (DOE) and Martin Marietta Energy Systems, Inc. (MMES). The goal of the project was to demonstrate a feasible and cost-effective process for closure and environmental restoration of the X-231B Solid Waste Management Unit at the DOE Portsmouth Gaseous Diffusion Plant located in southern Ohio. The X-231B Unit was used from 1976 to 1983 as a land disposal site for waste oils and solvents. Silt and clay deposits (K $<10^{-6}$ cm/s) beneath the unit were contaminated with trichloroethylene, 1,1,1-trichloroethane, and other volatile organic compounds (VOCs) (ca. 1-100 ppm range) and low levels of uranium and technetium. The shallow ground water (water table at ca. 4 - 5 m (12-14 ft.) depth) was also contaminated, with some contaminants at levels well above drinking water standards.

2. APPROACH

After an initial technology evaluation and screening phase, the X-231B project focused on research and demonstration of in situ treatment strategies using physicochemical processes (i.e., vapor stripping, chemical oxidation, solidification) in subsurface soil reactors achieved with soil mixing technology (Fig. 1). The objectives of the X-231B project were to define process operation and performance, including VOC removal/destruction efficiencies, off-gas composition, effects on soil chemistry and microbiology, the fate of heavy metals and radioactive substances, and soil homogenization and translocation.

Since July 1991 extensive research activities have been conducted. Site characterization was accomplished using a hydraulic probe sampling system to collect nearly 200 soil samples with onsite laboratory analysis for target VOCs. These data were used for statistical simulation and 3-dimensional modeling of contaminant distribution. Numerous laboratory experiments were completed using bench-scale apparatus as well as a pilot-scale reactor system in which 20 cm diameter by 60 cm long soil cores from the site were treated. A full-scale field demonstration was completed at the X-231B site in June 1992. Replicated tests of each in situ process were made in soil regions, 3 m (10 ft.) in diameter by 5 to 7 m (15 to 22 ft.) deep. Tracer studies were also conducted. A computerized data acquisition system linked to nearly 60 sensors enabled near-continuous monitoring of process operation and performance (e.g., auger position, off-gas air flow rate and VOC content, soil vapor pressure and temperature, at recording intervals of 0.2 to 2 min). In addition, nearly 500 soil matrix and soil gas samples were collected before,

F. Arendt, G.J. Annokkée, R. Bosman and W.J. van den Brink (eds.), Contaminated Soil '93, 1447–1448.

during, and after in situ treatment for analyses of physical, chemical, and biological properties. Soil matrix, soil vapor, and off-gas VOC measurements were made by multiple methods.

3. RESULTS

The wealth of data generated during the project is still being reduced and synthesized. Highlights of the results follow, while a series of project reports and open literature publications are forthcoming.

The use of a hydraulic probe for soil sampling with onsite VOC analyses and 3-dimensional modeling and visualization yielded enhanced information compared to conventional sampling, offsite analyses, and routine data treatment. Conventional sampling and offsite analyses exhibited up to 90% negative bias for TCE and related chlorinated organics.

In situ mixing facilitated the treatment of dense clay soils by increasing permeability and mass transfer. Mixing also appeared capable of attenuating hot spots, with translocation of soil material inward and upward. The treatment of VOCs was effectively and rapidly accomplished (e.g., >85% reduction at >15 m^3/h) while controlling the fate of VOCs and radioactive substances. There were beneficial changes in subsurface properties following treatment, including elevated soil temperatures (e.g., 45°C) and increases in soil bacteria levels (e.g., increases of up to 10^4 organisms/g). Operation and performance did vary for the different processes evaluated. In situ vapor stripping and chemical oxidation were particularly successful. In situ treatment costs were considerably lower than available ex situ alternatives.

As a result of the success of the X-231B project, the in situ treatment processes demonstrated are being utilized for full-scale remediation of several contaminated areas. Moreover, additional focused research and demonstration projects are being initiated.

Fig. 1. Illustration of the in situ subsurface soil reactor process. (Note: treatment agents are delivered through the mixing blade with emissions captured in the shroud)

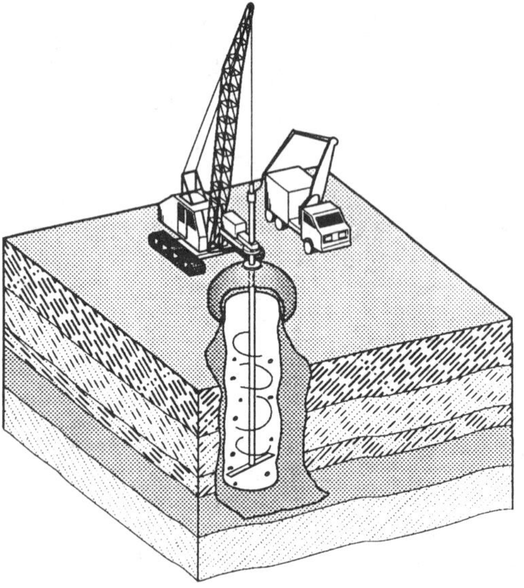

[1] Oak Ridge National Laboratory is managed by Martin Marietta Energy Systems, Inc. under contract DE-AC05-84OR21400 with the U.S. Department of Energy. The work described herein was funded by the U.S. Department of Energy's, Office of Environmental Restoration. The X-231B project was a cooperative research and demonstration effort benefiting from significant contributions of staff at ORNL, MMES, two universities (The University of Tennessee and Michigan Technological University) and six private industries (Chemical Waste Management, Millgard Environmental, Envirosurv, NovaTerra, IWT, and Lockheed).

BIOLOGICAL TREATMENT OF AN OLD DEPOSIT CONTAMINATED WITH CHLORINATED HYDROCARBONS: ELIMINATION AND BIODEGRADATION

DIPL.-BIOL. A. SIHLER, PD DR.-ING. W. BIDLINGMAIER
UNIVERSITY OF STUTTGART
INST. FÜR SIEDLUNGSWASSERBAU, WASSERGÜTE- UND ABFALLWIRTSCHAFT

1. SUMMARY

It is possible to compost deposits of greasy sludge contaminated with per- and trichloroethylene by mixing in organic additives. The toxic content of the sludge is significantly reduced by decomposition and elimination processes (**PCE up to 98%, TCE 30 – 80% in 5 – 7 weeks**). The addition of N-sources considerably accelerates the processes. Composting thus represents a system which is suitable for use on a large scale for the cleansing of this type of contaminated soil material.

2. INTRODUCTION

Chlorinated hydrocarbons (CHCs) are found throughout our environment. Their persistency and carcinogenic effect on human beings mean that they are an environmental hazard. CHCs continue to be used in industry in large quantities, for example as solvents and cleaning agents. The associated hazards and problems are thus constantly increasing. The yearly production of PCE world-wide is still around 1.1 million tonnes.

The technologies to be employed for the cleansing of contaminated soil should be selected from the ecological point of view with regard to local conditions, economic factors and the purpose for which the material is to be used after treatment.

The objective of the experiments which have been conducted is to demonstrate that biological aerobic treatment and cleansing of the contaminated material is possible.

3. EXPERIMENTS

On the basis of a comparison with other methods, composting is an economic cleansing method.

The following processes are conceivable:

- Decomposition of the toxic organic material
- Biothermal expulsion (elimination)
- Elution

F. Arendt, G.J. Annokkée, R. Bosman and W.J. van den Brink (eds.), Contaminated Soil '93, 1449–1450.
© 1993 Kluwer Academic Publishers. Printed in the Netherlands.

The test material consisted of greasy sludge contaminated with perchloroethylene (PCE) and trichloroethylene (TCE) from an animal-rendering plant which had been stored on a dump. The concentrations of toxic materials were 5 – 30 mg PCE/kg dry material and 0.1 – 11.5 mg TCE/kg dry material. It can be assumed that bacteria and fungi are able under favourable conditions to use CHCs as a C-source or to break these down cometabolically in accordance with the "Principle of Microbial Infallibility" of H.G. Schlegel.

The contaminated material was mixed with various organic materials such as hay, sawdust, paper, chopped leaves and bark and with various N-sources and composted on a laboratory scale over 5 – 7 weeks in Dewar vessels.

4. RESULTS

The mixtures displayed exceptionally good growths of various types of fungus mycelia. It was possible to achieve good aeration of the greasy sludge by mixing in the highly-structured additives.

Considerable biological self-heating and temperatures of 50 – 70°C over a period of approx. 20 days indicate a high level of metabolism during the composting process. The addition of ammonium sulphate, ammonium nitrate and urea increased the temperature in the Dewar vessels to over 70°C.

Gas chromatographs revealed the following:

- The PCE content was reduced by 80 – 98 %.
- The TCE content was reduced by 30 – 85 %.
- At high temperatures, the PCE content was reduced by over 95 %, while the TCE content was reduced by only 30 – 80 % despite its lower boiling point.
- PCE and TCE passed in part via the exhaust air into activated carbon filters, in which concentration took place (elimination).

The experiments showed that there was considerably less change in TCE content in comparison with PCE content in Dewar vessels without additional aeration. In certain cases, it was even possible to measure an increase in the TCE content.

The reasons for this TCE behaviour may be a degradation of PCE accompanied by no degradation or only a very slow degradation of TCE or an increased affinity of TCE for the material. The results of individual experiments indicate strongly that decomposition of PCE took place and that PCE and TCE were expelled thermally from the system.

It can be concluded that composting is a suitable method for the conversion of chlorinated hydrocarbons such as PCE and TCE and the elimination of these from contaminated soil material.

REMEDIATION EXHAUST AIR PURIFICATION BY MEANS OF AN ABSORPTION PROCESS

STEPHAN H. SINZ

SAN SANIERUNGSTECHNIK FÜR DEN UMWELTSCHUTZ GMBH

Physical absorption is a basic technique employed in process engineering, whereby a vapor phase substances (absorbate) is reversibly bound by a scrubbing liquid (absorbent) and thereby removed from the air stream, or exhaust air. This process, known in the chemical industry for many years, was first successfully employed in the field of purification of exhaust air from soil contamination remediation procedures by SAN Sanierungstechnik in 1991.

The gas phase components, which are to be separated, are reversibly bound to an absorbent in an absorption process taking place at atmospheric pressure and room temperature. They are thus removed from the exhaust air flow. The absorbent is continuously regenerated in a desorption column operating at elevated temperature and reduced pressure. The thus regenerated absorbent is recycled to the scrubber in a closed system without being consumed or depleted.

By means of judicious selection of high boiling absorbents, specifically chosen to match the contaminants present, it is possible to operate the remediation process within the legally established contaminant emission levels. It is even possible to vary the operating parameters of an already functioning system to optimize, or fine tune, the separation/purification performance. The scrubber liquids which we employ have been especially developed to optimize adsorption of organic compounds such as aliphatics, simple and polyaromatics, chlorohydrocarbons, chlorofluorohydrocarbons and alcohols. Should the contaminant concentration be high enough to form an explosive mixture in the air, the absorption scrubber system can be constructed in an explosion-proof form.

The purification of exhaust air by means of the absorption process is primarily applicable for larger mass feed rates of contaminant with low to moderate air flow rates. Under these conditions, the absorption process is considerably more economical than the comparable adsorption process. The efficiency of the system increases with increasing contaminant concentration in the air to be purified, since the operating costs per mass unit of revoerred solvent/contamination decrease. Rigorous implementation of heat recovered systems and a programmable control unit for the electrical system lead to minimization of energy use. Depending on the

F. Arendt, G.J. Annokkée, R. Bosman and W.J. van den Brink (eds.), Contaminated Soil '93, 1451–1452.

purity of the solvent in the feed air stream, it may be possible to recover the solvent in highly concentrated form, without any decomposition products, and to reuse it after suitable purification or preparation. Containerized construction enables the absorption system to be set up outdoors as well as indoors.

The range of applications for this system is quite broad:
- Treatment of contaminated soil-air from remediation (soil-vapor extraction) systems
- Treatment of exhaust air from ground water remediation (stripping)
- Treatment of landfill gas
- Treatment of exhaust air from industrial plants

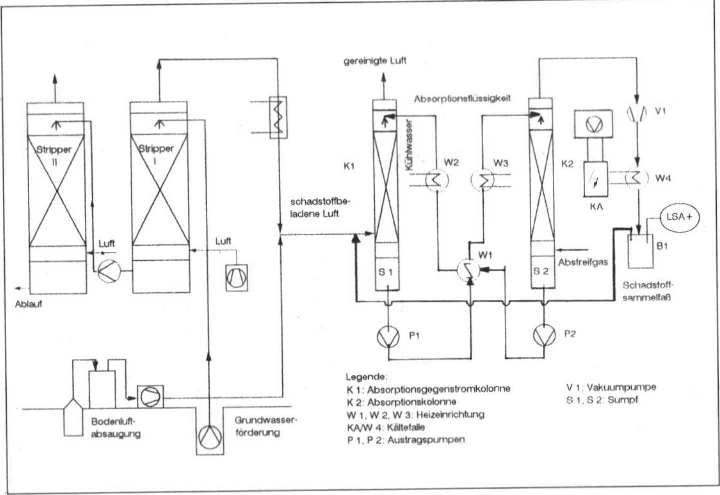

Fig. 1: Process Schematic

This technologiy will certainly experience increasing utilization stemming from the positive experiences gathered in successfully concluded exhaust air purification projects. In view of the problems associated with waste-generating processes, I believe that regenarative processes will rapidly become the technology of choice on both economic and ecological grounds. We believe that our experiences gathered to date warrant both utilization of absorption technology in present applications, as well as provide fundamental insights, which will lead to exciting future developments.

Fig. 2: Absorption Facility

THE POTENTIAL USE OF SURFACTANTS IN THE FIELD OF PHYSICOCHEMICAL AND BIOLOGICAL SOIL CLEAN UP

T. Sobisch, G. Reinisch, Heike Hübner und Anette Karg
Adlershofer Umweltschutztechnik- und Forschungs- GmbH, Rudower Chaussee 5, 0-1199 Berlin, Germany

H. Niebelschütz
ARGUS Umweltbiotechnologie GmbH, Reuchlinstraße 10 - 11, W-1000 Berlin 21, Germany

One of the main restricting factors in the remediation of soils is the low aqueous solubility of several organic compounds. This is connected with a resistance to mobilization by conventional pump and treat measures and a low bioavailability.
Therefore the residual saturation concentrations of mineral oils in sandy soils are in the range of 3-20 g/kg soil [1]. Higher molecular weight PAH's susceptibility to biodegradation is very low.
The application of surfactants to resolve these problems has been in discussion in recent years. The oriented adsorption of surfactants at the interfaces of the system organic contaminant/water/soil/air leads to a reduction in interfacial tension and enhances the wetting processes. The reduction of the interfacial tension provides a possibility to increase the interface by orders of magnitude via emulsification. This is important for physico-chemical mobilization but also for enhanced bioavailability of contaminants and metabolites. The solubilization of the hydrophobic compounds by micelles leads to a significant enhancement of their aqueous concentration.
The use of surfactants in on- and off-site soil washing [2] is quite common. A significant enhancement of oil mobilization was found under lab-scale conditions [3,4]. Surfactants have been successfully applied in bioremediation [5]. In the process of biodegradation the addition of surfactants can help to overcome the inhibitory effect of metabolic compounds [6]. This leads to a higher degradation rate and a higher degree of mineralization. Table 1 comprises some of the benefits of surfactant application.

TABLE 1
Benefits of surfactant application

	PCB/petroleum % removal	tetrachloroethylene days necessary for extraction	PAH mineralized %	model oil mineralized %
without surfactant	-	5000	5	17
with surfactant	92/93	3	50	39
ref.	[3]	[7]	[8]	[6]

However, these results have found limited application in the field because a number of field tests were quite unsuccessful [9-11]. A key problem is the formation of viscous emulsions which leads to a large reduction in the permeability of contaminated zones.

The purpose of the present work was to develop a surfactant formulation for a hydraulic in-situ measure to clean a diesel fuel contaminated area. It had to fullfil the following requirements:

- high oil removal rate
- low surfactant sorption losses
- no significant reduction of soil permeability

F. Arendt, G.J. Annokkée, R. Bosman and W.J. van den Brink (eds.), Contaminated Soil '93, 1453–1454.

With regard to physico-chemical aspects a mixture of two commercially available nonionic surfactants has been chosen for this purpose.

Two sets of laboratory column experiments have been carried out using sand which has been dried prior to artifical contamination.

In the first set the oil (30 g/kg) contaminated sand columns were washed - with 1) tapwater 2) surfactant solution of 1 w% (1 g/g oil) 3) tapwater - under constant hydrostatic pressure (1 m height).

In the second set oil (30 g/kg) was added on to the top of small sand columns and subsequently flushed with portions of 1) 1 w% surfactant solution and 2) water.

The results are summarized in Table 2 and 3. With surfactant A and the surfactant mixture A/B (9/1) a drastic reduction in the permeability was observed. Using a higher content of surfactant B this problem can be avoided. In addition, the residual surfactant sorption is markedly reduced (Table 2). From Table 3 it is evident that reducing the ratio of surfactant/oil the efficiency is reduced proportionally. It appears that in the second set of experiments the surfactant solution can bypass the contaminated zone, demonstrating thereby that mixture A/B (7/3) exhibits the highest efficiency for the removal of oil (Table 3). Surprisingly, mixture A/B (1/1) is less effective than water (Table 3).

We feel that the underlying physico-chemical principle is very promising for in-situ measures in general.

TABLE 2
Results of set 1

ratio surfactant A/B	residual oil g/kg	residual surfactant g/kg	permeability reduction
no surfactant	28	-	no
1/0	3.75	8.35	a
9/1	1.27	c	a
7/3	5.59	4.71	no
1/1	6.88	4.4	no

[a] clogging
[c] not determined

TABLE 3
results of set 2

ratio surfactant A/B	ratio oil/ surfactant	residual oil g/kg
no surfactant	1/0	12.7
1/0	1/1	b
1/0	5.49/1	7.68
9/1	1/1	0.83
9/1	5.31/1	2.53
7/3	1/1	0.60
7/3	2/1	1.42
7/3	5/1	2.36
1/1	1/1	18.8
1/1	5/1	19.5

[b] unpermeable

References

[1] K. Schwille; Deutsch. Gewässerk. Mittl. 10 (1966) 194
[2] B. Spei; Wasser, Luft, Boden 1991 7/8 77
[3] W.D. Ellis, J.R. Payne, G.D. Mc Nabb; EPA/600/2-85/129, U.S. Environmental Protection Agency, 1985
[4] C.C. Ang, A.S. Abdul; Ground Water Monit. Rev. 11 (1991) 121
[5] B. Ellis, M.T. Balba, P. Theile; Environ. Technol. 11 (1991) 443
[6] R. Müller-Hurtig, A. Oberbremer, R. Meier, F. Wagner; in "Altlasten'90", 3.Internationaler TNO/BMFT-Kongreß, Hrsg. Umweltbundesamt, Band I, 1990, 569
[7] J.C. Fountain, A. Klimek, M.G. Beikirch, T.M. Middleton; J. Hazard. Mat. 28 (1991) 295
[8] B.N. Aronstein, Y.M. Calvilio, M. Alexander; Environ. Sci. Technol. 25 (1991) 1728
[9] American Petroleum Institute; API Publication No.4390 (1985)
[10] J.H. Nash; EPA/600/2-87/110, PB-146808, Dez.1987
[11] J.H. Nash, R.P. Traver; in "Petroleum contaminated soils", Vol.1, Ed. P.T. Kostecki, E.J. Calabrese; S.157ff., Lewis Publishers 1989

Microbial Degradation of TNT (Trinitrotoluene) in Soil

Heiko Stoffers, Andreas Grothkopp, Ralf Winterberg
Contracon GmbH
Peter Henlein Str. 2-4
2190 Cuxhaven

In Germany at least 150 factories produced ammunition during World War II. Most of these ammunition factories caused serious environmental problems in the following decades. Explosives and other chemicals pollute the soil as well as the ground- and surface-water of large areas. Many of the substances found at these warfare related sites are known to be biohazards, which may cause human diseases. The importance of soil remediation has been realized and first attempts have been made to clean up soil and water on a laboratory and pilot scale. A very promising technique for the remediation of contaminated soils is microbiological treatment. A model substance which has been used for several laboratory studies is TNT (trinitrotoluene), an explosive found to be a major xenobiotic at many sites of explosive-production. For the investigations published up to now either pure cultures and well defined liqiud culture systems or very complex composts have been used. Aiming at a biological remediation system which is successful without the addition of organic additives in high concentrations, we investigate the degradation of TNT by mixed bacterial cultures in a model soil. Since recent work has shown that anaerobic and aerobic steps are necessary for the degradation-reaction and that co-substates might be necessary for some reaction steps, we especially focus on various culture conditions. Another important aspect is the detection of the degradation products, because the metabolic fate of microbiologically degraded TNT is still not completely known. The chemical analysis is completed by a toxicity assessment during micro-biological breakdown of the explosive.

F. Arendt, G.J. Annokkée, R. Bosman and W.J. van den Brink (eds.), Contaminated Soil '93, 1455.
© 1993 *Kluwer Academic Publishers. Printed in the Netherlands.*

BIOLOGICAL DECONTAMINATION OF GASWORKS SOIL IN LABORATORY EXPERIMENTS WITH UNDISTURBED SAMPLES

Dr. Ing. J. Swinianski, Dr. Ing. Ch. Lund[1], o. Prof. Dr.-Ing. G. Gudehus
Dept. of Soil Mechanics and Foundation Engineering, University of Karlsruhe (FRG)

Introduction

Laboratory experiments with large undisturbed samples ($\phi = 0.6$m, $h = 1.0$ m) were conducted in the framework of research program to develop and optimize an in situ-remediation method for contaminated sandy and gravelly soils.[2] When simulating in situ conditions, it is necessary to work with undisturbed samples because the natural soil structure, layering and porosity are important factors limiting in situ biodegradation. During the biological treatment the samples were aerated and moistened and the O_2 and CO_2 concentrations in the spent air were continuously monitored. The samples were regularly flushed with water in order to obtain a better distribution of nutrients.

Limitation of Biodegradation due to Oxygen Content

A study of the influence of oxygen content on biological activity was carried out to determine the critical O_2 concentration in the coarse pore system, which is sufficient for microorganisms. In this experiment, the active aeration of the sample was interrupted for various time intervals. After restarting the aeration, the amount of O_2 used up and the amount of CO_2 produced during the interruption was estimated. A shortage of oxygen was identified when a reduction in O_2 consumption rate or CO_2 production rate was found. The CO_2 and O_2 concentration in the spent air measured in the 16th and the 56th test weeks having different biological activities are presented in Fig. 1.

The increase of the CO_2 concentration was linear upto an oxygen content of about 13 Vol. % (for the 16th test week) and 11 Vol. % (for the 56th test week). Corresponding to this, no decrease in the oxygen consumption rate (0.70 Vol. % O_2/h and 0.19 Vol. % O_2/h) was observed. Below the O_2 concentration of 13 Vol. % (or 11 Vol. %) a reduction in the CO_2 production rate and thus biological activity occurred.

It can be concluded that, unfavourable conditions of diffusion can arise in the undisturbed soil samples which can hamper the biological activity even at relatively high oxygen concentration levels (over 10 Vol. %) in the coarse pore system of the soil.

Ozonization of the Soil

Ozone was injected in the large undisturbed sample in order to increase bioavailability of gasworks hydrocarbons by preoxidation. Two injections with gaseous ozone at a concentration of 30 g O_3/m^3 were conducted at the beginning of the test and after 70

---

[1]Trischler & Partner Consultants in Potsdam

[2]A comprehensive description of the laboratory experiments was presented in the paper "Biological In Situ-Remediation of Sandy Gravelly Gasworks Subsoils" by G. Gudehus et al. during the KfK/TNO Conference'93

1457

F. Arendt, G.J. Annokkée, R. Bosman and W.J. van den Brink (eds.), Contaminated Soil '93, 1457–1458.
© 1993 *Kluwer Academic Publishers. Printed in the Netherlands.*

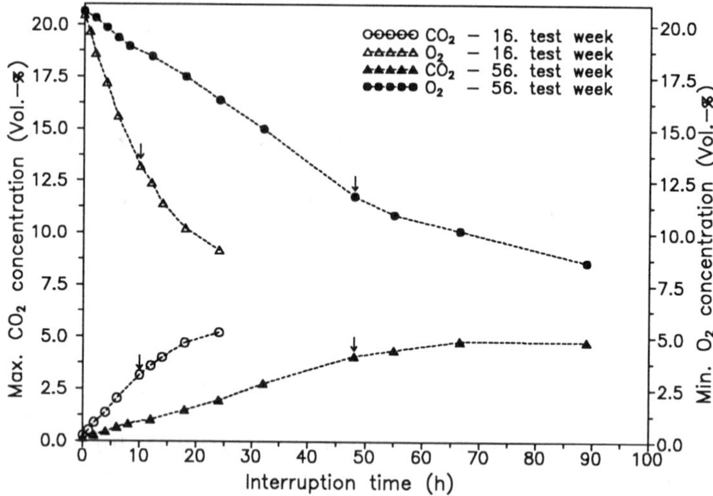

Fig. 1 Maximal CO_2 und minimal O_2 concentration in the spent air of the sample in dependence on the interruption time of aeration

days of biodegradation. Continuous CO_2 monitoring in the spent air was used to estimate the efficiency of ozone treatment and the amount of mineralized organic carbon. A total of 1930 g of O_3 (1st Ozonization) and 810 g of O_3 (2nd Ozonization) were delivered into the sample and 890 g of CO_2 and 520 g of CO_2 were produced during the ozone injections, respectively. Directly after the 1st ozone treatment, the sample was flushed with water. Chemical analyses of the flushed water showed, that the concentration of dissolved organics was significantly increased due to the ozonization (by a factor 4). The excess amount of the dissolved organic carbon found after the ozone treatment in the flushed water was only about 40 g. This amount of carbon would have been already mineralized in 4 days of biological remediation. A comparison between the totally mineralized carbon (about 240 g) and the dissolved organic carbon shows that, the increase of bioavailability was a secondary effect of the Ozonization. This could be an explanation for the fact, that no increase in the biological activity was observed after the ozone injection.

Though a relatively large amount of ozone was injected into the soil, no negative effects of ozone on the biological activity were observed. After the ozonization, it was not necessary to inoculate microorganisms into the soil sample.

The quotient of the produced CO_2 and the delivered amount of O_3 defines the efficiency of ozone treatment. This quotient was equal to 0.5 in the first and 0.7 in the second ozone injection (theoretical stochiometrical values: 0.75 to 1.3). The efficiency of ozone during the second injection was increased probably due to the fact that in the first 70 days of the test, a portion of the hydrocarbons was biologically preoxidized.

INVESTIGATION OF THE CONTAMINATION AND TREATMENT TECHNIQUES FOR CATCHER SAND IN A MILITARY FIRING RANGE

A. Thomas
Delft University of Technology, Faculty of Mining and Petroleum Engineering
Department of Raw Materials Technology, Mijnbouwstraat 120, 2628 RX DELFT, The Netherlands

1. INTRODUCTION

The emergence of a united Europe and trends of change in the world climate have led to a shift in military activities. In the Netherlands, the Ministry of Defense has planned to close several military firing ranges. Remedial actions are necessary because the bank of sand used to catch ammunition in a firing range is considered to be an environmental risk due to a high heavy metal content. Moreover, the presence of fine particles of heavy metals (primarily lead), which may become airborne, also poses a health risk, especially to those using the range.

Characterization of the sand provides important fundamental information regarding the condition of the sand and contamination. Expanding the scope of characterization to include results of various separation methods, it is possible to specify the contamination and gain insight into the viability of various processing methods.

2. CHARACTERIZATION

The sand is mainly quartz (80-85% by weight) with some feldspar and other minor minerals. It is contaminated by the metals used in the ammunition: Pb, Sb, Cu, Zn, and Fe. On an elemental basis the metal content is typically higher than 6% by weight. Lead is considered the primary contaminant, therefore the lead concentration is normally used as the reference to the level of contamination.

Due to the large difference between the specific gravity of the heavy metals and quartz, a concentration of the metal by conventional gravity separation methods is obtainable. However, the remaining sand is not devoid of contamination. Numerous compounds of the metals resulting from exposure to high temperature, friction, and atmospheric corrosion present an array of contaminant associations. The application of other standard mineral processing techniques shows that further decontamination is possible.

Outside of the contaminant as it exists in its liberated metal form, the next significant concentration of the contaminant is found in the slimes fraction (< 0.063mm). This fraction consistently contains higher than 50,000ppm Pb. Inspection of the fines with the electron microprobe revealed the ubiquitous nature of the metal in this fraction and the difficulty it presents in further cleaning of the slimes fraction.

Fines have also been identified as a coating layer on particles in larger size fractions. It appears as agglomeration and perhaps physical adhesion as a result of cooling of hot metal particles released upon the impact of a bullet. The implementation of a hydrocyclone at a high slurry density (50% solids by weight) for a scrubbing effect yields a new fraction of highly contaminated fines.

Organic material, primarily wood, is present in all size fractions. It represents about 1-2% by weight in total and is highly contaminated (>50,000ppm Pb). With microscopy it was possible to identify that microfine particles of metal are held within the fibrous matrix of the organic material. In the gravity separation tests the organic material reported to the light fraction as expected, usually floating on the water medium.

It is important to look beyond the elemental analysis alone, and to incorporate mineralogical, microscopic evaluations and the behaviour of materials in certain processes. In the gravity separation tests, for example, the light fraction also contained a significant amount of metal contamination:

F. Arendt, G.J. Annokkée, R. Bosman and W.J. van den Brink (eds.), Contaminated Soil '93, 1459–1460.
© 1993 *Kluwer Academic Publishers. Printed in the Netherlands.*

primarily iron, but also lead. Some of the products of corrosion, namely the goethite (FeOOH) and cerussite ($PbCO_3$), tend to have a porous structure and behave as a material of lighter density than the quartz sand middlings.

An association of iron and lead due to the displacement of metal during firing and impact provides a novel method of cleaning: high intensity magnetic separation. Tests conducted on a 0.125-0.250mm size range showed removal of over 60% of the lead in the paramagnetic fraction representing only 10% by weight of the feed. X-ray diffraction analysis of the paramagnetic fraction showed that it contained iron as goethite, magnetite, and metal, and lead as cerussite and metal. High intensity magnetic separation of the middlings fraction (sand) of the shaking table achieved further decontamination, thus removing contamination in a form that was evasive in gravity separation.

3. PROCESS POSSIBILITIES

Effective combination of the physical separation techniques which decontaminate the sand offers products of new specifications. In general such a treatment process achieves a concentrate of high metal content, a sand of reduced contamination, and a slimes fraction of high contamination.

An initial wet sieving step would allow an early removal of slimes and provide a rough size classification prior to gravity separation. Effective use of gravity separation requires a minimum influence of particle size difference. Due to the broad size range of the sand, different fractions are more effectively treated with different techniques. Moreover, for economic reasons it is more favourable to use high capacity methods wherever possible. The jig, fluidized bed and shaking table were applied to respectively smaller size fractions, attaining removal of 60 to 90% of the contamination by rough separation.

A rough gravity separation can remove most of the liberated metal portion as a heavy fraction and collects the organics in a highly contaminated light fraction. The sand middlings slurried to a high percent soilds ($\sim$50%) and scrubbed in a hydrocyclone frees a further highly contaminated fine fraction. Reduction of the slurry density then allows efficient removal of the fines and narrower size classification by hydrocyclones preceding a secondary gravity separation. Processing the sand middlings from the secondary gravity step by high intensity magnetic separation removes another metal fraction, yielding a considerably decontaminated sand product.

The scheme described is general and has of course several other options. For example, using hydrocyclones instead of wet sieving would also remove most of the organics with the slimes because the density of wood is lighter than water. The degree of size classification necessary depends on the extent of its influence on the efficiency of the gravity separation steps. Overall, the economic ramifications of different flowsheet options must be considered.

4. CONCLUSIONS

Through the use of physical separation techniques it is possible to achieve a certain level of decontamination and improve the feasibility of further treatment methods and ultimate reutilization of the material.

The concentrate of the contamination may be able to be used for its metal content. The sand fraction of reduced contamination may be able to be considered for reuse or safe disposal, and a fines fraction, which although may have to be disposed of as contaminated waste or treated with other methods, is of a significantly lower quantity than the total sand.

The applications of the product streams obtained by physical separation techniques constantly change due to changing environmental regulations. Depending on what type of specifications are necessary the decontaminated sand and the slimes may be further treated by leaching, flotation or other methods.

Although this investigation was conducted on sand of firing ranges that have been used for a long time, prevention of contamination is also important. If the catcher sand was treated on a periodic basis to remove at least the bullets and fines, for example, immediate environmental risk would be kept to a minimum and the future situation improved.

APPLICATION OF THE SULPHUR CYCLE FOR BIOREMEDIATION OF SOILS POLLUTED WITH HEAVY METALS

R. Tichý, J.T.C. Grotenhuis, A. Janssen, R. van Houten, W.H. Rulkens, G. Lettinga

Wageningen Agricultural University, Department of Environmental Technology, Bomenweg 2, P.O.Box 8129, 6700 EV Wageningen, The Netherlands

INTRODUCTION

Heavy metals belong to the most difficult compounds for soil remediation. Due to their recalcitrant character and complex behaviour in soil, the technologies removing metals are expensive and the applicability at large areas is limited. Moreover, the removal of heavy metals from the extractant and the recycling of process water contribute significantly to these problems.

Biological extraction processes using changes in the oxidation state of sulphur may be profitable. Processes of sulphur oxidation, sulphate reduction and sulphide oxidation can be combined, allowing for the full recycling of sulphur inside the treatment system (sulphur cycle). This paper aims at a summarization of the most important data about the individual processes which are of importance once the concept of the use of the sulphur cycle is considered.

SULPHUR CYCLE

Biological processes participating on the sulphur cycle are schematized in Figure 1.

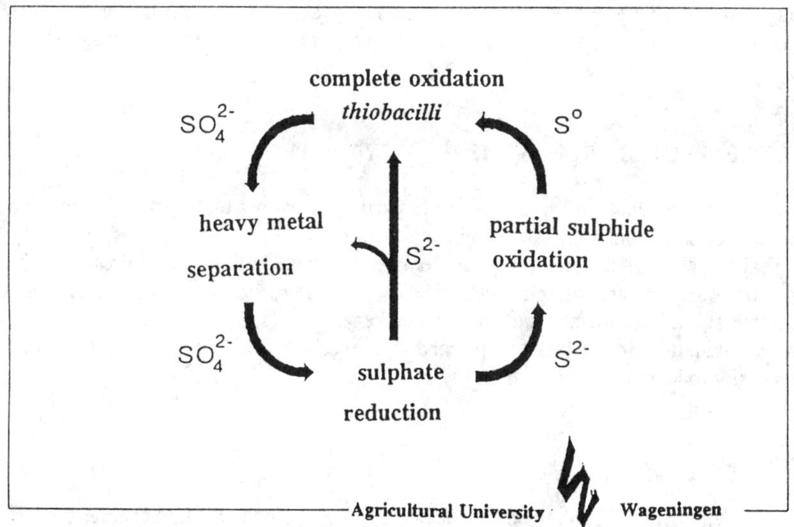

Figure 1-Processes within the sulphur cycle

Complete oxidation is a bioleaching process using acidophilic *thiobacilli* which utilize reduced sulphur compounds as a source of energy. This process is leading to (a) production of sulphuric acid, extracting metals from soil, or (b) direct solubilization of insoluble metal sulphides produced from the sulphate reduction step [van der Steen *et al.* 1992].

F. Arendt, G.J. Annokkée, R. Bosman and W.J. van den Brink (eds.), Contaminated Soil '93, 1461–1462.
© 1993 *Kluwer Academic Publishers. Printed in the Netherlands.*

Sulphate reduction is mediated by anaerobic auto- or heterotrophic organisms. As a substrate, hydrogen gas or reduced organic compound like acetate, ethanol, or more complex organic substrates can be used. Sulphide produced in a system can either be stripped as hydrogen sulphide gas, or it can form with heavy metals highly insoluble sulphides. This sludge containing very high heavy metals concentrations can be subsequently removed from the main stream [Barnes *et al.* 1991].

Partial sulphide oxidation process can biologically convert sulphide to the elemental sulphur in conditions of limited oxygen supply and/or sulphides overloading. Elemental sulphur produced by this process, "biological sulphur", can be utilized for the feeding of bioleaching step. Compared to hydrophobic sulphur flower, biological sulphur has a hydrophillic character allowing for progressively improved homogeneity of sulphur suspension in an aqueous medium [Steudel 1989].

Summarization of the main parameters for the three processes is given in Table 1.

	Complete oxidation	Sulphate reduction	Partial S^0 oxidation
pH optimum	1-4	6-10	7-8.5
pH trend	acidifying	alkalizing	alkalizing
redox-conditions	highly aerobic[*]	anaerobic	limited aerobic
substrate (electron donor)	elemental sulphur or sulphide	additional substrate[**] required	sulphide

[*] Although *thiobacilli* are not strictly aerobic, intensive aeration is required for the bioleaching
[**] Additional, non sulphur-containing substrate like organic compound or hydrogen gas

BENEFITS EXPECTED FROM A SULPHUR CYCLE PROCESS

Following items can benefit from the processes participating on a biological sulphur cycle applied for the treatment of soils contaminated with heavy metals:
- Biological oxidation of elemental sulphur mediates soil remediation by mobilization of heavy metals.
- Immobilization/concentration of heavy metals by sulphate reduction promotes handling with a waste liquid stream and allows metals and sulphur recovery.
- Partial S^{2-} oxidation neutralizes highly corrosive excess sulphide and produces elemental sulphur which is easily oxidized in a bioleaching step.

REFERENCES

Barnes, L.J., Janssen, F.J., Sherren, J., Versteegh, J.H., Koch, R.O., Scheeren, P.J.H. (1991) 'A new process for the microbial removal of sulphate and heavy metals from contaminated waters extracted by a geohydrological control system' Trans. I. Chem. Eng. 69A; 184-186

van der Steen, J.J.D., Doddema, H.J., de Jong, G. (1992)'Removal of heavy metals from waste streams using *thiobacilli* (in Dutch)' Dutch Ministry VROM Report 1992/10, The Hague, The Netherlands

Steudel, R. (1989)'On the nature of the elemental sulfur produced by sulfur-oxidizing bacteria- A model for S^0 globules' in Schlegel, H.G., Bowien, B. (eds.):'Autotrophic bacteria', Springer Verlag, Berlin, Germany; 289-303

SOIL VAPOR EXTRACTION (SVE): LIMITATIONS

K.WEHRLE

INSTITUTE FOR SOIL MECHANICS AND ROCK MECHANICS,
UNIVERSITY OF KARLSRUHE, GERMANY

In recent years soil vapor extraction (SVE) has been successfully used cleaning the unsaturated zone, mainly in cases involving contamination by chlorinated hydrocarbons (CHC). A schematic diagram of a SVE-system with a single extraction well is shown in *FIGURE 1*. The success of remediations by SVE can be demonstrated by the recovery of contaminants (measured as the load retained by a filter) and by analysis of the soil vapor. Recently soil clean-up target levels are set as an allowable soil vapor concentration after a certain shut down period (eg. total CHC $< 0.5 \text{ mg/m}^3$), but these targets are often not reachable. This observation, combined with a decrease in air effluent concentrations which often takes longer than anticipated means that the operating period and remediation costs may exceed expectations. What are the limitations for the extraction?

The basis for SVE is the removal of soil vapors containing highly volatile contaminants from the subsurface. As clean air flows through the subsoil and gets loaded on its way to the extraction well, a continuous extraction process is established. The following factors control the recovery of the contaminants, the operating period and the amount of contamination which may remain:

- the subsoil (soil stratigraphy, mineralogical composition, organic content, water content),
- the contaminants (distribution in the subsoil, density, vapor pressures, solubility, partitioning coefficients),
- operating conditions (construction and spacing of the extraction wells, flow rate, permanent or intermittent operation),
- ambient conditions (temperature, humidity, precipitation).

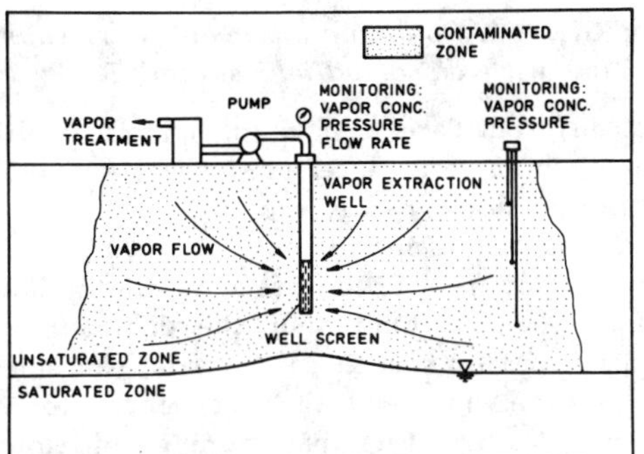

FIGURE 1: Soil vapor extraction system: Main devices and elements of the monitoring system

FIGURES 2 and 3 show the effluent vapor concentrations obtained during two column experiments. The main difference between the two experiments is the contamination level. The experiment depicted in *FIGURE 2* (after Mackay et al., 1990) was carried out with dry sand (0.25 - 0.5 mm) contaminated with liquid-phase trichloroethylene (TCE). The column was vented with air at a specific discharge of v = 15.9 cm/min. The effluent concentration remained close to the saturation vapor concentration until almost all of the total TCE mass was recovered. With regard to the total TCE mass the curve shows only a little tailing caused by the

F. Arendt, G.J. Annokkée, R. Bosman and W.J. van den Brink (eds.), Contaminated Soil '93, 1463–1465.
© 1993 *Kluwer Academic Publishers. Printed in the Netherlands.*

absence of the water phase. The correlation between tailing and soil moisture was illuminated by the experiments of McClellan and Gillham (1990). They also used Borden Sand and liquid-phase TCE but in the presence of water, and they observed a pronounced tailing. The experiment depicted in *FIGURE 3* (after Rauh, 1992) was carried out with moist sand (0.1 - 0.6 mm, f_{oc} = 0.14 %, inorg. carbon approx. 10 %) which was contaminated by flushing with TCE-loaded water (dissolved and sorbed contamination, no liquid phase). Afterwards the column was vented with air at a specific discharge of 21.1 cm/min. After starting the experiment the TCE concentration rapidly decreased and thus the removal rate also dropped down. During this experiment venting was interrupted for 3 periods of 24 hours. Each time, soil vapor concentrations peaked at levels more than 100 times above the starting concentration, although equilibrium between the water phase and the vapors was not reached after 24 hours. At the end of the experiment the air-filled pore volume has been exchanged more than 40,000 times and 97 % of the TCE mass was recovered, however the target mentioned above had not yet been reached.

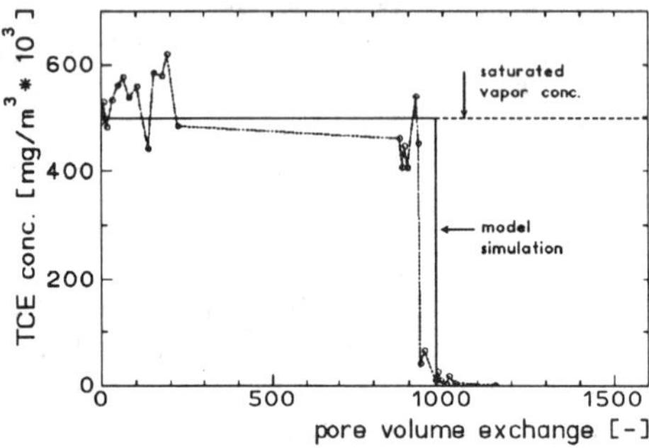

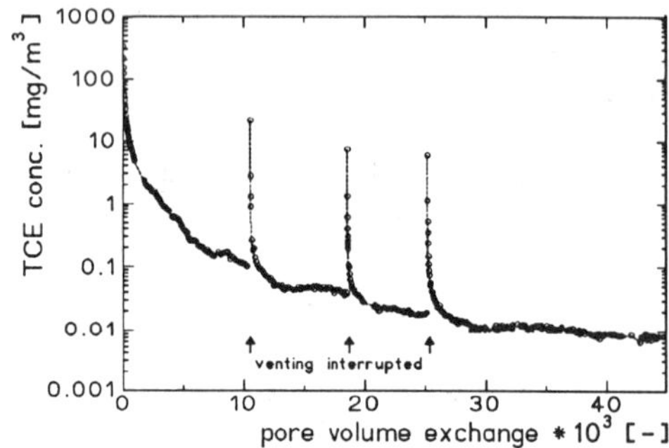

FIGURE 2: Column experiment, Borden Sand with TCE liquid phase

FIGURE 3: Column experiment, Karlsruher Sand with dissolved and sorbed TCE

Subsurface contamination appears in several different forms. After infiltration (spill) into the unsaturated zone liquid contaminants have large interfaces to the mobile air, so volatilization takes place quickly. But in some cases, especially long time after infiltration, liquid contaminants appear as little drops in residual form, which may be surrounded by water and therefore do not have an interface to the mobile air. In this case the transfer into the vapor phase is limited by the diffusion in the water. If contamination is caused by dissolved or sorbed compounds, the subsurface organic matter (organic content, particle size) and the micropores (pore radius, tortuosity, particle size) of the particles control their partitioning behaviour in the subsoil. Investigations by Grathwohl et al. (1992) showed these correlations. As an example, TCE took more than 100 days to reach equilibrium in a sand with high inorganic carbon content. The reason given for that was the very slow diffusion through mircopores in the soil particles. The soil at remediation sites has often been in contact with the contaminants over a very long period. The recovery of these contaminants is therefore often limited by diffusion in micropores or interactions between the organic matter and the contaminants.

The subsoil has to be sufficiently permeable to air if the application of SVE is to be successful. The recovery of the contaminants is dependent not only on the average permeability which can be determined by in situ air permeability (pump)

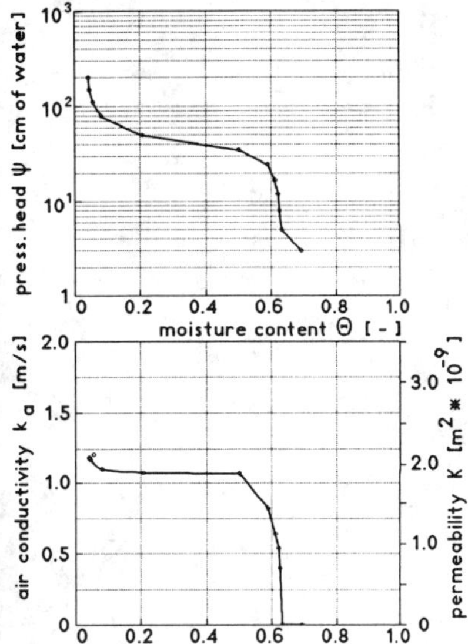

FIGURE 4: *Results of a combined measurement of capillary pressure and permeability vs. moisture content*

tests, but also on the homogeneity of the air flow in the soil. The permeability may vary greatly for reasons of soil stratigraphy (grain–size curves) and geological irregularities (fissures, pathways), as well as due to differences in water content (see *FIGURE 4*, after Brauns and Wehrle, 1992). The pore volume exchange of vapor is naturally slower in low permeability areas than in highly permeable areas. At a large scale, low permeability areas occur as silt and clay layers or lenses. At a smaller scale they can occur in the form of lokal heterogeneities. Under the pressure fields established during SVE it is possible that the air flow does not reach such low permeability areas at all, but flows around them. This leads to the fact that the contaminants inside this areas must diffuse a very long way to reach the mobile air and get out of the subsurface. The permeability of the soil and the diffusion time in the soil water limit the rate of transfer into the moving air.

During design of SVE systems in situ air permeability (pump) tests are often carried out. They give us important information about the flow rates which can be achieved and about the vapor concentrations in the initial stage of the remediation. This data often then gets used to predict the further course of the remediation. The reliability of such effluent concentration predictions depends on whether or not the potentially limiting processes have been observed. Therefore permeability test should be carried out until the vapor in the contaminated subsoil has been exchanged several times. Only then can the effects of the subsoil, the nature of the contaminants, the operating conditions and the ambient conditions express their influences, and allow some measure of prediction regarding SVE performance.

References:

Grathwohl, P., G. Einsele, T. Gewald, Ch. Schüth (1992): Untersuchungen zum Austrag organischer Schadstoffe aus kontaminierten Böden in das Grundwasser. In: KfK – PWAB 12, Kernforschungszentrum Karlsruhe (publication in work).

Mackay, D. M., M. B. Bloes, K. M. Rathfelder (1990): Laboratory studies of vapor extraction for remediation of contaminated soil. In: Weyer (ed.), Proc. IAH – Conf. on subsurface contamination by immiscible fluids. A.A. Balkema Publishers, Rotterdam (publication in work).

McClellan, R.D., R.W. Gillham (1990): Vapour extraction of TCE under controlled field conditions. In: Weyer (ed.), Proc. IAH – Conf. on subsurface contamination by immiscible fluids. A.A. Balkema Publishers, Rotterdam (publication in work).

Rauh, H. (1992): Untersuchung des Einflusses der organischen Substanz eines Bodens auf die Dekontamination von LCKW durch Bodenluftabsaugung. Thesis for diploma at the Institute for Soil Mechanics and Rock Mechanics, University of Karlsruhe (not publ.).

Wehrle, K., J. Brauns (1992): Bodenluftabsaugungsverfahren bei CKW-Schadensfällen – Einfluß maßgebender Bodenparameter auf die Luftströmung und den Schadstoffaustrag. In: KfK – PWAB 12, Kernforschungszentrum Karlsruhe (Publication in work).

METHODS OF TREATING SOIL CONTAMINATED WITH CHROMIUM

C WERNICKE[1],
R WIENBERG[2],
J GERTH, M WILICHOWSKI, U FÖRSTNER, J WERTHER[3]

[1] ENVIRONMENTAL AUTHORITY, HAMBURG
   Amelungstraße 3, D-2000 Hamburg 36, Germany
[2] BÜRO UND LABOR DR R WIENBERG – Consultants and laboratory
   facilities specialising in the chemical and biological
   aspects of contaminated sites, Gotenstraße 4, 2000 Hamburg
   1, Germany
[3] TECHNICAL UNIVERSITY OF HAMBURG-HARBURG
   Eißendorfer Straße 42, D-2000 Hamburg 90, Germany

1. INTRODUCTION

Suitable methods of cleaning up 2 sites of former
electroplating plants in Hamburg are to be found. Both soil
and groundwater are contaminated with chromium in both cases
(average soil contamination 800 mg of Cr/kg and 1500 mg of
Cr/kg respectively). Since little knowledge has been
accumulated to date about remedial action techniques for
treating soils contaminated with chromium, various methods
were subjected to laboratory-scale tests on behalf of the
Hamburg Environmental Authority.

2. SCOPE AND RESULTS OF TESTING

Extraction and speciation tests

Extraction and speciation tests (2,3) were carried out to
determine the form in which the chromium is present
underground and the soil fractions to which it is
predominantly bound. These tests were also used for risk
assessment and evaluating the soil to find which treatment
techniques are applicable. The most important result was that,
contrary to expectations, a certain proportion of even the
relatively insoluble Cr(III) can pass into the soil solution
and therefore be transported underground.

Tests were then performed to discover the extent to which the
treatment methods mentioned below are suitable for the soil
contaminated with chromium.

Washing soil with acid (laboratory tests)

The extraction tests (2,3) showed that washing the soil with
acid does not clean it satisfactorily. Cr(III) was only
partially eliminated, even when 1 M HCl was used and the
extraction period extended to several hours. It was only

F. Arendt, G.J. Annokkée, R. Bosman and W.J. van den Brink (eds.), Contaminated Soil '93, 1467–1468.
© 1993 *Kluwer Academic Publishers. Printed in the Netherlands.*

eliminated where it is bound to carbonate constituents of the soil. The residual chromium contents in the treated soil lay well above the target value of 100 mg of Cr/kg.

Washing soil with water (quasi full-scale tests)
These tests (3) showed that only a small proportion of the chromium passes into the water phase during washing. However, by using the washing to break down the soil structure and then fractionating, it proved possible to separate a coarse fraction (>0.3 mm) whose chromium content is low enough to make unrestricted use of this part of the soil a possibility. Due to its high eluate values, the fine residual fraction from this treatment must be conditioned before disposal.

Binding the chromium in soil-stabilisation grout (laboratory tests)
Testing based on in-situ and conventional on-site soil stabilisation techniques was carried out to determine the extent to which not only Cr(III), but also Cr(VI), can be permanently immobilised in a soil-stabilisation grout. The tests (2) were performed on 4 mixes using "Hochofenzement", "Heidelberger Schnellzement", "Sulfathüttenzement" and "Tonerdeschmelzzement" as the binding material. Both standard sand artificially contaminated with Cr(III) and Cr(VI) and soil contaminated on site with chromium were tested. Both Cr(III) and Cr(VI) were bound particularly effectively when "Sulfathüttenzement" was used. This result was also confirmed after the stabilised samples had been stored for over a year.

Binding the chromium to iron oxyhydroxide (laboratory tests)
On-site conditioning of the soil contaminated with chromium by binding the chromium to iron oxyhydroxide (2,3) was investigated as an alternative method to immobilisation. This involved testing the reduction and precipitation of chromium from artificially contaminated soils (sand, clay, peat, marl and loam), and from the soil contaminated on site with chromium. The tests showed that Cr(VI) is firmly bound to the iron oxyhydroxide phase that arises after reduction with iron(II) sulphate. Easily mobilised Cr(III) is also bound to freshly precipitated iron oxyhydroxide and immobilised. The treatment of a soil suspension can also be regarded as a suitable method of conditioning soil contaminated with chromium.

3. CONCLUSIONS

Decisions on measures for cleaning up sites contaminated with chromium are based on the results of the tests. It is advisable to separate a clean coarse fraction by washing the soil, and condition the fine fraction by binding it to soil-stabilisation grout or iron oxyhydroxide.

POSSIBILITIES FOR CYBERNETIC CONTROL OF FERMENTIVE ON-SITE REGENERATION OF SOILS CONTAMINATED BY ORGANIC COMPOUNDS

HERMANN WOLLNER

INNOVATION BODEN UND UMWELT

Considering the vast pollution of soils on all continents and the increasing ground and well-water pollution caused by this, the technological improvement of on-site treatment of contaminated soils in order to regenerate these is highly desirable.

This technological improvement will be achievable, if soil - regardless of the kind of invalidity - is considered as part of the natural transformation cycle of the substances of life (C, H, O, N, S, P) and the course of this cycle is interpreted on the basis of a cybernetic model.

Current soil treatment technologies follow linear models and tend to "centrifuge" results - contrary to nature.

Nature integrates rather than divides. Hence, the proposed model conceives soil regeneration as the appropriate remedy - and not merely soil treatment. The remedy process is the nutriment intake and the digestion process performed by biocenosae of micro-organisms, settling on mineral-organic substances, building up structured particles of humus, excreting gases and energy. This process is disturbed by anthropogenic organic compounds or xenobiotica.

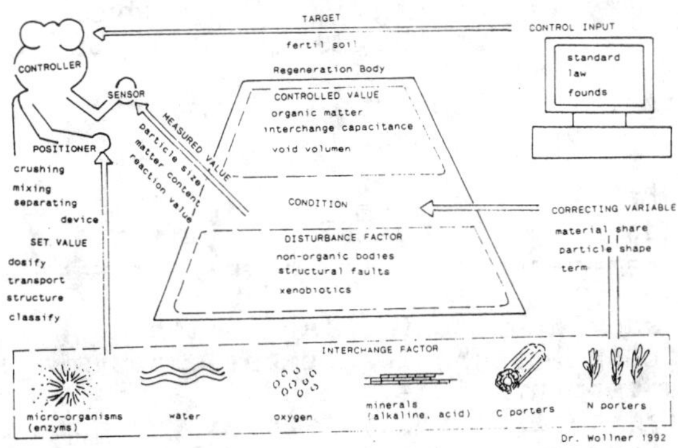

F. Arendt, G.J. Annokkée, R. Bosman and W.J. van den Brink (eds.), *Contaminated Soil '93*, 1469–1470.
© 1993 *Kluwer Academic Publishers. Printed in the Netherlands.*

The substances joined in the "Regeneration Body" (RB), mostly a heap, have to be converted into re-cultivable, vegetation bearing soil. These substances are considered neither as a sum of chemical elements or compounds nor as components of building sites - but as substances of the life cycle. Thus, all non-toxic (non-anthropogenic) substances found in polluted soils are considered as "soil content matter", if not as nutrients.

The value of soils should be characterized predominantly by physical and nutrimental parameters and not primarily as mere phenomena of non-integratable, disturbing matter.

The task of microbiological treatment of polluted soils is to convert disturbing compounds into useful ones and to exchange non-integratable radicals (e. g. halogens) for integratable ones (e. g. H_2, O_2). This task could be achieved through a liquid stream within the structured, well-homogenized RB. The stream gravitates downward and adsorbs increasingly the polluting radicals. The separation of disturbing elements on the gaseous path seems unnatural. This would lead to uncontrollable emissions sooner than does the conduction by the pollutant-adsorbing liquid to a well-sealed recipient in the bottom of the RB.

The essence of the cybernetic control of this process lies in
- the composition of a interchange-furthering structure of the RB
- the stockpiling and in-time-supply of such interchange elements - like micro-organism mass or microbial enzymes, water, oxygen and correctors of acidity (minerals).

The following elements of regeneration process control should be considered to be most important:

interchange element	command	measuring data	functional link	controlling exactitude
microbial enzymes	dosify	velocity of reaction	mixer	fuzzy
water	dosify	content in the RB	conveyor-distributor	fuzzy
oxygen	dosify	content in exh. air	mixer	fuzzy
pH	set/tune	content in the RB	distributor (mixer)	nearly exact

These elements should be controllable with much less measuring data than traditional technologies use. Fuzzy logical algorithms should substitute linear functions in the control system. The measure of control efficiency should be the rate of matter transformation. The economic importance of its improvement cannot be understated.

INVESTIGATION, IN PARALLEL WITH REMEDIAL ACTION, OF THE BALANCE OF MATERIALS AND FORMATION OF METABOLITES INVOLVED IN THE APPLICATION OF A WHITE ROT CLAMP TECHNIQUE FOR CLEANING UP SOIL CONTAMINATED WITH PAHs

MARTIN ZARTH[1], RENATE ELVERS-KÖHLER[1], REINHARD WIENBERG[2]

[1] ENVIRONMENTAL AUTHORITY, HAMBURG
Amelungstraße 3, D-2000 Hamburg 36, Germany
[2] BÜRO UND LABOR DR R WIENBERG – Consultants and laboratory facilities specialising in the chemical and biological aspects of contaminated sites, Gotenstraße 4, D-2000 Hamburg 1, Germany

1. OBJECTIVE

A white rot clamp technique was used in summer 1992 to clean up a fill contaminated primarily with polycyclic aromatic hydrocarbons, on which a depot for drums of tar and oil was originally constructed. Even after an extensive literature review (1), there are still some unanswered questions about the overall toxicological effectiveness of this technique developed by the contractor. In particular there is still little known about the formation of any toxic metabolites and about the immobilisation of PAHs and metabolites that have not degraded in the soil matrix. These questions are being pursued in parallel with remedial action as part of a combined white rot investigation project (WIP), in which 8 working parties are involved on behalf of the Environmental Authority, Hamburg.

2. INVESTIGATION METHODS

The investigation is being carried out:
1. As full-scale tests on two of the total of 15 bio-beds
2. As quasi-full-scale tests on four 100-litre reactors and
3. As laboratory-scale tests on six 6-litre and six 1-litre small reactors.

This approach should show up any **effects of scale** and allow valid extrapolation of the laboratory- and quasi-full-scale results to full scale.

To allow the residue of PAHs to be traced throughout the remedial action, the PAH concentrations and their changes as a result of mixing operations and inhomogeneities were **analysed rapidly with the mobile mass spectrometer** (MM1) during the initial excavation stage and construction of the clamps.

After being mixed with 25 per cent by weight of a substrate of mushrooms and straw, a total of 4800 m^3 of soil was incorporated in 2 m high and 4 to 8 m wide, trapezoidal, covered bio-beds, which were supplied indirectly with oxygen through air extraction pipes within the beds. The first WIP bio-bed with a length of 65 m was constructed at the beginning of July 1992 with sand containing clay, which is contaminated with 50 to 400 mg of PAHs/kg of solids. The WIP bio-bed with a length of 30 m was constructed at the end of August 1992 with clay contaminated with up to 230 mg of PAHs/kg of solids.

F. Arendt, G.J. Annokkée, R. Bosman and W.J. van den Brink (eds.), Contaminated Soil '93, 1471–1472.
© 1993 Kluwer Academic Publishers. Printed in the Netherlands.

The two WIP bio-beds and the large reactors are to be put to the test on 16 different dates over the course of the 2½-year period planned. The laboratory reactors are only being operated for 6 months for each trial, allowing 4 series of tests to be carried out one after the other over the same period of time.

Since the marked inhomogeneities in the soil contamination lead us to expect considerable scatter in the results of the analyses from clamps and large reactors, a statistically significant number of **rapid PAH analyses** are also being performed with the MM-1 on the individual mixed specimens provided as well as various soil fractions.

The metabolites are to be identified and quantified as comprehensively as possible with **special analytical methods**. To obtain indications of gentoxic metabolites possibly present, **gentoxic activity tests** are also being carried out on selected specimens. To achieve mass balancing along all residue paths that is as comprehensive and quantitative as possible, **substances marked with C-14** are being used in the laboratory. Qualitative and quantitative data that is as full as possible is to be acquired for contaminants entrained in the atmosphere, with new methods to be developed using a mass spectrometer.

To allow a comparison of the processes in the laboratory and large reactors as well as the bio-bed, it is necessary to regularly measure the **environmental conditions** (including moisture content, temperature, nutrient content, pH-value, redox potential, humus content, additives, soil characteristics), and the **microbiological activity** (including microorganism population, inhalation of CO_2 by the microorganisms, O_2, organic carbon, C/N ratio) in the WIP bio-beds and large reactors, and sometimes in the laboratory reactors. This data will also yield indications of the processes and parameters that affect the residual PAH concentration.

As a result of the natural factors and the operations of the contractor responsible for the remedial action, the parameters that determine the degradation processes vary to a certain extent with position and time. All of the data obtained should be evaluated systematically for this variation, in order to find out about the parameters determining the process. The parameters should not be changed deliberately in the bio-beds or in the large reactors by an amount going beyond this. Operation of the bio-beds is the sole responsibility of the contractor. Only in the laboratory reactors is such **variation of the parameters** planned to a limited extent in order to follow the effects of soil type, addition of straw, level of contamination and abiotic processes.

Since only a limited number of proven test methods are available for the above-mentioned investigation, in the course of the combined project some promising existing **test methods** and new ones required are to be developed and recommendations worked out as to which methods are suitable for monitoring future cases of remedial action.

3. REFERENCES

(1) Wienberg, R, Kästner, M, Mahro, B (1993). Biologischer Schadstoffabbau in kontaminierten Böden unter besonderer Berücksichtigung der Policyclischen Aromatischen Kohlenwasserstoffe. [Biodegradation of Soil Contaminants with Particular Reference to the Policyclic Aromatic Hydrocarbons]. Published by Economica-Verlag in Bonn.

DEGRADATION OF PCBs BY WHITE-ROT FUNGI IN SOLID PHASE SYSTEMS

Andreas Zeddel, Andrzej Majcherczyk, Aloys Hüttermann
Forstbotanisches Institut der Universität Göttingen, Büsgenweg 2, D-3400 Göttingen

Biodegradation, Polychlorinated Biphenyls, White-Rot Fungi

1. MATERIAL AND METHODS

Wood chips were coated with PCBs, extracted from an transformer-oil (Clophen A 30) or synthesised by chlorination of Biphenyl (^{14}C-PCB: [U][14]C-Biphenyl: 3,7 MBq, Sigma; [36]Cl-PCB: [36]Cl-NaCl, 3.7 MBq, Amersham Buchler) respectively. They were mixed with potato-pulp as a substrate and inoculated with white-rot fungi. The cultures were maintained in 700 ml glass tubes sealed with Teflon lids to allow ventilation with air from the bottom to the top. The bottom was filled with porous clay media which served as a water sink to adjust the water tension. The tubes were continuously ventilated with water saturated air. Analysis of PCBs was made by GC-MSD. Radioactivities were determined by liquid szintillation-counting.

2. RESULTS

2.1. Unlabeled PCBs on wood chips

An industrial PCB-isomer mixture, mainly tri- and tetra chlorinated biphenyls (concentration 2750 ppm sum of isomers) loaded on wood chips were degraded by more than 90% by *Pleurotus ostreatus* and *Trametes versicolor* in only 6 weeks. Penta- and hexachlorobiphenyls were also degraded by more than 50%, depending on their substitution pattern. Only 2,2′,4,4′,5,5′-hexachlorobiphenyl was resistant to degradation. The addition of lignosulfonate had no effect on degradation of PCBs. Mixing the wood-chips with garden soil did not affect the degradation process of PCBs. The direct application of the xenobiotics to the soil prior to the mixing with the wood chips, however, reduced the speed of their degradation to about 50%-60% of the rate obtained on wood-chips. No increase of bioavailibility of PCBs was achieved by the addition of tensides. PCBs with a highly symmetrically substitution pattern of chlorine, mainly 4,4′-, were more resistent to degradation.

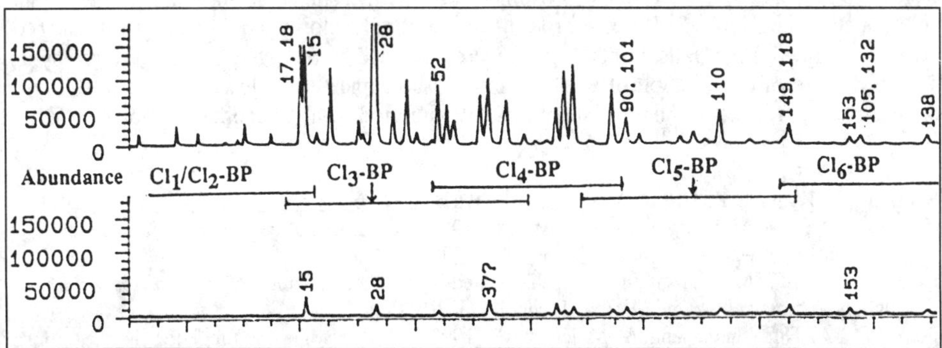

Figure 1. Comparison of GC-chromatograms of extractable PCBs in wood chips mixed with potato pulp and soil after incubation with fungi for 5 weeks. Control (upper) and degradation result (lower curve). The numbers indicate the PCB-congeners according to Ballschmitter and Zell 1980.

F. Arendt, G.J. Annokkée, R. Bosman and W.J. van den Brink (eds.), Contaminated Soil '93, 1473–1474.

2.2. Degradation products

Mineralisation of PCBs could be measured throughout the total incubation period. An example for the formation of $^{14}CO_2$ from labeled PCB´s is shown in Figure 2. The experiment demonstrates that differences between parallels were bigger than those between the different incubation systems on wood chips or on soil. The growth of the fungi correlates with the observed rates of PCB-mineralisation. After 4 weeks incubation 10% of initial ^{14}C-activity was found as CO_2 and CO_2-production continued for several months.

To investigate the fate of the chlorinated parts of the PCB molecule, the same experimental design was chosen for the monotoring of degradation of ^{36}Cl-PCB´s. After 4 weeks of incubation with *Pleurotus ostreatus ssp. florida* only 33 % of initial chlorine was detectable as unpolar extractable material, as seen in Figure 3. In this fraction no substances other than PCB´s were found by GC-MSD. The main activity was found as $AgNO_3$ precipitable chlorine ions, demonstrating total degradation of the PCB molecules. Another part of the activity was bound to the wood chip matrix. Only small amounts of the initially activity were found in the form of water or ethyl acetate soluble substances.

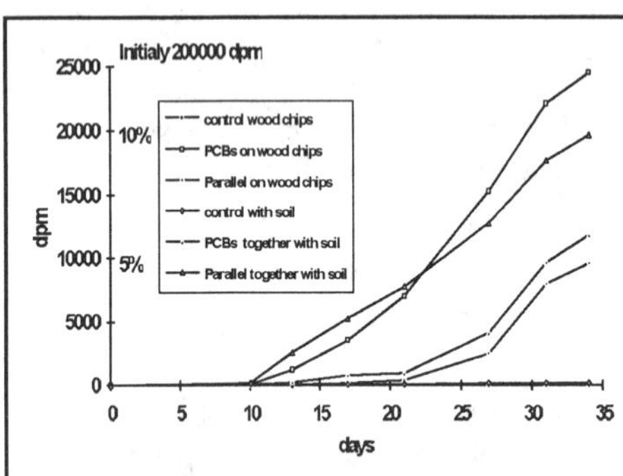

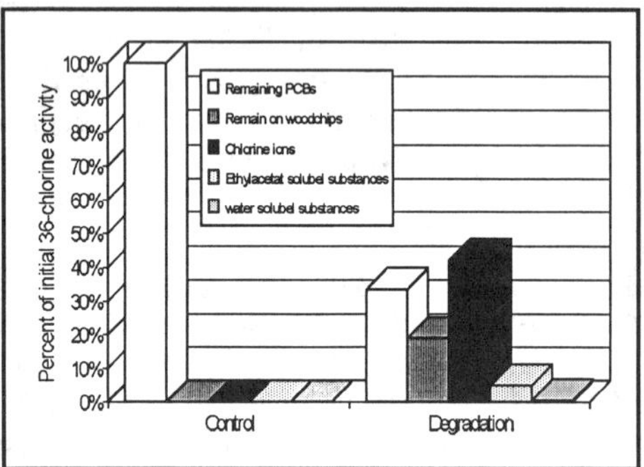

Figure 2: ^{14}C-CO2 formation from PCB- coated wood-chips by incubation with *Pleurotus ostreatus*.

Figure 3: Distribution of ^{36}Cl-activity from PCB-coated wood-chips after 4 weeks of incubation with *Pleurotus ostreatus*.

3. CONCLUSION

Incubation with white-rot fungi was found to be applicable for the degradation of PCBs in solid state fermantation. Opposit to the situation reported for bacteria, a single strain of the fungi were able to degrade a wide range of isomers of the environmental important tri- to penta-chlorinated biphenyls. Even PAHs, nitroaromatics and polychlorinated dioxins and furans (co-investigation with Prof Ballschmitter, Ulm) were degraded by the white-rot fungi tested by us. The lower rate of degradation of PCBs in soil can be explained by lowered bioavailability of contaminants. This fact requires a longer incubation period for the decontamination of soils.

4. LITERATURE

Ballschmiter, K.and M.Zell 1980. "Analysis of PCB's by GC" Fresenius Z. Anal. Chem. 302: 20-31.

Bumpus, J. A., Tien M., Wright D. and Aust S. D. 1985. "Oxidation of persistent environmental pollutants by a white-rot fungus" Science 228/4706: 1434-6.

Eaton, D. C. 1985. "Mineralisation of polychlorinated biphenyls by *Phanerochaete chrysosporium*: a lignolytic fungus" Enzyme Microb. Technol. 7/5: 194-6.

Loske, D., A. Hüttermann and A. Majcherczyk. 1990. "Use of white-rot fungi for the clean-up of contaminated sites". In: Coughlan M.P. and Collaco A. (Eds.) Advances of biological treatment of lignocellulosic materials, pp 311-322. London: Elsevier Applied Science.

Membrane-photo-reactor for microbial degradation of volatile and toxic substances

C. Zenneck, H. Märkl
Technische Universität Hamburg-Harburg, Arbeitsbereich Bioprozess- und
Bioverfahrenstechnik, Denickestr. 15, D 2100 Hamburg 90

Summary

BTX-Aromatics are exhaustless degraded by syntrophic algae- and bacteria cultures which are separated in a Membrane-Bioreactor. The photosynthetic production of O_2 and consumption of CO_2 by algae are conected with heterotrophic production of CO_2 and consumption of O_2 by bacteria for a closed loop circulation of the gases. This allows an exhaustless degradation of volatile or toxic substances. The degradation rate depends on the permeation rate of the used membrane. With the polypropylene membrane Celgard 2500 the degradation of Toluene amounts 6.96 mg/h*1 reactor volume. It can be concluded that the degradation rate can be increased with higher permeable membranes.

Introduction

Benzene,Toluene and Xylene (BTX-aromatics) represent a class of organic pollutants in industrial and municipal wastewater or landfill leachates.

These substances can be degraded succesfully by microorganisms [2]. Owing to the volatility of these substances the aerobic degradation is difficult. The aeration with air or oxygen and the production of carbondioxide during the process are stripping the volatile compounds out of the medium. The exhaust has to be cleaned with expensive filters and after use the filtermaterial has to be cleaned. With the Membrane-Photo-Reactor it is possible to degradate volatile sustances without stripping.

A schematic diagram of the reactor is shown in Fig. 1. It is composed of an outer chamber which is formed by a cylindrical polyamide foil (80 μm thick) and an inner chamber which is formed and separated from the outer chamber by a cylindrical membrane. Different types of membranes can be used, but in this study, a polyamide membrane (PA 6, 100 μm thick, Enka AG, Wuppertal, Germany) and a polypropylene membrane (Celgard 2500, 25 μm thick, Hoechst AG, Frankfurt, Germany) are used. A light-jacket with 36 light-tubes is installed around the outer chamber and supplies the phototrophic algae (*Chlorella fusca*) with light. Oxygen, produced from the algae by photosynthesis diffundates through the membrane in the inner chamber and is used by the heterotrophic bacteria (*Pseudomonas putida mt 2*) for degradation of organic pollutants. The carbondioxide from the degradation diffundates through the membrane in the outer chamber for carbon source of the algae. The exchange of gases between the cultures makes it possible to degradate volatile substances (e.g. BTX) without aeration.

Both chambers can be operated separatly with regard to the different grow-rates of the used organisms. In both chambers the dissolved oxygen is measured. For optimal working conditions in continous process a pH-regulation is installed. In continous process the outer chamber is working with a turbidostatic system which controlls the optical density and continously dillutes the culture with mineral-salt-solution. The whole process-parameters are measured and controlled by process-control-software.

Experiments

All experiments are realized with the following conditions:

Temp:	25 °C
pH:	6.8
Dry weight algae:	0.5 g/l

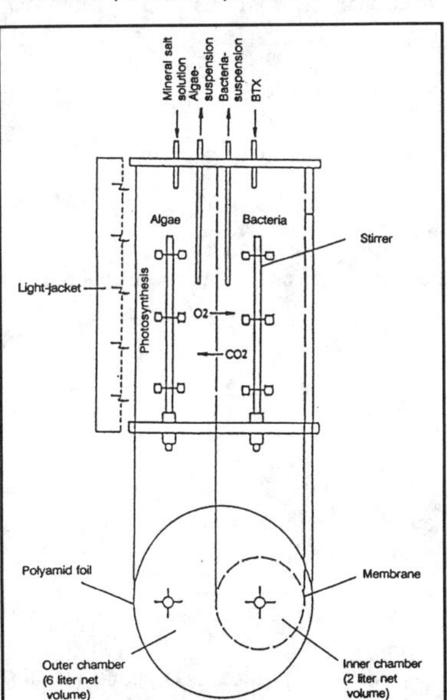

Fig. 1: Schematic diagram of the
Membrane-Photo-Reactor

1475

F. Arendt, G.J. Annokkée, R. Bosman and W.J. van den Brink (eds.), Contaminated Soil '93, 1475–1476.

1476

The inner chamber is working in fed-batch or continous conditions with different substrates.
After filling and sterilizing, the outer chamber was innoculated with *Chlorella fusca* and growing up to 0,5 g/l dry weight. Than the inner chamber was innoculated with *Pseudomonas putida mt 2*. The feeding rate of the substrates (BTX) depends by the oxygen consumption rate, see Fig. 2.

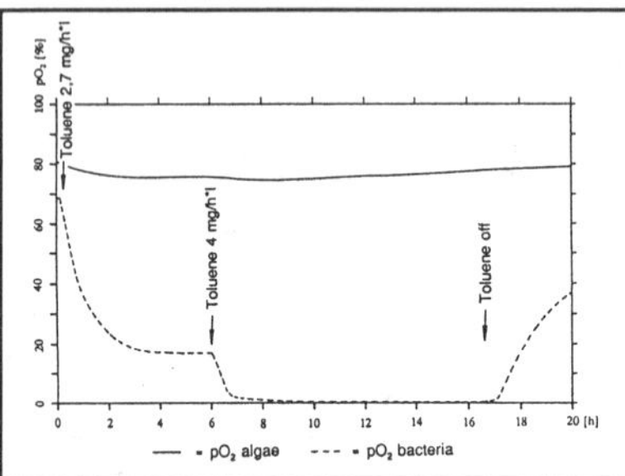

Fig. 2: Degradation of Toluene in continous culture

First the feeding rate of Toluene was 2.7 mg/h*l reactor volume. The DO in the inner chamber decreased from 70% to 21% in 3.5 h. After increasing the feeding rate to 4mg/*l the DO decreased to zero, the system was overcharged and needed 33 min after stopping the feeding to increase the DO in the inner chamber. A gaschromatographic control of samples from the inner chamber shows an continous rising concentration of Toluene to 1.93 mg/l. During optimal conditions (DO inside >2%) the substrate is not detectable. That means for the PA 6 membrane a degradation rate of 3.5 mg/h*l reactor volume. This value depends by the permeation rate of the used membrane, membranes with higher permeation rates are allowing higher degradation rates. The new tested polypropylene membrane Celgard 2500 with the double permeation rate of the PA 6 membrane shows a degradation rate of 6.96 mg/h*l. So it is necessary to search for membranes with higher permeation.

Fig. 3 demonstrates the gaschromatographic analysis of Xylene-mixture with 1,8% ortho-, 95,5% meta- and 2,3% para-Xylene before and after degradation in the Membrane-Photo-Reactor.

After a while the o-Xylene concentration is increasing. That means, o-Xylene is not or very slow degraded and accumulates in the reactor. This result concords to the results of [Worsey '75 and Chapman '81] who described only the degradation of meta- and para-Xylene for *Pseudomonas putida mt 2*.

Abb. 3: Xylene isomeres before and after degradation in the Membrane-Reactor

The purpose of the investigation is measuring experimentally the reaction dynamics of the system and comparing it with mathematical models. The data obtained from this investigations can be used for a scale up.

Conclusions

- With the Membrane-Photo-Reactor it is possible to degrade volatile or toxic subtances (e.g. BTX) exhaustlessly.

- The degradation rate depends by the permeability of the used membrane.

- The experimentaly measurement of the reaction dynamics of the system and analysis with mathematical models can be used for the design of technical plants.

This work is supported by the DFG.

References

[1] **Chapmann**, P.J.; Kunz, D.A.: Catabolism of Pseudocumene and 3-Ethyltoluene by Pseudomonas putida (arvilla) mt-2: Evidence for New Functions of the TOL (pWWO) Plasmid
J. of Bacteriol. 4: 1981

[2] **Clark**, P.H.: The metabolic versatility of Pseudomonads
Antonie van Leuwenhoek 48: 1982

[3] **Worsey**, P.A.; Williams, P.A.: Metabolism of toluene and xylenes by Pseudomonas putida (arvilla) evidence for the existence of new TOL-plasmids
Journ. of Bacteriol. 124: 1975

6. PREVENTION OF SOIL CONTAMINATION

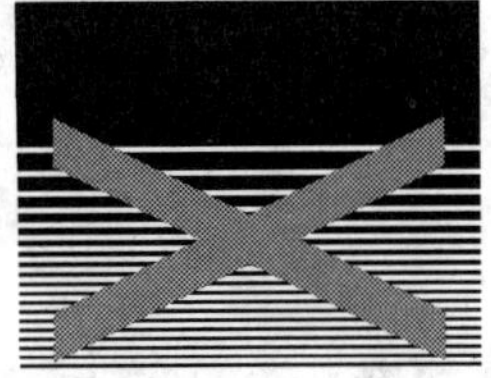

REQUIREMENTS FOR LANDFILLS IN NEW GERMAN TECHNICAL REGULATIONS
ACCORDING TO THE WASTE MANAGEMENT ACT
- A TRIAL TO AVOID FUTURE HAZARDOUS PROBLEM SITES -

KLAUS STIEF

FED. ENVIRONMENTAL AGENCY, BISMARCKPLATZ 1, D-1000 BERLIN 33,

1   INTRODUCTION
    The problem of abandoned hazardous waste sites in Germany
has been discussed many times (1,2). The majority of hazardous
waste problem sites are abandoned waste disposal sites or dumps
(3). Principally, abandoned waste disposal sites are classified
as suspected sites, according to the recommendations of a State
(Länder) working group (3).
    Future landfills should not be assessed as hazardous waste
problem sites, this will be achieved by establishing more
stringent federal standards for landfills.
    Federal nationwide standards for landfills have been esta-
blished in the second and third general administrative provision on the waste avoidance and waste management act [waste
management act] (4,5). The Federal Government presented
TA Siedlungsabfall on August 27th 1992 as proposal to the Bundesrat with the recommendation that it be approved. Until the
end of 1992 the Bundesrat has not yet reached a decision.
Therefore, all statements in this paper to TA Siedlungsabfall
refer to the Government proposal.
    In the first general administrative provision requirements
have been set for storing and landfilling of waste to protect
groundwater, which are also important for landfills (6).
    The general administrative provisions of the waste act ap-
ply to the State waste authorities responsible for licensing
landfill sites - they do not apply directly to the owners and
operators. The requirements on landfills have to be laid down
in the license documents so that they become legally effective.
    The cornerstone of the Federal strategy is "integrated
waste managment" where the following solid waste reduction and
management options work together to form an effective system:
source reduction, recycling, composting, combustion, and landfilling. The Federal strategy strongly encourages the reduction
of quantity and toxicity of materials and products entering the
waste stream (source reduction) followed by recycling,
including composting.

2   TA ABFALL AND TA SIEDLUNGSABFALL
    TA Abfall - among other things- sets requirements for land-
fills for waste requiring particular supervision (besonders
überwachungsbedürftige Abfälle), TA Siedlungsabfall sets requirements for class I and class II landfills for municipal
solid waste (MSW) to improve significantly the safety of existing and future landfills.

F. Arendt, G.J. Annokkée, R. Bosman and W.J. van den Brink (eds.), Contaminated Soil '93, 1479–1486.
© 1993 Kluwer Academic Publishers. Printed in the Netherlands.

For the purpose of this paper landfills for waste requiring
particular supervision will be defined as class III landfills.
TA Abfall was promulgated on April 1st 1991. The proposal
for TA Siedlungsabfall was issued by the Federal Government on
August 28th 1992.
Definitions for wastes requiring particular supervision
have been given in the Abfallbestimmungsverordnung (waste identification ordinance) (7). These types of waste are also listed
in appendix C of the TA Abfall. In this list recommendations
are given as to which are the most appropriate disposal paths.
When the recommendation for the disposal path "MSW landfill"
was determined, the assignment values for class I and class II
landfills were not known, so that in many cases the recommended
"priority 2 disposal paths: MSW landfill" will not be available
as the assignment values for class I or class II landfills are
lower.
Requirements in the TA Siedlungsabfall are set for: waste
medicine, waste and rest waste from conctruction activities,
demolition material, biological wastes, excavation waste, night
soil, night soil sludge, garden- and park waste, household
waste, sewage sludge, market place waste, residues from sewers,
bulky waste, street reconstruction waste, street cleaning waste
and water purification waste.
Furthermore, requirements shall be considered for waste from
industrial production and waste requiring particular provision,
which are suitable for joint or similar treatment or
landfilling as municipal solid waste.

3   PRINCIPLES FOR LANDFILLING OF WASTE
The principles of class I, class II, and class III landfills are similar: they specify that: (i) every landfill has to
have a bottom liner and a cap, (ii) uncontrolled leachate
discharge into the groundwater should be avoided, (iii)
leachate generation should be prevented and (iv) that monitoring and record keeping of landfill behavior should be obligatory during operation and in the post closure phase
The safety concept of future landfills is to avoid abandoned hazardous waste sites: landfill encapsulation alone is not
sufficient Therefore, special requirements have been set in
particular for the waste to be landfilled and for site selection.
Given the requirements for site selection, requirements for
geological and hydrogeological conditions are most important.
Wastes are organised in order of priority. The requirements
for class I landfills being the most stringent.
The problems of known abandoned hazardous waste disposal
sites are obvious. Landfilling of mixtures from several,
usually untreated, organic and inorganic wastes, causes
leachate and gas emissions with a variety of constituents impossible to identify analytically, as well as settlements not
quantifiable in the long term. Abandoned hazardous waste problem sites are usually not lined at the bottom. At the end of
the operational phase they are not capped, but just covered
with topsoil which is more or less water permeable , and then
possibly revegetated. Groundwater monitoring often reveals
heavy groundwater contamination. Very often it is difficult to

assess how far, and in what direction the polluted leachate has spread, because geological structures are fissured or have displacements. This means the site geology is inhomogeneous and has only limited pollutant attenuation capacity.

TA Abfall and TA Siedlungsabfall have established the necessary requirements to ensure that with future landfills the weak points of existing landfills and abandoned dumps will be avoided.

4   REQUIREMENTS FOR LANDFILL SITING

Abandoned hazardous waste problem sites are evaluated negatively if waste has been dumped into groundwater or if leachate has built up in the waste body.

At future landfills these negative issues will be avoided. The requirements are that:

The strata has to be 1 m above the groundwater table with unconfined groundwater, or 1 m above the groundwater pressure table with confined groundwater according to DIN 4049, part 1 (Ed. Sept. 1979), after settlements have ceased due to the load of the landfill body (4, No. 9.3.3; 5; 13.3.3)

and

landfills must not be sited in pits if discharge of leachate by gravity is not possible. (5; No. 13.3.1.d)

If these reqirements are met, leachate can only be generated from rainwater infiltrating into the landfill body which should be prevented by the cap system. These standards would improve future landfills significantly compared to abandoned hazardous waste sites.

The plume of polluted leachate in groundwater is determined by the geological barrier. With the worst abandoned hazardous waste sites the geological barriers very often have totally insufficient properties. At future landfills the pollutant transport in the subsoil (which is of course unwanted and in fact not to be expected in the near future due to the landfill encapsulation) should be hindered and predictable using groundwater models. In order to achieve this, particular requirements for the geological barrier have been set.

A geological barrier is that part of the natural subsoil of a landfill, that due to its properties and dimensions is determined for pollutant transport. Basically the geological barrier is of low to very low permeable unconsolidated material (Lockergestein) or hard rock (Festgestein) (DIN 18 130) of several meters thickness and high pollutant attenuation capacity, which is spread wide over the disposal area. Should these requirements for the geological barrier under or close to the landfill site not be completely fulfilled, it has to be proved, and if occasion arises ensured, that this will not lead to an increased risk to groundwater (5, Nr. 13.3.2).

The requirements for the geological barrier will strongly influence the siting of class II and class III landfills.

For class I landfills no particular Federal requirements for the geological barriers have been established. It is expected that, due to the more stringent assignment values for

class I landfills, the leachate pollution will be very low. It
will of course be necessary to consider the hydrogeological
conditions in the area of class I landfills.

In any case the requirements should make clear that a geological barrier must not be confused with the bottom liner
system. Therefore, an insuffient geological barrier can not be
compensated by a thicker mineral layer in the bottom liner
system. The safety concept for landfills requires that _in
addition_ to the requirements for the bottom liner system requirements for the geological barrier have to be met.

If the required geological barriers for class II and III
landfills are not available in a region, waste has to be treated to meet the assignment values for class I landfills.
Otherwise, cooperation with other authorities responsible for
waste disposal is necessary to use a landfill elsewhere.

In TA Abfall (4, No. 13.3.2) the requirements for geological barriers were prone to misinterpretation. Often it was
assumed that the geological barriers could be replaced by a
constructed support layer or by an additional mineral layer in
the bottom liner system. Such was not the intention.

5   REQUIREMENTS FOR THE WASTE TO BE LANDFILLED.

In negatively evaluated abandoned waste sites organic and
inorganic waste have usually been disposed of together. The
theory behind the idea of the co-disposal strategy (if it
really could be called a strategy at all) was that of an effective bio reactor. Within the waste body hazardous and toxic
constituents in the waste should be rendered harmless and de-
toxified by physical, chemical, and biological processes.

In Germany it is felt that this "Landfill strategy" cannot
not be proved to be effective. On the contrary, some of the
major important abandoned sites such as Hamburg-Georgswerder,
Gerolsheim, Malsch, Münchehagen came into existance as a consequence of the co-disposal of organic and inorganic MSW and hazardous waste at sites without bottom liners, and at locations
with no appropriate geological barriers.

Realistic approaches as to how in the long term to hinder
release of harmful constituents out of the landfilled untreated
waste and to keep the bottom liner systems effective against a
broad spectrum of chemical attacks from leachate, not predictable in composition, are unknown. Therefore, TA Abfall and
TA Siedlungsabfall require assignment values which have to be
met for waste to be landfilled at class I, II, and III landfills (Tab. 1).

Only solid waste is allowed to be landfilled. Solid waste
is defined by the assignment values for the required waste
strength. (Table 1, No.1)

Of particular interest are the assignment values No. 2
"Organic part of the dry mass of the original substance"
(Table 1). These assignment values restrict bio chemical
processes within the landfill to a minimum. As a consequence
the organic load of the leachate, and landfill gas generation
is expected to decrease dramatically compared with existing MSW
landfills, and furthermore, long term settlement, as result of
the decay of organic waste in the landfill, will be marginal.
Restricting the eluate concentrations should ensure that only

TABLE 1   ASSIGNMENT VALUES FOR CLASS I AND CLASS II LANDFILLS
           (TA Sie, AUG. 27.1992) und CLASS III LANDFILLS)

Nr.	PARAMETER	DIMENSION	LANDFILL CLASS I		LANDFILL CLASS II		LANDFILL CLASS III	
1	STRENGTH[1]							
1.01	shear vane strength	kN/m^2	$\geq$	25	$\geq$	25	$\geq$	25
1.02	axial strain	%	$\leq$	20	$\leq$	20	$\leq$	20
1.03	uniaxial shear strength	kN/m^2	$\geq$	50	$\geq$	50	$\geq$	50
2	ORGANIC DRY MATTER OF THE ORIGINAL SUBSTANCE [2] measured as							
2.01	loss on ignition	(Mass-%)	$\leq$	2	$\leq$	5	$\leq$	10
2.02	TOC	(Mass-%)	$\leq$	1	$\leq$	3 k A		
3	EXTRACTABLE LIPOPHILIC SUBSTANCES OF THE ORIGINAL SUBSTANCE	(Mass-%)	$\leq$	0.4	$\leq$	0.8	$\leq$	4
4	ELUAT CRITERIA							
4.01	pH-value			5,5 - 12.0		5,5 -12.0		4 - 13.0
4.02	conductivity	$\mu S/cm$	$\leq$	6,000	$\leq$	50,000	$\leq$	100,000
4.03	TOC	mg/l	$\leq$	20	$\leq$	100	$\leq$	200
4.04	phenols	mg/l	$\leq$	0.2	$\leq$	50	$\leq$	100
4.05	arsenic	mg/l	$\leq$	0.1	$\leq$	0.5	$\leq$	1
4.06	lead	mg/l	$\leq$	0.2	$\leq$	1.0	$\leq$	2
4.07	cadmium	mg/l	$\leq$	0.05	$\leq$	0.1	$\leq$	0.5
4.08	chromium VI	mg/l	$\leq$	0.05	$\leq$	0.1	$\leq$	0.5
4.09	copper	mg/l	$\leq$	1	$\leq$	5	$\leq$	10
4.10	nickel	mg/l	$\leq$	0.2	$\leq$	1	$\leq$	2
4.11	mercury	mg/l	$\leq$	0.005	$\leq$	0.02	$\leq$	0.1
4.12	zinc	mg/l	$\leq$	2	$\leq$	5	$\leq$	10
4.13	fluoride	mg/l	$\leq$	5	$\leq$	25	$\leq$	50
4.14	ammonium	mg/l	$\leq$	4	$\leq$	200	$\leq$	1,000
4.15	chloride	mg/l	$\leq$	500	$\leq$	5,000	$\leq$	10,000
4.16	cyanide, readily rel.	mg/l	$\leq$	0.1	$\leq$	0.5	$\leq$	1
4.17	sulphate [3]	mg/l	$\leq$	500	$\leq$	1,400	$\leq$	5,000
4.18	nitrite	mg/l	$\leq$	3	$\leq$	6	$\leq$	30
4.19	AOX	mg/l	$\leq$	0.3	$\leq$	1.5	$\leq$	3
4.20	watersolubles	(Mass-%)	$\leq$	3	$\leq$	6	$\leq$	10

1) 1.02 can be used together with 1.03 equivelant to 1.01. They are minimum requirements.
The strength of waste must be as calculated necessary for the landfill stability.
2) 2.01 can be used equivelant to 2.02; the assignment values are not required for contaminated soil, disposed of at a mono-landfill.
3) can be exceeded up to 1400 mg/l SO_4 for gipsum demolition waste and rest of gipsum construction material, as well as other demolition waste containing gipsum if the Ca-conzentration in the eluat is at minimum 0,43-times of the SO_4-conzentration.

waste from which the most important constituents will be
leached only in small concentrations will be landfilled. The
recommended leaching procedure is according DIN 38 414-S4.
    If the assignment values are exceeded, the waste has to be
treated before landfilling to achieve the required values.(Table 1)
    However, it will be possible to landfill waste even if the
assignment values are exceeded. Waste then has to be disposed
of at mono-fills or mono-sections at class I, class II or
class III landfills. For waste requiring particular supervision
it must be shown that the mono-landfill will not be more
disadvantageous than a landfill meeting the "normal"
requirements.
    With TA Siedlungsabfall the disposal of waste at a mono-
landfill is recommended if, due to the hazardous and toxic
constituents in the waste or the chemical binding of the
constituents in the waste, a mobilization of the constituents
and disadvantageous reactions of this waste with other waste is
not possible.
    It can be expected that the landfill behaviour at future
landfills will be much better than that of abandoned waste
dumps, and also much better than existing landfills. Landfill
gas emissions are not expected. Settlements will only be
marginal if the waste is highly compacted within the landfill
body. The organic load of the leachate will be low.

6   MONITORING LANDFILL BEHAVIOR
    For abandoned disposal sites in general, and also with many
existing landfills, it is usually not certain exactly , where,
what and when waste has been dumped or landfilled. The time dependent development of leachate and landfill gas generation as
well as of settlement is also usually unknown.
    For future landfills the operator or a commisioned consul-
tant has to prove that (i) the requirements for landfill behavior established in the landfill license document are fulfilled,
(ii) the landfill operation has been carried out as agreed, and
(iii) the effectiveness of the landfill bottom liner and cap as
well as the groundwater monitoring wells is guaranteed.
    Monitoring facilities have to be available and checked
regularily for their functionability. Required are:
    - groundwater monitoring wells
    - devices to monitor settlement and deformation of the
      landfill body
    - devices to monitor settlement and deformation of the
      landfill bottom liner system and the cap system
    - devices to measure meteorological data
    - devices to measure the quantities of water necessary to
      calculate water balances
    - devices to monitor temperature at the landfill bottom
      liner.
    Monitoring by the landfill operator or a commissioned con-
sultant has to be carried out during the operation period as
well as through out the post-closure care period of the
landfill.

*The landfill body behavior has to be documented by
record keeping of the leachate amount and composition
as well of the landfill gas emission and settlement, if
relevant. On the basis of the evaluation of the results
of measurement according to appendix G of the TA Abfall, a declaration for the landfill behavior has to be
written and annually presented to the responsible
authorities. In this declaration of the landfill behavior the time dependent development has to be shown and
to be compared with the assumptions for the landfill
body behavior ... and, if relevant, with the assumptions made in the license document for the amount and
composition of leachate and landfill gas.*

The annual declarations for the landfill behavior have to
be considered at the end of landfill operation in a final assessment, to decide upon the post-closure care measures on the
basis of reliable records compiled during the operational
phase.

The TA Abfall and TA Siedlungsabfall requirements on record
keeping during landfill operation and on the development of the
landfill body behavior - as well as the evaluated results of
landfill monitoring - should help to ensure that there is adequate information on (i) the landfill operation, (ii) the waste
landfilled, (iii) the encapsulation and (iv) the geological
barrier, which is necessary for environmental impact assessment
in future times.

7   FINAL REMARKS

If the requirements established in TA Abfall (4) and
TA Siedlungsabfall (5) are to be fulfilled in practice, the
result will be that future generations will not see our new
(future) landfills as hazardous waste problem sites, which
cause severe environmental hazards.

The most important condition for this assessment is that
the assignment values for waste landfilled on class I, II, and
III landfills will be met.(9)

The assignment values for many types of waste can be met
only after treatment before landfilling. To observe the
assignment value No. 2 of appendix D of TA Abfall, and No. 2 of
appendix C of TA Siedlungsabfall, it will be necessary to treat
many types of waste in thermal processes, usually incineration.
The opposition of groups of environmentalists and public
pressure groups against waste incineration plants is very
strong. The threshold values of the 17. BImSchV (8), however,
guarantee that there are no expected detrimental health effects
due to emissions from incineration plants.

For the disposal of household and other municipal solid
waste, biological treatment prior to landfilling (after separation of recyclable fractions) is proposed as an alternative
treatment process by groups in opposition to incineration. The
biologically treated rest-waste is said to meet the assignment
values for class II landfills, except for the parameter No. 2
"Organic part of the dry matter of the original substance". For
this parameter other more suitable criteria are proposed to
describe biological inactivity.

These concepts for biological pre-treatment of waste before
landfilling - if at all practicable - will generate landfills
with much better landfill behavior than that of existing
MSW landfills. However, it will be necessary to separate those
types of waste not suitable for biological treatment. The waste
separated has also to be treated to meet the assignment values
for the class I and class II landfills.

The categorical opposition against waste incineration is
seen to be not justified and unreasonable by the Federal Government. Many types of waste requiring particular supervision
with organic constituents must be incinerated to avoid environmental damage, and many types of waste generated or excavated
during remedial action on contaminated land can only reasonably
be incinerated. If there were no incineration capacity, thorough investigation and subsequent excavation of waste and of
contaminated soil would be unnecessary. Such sites would better
be encapsulated with slurry walls and caps, until the necessity
for decontamination becomes more urgent, or the opposition
against thermal treatment decreases. Our generation will
probably save money, but future generations will have to strug-
gle with the burdens we have left, and will have to spend at
least the money we have saved.

8   BIBLIOGRAPHIE

(1)   Der Rat von Sachverständigen für Umweltfragen (1989).
Altlasten. Sondergutachten Dezember 1989. Stuttgart: METZLER-
POESCHEL
(2)   Brandt, E. (Ed.) (1993). Altlasten: Bewertung,
Sanierung, Finanzierung. 3. neu bearbeitete und erweiterte
Auflage. Taunusstein: Blottner
(3)   LAGA (1989). Erfassung, Gefahrenbeurteilung und
Sanierung von Altlaten - Informationsschrift - .
Länderarbeitsgemeinschaft Abfall. Berlin: Erich Schmidt Verlag.
(4)   Anonym (1991). Gesamtfassung der zweiten allgemeinen
Verwaltungsvorschrift zum Abfallgesetz (TA Abfall), Teil 1:
Technische Anleitung zur Lagerung, chemisch/physikalischen,
biologischen Behandlung, Verbrennung und Ablagerung von
besonders überwachungsbedürftige Abfällen vom 12. März 1991,
GMBl, 42 Jg. (1991) Heft 8, Seite 139, Köln, Carl Heymanns,1991
(5)   Dritte allgemeine Verwaltungsvorschrift zum
Abfallgesetz, (TA Siedlungsabfall), Technische Anleitung zur
Vermeidung, Verwertung, Behandlung und sonstigen Entsorgung von
Siedlungsabfällen; Kabinettsbeschluß vom 27.08.1992,
Bundesratsdrucksache 594/92
(6)   Erste allgemeine Verwaltungsvorschrift über
Anforderungen zum Schutz des Grundwassers bei der Lagerung und
Ablagerung von Abfällen vom 31. Januar 1990. GMBl. S. 74
(7)   Siebzehnte Verordnung zur Durchführung des Bundesim-
missionsschutzgesetzes (Verordnung über Verbrennungsanlagen für
Abfälle und ähnliche brennbare Stoffe - 17. BImSchV) vom 23.
Nov. 1990 (BGBl.I, S. 2545, ber. S. 2832).
(8)   Verordnung zur Bestimmung von Abfällen nach Para 2
Abs. 2 des Abfallgesetzes (Abfallbestimmungs-Verordnung -
AbfBestV)
(9)   Stief, K. (1992). Siedlungsabfalldeponie - Quo vadis?
Abfallwirtschaftsjournal 4. Berlin: EF-Verlag für Energie- und
Umwelttechnik GmbH

LANDFILL RECONSTRUCTION INVESTIGATIONS AT THE SCHÖNEICHE AND SCHÖNEICHER PLAN LANDFILLS NEAR BERLIN

Dipl.Ing. Gregor Heckenkamp

ITU - Ingenieurgemeinschaft Technischer Umweltschutz GmbH, Berlin, Germany

1. ABSTRACT

This paper presents a comprehensive research project for landfill reconstruction investigations, funded by the Bundesministerium für Forschung und Technologie (BMFT). Two landfills in the vicinity of Berlin have been chosen in which various waste excavation and treatment methodologies will be tested. The conditions of the landfilled waste as well as the potential use of common methods for volume reduction and decontamination of hazardous materials will be evaluated. In parallel, occupational and environmental protection measures within the frame of this project will be tested in the field and evaluated.

2. INTRODUCTION

Available landfill space is a decreasing resource, especially in those that are near densely populated areas. The landfilling practices of the past have tended to squander this valuable resource. For example, reusable and recyclable materials, demolition waste, and hazardous waste were landfilled without control.

Current requirements regarding landfill technologies (such as gas and leachate collection systems) are a result of increasing knowledge of the danger and explosiveness of this type of uncontrolled mixed waste. These technologies were unthinkable in the 1970's. Landfills located in former clay and gravel pits as well as unlined landfills (which were common in the past) are in many cases contaminated sites.

Adequate amounts of hazardous materials exist in a household waste landfill that could render groundwater unfit for human use. Gases that are produced in the landfill are comprised of many toxic components and also contribute to Global Warming.

Decontamination measures that have commonly been undertaken on such problem landfills are in actual fact reactive security measures that do not necessarily deal with the source of the problem. The most common security measure is addition of constructed or hydraulic barriers. The prime purpose of these

F. Arendt, G.J. Annokkée, R. Bosman and W.J. van den Brink (eds.), Contaminated Soil '93, 1487–1495.

barriers is the minimization of water and atmospheric contamination. The results in unpredictable long term risks since decontamination of the landfill has not been carried out.

The landfills Schöneiche and Schöneicher Plan which have been chosen for this research project are located in the outskirts of Berlin, in the State of Brandenburg. Their histories provide an interesting and relevant framework for the project. The Schöneiche landfill has been operating since 1977 and accepts wastes solely from West Berlin. The Schöneicher Plan landfill on the other hand, contains waste from East Berlin and Potsdam. Thus both landfills are considered to be typical of the old and new German states. Furthermore, both landfills are not lined and as a result of their significant sizes are considered to necessitate decontamination and reconstruction.

3. PROJECT GOALS

The most important goals of the research project as seen in Figure 1 are:

- long term security of the landfills
- reduction of hazardous components
- reconstruction using current technology
- increase of available landfill space

A comprehensive decontamination program will be developed in which current technologies will be used for waste excavation, treatment and reintroduction.

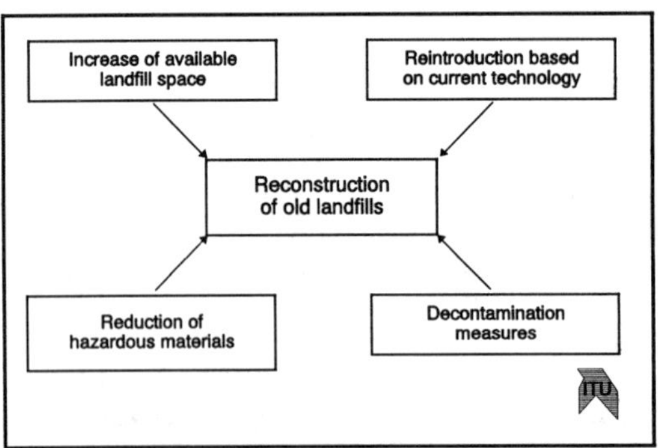

Figure 1: Project goals

The main purpose of treating the waste is to remove hazardous and recyclable materials. Special consideration will be given to creating a residual waste stream which may be reintroduced to landfill based on the upcoming household waste management guidelines for Germany (TA Siedlungsabfall). Landfill space is made available through volume reduction by sorting and treating previously landfilled waste.

All methods for treating residual waste currently discussed will be considered. These include: mechanical separation, incineration and biological treatment.

Security measures can be introduced into the landfills to bring them to current technical standards once the waste has been excavated.

4. PROJECT PROCESS

A multi-phase process was developed which is shown in Figure 2.

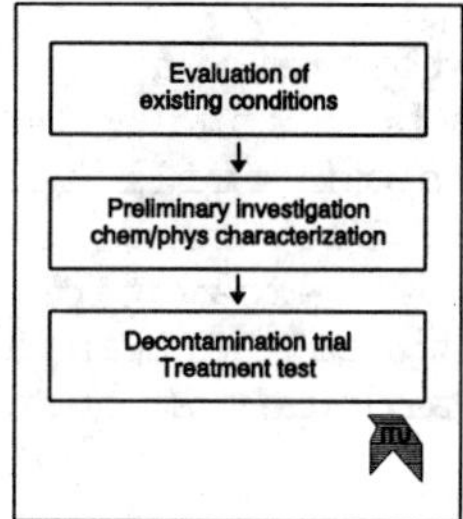

Figure 2: Program phases

Each of the phases is described in detail below:

4.1 <u>Analysis of historical data to evaluate the existing situation</u>

The goal of the historical evaluation is the reconstruction of the landfill process in terms of waste types and when they were deposited. This is especially important in determining the appropriate occupational health and environmental protection measures to be implemented during waste excavation. Figure 3 show the types of research that were carried out. The following issues were to be clarified in the evaluation:

- types of wastes that were deposited (quantities, composition)
- geographic concentrations of hazardous materials
- whether specific waste types (demolition wastes, sewage sludge) were deposited separately

1490

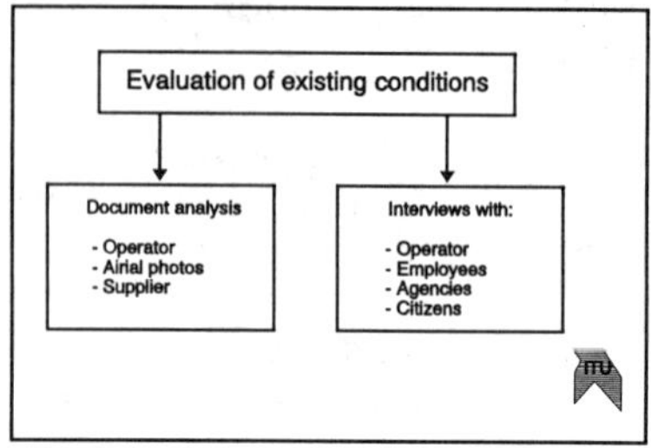

Figure 3: Evaluation of existing situation (Phase I)

4.2 Preliminary investigations - gas analysis and phys/chem characterization of materials

Based on the historical evaluations, preliminary investigations will be carried out. The primary goal of these investigations is the determination of appropriate areas for the decontamination trials. Furthermore, the landfilled materials will be characterized and an assessment will be made as to the degree of hazardous contamination. The results will be used in the development of a waste excavation and treatment plan. Appropriate occupational health and environmental protection measures will also be determined at this point. The steps to be undertaken in the preliminary investigations are shown in Figure 4.

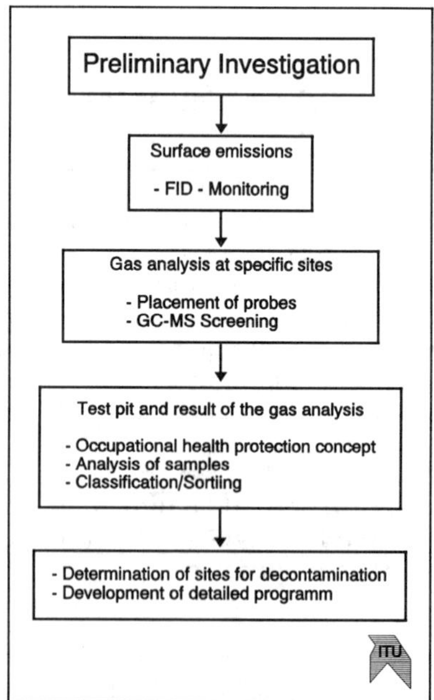

Figure 4: Preliminary investigations process (Phase II)

Landfill gas emissions concentrations are measured at the landfill surface. The results provide an indication of the biological activity in the landfill.

Gas samples will be taken in specific areas using specialized gauges. The results of the analysis of these samples will provide information regarding the landfill contents, biological activity and hazard potential of the landfill.

Following the evaluation of the results, comprehensive physical/chemical analyses will be carried out on samples of landfilled wastes that have been removed from specific areas.

4.3 <u>Decontamination trial / treatment tests</u>

Decontamination trials will be carried out in those areas of the landfill which have been preliminarily investigated. Special emphasis will be placed on the occupational health and environmental protection measures. Waste removal, monitoring, preparation and treatment processes will be tested. The goal of these tests is to produce a residual waste stream that meets the draft guidelines for household waste (TA Siedlungsabfall).

The excavated waste will be prepared to provide specific waste streams comprised of certain materials. These waste streams will be allocated to appropriate treatment measures based on a specialized on-site coarse analysis. Figure 5 illustrates a possible decontamination trial process.

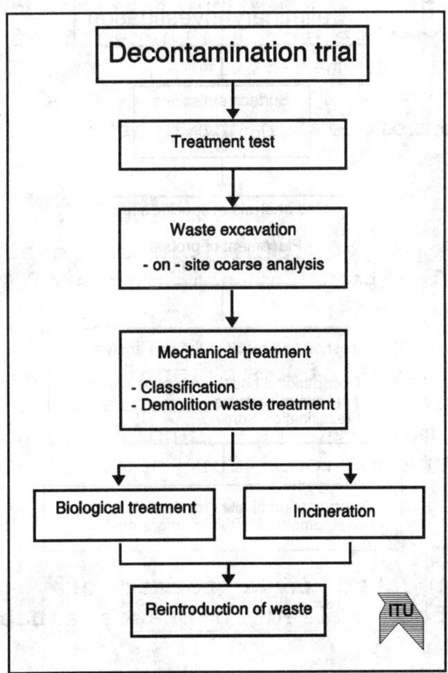

Figure 5: Decontamination trial process (Phase III)

The goal of this trial is the evaluation of the following points in relation to the preliminary design of a pilot treatment facility for excavated landfill wastes:

- waste excavation processes and equipment
- occupational health and environmental protection measures in landfill excavations
- physical/chemical conditions of the wastes
- mechanical on-site waste preparation
- use of waste treatment processes to produce residual waste able to meet guidelines (TA Siedlungsabfall)
- on-site coarse analysis of samples
- monitoring of decontamination process

The results of the decontamination trial will be used in the development of a preliminary design of a pilot facility for the treatment of previously landfilled wastes.

This research project is funded by the BMFT and the city-state Berlin. It is directed by the Umweltbundesamt and is being carried out by the following firms:

- **ITU - Ingenieurgemeinschaft Technischer Umweltschutz** is the scientific advisor of the project. ITU is also responsible for the analytical portions of the project and the composting trials.

- The **Gesellschaft für Umweltverfahrenstechnik und Recycling (UVR)** is responsible for all aspects of waste preparation and analysis of raw materials

- **Züblin** is responsible for the construction measures and the incineration trial.

5. RESULTS
Phase I of the project (historical evaluation) has been completed. The preliminary investigations of Phase II have begun at the time of writing.

The existing conditions of both landfills were evaluated based on in-depth research through documents and interviews. Aerial photos infra-red maps of the landfills were evaluated. The results of the evaluations for each of the landfills are described below.

Schöneicher Plan
The landfill is located in a former clay pit from the 19th century that has been filled in phases since the beginning of this century.

The landfill has been the central landfill for East Berlin and

Potsdam since the 1950s. Household wastes, ashes and demolition wastes are accepted for disposal in the landfill. Industrial hazardous wastes of a certain class[1] were allowed to be deposited based on special certification until the spring of 1990.

30 million m^3 of waste are contained in the landfill which has a total area of circa 120 ha. Based on the remaining landfill space, it is estimated that the landfill will close by 2010.

The phases of landfilling were able to be reconstructed based on the evaluation of aerial photos.

Much of the waste was delivered by rail, whereby the hazardous materials tended to be delivered by trucks. These hazardous wastes were deposited in "hazardous areas" which are shown in the existing landfill mapping.

The waste composition was difficult to determine, as the waste was not weighed upon delivery. Data regarding the commercial waste composition do not exist; in some cases, only partial handwritten lists were available. However, the waste composition is shown in Figure 6.

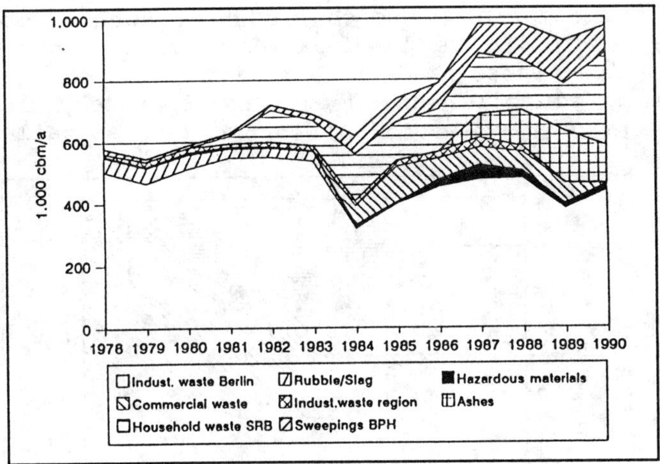

Figure 6: Time line of waste deposition at Schöneicher Plan

Most of the deposited waste was industrial (mainly from the rail construction industry) until the early 1980s. From then increasing amounts of household waste, sweepings and ashes from power plants were disposed of in the landfill. The delivered

---

[1] Differences were made between "toxic" and "other contaminant-containing materials in the pertinent law. The Schöneicher Plan landfill accepted wastes with lower hazard potentials.

1494

waste quantity increased from approximately 600 thousand m³ in
1980 to nearly 1 million m³ in 1990. Since much of the household
waste in the GDR was composed of ashes, the landfill is
considered to contain mainly ashes and inert materials.

Approximately 214,000 Mg of hazardous materials were deposited
between 1977 and 1989. This corresponds to 1 - 5 % of the annual
delivered waste quantity. Based on the existing data, it is very
difficult to classify the wastes according to the German waste
catalogue. However, according to the required waste management
practices from the TA Abfall, a large proportion of the waste
would need to be dealt with in hazardous waste management
facilities. Approximately half of these wastes consist of
contaminated soils, demolition wastes and aluminium slag. About
15 % are slags from the metal industry and a further 15% are
mineral oil wastes and wastes from the hydrocarbon industry.

Schöneiche
This landfill began operation in 1977. The base of the landfill
was covered with a 1 m layer of demolition waste prior to the
deposition of other wastes. A gas collection system was
installed on part of the landfill once the landfill was
operating.

Household and demolition wastes from West Berlin were disposed
of in the landfill. The landfill standard is typical of West
German municipal waste landfills at that time. Demolition wastes
and contaminated soils were deposited separately in a central
area. Sewage sludge was disposed of in so-called "sludge
basins".

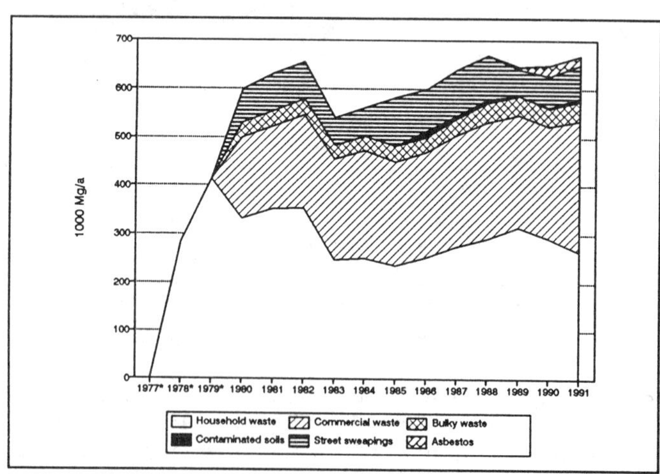

Figure 7: Time line of waste deposition at Schöneiche

A total of about 15 million Mg of waste were landfilled until
1990. The household waste quantities remained fairly constant

over this period at approximately 600,000 Mg/a, while the demolition waste quantities fluctuated between 200,000 and 900,000 Mg/a. The composition of the household waste could be determined quite accurately using the existing weigh scale data. However, problems may arise during excavation activities with regards to hazardous wastes that have been mixed illicitly with the household wastes and asbestos that may be contained in the demolition wastes.

6. EVALUATION AND FUTURE WORK

The historical evaluations provided an initial overview of the existing conditions of the landfills. With regard to the remaining steps in the project, the following issues need to be considered:

The potential contact with hazardous and gas-producing wastes needs to be considered seriously in areas of the Schöneicher Plan landfill. It is especially important to determine homogenous areas that are appropriate for the decontamination trial by means of the gas analyses. The excavation of a hazardous waste area is only possible if stringent precautionary measures are taken. The potential for biological activity in the landfill is small as a result of the large proportion of inert materials.

In contrast, the Schöneiche landfill consists of very homogeneous areas in which the biological activity will be related to the age of the deposited material. Hazardous materials are expected as point sources and in small quantities only.

At the time of writing, gas analyses were being carried out to determine specific sites for the decontamination trials. The decontamination trials are planned to commence in the summer of 1993.

ENVIRONMENTAL PROVISION AND DUMP PLANNING

LIEBER, M., STOLPE, H.

1. SUMMARY

When planning dumps security measures are considered with the goal to avoid that harmful substances will be spread out into environment. Dump engineering as well as dump operation are to be understood as part of the protection from danger. Before starting dump planning effective environmental provision should be taken into account with regard to waste avoidance and the reduction of harmful substances. Waste treatment and dump engineering are necessary as well as supplementary measures for an effective environmental provision. A reduction of ecological damages caused by dumping of non-avoidable and non-recyclable waste can be achieved by applying Environmental Impact Studies (EIS) to the entire planning procedure of a dump (dump site search, comparative EIS within the development planning procedure, project-related EIS within the planning permission procedure).

2. THE CONTRIBUTION OF DUMP PLANNING TO AN EFFECTIVE ENVIRONMENTAL PROVISION

Dumping is the last link in a chain of waste management measures. Therefore, it is useful and necessary to look into the question where are the main starting points for provision of pollution and/or for the environmental provision in the waste sector and which task dump planning will have under the headline "Prevention of Pollution".

2.1. Production and Waste Management

The main starting point for environmental provision in the waste sector is situated on the level of production as well as waste avoidance and waste recycling.

Entrepreneurial decisions on products, on waste caused by production as well as on the contents of harmful substances, the re-usability and recyclability of the products after beeing used will set decisively the course with regard to environment.

An ecological orientated restructuring will be necessary for an effective environmental provision. That is the only way to achieve an effective reduction of waste quantities and of the harmful substances to be encapsulated in dumps. Moreover, the increasing problem of emission as well as the consumption of ressources and energy of waste utilization and treatment plants can be counteracted. Both, product responsibility of the enterprises and legal measures to be taken as well as a detail waste management

---

Authors' address:
Dipl.-Ing. Manfred Lieber, Dr. Harro Stolpe, AHU - Büro für Hydrogeologie und Umwelt GmbH, Kirberichshofer Weg 6, D-5100 Aachen

F. Arendt, G.J. Annokkée, R. Bosman and W.J. van den Brink (eds.), Contaminated Soil '93, 1497–1502.

conceptof the cities and administrative districts which are obliged to take actions for waste disposal are essential conditions for an effective environmental provison in the waste sector.

Concerning the production detail lectures will be hold during the meeting, as referred herein.

2.2. Dump Sites

Also a society which is obliged to the principles of an effective environmental provision and to the economic principle of circulation will need plants for the storage of remaining waste. Even then dump sites are to be localized which should be comparatively small interventions in nature and landscape and should have possibly small risks and effects on human and environment.

As Germany is a quite intensive used and densely populated country dump sites will always be in competition with still existing uses and plannings. There is no dump site in Germany which can exclude a painful intervention in nature and landscape as well as effects on human - even if they are only of subjective manner.

Today, the principle of conflict minimization is persuaded to localize such undesired sites for dumps. This means that dumps may be located in low populated, low sensible areas considering ecological aspects. The result of such a planning procedure is the distribution of ecological damages on the entire area with regard to a long-term period.

Keeping this in mind, experts are discussing whether the principle of conflict minimization can be combined with an effective environmental provision.

2.3. Dump Engineering/Operation of Dumps

Protection against danger are the essential measures of dump engineering and the operation of dumps. The main tasks of dumps are to encapsulate the existing potential of harmful substances from environment as long as possible, as safe as possible by realizing the multibarrier principle.

The result of learning processes caused by heavy ecological damages due to inadequate former dump engineering is the realization of an essentially improved security standard for new dumps taking the multibarrier principle into account.

But, it should be borne in mind, that all new devolped dump sealing, controlling and security systems can only be used for a limited period of time. Taking into consideration that the effects of dumping are unlimited, the dumping-related environmental problems will be transferred to future generations. This can only be counteracted by regular controlling and repairing (e.g. surface sealing).

Today, waste treatment is the main emphasis of the reduction of risk potential of the waste to be encapsulated - be it thermical or mechanical-biological treatment. This reduction of risk potential of waste required by dump security is nevertheless combined with emissions and other environmental harness. But this can be counteracted by a technical high standard of filter and other controlling measures. That is why the target of waste treatment as well as the one of dumping is mainly the protection of danger.

3. ENVIRONMENTAL IMPACT STUDIES CARRIED OUT FOR ENVIRONMENTAL PROVISION

As to the law for the Environmental Compatibility Test (ECT) of February 12, 1990 the ECT should be regarded as instrument for an effective environmental provision. According to the ECT law Environmental Impact Studies (EIS) are to be performed for dumps. Experts perform Environmental Impact Studies (EIS) and/or Environmental Compatibility Assessments (ECA) as qualified basis for Environmental Compability Tests (ECT) conducted by official authorities.

The ECT law essentially refers to the project-related EIS e.g. to individual planned dumps. The explanation in para 2 shows that an effective environmental provision should be set up prior to the contemplation of a single plant.

The following should be mentioned:

- decree for waste avoidance 14 waste law,
- development and production of less harmful products causing less waste,
- further measurements for waste avoidance,
- a waste management concept should be made.

The ECT law does not include the conditions for an effective environmental provision with regard to the waste sector. Considering this, the EIS can only partly meet its requirements. In addition, the legislator requires for the EIS an orientation towards the legally related evaluation criteria (e.g. the limit values of the German TA Luft). But limit values are orientated towards the protection of danger. To conduct an effective environmental provision partly more severe standards are necessary.

Thus, the possibilities of the EIS refer mainly to the avoidance and reduction of effects and risks caused by dumps with regard to the dump site search as well as to planning, technical and operational measurements on a direct project planning level.

The systematic application of Enviromental Impact Studies with regard to the dump planning is described below. Furthermore, hints concerning the method and the investigations to be carried out are given.

4. SYSTEMATIC APPLICATION OF ENVIRONMENTAL IMPACT STUDIES FOR DUMP PLANNING

4.1 Overview on structure and procedure

Considering the area where waste disposal has to be performed a regional localization of possibly suitable sites is to be conducted. This should result in a choice of several possible sites which will be evaluated in the next step.

A comparative EIS will be performed in the second step. The goal of the EIS is to determine a suitable environmentally acceptable site for which an development planning procedure must be performed in accordance with § 6a of the development planning law.

Then planning permission documents for the determination of the site and a project-related EIS are to be prepared according to the ECT law (third step). On this basis the formal planning permission procedure should be carried out.

When the planning permission procedure has been completed the dump will be constructed. The construction should be supervised in accordance with the recommendation of the EIS. Moreover, it is essential to monitor the dump in operation and to monitor the after care measures (control). In order to observe possible effects on the environment, to detect defects and, if necessary, to take appropriate measures to minimize or avoid harzardous effects on environment.

The three steps (mentioned above) of the EIS-procedure will be discussed in further detail. They have different goals, serve different functions and require specific methods. Figure 1 shows a summary of the entire process and the essential methodically basis.

4.2. Environmental Impact Studies Performed within the Framework of Dump Site Search

The procedure conducted for dump site search will already in an early stage take environmental problems into account. In this first step the entire area where waste disposal has to be conducted will be investigated in several steps which refer to each other. Firstly the investigation refers to such areas which can be excluded for the construction of a dump. Then more detail investigations and evaluation with regard to existing restrictions are to be performed for the remaining areas. The result of this first step is a bigger number of possible suitable sites.

A map scale of 1:25.000 and 1:50.000 applies to the dump site search. At this stage existing geological and topographical maps and related documents are compiled. It is assumed that the site will have negative effects on the environment and that these spread out evenly in all direction. For this reason the distances stated in the relevant list of criterias must be equal in all directions from the investigated sites.

4.3. Comparative Environmental Impact Studies for Possible Suitable Sites

The Comparative EIS refers to such sites which are possible suitable sites. A site should be determined which on the one hand meets the minimum requirements of the dump sites (e.g. existing of a geological barrier). On the other hand, the site should be located comparatively suitable with regard to the risks and effects on human and nature.

The planning permission procedure will be initiated for such a site after legal action. According to the planning law 6a the EIS is part of the planning permission procedure. In case of an appropriated conception the documents of the comparative EIS can be used.

A map scale of 1:5000 and 1:25000 applies to the comparative EIS. Here, the information retrieved from maps and documents during the preceeding processing state are no longer sufficient.

The information basis necessary for the EIS requires the development of a first technical design of the dump as well as an estimation of the emissions to be expected. Further information only yield by field investigations are necessary. The investigation program only serves for orientationally purposes. Here, great importance is to be drawn to the investigation of the subsoil conditions.

On the basis (estimate of emission, orientational investigations) a rough statement can be made on the effects of dumps. The exactness of the statement is sufficient for a site comparison and the preparation of documents for the development planning procedure.

4.4. Environmental Impact Studies with Regard to the Planning Permission Procedure

The following tasks are to be met within the framework of the systemically application of Environmental Impact Studies with regard to dump planning on the level of the project-related EIS introduced herein:

- supplementary control whether the minimum requirements are met on a detailled information basis,
- extensive investigation, description and evaluation of the effects of the planned dump on human, nature, environmental media as well as on material and cultural values.

A map scale of 1:500 and 1:5.000 applies for the project-related EIS. In the project-related EIS detailed prospects with regard to the effects of the dump (emission, effects on nature) on human, environment to be expected as well as risk estimations are made on the basis of a direct dump planning and extensive local investigations. These refer on the one hand to the construction, operation and after care phase of a dump and on the other hand to the risks of normal operation and malfunctions.

In the project-related EIS a close agreement between the EIS experts and the dump experts is necessary in orientation towards the technical planning steps (preplanning, concrete planning, permission procedure). Goal is the optimization of the technical planning with regard to the avoidance and reduction of risks and effects on human and environment. At the end of the project-related EIS a statement can be made with regard to the remaining risks and effects of a dump on human and environment.

5. RESULT AND PROSPECT

- The explanations have shown that within the framework of dump planning a reduction of the risks and effects of a dump on human and environment can be achieved by applying Environmental Impact Studies to dump planning.

- The development of environmental quality goals which refer to special areas is the starting point for a higher consideration of the environmental provision in Environmental Impact Studies. Environmental quality goals can be used as a more improved basis for evaluation standards concerning EIS. The blueprint of the administration regulation of the ECT law contains starting points for the above-mentioned problem, but which are to be concreted and supplemented. The result could be a planning orientated towards the principles of keeping and developing of still existing quasi-natural or cultural landscapes.

- Due to the basic conditions of waste management and development planning the possible number of sites for dumps in Germany will soon be exhausted.

- The determination of necessary dump sites can be facilitated today and in future and can be more easily realized from a political and planning point of view if quantity and risk potential of waste are minimized.

- The main starting points for an effective environmental provision in the waste sector are situated prior to the dump site determination and dump planning. Effective environmental provision should above all start on the level of production and waste avoidance. Here, legal measures are also required as an ecological orientated restructuring of production. Without setting the course appropriately the environmental problems caused by waste cannot be managed.

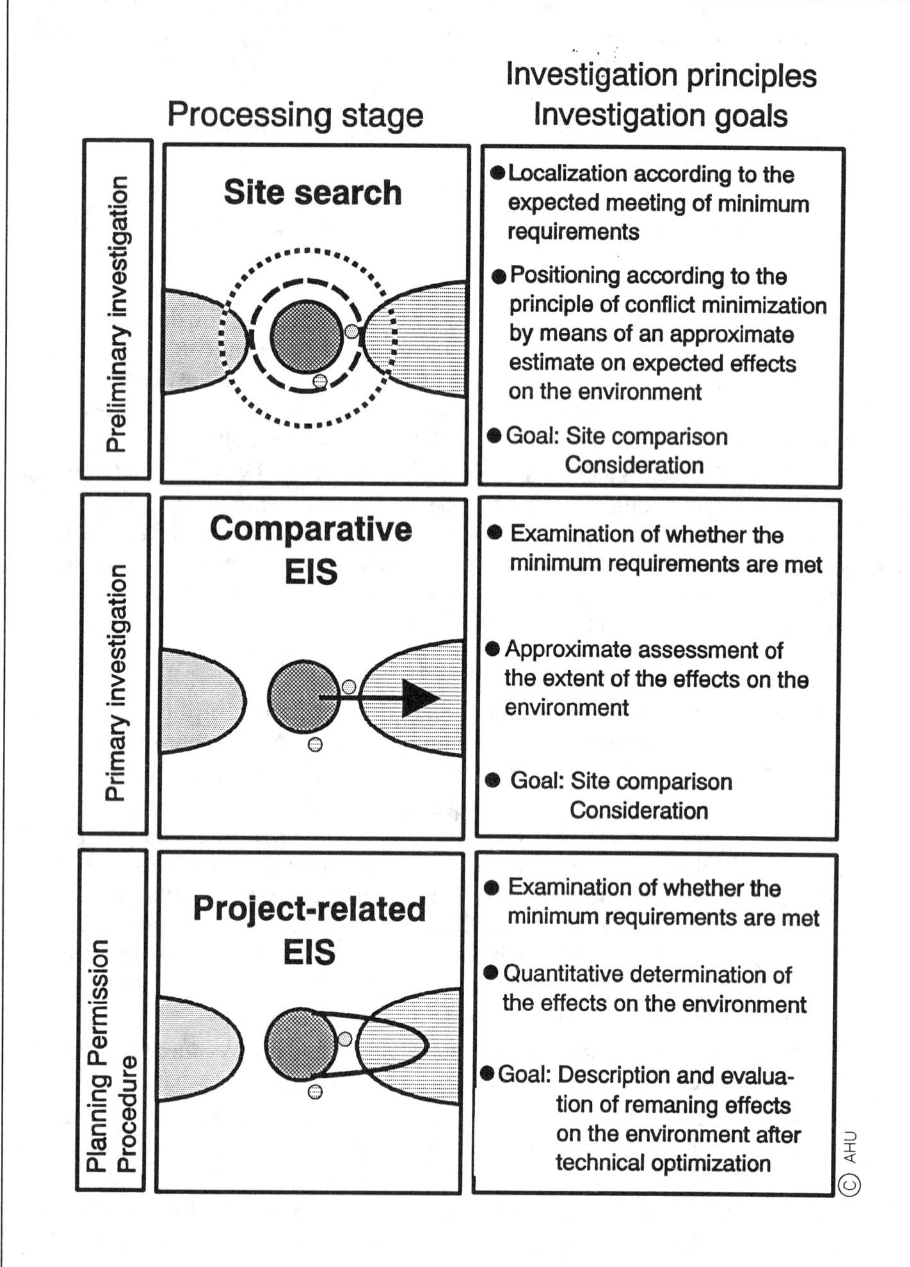

Figure 1: Systematic Application of Environmental Impact Studies for the Dump Planning

N-fertilization under consideration of ecological aspects

Prof. Dr. habil. Martin Körschens, Centre of Environmental
Research Leipzig-Halle,Division of Soils Research,
Hallesche Straße 44, O - 4204 Bad Lauchstädt, Germany

Summary

Nitrogen belongs to the most important but also to the
difficultest plant nutrients and simultaneous it is cause of
spreaded detrimental effects. In the last decades fertilizer
use as well as immission have been increased considerably.
The surpluses of N-balances in Germany amount about 100 kg.
$ha^{-1}.a^{-1}$. Insufficient possibilities to quantify the N-
release from the soil are among other things cause of raised
N-losses. There is a demand for research concerning the
further clearing up of the nitrogen cycle.

Nitrogen belongs to the most important but also to the
difficultest plant nutrients and simultaneous it is cause of
spreaded detrimental effects. The use of mineral N-fertili-
zer has been increased since 1950 from 26 $kg.ha^{-1}.a^{-1}$ to 134
$kg.ha^{-1}.a^{-1}$ and diminished afterwards only to 115 $kg.ha^{-1}.$
a^{-1}. In the same period the livestock increased, to some
extent on the basis of noticable feedstuff imports and as
consequence the reversion of organic fertilizer is grown.
The increasing concentration of the animal production caused
additional problems of organic manure distribution and led
to partial overdressing. As result of this development the
surplus of the N-balances in Germany amounts about 100
$kg.ha^{-1}.a^{-1}$ (FINCK, 1990; KÖRSCHENS, 1986; KÖSTER ..., 1988;
STURM ..., 1989). Thereby the N-losses are equivalent to
about 80 per cent of the quantity applied by mineral fertilization.

Comparable values were determined in other countries too. In

F. Arendt, G.J. Annokkée, R. Bosman and W.J. van den Brink (eds.), Contaminated Soil '93, 1503–1510.

Poland the N-surplus amounts 90 $kg.ha^{-1}.a^{-1}$. In the Netherlands the balance shows a surplus of more than 400 $kg.ha^{-1}.a^{-1}$ by including of all storage losses of the organic manure. The named figures represent average values which differ from one another in special case considerably. There are already high variations around between 30 and 170 $kg.ha^{-1}.a^{-1}$ among the several lands of Germany. However in the end the balance of a single farm or a single field is of importance for the environment. There the surplus can amount at least temporary and they cause high stresses for the environment. BECKER (1989) calculated for several farms in dependence of the land use system surplus balances between 137 and 440 $kg.ha^{-1}.a^{-1}$.

The extent of the gaseous N-losses and their dependence of physical, chemical and biological soil properties is up to now insufficiently known. Improper fertilizer use, unfavourable soil conditions and excessive humus content can cause gaseous N-losses up to 100 $kg.ha^{-1}.a^{-1}$.

The resulting stresses for the drinking water with nitrate and for the atmosphere with nitric oxides require also in view of soil protection compelling steps for the reduction of N-losses.

Causes of high N-losses are:
- Disregard of known principles of the N-fertilization especially by using organic fertilizer
- Excessive yield expections which inevitably lead to overdressing as well as unsureness to calculate the with yield connected N-uptake. In the Static Fertilization Experiment Bad Lauchstädt the N-content of a wheat grain varied dependend on fertilization and annual weather between 1,6 and 3,2 per cent.
- Fallow land in a crop rotation or continous fallow land if there is not provide for skiming of the grown crops, that means harvest and transport of the N-amounts taken up by the plants.

- Insufficient knowledge of the N-release from the soil organic matter and from the organic primary substance (crop and root residues, organic fertilizer etc.). This N- release strongly varies in dependence of the soil conditions and the annual weather and can amount between 50 and more than 200 $kg.ha^{-1}.a^{-1}$.

The figures 1 and 2 show examples for the N-release of different crops by omission of all fertilization. The amount of N-uptake varied considerably in dependence of the annual weather conditions and the cultivated crop. Even after several years without fertilization the N-uptake is still about 150 $kg.ha^{-1}.a^{-1}$.

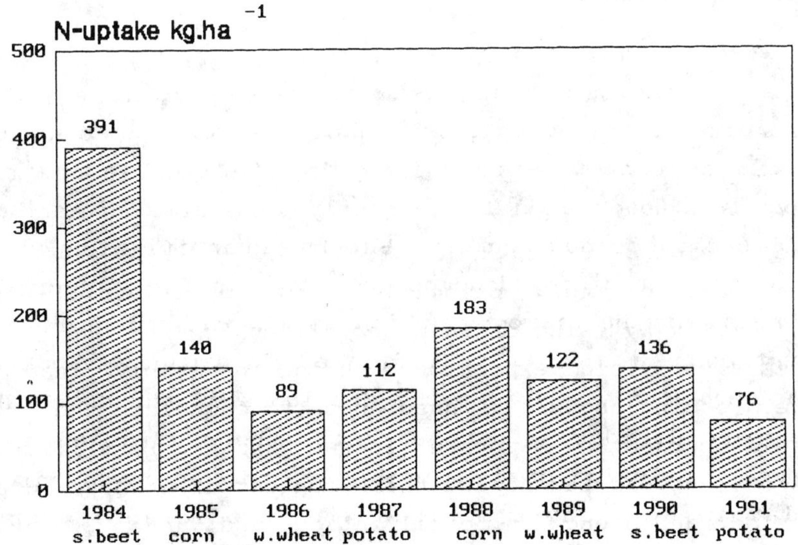

Fig.1:N-uptake without fertilization.
Long term experiment on black earth
in Bad Lauchstädt

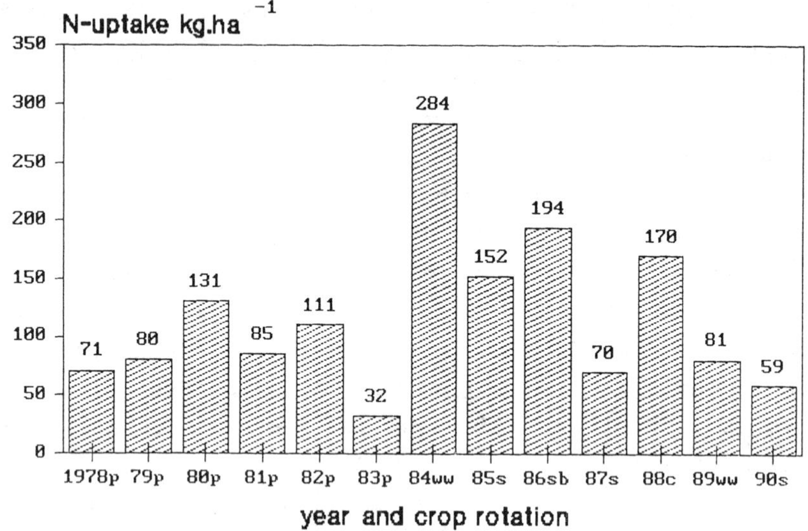

Fig.2:N-uptake without fertilization-
long term experiment on black earth
in Bad Lauchstädt.

This is cause of the highest unsureness with regard to
environment-conscious calculation of the fertilizer amount.
The soil content of anorganic N is determinable and this
method is already used for a fairly long time as help for
fertilization proportionment. But the quantifiation of the
nitrogen from the easy decomposable part of the soil organic
matter and of the nitrogen, which is mineralized during the
vegetation period is very difficult. The known methods are
still very costly. Besides the knowledge of the expectable
weather conditions is necessary for the prediction of N-
release. A solution of this problem can be achieved only by
simulation over short periods with predictable weather
conditions.

The N-input from "other sources" (immission, asymbiontical
N-linkage) mostly remains disregarded. These amounts can be
calculated with about 50 kg.ha^{-1}.a^{-1} (KÖRSCHENS, 1987;
WELTE and TIMMERMANN, 1987; TUDGE and FOWDEN, 1986; and

others). The N-uptake from control plots of long term experiments can be used as standard for the N- input from other sources (table 1).

Table 1:

N-uptake from unfertilized control plots of selected long term experiments

Place	Country	Soil type	years after initiation	N-uptake $kg.ha^{-1}$	
Bad Lauchstädt	Germany	loam	42.- 51.	35	
			52.- 61.	42	
			62.- 71.	49	
			72.- 81.	70	
			82.- 89.	56	
Halle	Germany	loam	31.- 36.	65	(STUMPE ..., 1990)
Rothamsted	GB	loam	137.-140.	50	(ROTHAMSTED REPORT,1986)
Praha	CSFR	loam	1.- 23.	60	(SKARDA ..., 1991)
Ascov	Denmark	loam	80.- 91.	46	(DAM KOFOED 1987)
		sand	80.- 91.	21	
Müncheberg	Germany	sand	21.- 27.	33	(SMUKALSKI ...,1990)
Groß Kreutz	Germany	sand	1.- 28.	34	(ASMUS,1990)

On loamy soils with low leaching danger for nitrat-N the N-uptake from the unfertilized plots amount between 46 and 70 $kg.ha^{-1}.a^{-1}$. On the basis of the Static Fertilization Experiment Bad Lauchstädt a considerable increase of the N-input from other sources could be ascertained over the last 40 years. The decline in the last examination period from 70 $kg.ha^{-1}$ to 56 $kg.ha^{-1}$ is attributed to the low yields as consequence of the uncommon dryness during the last 4 years. It can not be valued as result of lower N-immission.

The optimal arrangement of fertilizing scheme and a realistic yield assessment during the vegetation period (KOCHS and HEYLAND, 1990) are assumption for a considerable reduction of N-losses. On loess-black earth in a water short area a apperent utilization up to 100 % of the applied N can be achieved with application rates up to 200 kg.ha^{-1}.a^{-1}., that is the inevitable happend losses are overcompensated by the N from other sources (figure 3).

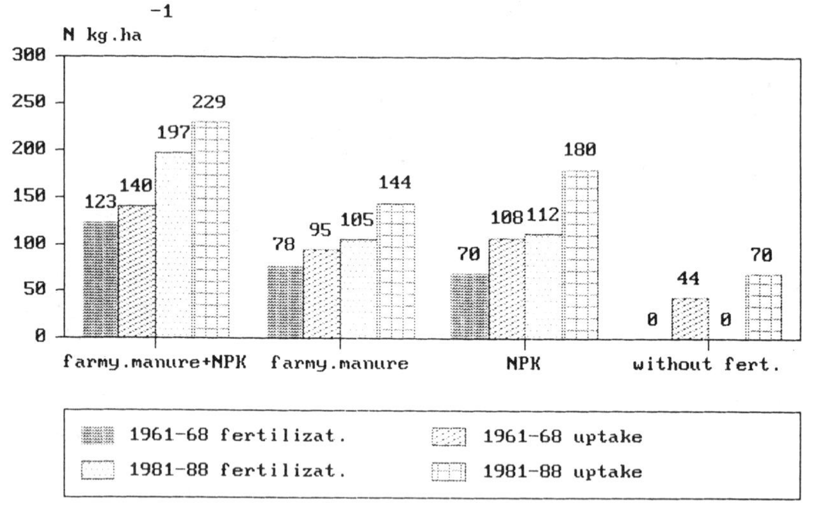

Fig.3:Nitrogen balances- Long term experiment Bad Lauchstädt on black earth in a dry region.

A improve of N-utilization also with increasing N-application rates, mainly established by improved fertilization methods. However there is to consider that this is a deep loess in a dry region with slight new formation of ground water. On sandy soil in areas of heavy precipitation the boundary among profit and loss is between 130 and 150 kg.ha^{-1}.a^{-1}. The consideration of soil humus content and the resulted N-release are also of high importance for the calculation of the N-application rates. The risk of N-release grows with the increase of humus content. The higher the humus content the more important is the utilization of the whole vegetation period to skim the mineral N by plants.

The presented results allow the conclusion, that the reduction of the fertilizer application rate is necessary and also possible without the disadvantage of having economically significant lower yields. The using of fertilizing schemes on the basis of modelation under consideration of the available N from soil organic matter can be of valuable support. Further research work is necessary to quantify the N-cycle and to improve the determination methods for the N-release.

<u>REFERENCES</u>

Asmus, F. (1990). Versuch P 60 Groß-Kreutz – Prüfung verschiedener Möglichkeiten der organischen Düngung. In: <u>Dauerfeldversuche</u>, Berlin AdL. 2. Auflage, 231-243.

Becker, K.W. (1989). Betriebsbedingte Stickstoffbilanzen. <u>Stickstoffbilanz.</u> Referate der gemeinsamen Tagung des Verbandes der Landwirtschaftskammern e.V. mit dem Industrieverband Agrar e.V. am 18./19. April in Würzburg. 47-51.

Finck, A. (1990). Umweltbelastung durch Düngung: Kritische Interpretation regionaler Daten. <u>102. VDLUFA-Kongreß Berlin (17.-22. Sept.)</u>, Kurzfassung d. Vorträge, S. 35.

Kochs-Heyland (1990, Mai). <u>Stickstoff-Prognose mit dem Computer.</u> Pflanzenbau Integriert. Mitteilungen der Fördergemeinschaft integrierter Pflanzenbau.

Kofoed, A. dam (1987). Staldgodningens betydning-danske erfaringer. (The significance of farmyard manure-Danish experiments.) <u>K. Skogs-O, Lantbr. akad. tidskr. Suppl. 19:</u> 37-63.

Körschens, M. (1986). Beitrag der organischen Substanz des Bodens und der organischen Dünger zur Stickstoffversorgung der Pflanzen. <u>Tag.-Ber. Akad. Landwirtsch.-Wiss. DDR, Berlin 248:</u> 125-132.

1510

Körschens, M. (1988). N-Bilanzen, abgeleitet aus Dauer-versuchen. Stickstoff u. Phosphor im System Boden-Dünger-Pflanze. Wissenschaftliche Jahrestagung 1987 Berlin, Tag.-Ber. Sektion Pflanzenproduktion der Humboldt-Univ. Berlin: 50-54.

Köster, W., Severin, K., Möhring, D. u. Ziebell, H.-D. (1988). Stickstoff-Phosphor- und Kaliumbilanzen landwirt-schaftlich genutzter Böden der Bundesrepublik Deutschland von 1950 - 1986. Landwirtschaftskammer Hannover, Landwirt-schaftliche Untersuchungs- und Forschungsanstalt Hameln.

Sapek, A. u. Sapek, B. (1992). The budget of nitrogen in Polish Agriculture. Poster. Churchill College Cambridge - Nitrate and Farming Systems: 29. June - 1. July.

Skarda, M., Zobac, J. and Marhová, J. (1991). Organické hnojeni bez zivocisne vyroby. Rostlinna Vyroba 37(11): 867-877.

Smukalski, M., Kundler, P. u. Rogasik, J. (1990). Der Müncheberger Nährstoffsteigerungsversuch. In: Dauerfeldver-suche, Berlin AdL, 2. Auflage, 251-259.

Stumpe, H., Garz, J. u. Hagedorn, E. (1990). Die Dauer-düngungsversuche auf dem Versuchsfeld in Halle. In: Dauer-feldversuche, Berlin AdL, 2. Auflage, 25-71.

Tudge, C. and Fowden, L. (1986). The future with Rotham-sted. Harpenden/Herts. Rothamsted Experimental Station.

Welte, E. and Timmermann, F. (1987). Effect and sources of nitrate pollution and possibilities for emission reducti-on. Proc. 5th International Symposium of CIEC 1: 14-33 Bala-tonfüred, Hungary. 1. - 4. Sept.

Wadmann, W.P. and Jeeteson, J.J. (1992). Nitrate leaching topes from organic manures - the Dutch experience. Nitrate and farming systems. Aspects of Applied Biology 30: 117-126.

Sturm, H. (1989). Überlegungen zu einer Gesamtstickstoff-bilanz. Stickstoffbilanz. Referate der gemeinsamen Tagung des Verbandes der Landwirtschaftskammern e.V. mit dem Indu-strieverband Agrar e.V. am 18./19. April 1989 in Würzburg, 19-34.

COPING WITH AGRICULTURAL WASTE PROBLEMS IN AN ECONOMIC WAY

Dr. J.P.M. Sanders, Triple A Agro Amino Acids, PO Box 1, 2600 MA DELFT, the Netherlands

Traditionally Agriculture did have the function of the Producer of primary products as food (and in some cultures also of secondary products like medicines, toxins for hunting, and stimulants like tobacco and flowers).

As with many other human activities the agriculture has been mechanized and rationalized mainly in the last century.
Irrigation techniques and chemicals to increase the harvest per hectare cultivated land, as well as growth inhibitors of undesired weeds as well as fertilizers increasing the yield of the crops wanted. Also the crops have been genetically improved by crossing and selection during many decades.
The combined technology developed enabled mankind to grow desired crops at places where they are required in large quantities and where naturally these products would not have any chance of being cultured in a more and more economical way.
As the (drawback?) result of our economical system at last also farmers had to become normal players in the system and since especially this century many countries liked to become selfsufficient in their food supply, an industrialization of agriculture really broke through. Farmers had to change their generalistic way of living in a specialistic profession like most other members of our society. Development of processes got a lot more emphasis than the products manufactured.
Economies of scale was a general applicable feature.

While nature had always been in equilibrium as far as elemental cycles are concerned such as carbon, nitrogen, sulphur etc. now at the concentrated processes in beer, wine, olives, sugar, potatoes, milk processing and slaughter-houses etc. high concentrations of compounds came available that have not been in the environment before, giving rise to a disturbance of the equilibrium.
Urbanization caused another impulse to this development.
Because of economic constraints agriculture has become more and more industrialized. Fertilization has been optimized for growth rate of crops as well as husbandry. This often has led to overfertilization that is leading to an enormous problem in surface water, in air and certainly by the accumulation in soils.

In general there are eight ways to cope with waste problems:

F. Arendt, G.J. Annokkée, R. Bosman and W.J. van den Brink (eds.), Contaminated Soil '93, 1511–1517.
© *1993 Kluwer Academic Publishers. Printed in the Netherlands.*

1512

1. Ignore   5. Dilute
2. Hide   6. Transform
3. Wait   7. Prevent
4. Cure   8. Improve

<u>Ignore</u> and <u>Hide</u> essentially mean leaving the problem for the next generation and should be considered unacceptable, although at this moment they seem to be the most widely practised.

<u>Wait</u> may be an instinctive human reaction to a seemingly insoluble problem but it is just about acceptable in those cases where an affordable solution is within reach.

<u>Cure</u> is an obvious action but a very costly appraoch.

<u>Dilute</u> is marginally acceptable and only in those situations where nature's recycling capacities can cope with the level of waste offered.

<u>Transform</u> a waste stream into valuable products is a combination of two things. Environmental requirements could be met while at the same time a product is manufactured resulting in the prevention of waste elsewhere.

<u>Prevent</u> definitely ends up in a clean environment but if it means a stop to production costs may be very high.

<u>Improve</u> means a clean environment and an improved product/process.

It will be evident that in the order they have been mentioned the efforts and technological input required will increase. From the point of view of a sustainable agriculture only the last three options are viable, although they may initially require considerable investment. However, compared to the environmental costs thay may be attributed to the other methods they will be more economic in the longer term. It is especially in those areas that biotechnology provides powerful tools for solutions. They agricultural nitrogen cycle in the Netherlands provides an excellent analysis as a starting point to design valuable technical solutions.

ANALYSIS HOW TO PREVENT WASTE PROBLEMS

From figure 1 it can be deducted that approx. 70% of the nitrogen is converted into <u>waste</u> products and thus causing enormous problems. It should be realized that most of the nitrogen is produced from nitrogen from the air at high energy costs and high CO_2 production. The energy value should enable us to recycle the fixed nitrogen back into an economic process. How should we proceed?

A more detailed picture (figure 2) unveils that nitrogen is mainly applied in animal husbandry where expecially milk and meat production from cows shows a very low yield

9%). The low yield may be explained by the fact that the production process is optimized on costs and does not reflect any environmental effects. This is further illustrated in biofarming where the same process is optimized on such parameters as fertilizer use, non-use of pesticides and no additional feeding. A yield of approx. 35% may be reached, however, at much higher product costs. Nevertheless, this may serve as an indication that there is room for improvement, from an environmental point of view.

SOME SOLUTION

Relatively simple mechanical means like injection of nitrogen into soil rather than dumping it on the soil in combination with a dosage scheme based on the real need of the crops, will provide considerable improvement. The latter principle applied to animal feed schemes rather than an ad libitum scheme also adds to a reduction of waste.

Biotechnology becomes prominent when the overall production efficiency of the animal itself is improved. This may be achieved through improvement of the digestibility of feed ie. by the addition of enzymes like Phytase resulting in better conversion and less waste.
Figure 3 shows that Phytase is preventing that phosphate should be added to the feed for pigs and poultry.

Another opportunity is by adding amino acids produced from fermentation to the feed increasing its nutritional value and reducing waste. A next step is the production of these amino acids from manure which in essence means: closing the circle.

THE MANURE PROBLEM

A Considerable part of intensive stock breeding, especially of pigs, poultry and calves and to a lesser extent of cattle, is concentrated in confined areas. Due to sufficient supplies of feedstuffs the amount of livestock has become independent of the available land area. As a result of this concentration problems arise due to a surplus of animal manure.

Although the Southern and Eastern parts of the Netherlands form the most extreme example of this problem, various other regions in Europe are known as well to have a significant excess of manure, such as Flanders (Belgium), Nieder-Sachsen (Germany), the Po region (Italy) and Bretagne (France). Outside Europe Taiwan is an example of a country with a serious surplus of manure.

Various solutions are used to reduce these problems, such as the reduction of those minerals in feed that specifically cause the problems when excreted, and the transport of manure from areas with a surplus to other areas. The environmental problems caused by the excess of manure, however, are huge, and measures are getting more severe in the course of the years. Finally, therefore, the inevitable choice has to be made between either the processing of the remaining surplus of manure or a manditory reduction of the number of livestock.

The government of the Netherlands started with various directives, supported by penalties on the overproduction of manure. Some time ago, however, the situation led up to a regulation, according to which a minimal amount of 6 million m^3 of manure must be processed yearly, from on January 1995. If not, livestock will have to be reduced. This surplus of 6 million is likely to grow to a surplus of 20 million m^3 in the year 2000.

Measures in other EC memberstates will get more severe over time as well. Among others the memberstates face implementation of the nitrate directive, allowing a maximum of 170 kg nitrogen/hectare/year. Before 1993 they have to submit a plan on how to reduce nitrate loading in excess areas. This will pressure governments to take actions. It will also increase the surplus of manure and the need for economically justified methods to process at least part of that surplus.

FIRST GENERATION SOLUTIONS

The processing of manure has been started up at several places in the Netherlands, and the first plant, with a capacity of 500.000 m^3 manure/year is under construction. Although based upon different principles, all of these processes are drying the manure and producing manure-pellets. These pellets are welcomed on lands, poor of organic substances. The payments for these pellets, however, are not sufficient to make these processes costeffective. In fact losses are made of ECU 8-16/m^3 of manure processed.

Some years ago, Gist-brocades started to investigate the role that modern fermentation know-how could play in a more profitable way of processing manure. This led to the development of a process to produce amino acids, like lysine, out of pig manure. Lysine has already been used as a feed additive for over 20 years, with an ever growing demand because of its potential environmental advantage. When adding the limiting amino acid lysine to pig and poultry diets, the total percentage in these diet of proteins, and thus of nitrogen, can be reduced, leading to a reduction of nitrogen excretion in manure.

It appeared that the added value of lysine, as produced with this proces, could well exceed the total operational costs of processing manure. The impact of this new and costeffective process on the solution of the National manure problem was widely recognized. This led to the foundation of Triple A, (Agro Amino Acids), a daughther company of Gist-brocades, The Rabobank and the PVVR (Board of Compound Feed Industry) the latter representing the complete Dutch Compound Feed Industry.

THE TRIPLE A PROCESS

The excess of ammonia that is present in the manure is converted to lysine by a fermentation process. The process can be schematically represented as given in figure 4. Actually the lysine technology is being completed to the existing first generation manure processing technology. A drawback of this first generation technology is that it can only produce fertilizer pellets of a given relation between nitrogen, phosphate and potassium. Since the fertilizer directives in the Netherlands but also in other countries are developing in the direction of equilibrium fertilization, many agricultural places only can absorb one or two of these elements but certainly not all three together.
The Triple A process enables the separation of all three of these elements as a bonus for the approach: the conversion of one of them to lysine. Because of this separation the excess of elements can be placed in the Netherlands as will be indicated by a diagram of the Agricultural business column.

Closing the elemental cycles in small loops closeby where these elements have been used, enables to prevent waste problems because closely users will then recognize the value of what faraway observers otherwise would see as an environmental problem.

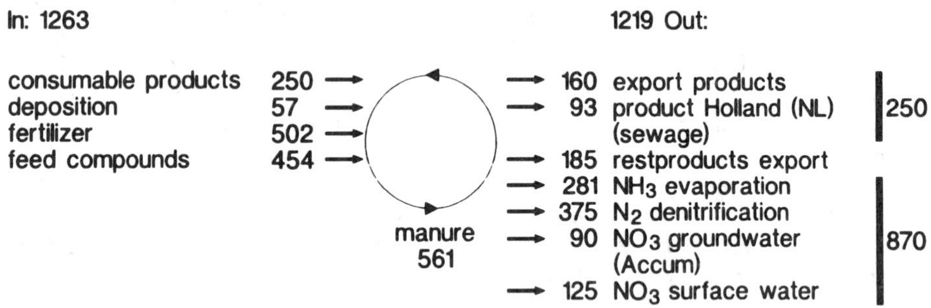

Fig.1

Nitrogen streams per year (millions kg N)

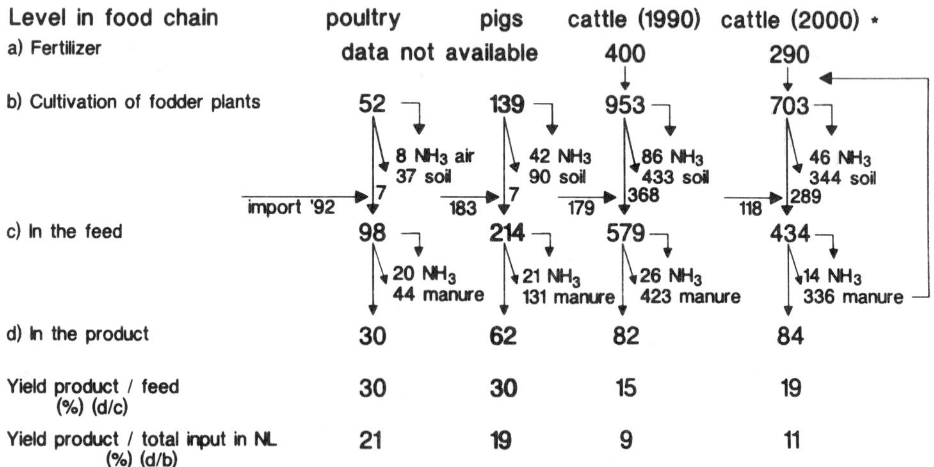

Fig.2

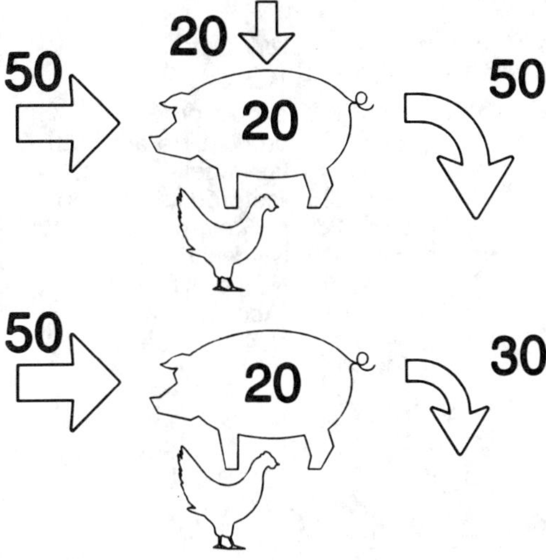

Fig.3

Process routes

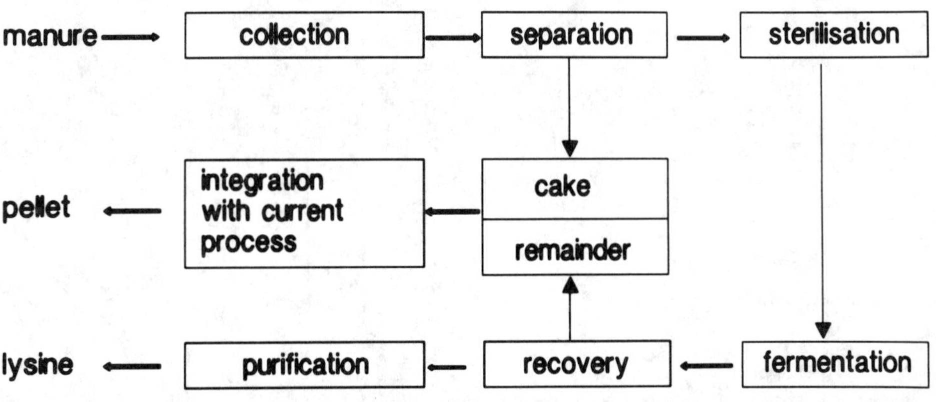

Fig.4

A SOUND AND PRACTICAL METHOD TO DETERMINE THE QUALITY OF STABILIZATION;

Merten Hinsenveld
Center Hill, USEPA Solid and Hazardous Waste Research Facility
5995 Center Hill Road, Cincinnati, Ohio 45224, USA

SUMMARY
The bulk diffusion model provides a curve fitting equation rather than a mathematical representation of the leaching process in cement based stabilization systems. Based on concentration profiles in the interior of the specimen and observation of the acid dependency of the leaching process, it is shown that the shrinking core model is a more appropriate model for cement stabilized waste. It is argued that the ANS 16.1 and the TCLP procedures do not provide adequate quality indicators. A new parameter the so-called <u>Stabilization Quality Index</u> is proposed to form the basis of regulatory decisions concerning the quality of stabilization. A simple test method outlines how this parameter can be obtained.

1. INTRODUCTION

An assessment of leach rates is often obtained from a standardized test procedure. The results of such a test may be decisive in whether or not a certain waste is accepted for disposal and can, therefore, strongly impact the cost of remediation. It is of utmost importance to have at least a basic understanding of what these tests actually measure. In this paper it will be shown that the two major standard tests available to date, i.e. the ANS 16.1 and the TCLP, are built on a rather poor fundament and do not measure what they claim they do.

The bulk diffusion model is widely used for modeling purposes and forms the basis of the ANS 16.1 leach test. We will start this paper with a brief introduction to this model to indicate why it is not appropriate for cement stabilized waste in Section 2. Recent research has indicated that matrix dissolution rather than bulk diffusion is the key factor in the leaching process. We will examine the response of cement to acid exposure in Section 3 to illustrate that a shrinking core model is a more appropriate model for acid leached specimens.

The concepts introduced in this paper are best illustrated with a mathematical treatment of the simplest case. The shrinking core model will derived for a simple geometry in Section 4, introducing the concept of exposure and a new parameter the so-called Stabilization Quality Index (SQI). The SQI is argued to be a better and more widely applicable quality indicator than the ones obtained from the TCLP and ANS 16.1 procedures.

Leaching experiments supporting the shrinking core model will be discussed in Section 5. It is shown that the model adequately incorporates the acid dependency of the leaching process as well as the parabolic leach rate and the concentration profiles in the interior of the specimen. Based on these findings, it will be indicated why the TCLP extract is not an adequate quality indicator for cement stabilized waste.

How the Stabilization Quality Index may be determined from the results of a simple test procedure is discussed in Section 6. We will evaluate this new test procedure and propose that the SQI serve as a basis for regulatory decisions concerning the quality of a cement stabilized waste in Section 7.

2. LIMITATIONS OF THE BULK DIFFUSION MODEL

The stabilized waste form in the bulk diffusion model is assumed to behave as an unreactive solid from which contaminants leach by a purely diffusive mechanism. The model equation can be obtained by applying Ficks law on a segment of the waste form and reads:

$$\frac{\partial c}{\partial t} = D_e \frac{\partial^2 c}{\partial x^2} \tag{1}$$

where c is the contaminant concentration in the waste form [kmole/m^3] and D_e is an effective diffusion coefficient [m^2/s].

For short leach times, the concentration in the center of the specimen remains close to the original concentration and we can use the boundary conditions of a semi-infinite specimen. With initial condition $c(x,o) = 0$, and boundary conditions $c(\infty,t) = c_o$ and $c(0,t) = 0$, we find the concentration profile in the specimen as:

$$c(x,t) = c_o \, erf(x/4D_e t) \tag{2}$$

where c_o is the initial concentration in the solid [kmole/m^3] and erf is the error function. The cumulative amount released into the leachant per unit exposed surface area follows from:

F. Arendt, G.J. Annokkée, R. Bosman and W.J. van den Brink (eds.), Contaminated Soil '93, 1519–1528.
© 1993 *Kluwer Academic Publishers. Printed in the Netherlands.*

1520

$$M''(t) = c_o \sqrt{4 D_e t/\pi} = \sqrt{4 D_e c_o^2/\pi} \cdot \sqrt{t} \tag{3}$$

The concentration profiles in the specimen, as given by equation 2, are depicted in Figure 1 for five times ($t_5 > t_4$, etc). Initially ($t_o = 0$) the solid has a uniform concentration, c_o, while at increasing exposure times the concentration decreases. The model is accurate up to t_5, where the assumption of a semi-infinite specimen ceases to hold (equivalent to about 20% leached). Usually, we do not carry the process up to that point. During the leaching process no dissolution or precipitation is assumed and the matrix remains intact, hence, the effective diffusion coefficient is assumed to be constant.

Figure 1: Concentration Profiles as Predicted by the Bulk Diffusion Model

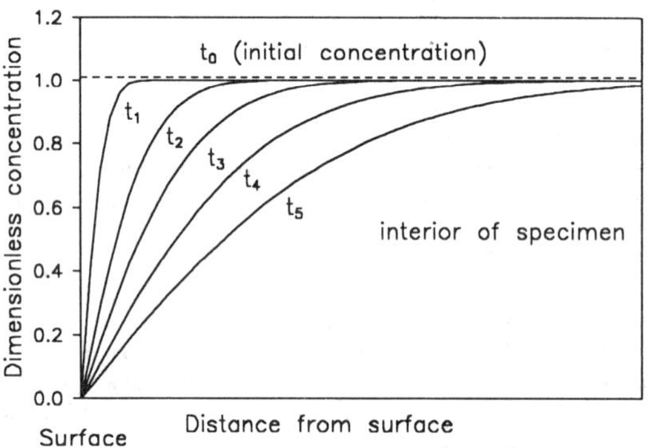

Initially, the model predicts the cumulative amount leached to follow a square root of time relationship as can be seen from equation 3. This is the reason that most researchers have plotted their leaching results on a t-axis scaled by a square root to obtain a straight line; that is, $M''(t) = K \cdot t^{\frac{1}{2}}$. Obtaining this straight line is usually considered a justification of the model. It is important to note, however, that (1) many models can be fitted to a square root of time relationship, e.g. the shrinking core model described hereafter and (2) the model predicts concentration profiles in the interior of the specimen as indicated in Figure 1. Although a comparison of the predicted concentration profiles and the measured concentration profiles is a powerful indicator of the rightness of the model, this comparison is only seldom made.

The boundary conditions used in the derivation of the bulk diffusion model do not include an explicit statement about the leachant. As far as this model is concerned, the leachant could be any medium, provided that the contaminant is soluble in the leachant and removed by leachant renewal (to provide the zero boundary condition). Since most inorganic contaminants of concern (e.g. Cd, Pb, As) are more soluble at pH values below pH = 7 than they are at the pH value of normal cement (about 12.5) and the L/S ratio is taken to be 10, the zero boundary condition (for the contaminants) is easily obtained. Increasing acid strengths from pH = 7 downward, therefore, should not have any influence on the leaching process according to this model, e.g. similar leaching results would be expected for a leachant of pH = 7, 6, 5, 4 etc (it should be realized that acid refers to acid in comparison to the situation in natural cement; deionized water in this respect is also an acid).

Figure 2: Leaching Rates Observed for Various Acid Strengths

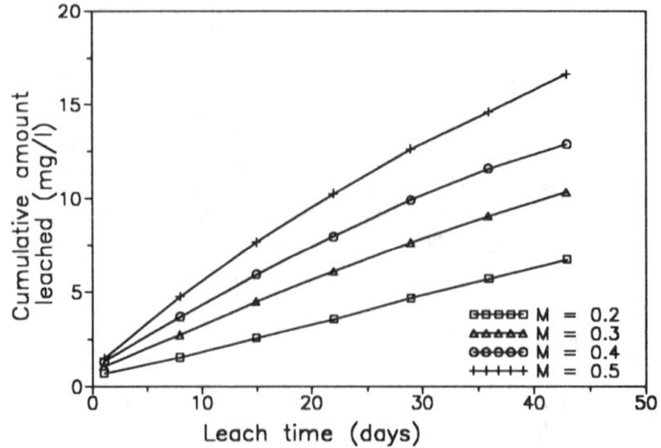

Comparing the leach rates for different acid strengths reveals immediately the inadequacy of the bulk diffusion model. The leach rate is not independent of the acidity but, moreover, it is a strong function thereof. To illustrate this point, I have plotted the leaching results of a lead sludge stabilized with cement for various acid strengths in the same graph, Figure 2. Instead of the one curve we would have expected, we have obtained four curves, one for each acid strength, although we have changed neither of the variables in the bulk diffusion model.

The observation of increasing leach rates with increasing acid strengths suggests that the acidity in the leachant, rather than the bulk concentration of the contaminants in the cement matrix, is the driving force in the leaching process. A model that does predict this acid dependency and gives a parabolic leach rate for a flat surface (at constant pH values) is the shrinking unreacted core model.

3. RESPONSE OF CEMENT STABILIZED WASTE TO ACID EXPOSURE

In cement stabilized wastes, the major acid consuming species is free calcium hydroxide. The result of acid leaching is that the free calcium is released into the leachant, leaving a friable outer shell at the surface, while the core remains unaltered [Van Dijk, et al., 1986; Chandra, 1988; Fattuhi and Hudges, 1988; Cheng, 1991; Buil, et al. 1992]. The leached shell in hydrochloric acid leached specimens, investigated by Chandra (1988) seemed to form an effective barrier against further acid attack, as was concluded from the fact that chloride ions were only present up to unleached core of the specimen. The same phenomenon was observed by Cheng (1991) for acetic acid leached specimens. In addition the concentration of heavy metals in the core was found to remain unchanged during the leaching process. Even cement blocks exposed for a number of years to seawater (pH $\approx$ 8.5) where found to form the friable surface layer, while leaving an unaltered core [Hockley and van der Sloot, 1991]. These observations strongly support the adequacy of a shrinking core model as a mathematical representation of the leaching process.

In addition to the formation of a leached shell, some researchers observed a remineralization region at the core boundary [Fattuhi and Hudges, 1988; Cheng, 1992; Miller and Isenburg, 1992]. The occurrence of this zone was theoretically explained by Cussler. (1984). He stated that a rapid dissolution of calcium hydroxide at the core boundary creates a high local calcium concentration. The resulting gradient works in two directions: (1) outward, leading to a release of calcium into the surrounding leachant and (2) inward, leading to precipitation of calcium at the core boundary. A simplified picture of the various concentration profiles is given in Figure 3 (I have omitted the Ac⁻ ions in this picture). A zone of remineralization was not observed in the hydrochloric acid leached specimens. It would, however, lead us onto a side track to discuss the conditions at which the remineralization zone occurs. An explanation can be found in Hinsenveld (1992b).

Figure 3: Remineralization at the Core Boundary

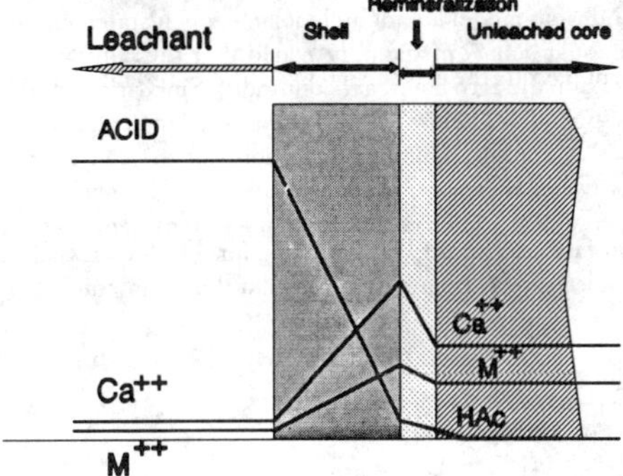

4. CONCEPT OF THE SHRINKING CORE MODEL

It is well known that acid attack on concrete can be described as a shrinking core process [Neville, 1981]. The shrinking unreacted core model also finds wide application in studying leaching of metals from mineral ores [Lasaga, 1988] and in studying glass corrosion [Murakami and Banba, 1984]. Buil, et al. (1992) used the model to describe the leaching of calcium from cement in a neutral leachant. They were concerned with the release of different forms of calcium from specimens leached in neutral leachants. In their model, which required numerical solution, the calcium concentration was assumed to be the driving force. Modeling the release of contaminants along with the free calcium from cement stabilized specimens by using an analytical version of the shrinking core model was only recently proposed by Hinsenveld (1991).

The aim of this section is not to give a rigorous derivation of the model, but to evaluate some of its characteristics as well as to state some of the important assumptions underlying it. The use of the model for various geometries and limiting regimes is extensively described in Hinsenveld (1992a,b). Here, we will derive the model only for the simplest situation of a flat specimen for which the leaching process is limited by the diffusion of acid into the specimen.

Figure 4: Principle of the Shrinking Core Model

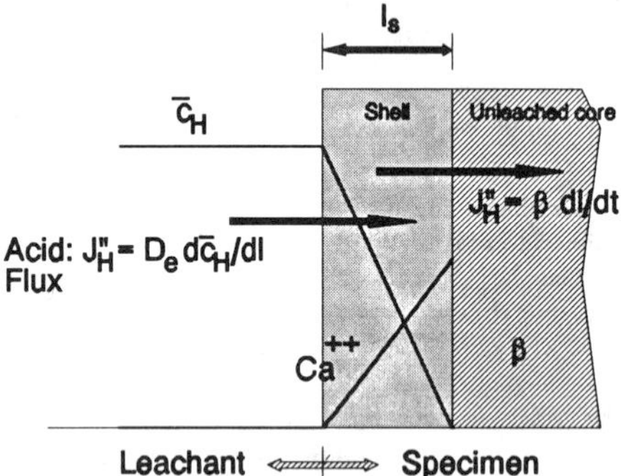

The specimen as perceived in the unreacted shrinking core model consists of an unreactive matrix, accommodating a reactive substance (e.g. $Ca(OH)_2$ and $Pb(OH)_2$). The model assumes the reaction to proceed at a sharp interface that gradually moves inward, leaving a reacted (leached) shell, while the core remains unaltered, Figure 4.

In the derivation, a pseudo-steady state approximation will be used (i.e. the movement of the boundary does not lead to non-linear concentration profiles in the shell). In the simple case discussed here, we do not really need this assumption, but it makes the derivation a lot easier. A discussion of the validity of the pseudo-steady state approximation can be found in Bischoff (1963).

The advancement of the leaching boundary is determined by the acid diffusing through the leached shell and the acid neutralization capacity of the specimen. The diffusive flux through the leached shell is given by:

$$J_H'' = D_e \frac{c_H}{l_s} \tag{4}$$

where we have used the pseudo-steady state assumption. It is coupled to the advancement of the core boundary by:

$$J_H'' = \beta \frac{dl_s}{dt} \tag{5}$$

In these equations we have written c_H for the acid concentration in the leachant [kmole/m^3], l_s for the thickness of the leached shell [m] and β for the acid neutralization capacity [kmole-eq/m^3]. Eliminating the acid flux from equations 4 and 5, gives us the thickness of the leached shell:

$$l_s = \sqrt{\frac{2D_e}{\beta}} \sqrt{\overline{c_H} \cdot t} \tag{6}$$

Since the leachable (mobile) fraction of metals, f_{mo}, is assumed to leach along with the acid neutralization capacity, it is proportional to the amount of ANC leached and, hence, the thickness of the leached shell. We thus find for the cumulative amount leached:

$$M''(t) = l_s \cdot c_o \cdot f_{mo} = \sqrt{\frac{2D_e c_o^2 f_{mo}^2}{\beta}} \sqrt{\overline{c_H} \cdot t} \tag{7}$$

The acidity in the leachant decreases as the leaching process proceeds and we need to extend the expression in equation 7 to incorporate this time dependency. For similar reasons as stated earlier for the pseudo-steady state approximation in the shell, we may assume that the response time to slowly changing leachant conditions is small compared to the rate of these changes. This assumption enables us to integrate the concentration with respect to time to replace $c_H \cdot t$ as was proposed earlier by Van der Zee, et al. (1989). Therefore:

$$M''(t) = l_s \cdot c_o \cdot f_{mo} = \sqrt{\frac{2\,D_e\,c_o^2\,f_{mo}^2}{\beta}} \cdot \sqrt{I} \qquad (8)$$

where we have defined the so-called exposure integral (I) as:

$$I(t) = \int_0^t c_H\,dz \qquad (9)$$

Equation 8 gives us the leaching rate as a constant times the square root of the exposure. This constant is very important, since it determines the resistance of the specimen to acid leaching. I have chosen to call it the <u>Potential Release Factor</u> (PRF), hence:

$$M''(t) = PRF \cdot \sqrt{I} \qquad (10)$$

with:

$$PRF = \sqrt{\frac{2\,D_e\,c_o^2\,f_{mo}^2}{\beta}} \qquad (11)$$

Because we usually want to qualify (rank) the stabilization quality by an easy to use qualifier, we will define a <u>Stabilization Quality Index</u> (SQI) as:

$$SQI = -\log(PRF) \qquad (12)$$

to obtain a parameter similar to the ANS 16.1 Leachability Index. It is this parameter that we aim to determine in the test procedure described hereafter. For a properly stabilized product, we would like a high SQI. It is important to note that, unlike the Leachability Index, the SQI is independent of the acidity of the leachant. Targeting of a specific pH, a common trick to pass the TCLP, therefore, can not be used to improve the "quality" of the stabilization. Also note that at increasing solid concentrations of the contaminant, the SQI decreases. Keeping the solid concentration of the contaminants in the index seems natural. For increasing solid concentrations to give similar leaching results, we either need a denser shell or a higher acid neutralization capacity of the binder (these two may be conflicting requirements) to remain below a predefined number.

Figure 5: Theoretical and Experimental Leaching of Lead as a Function of Exposure

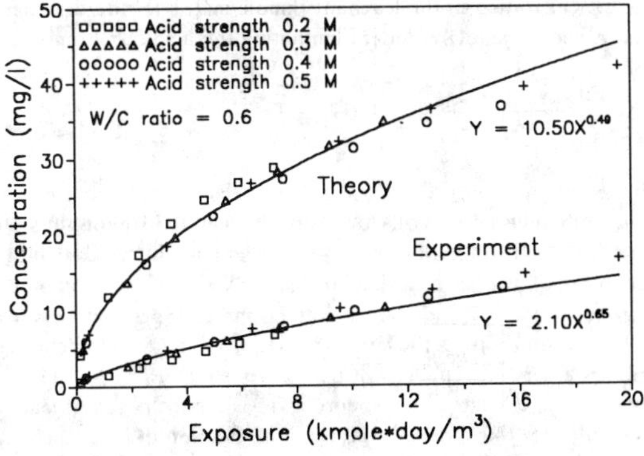

5. EXPERIMENTAL VERIFICATION OF THE MODEL

In this section we will use a small part of the experimental results provided by Cheng (1991); the data has been taken from the appendix of his dissertation. This researcher used the data to fit the bulk diffusion model discussed earlier. Since the leaching values were different for each condition, about 30 models were needed to fit the leaching data (for each condition one). It clearly shows that the bulk diffusion model is not the proper model in this case. However, this research project has provided me with a very useful set of data to validate the shrinking core model.

In the experiments, a large number of small cement stabilized spherical specimens (diameter 37 mm), contaminated with cadmium, lead and arsenic, were leached in acetic acid of different strengths (0.2, 0.3, 0.4 and 0.5 M, respectively). For reasons of brevity we will only use the data for lead in this paper (the data for the other contaminants equally well supports the conclusions reached at the end of this section). The leaching experiments were conducted in seven leach cycles, the first being 1 day and the following each 7 days long (1, 8, 15,..,43 days). Upon conclusion of each of the seven leach cycles, the leachant was removed for analysis and a subsequent leach cycle was initiated with new leachant of the original acid strength. A small number of specimens was investigated internally by cracking the specimens in half and determining the pH profile in the interior by using color indicators. The composition of the specimens is summarized in Table 1.

TABLE 1: COMPOSITION OF THE SPHERICAL SPECIMENS

Parameter	Value
W/C ratio	0.6
Initial lead concentration (kg/m^3)	1.43
Initial calcium concentration (kg/m^3)	510.7
Acid neutralization capacity $(kmole\text{-}eq/m^3)$	25.7

From the data, it was concluded that the leaching process was diffusion limited for two reasons (1) from the internal pH measurements the acetic acid concentration at the core boundary could roughly be estimated; it was found to be reduced to about 15% of its original value, and (2) the theoretically expected leach rates for a diffusion limited regime correlated best with the measured data.

The acid neutralization capacity leached (ANC) is directly related to the reduction of the core as was obtained from comparing the cumulative amount of ANC leached with the physical measurement of the leached shell thickness. The thickness of the shell was found to increase as if the solid ANC (β) was 22.6 kmole-eq/m^3. This ANC deviates somewhat from theoretical ANC (25.7 kmole-eq/m^3), which may be due to the fact that a small fraction (12%) of the ANC is bound in insoluble form or not immediately available at the mild leaching conditions used in the experiments.

In the diffusion limited regime, the total concentration of hydrogen ions and the undissociated acid forms the driving force. Effectively this driving force is equal to the bulk concentration of undissociated acetic acid, since the concentration of its dissociated form is orders of magnitude lower ($\alpha \approx 10^{-3}$). The HAc concentration in the leachant, however, was not measured and I have used the calcium concentration in the leachant as an indicator of the HAc concentration, assuming that it originated from its reaction with acetic acid (equivalent to a titration of a weak acid with a strong base). Since only initial and final values where available, the exposure integral was obtained by using the average acetic acid concentration and adding the contributions of each leach cycle (i). Hence:

$$I(t_n) = \sum_{i=1}^{n} \frac{c_{H,initial,i} + c_{H,final,i}}{2} \cdot (t_i - t_{i-1}) \tag{13}$$

The cumulative amount of lead leached as a function of the exposure integral is given in Figure 5. The curve designated as "experimental leaching" comprises the data actually measured, while the curve designated "theoretical leaching" comprises the data that would have been obtained if all lead would have been present in a readily leachable form. It is clear that the use of the exposure integral strongly correlates the leaching data. Instead of the four curves obtained earlier in Figure 2, we now have only one function for all acid strengths (since we have used spherical specimens instead of flat ones the curve is not entirely parabolical). The effective diffusion coefficient calculated from the data indicated that the tortuosity in the leached shell was about 3, which is a very reasonable value for porous materials. The power of the measured leach rate is slightly higher than we would have expected on a theoretical basis and the actual leaching rate is a fraction of the expected one. We can understand what this means by plotting the fraction of lead leached, Figure 6. We can draw two conclusions from Figure 6 (1) there is a slight dependency of the leach rate on the acidity in the leachant and (2) there is a slow kinetic release from the shell as can be seen from the positive slope of the fractions leached.

As indicated in the previous sections, the release of heavy metals is still insufficiently understood, but it is clear that metals release involves kinetic as well as a compositional factors. It seems that one part of the contaminants dissolves readily along with the free calcium (probably hydroxides); another part dissolves slowly and a third part remains fairly

Figure 6: Fraction of Theoretical Lead Leached

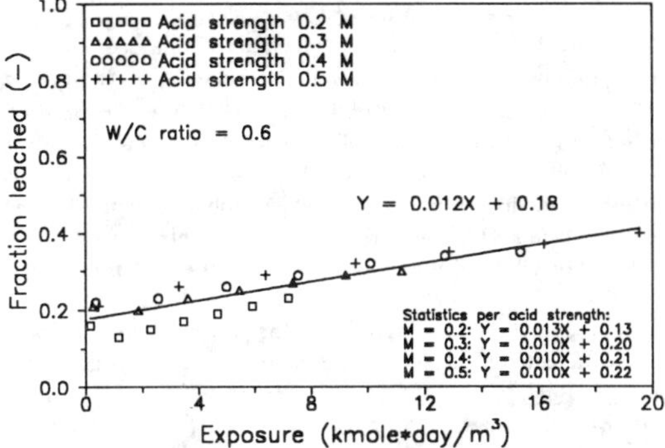

fixed in the matrix. We can picture this behavior as indicated in Figure 7. In this figure, the insoluble part is indicated as a layer along one of the pores, while the second layer on top of it comprises a part that is kinetically released. The contaminants present as readily leachable hydroxides are indicated along with the free calcium in the middle of the pore (for clarity, I have omitted open pore space). Although these layers do not exist in this manner, the picture provides an adequate model for the leaching results described in the previous section.

Ortego, et al. (1991) indicated that the kinetically released metals, that we have indicated as the second layer, actually comprises metals incorporated within the calcium silicate hydrates. The release of these metals coincides with the transition of these hydrates into something like amorphous silica upon contact with the leachant.

This brings us to the TCLP extract as a quality indicator of the stabilized waste form. When a waste form is leached in a fixed amount of acid, i.e. in the TCLP, the readily available hydroxides will rapidly neutralize the acid. The theoretical maximum amount of metals that could be present in the TCLP extract is the amount present in the small

Figure 7: Picture of the Various Leachable Fractions as Supported by the Shrinking Core Model

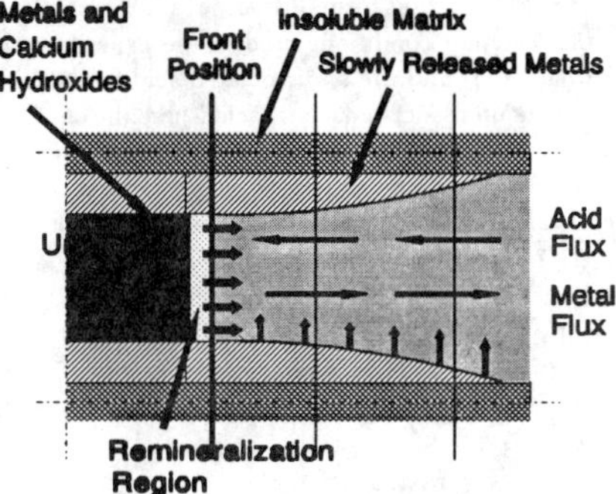

volume of waste that the TCLP leachant digests (i.e. the volume of leached shell it produces). To give some idea, the TCLP protocol prescribes 2 equivalents of acid per kg of wet waste. A 50/50 cement waste mixture has an ANC of about 10 eq/kg. The TCLP, therefore can extract a maximum of only 2/10 of the contaminants present in the waste. Since there is virtually no acidity left in the final extract, the TCLP will measure primarily the faster leaching hydroxides from the shell (which was less than 20% of the total leachable fraction even at the fairly constant pH values used in the experiments we discussed), the metals present in the silica hydrates will virtually not leach.

After conducting leaching experiments on zinc and lead from cement, Ortego, et al. (1991) wrote similar concerns. They stated that the leach rate in buffered systems was extremely high compared to the TCLP protocol and they severely questioned the usefulness of the TCLP as an indicator of the leaching process. In addition to the kinetic aspects, the high pH value in the final TCLP extract will severely limit the solubility of initially released metals, causing them to remineralize on the solids as the leaching proceeds.

6. A NEW TEST METHOD TO DETERMINE THE QUALITY OF STABILIZATION

In setting up a leach test, we may pursue different goals (1) a simple test that quite accurately gives us the reduction of the core, while (over)estimating the release of contaminants; (2) a more involved test that gives us the leaching rate of the contaminants as accurately as possible, while still using the shrinking unreacted core model and the concept of exposure. Such a test would require definition of the "leachable fraction"; and (3) a scientific test, in which the various parameters are carefully determined and factors such as kinetic release from the shell are incorporated. This approach requires use of numerical methods.

From a practical standpoint, the first type of test is preferred. That is, we are willing to sacrifice some of the accuracy of the results if it would simplify the test or reduce the amount of information needed. To avoid the problems in conjunction with the "leachable fraction", the straight forward assumption that all hazardous compounds are released from the shell is preferred. It is a conservative estimate and it gives more accurate results if long term effects are taken into account. In the experiments we have discussed in section 4, this assumption initially results in a factor 5 overestimate for lead, but after 43 days this is already reduced to a factor 2. We will retain the "leachable fraction" (f_{mo}) in the correlations to be discussed hereafter, but in the calculation of the SQI we will assume that the shell is entirely depleted of contaminants ($f_{mo} = 1$).

Procedure. With the general outline given above we can describe a simple test procedure that gives us sufficient information, yet requires a minimum of measurements. In fact, we will give the minimum procedure needed to determine the <u>Potential Release Factor</u> (PRF) and the <u>Stabilization Quality Index</u> (SQI), and to estimate the leach rate. In practice we would entirely be satisfied by this procedure if the general response of the waste to acid exposure is known (i.e. does it behave more or less as a shrinking core). Since we are interested in the principles of the test, we will omit details. In this simple procedure, only two parameters need to be determined prior to the leach test. These are:

(1) The solid ANC of the specimen (β) in kmole-eq/m^3 (titrate specific waste volume to neutral pH); and
(2) The solid concentration of the contaminants (c_o) in kmole/m^3 (digest specific volume of the waste form).

During the leach test, the amount of acid needed to keep the leachant at a constant pH must be recorded. Since the test is conducted in a closed system, we could alternatively determine only the ANC leached each day (without pH control), which would give us the SQI directly (although initial effects may interfere). However, the release of metals from the shell is pH dependent and each waste would influence this release differently, leading us into similar interpretation problems as mentioned for the TCLP. We may, however, allow for small variations in the leachant concentration and intermittent acid addition may be considered. In all cases, it is advisable to determine the ANC leached regularly (it is very easy to determine).

In order to use our measurements we will derive a simple equation that can be used to determine the PRF. The equivalents of acid added ($V_a \cdot N_a$) to keep the pH in the leachant constant is equivalent to the ANC leached. Since the solid ANC is given in equivalents per volume of leached shell, the thickness of the leached shell is given by:

$$l_s = \frac{(V_a \cdot N_a)}{A\,\beta} \tag{14}$$

where A is the exposed surface [m^2]. Combining equation 14 with equation 8, we find for the PRF [kmole$^{0.5}$/(m$\bullet$s)$^{0.5}$]:

$$PRF = \frac{(V_a \cdot N_a)\, c_o\, f_{mo}}{A\,\beta\,\sqrt{I}} \tag{15}$$

Since all parameters except the total amount of acid added and the exposure are constant, we can write:

$$PRF = H\,\frac{V_a(t)}{I(t)} \tag{16}$$

with:

$$H = \frac{N_a\, c_o\, f_{mo}}{A\,\beta} \tag{17}$$

The parameter H in this equation can be calculated from the data (it has no meaning in itself). The PRF can now be obtained by correlating $V_a(t)$ and $I(t)$ or graphically from the slope of a plot of $V_a(t)$ as a function of $I(t)$.

Example calculation. Suppose that we have a cement stabilized waste containing cadmium as the main contaminant. The solid ANC has been determined and was found to be equal to 22 kmole-eq/m^3, while the concentration of cadmium in the stabilized waste form is equal to $7 \bullet 10^{-3}$ kmole/m^3. The specimen has an exposed surface area of 0.01 m^2, is subjected to an acetic acid leachant of 0.1 M strength and leached for 5 days (both arbitrary). The pH is kept at the original value (pH-stat). The L/S ratio may be any value, e.g. L/S = 20 m, the choice being based on practical considerations; the calcium concentration should not be allowed to build up to the point where its influence on the H^+/HAc equilibrium gets notable. In short tests, there may be no need for leachant renewal.

Summarizing the data, we have: (1) $\beta = 22$ kmole/m^3 (solid acid neutralization capacity), (2) $c_o = 7 \cdot 10^{-3}$ kmole/m^3 (solid contaminant concentration) (3) $A = 0.01$ m^2 (exposed surface of the specimen) (4) $M_a = 0.1$ kmole/m^3 (molarity of acetic acid leachant) (5) $N_a = 5$ kmole/m^3 (normality of an acid for pH control) (6) $H = 0.16$ (kmole/m$\cdot$s)$^{0.5}$ (calculated from the data).

Since the pH is constant (adjusted), and hence the HAc concentration in the leachant is constant, the exposure is a linearly increasing variable. If we choose to daily record the total volume of (continuously) added acid, the exposure increments will be (0.1 kmole/m^3)*1 day = 0.1 kmole*day/m^3 per day = 8,640 kmole$\cdot$s/m^3 per day. Suppose that we have found the results indicated in Table 2.

TABLE 2: VOLUME OF ACID ADDED AND CALCULATED EXPOSURE

Day	Volume acid added (10^{-3}m^3)	Total volume acid (10^{-3}m^3)	Exposure (kmole$\cdot$s/m^3)	Exposure$^{1/2}$
0	0	0	0	0
1	1.29	1.29	8,640	93
2	0.54	1.83	17,280	131
3	0.41	2.24	25,920	161
4	0.35	2.59	34,560	186
5	0.30	2.89	43,200	208

Using the data from cycle 3 (day 3) and the H-value given in the data listing gives us: PRF = [0.16 (kmole/m^5)] $\cdot$ [2.24$\cdot$10^{-3} m^3)/(161 (kmole$\cdot$s/m^3)$^{0.5}$] = 2.22$\cdot$10^{-6} (kmole/m$\cdot$s)$^{0.5}$. Since we have constructed the data, the PRF value is the same for each test cycle and we can omit a least square fit (the values chosen, however, are very close to the ones obtained in the experiments discussed earlier). The SQI can be calculated from the PRF and is equal to SQI = - log (2.22$\cdot$10^{-6}) = 5.65.

7. CONCLUSIONS

We have seen that the example leach test is not more difficult than the TCLP or the ANS 16.1 test. The new test gives a good and conservative estimate of the amount of contaminant released from the specimen, as long as the shrinking core model is a reasonable representation of the leaching process. The exposure correlations have a sound mechanistic background, and although the response of cement stabilized specimens may not entirely be as simple as we have described, the mechanisms incorporated in the model are essentially present. The SQI, therefore, is a simple and adequate qualifier that can be used by regulatory agencies as part of the procedure by which a certain waste is accepted for disposal. In the test procedure, we have assumed that all metals were released from the shell, hence we have put $f_{mo} = 1$ to determine the SQI. It is up to the regulatory agencies to decide to what extent vendors of stabilization technology are allowed to increase the SQI by assuming (or measuring according to a certain procedure) a fraction to be completely immobilized.

Note that we did not need to measure the concentration of the contaminants in the leachant. It will be clear that regulatory agencies will require this measurement. It is recommended to digest the core and the shell separately after the leaching process has been completed as one of the checks of the adequacy of the shrinking core for that particular case). In addition, cracking a few specimens open and applying a pH indicator prior to the leach test may easily reveal the presence of a leaching boundary. If there is no boundary present, the test method proposed here may or may not apply and further research is necessary.

Despite the fact that it was established with some confidence that the leaching process in the case study specimens was diffusion controlled, this need not necessarily true for all specimens subjected to acid leachants. However, the procedure can easily be adapted for reaction limited regimes (leading to a linear function for the cumulative amount of calcium leached as a function of the exposure; this is not often observed in practice). If the leaching rate is reaction limited, the SQI as given in equation 12 may be more conservative than necessary. Defining a separate SQI for reaction limited regimes (SQIR) by introducing a reaction constant then may be useful (the interpretation of the leach test given before changes if the leach rate is diffusion limited!). The SQI as defined in equation 12, however, is the most conservative and reliable parameter. Furthermore, any system will eventually reach a stage where the leaching process becomes diffusion limited.

Although we have continued to use the exposure integral as a variable, it is clear what the advantage is of keeping the acid concentration in the leachant constant. We can then take equation 7 to describe the leaching process and rewrite it as:

$$M''(t) = \sqrt{\frac{2 D_e c_o^2 f_{mo}^2 c_H}{\beta}} \sqrt{t} = K \sqrt{t} \tag{18}$$

where K is some constant. As with the bulk diffusion model, the shrinking core model then predicts the leaching rate to be proportional to the square root of time at each constant pH level. It demonstrates once more why researchers usually find the cumulative amount leached to be parabolic in time ("square root of time relationship"). Square root of time relationships can also be fitted to leaching results from spherical and cylindrical specimens, but are then an approximation. In general, it is likely that we want to correct for small changes in the acidity of the leachant (e.g. when we use intermittent acid addition). Use of the exposure integral as a variable, therefore, is preferred.

Notice that from equation 18 it seems that the ANS effective diffusion coefficient is an uncorrected potential release factor. This is, however, coincidental. The ANS 16.1 model has no similarity with the model described in this paper. The parabolic leach rate (for constant acid strength) is a consequence of the parabolic movement of the core boundary and not a consequence of bulk diffusion.

Acknowledgement:
This research was funded by Center Hill Research Facility. Dr. Paul Bishop chaired the PhD committee.

8. LITERATURE

American Nuclear Society (1981) Measurement of the leachability of solidified low-level radioactive wastes.

Bischoff, K.B. (1963) Accuracy of the pseudo steady state approximation for moving boundary diffusion problems, *Chemical Engineering Science*, Vol.18, pp 711-713.

Buil, M., E. Revertegat, J. Oliver (1992) A model of the attack of pure water undersaturated lime solutions on cement, T.M. Gilliam, C.C. Wiles (eds) Stabilization and solidification of Hazardous Radioactive and Mixed Wastes, ASTM STP 1123, pp 227-241.

Chandra, S. (1988) Hydrochloric acid attack on cement mortar - An analytical study, *Cement and Concrete Research*, Vol.18, pp 193-203.

Cheng, K.Y. (1991) Controlling mechanisms of metals release from cement-based waste form in acetic acid solution, PhD Thesis University of Cincinnati, 1991.

Cussler, E.L. (1984) Diffusion - Mass Transfer in Fluid Systems, Cambridge University press, Cambridge, 1st edition.

Dijk, J.C. van, P.J. de Moel, W.F.J.M Nooyen, P.C. Nuiten (1986) Diffusiemodel voor zure aantasting van cement gebonden materialen, H2O, Vol.19, No.20, pp 482-487 (in Dutch).

Fattuhi, N.I., B.P. Hudges (1988) The performance of cement paste and concrete subjected to sulfuric acid, *Cement and Concrete Research*, Vol.18, pp 545-553.

Hinsenveld, M. (1991) Towards a new approach in modeling leaching behavior in J.J.J.R. Goumans, H.A. van der Sloot, Th. G. Aalbers, Waste Materials in Construction, Proceedings WASCON Conference, Maastricht, The Netherlands, 11-14 November 1991, pp 331-340.

Hinsenveld, M. (1992a) A shrinking core model as a fundamental representation of leaching mechanisms in cement stabilized waste, PhD Thesis University of Cincinnati, 1992.

Hinsenveld, M (1992b) Leaching Behavior - Mechanisms and Interactions, Kluwer Academic Publishers, 1993.

Hockley, D.E., H.A. van der Sloot (1991) Long-term processes in a stabilized coal-waste block exposed to seawater, *Environmental Science and Technology*, Vol.25, No.8, pp 1408-1414.

Lasaga (1981) Rate laws of chemical reactions in Reviews in Mineralogy, Vol. 8: Kinetics of Geochemical Processes, Mineralogical Society of America, pp 1-68.

Miller, P., J. Isenburg (1992) Durability of stabilized waste forms to leaching and weathering - Leaching of precracked samples, Draft report, EPA contract No. 69-C9-0031.

Murakami, T., T. Banba (1984) The leaching behavior of a glass waste form - part I: the characteristics of surface layers, *Nuclear Technology*, Vol.67, pp 419-428.

Neville, A.M. (1981) Properties of Concrete, Longman, New York, 1981, 2nd edition.

Ortego, J.D., Y. Barroeta, F.K. Cartledge, H. Akhter (1991) Leaching effects on silicate polymerization - An FTIR and ^{29}Si NMR study of lead and zinc in portland cement, *Environmental Science and Technology*, Vol. 25, No. 6, pp 1171-1174.

Zee, S.E.A.T.M. van der, W.H. van Riemsdijk, J.J.M. Van Grinsven (1989) Extrapolation and interpolation by time-scaling in systems with diffusion controlled kinetics and first order reaction rates, *Netherlands Journal of Agricultural Science*, Vol.37, pp 47-60.

PREVENTIVE SOIL PROTECTION AT INDUSTRIAL SITES

Pieter A. Ruardi

Ministry of Housing, Physical Planning and Environment, Department of Soil Protection, The Hague, The Netherlands

1. ABSTRACT

The paper presents the preventive soil protection policy for industrial sites in The Netherlands.
First, the general preventive soil protection policy is introduced with reference to EC-policy and EC-Council directives. Then the problematic nature is mentioned of the short term implementation of soil protection at existing industrial sites. Some attention is given to the soil remediation procedures and the relationship with the preventive soil protection policy. Next, those aspects of soil protection where problems are encountered are explained. For these problems, solutions are given by means of using available policy-instruments. The paper concludes with a program of knowledge development and transfer, and quality control by guidelines that should lead to an adequate preventive soil protection.
A list of references has been added.

2. THE GENERAL PREVENTIVE SOIL PROTECTION POLICY

Soil protection measures for localized sources of potential contamination include emission preventing (installation related) measures, containment (isolation) measures and a combination of both kinds of measures.
When soil-endangering emissions can be prevented, normally isolation measures will not be necessary. These process-, installation-, or equipment-related measures, although of utmost importance, are not primarily included in the soil protection policy. Those measures are pursued within the more integral environmental and safety policies.
A distinction is made between structural emissions, in principal foreseeable, and incidental emissions (during calamities), mostly not foreseeable.
As far as structural emissions can not (yet) be prevented, containment measures have to be taken.
The Netherlands' basic principle for the preventive soil protection by containment is the avoidance of unacceptable spreading of soil endangering substances by application of the ICM-criteria (Isolation, Control and Monitoring). These criteria should guarantee a lasting good soil quality.
Some information about the ICM-criteria:
- Isolating measures, such as impermeable layers, are necessary where spreading of soil endangering substances is expected.
- Control measures can be technical or organizational and must ensure that the conditions are maintained with which those

F. Arendt, G.J. Annokkée, R. Bosman and W.J. van den Brink (eds.), Contaminated Soil '93, 1529–1536.
© 1993 *Kluwer Academic Publishers. Printed in the Netherlands.*

substances are handled responsably.
- Monitoring measures are necessary to enable the control of
either the isolating measures, or the quality of the soil (or
both).
With the EC-treaty the legal grounds have been established for
an European environmental policy. Until now 3 EC-Council directives have been produced which are important for the soil
protection policy, such as the Groundwater Directive.
The participants on the EC Ministerial Seminar on Groundwater,
The Hague 1991 agreed a.o. - A system of permits, general rules
and codes of practice, including, where appropriate, possible
sanctions, to be reviewed on a regular basis for those activities which could lead to pollution of groundwater or soil,
and including general rules for facilities in which substances
hazardous to groundwater are produced, used, stored, treated or
transported; and the establishment of general rules for the
disposal of waste -. The discussions led to a declaration with
a EC groundwater action program to be implemented by the year
2000, approved by the EC Environmental Council. RIVM (National
Institute of Public Health and Environmental Protection) and
RIZA (Institute for Inland Water Management and Waste Water
Treatment) prepared a substantial contribution to this seminar
with a report on the actual threats to groundwater systems
(lit. 6.).
The Dutch National Environmental Policy Program (NEPP) 1990-
1994 indicates a number of tasks on preventive soil protection
which gradually are being performed.
The General Administative Orders (Governmental Directives) on
soil protection announced and partly already in effect focus on
landfills, storage activities, recycled materials for building
purposes, underground tanks, direct discharges into soil aquifers, dredged mud depots and obligatory soil-quality investigation.
Soil protection regulation has, because of the necessity of
knowledge transfer for effective soil protection, the character
of a multi-staged platform as follows:
1. General Aministrative Orders.
2. Ministerial directives.
3. Guidelines.
4. Handbooks and protocols.
5. Technical/scientific literature.
These related regulations and documents give information about
the practical design, construction and management of soil
protection. The philosophy behind such harmonizing regulating
instruments is quality assurance by knowledge transfer.
The soil protection strategy is based on the state of the art
of soil protection, according the ALARA-policy (emissions as
low as reasonably achievable) and the monitoring of protection
effectivity.

3. SOIL CONTAMINATION PREVENTION AT INDUSTRIAL SITES
As has been shown in the soil remediation program, potential
sources of soil contamination are :
        - waste dumps at industrial sites,
        - storage and transfer,
        - transport,

- handling,
- sewerage.

The soil has functioned mostly up till now as a sink for contaminating substances sprung from industrial activities. A number of companies has been confronted already with the consequences of such a behaviour. In a relatively short period the score will be enlarged with a program of voluntary site-remediation by the industry, which resulted from an agreement between the authorities and the industry. This operation is also important for preventive soil protection activities. A remediation procedure is only effective when at the same time preventive measures are taken according the state of the art. Companies that are permitted, in line with the remediation program, to postpone their cleaning-up for several years will be obliged to take preventive measures in the meantime. Depending on the delay of the remediation start-up and the life of soil protective constructions, for example temporary floors or covers could be installed.

Crucial knowledge gaps in soil protection are indicated below.

3.1. Assay-criteria.

The measurable limits of admissible spreading of soil endangering substances are yet to be determined. This is a complicated matter, for instance at landfills with their endless existence and dependancy on local geo-hydrological aspects. State of the art techniques and monitoring results could demand extra control measures and eventually remediation if limits have been exceeded.

3.2. Design requirements.

For the design and the maintenance of preventive measures, requirements have to be determined for permeability, sustainability and security. The problem is that they have to be in accordance with the assay-criteria and the (potential) state of the art. The bigger the risks or effects of soil endangering activities the more restrictive measures are in demand to enable the same efficacious protection level.

3.3. Protection at existing sites.

Preventive measures are far more cost effective than remediation after omission of prevention and definitely better budgettable (lit. 2.).

At existing industries with their fixed lay-out and installed facilities a cost-effective soil protection is more complicated than for new industrial areas. At most companies the necessary protection techniques are lacking and control measures too (lit. 3.). Impermeable flooring is often prescribed by authorities, but a technical translation "down to earth" is seldom given, by lack of criteria.

3.4. Monitoring.

Because of important knowledge gaps in monitoring technology the effectivity of monitoring and related control measures is far too low for a sustainable soil protection. Especially (semi-) automatic indicative monitoring for small industrial activities (like petrol pumps, tanks and filling stations) should be developped.

3.5. Aftercare of landfills.

By implementing waste management and soil legislation, the necessity of a lasting responsability at landfills has become

the centre of focus. The technical, organizational, financial
and administrational aspects of post closure management of
landfills have been made subject of study. Aftercare is influenced by the construction, the exploitation of the disposal
facility and the eventual exploitation of the closed facility
for other purposes, often of a recreational nature. However for
landfills exploited by industrial companies on their own sites
the responsability for a lasting soil protection has not yet
been established.

4. SOIL PROTECTION POLICY INSTRUMENTS
Several factors contribute to the solution of the indicated
problems. Feasible regulation-related policy instruments are
mentioned hereafter.
4.1. Risk-assessment.
Risk assessment should get an important place in a cost-effective soil protection. An investigation of risk-assessment
methods for soil protection and the needs of industry and
authorities is in a conclusive phase. A method developped by
TNO-IMET will be used more frequently (lit. 11.). Such a risk
assessment document, with which the risks and hazards of a
potential soil endangering activity are estimated, can be used
as an environmental audit instrument.
4.2. Environmental management.
With a system of (company-)internal environmental management,
adopted by (part of) the industry, the environmental responsibility will be delegated by the authorities to the industrial
companies. The principal aspects are: an environmental declaration, environmental program, environmental coordinator,
measurements and registration, internal control, internal
education, internal and external reporting and an environmental
audit.
4.3. Quality assurance.
Quality assurance aims on the realization of the quality policy
of an organization. Building authorities use an external quali-
ty-assurance for ordered products and services. Such an organized care enables efficacious soil protection without trespas-
sing authorities' capacities. The development of clear standardized provisions and measures will be promoted in cooperation
with the building industry, according the ISO-9000-series.
4.4. Infrastructure knowledge transfer.
As yet the knowledge on preventive soil protection is too
fragmented, sparse, insufficient, controversial and inaccessa-
ble. In order to obtain an adequate soil protection, scientists
and technologists need infrastructures for information and
knowledge transfer. For the scientific development the government created and subsidizes a "Speerpuntprogramma Bodemonder-
zoek" (Program of soil research). For technologists a Soil
Protection Expertise Assessment Network is engaged with the
following activities:
- Surveying, analyzing and selecting information about the
technical and organizational preventive soil protection measures at sources of potential soil contamination.
- Knowledge transfer to authorities,industry and others.
- Investigation of knowledge gaps and stimulation of research.
The National Institute of Public Health and Environmental

Protection (RIVM) is the centre of a network with 3 other institutes TNO, KIWA and Staring Centre. Such a balanced combination of expertise is considered to meet the needs of a decisive instrument for an adequate preventive soil protection.

4.5. <u>Guidelines on soil protection</u>.
For the implementation of soil protection, guidelines or handbooks are necessary in which criteria have been laid down and knowledge has been assembled. Such knowledge documents are produced in cooperation with research institutes together with the construction industry. Often those documents enable to carry into effect governmental directives.
Especially in case of EC-directives it is necessary to produce translations on technical, juridical and administrative level. Two international groups of experts (ISWA and ETC 8) are developing such information for the design, construction and management of landfills and depots. The European Technical Committee No.8 is producing knowledge documents on geotechnical aspects of landfills and isolation. The first version was published in 1991 (lit. 9.). This information deals with the geo-technical design, including a design report, a quality assurance plan and risk assessment.

4.6. <u>Development of soil protection techniques</u>.
Knowledge gaps in soil protection technology will be detected in cooperation with the before mentioned expertise network, as was decribed in a RIVM study (lit. 7.). Knowledge development will be stimulated by policy supporting research and production of guidelines and handbooks.

5. IMPLEMENTATION-PROGRAM OF SOIL PROTECTION POLICY
The described policy instruments are used in a program to be implemented for soil protection at industrial sites.
The knowledge transfer by guidelines, based on policy supporting research and partly achieved in cooperation with the construction industry, has been planned as follows:

5.1. <u>The Soil Protection Expertise Assessment Network</u>.
The extension of the activities of the expertise network (Expertisenetwerk Bodembescherming) is planned for the next two years to go forward. In 1996 the target groups, the participants and the principals of the project will decide if continuation of the network is significant and possible.

5.2. <u>The Guideline on bottom lining of landfills</u>.
Supplementary to the already produced - Guideline for the construction of impervious covers for waste-poduct storing sites and landfills - (lit. 8.) a guideline will be produced on the bottom sealing (lining) of landfills. This guideline is based on a study on this subject. Also a guideline for the bottom lining of depots for bulk products or waste will probably be made.

5.3. <u>Guidelines on impermeable flooring</u>.
For the application of concrete and asphalt for impermeable industrial flooring guidelines/handbooks will be made. The - Handbook concrete and environment - made on behalf of the Betonvereniging will give up-to-date information on the application of concrete and related materials. Combination materials will also be dealt with. The VBW Asfalt is going to produce a guideline for soil protection with asphalt. CUR recommendations

will inform about quality control of sand-bentonite liners. For
the application of bentonite composites with geotextiles, with
which easy assembling on site seems advantageous, a handbook or
manual will be needed. As those knowledge documents (made by
the construction industry) are promoted by the Ministry, a
corresponding budget will be necessary. Adequate Ministerial
funds for thes purposes are difficult to obtain as is the
experience in recent years, The end of the year 1994 is the
goal for these documents.
The development of the application of adequate materials and
constructions will go forward in the meantime. This is not a
really new technology as construction protection techniques (a-
gainst acids etc.) are being used at least 30 years already.
For soil protection new aspects, criteria and substances are to
be dealt with. A Guideline on the application of geomembranes
for soil protection has been made available (lit. 12.). Protocols on materials and construction follow.
The before mentioned aspects are also valid for basins and
other containment as far these materials are applicable. For
tanks out of steel and reinforced plastic the socalled CPR-
guidelines (CPR = Committee Prevention Catastrophs) are being
used and governmental directives are planned.
5.4. Guidelines on sewerage.
The first 2 parts of a Handbook on sewerage techniques is available. These information documents are being produced by the
Vereniging van Producenten van Betonleidingsystemen (VPB). For
industrial sewer systems made of concrete these booklets could
be used. For the application of other more chemical resistant
materials (plastics and ceramics) complementary information
will have to be gathered. For other waste water transport pipe
lines the same materials can be used. Monitoring of the effectiveness of these drainage systems is needed.
5.5. Guidelines on monitoring.
Signalizing soil quality changes is of utmost importance for
soil protection. The - Survey development of soil protection
techniques - (lit. 7.) gives an exposition of monitoring systems and techniques:
- Level/sampling tubes establish in the subsoil an own geohydrological and (geo-)chemical system, out of which (with help
of sampling and analysis protocols) samples are taken for
determination of the soil quality. The system influences the
results of sampling and analysis.
- Drainage monitoring tubes have advantages over level/sampling
tubes with the possibility of a closer horizontal measuring
network, but their response time is longer (several years)
because of their positioning.
- With subsoil pushable sample probes a swifter and more accurate investigation on specific parameters is possible. With
such in situ measuring techniques (for pH and conductability)
no changes occur in the chemical equilibriae, because no gro-
unddwater is withdrawn and analysed elsewhere.
- Geophysical measuring techniques are geo-electrical (resistance measurement), electro-magnetical (groundradar) and (for
soil protection seldom used) seismic techniques. Besides equipment to be used at the surface, subsoil pushable equipment
exists.

The state of the art of monitoring is described in the - Guideline of drainagesystems and monitoring systems groundwater for waste-product storing sites and landfills - (lit. 13.). For industrial sites the technical development of (semi-)automatical monitoring systems is needed. Afer an extensive inventarisation will been done in 1993, a guideline/handbook for all monitoring systems will be produced. In that document the connection has to be established with an intended General Administrative Order for soil quality investigation.

5.6. <u>Aftercare</u>.
The post-closure control of landfills, exploited by industries on their own premises will have to be dealt with in near future. In accordance with the industry a policy will have to be developped for (eventual) continuation and improvement of these dumps and the technical, juridical, administrational and financial aspects of aftercare.

5.7. <u>Provisions for new industrial sites</u>.
A basic packet of (standardized) soil protection provisions for (new) industrial sites could be developped. A study indicates the possibility of such basic provisions and measures (lit. 1.). Soil protected new industrial sites tend to be more expensive for settlers than those available without protection provisions, but those investments will pay off.

5.8. <u>Risk and hazard assessment approach</u>.
A inventarising study on risk-analysis and hazard assessment methods will be concluded 1992/93. When knowledge gaps are ascertained, as is to be expected, further research for the development of these methods should be necessary.

6. CONCLUSION
The implementation program of soil protection policy for industrial sites, having started in 1992, is expected to be continued during another 3 years. The length and the quality of the program is strongly effected by the budgetary possibilities of the Ministry and its partners.
The development and transfer of knowledge and the quality control, necessary for adequate preventive soil protection, seem to be guaranteed with this policy program.

7. REFERENCES

1. Soil-safe lay-out of industrial sites. DHV (1986), VROM Bodemreeks no. 70 (in Dutch). Bodemveilige inrichting van industrieterreinen.
2. Colon et al. (1987). Prevention, control, remediation, the control-mix, paper for the BMRO-symposium - soil control industrial sites - (in Dutch). Preventie, beheersing, sanering, de beheersmix.
3. Blenkers, J., and Swinkels, F. (1990). Potential soil endangering industries. Environmental state of affairs at 21 industrial companies in Noord-Brabant, RIMH N-B. (in Dutch). Potentieel bodembedreigende bedrijven. Milieuhygiënische stand van zaken bij 21 bedrijven in Noord-Brabant.
4. Kaltenbrunner et al. (1990). Inventory of technical and control measures for soil protection. RIVM reportno. 736102004 (in Dutch). Inventarisatie van technische voorzieningen en beheersmaatregelen voor bodembescherming.

5. Ruardi, P.A., (1991). Project Soil expertise assessment network Endreport. RIVM reportno. 738708009 (in Dutch). Project Bodemexpertisenetwerk Eindrapport.
6. Sustainable Use of Groundwater. Problems and threats in the European Communities. RIVM/RIZA (1991) Report no. 600025001.
7. Meeder et al. Survey development of soil protection techniques (1992) RIVM reportno. 736101015 (in Dutch). Overzicht ontwikkeling bodembeschermingstechnieken.
8. Guidelines for the construction of impervious final covers for waste-product storing sites and landfills. VROM/Heidemij Adviesbureau (1991).
Richtlijn dichte eindafwerking op stortplaatsen en afval- en reststofbergingen VROM reeks bodembescherming 1991/2.
9. Geotechnics of landfills and contaminated land. Technical Recommendations "GLC"/ ed. by the German Geotechnical Society for the International Society of Soil Mechanics and Foundation Engineering. - Berlin: Ernst (1990).
10. Definition of impermeable floors/covers. Tebodin (1991 in Dutch). Definiëring van het begrip vloeistofdichte vloeren.
11. Van Deelen, C.L., (1990). A method for the assessment of the risk of soil pollution by industrial companies. TNO-MT (in Dutch). Een methodiek voor de bepaling van het risico van bodemverontreiniging door bedrijven.
12. Guideline for application of geomembranes for environmental purposes. KRI-TNO (1991 in Dutch). Richtlijn voor het toepassen van geomembranen voor de bescherming van het milieu. VROM reeks bodembescherming 1991/5.

Safe Management of Mega Amounts of Inorganic Residues from Energy Production and Waste Treatment

H. Pentinghaus, B. Kienzler
Kernforschungszentrum Karlsruhe GmbH
Institut für Nukleare Entsorgungstechnik (INE
P.O. Box 3640, D-7500 Karlsruhe 1 FRG

INTRODUCTION

The primary residues of solid fossil fuel for energy production and of waste incineration will yield in amounts of geological dimension (km^3) in Germany during the next 100 to 150 years. The dimension remains almost the same even if we take into account a possible future reduction in fossil fuel consumption.

German legislation strictly demands reusing of those residues for which it can safely be done. Therefore we need reliable criteria to distinguish between safely reusable residues and those which have to be disposed of. This also holds for those residues which have already been used as building materials like gypsum, and will arise again after some 60 to 80 years as building rubbish. In order to re-use and to dispose of residues we have to be able to predict their geochemical behaviour under a variety of conditions in order to avoid future contaminations with respect to releasing of toxic metals and anions like sulphate into the regional hydrology. The result might be a lessening of the quality of drinking water or even its toxicity.

CONCEPT OF FINAL DISPOSAL

To solve the problem of final disposal of those residues which cannot be reused we propose a final repository in ground water without technical barriers. This is in contrast with the present ideas of repositories above the groundwater table with complex technical barriers and covering systems. Although their technical realisation would be possible even for the necessary dimensions we cannot predict an unlimited lifetime.

Stable deposit bodies composed of thermodynamically stable phase assemblages can only develop in a closed system where the residues are permanently kept in water. This is based on a number of already known reactions betweeen the residues with water as well as with its inventory but also on those reactions which have to be studied in the future. These reactions lead to the formation of

F. Arendt, G.J. Annokkée, R. Bosman and W.J. van den Brink (eds.), Contaminated Soil '93, 1537–1541.
© 1993 *Kluwer Academic Publishers. Printed in the Netherlands.*

a number of synthetic minerals [1] which can incorporate toxic elements and contaminating anions via solid solubilities. We have to distinguish between fast hydraulic reactions and other much slower hydrolytic interactions between reactive glasses in the residues and the aquatic system. Both kinds of reactions will finally lead to a complete stable phase assemblage. Very slow flow of ground water favours homogenization of the deposit body. Characteristic natural examples of such processes are the formation of sedimental rocks from sediments and the crystallization of pyroclastic rocks through hydration reactions [2]. The kinetics of these diagnetical alterations*) which result in a solidification of the loose residues strongly depend on their composition and the composition of the aquatic system.

RESIDUES

The present annual yield of residues from non-nuclear energy production and waste treatment is about 10^7 tons. Though the amount of partial waste streams may vary with time we can expect a total annual amount of this order of magnitude for the next 100 to 150 years.

The essential task of safely managing these amounts of residues consists in the first step, of characterizing them with respect to chemical composition, phase assemblages, texture and chemical reactivity. The contaminating potential of these residues is given by their inventories of toxic metals, polluting anions and their chemical forms [3]. So their determination is an important task. Furthermore there is a definite correlation between the chemical composition and grain size of the residues [4]. This is a further property the knowledge of which enables a sorting of the residues with respect to specific hazard indices. Due to this knowledge fractions of residues can be separated and specifically reused. Chemical reactivity of the waste is determined by the content of phases showing hydraulic properties and the amount of active glasses [5]. These inventories have to be measured.

LARGE FINAL REPOSITORY IN GROUNDWATER

The scenario of such a repository is shown schematically in figure 1. The deposit body permanently lies in the groundwater. Potentially we may expect material transport out of the groundwater and surface water into the repository (import) but also vice versa (export). The situation is analogous to early diagenesis, e.g. the cementation of a sediment. Transport of matter is mainly influenced by the porosity of the sediment (loose residues) and the ratio of the surface area of the solid phases to the volume of the solution. In contrast to natural sediments the materials considered in this context show different properties because they contain phases which give rise to hydraulic as well as to hydrolytic reactions. Fast hydraulic reactions my reduce the open porosity of the deposited body substantially. A further reduction of porosity may be caused by the mass of the repository.

*)Diagenesis comprises geochemical processes in water which finally lead to a cementation of loose sediments.

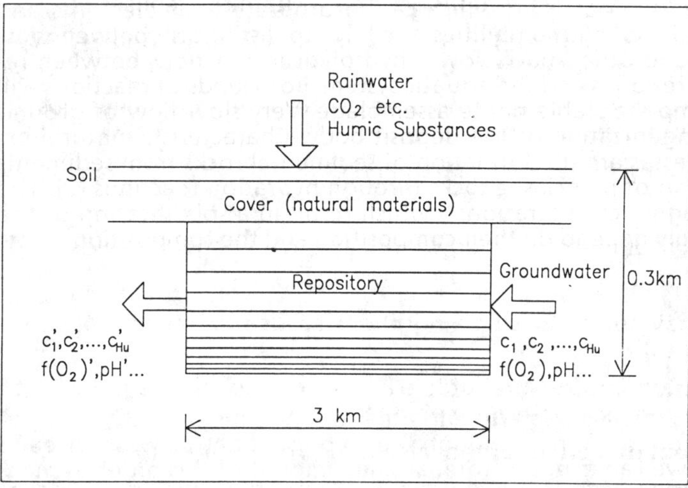

Fig. 1: Scheme of a large repository in groundwater

The geochemical interactions in the deposit body occur between the aquatic inventories and between the solution and the surfaces of the solid phases within the pore volume. Formation of new phases consists of primary reactions (hydraulic) and of dissolution and precipitation processes. They are comparable with authigenic mineral formation. Individual reactions are controlled by different kinetics. It is known that authigenic quartz and feldspar crystals can be formed at ambient temperatures within a couple of weeks. Reaction with reactive glasses will most probably proceed somewhat more slowly. Therefore it is expected that the phase assemblages formed first will change in due course of time [6] until equilibrium will be reached. So also porosity may change in time. This has to be determined. The final goal is to balance the total material transport within the repository and to geochemically model the exchange processes involved [7,8].

R & D Work

The work necessary is partly research on the crystal chemistry of the phases which can incorporate toxic metal cations and contaminating anions /9/ and their textural development in individual experiments and development work on instrumented model repositories on pilot scales. In order to quantify the transport of matter within such a model repository the occurring reactions have to be deciphered and the observed sequence of phases formed has to be geochemically modelled with respect to their stability.

Therefore we propose a R&D-program to study the long-term behaviour of disposed of inorganic residues in groundwater. The aim of this study is the safe management of mega amounts of inorganic residues from non-nuclear energy production and waste treatment. So far we can distinguish between the following objectives:

- Quantification of the mineral reactions occurring within the repository,
- Establishing obligatory regulations for the final disposal of primary inorganic residues which cannot be reused,
- Evaluation of secondary residues (those which have already been reused) with respect to further reuse or final disposal,
- Proof of safely reusing of fractions of primary residues.

Based on these objectives one can already define the following R&D fields:

Research:

I Geochemical behaviour of specific amorphous solids (glasses as main constituents of residues)

  a) Determination of structural properties of reactive glasses in order to make their chemical behaviour predictable.

  b) Development of models for the release of toxic metals etc.

  c) Determination of reaction kinetics with respect to the formation of crystalline phases.

II Geochemical behaviour of the crystalline phases involved

  a) Study of the dissolution of crystals

  b) Study of the formation of secondary phases and their incorporating capacity of toxic and polluting cations and anions

III Evaluation of natural analogues (e.g. altered pyroclastic rocks)

IV Statistical evaluation of properties regarding texture, phases, and chemical reactivity

V Geochemical modelling

  Coupling of transport and geochemical equilibrium models.

Development 1

Disposal of residues

I Characterization of residues with regard to

  a) kinds and properties

  b) evolution of yields in time and region

II Evaluation of repository sites with regard to regional geology, soil and hydrology

III Pilot scale repository (1-1000 m^3)

IV Study of existing deposits, their inventories and chemical states

V Instrumentation of a large final repository for long-term observation

VI Proposals for obligatory regulations for the disposal of residues

Development 2

Contributions to reusing residues

I   Study of long-term behaviour in different fields of application

II   Evaluation of polluting potential under the conditions of application

These studies can only be completed after longer periods of time because the activities necessary are very time-consuming. They cannot be replaced bei engineering operations. As the study will lead to obligatory legal regulations it is inevitable to cooperate confidentially with the electric power suppliers generating the residues, universities and other research institutions.

Due to the public importance of the studies and their interdisciplinary character it is an adequate task for a public research centre.

LITERATURE:

[1]   BAMBAUER, H.U., "The application of mineralogy to environmental management - an overview", Proc. Int. Conf. Appl. Min., Paper 3, Vol. 1, Pretoria, (1991).

[2]   RICE, S.B., PAPKE, K.G., VAUGHAN D.E.W., "Chemical controls on ferrierite crystallization during diagenesis of silicic pyroclastic rocks near Lovelock, Nevada", American Mineralogist, Vol 77, 314-328, (1992).

[3]   WADGE, A., HUTTON, M., "The leachability and chemical speciation of selected trace elements in fly ash from coal combustion and refuse incineration", Environmental Pollution, Vol. 48, 85-99, (1987).

[4]   KEYN, J., SCHREITER, P:;"Zum Phasen- und Gefügeaufbau von Braunkohlenfilteraschen im Feinstkornbereich", Silikattechnik Vol. 36, Heft 11, 341-343, (1985).

[5]   ZSCHACH, S., "Mineralogical properties of fly ashes", Chem. Erde, Vol.87, 330-355, (1978).

[6]   KRONIMUS,B., PENTINGHAUS, H.J., "Niedrighydrothermale Umwandlung von Na-Borosilikatgläsern zur Verfestigung von HAW", Europ. Journal of Mineralogy, Vol. 3, 156, (1991) und Report KfK 5033, 10-11, (1992)

[7]   KIENZLER, B., KÖSTER, R. "Experimental and theoretical investigations of corrosion mechanisms in cemented waste forms", Nuclear Technology 71 , 590-596, (1985).

[8]   BERNER, R.A., "Early diagenesis", Princeton Series in Geochemistry, Princeton University Press, Princeton, New Jersey, (1980).

[9]   PÖLLMANN, H., " Mineralogisch-kristallographische Untersuchungen an Hydratationsprodukten der Aluminatphase hydraulischer Bindemittel", Habilitationsschrift, Erlangen, (1989).

Hydrological, geophysical and geochemical investigations for a future waste deposit site in Saxony

MATTHIAS BEUSHAUSEN, GERHARD LANGE, HEINRICH KRUMMEL &
WERNER HILTMANN
Federal Institut for Geoscience and Natural Resources (BGR)

In the joint research project "Methods on the investigation and description of the underground of operating and abandoned waste deposit sites" that is sponsored by the Federal Ministry of Research and Technology, hydrogeological, geophysical and geochemical investigations at different testing sites in Germany are being carried out. The results of these investigations, along with those of 37 individual research projects of geophysics, hydrogeology and geochemistry, will contribute to a manual of methods which will serve as a working support for experts on the field of waste management and remedial actions.
As an example a testing site west of Chemnitz has been selected, the area of a potentially new waste deposit site. The geological set up is characterized by north-south striking and east dipping metamorphic slates with greywackes of lower Carboniferous age adjacing to the east.
The investigations at this site are aimed to:
- Position and extension of the rock sequences
- Location of faults and cleavages
- Thickness and chacter of the zone of weathering
- Thickness of the loess cover
- Permeability of the geological entities
- Local and regional groundwater system
In the first phase of the study geoelectric, electromagnetic and magnetic measurements have been conducted which along with a detailed geological mapping resulted in a preliminary model of the geological structure of the area. Basing on the first results bore hole sections and hydrological test will contribute to a better understanding of the lithology and the local groundwater system. The investigations are designed in a way that apart from obtaining new information on hydrogeo-logic parameters of the study area a direct comparison of the methods applied is facilitated. Furthermore geochemical, mineralogical and soilmechanical investigation shall provide information on the background level and the underground's pollutant retaining capacity.

F. Arendt, G.J. Annokkée, R. Bosman and W.J. van den Brink (eds.), Contaminated Soil '93, 1543.
© 1993 *Kluwer Academic Publishers. Printed in the Netherlands.*

Estimation of emission for dumps according to the draft of the administration regulation "TA Siedlungsabfall"

Dr.-Ing. Klemens Finsterwalder, Dipl.-Ing. Ulrich Mann,
DYWIDAG Umweltschutztechnik GmbH, Erdinger Landstr. 1,
D-8000 München 81

The German administration regulation "TA Siedlungsabfall" lays down minimum values for the dimensions of surface and base sealings resp. as well as minimum characteristics for the applied materials. Furthermore certain requirements are made on the eluate of the waste to be deposited. This is based on the idea that all aspects of the environmental protection are kept in, when these demands are fulfilled.

No information about permissible emission of individual substances are existing in the a.m. administration regulation although it is indisputable that dumps emit substances in the long-term. As the toxic effect of a substance on flora and fauna each depends on its concentration, the knowledge of the emission and the tolerable ultimate emission resp. is of great importance.

The emission of a dump can be estimated for different sealing systems and materials by means of mass transport calculations. For the examination of the efficiency of a dump design according to the minimum requirements of the TA Siedlungsabfall with regards to emissions of contaminants, emission calculations under consideration of safety factors are executed, which include both installation and the sensitivity of the used building materials.

The mass transport parameter of the used sealing materials are, as input values for the calculations depending on the material composition and the installation method, surcharged with safety factors. So e.g. material treatment and sensitivity with regard to the drying out are considered (Table 1).

For lack of existing limit values for dump emissions, limit values for concentration like the are proposed in the "special report for contaminated land" in the field of soil purification are added, so that an assessment criteria for the calculated emissions is existing.

The emission calculations, executed for six different sealing materials and installation methods resp. show that with regard to the given limit values only insufficient safeties are exi-

F. Arendt, G.J. Annokkée, R. Bosman and W.J. van den Brink (eds.), Contaminated Soil '93, 1545–1546.
© 1993 *Kluwer Academic Publishers. Printed in the Netherlands.*

sting (Table 2). Only two methods can with regard to certain contaminants reach safeties > 1.

These examinations shall at the tendering authorities achieve a higher safety thinking referring to material and installation methods. At last the have the possibility to increase the requirements with regard to the administration regulation.

Furthermore it is also senseful to discuss the ultimate emissions of dumps to avoid a new formation of waste loads as reason of an insufficiently controlled dump technology.

	Wet installation according to proctor						Dry installation	
Safety coefficient for the k-value	A without homogenization		B with homogenization		C with controlled grain size distribution		D with controlled grain size distribution	
	O˙	B˙	O˙	B˙	O˙	B˙	O˙	B˙
V_1 chemical resistance	1,5	1,5	1,3	1,3	1,15	1,15	1,15	1,15
V_2 material properties	3,0	3,0	2,0	2,0	1,15	1,15	1,15	1,15
V_3 sensitivity after installation	3	1,5	3	1,25	2	1,25	1	1
$V_k = V_1 \cdot V_2 \cdot V_3$	13,5	6,75	7,8	3,25	2,65	1,65	1,32	1,32

Table 1: Development of safety coefficients for the k-value depending on the treatment method

˙ O = surface sealing   B = base sealing

	Toxic substances	Safeties against inconsistent soil contaminations					
		A	B	C	D	C1	D1
1	Lead	0,13	0,14	0,16	0,19	0,44	4,8
2	Arsenic	0,26	0,23	0,31	0,38	0,88	9,6
3	Chromium	0,54	0,55	0,62	0,77	1,7	20
4	Cadmium	0,05	0,06	0,06	0,08	0,17	2,0
5	Mercury	0,05	0,05	0,05	0,06	0,08	0,172
6	Nickel	0,14	0,14	0,16	0,19	0,44	4,8
7	Fluoride	0,08	0,09	0,11	0,16	0,64	1,89
8	Nitrate	18	20	28	38	167	500
9	Nitride	0,02	0,02	0,03	0,04	0,15	0,45

Tabelle 2: Safeties of the porewater concentrations with regard to the EG-regulations for drinking-water recovery

LEACHING OF CHROMIUM FROM A DISPOSAL OF THE LEATHER INDUSTRY

Dr.F.Glombitza[1], Dr.J.Ondruschka[2]

1 DFA GmbH Consulting und Engineering Jagdschänkenstr.52
Postfach 89 0-9030 Chemnitz Germany
2. KAI e.V. WIP-AG "Laugung " Permoserstr.15, 0-7050 Leipzig

1.Abstract

Investigations of a chromium containing sludge from a tanning manufactory showed a growth of the used population of the Thiobacillus strains up to a concentration of chromium of about 800 mg/l of chromium III -ions and the possibility to realize a relatively highly chromium concentration in the water phase combined with a removal of about 80 % of chromium from the sludge with a chromium concentration of 3.5 % by means of a partial leaching or drainage water recycling.

2.Introduction

The deposited sludge of a tanning manufactory contains organic waste besides sulfides and chromium salts. Investigations on a disposal demonstrated a mobilization of chromium in the drainage water and a decrease of the pH-value. The reasons for these are assumed in a microbial oxidation of the sulfides and a dissolution of the chromium III hydroxide. (1.2.3.) That requires the presence of chromium resistant and organotolerable chemolithoautotrophic microorganisms.

3.Tasks

1.) Investigation of the ability of leachable microorganisms to grow in a chromium ions containing solution.
2.) Determination of the reachable maximum chromium ion concentration.
3.) Treatment of the waste product for detoxification by leaching and removal of chromium.

4.Methods

Microorganisms strains of Thiobacillus ferrooxidans and non characterized chemolithoautotrophic strains were cultivated in a chromium III – ions containing solution with 9 K – medium and iron II ions as energy sources after adaption. The chromium containing wastes were treated by means of column leaching methods with a content of 1 kg material and a pH-controlled as well uncontrolled continuous process. Iron II ions were used as energy sources but a leaching without iron II – ions was carried out further. An using of S^{2-}-ions from the sulfides of the sludge as well as Cr $^{3+}$-ions is assumed in this case.

5.Results

The growth of the Thiobacillus microorganisms in a 9 k iron II – ions containing solution demonstrated the ability of the microorganisms to grow up to a concentration of 900 – 1000 mg Cr^{3+} / l. (Fig. 1)

F. Arendt, G.J. Annokkée, R. Bosman and W.J. van den Brink (eds.), Contaminated Soil '93, 1547–1548.

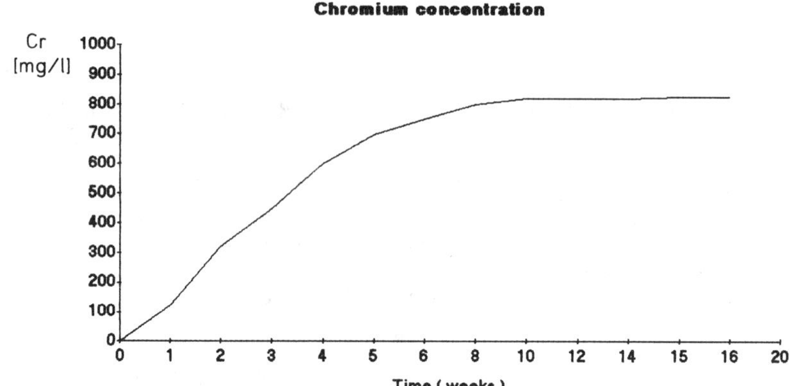

Fig.: 1: Leached chromium ions as a function of time

A treatment of the chromium containing dried sludge in a column leaching process with and without iron – II – ions demonstrated the possibility of a separation of chromium.
The concentration increases in the irrigation liquid to 800 – 1000 mg / l. An exchange of the leach solution and a decrease of the chromium concentration gives a further separation of chromium. A separation of chromium is possible by a repeated exchange of the solution by the way.(Fig.2)

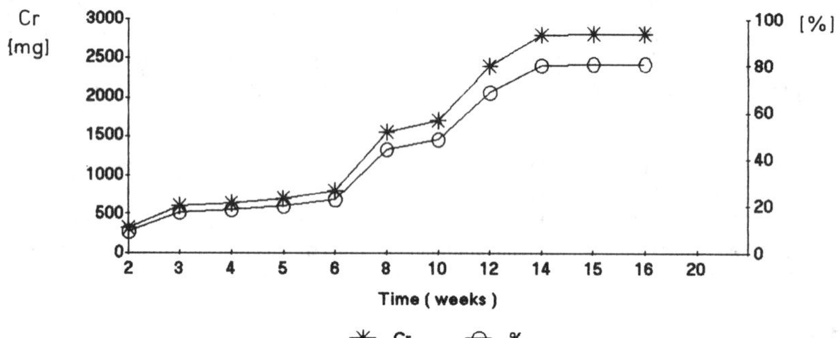

Fig. 2: Leaching efficiency (%) and leached amount of Chromium

A removal of about 80 % of chromium from the sludge by means of a partial leaching or drainage water recycling was possible.

6.References

1.) Rossi G."Biohydrometallurgy" Mc Graw Hill Book Company G.m.b.H. 1990 Hamburg
2.) Glombitza F.,Iske U. et.al. " Biotechnology based opportunities for environmental protection in the uranium mining industry " Acta Biotechnol.12 (1992) 2 79–85
3.) Puhakka J.,Tuovinen O.H.," Biological Leaching of sulfide minerals with the use of shake flask, aerated column, airlift reactor and percolation technique" Acta Biotechnol.6 (1986) 4 345–54

STUDIES OF THE IMMOBILIZATION OF HEAVY METALS FROM WASTE
MATERIALS BY SOLIDIFICATION WITH HYDRAULIC BINDERS

K.P.Großkurth, W.Malorny

Institut für Baustoffe, Massivbau und Brandschutz
Technical University of Braunschweig, Germany

Despite intensified efforts with regard to the reuse of waste
components there is still a lot of waste materials, for
example residues from incineration processes, that have to be
disposed of in landfills. Because of their potentially toxic
effect on living beings heavy metals belong to the harmful
waste elements as - contrary to organic compounds - they are
not destroyed by thermal treatment. One requirement for the
ultimate disposal of heavy metal-containing wastes is that no
risks for the environment arise from them in the long run.
Therefore they often have to be solidified by means of an
appropriate binder. Outlined by the terms "encapsulation",
"stabilization", and "fixation", the aim is to minimize the
mobility of hazardous waste elements.

As far as hydraulic binders are concerned experience reaching
back to history shows that their long-term durability is at
least given in principal. Our knowledge on the physico-
chemical reactions between binder and waste, however, is not
fully satisfying yet. This is holding for solidfication
procedures already established on the market, too.

The complexity of waste-binder interactions demands the study
of the fundamental mechanisms of immobilization on simple
well-defined model systems consisting of both single cement
clinker minerals and heavy metal compounds. On the analogy of
waste water treatment methods that are based on lye addition,
cement-containing binders stabilize dissolved heavy metal
salts as metal hydroxides or basic carbonates of low solubility. This is due to their high alkaline buffering capacity.
Additionally, hazardous cations as well as anions may undergo
chemical fixation by incorporation into the hydrates of the
calcium silicates and calcium aluminates of cement-based or
pozzolanic binders. In this context the cement mineral
ettringite has to be mentioned particularly, being a potential
candidate for the fixation of toxic heavy metals, for example
the chromate ion in its crystal structure. This is shown by
means of scanning electron microscopy in combination with

F. Arendt, G.J. Annokkée, R. Bosman and W.J. van den Brink (eds.), Contaminated Soil '93, 1549–1550.
© 1993 *Kluwer Academic Publishers. Printed in the Netherlands.*

energy-dispersive x-ray spectroscopy and x-ray diffraction.

Leaching tests confirm the effectiveness of the immobilization of waste components by means of hydraulic binders. They were performed on mixtures of flue gas clean-up residues and ordinary portland cement either according to the German standard method DEV S4 or - in a more realistic simulation of disposal site conditions - by means of triaxial cells designed for the evaluation of the permeability of solidified waste products.

Obviously the solidification of heavy metal-containing waste by means of hydraulic binders lends itself to more than a simple improvement of the mechanical behaviour of wastes for land disposal. It is better characterized as a physico-chemical treatment, regarding the generally accepted multi-barrier concept for hazardous waste disposal, saying that the waste product itself must be an efficient barrier against the liberation of toxic substances.

Numerous possibilities for optimization result from adaption to special requirements, for example by an adequate choice of the cement type used or by admixture of additives. Economic interests will be taken into account best by combining several kinds of wastes such as waste water from the clean-up of flue gases and incineration residues to solidify so to speak waste with waste.

PHOSPHATE SATURATION AND GROUNDWATER QUALITY IN SAND SOILS IN THE PROVINCE OF FRIESLAND

H. HIDDING, P.O. DE VRIES, 'Oranjewoud' BV, P.O. Box 24, NL 8440 AA Heerenveen. A. HAHN, R. VEENINGEN, Province of Friesland.

1. INTRODUCTION

In a number of areas in the Netherlands applications of phosphate by means of fertilisation and manuring exceed the removal of this element by crops and leaching. As the phosphate bonding capacity of the soil is limited, when over-fertilisation continues, saturation will eventually occur and phosphate may leach into surface waters and contribute to their eutrophication.

In sand soils the phosphate bonding capacity consists of the reactive Fe and Al as determined by oxalate extraction. In the Netherlands phosphate saturation measurements are based on the oxalate-extractable concentrations of phosphate, iron and aluminium in the soil layer between ground level and the average highest groundwater level. Consequently the sensibility of the soil to phosphate saturation is closely connected with the soil type and the groundwater fluctuation classes, determined according to the standard classification in the Netherlands. According to the standards set by Van der Zee and Riemsdijk, sand soils are considered Phosphate-saturated when:

Degree of P-saturation = (Mole oxalate extractable P)/ 0.5(Mole oxalate extractable iron and aluminium) ≥ 0.24

This level is based on an expected phosphate concentration in surface water of 0.1 mg/l ortho P, or 0.15 mg/l total P at the above mentioned ratio of 0.24 between on the one hand oxalate-extractable phosphate and on the other hand half of the oxalate-extractable iron and aluminium in soil.

2. INVENTORY

In the spring of 1992 'Oranjewoud' BV carried out an inventory of the occurrence of phosphate saturation in the province of Friesland. The soil, groundwater and surface water were sampled and analysed at 68 sites (pastures), having different, frequently occurring soil types. The standard error of sampling and analyses appeared to be such that a parcel with a measured average P-saturation of 0.305 is to be considered to have a true level of saturation higher than 0.24 with a reliability of 97.5 %.

In Table 1. the phosphate bonding capacities, as measured in podsols in sand and loamy sand, differing in thickness of the A-horizont and groundwater regime (37 parcels sampled) and the occurrence of P-saturation ≥ 0,305, are presented. As expected, the phosphate bonding capacity on loamy sands is larger than on the sand soils.

It was found that soils consisting of fine sand and having a high groundwater level are relatively sensible to saturation. 24 % of the sands and loamy sands investigated appeared to be clearly Phosphate-saturated, 32 % of the parcels were possibly Phosphate-saturated (0.24 ≤ P-saturation ≤ 0.305).

The degree of phosphate saturation and the log-transformed concentration measured in the groundwater are correlated on a 95 % significance level (r=0.33, n=37).

F. Arendt, G.J. Annokkée, R. Bosman and W.J. van den Brink (eds.), Contaminated Soil '93, 1551–1552.
© 1993 *Kluwer Academic Publishers. Printed in the Netherlands.*

TABLE 1. Phosphate bonding capacity F (tonnes/ha) and the occurrence of Phosphate-saturated soils, in relation to the average highest groundwater table MHG (m below ground level) and loaminess in sand soils.

MHG	Sand (0 - 10 % silt) F			Loamy sand (10 - 32.5 % silt) F		
0.2	4.2	(1.9–9.3)	1/2*	–		
0.3	6.9	(3.1–15.2)	0/1	10.4	(7.0–15.3)	1/5
0.55	8.5	(5.3–13.6)	1/2	16.6	(11.3–24.5)	0/2
0.6	14.0	(8.8–22.4)	5/12	22.4	(15.2–33.1)	0/9
1.0	26.6	(16.6–42.5)	1/4	–		

* 1 parcel with degree of phosphate saturation ≥ 0.305 out of 2 investigated.

The phosphate concentrations observed in groundwater are (still) relatively low however. This can be explained by an overestimation of the phosphate saturation; the relatively low degree of phosphate saturation of the lower soil horizonts could not be taken into account fully. The current groundwater regime is drier than the regime used in the calculation of the phosphate saturation.
A correlation between phosphate saturation and the high phosphate concentrations in surface waters bordering the sampled parcels could not be established. The concentrations in groundwater are determined by many other factors than groundwater quality and the variability of phosphate concentrations in time, especially in surface waters is very high.

3. PHOSPHATE SATURATION AND SOIL FERTILITY

Based on the correlation between phosphate saturation and the phosphate nutrient state for plants (as determined by extraction with ammoniumlactate/acetic acid at pH 3.7)(Schouwmans et al), it can be concluded that parcels having a 'sufficient' or 'low' phosphate status from the point of view of soil fertility, are not provably saturated with phosphate, while parcels classified as having an 'amply sufficient' or 'high' phosphate nutrient status are most certainly saturated with phosphate. So, when following the prevailing fertilisation recommendations, a high degree of phosphate saturation should not be expected.

4. REFERENCES

Ingenieursbureau 'Oranjewoud' BV., 1992. Inventariserend onderzoek naar fosfaatverzadiging in Friesland.

Schouwmans, O.F., A. Breeuwsma, A. El Bachrioui-Louwerse en R. Zwijnen, 1991. De relatie tussen de bodemvruchtbaarheidsparameters Pw-en P-Al-getal en fosfaatverzadiging bij zandgronden. Staringcentrum-DLO Wageningen, rapportnr. 112.

Zee, S.E.A.T.M. van der, W.H. van Riemsdijk en F.A.M.de Haan, 1990. Het protocol fosfaatverzadigde gronden, deel I: toelichting. Vakgroep Bodemkunde en Plantevoeding, Landbouwuniversiteit Wageningen.

Zee, S.E.A.T.M. van der, W.H. van Riemsdijk en F.A.M.de Haan, 1990. Het protocol fosfaatverzadigde gronden, deel II: technische uitwerking. Vakgroep Bodemkunde en Plantevoeding, Landbouwuniversiteit Wageningen.

Application of Geophysical and Hydrogeological Methods at the
Abondoned Waste Disposal Site "Eulenberg" Near Arnstadt,
Thüringen

HEINRICH KRUMMEL, KLAUS KNÖDEL, MATTHIAS BEUSHAUSEN
Federal Institute for Geoscience and Natural Resources (BGR)

As part of the current joint research project "Methods for the
Description and Investigation of the Subsurface of Operating
and Abandoned Waste Disposal Sites" (sponsored by the Federal
Ministry of Research and Technology) the BGR is carrying out
extensive geoscientific investigations at different testing
sites. Further development and testing of methods for the
evaluation of the subsurface below waste disposal sites is the
aim of this interdisciplinary joint research project. The
scientific interest is focused on the formation's retention
regarding pollutants.
The abandonned waste disposal site "Eulenberg" is one of the
testing sites where different methods are tested in the field.
800.000 m^3 of unspecified domestic and industrial waste that
had been disposed of until 1979 forms the waste body. A bunker
system constructed during the second world war is situated in
the core of the site. Today the waste completely covers this
area. Neither a base nor a surface sealing exists.
The following three problems are to be solved:
- mapping and description of the subsurface regarding permeability and retention
- tracking of possible contamination plumes
- locating of the constuctions below the waste body.

Geophysical and hydrogeological methods have been applied,
evaluated and compared to each other. In addition the cost
efficency of these methods has been taken into consideration.

F. Arendt, G.J. Annokkée, R. Bosman and W.J. van den Brink (eds.), Contaminated Soil '93, 1553.
© 1993 *Kluwer Academic Publishers. Printed in the Netherlands.*

ECOLOGY-MINDED PRODUCTION AND WASTE UTILIZATION - NECESSARY TO SAVE OUR WASTE DISPOSAL CAPACITIES

H. Lehmann
Prof. Dr.-Ing. habil.

Universität Hannover
Institut für Wasserwirtschaft, Arbeitsbereich Wassergüte und Umweltschutztechnik

Due to handling of materials harmful to the environment, dangerous waste is rising constantly. Up to now this waste is still deposited on free waste deposit areas. We have to find new ways to handle the still enormous amount of waste in the West and the even increasing amount in the East, since the current way will lead to a dead end in waste economy.

Waste has to be both avoided and disposed ecology-minded, though its avoidance and recycling has to have absolute priority above treatment and depositing. On one hand this can be achieved by retrieving all arising residues, process materials and additives into a closed material cycle. On the other hand the production process has to consider the regaining of raw materials and recycling. This means an increasing use of secondary material instead of raw material.

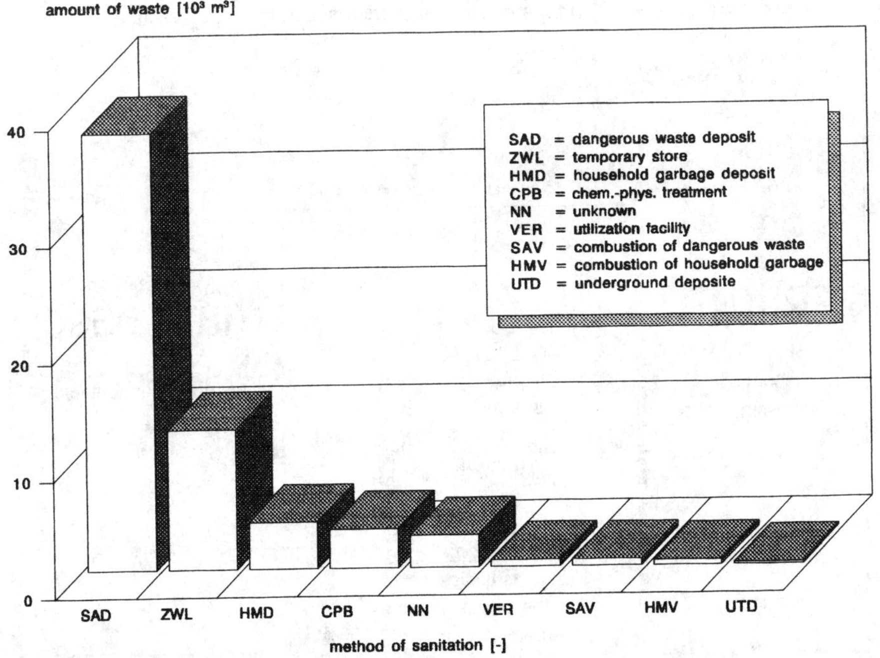

Fig.1   Methods of sanitation (example area)

F. Arendt, G.J. Annokkée, R. Bosman and W.J. van den Brink (eds.), Contaminated Soil '93, 1555–1556.
© 1993 Kluwer Academic Publishers. Printed in the Netherlands.

Data from research projects included in the integrated BMFT-project "Residue streams in the metal industry" show the current ways and possibilities. Fig. 1 shows how the metal industry gets rid of its dangerous waste. Approximately 87.5 % of all arising dangerous waste is deposited.

Usually, however, the factories do not have the possibility to utilize residues in their own processing. Capacities for charging of residues are at hand in other factories, but they cannot be used because of lacking information or due to necessary intermediate treatments to reach the limiting value eligible for further processing. This information is especially important for small and average enterprises because they cannot afford to build a facility of their own. The data gained out of the factories and facilities of an example area and out of the general examination of technical possibilities are collected and process in an information system, which is able to point out arising, avoidance and utilization possibilities of residues in different factories. In addition, this kind of information can be very useful for authorities to compare and evaluate the facilities, and thus recognize whether a facility meets the state of the art. An ecology-minded production first consists of inventoring all material streams including storing, raw material, production process, product and finally waste. In the following a balance of both total emission and energy, together with an evaluation of employed materials and technologies, enables a thorough assessment of the facility.

Fig. 2 shows the structure of this information system, consisting of two units. The first unit is the data collecting subsystem EPIKUR which enables input of all data in an easy and flexible way. All characterizing data important for assessment of a facility are then transferred into a data base, the second unit. Since the two subsystems communicate the pool of all collected data can be interpreted focussing on the development of characterizing values to evaluate a facility in view of its ecological effects. The data analysis takes current laws and order into account, and the output of the results takes place in EPIKUR.

So the impeding tight structures of a common data base system are avoided, and extension and improvement of this system is possible without much additional effort. On the basis of a complete data acquisition the program delivers a highly valuable tool in estimating the current arisings of contaminants, especially in view of further avoidance and decrease of industrial residues.

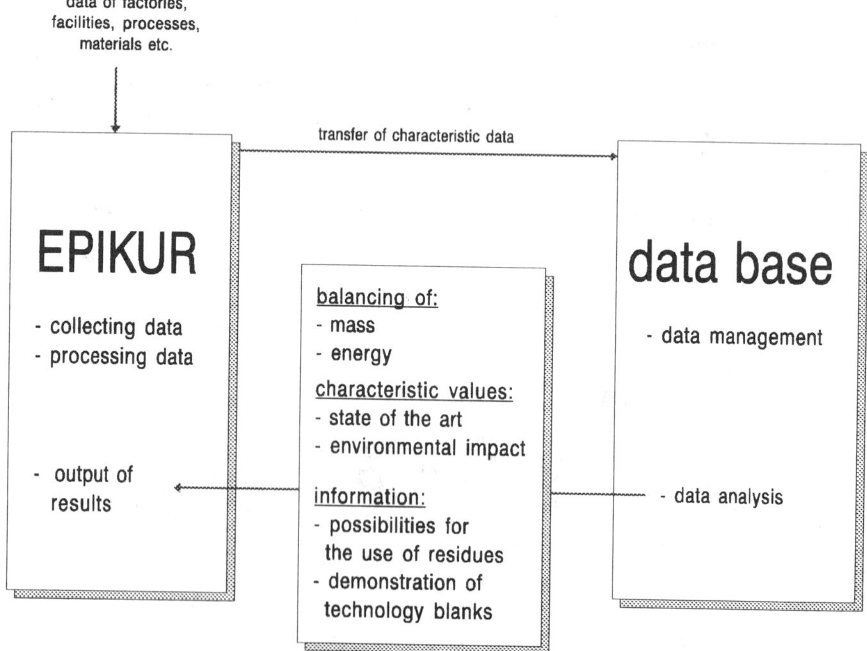

Fig. 2   Structure of the information system

Comparison of Basic Conceptions for Repositories of Radioactive Waste and Toxic Waste

Annette Rolle, Bernd Schulz-Forberg, Bernhard Droste
Bundesanstalt für Materialforschung und -prüfung,
Unter den Eichen 87, 1000 Berlin 45

Introduction

Fundamental safety features for the disposal of highly hazardous waste are the protection of health and environment and the responsibility for future generations. The safety considerations for the final storage of high active radioactive waste have to be extended over a long time into the future because of the partly very long half-life periods. The same aspects apply also to stable chemical wastes of high toxicity. A comparison of existing conceptions for the disposal of radioactive and toxic waste shows that there exist similar fundamental principles of safety measures.

Safety Analyses

The safety analyses for a storage facility have to consider the operation phase as well as the post-operation phase. Essential aspects for the evaluation of a disposal facility are the need for a waste immobilization as a first barrier against a release of hazardous substances and the sound multi-barrier conception i.e. the conception of technical and natural barriers whose efficiency sums up to a general protection of the biosphere. It is one objective of the safety analyses to prove that there are no dangerous impacts for the workers and for the public during the normal operating phase of storage systems and in accident situations, respectively.

Waste Requirements

In analogy to the usual radiation risk analyses the environmental risk caused by stored chemical materials should also be evaluated taking into account all possible pathways of impact. In order to characterize the radioactive waste, information about radiotoxicity of included nuclides, the residual heat generation (e.g. of spent reactor fuels) and about the soluability and there compounds are nessesary. Information about toxicity and of secondary risks are nessesary for evaluation of chemo-toxic waste. In order to minimize the health risks especially high-active-waste will be generally stored insoluable and compacted form. For the same reason toxic waste is only foreseen for disposal in dry and inactiv condition. With respect to the radioactive waste (one has to distinguish between high-active (with higher heat generation) and low-active materials (with nearly no head generation).

F. Arendt, G.J. Annokkée, R. Bosman and W.J. van den Brink (eds.), Contaminated Soil '93, 1557–1558.

The Multi-Barrier System

The storage of hazardous waste in deep and stable geological formations is considered as very safe. In Germany salt domes are the most suitable sites for respositories. The multi-barrier system as basic conception for a long-time storage of radioactive and chemotoxic waste includes:

1. Technical barriers (waste conditioning, waste container)

2. Geo-technical barriers (enclosure of the emplacement room, drill hole closure, dams, mine filling)

3. Geological barriers (salt domes and other geological formations)

Taking into account the properties of waste, the function and efficiency of the several barriers in the general system has to be considered.

The operational safety of storage facilities, especially the protection of the workers during their above-ground and underground activities must be quantified by the containment and the shielding efficiency of the technical barriers. In this context, questions of suitable and tolerable containment materials and containers (for chemical waste) and of the durability of packaging systems (for radioactive waste) are of importance.

Monitoring and Inspection

The examination of all components and systems of a repository for radioactive and high-toxic waste during the phase of it´s manufacturing and before taking into operation the testing and licensing of storage containers and the inspection of each single waste package are the most important quality assurance measures to be used. Before the disposal of the container a correct and complete documentation of the quantity of waste, the composition of the waste, and of the results of the container function tests have to be delivered to the respository. There is included once more a control of the packaging with respect to the compliance of the safety and design requirements. During the operational phase also the parameters of storage fields (temperature, air composition) in districts behind dams shall be controlled with the objection to quaranteee it´s operational safety and to exclude undue accidents. Post-operational monitoring of a repository shall also be apllied.

Literatur

/1/   B. Schulz-Forberg, P. Blümel: Verträglichkeit von Füllgut und Wandungsmaterial als Voraussetzung für die Langzeitlagerung; Symposium "Die Deponie - Ein Bauwerk?", Aachen, 18.-19.9.89, IWS-Schriftenreihe 1/87, S. 297 pp

LEACHATE OF WASTE IN STANDARD LABORATORY TESTS,
IN TESTS UNDER LANDFILL CONDITIONS AND IN A LANDFILL

J.J. STEKETEE*, W.F.T. TROMP** & L.G.C.M. URLINGS*

TAUW INFRA CONSULT B.V., P.O. BOX 479, 7400 AL DEVENTER, THE NETHERLANDS
*   RESEARCH & DEVELOPMENT
** LIFE CYCLE ANALYSIS AND ENVIRONMENTAL MANAGEMENT

The contamination of the soil and groundwater underneath a landfill occurs
via the transportation of soluble contaminations in the percolated water.
Therefore, the leachable fraction of the contamination present in the waste
generally poses a greater threat to the environment than the entire
concentration. It is desirable to determine the potential amount of leaching as accurately as possible. To that end, various tests are available,
e.g. the German S_4 test (prescribed in the proposed EC Council Directive on
the Landfill of Waste) and the Dutch tests in accordance with the NEN 7343
method (column and shake tests). It is generally known that similar tests
cannot accurately predict the amount of leaching under landfill conditions.
The tests will be carried out individually for each waste, using demi-water
(or comparable) as a leaching medium, under aerobic conditions. In a landfill, under practical conditions, there is a complex mixture of waste, a
leaching medium with large concentrations of salt and organic complex
formers and anaerobic conditions. Furthermore, the waste in the landfill
undergoes transformations, whereby the leachability can change drastically.

In this poster the leaching of various types of waste, under standard laboratory conditions, will be compared with leaching under landfill conditions. Four individual waste streams are examined, namely fly ash from a
waste incinerator, spent blasting grit, waste water sludge and the residue
of a soil cleaning installation. The effect of mixing, anaerobic conditions
and the composition of the leaching medium is illustrated in Figure 1.
These parameters have a great individual effect on the leaching behaviour
of heavy metals. By summation it seems that the leaching of certain elements, under landfill conditions, has been reduced till nil (compared with
standard leaching tests). Other elements have sharply increased in leaching, maximum a factor of 100. (See Figure 1).

F. Arendt, G.J. Annokkée, R. Bosman and W.J. van den Brink (eds.), Contaminated Soil '93, 1559–1560.

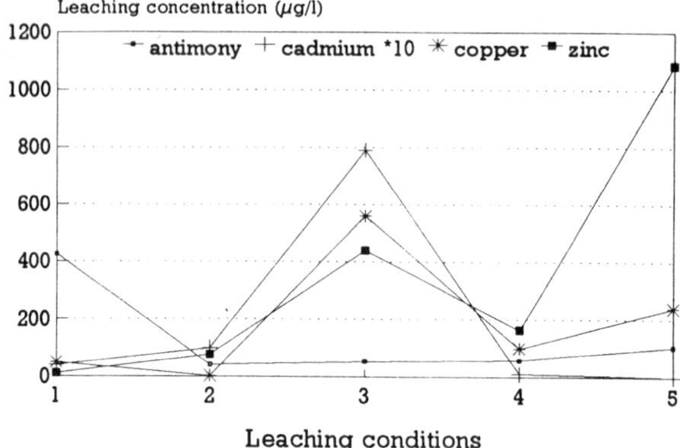

FIGURE 1. Effect of mixing, leachant composition and redox potential on the leachate concentrations of fly ash, blasting grit, waste water sludge and soil cleaning residue. Shake test L/S 20; 1: demi-water, individually; 2: demi-water, mixed waste; 3: acid landfill percolate, Eh = 17 mV, mixed; 4: acid landfill percolate, Eh =-183 mV, mixed; 5: methanogenic landfill percolate, Eh =-178 mV, mixed.

Comparison of the practical leaching (percolate concentrations) of a land-fill for industrial waste with the leaching of the input, such as that measured with standard laboratory tests, leads to the conclusion that the laboratory tests grossly overestimate the leaching (see Table 1). This effect cannot be completely explained by the factors previously mentioned (anaerobic conditions, mixture of waste and composition of the leaching medium). It can be concluded that the transformations in the material also plays an important role.

TABLE 1. Theoretical and practical leachate concentrations at a landfill for industrial waste.

	Calculated, Leaching tests	Calculated, Leaching tests and corrections[1]	Measured
Antimony	230	28	3
Copper	3370	90	2
Lead	2100	37	6
Zinc	3900	830	110

(1) Corrections for mixing, anaerobic conditions and leachant composition (according to Figure 1).

It can be concluded that the standard laboratory tests cannot predict the leachability under landfill conditions. Improvements are possible by taking the anaerobic conditions, mixing and the composition of the percolate water into account. If transformations in the waste do have an important in-fluence on the leaching behaviour, these effects have to be determined under semi-practical conditions, by means of prolonged tests.

SETTING STANDARDS FOR THE QUALITY AND APPLICATION OF BIOGENIC WASTE COMPOST

Koos J. Verloop

Technical Soil Protection Committee, P.O. Box 30945, 2500 GX Den Haag, The Netherlands

INTRODUCTION

In the Dutch soil protection policy the starting point with respect to diffuse sources of pollution is that the input of pollutants to the soil system should be balanced by the output. This balance should result in an equilibrium system where the contents of pollutants in all relevant compartments such as soil, crops and groundwater, do not exceed the quality standards. Composts made of yard waste or organic household waste contain considerable amounts of heavy metals. In an advise to the Minister of Housing, Physical Planning and Environment the Technical Soil Protection Committee evaluated quality and application standards for compost (1). The main lines of the evaluation and some conclusions are presented here.

USING THE BALANCE EQUATION

Creating a balance situation for input of persistent pollutants such as heavy metals to soil and output to adjacent compartments, results in maintenance of the content of compounds in the soil. Figure 1 shows that knowledge of the amount of leaching of heavy metals to groundwater, the atmospheric deposition and uptake by crops as well as input of heavy metals by application of fertilisers is needed for the calculation of the standards for accep-

table load of persistent substances from compost and other fertilizers. It will be very complicated to assess at what standards for compost, the balance situation will be established, because most processes are poorly quantified. However, both fertilising with compost and harvesting of crops can be regarded as one agricultural activity of which the effect on the mass balance of the soil should be neutral. If other fertilisers are used along with compost, the total amount of compounds added to the soil by fertilising, should be taken into account.

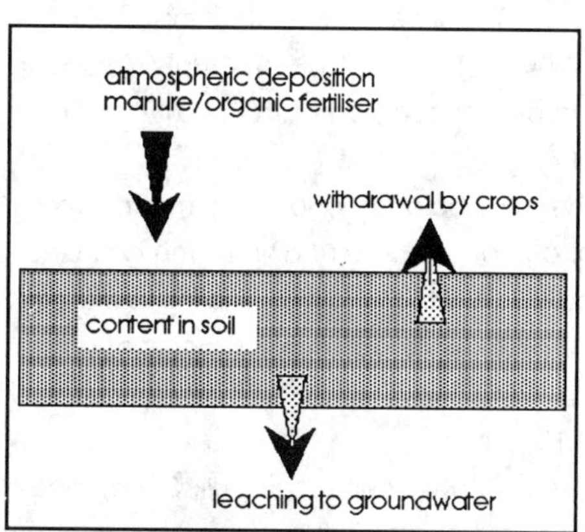

FIGURE 1: Scheme of input and output factors of heavy metals to the soil system.

F. Arendt, G.J. Annokkée, R. Bosman and W.J. van den Brink (eds.), Contaminated Soil '93, 1561–1562.
© 1993 Kluwer Academic Publishers. Printed in the Netherlands.

For the evaluation of standards according to this approach a simple balance equation can be used:

$$F_c \times C_c = H \times C_h$$

where: F_C = the amount of composts applied (kg/ha.yr), C_C = the heavy metal content in compost (mg/kg), H = amount of biomass carried off by harvest (kg/ha.yr) and C_h = heavy metal content in crops (mg/kg).

THE BASE CONTENT

In biogenic waste compost a considerable amount of soil particles occur. The mass fraction of the anorganic soil material in household waste and yard compost is estimated at 70% dry matter. A part of the heavy metals in compost can be ascribed to the soil fraction. The content of heavy metals in compost at which the soil fraction in compost would meet the dutch target values (2) for good soil quality is called the base content.

The base content is taken into account in the calculations with the balance equation. Only the amount of heavy metals that exceeds the base content must be balanced by plant uptake. Thus the balance equation can be elaborated:

$$F_c \times (C_c - B_c) = H \times C_h$$

where: B_C = base content (mg/kg).

With this equation it can be calculated whether combinations of quality standards (C_C-B_C) and application standards (F_C) as proposed in a draft General Administrative Order result in a balance of input and uptake of heavy metals. Data of the withdrawal of biomass (H) and concentration of heavy metals in corps (C_h) where collected by Ferdinandus (3).

CONCLUSIONS

The application of compost that meets the quality standards evolving from the calculation of the base content (C_C-B_C=0), is not restricted by the mass balance of heavy metals. For compost of this quality the limiting factor is the phosphate application standard of 70 kg P2O5/ha.yr.

The mass balance of organic matter would be established in most cases at an application of compost of 3 ton/ha.yr. At the combination of this application rate with most of the proposed quality standards, the influx of heavy metals would be balanced by the uptake by corps. In some cases adjustments of quality standards where recommended.

REFERENCES

1. Advies "Kwaliteit en gebruik GFT-compost", 1991. Leidschendam, TCB, A90/04.

2. Environmental Quality standards for soil and water, 1991. 's Gravenhage, VROM.

3. Rapport "Berekening van zware-metaalbalansen voor de bodem", 1989. Leidschendam, TCB A89/01-R.

THE MINERAL BASE LINER - A HYDRAULIC AND/OR A GEOCHEMICAL BARRIER?

Jean-Frank WAGNER

Department for Applied Geology, Karlsruhe University
Kaiserstr. 12, W-7500 Karlsruhe 1, Germany

INTRODUCTION

The general objective of all base liners consists in the highest possible retention of landfill leachate and of the noxious substances contained therein. This is why a sealing system has to meet the requirements of the lowest possible permeability and the highest possible sorption capacity.

For a mineral liner of a swelling clay mineral mixture (f.ex. bentonite) there is, in addition to the very low permeability, also a very high sorption capacity. For this sealing material however, the stability required and to be required over historical periods so as to ensure a safe operation of the landfill in the case of a chemical attack can not be definitely proved. For a mineral base liner of bentonite or bentonite-like clay mixtures, permeability can slowly increase as a result of the sorption processes. On the one hand this has been proved in practice through clay mineral changes appearing after several years, on the other hand it can also be predicted by theoretical consideration. The increase in permeablility of bentonite sealing materials that had been treated with organic as well as inorganic substances is in many cases attributed to a crack formation, for which the following causes may be possible:

a) *Clay mineral alteration:* f.ex. collapse of the silicate layers by inclusion of pollutants into the interlayer space of clay minerals.
b) *Mineral transformation or formation of new minerals:* f.ex. dissolution of calcite and precipitation of a new mineral phases.
c) *Decrease of the "Diffuse Double Layer"* by an increase in electrolyte concentration.

DOUBLE MINERAL BASE LINER

Finally, the long-term stability of the smectite and thus the function as a hydraulic barrier of the mineral bentonite liner as well do not seem to be given because of the ever proceeding interaction with the leachate components. This is why mineral sealing materials of more stable clay mineral mixtures (f.ex. kaolinitic clays) are used increasingly. These however only show minimal sorption capacities, so that they can only extract a minimum of pollutants from the leachate.

This led to the consideration that the highest possible elimination of pollutants in landfill leachate as well as a low long-term permeability can not be obtained by one sealing layer only. Therefore a combined installation of two separate mineral layers (Double Mineral Base Liner; Fig. 1), one of which assuming the sorption function ("geochemical barrier"), the other one ensuring a low long-term permeability ("hydraulic barrier") is proposed in the papers of Wagner (1991a, 1991b).

F. Arendt, G.J. Annokkée, R. Bosman and W.J. van den Brink (eds.), Contaminated Soil '93, 1563–1564.
© 1993 *Kluwer Academic Publishers. Printed in the Netherlands.*

An upper, highly "active" layer (geochemical barrier), which is orientated towards the waste should extract as much pollutants as possible from the leachate. This filtering layer can be composed of minerals having a very high sorption capacity and/or very reactive mineral phases. The reactions (adsorption, dissolution/precipitation, etc.) within the active layer can however lead to an increase in permeability. A constantly low permeability of the sealing system will be ensured with the installation of a second layer, the so-called "inactive" layer (hydraulic barrier), which is not reacting with the pollutants. The lower layer consists in 'inactive', chemically very stable sealing materials. The permeability of the inactive layer can be kept very small, given an adequate granular grading and an appropriate compaction, so that the advective solute transport will be in any case lower than the diffusive solute transport (s. Wagner & Egloffstein 1990). The installation of an inactive layer underneath the sorption layer allows the leachate to be held back as long as possible and thus maximizes the reaction time between active layer and pollutants.

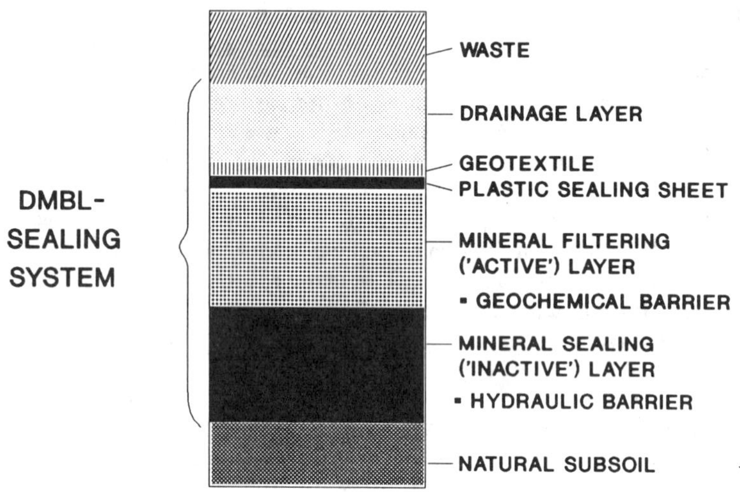

Fig. 1: Schematic of a vertical section through a Double Mineral Base Liner (DMBL).

REFERENCES

WAGNER, J.-F. (1991a): Die Doppelte Mineralische Basisabdichtung - ein Deponieabdichtungssystem mit zwei unterschiedlichen Funktionsweisen. - Ber. 8. Nat. Tag. Ing.-Geol., 75-81, 2 Abb., Berlin.

WAGNER, J.-F. (1991b): Concept of a Double Mineral Base Liner. - Proc. Sardinia 91, Third Int. Landfill Symp., vol. 1, 539-545, CISA, Cagliari, Italy.

WAGNER, J.-F. & EGLOFFSTEIN, Th. (1990): Advective and/or Diffusive Transport of Heavy Metals in Clay Liners. - Proc. 6th Int. Congr. Int. Assoc. Eng. Geol., 6-10 August 1990, Amsterdam, Vol. 2, 1483-1490, Balkema, Rotterdam.

OPTIMIZATION OF BIOLOGICAL PRETREATMENT OF RESIDUAL WASTE

Dr. Ing. Ulrich Wiegel, Dipl.Biol. Gabriele Janikowski

ITU - Ingenieurgemeinschaft Technischer Umweltschutz GmbH,
Berlin, Germany

1. ABSTRACT

Only materials producing insignificant quantities of gases and leachates over the long term will be accepted at landfills in the near future. Therefore, the biological pretreatment of waste is being considered as an alternative to thermal oxidation (incineration). This process is an optimized biological decomposition of the organic components with the goal to produce materials that are inert and able to be landfilled. To ensure these high standards, the decomposition process must exceed the level reached by conventional composting processes.

2. GOALS

The following goals were identified for this composting optimization process:
- Obtain the greatest possible concentration of organic content in specific waste fractions
- Target the extraction and treatment of the organic content using specialized procedures
- Optimize the decomposition process for the high organic content fraction, parallel to ensuring operational safety (optimal operating conditions).

3. CONCEPT

The main proportion of organics in the waste stream lies between 8 - 80 mm in size. The first step in material preparation is a two level size screening to concentrate the organic material in Fraction II (Figure 1). Fraction I (> 80 mm) will still contain large quantities of paper, even though recyclables will have been previously hand-sorted. Based on experience, these large quantities of paper will not be decomposed during a "normal" composting process. The remaining fine material (Fraction III, < 8 mm) still contains a significant proportion of organics (> 15 %) which must be extracted before this fraction can be landfilled.

F. Arendt, G.J. Annokkée, R. Bosman and W.J. van den Brink (eds.), Contaminated Soil '93, 1565–1567.
© 1993 *Kluwer Academic Publishers. Printed in the Netherlands.*

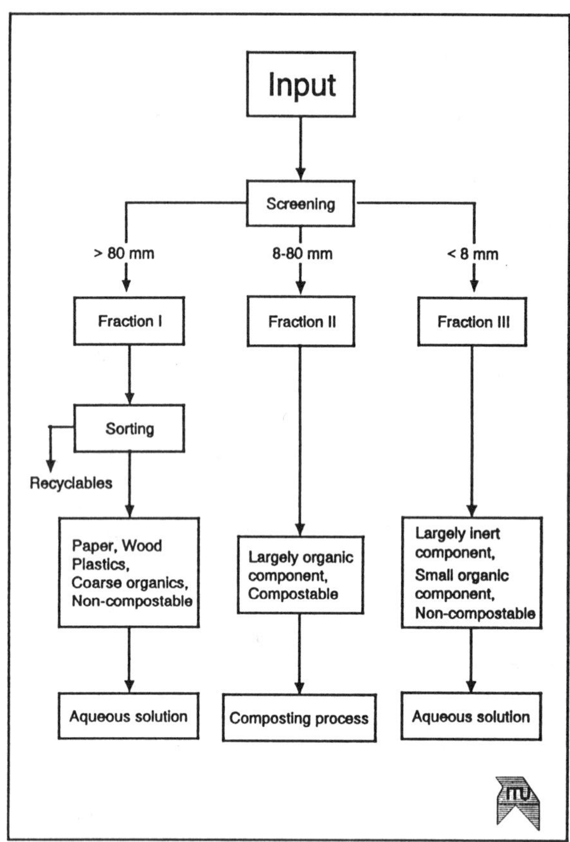

Figure 1:   Distribution of waste over two step screening for the
             preparation of treatment processes

Therefore, the large and small fractions (I and III) will be
treated using a specialized washing/dissolving process. The paper in
Fraction I will be dissolved, while the organics in Fraction III will
be washed out. Both of these liquidized fractions will then be treated
by anaerobic digestion.

Fraction II which contains most of the organics will be treated
using a conventional aerobic composting process. The main task of this
process is to optimize the decomposition in such a way that microbial
conditions exceed those in conventional waste composting. Figure 2
shows the decomposition with varying degrees of optimization -
partially through an increase in initial biological activity through
optimized aeration and moisture content, and partially through process
reactivation with the addition of new material in dosages.

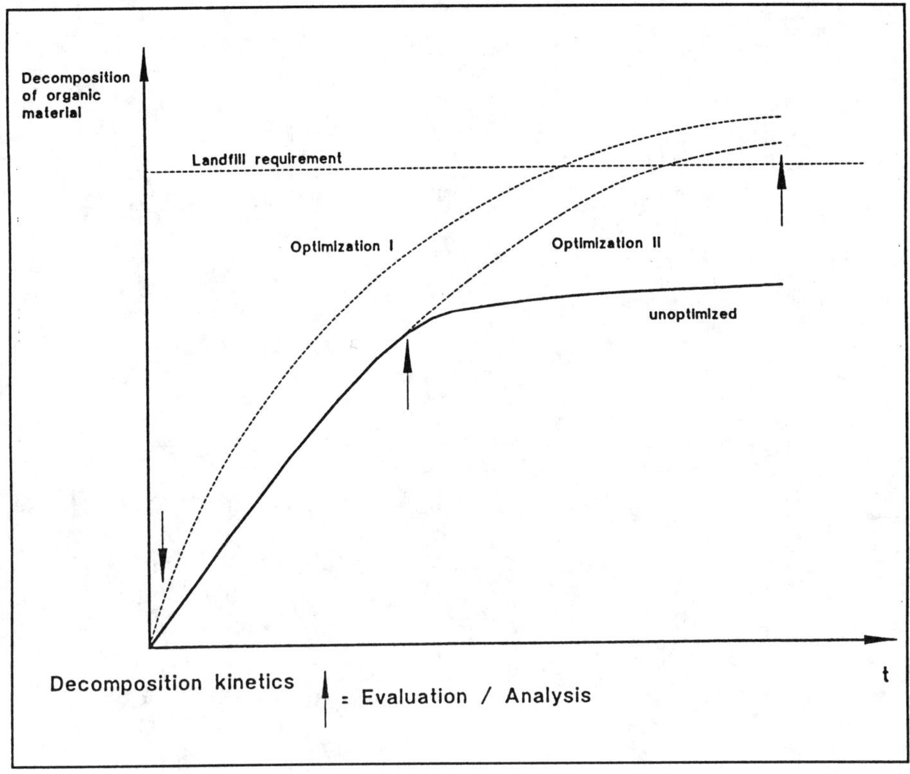

Figure 2:   Targeted increase of decomposition during optimal processing

4. REALISATION

The investigations are planned to take place over 1 year in a composting container system with three separate controllable 25 m^3 reactors. The three fractions will be separated using a mobile sorting screen. Each of the fractions will be evaluated for their organic contents and availability of organics. The composting process will be monitored continuously. The decomposition progress will be evaluated by testing the material and CO_2 emissions. The ability for separation and decomposition of Fractions I and III will be evaluated in the lab and later at a pilot scale. The optimal size screening into fine, medium and large fractions will be optimized within the investigations.

The results will be used in developing optimal operating conditions for decomposition as well as obtaining residual waste with the required quality for landfilling.

7. WORKSHOPS

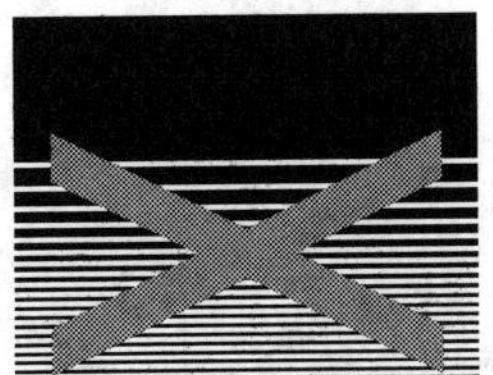

WORKSHOP IN-SITU SOIL REMEDIATION

E.R. SOCZÓ[1], T.A. MEEDER[1], H. EIJSACKERS[2]

1   National Institute of Public Health and Environmental Protection, P.O.Box 1, 3720 BA Bilthoven ,The Netherlands
2   The Netherlands Integrated Soil Research Programme, P.O.Box 37, 6700 AA Wageningen, The Netherlands

For the remediation of contaminated sites, ex-situ treatment in thermal and soil washing plants and solidification/stabilization are the techniques most used in practice. While, however, biological and in-situ treatment have been increasing. Several in-situ soil remediation techniques, like soil vapour extraction, in-situ bioremediation, in-situ extraction (soil flushing), in-situ vitrification and electroreclamation, are applied at the moment to soil and groundwater. This in contrast to the so-called "pump and treat" techniques, used for the treatment of contaminated groundwater. A recent international study [1] on the applicability of in-situ techniques concluded:

- Soil Vapour Extraction is a viable technology for the unsaturated zone remediation of (semi)volatile contaminants; off gasses can be treated by conventional technologies.
- Permeability is a key factor in applying in-situ bioremediation. A combination of venting and a water recirculation system improves the biodegradation of the contaminants and can be cost-effective.
- The use of extraction techniques in-situ is limited to sandy soils, which have limited absorptive capacity.
- Electro-reclamation is a new technology, which can be applied to clean clayey soils contaminated with heavy metals.
- In-situ vitrification is a promising technique for highly contaminated, mixed organic and inorganic wastes. Vitrification combines the advantages of the thermal process (destroying organic compounds) with immobilization of the inorganics.
- The pump and (above-ground) treatment techniques form a limited technology for remediating aquifers.

There is much interest in the application of in-situ techniques, especially for the remediation of industrial estates, where the cost of excavation is high and immediate removal is not necessary. However, a few limitations prevent large-scale application of in-situ techniques. Soil heterogeneity leads to high spatial variations in soil cleanup, and high local residual concentrations of the pollutant. Soil heterogeneity, together with the heterogeneous distribution of the contaminants, hamper monitoring of the remediation process, prediction of time needed for the cleanup and the assessment of the final result. To achieve better control, sampling strategies and procedures for analyses should be developed and/or standardized. Advanced system design and the use of simulation models could help to optimize the process conditions.

F. Arendt, G.J. Annokkée, R. Bosman and W.J. van den Brink (eds.), Contaminated Soil '93, 1571–1573.
© 1993 *Kluwer Academic Publishers. Printed in the Netherlands.*

The aim of the workshop is:
- to give a critical review of the main aspects of in-situ remediation;
- to define the bottlenecks in technology development and impediments in application on a large scale;
- to show the interlinkage of fundamental research and practice;
- to discuss the desired (research) strategy.

The workshop will progress from fundamental aspects of the in-situ remediation process to the application of the techniques in the field. Modelling can be considered as the link between research and practice and, consequently, will form the link between the three presentations. Simulation models (will) play an important role in the optimization of the treatment process, as well as in the prediction and evaluation of the cleanup results. Three introduction papers, each of which will be concluded with a few challenging theses, will be followed by a discussion.

Dr. M. Loxham (Delft Geokinetics, The Netherlands) will present the geohydrological and mass-transfer aspects of in-situ soil remediation, including modelling the processes. Several factors which determine the transfer of contaminants on micro and macro-scales will be discussed. On the macro-scale the transfer of reactants, solvents and carriers into the contaminated soil volume (and normally their subsequent removal) is an advective process following the principles of geohydrology. On the micro-scale of the process, in the soil itself, several other factors may determine the overall rate of the cleanup. These include diffusion, both in the fluid phase and along the particle surfaces, dissolution kinetics, adsorption-desorption rates and reaction rates. Ostwald effects also play a role.

These processes can be modelled in one or two theoretical frameworks. The starting point of the first is the convection-diffusion equation, with emphasis on the spatial distribution of the various concentrations, whereas the second, based on unit operation concepts borrowed from chemical engineering principles, highlights the mass-transfer rates and the chemical processes taking place.

Mr. A. Bachmann (MBT Environmental Europe, Zurich, Switzerland) will focus on the effect of mass-transfer on in-situ (bio)remediation. Data will be presented to show the effects of desorption from soil aggregates on the bioremediation kinetics of pesticides. Two models have been applied to the desorption and biodegradation kinetic data, a first-order model (FOM) and a sorption-retarded radial diffusion model (RDM). Only RDM could explain the effect of aggregate size on pesticide desorption and bioconversion rate. RDM also yielded the best fit to the desorption kinetic data. The practical consequences for in-situ remediation techniques will be explained by means of data interpretation.

The third speaker (a representative from the U.S. Environmental Protection Agency) will discuss how to integrate fundamental knowledge into a site-specific approach for in-situ remediation. A brief review will be given of results in cleanup using conventional in-situ techniques and the improvement through advanced system design. Special attention will be paid to the heterogeneity of the soil and its effect on the design of the treatment system. The role and implementation of computer models in practice (to help predict and evaluate the results of the remediation) will also be discussed. The final discussion point will be how to achieve cost-effectiveness in in-situ remediation.

References:

1. Olfenbuttel, R.F. (ed., 1992). Demonstration of Remedial Action Technologies for Contaminated Land and Groundwater, Final report NATO/CCMS Pilot Study, Committee on the Challenges of Modern Society, Number 190, Brussels, Belgium.

PRESENT STATE AND ASSESSMENT OF ENVIRONMENT AND SOIL
CONTAMINATION IN CZECH REPUBLIC

MEČISLAV KURAŠ[1] and JAN MIKOLÁŠ[2]

[1] INSTITUTE OF CHEMICAL TECHNOLOGY, PRAGUE, 166 28 PRAGUE 6
[2] ENVIRONMENTAL MANAGEMENT OFFICE, 120 00 PRAGUE 2,
CZECH REPUBLIC

1. ABSTRACT

This paper deals with the present state and the causes of
enormous environmental contamination in Czech Republic. Some
data concerning the landfills of different wastes and the contaminated soil in military areas are presented. Starting points
for improvement the environmental conditions and for soil decontamination are briefly discussed.

2. CAUSES OF CRITICAL ENVIRONMENT CONTAMINATION

Czech Republic belongs to the countries with the most contaminated environment in Europe and consequently with maximum
health, ecological and cultural damages and consequences. In
the north-west Bohemia the environment is in the critical state.

Czechoslovak economy was oriented expecially to the heavy
industry with high demand on primary sources. The specific consumption of more than 7 tons of specific fuel per habitant puts
us on the fore place in the world. The share of the industry in
power consumption is much higher than in the neighboring countries. In Austria it amounts only 30%, in Germany 32%, but in
former Czechoslovakia 54 %. The main energetic source is still
brown coal with a high content of ash and sulphur. The content
of sulphur is in average about 1.5%. In sixties the share of
coal in the energy production reached 90% and remains still
high (Moldan et al. 1990).

The World Bank recently estimated that it would take a total investment of 50 billions US dollars to help the former
Czechoslovakia achieve acceptable environmental standards, and
even with this, accompanied by strenghtened regulations and
marked pricing, it will be decades before the main environmental problem in this country are mitigated.

3. ATMOSPHERE CONTAMINATION

Especially dangerous from the point of view of health consequences is the atmospheric contamination. The maximum deposition of total sulphur (above 10 g per year and quadrate meter) occurs in the border region of north Bohemia which makes
a part of so caled Black Triangle. Czechoslovakia was on the
6th place in Europe with the total amounts of SO_2 emissions, on
the 8th place with the NO_x emissions, but on the first place
with regards to the state area. The total emissions of SO_2 in

F. Arendt, G.J. Annokkée, R. Bosman and W.J. van den Brink (eds.), Contaminated Soil '93, 1575–1579.
© 1993 *Kluwer Academic Publishers. Printed in the Netherlands.*

former Czechoslovakia amounts more than 3 millions tons per year, that of solid matter 1.7 millions tons per year. The annual emissions of other noxious compounds amount to about 1 million tons of NO_x, 1.3 millions ton of CO and 0.2 millions tons of hydrocarbons.

In some regions of north Bohemia the highest average acceptable SO_2 concentration in atmosphere, 150 $\mu g.m^{-3}$, is exceeded for more than 100 days in a year. These values are frequently exceeded in the extreme cases during the inversion emergency – the maximum measured value was as high as 2974 $\mu g.m^{-3}$. In Prague at the same time the maximum daily concentration of SO_2 was more than 3000 $\mu g.m^{-3}$ which points to the fact that the emissions from the extensive complex of local heatings with low chimneys have during inversion situation the same consequences as the emissions from large point sources emitted in the height of several hundreds meters.

The SO_2 emissions from traffic increased from the year 1971 by 19% and NO_x emissions by 50% and a further increase is expected. Only the Pb emissions decreased from 1100 tons in 1984 to 300 tons in 1989 owing to the decrease of lead in petrol from 0.6 g to 0.15 g per litter.

According to the model calculations it can be supposed that about 70% of SO_2 emissions and 90% of NO_x emissions enter into distant transfer. From this amount more than 1/3 is deposited in our territory and remainder falls in other European countries and adjacent seas. The north-western part of Czech Republic is loaded by a total deposition of sulphur of 15 $g.m^{-2}$, whereas the clean part of south Bohemia in Šumava mountains only by 2 $g.m^{-2}$ (Viturka et al. 1992).

Especially polluted north-west Bohemia makes only 2% of the total areas of the former Czechoslovakia but gives rise to 45% of SO_2 emissions and 40% of NO_x emissions. The emissions from the thermal plants in this region are as high as 1.2 millions tons.

In Czech Republic there are more than 600 stations measuring the SO_2 content in atmosphere by West Glaecke method. In some of them other noxious compounds are also measured. For the assessment of the total atmospheric toxicity in different places in Prague we have succesfully used the bioindicator containing the conidia of the mould strain Aspergillus niger (Bomar et al. 1992, 1993). The toxicity evaluation is based on the measurement of inhibition of conidia growth of this strain. For this year, the Prague Municipal Office prepare the vast programme for complex measurement of noxious compounds in atmosphere and soil in Prague in which the Institute of Technology, Prague also takes part.

4.   SOIL CONTAMINATION
The total area of former Czechoslovakia was 12.79 millions hectars from which 53% is agricultural soil and 36% forest soil. It represented 0.44 ha of agricultural soil and 0.30 ha of arable soil per habitant. Due to the unreasonable cultivation in the former years a considerable part of soil is subjected to erosion. On the heavy eroded soils, expecially on the slopes, the yield of agricultural products are decreasing by as

much as 70%. It is equal to the loss of agricultural production
from 330 000 ha.

More than 700 000 ha of soils in former Czechoslovakia are
damaged by the emissions. Due to this fact the soil continue to
be more and more acid (pH value 3 to 4) and cause the yield decreasing of agricultural products. Under the direct influence of
SO_2 and NO_x emissions is more than 10% of soil. That has evi-
dently a negative influence on the metabolism of plant and food
chain.

The worst affected are the forest massives, especially all
boundary mountains in the north Bohemia and the speed of their
withering away increase rapidly. It seems that beside the SO_2
and NO_x emissions the other harmful substances, such as ozone
and fluorine influence the devastation of the forests. Even if
a substantial decrease of emissions occurs, the destruction of
forests in this region (and in other regions as well) will continue at least to the year 2010 to 2020.

Especially susceptible to the threat of acidification are
the unsaturated soils of the mountains in northern Bohemia, of
the north Bohemian sandstones and the soils of gravel sand terraces in central Bohemia, where the considerable influence of
contamination from large sources is observed. The soil relatively most endangered by acidification can be found in the northern Moravia in the Ostrava region.

The great threat to the soil and especially to the ground
and underground water contamination presents the enormous application of industrial fertilizers in agriculture. The present
-day consumption of nitrogenous, phosphorous and potash fertilizers per hectar of agricultural land place our country to the
fifth place in Europe. Maximum amount of pure nutrient applied
per hectar of agricultural land is in some regions higher than
350 kg.

Erosion damages occur mainly in hilly land areas and on the
edges of floodplain. The largest denudation occurs in soils in
which a considerable part of national plant production is realised. The intensity of soil denudation in some areas reaches
millimeters to centimeters every year, thus exceeding by at
least an order the rate of soil forming proceses.

5. HEAVY CONTAMINATED LOCALITIES

There are 150 localities in Czech Republic, which were un-
till recently used by Soviet army. The majority of these localities are heavily contaminated mostly by petroleum products
and by other noxious compounds. In many localities the layers
of hydrocarbons, more than 1 m deep on the surface of underground water are not exception. In an airport in northern Bohemia the layer of petroleum products on the surface of under-
groundwater was as high as 4 m. In an airport in Moravia several thousands tons of fuel were extracted from the underground
level during the exploration works.

In several territories the concentration of hydrocarbons in
soil were found to be as high as 10^4 to 10^5 mg.kg^{-1}. In Czech
Republic, there are more than 1800 legal landfills and about
10 000 illegal landfills of different types of wastes. These
illegal ones whose existence is mostly unnoticed and often kept

secret contain numerous toxic pollutants such as chlorinated
hydrocarbons, heavy metals and other noxious compounds.

6. MEASURES FOR SOIL DECONTAMINATION

In Czech Republic the works on the decontamination of soils
contaminated especially by petroleum products has started several years ago. Recently, several industrial units for the soil
decontamination by solidification have been established. Bacterial soil decontamination is now provided by numerous companies
utilizing the different bacterial strains. Especially succesful
proved to be the strain Pseudomonas putida. Its industrial preparation is based on bacteria isolated from natural sources. In
our Institute we are now investigated the influence of the
structure and amount of hydrocarbons and chlorinated compounds
on the speed of their bacterial degradation.

7. STARTING POINTS TO ENVIRONMENT IMPROVEMENT

Despite given problems the environmental situation in
Czech Republic is not without promise. It can be seen from the
experience of Western countries. Thanks to the developed market
economy, and thanks to the existing democratic system, these
countries have been able to improve the state of environment
considerably during past twenty years.

Also in Czech Republic basic political and economic pre-
conditions for the gradual improvement now exist. Bad economic
policy in the past was one of the main causes of the present
state of affairs; a reformed economic policy can thus have immediate positive effects on the levels of pollution. Market
conditions should bring about the closure of industries which
are obsolete and often the biggest polluters of the environment.

Some estimates state that the economic reform will result
in the cutting of SO_2 emissions by at least 30% of the present
levels within the next two years, and the substantial decrease
in the consumption of energy was already recorded. More than
1.6% of the Gross Domestic Product was dedicated to protecting
the environment in 1991 which is double that of the amount in-
vested in the 1980´s. A law on tax reforms was passed which
harmonizes the tax system in the Czech Republic with the systems of the developed nations, and in which support for the
protection of the environment has been formulated. The Federal
Act on Large-Scale Privatisation was amended in February 1992
to require environmental audits for properties undergoing transformation. New options for incorporating environmental concerns into the mass privatization program are under discussion,
including possible indemnification of buyers for third party
liability and off-site effects, along with a sharing of liability between the government and the buyer for on-site contamination from past industrial activities.

Also new environmental laws have been adopted. We have a
new Clean Air Act, Law on Waste, Law on Environment, Law on Environmental Impact Assesment. Other new laws on soil and forest
protection, on water pollution etc. are under preparation. New
laws and standards are or will be consistent with the European
Communities laws which is very important for foreign investors.

8.   PRIORITIES AND INTERNATIONAL CO-OPERATION
    All environmental problems are top priorities in Czech
Republic. The greatest problems facing us are air pollution
and the contamination of the food chain, followed by the problems of the treatment of water pollution and solid wastes.
    The immediate task to be undertaken is to decrease SO_2
emissions, with an emphasis particularly on regional air pollution problems. The decontamination of the food chain is closely connected to the reducing waste at the source, particularly hazardous wastes, and similarly, with the proper treatment
of these wastes in incinerators or controlled waste landfills.
The construction of municipal waste water treatment plants
must continue so that we can assure clean water supplies.
    Foreign co-operation plays an important role in the at-
tempt to solve the environmental problems in Czech Republic.
We are aware, however, that it is citizens of Czech Republic
and the Czech enterprises who play a decisive role in cleaning
up the environment.

9.   REFERENCES

    Moldan B. et al. (1990). Environment in Czech Republic.
Prague: Academia Press.
    Viturka M. et al. (1992). Atlas of the Environment and
health of the population of the CSFR. Prague: Federal Committee
of Environment.
    Bomar M.T., Punčochářová J., Kuraš M. (1992). Biomonito-
ring of contaminated atmosphere from different sources by mould
strain Aspergillus niger. Proceedings of the conference Envirotech Vienna 1992 (W.Pillmann, Ed.) Vienna: International Society for Environmental Protection.
    Bomar M., Bomar M.T., Punčochářová J., Kuraš M. (1993).
Untersuchung der Umweltbeschaffenheit im Gelände der Technischen Universität, Prag. Staub-Reinhaltung der Luft, in press.

SOIL CONTAMINATION IN POLAND

Henryk Greinert

Higher College of Engineering, Department of Environmental
Restoration, 50 Podgórna Street, 65-246 Zielona Góra, Poland

ABSTRACT
   The majority of soils in Poland developed from sands
various origin. Most of them are acid and have low buffer
capacity. Therefore the soils are sensitive to contaminants.
The greatest amount of pollutants is emitted by the coal
power plants. The large doses of heavy metals in some soils
come mostly from the smelters of non-ferrous metals. The best
way for improving the environmetal condition is the change of
the structure of industry and to install cleaner
technologies. Several decontamination measures and
recultivation in wider scope will be also necesarry.

INTRODUCTION
   Poland is situated in the zone of moderate climate,
between 49° and 55° northern geographical latitude. The mean
temperature for the country varied from 6,5^O C in the east to
8,5^O C in western Poland. The average annual precipitation
rate amounts in most cases 500-700 mm. 75,1 % of the
territory has the absolute height below 200 m.
   The area of Poland is 312.68 thousand sq km. The land use
is presented in Table 1.

TABLE 1. The land use in Poland (1).

	Thousand hectars	Distribution %
Total area	31268	100
Agricultural area	18784	60,1
Forests	8884	28,4
Inland waters	826	2,6
Mine areas	42	0,1
Traffic sides	989	3,2
Settlements	952	3,0
Waste land	504	1,6

F. Arendt, G.J. Annokkée, R. Bosman and W.J. van den Brink (eds.), Contaminated Soil '93, 1581–1591.
© 1993 Kluwer Academic Publishers. Printed in the Netherlands.

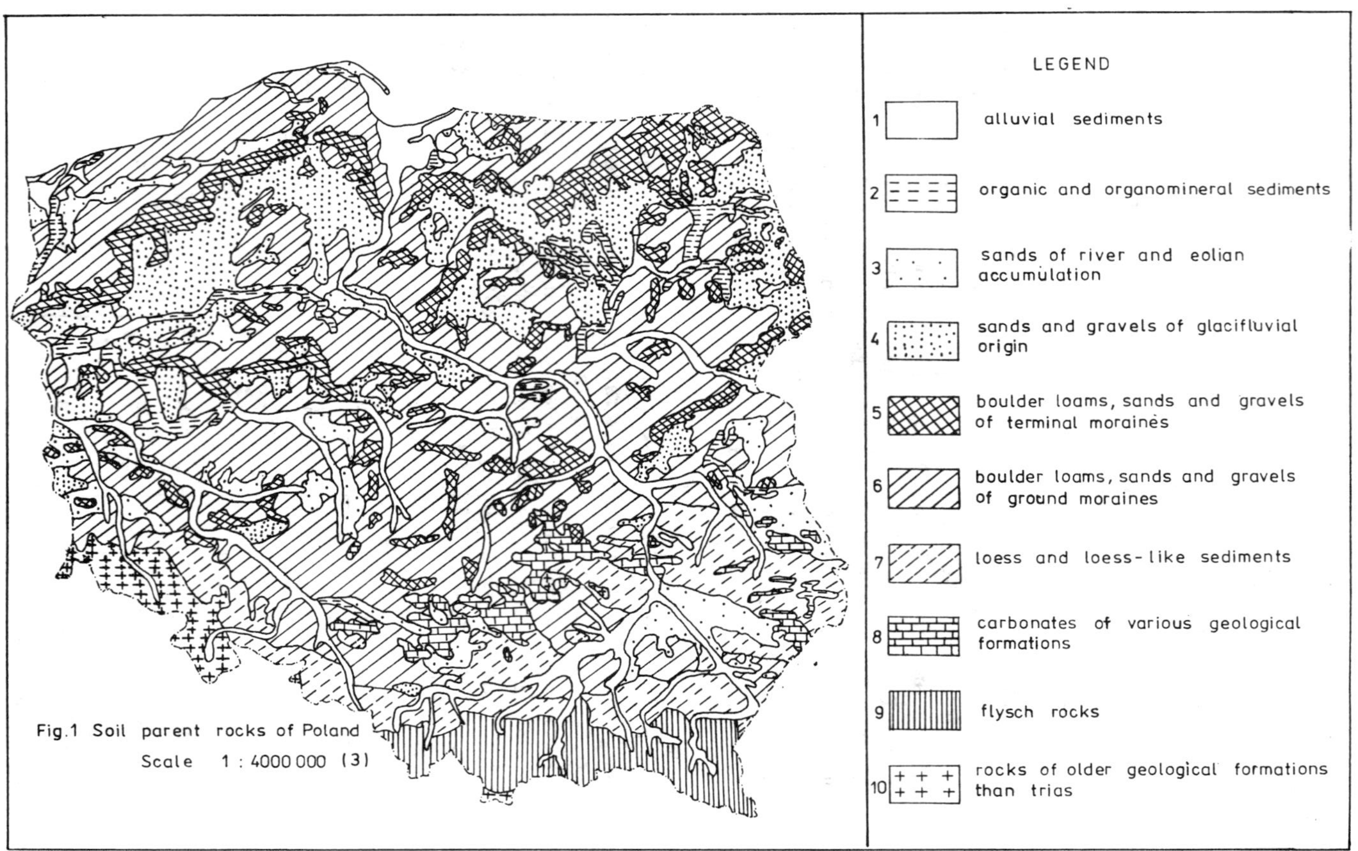

Fig.1 Soil parent rocks of Poland
Scale   1 : 4000 000 (3)

About   75 %   of polish soils developed from glacial
sediments.   Only   in   the   southern part of the country occur
rocks   of   older origin. The alluvial sediments and peats are
the youngest soil parent rocks (fig. 1).

From the mechanical composition point of view the largest
group are sandy soils of various origin (60 % of mineral
soils).   In   the   lowland   part   of Poland most sandy soils
developed   from   the   sands   of   the old fluvioglacial
accumulation terraces and outwash plains. These are the
podzols, brown acid soils, pseudogleys (fig. 2). As a rule,
this soils are acid, what forms favorable conditions for
heavy metal solubility and availability of these elements
(5).

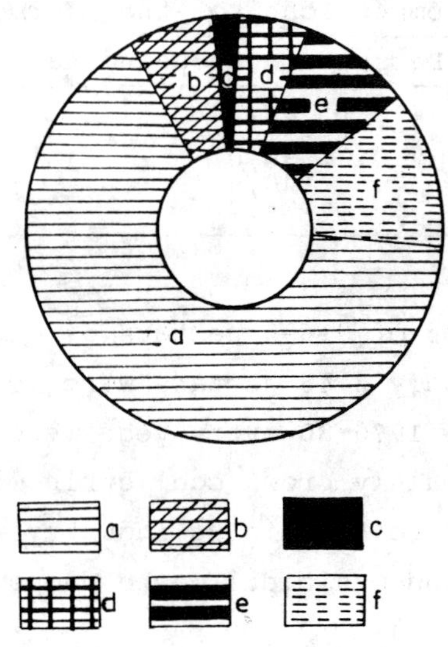

FIG.2. Soils of Poland.

Fig.2. Soils of Poland.
The genetic classes of soils, expressed in the percent of total area.
a). podzols, pseudogleys and brown acid soils, b). typic brown soils and lessive, c). chernozems, d). alluvial soils, e). mountain soils, f). others.

THE DEVASTATION AND CONTAMINATION OF SOILS IN POLAND

1. Decrease of the area of soils in Poland.

In the year 1946 in Poland was 20440 thousand hectars in agricultural use. In 1991 the area decreased to 18780 thousand hectars. Table 2 give some informations about this process.

TABLE 2. The decrease of the land in agricultural use in Poland in comparison to the former year (2).

Year	Decrease - thousand ha
1980	51,3
1985	31,1
1989	30,0
1990	20,9
1991	24,3

In the year 1991 area of 91695 hectars of devastated land needed recultivation. Only 2146 hectars were recultivated in this year. In the decade 1970-80 every year were recultivated 4000 - 5000 hectars, mainly brown coal strip mine dumps. 42 thousand hectars belongs to the mining industry, and 11725 ha is under the industrial and municipal waste dumps.

2. Soil erosion

According to the evaluations done by the Institute of Plant Cultivation, Fertilization and Soil Science in Puławy (IUNG), 27,9 % of the soils in Poland are affected to water erosion (12,6 % of them strong, 38,7 % - medium and 48,7 % - weak). The wind erosion threated 28,2 % of arable land.

The most intensive soil erosion appear in the mountains, the loess area and on the marginal moraines. Also the strip mine dumps, wastes dumps and construction sites are influenced very much by erosion. A lot of contaminats from this sides spread over the environment.

3. Soil contamination with industrial dusts.

Although the amount of industrial dusts decrease from year to year, what is the effect of electrofilter installation with better effectivity, still in many areas the dust pollution overruns the upper limit 250 tons/sq km. To the very polluted soils with heavy metals (Cd, Zn, Pb, Cu, Ni, As) belongs the Silesia area (fig.3). There are many sources of air pollution, like iron and non-ferrous smelters, foundries, cokeries, power plants fired with pit coal. In many cases the concentration of pollutants is higher than the C-values of the Dutch list. Other heavy polluted with Pb and Cu region is the industrial complex in Legnica voivodeship. Both above mentioned regions were officialy classified as areas of ecological catastrophe. Near Tarnobrzeg the sulphur dust acidified the soils in the vicinity of the storing and reloading places to the pH 1.

4. Soil pollution with gaseous contaminants.

Power plants in Poland are fired in over 90 % with pit coal and brown coal. As a result of this about 4 milion tons SO_2/year were emitted every year. Nearly the same amount of lime was applied to the arable soils by the farmers. The sulphur dioxide reduced the effect of liming on the arable soils and worsened the growing conditions in forest soils, where the additional acidity was not neutralized. This may be one of the important reason of drying out of the pine and spruce trees in some regions of Poland.

It is necessery to mention, that quite large amount of

SO_2 is blowed by wind mainly from the west.

In the vicinity of nitrogen fertilizer plant in Puławy the air pollution with nitrous oxides caused the completely drying up of the pine forest.

5. The influence of the industrial and municipal waste materials on the soil quality.

The amout of stored waste materials increased from year to year:

in   1980 was stored   914 347 thousand tons wastes
  "   1985   "   "   1324 264   "   "   "
  "   1990   "   "   1637 900   "   "   "

The largest amount of wastes produces the mining industry. Most of this wastes, especially from the strip mines, can be relatively easy recultivated. The exception are the pyrite containing wastes, which by contact with water and oxygen forms sulphuric acid, and their pH dropped to low, phytotoxic level (6). Many difficult problems create by the recultivation dumps of flotations slurries, flying ash from power plants, phosphogypsum, mineral wool wastes, and salt-containing wastes.

Many waste substances are used in the agriculture as fertilizers or soil conditioners (lime from many processes, organic matter). Not always the result are good, because the wastes contain admixtures of toxic substances, like heavy metals (7). Therefore before applying such materials chemical analysis are always necessery.

The largest amount of industrial wastes is accumulated on the area of Katowice voivodeship. On this territory is 1/3 of all wastes in Poland (in Poland are 49 viovodeships). Additionally, the wastes are often very toxic.

On the area of copper mines and copper smelters complex Legnica-Głogów are also large waste dumps.

6. Other kinds of soil contamination or devastation in
   Poland.

   In many parts of Poland were noted also other negative
changes in the soils, caused by industrial or agricultural
activity. On the areas and various construction sites the
soils the are usuelly geomechanically disturbed. A_1 horizon,
as a rule, is removed, the soil layers are dislocated. Much
wastes from the construction sites were mixed with the soil.
Heavy machinery compact the whole soil profil. The changes are
often so deep, thet for the establishing of greenland soil
material is taken from outside, in many cases it is expensive
peat.

   In the vicinities of strip mines, other deep excavations,
intensive exploited wells, the water table decreases. This
phenomenon has the most negative influence on hydromorphic
soils and sandy soils with shallow water level.

   In many parts of the country the water resources diminished
as a result of improper amelioration. The water is not
accumulated in spring, but fast carried away from the drainage
area, and later is not enough water in soil for the plants.
Many authors defined this result of wrong amelioration as a
change to the steppe conditions.

   Around the large industrial cattle and swine farms the
soils are often overfertilized with this material. The plants
show the symptoms of overdosage of nitrogen, and the soil
looses the proper structure. These events occur only than,
when the fertilizer distribution is not under control.

   The mineral fertilizer use in Poland was in year 1979/80
192,9 kg NPK/hectar, and in 1989/90, 163,9 kg NPK/hectar. The
level of nitrogen did not change (69,6 kg N/ha in year 1979/80
and 68,9 kg N/ha in 1989/90), the decrease of NPK level is
mainly in potassium and phosphorus.

   The pesticides use in 1989/90 was low. The average of
active substances used pro hectar was 0,4 kg (in 1979/80 - 1.1
kg/ha).

   The data presented above show, that the chemicals use is
generally low. For still many small farmers the method of
organization of the storage cemeteries for overdues substances
and empty containers of pesticides is not convenient. The best
way to improve the situation is to sample this materials by
the pesticides salesman.

THE DISTRIBUTION OF CONTAMINATED SOILS IN POLAND
   The long-time experience and results of many research works

showed, that soils located inside the larger towns are
contaminated in lesser or greater degree. The most frequent
pollutants are sulphates, heavy metals, phenols, mineral oils.
The contamination of soils in some towns is even larger, than
around many industrial plants. Usuelly the most polluted parts
of towns are the downtowns and the industrial and storage
quarters.

Fig. 3 shows the localization of 27 regions in Poland,
which have been oficially recognized as areas of ecological
emergency. The regions are including territories, in which at
least two kinds of pollution overruned the trigger values, or
one pollutant exceeded the limit more times. Not in all of the
ecological emergency regions occur high soil pollution. Very
often the water pollution is the most important problem.
The short review ilustrates the soil contamination in this
regions.

1. Szczecin. The largest area of polluted soil are the dredged
   bottom sediments from the water-way, which contain high
   amounts of Zn, Cd, Pb, Cu and Mn (8). According to the
   Dutch list it is a category B and C. In addition 2 large
   industrial waste dumps (phosphogypsum and fly ash), about
   100 ha each occur here. Around the fertilizer complex in
   Police higher level of fluorine exist in soils.
2,3,4 and 5. No larger areas of contaminated soil occur in
   this regions, but smaller spots of very contaminated soils
   exists around some chemical plants.
6. Konin. Large area degradated by brown coal strip mining
   and emission of 3 large power plants, using brown coal.
7 and 8. No larger soil contamination.
9. Legnica-Głogów. Soils of good quality contaminated on large
   area with Cu and Pb (category B and C). Additionally very
   large amount of mine and flotation wastes.
10. Wrocław. Many contaminated area of smaller size.
11. Bełchatów. Soils degradated by brown coal strip mining and
    contaminated with emission from power plant, brown
    coal-fired.
12. 12 and 13. Spots of contaminated soils smaller size.
13. 14. Puławy. The forest soils around the nitrogen
    fertilizer plant are continually polluted from the air by
    very high, toxic concentration of nitrates. The
    remediation of soil can be possible after stopping the
    emission of N commpounds.
15. Emission not harmful to the soil (cement dust).
16. Turoszów. Soil degradated by brown coal strip mining and
    emission from the power plants, brown coal-fired.
17. No larger soil contamination.

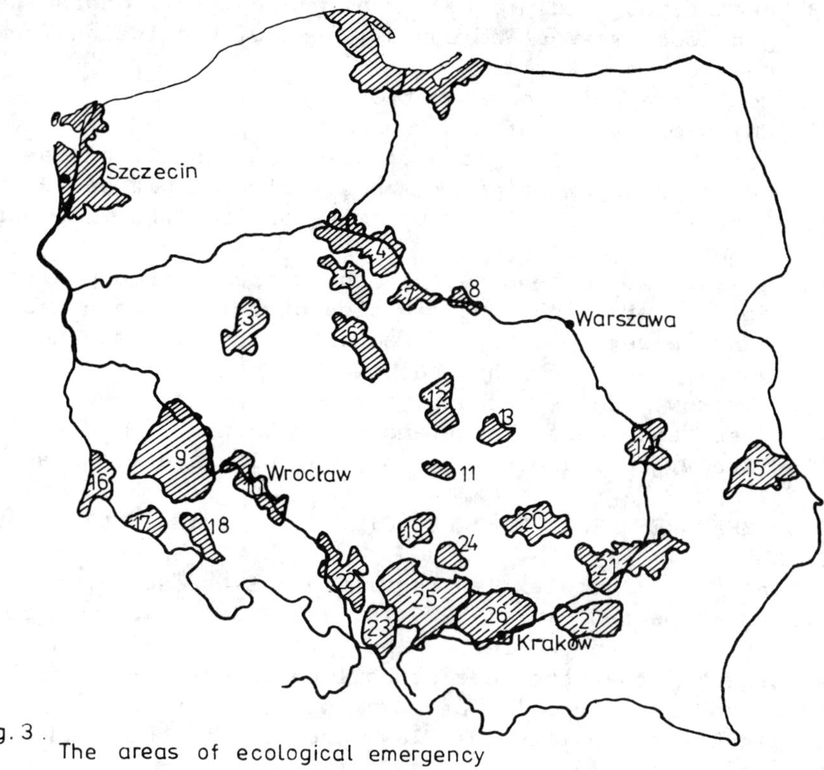

Fig. 3 . The areas of ecological emergency in Poland (2)

18. Wałbrzych. Old pit coal mining area. Soil degradated with pit coal wastes and contaminated with emission from power plants and cokeries.
19. Not large soil contamination.
20. Emission not harmful to soils.
21. Tarnobrzeg. The acidification of soils by the sulphur dusts. Also emission from power plants, pit coal-fired.
22. Opole. Soils polluted by emission from the iron works and chemical plants.
23,24 and 25. The pit coal mining area. Large mining wastes dumps. Very high contamination of soils with emission from many power plants, fired with pit coal, smelters, steel works, cokeries, chemical plants.
26. Kraków. Soils around the steel works are polluted with dusts and gases. Large amount of wastes.
27. Tarnów. Soil around the chemical plants are contaminated.

THE MEASURES FOR THE REDUCING THE SOIL CONTAMINATION IN POLAND.

The most complex law regulation of the soil protection in Poland appeared in 1982. It was the law concerned the agricultural and forest land protection. There was a clear statement, that the doer of soil contamination or devastation have to retrieve it on own cost. Practically it was very difficult to realize this law, because the doer and owner was the same.

The proclamation of the 27 regions of ecological emergency in 1983 was also a step in direction of focussing the efforts on these areas. Construction of new and extension of old burdensome factories were prohibited on this areas.

The payments for the pollution disposal to the environment increased sharply from year to year. Lately the Minister of Environmental Protection had to lower the payments, because the plants were not able to pay so much.

The changes of the polish economy create the possibility for improving the environmental protection measures, including soil protection and remediation. The production decrease of our industry caused decrease of SO_2 emission from 4 mil. tons to 3,2 mil. tons/year. The hard coal output diminished from 190 mil. tons to 135 mil. tons/year. Many old, energy-consuming factories are shutted down. In the near future the old steel works will be closed, or modernized.

By the planning of new investments or purchasing the EIA analysis are provided.

We can resume, that the climat for soil protection and remediation is much better, than before, but the money resources problem for these action is very difficult.

CONCLUSIONS.
1. The most contaminated soil resources in Poland are located in the southern part of the country.
2. The main soil pollutants come from the steel works, smelters and coal-fired power plants.
3. The changes of the structure of industry and technology will decrease the soil pollution and devastation.
4. Conditions for soil remediation are better from the law regulations point of view, but the economical problems make it difficult.

REFERENCES.
1. Główny Urząd Statystyczny: Rocznik Statystyczny 1992. (Statistical Yearbook). Warszawa.
2. Główny Urząd śtatystyczny: Ochrona Środowiska (Environmental Protection). Warszawa 1991.
3. Dobrzański et. al.: Zarys charakterystyki gleb Polski (The general characteristics of polish soils). Wyd. Geol. Warszawa 1973.
4. Musierowicz A.: Gleboznawstwo szczegółowe PWRiL. Warszawa 1968.
5. Greinert H.: The solubility and plant availability of heavy metals in soils contaminated with industrial dusts and sludges. Euro sol 1992 (in press).
6. Greinert H.: Recultivation of phytotoxic miocene sands. Contaminated Soil 90, 303-310.
7. Greinert H.: Heavy metal contamination of reclamated soils as a consequence of liming with Ca-Mg oxides from a zinc smelter. Euro sol. 1992 (in press).
8. Niedźwiecki E.: Zagrożenia ekologiczne w strefie ujścia Odry ze szczególnym uwzględnieniem metali ciężkich (Ecological problems in the mouth of the Odra river with a special emphasis on heavy metals). "Pollutants in Environment". ART Olsztyn 1991, 3-6.

Agro-environmental assessment and remediation. Case study in Hungary

dr. Péter Szabó

Comco Martech Hungary, Budapest

Abstract

Comco Martech Hungary carried out a comprehensive agro-environmental assessment and remediation of a large-scale Hungarian farm. Soil and ground water contamination was detected close to agricultural fields originated from improper handling and storing of different wastes - paint waste, solvents, lacquer resin and oily waste. Heavy metal /Zn, Cr, Ni, Pb/ and solvent concentration in the soil exceeded the limiting value in some places. Remediation of the contaminated soil was implemented by removal and transportation of the most polluted surface layers to safe off-site landfill and by in-situ treatment of the layers with less pollutants. Clearance Certificate was made to guarantee the non-contaminated status for 1106 ha of planned vegetable cultivation area based on detailed soil and water tests.

1.0. Introduction

The former East European block countries, including Hungary inherited a large number of environmental problems partly because of the forced heavy industry development partly as a result of neglecting of environmental concerns. Hungary however never put in the practice so drastic industrial development like the other former socialist countries having stronger industrial tradition. In spite of this newer and newer contaminated "hot spots" are being revealed in Hungary too.

During the privatization process taking place in Hungary an agribusiness oriented English-Hungarian joint venture was formed having 9000 ha of agricultural area near Apaj (between Danube and Tisza river). The company was particularly interested in elimination of the environmental problem occurring there - soil and ground water contamination -which was endangering their intention to grow quality vegetables for export and made a contract with an American-Swiss Environmental Consulting Firm, Comco Martech Hungary (CMH) to carry out the remediation work.

2.0. History of the site

State Farm "Kiskunsági" started different non-agricultural activities in this relatively poor area in the past to produce sufficient profit to maintain the old state owned large scale farm structure. Chemical industrial branch was created in 1981 for producing putty and for temporary storage, recycling and incineration of different wastes coming from paint factories, railway company, telephone factory, ship and crane factory and so on. This

F. Arendt, G.J. Annokkée, R. Bosman and W.J. van den Brink (eds.), Contaminated Soil '93, 1593–1601.
© *1993 Kluwer Academic Publishers. Printed in the Netherlands.*

branch was located near the village "Apaj" among agricultural fields on an area with the size of 5000 m^2 . More and more barrels containing hazardous wastes arrived and were stored here. Some of the materials were burned, some of them recycled but the most were piled up and the temporary storage site became full. Since the waste business was very profitable the farm leaders decided to take off more waste and eliminate them on an illegal way: putting the waste with the barrels into excavated holes, abandoned sand mines, hidden in small forests or pour the liquid wastes on the soil surface nearby.

Because of the illegal on-purpose and improper storage and disposal of the waste significant soil and ground water contamination followed. This criminal action came to light forming probably the first Hungarian environmental scandal. The State Farm was ordered by the authorities to decontaminate the waste and to remediate the area. The incineration and the acceptance of further wastes were banned in 1988 and the barrels were stored in 14 newly constructed concrete platforms. The Farm could transport half of the waste to registered landfill and incinerator for safe decontamination, revealed and remediate 8 illegal dumpsite in the surroundings. Then the farm went to bankruptcy and was put on a list of farms to be privatized. Privatization process made the financial background for completion of the agro-environmental remediation.

3.0. Project structure

Project consists of three parts:

Part I.: Onsite treatment, preparation, necessary packing and safe off-site disposal of hazardous wastes including mixed solvents, paint sludge, lacquer resins, paint containing wastes and oily wastes approximately 2300 tons stored in barrels on the barrel storage area.

Part II.: Remediation of the barrel storage area

Sampling and analysis of the barrel storage area, compilation of the remediation plan. 300 soil and water samples were taken from different sites (from and under the concrete and from the soil between the concrete units) and depth according to the authorities' requirements. The analytical results show no contamination at he back part of the area while some contamination was detected at the front part (6 concrete units from the total 14 and intervals). Removal of the surface layer of the contaminated concrete and soil and onsite treatment of the deeper soil layers was necessary.

Part III.: Clearance Certificate based on detailed soil and water sampling and analysis of the agricultural fields of the farm. Sampling and analysis of the fields planned to be used as vegetable growing area determined by the client (1106 ha) has been completed and issued.

4.0. Part I.:

Based on the remediation plan CMH carried out the necessary treatment, repackaging and transportation of hazardous waste between July 13, 1992 and October 20, 1992. 9000 barrels were stored on wooden supports on 14 concrete platform. Because of the improper storing, some barrels were damaged and leaked continuously polluting the environment. Due to the low flammable point and the high and dry summer weather condition water cooling and night shift waste treatment proved to be necessary.

The waste types were investigated and treated separately and transported to different safe decontamination place. /Table 1., 2., 3. and 4./

Table 1. Composition of the wastes

mixed solvents	400 t
paint sludge	900 t
lacquer resin	700 t
paint waste	200 t
oily waste	100 t
Total	2300 t

The drums were separated according to the kinds of wastes, they were marked with identification numbers and the weight was painted on the drums.
Landfill disposal needed additive /soil, sand, perlite/ usage. The total amount of waste was 2300 tons.
In addition to this 500 m^3 contaminated rainwater collected in 6 ditches had to be transported to a sewage purification plant.

Table 2. Average composition of solvents

ethanol	0.51%
i-propanol	0.33%
n-butanol	0.73%
i-butanol	10.54%
methyl acetate	0.89%
iso-butyl acetate	3.91%
acetone	4.29%
methyl-isobutyl-ketone	3.57%
methyl-ethyl-ketone	0.80%
toluene	24.16%
ethyl-benzene	0.36%
1,3,5 trimethyl benzene	0.26%
xylol	34.24%
petrol	6.54%

Table 3. Metal content of the wastes

Lead	max. 600 mg kg^{-1}
Zinc	max. 800 mg kg^{-1}
Copper	max. 980 mg kg^{-1}
Cadmium	max. 28 mg kg^{-1}
Chromium	max. 70 mg kg^{-1}
Nickel	max. 31 mg kg^{-1}

Table 4. Decontamination places

Incinerator, Dorog	paint waste, mixed solvent and lacquer resin
Incinerator, Győr	lacquer resin
Incinerator, Székesfehérvár	oily waste
Incinerator, Szombathely	mixed solvent and oily waste
Recycler, Kápolnásnyék	mixed solvent
Landfill, Aszód	paint waste, lacquer resin

5.0. Part II. Remediation of the barrel storage area

After the removal of the hazardous waste environmental remediation plan was made based on soil and ground water sampling and analysis. The plan was approved by the relevant authorities.

5.1. Geological properties of the storage site

The barrel storage area can be found between the Danube and Tisza rivers on the Northern part of "Solt" plain. The relief is affected by the former river beds and quick sand hills. 2-5 m deep fine alluviums - silt, fine sand and clay - deposited on 10-20 m deep gravel, gravel-sand and coarse sand laying on pannonian impermeable clay. The depth of gravel layer in the questionable area can be estimated as 17 m being covered by 2-3 m deep cover layer. The hydraulic conductivity /K factor/ of the grave is about 4×10^{-3} m/s, that of the cover layer is less than 5×10^{-5} m/s.

The typical soil types of the area are calcareous meadow-chernozems, calcareous meadow soils, calcareous chernozem-meadow soils, in the deeper sites calcareous solonchaks, calcareous solonchak-solonetz soils, meadow solonetz, solonetz-like meadow soils and bogy-meadow soils.

5.2. Sampling plan and implementation

The sampling sites were determined based on the results of the previous samplings and the history of the area. 57 sampling sites were determined: 25 between the concrete platforms, 24 on the platforms, 6 in the water collecting ditches, 2 outside the barrel storage area. Drillings were made down to 260 cm, where the ground water appeared. Topsoil and 50 cm deep soil layers were separately sampled. Two samples were taken from the 30 cm deep concrete platform too.

10 ground water monitoring wells - established earlier - were also sampled.

5.3. Soil and ground water analysis

Soil and concrete analysis concentrated on two different parameter groups: heavy metals and solvents.

Applied analytical standards:
Total heavy metal content:
MSZ-08-1933/1-2-3-4-5-6-7-8-9-10-17-19/1984.
Total arsenic (As) and mercury (Hg) content: MSZ-08-1944/17-1988 and MSZ-08-1944/20-1988.

Solvents: Phenols and Phenol Derivatives:
The EPA 604 method, adapted to soil samples, was used.
Benzene Homologues and Aliphatic Chlorinated Hydrocarbons:
The EPA 3550 soil measurement was employed, as adapted by the technique developed in the laboratory of Plant Health and Soil Conservation Station (PHSC), Borsod county. 10 g sample of solid-phase extract is mixed with 5 ml of methanol, which dissolves all of the extractable substances from the soil.

5.3.1. Heavy metal content
Total Zn, Cr, Ni and Pb content exceeded the limiting value while total Cu, Cd, Hg, As and Sn content proved to be normal (Table 5., 6., 7.).

Table 5. Threshold values of different elements and toxic substances in the soil (mg kg^{-1})
(MI-08-1735-1990 Hungarian Technical Guideline)

Element	Cation	Exchange	Capacity	Remark
		me 100^{-1} g soil		
	5-15	15-25	25-35	
As	7	10	15	
B	100	100	100	
Be	10	10	10	++
Cd	1	2	3	++0
Co	50	50	50	
Cr	75	100	100	xx0
Cu	75	100	100	xx0
F	500	500	500	
Hg	1	1	1	++0
Mo	10	10	10	
Ni	50	50	50	xx0
Pb	100	100	100	xx0
Se	10	10	10	
Zn	200	250	300	xx0
PAH	1	1	1	
TPH	100	100	100	

Remark:
++   special care needed
xx   in hop garden and vineyard as well as in case of soils having more than 5% CaCO$_3$ higher value by 25% is acceptable
0   in case of grassland and below 6.5 pH half of the value in the table is valid.

Table 6.:Existing and permissible concentration of various elements in the soil with regard to the tolerance of plants /based on Tietjen's work/

Elements	Deviation	Total content of elements (mg kg^{-1}) in healthy soils Most frequent (mg kg^{-1})	Permissible
Be	0.1-10	1-5	10[++]
B	2-100	5-30	100
F	10-500	50-250	500
Cr	1-100	10-50	100
Ni	1-100	10-50	100
Co	1-50	1-10	50
Cu	2-100	5-20	100[+]
Zn	10-300	10-50	300[+]
As	1-50	2-20	15[x]
Se	0.1-10	1-5	10
Mo	0.2-10	1-5	10
Cd	0.01-1	0.1-1	5[++]
Hg	0.01-1	0.1-1	2[x++]
Pb	0.1-10	0.1-5	100

Note:
[+]   In hop and grape plants higher values are also permissible.
[++] Special care is needed
[x]   More stringent values based on the investigations of the National Public Health Institute

High Zn content was measured at the platforms of I., II., III., V. and VI. in the upper 2cm of the concrete and beneath in the 10 and 60 cm deep soil layers.
High Cr content was found at the platforms of II., III., V., VI. and their intervals.
High Ni content was analyzed between the platforms of III. and V.
High Pb content was measured at the platforms of I., II., III., V. and VI. and in their intervals /Table 7./.

Table 7. Analytical results of heavy metals in the most contaminated part under the platform of the barrel storage area

Sample code	Sampling depth (cm)	Zn mg kg^{-1}	Cr mg kg^{-1}	Ni mg kg^{-1}	Pb mg kg^{-1}
Limiting value for soil		250	100	50	100
Limiting value for sewage sludge		3000	1000	200	1000
Detectable limit (KH)		0.1	0.1	0.1	0.1
I/1a	0-2	58.8	17.3	10.0	8.9
I/1b	0-10	594.0	42.3	7.6	169.7
I/1c	60	1075.0	94.5	9.0	465.3
I/1d	110	105.1	6.3	7.9	22.7
I/1e	160	17.6	1.4	8.2	0.4
I/1f	210	20.1	1.6	6.7	2.0
I/1g	260	11.7	1.3	5.6	0.6
I/2a	0-2	241.9	39.9	17.0	49.0
I/2b	0-10	260.0	18.3	9.4	62.4
I/2c	60	576.1	31.8	7.2	139.0
I/2d	110	82.5	6.8	14.8	12.2
I/2e	160	21.5	1.5	7.7	0.9
I/2f	210	16.2	0.9	5.8	0.6
I/2g	260	17.0	2.0	7.3	0.9
II/1a	0-2	218.6	9.7	30.5	165.1
II/1b	0-10	68.8	6.7	8.1	14.0
II/1c	60	5478.0	430.5	10.3	2123.0
II/1d	110	122.2	12.0	12.2	38.1
II/1e	160	49.7	8.2	18.2	3.5
II/1f	210	32.6	5.3	15.4	2.5
II/1g	260	12.9	2.2	6.0	1.7
II/2a	0-2	262.6	61.3	28.7	103.1
II/2b	0-10	50.6	5.6	9.0	6.7
II/2c	60	2472.0	50.9	12.4	198.6
II/2d	110	152.5	6.5	11.9	10.1
II/2e	160	49.6	7.6	18.9	1.7
II/2f	210	42.5	7.2	16.0	9.0
II/2g	260	14.1	1.6	6.1	1.4

Based on the monitoring wells analytical data no significant heavy metal contamination could be detected in the ground water: the alkaline soil /$CaCO_3$ = 10-20%, OMC in the upper 80 cm = 1.5-2.5%/ hindered the vertical movement of the heavy metals.

5.3.2. Solvent content

Evaluating the solvent content the Dutch Standard was taken into consideration /benzene 0.5 mg kg^{-1}, toluene 3.0 mg kg^{-1}, total aromatics CH 7.0 mg kg^{-1}/. Benzene, toluene and xilene concentration proved to be high at the upper 2 cm of the concrete platforms and at the depth of 10 and 60 cm below the platforms in the front part of the storage area but in deeper layers also could be measured high values since the absorption properties of the soil can not hinder the vertical movement of the solvents. Solvents were found in the ground water representing a relatively fast solvent migration in the case of shallow ground water table. Since the barrel storage area is relatively close to a village detailed calculations were made for modeling the actual status and the further vertical and horizontal movement as well as the expectable dilution of the contaminants. The ground water slowly flows from North-West to South-East with an average speed of 3.89 m/year, the pollution source has been eliminated and the measured solvent concentration can be totally decomposed and diluted within 10-15 years while the polluted ground water covers only 60 m. So the contaminated ground water can never get the 300 m far village.

Table 8. Analytical results of solvents in the most contaminated part under the platform of the barrel storage area

Sample code	Sampling depth cm	Benzene mg kg^{-1}	Toluene mg kg^{-1}	o-xilene mg kg^{-1}	m-xilene mg kg^{-1}	p-xilene mg kg^{-1}
Detectable limit (KH)		0.2	0.2	0.3	0.3	0.3
I/1a	0-2	<KH	1.9	0.4	<KH	3.7
I/1b	0-10	<KH	<KH	<KH	<KH	<KH
I/1c	60	<KH	<KH	2.1	11.5	6.4
I/1d	110	<KH	<KH	<KH	<KH	<KH
I/1e	160	<KH	<KH	<KH	4.8	<KH
I/1f	210	<KH	<KH	<KH	<KH	<KH
I/1g	260	<KH	<KH	0.4	1.2	0.7
I/2a	0-2	<KH	<KH	0.4	<KH	2.2
I/2b	0-10	<KH	1.0	3.2	2.6	<KH
I/2c	60	<KH	<KH	1.3	2.3	1.0
I/2d	110	<KH	<KH	<KH	<KH	<KH
I/2e	160	<KH	<KH	<KH	<KH	<KH
I/2f	210	<KH	<KH	<KH	<KH	<KH
I/2g	260	4.3	<KH	<KH	<KH	<KH
II/1a	0-2	<KH	4.8	4.0	19.6	10.3
II/1b	0-10	<KH	0.2	0.5	0.5	<KH
II/1c	60	<KH	0.8	1.9	3.9	2.4
II/1d	110	<KH	<KH	<KH	<KH	<KH
II/1e	160	<KH	0.8	<KH	<KH	<KH
II/1f	210	<KH	<KH	<KH	<KH	<KH
II/1g	260	<KH	<KH	<KH	<KH	<KH
II/2a	0-2	<KH	0.8	0.3	0.8	0.5
II/2b	0-10	<KH	<KH	<KH	<KH	<KH

II/2c	60	<KH	0.5	4.5	23.5	6.8
II/2d	110	<KH	<KH	<KH	0.3	<KH
II/2e	160	<KH	0.4	<KH	<KH	<KH
II/2f	210	<KH	0.3	<KH	<KH	<KH
II/2g	260	<KH	<KH	<KH	<KH	<KH

5.4. Remediation plan and implementation

First the contamination source had to be eliminated. During this process the most polluted upper 20 cm of the open soil surface was removed and added as a stabilizer to the waste. Special attention was devoted to avoid secondary contamination during the handling of the barrels. The heavily contaminated upper 2 cm of concrete platforms I., II., III., V. and VI. was also removed and transported to a off-site landfill. In order to increase the buffer capacity of the soil clay minerals - bentonite - mixed with lime sludge was injected into the 60 cm deep soil layer.

6.0. Clearance Certificate for cropland

After completion of the remediation work CMH started an investigation program for screening the contamination status of certain agricultural fields /1106 ha/ as well as to prove that the area is suitable for agricultural production for human consumption. In the frame of this job detailed sampling system was made:

V shaped sampling lines were determined. Samples were taken from 1-1 site /point-samples/ from various depths: in each sampling site from the ploughed layer /0-25 cm/, in certain sites from 50 and 100 cm depth /"deep samples"/. Deep samples were taken from soil profiles. Samples for solvent analysis were put into cooler and transported for immediate analysis.

Part of the samples was put into a storeroom, in case of solvent analysis into refrigerator, being stored till the end of the project making possibility to reanalyze certain questionable samples.

6 parameter groups were created:

agronomy, heavy metals, solvents, pesticide residues, radiation, oxygen demand and microbiology/ only for water/.

Agronomy parameters /texture, pH, organic matter, $CaCO_3$, EC, total and available macro and meso elements, available micro elements/ represented fertile, cultivable land.

Total heavy metal content /incl. Pb, Zn, Cd, Cr, Cu, Ni, As, Sn/ was found to be typical for normal soils.

No significant solvent contamination was measured in the soil, while some phenol was analyzed in the irrigation channel water coming from the water source, Danube river, not exceeding the irrigation water standard.

Chlorinated hydrocarbons, carbamide derivatives and triazine were investigated and no harmful residues were found

The radioactivity was everywhere normal: no Chernobyl effect was detected.

The chemical and biological oxygen demand proved to be acceptable.

No coliforms were found in the microbiological analysis.

On the basis of the investigation our Clearance Certificate guarantees the contaminant-free status of the investigated area.

CONTAMINATED SITES IN HUNGARY

Attila Takáts
Technical University of Budapest
H–1111 Budapest, Müegyetem rkp. 3, Hungary.

ABSTRACT

The problem of contaminated sites in Hungary is relatively severe, having regard to the development and volume of the economy. Work on the discovery, investigation and remediation of such sites began only 2–3 years ago. There is no legal regulation, so the initiatives taken by the authorities have little effect. The expensive and extensive remedial work required cannot be financed, either by government or by companies. There is no expert training available. Despite this, however, expertise is available in specialized companies, and remedial activities in Hungary are at a satisfactory level.

1. Origin of Contaminated Sites

In the 40 years that followed World War II Hungary experienced two periods of forced industrialisation, the underlying aim of which was to fulfil the objectives set by the country's leaders in accordance with goals set by COMECON. Agriculture, too, experienced extremely rapid, enforced industrialisation at the same time.

Two factors were either partly or totally neglected during these 'development' processes: cost efficiency and environmental protection. According to the general practice in the socialist countries, the costs of investment were not completely allowed for in the planning process, leading to results that were of poor quality, as well as outdated. The situation was frequently exacerbated by waste and negligence. During the later industrialisation phases it was obligatory to take account of environmental protection measures in the planning phase. The underplanned investment costs were at that time supplemented by investments from a fund for environmental measures. But environmental protection received only any money that might be left over after all other requirements had been met. This is the notorious 'principle of the remainder' which governed the protection of the environment throughout the whole of the socialist bloc.

If there was any money left over for environmental protection, the enterprises used it for wastewater treatment (without sludge treatment) and air quality protection. Investments for solid waste treatment were only made at the end of the second period of industrialisation. With some honourable exceptions, investment in the prevention of soil contamination is absent, even today.

Under such circumstances, solid production waste was disposed of on garbage dumps, having no regard to the toxicity of the waste, or else tipped on carefully concealed 'wild' dumps.

To illustrate this situation, an example from the 1970s may be instructive. In a Budapest chemical factory, solid and liquid toxic wastes were collected in drums inside the factory. The factory site was very cramped, so it was decided o make the wastes 'disappear'. There was an open sandpit close to Lake Balaton. Mining activities there had been terminated, and the mining enterprise was obliged to restore the landscape. The chemical factory and the mining enterprise agreed to fill the pit together. The chemical factory transported its toxic wastes to the pit, 1.5 km away from Lake Balaton, where the mining enterprise planned to

F. Arendt, G.J. Annokkée, R. Bosman and W.J. van den Brink (eds.), Contaminated Soil '93, 1603–1607.

cover the drums with sand and grass.

A policeman discovered this dangerous plan when the drums were transported to the pit. The drums were taken back, then taken further on to a little city near Budapest. After two years the drums had rusted and the load of chemicals percolated into the soil and then into the groundwater. The city was supplied with water from this polluted groundwater. No grave poisoning occurred.

Although some improvements occurred between 1983 and 1990, the consequences of this approach to waste disposal is about 4000 (four thousand) suspect sites. A further 10,000 to 15,000 suspect sites may be contaminated with mineral oil and oil derivatives. Hungary, even today, buys its oil from the former Soviet Union. It is transported through the 'Friendship' pipeline. One section of this pipeline, in Eastern Hungary, ruptured frequently, releasing about a hundred cubic metres of mineral oil to the environment in each accident. There are 20,000 to 30,000 underground tanks, each of over 5 m^3 capacity, for the storage of mineral oil products. These are unprotected. There are two oil refineries in Hungary, etc., etc.

The number of suspect sites caused by the military activities of both the Soviet army and the Hungarian National army can be estimated as some few hundred. The majority of these are polluted by mineral oil derivatives (aircraft fuels, etc.). There is a similar number of underground mines and factory sites. Among these are some factory sites which are filled with toxic waste (heavy metals, cyanide, materials containing PAH, etc.). Some of the factory buildings have been erected on this layer.

Finally, some issues should be mentioned which have not yet been estimated.
– Traffic problems (airports, railroads, bus stations, transport repair workshops, etc.).
– Agricultural problems (pesticide residues, etc.).
– 'Environmental protection' problems.

In summary, the Hungarian situation may be characterised in terms of a minimum of 25,000 contaminated sites within the 93,000 km^2 territory of Hungary. This is a very large number in regard to geological contamination. About 80% of the drinking water supply uses underground water. The bases of this drinking water supply are not geologically protected in about 1/3 of the country. It is regrettable that this unprotected part has the highest industrial density and is the most densely populated, and consequently is the most polluted.

It is obvious that immediate measures are necessary to remedy this situation; measures that will also have to be continued for some time to come. In the following paragraphs I shall examine the conditions at our disposal for taking those measures that are considered to be indispensable.

2. Administration

The present author raised the issue of contaminated sites many times in Hungary in the 1980s. An attempt was made in 1988–1989 to achieve some kind of regulation. This attempt failed. No new attempt has yet been made because of the difficulties in composing a Law on Environmental Protection.

As a consequence of the lack of legal rules, the measures taken by the authorities are open to dispute. The authorities can only work within an existing legal framework, such as the Water Protection Law or an order on the treatment of hazardous waste. In other cases, any measures taken may be disputed, or they have to be arrived at by agreement.

Another consequence of this problem is the lack of any official inventory of suspect sites and no classification system for specifying the urgency with which suspect sites have to be tackled. This latter problem means that there is no idea of cost effectiveness in the case of any single clean-up process.

Let me make this clear by giving a widely known example. In a large Hungarian chemical factory, where one of the most important production processes is chlorine-alkaline electrolysis using a mercury cathode, mercury was lost from the electrolysis equipment over a period of about ten years. The major part of the mercury passed though the ground floor of the building and percolated deep into the soil. The total quantity of mercury lost amounted to some hundreds of tons (metric). An NGO working in the region discovered this problem and, together with the local inhabitants, exerted pressure on the local authorities and the factory. A large amount of money (some ten million forints) was spent on delimiting and examining the mercury and its status. Finally, in a very important examination, it was found that the mercury was in a passive state and that the immediate risk was very low.

At the same time, the factory still has some soil basin reservoirs for water containing mercury and for sludge. These are really very hazardous, and they have not been examined to this day.

3. Technical and Economic Conditions

3.1. PHARE Projects

The PHARE1 projects on clean up issues have been in existence for two years now. Project number 121 is related to the Waste Management Division of the Ministry for the Environment and Regional Planning. The aim of this project is to create a 'Geoinform' system about that part of Hungary where the underground water reserves used for the supply of drinking water are in danger. The Geoinform system contains geological structures, data on water bases, and records of suspected sites. The work is being executed by the Dutch firm BHV.

A second PHARE1 project, number 134, is in progress under the control of the same ministry. Researchers in this project are using a model field to examine the different tracing, discovery and investigational methods under practical conditions. The model field is Csepel Isle, an island in the Danube built up from alluvial deposits (mostly gravel). The island is utilised for the extraction of gravel-layer-filtered Danube water. The work is being performed by the Hungarian Geophysical Institute, the Environmental Management Institute, and the Water Management Research Institute.

3.2. Expert Staff

In order to discover and clean up the contaminated sites that are increasingly coming to light it will be necessary to have available a considerable number of well trained specialists and specialist companies. At the present time there is no specialist training available in this field in Hungary. There are only a few real experts, who have either learned abroad or have studied the foreign literature on their own.

Since the problems of the contaminated sites are far more urgent to allow the training of experts, and the tasks of dealing with them are recurrent, certain environmental enterprises (working in the areas of water protection and waste management) started to contract work in this field. Their solutions are very often objectionable, either because of their provisional nature or their poor cost effectiveness.

Despite that, I believe that the approach to a solution by such enterprises, using teamwork (geologists, chemists, biologists, etc.), may be able to form a Hungarian clean up industry within a short time.

3.3. Tools, Technologies

The tools and technologies used by the Hungarian clean up enterprises are not very sophisticated. The main preference is for two technology groups. One is different types of immobilization, which is not very favourable since it tends to be used instead of a final solution. The second technology group is biodegradation, either _in situ_ or _ex situ_. The main field of application for this technology are sites contaminated with mineral oil and oil derivatives. There are about ten different microorganism cultures available which can be used for this purpose. Bacteria can also degrade other materials (e.g. PACs, etc.). This is important since, in the south part of the country, there is a deposition site which contains about 15,000 tons of polychlorinated waste, and under that soil (very good quality clay) there is very heavy pollution by this material. This task would be insoluble without microbiological treatment, but a small Hungarian enterprise has obtained the appropriate culture for cleaning up the soil.

We can say that the level of our technology is far behind that of the Dutch or the Germans. We have no large, mobile instruments (incinerators, soil washing equipment); in general, Hungarian enterprises use the equipment and machinery of civil engineering. Despite that, the Hungarian clean up activities require only one type of technologically developed tool: a mobile washing installation which is capable of washing heavy metals out of soils.

3.4. Economic Background

In Hungary now the only factor that hinders us from overtaking the developed countries is our almost complete lack of economic activity. We are at present in a the midst of a deep recession. The volume of industrial production has decreased by 50%. Those firms that remain in production have been hit by the high degree of centralisation in the incomes policy, so no amount of pressure from the authorities can force the firms to participate effectively in clean up activities (with the exception of some well-run enterprises). On the contrary: they ask for government support in solving this problem, stating that the centralised incomes policy has made it impossible for them to find a solution. But the Central European Protection Fund contains only a small sum which is earmarked for this purpose.

There are two possibilities to escape from this trap. First is the 'debt-for-nature' swap, which the OECD countries applied to Poland. But since Hungary, even now, has the ability to pay off its debts, this option is not all that likely. The second possibility, which may work now, is privatisation. Foreign firms are willing to buy Hungarian state-owned firms if the environmental problems can be fully settled, including the problems of contaminated sites. The money which is necessary to settle the pollution problems originates from the sales value of the enterprise. For example, when a Hungarian refrigerator factory was acquired by foreign firm, a large land area (many hectares) was investigated and cleaned up by different firms, both Hungarian and foreign. The costs of the clean up activities exceeded 1 billion forints.

4. Brief Case Studies

A very old problem is that of the so-called Lux mass. This is a special filter material which retains sulphur in coal distillation plants. This mass was changed regularly in the gasworks in days gone by, and deposited anywhere (before the hazardous waste decree came into force).

At the present time, six such places have come to light. The latest is quite interesting because at that time (about 30 years ago) it was used for liquidation flats in artificial caves. Today the sulphur content of the mass has destroyed the caves, the hill, the channels, etc., so it is now necessary to extract the rest of the mass from a densely populated area.

Another old problem is the very old (almost 100 years) nonferrous metallurgy works mentioned above. Apart from the 1,000,000 tons of metallurgy slag deposited in the factory's grounds, a circle of about 1.3 km diameter is heavily polluted with lead. The measured lead concentrations in the upper soil, in plants, even in the blood of the inhabitants, are over the limit values. At present nobody, not even the Dutch experts, can think of a solution.

A more recent problem is the quality of the soil and the underground water on the former Soviet military bases. Thousands of cubic metres of aircraft fuel leaked from underground tanks. There are about 100 such fields.

I have to state that all of the contaminated sites (and even more) mentioned in this paper are under intensive investigation, and some of them are being remediated. The larger problem is what we do not yet know.

SUSTAINABLE SOIL USE
A TNO-concept

A.J. Palsma (ed.), M.J. Diependaal (ed.),
F.G. Aelmans, E.J. Hoekstra, A.C. Leget, H.J. van Veen

Institute of Environmental Sciences TNO, P.O. Box 6011, 2600
JA Delft. *The Netherlands.*

1. Introduction

In the past few decades the authorities in many countries have been confronted with soil pollution. Most soil pollution originates in the past. Soil pollution was only recognized as an
important environmental problem in the '80s. To deal with these problems soil clean-up policies were rapidly drawn up in several countries. A few years later these were followed by prevention policies.

The current soil use in various countries in the west is sketched in figure 1. Clean soil is used
by man. The soil may be used either as a carrier or as a raw material. Part of the soil will be
polluted by accidental or regular actions. In many cases this pollution is so serious that the soil
has to be removed and stored in a landfill deposit as waste. The authorities are now trying to
effect the clean-up of polluted soil through legislation. However, there is a major problem. The
pollution of clean soil, its excavation and storage in landfill deposits, is much cheaper than
remediation of polluted soil. As a consequence the flow sheet in Figure 1 results in a reduction
of the volume of clean soil and continued storage in landfill deposits of polluted soil. In other
words, the flow from left to right in the figure exceeds that from right to left. Hence, clean soil
is consumed and soil clean-up is delayed, i.e. such soil use is not sustainable.

Recent discussion within TNO has resulted in the development of a concept for the
implementation of sustainable soil use. This concept is based on two premises:
-    the availability of good quality soil is limited and the resources should be maintained;
-    soil quality has an economic value.

The "TNO concept", a proposal for sustainable soil use, was developed on the basis of these
premises. This concept will provide the direction to a discussion at a workshop during the
TNO/KFK Soil Congress to be held in Berlin in May 1993.

2. Current situation

In the early 1980's a number of countries in the west were confronted with soil pollution.
Policies to deal with this issue were drawn up in a relatively short time. At that time the
Netherlands, one of the first countries confronted with the problems, set the direction for the
clean-up policies.

Besides policies, clean-up technologies were developed rapidly. Initially the approach was to
deal with the problem by removing all polluted soil and processing this on an industrial basis.
Soon, however, it became clear that the scale of the soil pollution was far greater than first
anticipated. At the international level the soil-pollution issue was also huge and the cases on
the record are becoming more numerous.

1609

F. Arendt, G.J. Annokkée, R. Bosman and W.J. van den Brink (eds.), Contaminated Soil '93, 1609–1615.
© 1993 *Kluwer Academic Publishers. Printed in the Netherlands.*

Current remediation policies are largely based on the assumptions and estimates made in the early days. As available remediation technologies are often too expensive or unable to reduce the pollutants to the required level, polluted soil is often stored in a landfill deposit.

One can observe that in many countries the large-scale remediation of polluted soil is slowing down, while the development of remediation technology is also lagging behind. The prevention of new cases of soil pollution is also insufficient and the number of cases continues to increase.

The overall effect of this process is the consumption of clean soil.

3.   TNO-concept

To implement the sustainable use of soil the use and quality of the soil will have to be integrated in an economic structure. In this way the prevention of soil pollution may be promoted and funds for remediation can be set aside. At present land prices are determined largely by the location and options for use within the existing planning framework. Soil quality and associated implications for use are less important. Furthermore, a functional use of the soil should be implemented. The proposed structure is outlined in Figure 2. Like Figure 1, this figure includes soil use and remediation of soil pollution. The TNO concept is characterised by:

Clean soil as a limited resource

Clean soil is a limited resource which has to be used and managed carefully. Different countries apply different standards to cleaned soil. In the Netherlands for example cleaned soil should be fit for all functions which were originally possible (multi-functionality). A wide range of activities, associated with actual or potential pollution is often undertaken on the soil. In many cases the resulting soil pollution is so great that it cannot be reinstated to multi-functional quality. Hence, the stock of clean soil is being reduced.

Functional application

Soil is used, now or in the future, to fulfil a function such as agriculture or residential area. An acceptable maximum pollution level for a given plot is determined, for the period of the functional use. Soil use is sustainable if a function can be maintained at a location for an indefinite period. Adequate prevention is an important element in this. Prevention can be promoted by putting a price on each function. These prices should depend on the potential pollution associated with the function. When the function is changed, the soil should be reinstated to a quality appropriate to the new function. Remediation costs can be set aside through the payments referred to above. When the level is exceeded, remediation activities will have to be carried out or the functional use should be changed.

Setting Standards

If function-dependent standards were introduced, soil quality could be defined as the fitness for use. Toxicological and ecotoxicological risks and effects could be evaluated for a given function or area which may be polluted. In this way a set of standards can be defined for each function, depending on the type of soil. When setting standards, the risk of pollution transport (through water or atmosphere) should be considered as well as the toxicological and ecotoxicological aspects. Protective measures, such as hydrological isolation, should be taken if the toxicological or ecotoxicological risk is acceptable but the risk of pollution transport through the groundwater is too high.

Such a quality criterion may introduce problems when the use of soil is changed. Soil suitable for agriculture may not be suitable for an area of natural interest or residential areas. To determine the suitability of soil when its function is changed, it is not enough just to set standards for potentially toxic substances. The standards will have to cover a wide range of chemical, (physico)chemical, hydrological, mineralogical and ecological variables. Standards will also have to be established for clean soil (good soil quality). In principle such standards will be independent of toxicological, ecotoxicological and hydrological risks. Even when dealing with clean soil, it is not enough to specify the levels of potentially toxic substances. In most parts of the Netherlands soil properties have changed as a result of the use made of it (e.g. agricultural soils enriched by human activity, areas where the groundwater level was lowered, sand brought on site for construction works). For soil whose properties have undergone major changes the question arises to what extent it is useful and practical to define a "clean original situation".

Remediation measures chosen on the basis of environmental yield
Remediation may be undertaken to reinstate a functional level of soil quality. Historical pollution is particularly relevant to remediation policies. The concept described here interprets sustainable use as an equilibrium situation. However, given the historical pollution, this equilibrium can only be obtained by making up for past problems. Environmental yield (return on investment) should be the main criterion for a choice from possible alternatives when considering remediation activities. Figure 3 defines "environmental yield" as the "cost effectiveness of a measure". The environmental benefits or environmental effectiveness are subcriteria in which location, process and product aspects are combined.

In this context location aspects refer to the effects of the remediation operation on the site to be treated. This includes the environmental impact of the process to be used and the effects of the operation on the soil as a product. Other factors besides environmental efficiency affect the feasibility of the measures, for example administrative aspects and social desirability.

Technology
The concept presented here will affect the development of soil-remediation technology. Current techniques aim to reduce the level of contamination to the standard for clean soil in a short time. In view of the above considerations, future techniques should not only reduce the level of toxic substances, they should improve the overall quality of the soil. As a result the contamination level may be acceptable even if it exceeds the current reference values used in the Netherlands, provided that the soil is suitable for its function.
After the clean-up a risk analysis should be carried out to determine the purposes for which the soil (i.e. the product) can be used. This will make remediation techniques which do not affect the properties of the soil more feasible. In situ, non-intensive remediation techniques would appear to be particularly appropriate. However, such long-term remediation will only be acceptable if the risk of pollution transport during the process is monitored and remains within limits.

Soil remediation will not be limited to those cases where contamination levels are so high that the land cannot be used for any function. Non-intensive techniques may also be used at lower pollution levels as the economic structure proposed here guarantees that the value of the land will increase if the soil quality is closer to the standards for clean, good quality soil.

Soil use within the economic system
Sustainable soil use will only be feasible if the soil is part of an economic system. Although land is already included in the economic system, the price is largely determined by its location

or the activities which are feasible or permitted. At present quality and scarcity of soil as a resource are less important.

Remediation of polluted soil can be encouraged by an "economic drive". Such an economic drive can be implemented in a variety of ways, for example, a fund could be established. Such a fund could be built up through annual contributions from those using the soil. The level of their contributions could depend on the use of the soil, its quality and the measures taken against soil pollution. For long-term responsible use a no-claim discount could be installed. Whatever its structure, the essence should be that a cost would be incurred in the event of pollution and that soil remediation should be rewarded. The reward should meet the costs required to obtain the result.

Figure 4 illustrates an estimate of the risks and relative costs of various functions. Naturally the costs per area unit for a given functional area correspond with the risk of pollution. The costs of a given function will depend on the risk and preventative measures taken. Figure 5 illustrates a possible differentiation scheme for landfill sites.

Transitional situation
As referred to above, the proposed concept is based on a steady state. However, at present the large number of soil pollution sites due to past activities amount to a considerable burden on the economy and the environment. Clearly sustainable soil use will require remediation of serious cases. Less serious cases can be incorporated in the system. The proposed criteria can then be applied to them. Those responsible for such problems created in the past will have to fund the remediation operations if they can be identified. Other cases could then be financed from the soil remediation fund.

4.   Comprehensive package
The idea presented here as the TNO concept is essentially simple. It not only provides soil users with the facility to implement functions but it also makes users financially responsible for any pollution associated with that use. The proposed system will only lead to the desired self-regulating, sustainable soil use if it is implemented completely, as a comprehensive package.

5.   Recommendations
If the TNO concept is accepted as a feasible option, it will lead to a large-scale operation to develop certain aspects of the concept, for example:
- defining functions and the associated standards;
- developing standards to define the concept of "good soil quality" in broader terms (chemical, physical, hydrological and ecological aspects);
- developing and implementing remediation techniques, including non-intensive techniques;
- selecting the most suitable form of "economic drive";
- developing and implementing soil and groundwater quality monitoring systems.

6.   Summary
Theses:
- Soil is a limited resource; soil pollution results in an irreversible reduction in the stock of high-quality soil.
- Soil quality should be associated with an economic value.

Conclusions:
- Soil is currently being consumed.

- Remediation of polluted soil is slowing down.
- There is extensive, and still increasing, soil pollution.

Solutions:
- The policies will continue to aim to protect clean, high-quality soil.
- Functional soil use associated with an acceptable pollution level, depending on the function (areas where pollution is acceptable).
- Remediation activities should be selected on the basis of environmental efficiency.
- Economic drive: a "deposit" or "insurance" for the use of soil.
- Creation of a national soil fund. This will have a preventative effect and set aside resources for remediation activities.

Sustainable soil use
The TNO concept will lead to sustainable soil use if both the functional use and the economic engine are implemented.

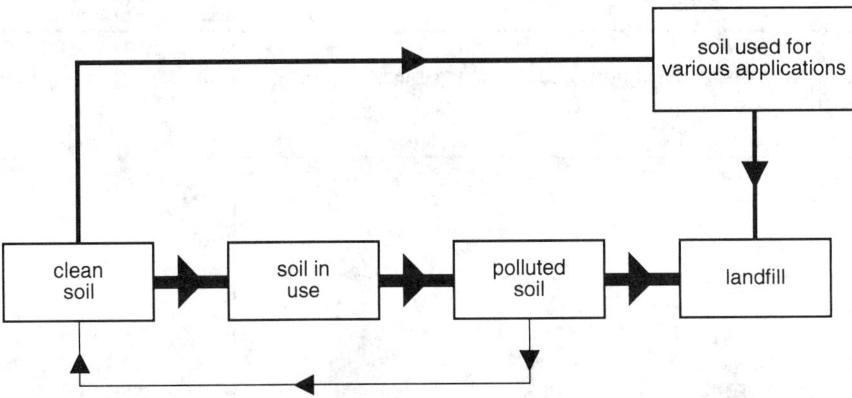

FIGURE 1.   Soil use

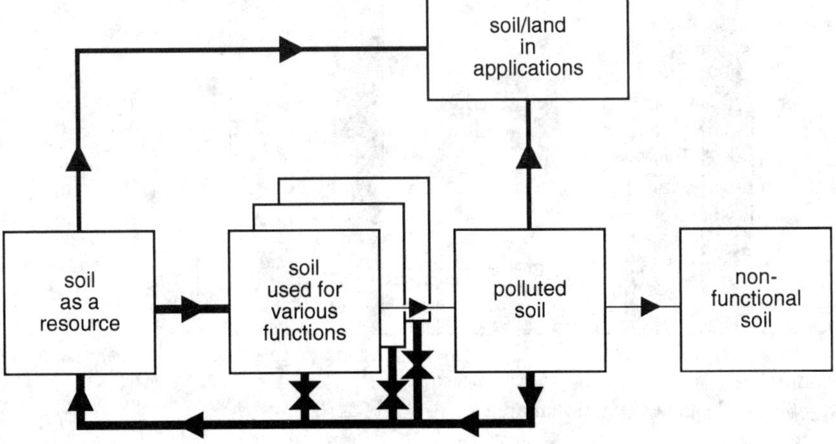

FIGURE 2.   The TNO concept

PARAMETER	SUBCRITERION	CRITERION	MAIN CRITERION	
Actual risk Potential risk	Location aspects	Environmental benefit or Environmental effectiveness	Environmental yield	P R I O
Waste disposal residues Energy, emissions	Process aspects			R I T
Structure Organic C content pH Ecological recovery level of contaminants	Product aspects			I E S
Costs/cost effectiveness				
Social desirability				
Administrative and legal aspects				

FIGURE 3.   Suggested definitions and structure of terms associated with the "environmental yield" concept.

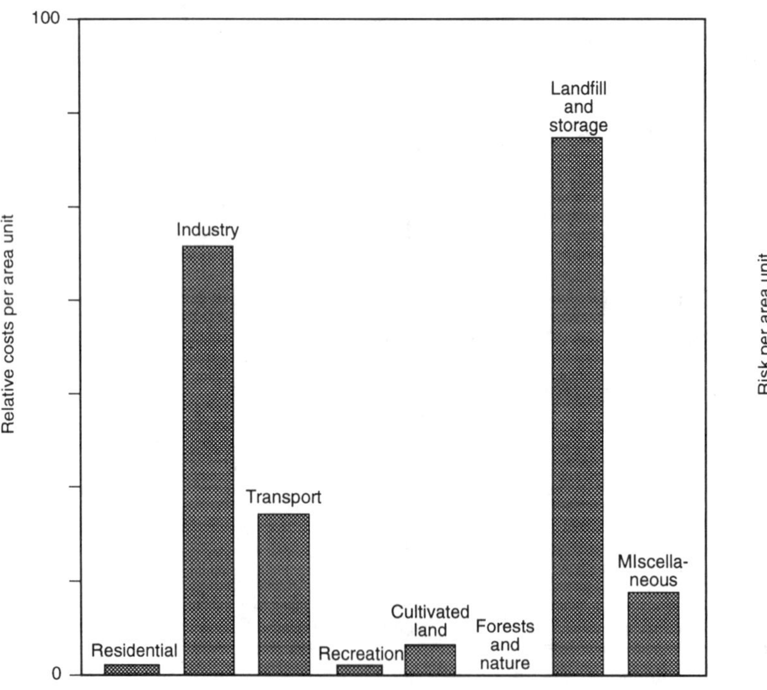

FIGURE 4   Costs per functional area

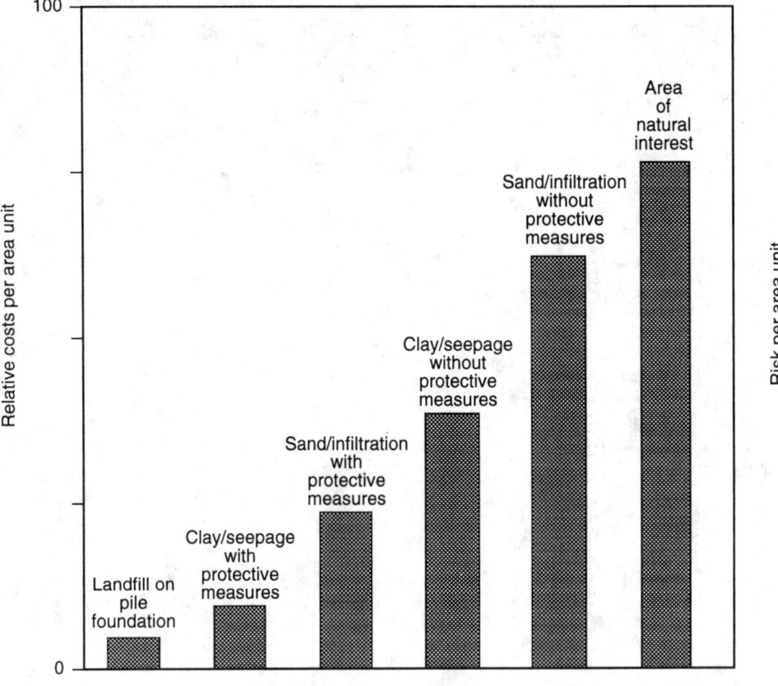

FIGURE 5   Cost differentiation of landfill sites by location and protective measures

MANAGEMENT FRAMEWORK FOR THE SURFACE CLEANUP OF URANIUM MILLING WASTES

Ned B. Larson[1]
Ralph G. Lightner[2]
Denise Bierley[3]

ABSTRACT

The cleanup and disposal of low-level radioactive wastes from the former milling of uranium in the United States was legislated in 1978 with the passage of the Uranium Mill Tailings Radiation Control Act (UMTRCA). This act required the Environmental Protection Agency (EPA) to set the cleanup standards that would be used. UMTRCA also required the Department of Energy (DOE) to fund and carry out the remediation activities at designated inactive tailings sites, and the Nuclear Regulatory Commission (NRC) was to review and concur with the DOE plans to ensure that EPA standards were met. In addition to the involvement of these Federal agencies, the States were required to fund 10 percent of the remedial actions so the States would also participate in the planning and implementation of the remedial actions. Because the remedial actions affected local communities and counties, they were consulted in most aspects of the remedial action efforts. This paper is an overview of the management framework that is used to implement remedial actions while involving all the cooperating agencies.

INTRODUCTION

The Uranium Mill Tailings Remedial Action (UMTRA) project's objective is to stabilize and control the cleanup of the uranium mill tailings, contaminated materials at nearby sites, and other residual radioactive materials at 24 sites and over 5,100 vicinity properties in accordance with EPA standards. The remedial solution must insure that the disposal option chosen is safe and environmentally sound to eliminate, or significantly, reduce radiation and other potential health hazards to the public.

During the early years of uranium milling, the U.S. Government purchased uranium from a number of suppliers for research and for the production of nuclear weapons. At the time that UMTRCA was passed, these mills had produced a total of 41 million cubic yards of tailings and other radioactive wastes. As the mining and milling of these radioactive materials became uneconomical, the mining companies closed operations and

---

[1] Deputy Director, Office of Southwestern Area Programs, Environmental Restoration, Department of Energy, Washington, DC

[2] Director, Office of Southwestern Area Programs, Environmental Restoration, Department of Energy, Washington, DC

[3] Senior Project Manager, Roy F. Weston, Albuquerque, New Mexico

F. Arendt, G.J. Annokkée, R. Bosman and W.J. van den Brink (eds.), Contaminated Soil '93, 1617–1622.
© 1993 *Kluwer Academic Publishers. Printed in the Netherlands.*

the radioactive materials were left and scattered by natural weathering forces and by man's activities. About 2,000 acres were contaminated by wind and water-born materials, and nearly 300 acres were contaminated by man's activities.

Remediation activities for these sites began in 1981 with planned surface activities continuing through 1996. Currently, 10 of the original 24 sites have been cleaned up and construction is being performed on 6 other sites. Vicinity properties are sites where tailings were used as construction materials or where contamination was spread from tailing sites by natural phenomona. Of these, more that 4,400 sites have been remediated. The total cost for performing all surface remediation is expected to reach approximately $1.4 billion.

<u>COOPERATING AGENCIES</u>

The act requires that a number of different agencies participate in the remediation of these former uranium mills and tailing facilities. The principal agencies are EPA, NRC, and DOE. Below is a brief description of how each of these agencies cooperates in the final remedial actions:

- <u>EPA</u>
  EPA performs the following actions:
  - Promulgates the standards for cleanup and disposal at UMTRA sites
  - Consults with DOE on the definition/interpretation of the standards, as needed
  - Concurs in the establishment of alternate standards where they are desirable

- <u>DOE</u>
  DOE performs the following actions:
  - Establishes the funding levels necessary to perform all remedial actions and obtains that funding from Congress
  - Establishes cooperative agreements with the affected States and Indian tribes
  - Cleansup and disposes of the radioactive materials
  - Provides for public information and participation in the performance of remedial actions at the contaminated areas
  - Certifies to NRC that the tailings remediation activities meet the requirements of EPA standards
  - Obtains a general license for all of the sites from the NRC, providing for Federal custody, long-term surveillance, and monitoring

- <u>NRC</u>
  NRC performs the following actions:
  - Enforces EPA standards
  - Concurs in the selection and performance of remedial action for each site
  - Issues a general license for long-term site surveillance and monitoring
  - Concurs in land acquisition and disposal decisions

- **<u>STATE/INDIAN TRIBE</u>**
  The affected State or Indian tribe, through cooperative agreements with
  DOE, has the following responsibilities:
  - Recommends alternative disposal sites and remediation technologies
  - Acquires processing and disposal sites where deemed appropriate
  - Consults on "Environmental Compliance Documentation"
  - Consults on the protection of cultural resources
  - Concurs in Remedial Action Plans (RAPs)
  - Encourages public participation
  - Shares 10 percent of the remedial actions within the State (the
    Indian Tribes do not cost share)
  - Conducts monitoring during remedial action

<u>DOCUMENTS</u>

The UMTRA Project encountered significant problems that needed to be
addressed early in the process. Therefore, the cooperating agencies
established programmatic documents to identify the areas of major concern
and to define interaction and participation between the agencies.

- **<u>Memorandum of Understanding (MOU)</u>**
  To provide an orderly process for executing their respective statutory
  responsibilities under the Act, DOE and NRC entered into a MOU. This
  MOU has been revised at times to account for changes in the project and
  to better reflect the responsibilities and time limitations to perform
  the required actions.

- **<u>Cooperative Agreements</u>**
  The Act requires DOE to enter into cooperative agreements. These
  agreements are between DOE and the respective States/Tribes. These
  documents outline the roles and responsibilities of DOE, the States,
  and Tribes as they interact for the completion of the UMTRA Project.

- **<u>Technical Approach Document (TAD)</u>**
  The standards established by EPA call for disposal facilities that must
  exist relatively maintenance free for at least 200 years and, where
  possible, for 1,000 years. Design methods to be used predict the
  disposal facility performance this far in the future had to be agreed
  upon by the cooperating agencies.

  At the beginning of the UMTRA Project, there was little or no
  experience in the design of facilities that were to have maintenance-
  free designs which approached 1,000 years. Because of this, NRC and
  DOE worked together early in the program to establish design techniques
  and procedures. To address some of these problems, near the beginning
  of the Project, NRC issued a Standard Review Plan which later evolved
  into the Standard Format and Content (SFC) Guide. This document
  outlines the items that NRC wants to see addressed in the site-specific
  Remedial Action Plan (RAP), and the format in which RAP should be
  designed. Some of the items listed in the SFC that are to be addressed
  in the RAP are erosion potential, earthquake protection, slope
  stability, cover performance, and ground water protection.

Since these items could be addressed in a large variety of ways, NRC and DOE met and determined the best methods to address these issues. All these methods were then compiled into the TAD. By having NRC and DOE agree on these methods prior to developing site-specific remedial actions, DOE had a clear understanding of the requirements regulated by NRC. This eliminated the "bring me another rock" syndrome that is sometimes encountered when dealing with regulatory agencies. NRC also found that it reduced the amount of time needed to perform a review since the methods to be used to design the disposal facility were already defined.

The TAD is not a "cookbook" type document but provides the methods and approaches which, if used correctly, would be more readily approved. The implementation details are still left to the designers so that site-specific problems can be addressed and accounted for. There still remains the flexibility that if an agreed-upon method to solve a problem was not appropriate for a site, then other methods can still be used but, the new method must be thoroughly documented for regulatory evaluation.

SITE-SPECIFIC DOCUMENTS

In addition to programmatic documents, site-specific documents are produced and given to the regulators for information and comment and for concurrence, where appropriate. The main goal of these documents is to provide an early exchange of information and ideas between the design teams, the other cooperating agencies, and public so problems can be addressed. In this way costly redesign and associated delays can be avoided. A brief description of these documents is given below along with their purpose.

- **Comparative Analysis of Disposal Site Alternatives Report (CADSAR)**
  When beginning an initial site assessment, a large number of alternatives for remediating the site are identified. Solutions for disposing of the contaminated materials can range from stabilizing it in place, picking it up and moving it to a separate location, or some combination of these alternatives. A very methodical process was developed to look at the different alternatives to determine which would be considered the preferred option. The CADSAR provides the basis for agreement with the affected State or Indian tribe on a preferred alternative for a sites's remedial action and with NRC that the preferred alternative will meet EPA standards.

  Although this is not an NRC concurrence document, it is given to them for comment so that if they see something which concerns them early on, they can comment on it and their comments can be addressed before significant amounts of time and money are spent on characterization or design.

- <u>National Environmental Policy Act Documents (NEPA)</u>
  Environmental Assessments (EA) or Environmental Impact Statements (EIS)
  are completed for each site to meet the requirements set forth by NEPA.
  The major purpose of the documents is to demonstrate compliance with
  the law and other associated standards, to get information to the
  public on the proposed solution, and to allow them to comment on it.
  This is considered to be a decision making document and does not
  require NRC approval.

- <u>Remedial Action Plan (RAP)</u>
  The RAP is a design document. It is the primary concurrence document
  for NRC, States, Tribes, and DOE. Because of its importance, there are
  several review stages before final concurrence is reached by all
  cooperating agencies. Once a preferred alternative is selected, all
  remaining characterization is completed and design begins.

  When the design reaches the 60 percent completion stage, a copy is
  forwarded to NRC for informal review and comment. Here again, NRC is
  given a chance to have their concerns addressed early. Once the
  comments are received, they are incorporated and the design process
  continues until the 90 percent point. Here NRC is given another change
  to review the RAP, to make sure that previous concerns were addressed,
  and that the RAP meets the requirements of NRC's Standard Format and
  Content. The same process is continued through to 100 percent design
  completion phase. Although it may seem time consuming and cumbersome
  to allow for such a large number of reviews, the time involved is
  actually less than waiting, forwarding it all to NRC at one time,
  receiving their comments, resolving the comments, and then
  incorporating the comments into the final design. Experience has shown
  as would be expected that as the RAP goes through the review process,
  fewer and fewer comments are received; and, as the design approaches
  completion, all issues have been identified and resolved while
  eliminating a large amount of rework or wasted effort and additional
  characterization.

- <u>Project Interface Document (PID)</u>
  Although the final RAP is concurred upon, before construction begins,
  changes from the plan may be required due to unforeseen field
  conditions. When design changes need to be made on the remedial
  actions, then again, NRC is brought in early to explain the problem and
  a PID is completed which explains the problem and how the design will
  be modified to address the problem. This document is incorporated into
  the RAP as an addendum so all cooperating agencies are given the chance
  to review and approve the changes before implementation.

- <u>Completion Reports</u>
  Once remedial action at a site is completed, a completion report is
  submitted to NRC. The main purpose of the completion report is to
  document to NRC that the RAP was followed during construction. This
  includes everything from showing that the disposal cell was constructed
  as shown on the as built drawings, to documentation showing that the
  correct number of engineering, geotechnical and radiological tests were
  made, and that design specifications were met. Before final
  concurrence on this document, all cooperating agencies send
  representatives to the site for a walk over and assessment of the

remedial actions. After the walk over, NRC completes reviewing the completion report and formally concurs with its content. Since NRC has already seen and approved all design changes as they occurred, this document is concurred upon readily by NRC.

* <u>Long Term Surveillance and Maintenance Plan (LTSP)</u>
  After construction is completed and documented, DOE prepares an LTSP Plan. The LTSP describes the surveillance and monitoring requirements for the disposal site after remedial actions have been completed and provides the basis for obtaining the site license. The draft is reviewed and commented on by all the reviewing agencies. Issues and comments are then resolved and incorporated into the final plan.

<u>CONCLUSIONS</u>

Although there is no such thing as a perfect management framework, the management framework that was devised for the UMTRA Project has been successful. The cooperating agencies have well-defined roles, yet there is still enough flexibility that site-specific problems can be addressed and resolved without stopping work. With a system that has these qualities, work has been planned, approved, and executed efficiently. To date, remedial actions at 10 of the UMTRA sites has been completed and work is continuing on the remaining sites. The fundamental principle that allows for the required flexibility within the management framework is the close interaction and frequent communication between the cooperating agencies. With this communication, work is planned and completed on a reasonably consistent basis.

CHARACTERIZATION AND PHYSICAL SEPARATION OF RADIONUCLIDES FROM
CONTAMINATED SOIL

M. Misra, C. Neve and A. Raichur

Department of Chemical and Metallurgical Engineering
Mackay School of Mines
University of Nevada, Reno, NV 89557, U.S.A.

ABSTRACT
     The physical and chemical characterization of soils contaminated
with low-level uranium and transuranic elements is essentially required
for the development of contaminated soil remediation and clean-up
technology. The radionuclides are invariably present in trace amounts
and they are either attached to the soil particles due to physical
entrapment or chemically bonded to the soil fractions. Ongoing work with
plutonium contaminated soils from the Nevada Test Site and other
contaminated soils have shown that most of the radionuclides are present
in the finer size fractions of the soil. However, some of the
radionuclides are also present in the coarse size fractions. In some
instances the radionuclides are bonded to magnetite and clay particles.
SEM/EDX and petrographic analyses have shown that the major
mineralogical constituents of the soil are clay, quartz, magnetite,
titanomagnetite and limestone. The surface chemistry and physical
properties of radionuclides are significantly different from that of the
host soil. For example, the density, magnetic susceptibility, surface
wettability and electrical charge of certain transuranic elements are
markedly different from that of quartz, clay and other minor
constituents. In view of the above differences in surface properties,
several physical separation processes have been proposed and tested for
the separation of radionuclides from contaminated soil. These processes
will not only reduce the volume but will also be economically attractive
in remediating high volume of contaminated soil. The potential processes
are size classification, gravity based separation, flotation and
magnetic separation. In addition to the physical separation processes,
this paper also discusses several potential materials (having surface
properties close to that of plutonium and uranium) which can be spiked
with the clean soil for testing in an unrestricted environment as
surrogates for radionuclides.

INTRODUCTION AND BACKGROUND
     The physical separation processes have strong potential for
significantly reducing the volume of transuranic contaminated soil
fractions so that it can conveniently be disposed of by shipping to an
appropriate repository. It has been established that most of the
radionuclides are dispersed over a wide area containing low-level
uranium and transuranic elements. In the Nevada Test Site, it is
estimated that the top 5 cm of the soil was contaminated with plutonium
during the Safety Shot testing in early 1950-1960's (1-3). The total

F. Arendt, G.J. Annokkée, R. Bosman and W.J. van den Brink (eds.), Contaminated Soil '93, 1623–1631.

volume of that contaminated soil is approximately 1,600,000 m³. Assuming a disposal cost of $300/ton, it will cost around 500 million dollars to cleanup the test site (1). In Rocky Flats, it is estimated that 13,600 m² of soil has been contaminated with radionuclides (4-5). The contamination depth varies from 2 to 20 cm of the top soil (2). At the Fernald site, it is reported that the top 10 cm. of soil is contaminated with 10 to 280 pci/gm of uranium.

Previous analyses have shown that a significant amount of radionuclides are present in the fine size fractions of the host soil. Table 1 summarizes the approximate distributions of radionuclides in the soils from different DOE test sites.

As can be seen from Table 1, the occurrence and distribution of uranium and transuranic elements are invariably in the fine soil fractions. In order to physically separate these radionuclides, an efficient process (or processes) uniquely suitable for fine particle processing is (are) required.

Table 1:   Approximate Distributions of Radionuclides in Fine Soil Fractions (1-5)

Selected DOE Sites	Percentage Radionuclides in Fine Size Fractions of Soil
Nevada Test Site (NTS)	minus 20 microns size ~ 31%
	minus 125 microns size ~ 95%
Rocky Flats	minus 100 microns size ~ 95%
Hanford	minus 100 microns size ~ 82%
Fernald Site	minus 75% microns size ~ 48 to 79%
INEL	minus 100 microns size ~ 84%

The objective of this paper is to elucidate the physico-chemical characterization of soil and radionuclides present in the soil. The second objective is to develop some potential physical separation processes designed for selective separation of radionuclides. Because of the inherent radioactive nature of the contaminants, some potential surrogates (for plutonium) have been identified to be tested in an unrestricted environment to establish the feasibility of the process.

SOIL CHARACTERIZATION AND SIZE DISTRIBUTION
    In this paper, most of the attention is focused on the Nevada Test Site (NTS) Soil. It is anticipated that the soils from other DOE sites will be different. Nevertheless, general applicability of the process will be the same. The size distribution of the NTS soil (from area 11) is given in Figures 1 and 2 for both dry and wet screening. It can be seen that the amount of fine particles (passing 38 microns, 400 mesh)

increased from 3 % (dry screening) to 27 % (wet screening). It is apparent that mixing followed by wet screening disintegrates the lumped clay particles and increases the amount of fines present in the soil. The wet screening is rather a good indication of the true size distribution. Recent work by Murarik (2) has shown that activity of plutonium increased in the wet size fractions as compared to the dry size fractions. This fact reinforces the notion that the transuranics are finely dispersed and loosely bound to the clay particles which can be easily separated by soil pretreatment techniques. The complete size distribution of the NTS soil with approximate distribution of plutonium activity (as per Tokamura) is given in Figure 3. From the figure, it can be seen that a significant amount of soil is in the fine size fractions containing appreciable amounts of plutonium.

SEM AND PETROGRAPHIC CHARACTERIZATION

The Scanning Electron Microphotograph of NTS soil (- 400 mesh) is shown in Figure 4. The attachment of fine clay and silt to relatively large particles and agglomeration of fine clays are quite evident. This photograph along with the wet screening results suggest that the soil pretreatment steps are necessary to disintegrate the lumped particles for effective separation. The mineralogical analysis of coarse size NTS soil reveals the presence of magnetite, sanidine, biotite and titanomagnetite (Figure 5). The mineralogical composition of the fine fractions showed the presence of clay, quartz and feldspar.

SURFACE CHEMISTRY OF PLUTONIUM AND POSSIBLE PHYSICAL SEPARATION PROCESSES

The effective separation of plutonium from the associated soil depends on the soil/radionuclide associations, surface chemistry and physical properties of plutonium. From the preliminary investigations, three different possible physical separation techniques have been envisioned for soil decontamination.

Gravity Based Process

The gravity based process uses Stokes' law for differential settling of heavy fractions from the light ones. The density of different radionuclides and selected minerals are given in Figure 6. Using a concentration criterion, the effective separation of radionuclides from the host soil can be estimated. The gravity based separation has been widely used for different minerals including gold. As can be seen, the separation of plutonium from NTS soil can be accomplished with good success provided radionuclides are liberated. However, the gravity separation process has some limitations with respect to particle size. It is not very effective for fine particles. Figure 6 also shows that bismuth and WO_3 can be used as surrogates for plutonium with respect to gravity separation.

Magnetic Separation

The volume magnetic susceptibility of selected radionuclides and associated materials is given in Table 2.

Table 2:   Volume Magnetic Susceptibility

Compounds	Magnetic Susceptibility (CGS units)
Pu	707×10^6
PuO_2	384×10^6
U	411×10^6
CuO	242×10^6
SiO_2	-44.0×10^6
Al_2O_3	-18.0×10^6

Because of significant differences in magnetic susceptibility it is quite possible to separate radionuclides from the host soil. NTS soils and other soil samples do contain some magnetite, titanomagnetite and iron oxides. Experiments with CuO surrogates show that it is possible to separate finely dispersed CuO surrogate from a well mixed NTS soil sample.

Flotation
        Froth flotation is an established process which has been traditionally used for the concentration of mineral values for ores. The process uses selective attachment of an air bubble to a desired constituent in a slurry and subsequent levitation of particle/bubble aggregate to froth phase. A schematic of the flotation process is given in Figure 7. The process can be made very selective by appropriate selection of surfactants, activator, and frother. This process is uniquely suitable for very fine sized materials. The selective attachment of surfactant to radionuclides in order to make them hydrophobic is dependent upon the surface charge of the particle, oxidation state and the pH of the system. The zeta-potential of the NTS soil and PuO_2 is given in Figure 8. As can be seen, the point of zero charge (PZC) of the soil is around 1.8 whereas PuO_2 has a PZC around pH 6.7. The distribution of plutonium species for different oxidation potential and pH is given in Figure 9. From the zeta potential study and potential-pH diagram it can be observed that it is possible to separate PuO_2 from soil by using anionic or cationic collectors at an appropriate pH. Experimental results using two surrogates (i.e. CeO_2 and TiO_2) with NTS soil has shown that more than 90 percent of the surrogates can be separated from the soil using oleic acid at an alkaline pH. Experiments are in progress with conventional and column flotation using contaminated soils with different collectors.

CONCLUSIONS
        The distributions and association of radionuclides with the contaminated soil matrix are complex. The radionuclides are present in

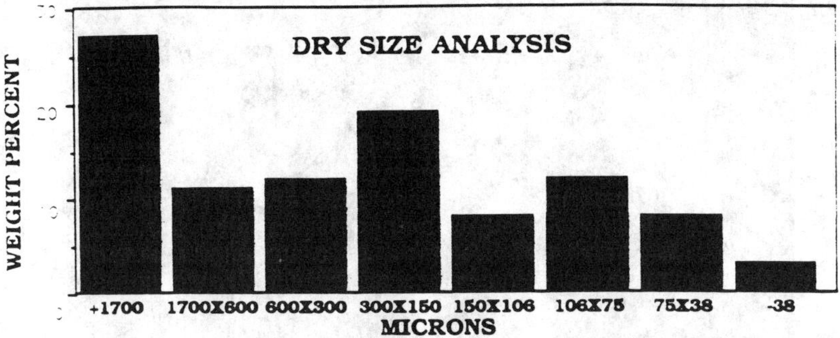

Figure 1. Dry Size Analysis of Nevada Test Site (NTS) Soil

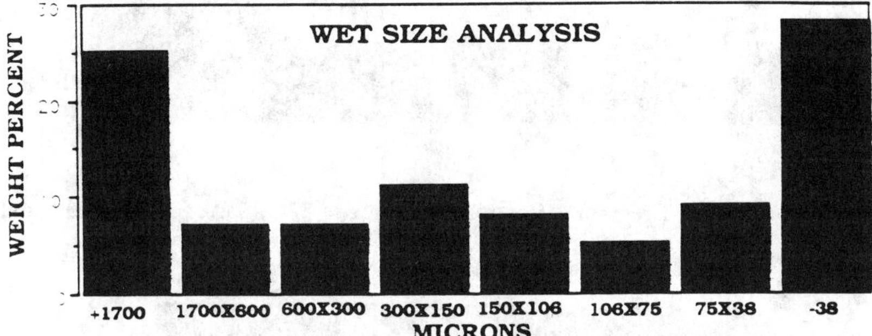

Figure 2. Wet Size Analysis of Nevada Test Site (NTS) Soil

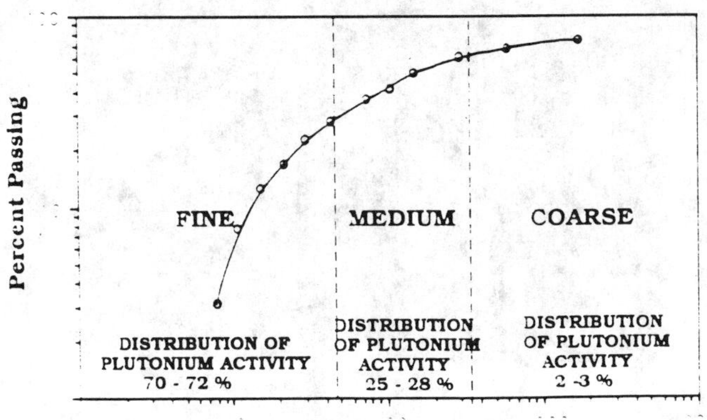

Figure 3. Complete Size Distribution of NTS Soil Indicating Approximate
Plutonium Distribution of Activity (Taken From Reference 3)

Figure 4. Scanning Electron Microphotograph of NTS Fine Soil (-400 Mesh)

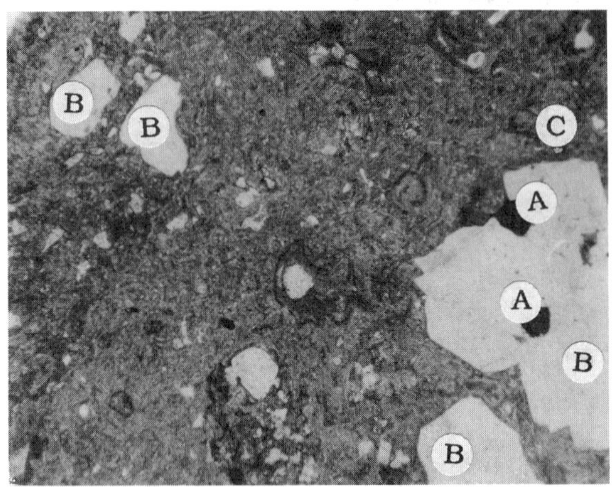

A.   Titanomagnetite
B.   Sanidine
C.   Biotite

Figure 5. Petrographic Analysis of NTS Soil

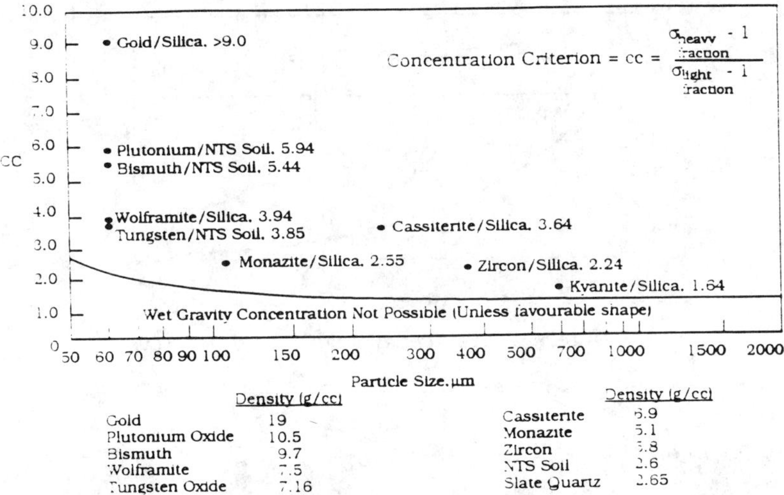

$$cc = \frac{\sigma_{heavy\,fraction} - 1}{\sigma_{light\,fraction} - 1}$$

Figure 6. Potential of Gravity Separation For Plutonium And Other Heavy Minerals. Concentration criterion> 9 Indicated Excellent Separation

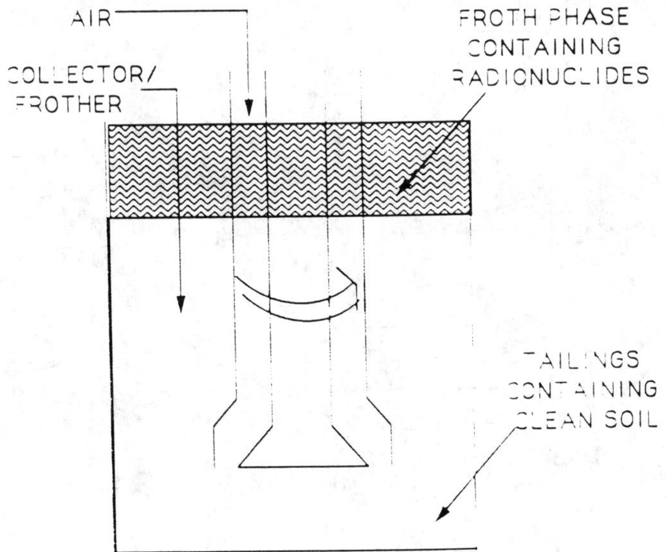

Figure 7. Schematic of Flotation Separation of Radionuclides Using Conventional Flotation Cell

1630

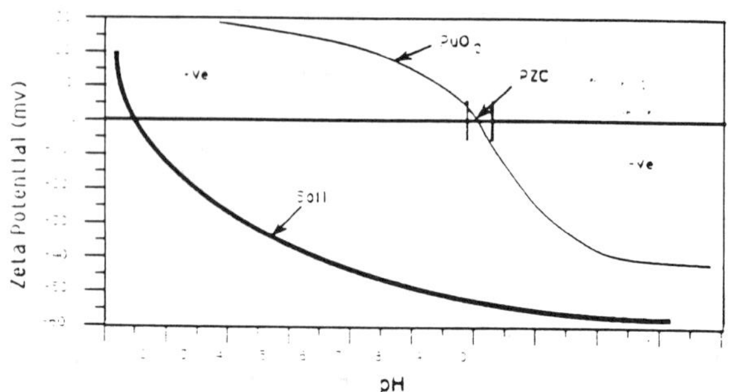

Figure 8. Zeta Potential of PuO2 And NTS Soil

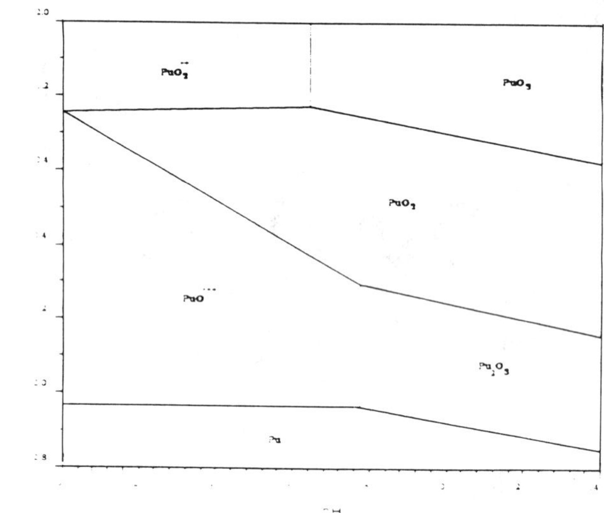

Figure 9. Eh-pH Diagram of Plutonium - Water System

the fine fractions of the soil due to physical or chemical bonding. The surface chemistry and physical properties of radionuclides are different from the soil. The density, magnetic susceptibility, electrical charge, and wettability of plutonium is different from that of the associated soil. As a result, it is possible to separate plutonium and other radionuclides from the soil using one or a combination of physical separation processes. Experiments using surrogates showed that good selective separation can be accomplished at appropriate conditions.

ACKNOWLEDGEMENTS

The authors wish to acknowledge the financial support from DOE through DRI. Technical support provided by REECo, EG&G, and MSE, Inc. is gratefully appreciated.

REFERENCES

1.   Ebeling, L.L., Evans, R.B., and Walsh, E.J., "Land Surface Cleanup of Plutonium at the   Nevada Test Site", <u>Reynolds Electrical & Engineering Co. Inc. Report</u>, 1991.

2.   Murarik, T.M., Wenstrand, T.K., and Rogers, L.A., "Characterization Studies   And Indicated Remediation Methods For Plutonium Contaminated Soils at the   Nevada Test Site", <u>Spectrum '92</u>, Boise, Idaho, August 23-27, 1992.

3.   Tamura, T., "Characterization of Plutonium in Surface Soils Area 13 of the   Nevada   Test Site", <u>The Radioecology of Plutonium and Transuranics in Desert   Environments</u>, HVO, 153, 1975.

4.   Lee, S.Y. and Marsh, J.D. Jr, "Characterization of Uranium Contaminated Soils   from DOE Fernald Environmental Management Project Site: Results of Phase I   Characterization, <u>Oak Ridge National Laboratory</u>, Publication no. 3786, 1992.

5.   Stevens, J.R., Kochen, R.L., and Rutherford, D.W., "Comparative Scrub-Solution   Tests for Decontamination of Transuranic Radionuclides From Soils", <u>Rockwell   International</u>, (Report) RFP-3161, DOE/TIC-4SW, 1982.

6.   Garnett, J., Mitchell, D.L., and Faocim, Peter, "Initial Testing of Pilot Scale   Equipment For Soil Decontamination", <u>Rockwell International</u>, RFP 3072, 1980.

7.   Nishita, H, "A Review of Behavior of Plutonium in Soils and Other Geological Materials", <u>Report to U.S. Nuclear Regulatory Commission</u>, NUREG/CR 1056,   1979.

A New Fiber-Optic Sensor Technology for Rapid and Inexpensive
Characterization of Soil Contamination

Fred P. Milanovich, Steve B. Brown, Billy W. Colston, Jr., and Paul F.
Daley

Lawrence Livermore National Laboratory, Livermore, CA 94550

Joe Rossabi

Westinghouse Savannah River Technology Center, Aiken, SC

1. ABSTRACT

The extent and complexity of worldwide environmental contamination
are great enough that remediation will be extremely costly and
lengthy. There is an urgent need for characterization techniques that
are rapid, inexpensive, and simple and that do not generate waste.
Towards this end LLNL is developing a fiber-optic chemical sensor
technology for use in groundwater and vadose-zone monitoring. We use a
colorimetric detection technique, based on an irreversible chemical
reaction between a specific reagent and the target compound. The
accuracy and sensitivity of the sensor (<5 ppb by weight in water,
determined by comparison with gas chromatographic standard
measurements) are sufficient for environmental monitoring of at least
trichloroethlyene (TCE) and chloroform.

2. BACKGROUND

The U.S. Department of Energy (DOE) has begun an ambitious program
to clean up its weapons complex facilities and to bring all continuing
activities into compliance with United States environmental
regulations. The physical magnitude of this effort is unprecedented in
modern history. Recent estimates[1] indicate that the complex has in excess of
1.4 million drums of buried and stored high- and low-level waste; 300,000 m³ of
transuranic waste; 2.5×10^6 m³ of uncontained low-level radioactive waste;
3700 hazardous, radioactive, and mixed waste sites; 5200 mill-tailing
sites; and hundreds of facilities requiring decontamination and
decommissioning. The ultimate cost of this remediation will be
strongly influenced by the number and cost of environmental samples.
It is estimated that these costs will reach into the tens of billions
of dollars; over 50,000 individual samples may have to be analyzed.
There is therefore an urgent need for new analytical techniques that
can provide rapid, inexpensive, and credible environmental sample
analyses.
One area of particular importance is in-situ characterization of
soils. Here a rapid analytical technique could greatly reduce costs by
significantly reducing sampling and analysis costs and by permitting
optimum placement of monitoring locations (monitoring wells or soil

F. Arendt, G.J. Annokkée, R. Bosman and W.J. van den Brink (eds.), Contaminated Soil '93, 1633–1637.
© 1993 *Kluwer Academic Publishers. Printed in the Netherlands.*

vapor extraction points). Towards this end we are developing and field-testing a fiber-optic sensor technology for monitoring chlorinated organic solvents such as trichloroethylene (TCE) and chloroform. In this paper we describe the sensor technology, emphasizing a design that permits the placement of the sensor into the subsurface solely with hydraulically driven penetrations (hereafter referred to as cone penetrometry).

3. EXPERIMENTAL

The sensor operates on colorimetric principles. In the basic sensor design, optical fibers are used to remotely monitor a quantitative, irreversible chemical reaction that, upon exposure to various target molecules, forms products that absorb visible light. The primary chemical reagent is an outgrowth of the work of Fujiwara[2], who first demonstrated that basic pyridine, when exposed to certain chlorinated compounds, develops an intense red color. We have shown[3] that the absorption of 560-nm light by these colored products is directly related to the vapor-phase concentration of the contaminant and that some selectivity of target molecules is possible through modifications of the reagents and of their relative proportions. We have evaluated the sensor against gas-chromatographic standards; its accuracy and sensitivity (<5 ppb by weight) are sufficient[4] for the monitoring of the environmental contaminants trichloroethlyene (TCE) and chloroform.

The sensor system has three major components: a miniature pumping and delivery system for renewing the chemical reagent, an electro-optic readout device for monitoring the transmission of the sensor, and the sensor itself. The system design allows consecutive measurements to be taken at short intervals in an on-demand basis; control and monitoring are done remotely. The small reaction volume of the sensor allows a large number of measurements to be made before the reagent reservoir must be recharged. This allows the sensor to stay in place for a long time before it must be serviced.

Figure 1 shows the most versatile version of the reagent delivery system, which was designed to fit into a standard penetrometry cone. The high-aspect-ratio design was achieved by placing the reagent supply into a gas-tight 5-mL syringe that is opposed to an identical, empty syringe. The syringe plungers are fixed with respect to one another and are simultaneously driven by a motor and drive screw. Therefore as fresh reagent is delivered to the sensor, spent reagent is captured in the empty syringe, making recovery and disposal easy. The motor is encoded to permit monitoring of the remaining reagent volume.

A custom penetrometer cone, also shown in Figure 1, was designed by S. Cooper of the U. S. Army Corps of Engineers Waterways Experiment Station (COE/WES) and S. Brown of Lawrence Livermore National Laboratory (LLNL) and was constructed at LLNL. The cone (495 mm long, 44.5 mm o.d., 33.75 mm i.d.) is threaded to receive standard 1 m x 33.75 mm COE/WES push rods. The reagent delivery system and sensor are housed in the cone and are isolated from the environment until the measurement depth is reached. The TCE sensor was exposed to soil vapors by means of a removable and disposable ground-penetrating tip. The cone is driven to the desired depth and then retracted by 15 cm, thereby leaving the removable tip at depth while exposing the sensor to soil vapors and creating a void or "probed volume" of approximately 200 mm³. Communication from the ground surface to the sensor and support equipment housed in the cone is accomplished with a custom-built 60 m long umbilical cord containing two optical fibers, a tube for vacuum evacuation of the probed volume, and a six-wire electrical cable for electronic instruction delivery and data acquisition.

4. RESULTS AND DISCUSSION

The DOE Office of Technology Development Integrated Demonstration at the Savannah River Site (SRS) was chosen as the initial location for the field demonstration and evaluation of the sensor technology. The objectives of the work reported here were to deliver the sensor, using cone penetrometry, to multiple depths at multiple locations; to expose the sensor to a range of TCE concentrations; and to observe the effect of pumping, as in a vadose zone "grab sample" operation, on the measurement process.

Before the experiments the umbilical cord was threaded through 47 one meter long push rods, allowing us to reach depths of at least 41 m. Experiments were controlled from the surface by means of a portable optical sensor readout device and portable computer. All "pushes" were carried out by COE/WES participants. The average push rate was 1 m/min.

Figure 1. Above right is an assembly drawing of the the reagent delivery system and penetrometer cone. The reagent is connected to the sensor by means of micro-tubing. The syringe reservoirs are held in place with quick disconnect clamps for easy servicing. The cone is made from corrosion resistant steel tubing with 6.35mm wall thickness.

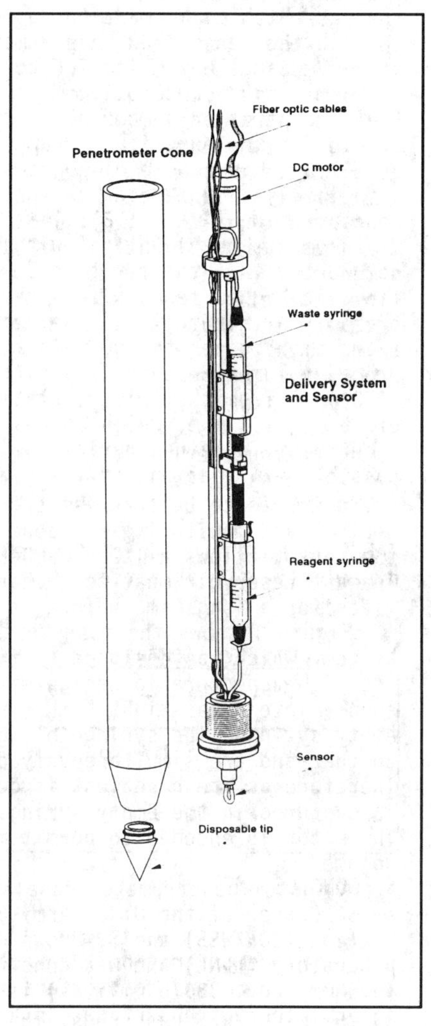

Four separate pushes (to depths of 9, 14, 41 and 33 meters) were accomplished over two days. Measurements were made in both a quiescent ("static") and vacuum extraction ("dynamic") mode. A total of 121 individual TCE concentration measurements were made during the four pushes; measured concentrations ranged from ~2 ppm by volume (static measurement at 9 m depth) to ~120 ppm by volume (dynamic measurement at 14 m). It is of particular note that the time to depth for the 41 m push was only 50 min.

Table 1 gives the results of an extended experiment performed after the sensor was pushed to 33 m, which placed it approximately 6 m above the water table. We made three measurements with the cone tip in place to zero the sensor and then removed the tip and made 15 consecutive static measurements. The TCE concentration was 18.8 ± 5.0 ppm. We then performed nine consecutive measurements under dynamic conditions. Here the concentration jumped abruptly into the 30-ppm range and remained stable for the duration of pumping, yielding a TCE concentration of 34 ± 1.3 ppm. When we discontinued pumping, the concentration almost immediately dropped back to the previous static values (average concentration 14.4 ± 1.8). This is slightly less than the initial measurements but identical within experimental uncertainty. It is noteworthy that the first three measurements after retraction of the tip gave higher results than the next eight measurements. This is probably attributable to the vacuum, and the resulting greater soil extraction, generated upon initial retraction of the cone to form the probed volume. Removing the first three measurements (29.8, 26.9 and 20.0 ppm) from the initial static set yields a mean TCE concentration of 16.1 ± 1.4 ppm, which agrees very well with the mean and uncertainty obtained during the second static set of measurements. One possible explanation for the higher values obtained during pumping is increased volatility of the TCE during flowing conditions.

TABLE 1. Measured TCE concentrations under static (unpumped) and dynamic (vapor extraction) conditions.

[TCE] (ppm, by volume)											Pump Average
On	29.8	26.9	20.2	18.2	18.0	16.6	16.7	15.8	15.2	15.0	14.6	18.8
Off	35.2	33.7	32.1	33.6	33.3	32.5	36.2	35.2	34.3			34.0
On	23.7	18.4	16.0	15.5	14.2	13.2	14.4	12.9	13.0	13.0	13.2	14.4

5. CONCLUSIONS

All objectives of this initial penetrometer-mediated sensor-placement demonstration were accomplished. This technology represents the most rapid and cost-effective means to obtain initial survey information on subsurface contamination distribution for a variety of chlorinated solvents. Its impact on optimizing the siting of monitoring wells will significantly reduce remediation costs.

This work is supported by the DOE Office of Technology Development and performed under the auspices of the U.S. Department of Energy by Lawrence Livermore National Laboratory under Contract W-7405-Eng-48. We are indebted to Dr. Lloyd Burgess and the Center for Process Analytical Chemistry, University of Washington, for collaboration that led to the design and demonstration of the continuous sensor. We acknowledge the invaluable contributions of the COE/WES staff during the course of the field experiments.

6. REFERENCES

1. *United States Department of Energy Environmental Restoration and Waste Management Five-Year Plan, Fiscal Years 1992-1996*, June 1990.

2. K. Fujiwara (1916), *Sitzungsber. Abh. Naturforsch. Ges. Rostock* **6**, 33.

3. S. M. Angel, P. Daley, K. Langry, R. Albert, T. Kulp, and I. Camins (1987, Jun), *Mechanistic Evaluation of the Fujiwara Reaction for the detection of Organic Chlorides*, Lawrence Livermore National Laboratory, Livermore, CA, UCID-19774.

4. B. W. Colston, Jr., S. B. Brown, P. F. Daley, K. Langry, and F. P. Milanovich (1992, August), "Monitoring Remediation of Trichloroethyene Using a Chemical Fiber Optic Sensor: Field Studies," *Proceedings of the International Topical Meeting on Nuclear and Hazardous Waste Management*, Vol. 1, 393–396.

CONTAINMENT AND STABILIZATION TECHNOLOGIES FOR MIXED HAZARDOUS AND
RADIOACTIVE WASTES

James L. Buelt

Pacific Northwest Laboratory'
Richland, Washington

1. ABSTRACT

A prominent approach to the cleanup of waste sites contaminated
with hazardous chemicals and radionuclides is to contain and/or
stabilize wastes within the site. Stabilization involves treating the
wastes in some fashion, either in situ or above ground after
retrieval, to reduce the leachability and release rate of constituents
to the environment. This approach is generally reserved for
radionuclides, inorganic hazardous contaminants such as heavy metals,
and nonvolatile organic contaminants. This paper describes the recent
developments in the technical options available for containing and
stabilizing wastes. A brief description of each technology is given
along with a discussion of the most recent developments and examples
of useful applications.

2. INTRODUCTION

Containment and stabilization technologies are the most common
method of treating inorganic contaminants such as heavy metals and
radionuclides. Volatile and semi-volatile organic contaminants are
more effectively treated through recovery or in situ destruction
techniques. New reagents and techniques for immobilizing these
contaminants that cannot be destroyed are evolving as land disposal
restrictions are enforced. This paper is intended to provide a brief
summary of the containment and stabilization techniques currently
available.

3. TECHNOLOGY DESCRIPTION

This paper briefly describes the following containment and
stabilization technologies:
- vitrification, both in situ and above ground, for immobilizing
  contaminants and any associated contaminated medium such as
  soils, sludge, or sediments in a durable glass and crystalline
  waste form;

'Pacific Northwest Laboratory is operated by Battelle Memorial
Institute for the U.S. Department of Energy

F. Arendt, G.J. Annokkée, R. Bosman and W.J. van den Brink (eds.), Contaminated Soil '93, 1639–1645.
© 1993 *Kluwer Academic Publishers. Printed in the Netherlands.*

- solidification technology, whereby a reagent such as grout is mixed with wastes to form a solid monolith to reduce leachability of waste and improve its physical handling characteristics; and
- bitumenization, the process of microencapsulation of waste materials in asphalt.

3.1 Vitrification

3.1.1 Process Description. Vitrification is a class of thermal treatment processes that convert contaminated waste materials, usually solids such as soils, sediments, and sludge, to a glass or glass and crystalline product. Heavy metal and radionuclide contaminants are actually incorporated into the random structure of the glass product, rather than simply encapsulated. Consequently, the product is relatively durable and resistant to leaching and is capable of withstanding disposal conditions for thousands of years.

3.1.2 Electrical Heating. Although several types of vitrification processes exist, the most common are electrically heated. These processes, which include refractory lined melters, earth melters and In Situ Vitrification operate on the principle of joule heating. Electrical energy is imparted to the glass. Normally glasses are not electrically conductive, but when in the molten state, the alkaline elements within the glass, such as sodium and potassium oxides, ionize and conduct electrical current. Submerged electrodes within the molten glass impart an electrical voltage at either end of the melting cavity, which generates electrical current and heat within the molten material being vitrified. As more heat is generated within the molten pool, the glass becomes more fluid and more conductive. Convective currents are established within the molten pool, which help carry heat to the material being vitrified and make the product homogenous.

3.1.3 Types of Processes. Of the electrically-heated vitrification processes, there are two basic types: refractory-lined melters and In Situ Vitrification. In refractory-lined melters, the molten glass is contained within a high temperature, refractory-lined cavity, and material being processed is fed to its surface. In Situ Vitrification (ISV) is generally initiated at the surface of the contaminated soil or sludge being processed, and proceeds downward, thereby vitrifying the contaminated material in place.

3.1.4 Refractory-Lined Melters. Figure 1 depicts the refractory-lined melter. As contaminated material is fed to the surface, molten glass product is drained either through a bottom freeze valve or through and overflow weir. Off gases, which consist primarily of water vapor and air in-leakage, are collected in the plenum area above the molten pool, which is maintained at a negative pressure of about 10 to 20 cm of water. These gases are then directed to an off gas cleanup system consisting of wet scrubbers and or dry filters, depending on the hazard and nature of the waste constituents. This process was originally developed for high level radioactive wastes for operation in a remote canyon or hot cell facility. The most recent developments of this process extend its operation to combustible materials, allowing a diversity of mixed hazardous and radioactive

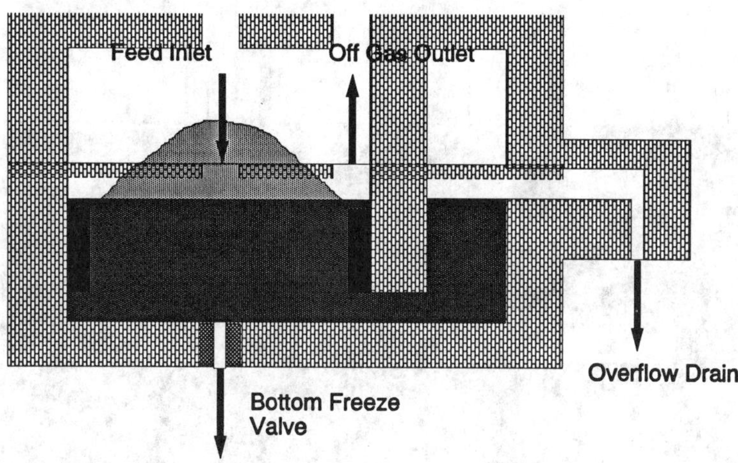

Figure 1. Graphical Depiction of a Refractory-Lined Melter

wastes to be treated by this process. Successful tests have been
completed on use of the melter on hazardous chemical wastes (Freeman
1988) and municipal solid waste incinerator ash (Chapman 1991).
 3.1.5 <u>In Situ Vitrification</u>. ISV is generally performed as a batchwise process without disturbing the contaminated soil or sludge being
processed. As shown in Figure 2, a mixture of graphite flakes and
ground glass is laid among four vertical electrodes on the surface of
the contaminated soil. These electrodes are 30-cm diameter cylinders
of graphite. An alternating electrical voltage imposed on these four
electrodes induces electrical current and heat to the graphite path,
which in turns melts adjacent soil. Once molten, the soil begins to
conduct electricity and generate heat within it. The graphite is
eventually consumed by oxidation, and the molten soil grows outward
and downward until the desired treatment depth is attained. Although
most contaminants (generally 90 to 99.99 wt%) are either immobilized
in the vitrified product or destroyed by the process heat, off gases
are contained in a hood over the site being vitrified and treated by
wet scrubbers and dry filters. ISV has been demonstrated on mixed
hazardous and radioactively contaminated soils at two sites at Hanford
near Richland, Washington (Luey et al 1992). Although the target
depths of 7 m at these sites were not achieved (depths of 4.5 m and
5.2 m were obtained) the demonstrations proved successful in
immobilizing radioactive and hazardous chemical contaminants in a
glass and crystalline waste form. Recent developments include the

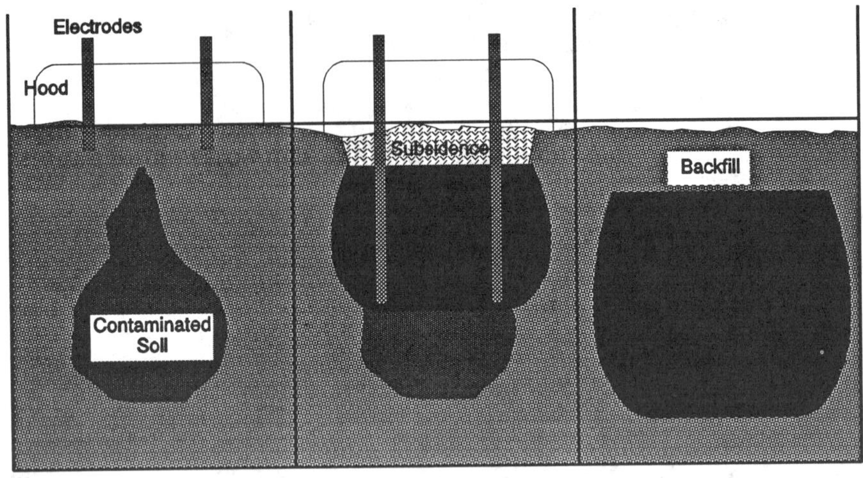

Figure 2. Three stages of the In Situ Vitrification Process

demonstration of an electrode feeding device that eliminates the need
to install electrodes and their casings into the contaminated soil
prior to treatment (Buelt and Farnsworth 1991). The electrodes are
instead fed downward as the soil is vitrified. This development also
allows high concentrations of scrap metals to be processed; up to 25
wt% scrap metals have been demonstrated with the electrode feeding
concept.

3.2 PHYSICAL SOLIDIFICATION AND STABILIZATION TECHNOLOGY
   3.2.1 <u>Process Description</u>. Solidification and stabilization
technologies are treatment processes designed to improve the physical
handling of a waste or improve its resistance to leaching. Process
options vary from simply adding absorbing agents to free liquids to
reducing the surface area to mass ratio through addition of a physical
solidification reagent such as grout to limiting the solubility of
hazardous constituents by addition of a sorptive reagent. Both in
situ and above ground processes are available. Examples of the
various process options available include the following:
   • in situ solidification or deep soil mixing,
   • continuous or batch mixing, and
   • macroencapsulation (Pacific Northwest Laboratory 1991).
   3.2.2 <u>Deep Soil Mixing</u>. Deep soil mixing is an in situ stabilization
process capable of stabilizing selected soils and sludge in place,

thereby improving the stability and resistance to leaching of the contaminated soil. Without having to excavate the process mixes solidifying reagents such as grout with the soil to reduce its penetration to water as well as other geotechnical improvements. The process is accomplished through the use of hollow-stemmed augers and mixing paddles. As the auger penetrate downward through the soil, cement-based or pozzolanic stabilizing agents are injected through the hollow-stemmed auger. The mixing paddles mix the reagents with the contaminated soil. Although this technique has been proven as a technique to stabilize soils for containment and foundation applications (treatment depths of 30 to 40 m have been demonstrated), it has only recently been demonstrated as a contaminated soil stabilization technique.

  3.2.3 <u>Continuous or Batch Mixing</u>. In this process option, contaminated materials are mixed with stabilizing reagents in commercially manufactured treatment/mixing plants. The treated mixture is then delivered to a truck for off site transport and disposal or stockpiled for on site disposal. Inorganic or organic binding agents can be used. Inorganic materials consist primarily of cement-based materials or pozzolanic-based materials. Organic binding agents consist of thermoplastic materials such as polyolefins and epoxies; organic polymers such as urea formaldehyde; and activated carbon. Cement-based materials consist of soluble silicates and portland cement that bind with water to form a cementatious matrix. They are best suited for inorganic contaminants, such as radionuclides and heavy metals. Pozzolanic materials consist of fine silica originating from fly ash, kiln dust, and lime. They behave similarly to cement-based materials, but react more slowly and may offer economic advantages. Thermoplastic materials and organic polymers are designed to microencapsulate contaminants by encasing them within their solidified matrix. These materials are generally better for stabilizing organic contaminants.

  3.2.4 <u>Macroencapsulation</u>. Macroencapsulation is a process by which a volume of wastes are enclosed within an inert jacket or container. This process can take the form of a plastic jacket fused onto the surface of solidified waste, a plastic jacket sprayed or fused around a fiberglass container that houses containers of solidified waste, and overpacks of waste drums (Lubowitz and Telles 1981a and Lubowitz and Telles 1981b). In the United States, the encapsulated material must be able to withstand compressive forces at burial depths of 15 m. The encapsulation material must be compatible with the wastes it contains and be able to withstand bacterial and fungal decay, freeze/thaw cycles, internal pressures, and ultraviolet radiation exposure. This technique has been used for both on site storage and landfill disposal.

3.3 BITUMENIZATION

  3.3.1 <u>Process Description</u>. Bitumen or asphalt is a relatively leach resistant thermoplastic substance. The process is designed to embed contaminants within molten bitumen, which when it cools provides microencapsulation of these residues. As shown in Figure 3, the

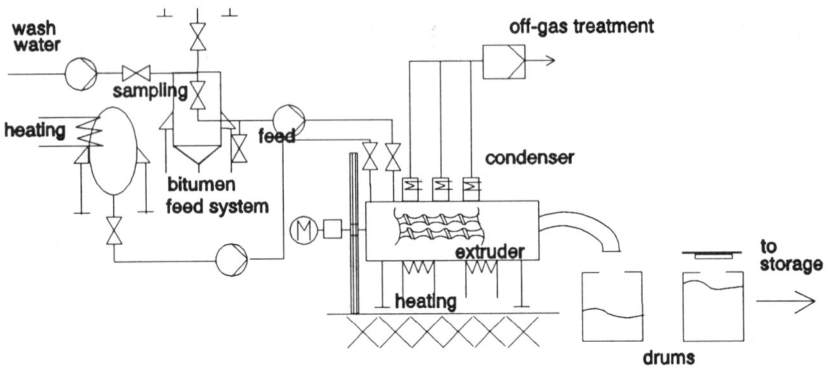

Figure 3. Graphical Depiction of a Bitumen Extruder for Radioactive Slurries

process operates by extruding heated bitumen from a storage tank at approximately 10 to 50 liters per hour into an extruder. Simultaneously, the contaminated material, usually in slurry form, is pumped into the extruder. Water is evaporated to a residual moisture content of about 0.5 wt% (Nukem 1984). The final product is a homogenous mixture of extruded solids and bitumen. The product is discharged from the extruder into a waste drum. Bitumenization has been developed and applied primarily for low and medium level radioactive wastes. Simpson et al (1988) have evaluated the process for possible use on mixed radioactive and chemically hazardous wastes. They postulated potential applications for sludge from metal plating operations, laboratory sink drains, and decontamination solutions. It was targeted in this evaluation at such contaminants as nitrates, cyanides, nickel, cadmium, and radionuclides.

4. REFERENCES

Buelt, J. L. and R. K. Farnsworth. _In Situ Vitrification of Soils Containing Various Metals_. *Nuclear Technology*, **96**, 178-184, 1991.

Chapman, C. C. _Evaluation of Vitrifying Municipal Incinerator Ash_. 1991. PNL-SA-18990, Pacific Northwest Laboratory, Richland, Washington.

Freeman, H. Standard Handbook of Hazardous Waste Treatment and Disposal, McGraw-Hill Book Company, New York, New York, 1988.

Lubowitz, H.R. and R.W. Telles 1981a "Securing Containerized Hazardous Wastes by Encapsulation with Spray-on/Brush-on Resins." US Environmental Protection Agency, Cincinnati, OH

Lubowitz, H.R. and R.W. Telles 1981b "Securing Containerized Hazardous Wastes with Polyethylene Resins and Fiberglass Encapsulates." US Environmental Protection Agency, Cincinnati, OH

Luey, J. et al 1992 _In Situ Vitrification of a Mixed-Waste Contaminated Soil Site: The 116-B-6A Crib at Hanford_ PNL-8281 Pacific Northwest Laboratory, Richland, Washington

NUKEM Co. 1984 "Volume Reduction and Solidification Systems for Radwaste."

Pacific Northwest Laboratory, _Technology Modules for the Remedial Action Assessment System (RAAS)_. 1991, Richland, Washington.

Simpson, S., H. Vidal, and M. Morris. _Mixed and Chelated Waste Test Programs with Bitumen Solidification_. Paper presented at Spectrum '88, Nuclear and Hazardous Waste Management International Topical Meeting, September 11-15, 1988, Pasco, Washington.

RADIOLOGICAL EXPOSURE ASSESSMENT OF THE FERNALD ENVIRONMENTAL MANAGEMENT PROJECT

Dr. Roy E. Eckart
Raymond P. Wood
Talaat Ijaz

Department of Mechanical, Industrial, and Nuclear Engineering
University of Cincinnati
Cincinnati, Ohio

ABSTRACT

This case study estimated the potential human exposures from the Fernald Environmental Management Project located outside Cincinnati, Ohio. Three human activity scenarios were included; an onsite resident farmer, an offsite resident farmer, and the 50 mile offsite population group. The onsite resident farmer scenario estimates the maximum individual exposure after free release of the site under a variety of site cleanup levels. The offsite farmer and 50 mile population scenarios provide estimates of maximum individual and population group exposures assuming continued administrative control of the site. The results for the various scenarios show onsite resident farmer exposure estimates ranging from nearly 60 Rem CEDE under worst case no cleanup assumptions to about 30 mRem CEDE under best estimate assumptions for the site as it currently exists. Dose to source ratios and soil cleanup guidelines under the resident farmer scenario were also calculated for each of the isotopes present. Offsite farmer and population group exposures were estimated from measured radionuclide concentrations and estimated site releases, respectively. The average offsite farmer exposure was estimated at 25 mRem, and the 50 mile population group exposure was estimated at 22 person-Rem. Dose to source ratios were also calculated for each of these groups.

INTRODUCTION

This paper summarizes the results of an exposure assessment done for the Fernald Environmental Management Project (FEMP). The FEMP is located about 20 miles northwest of Cincinnati, Ohio on a 1050 acre site in a highly productive agricultural region. The site is a contractor operated U.S. Department of Energy facility whose principal objective traditionally was production of pure uranium metals for the military. The FEMP site also served as the U.S. repository for thorium, and consequently has large amounts of thorium stored on-site. The facility has been decommissioned as a production site since 1989, and now is operated as a prototype site for cleanup. The historical production unit of the facility is located on 136 acres near the center of the site.

The goal of this analysis was to estimate human exposures to the general public from radiological contamination at the FEMP under a variety of scenarios. Exposures from radiological contamination present in soils and groundwater, and from releases to the environment are dependent upon the scenarios which are considered. Each scenario will activate different means of radionuclide transport, exposure settings, and mechanics of uptake. Three credible scenarios are considered; these envision either unrestricted site use after institutional control (the onsite resident farmer) or continued control with lesser cleanup (offsite farmer and population). The first scenario is concerned with exposure to a resident farming family on the site itself from which all exposures originate. The second scenario analyzes exposure to the nearest off-site farmer who is subjected to existing radionuclide concentrations in the three basic environmental media (air, soil, groundwater). A third scenario calculates exposure in person-Rem to the 50-mile population from the latest available measured releases.

Results for the scenarios are presented as 1 year Committed Effective Dose Equivalent (CEDE) (ICRP 1977), which is defined as the 50 year committed effective dose equivalent from a one year exposure period.

Other potential sources of radiological exposure such as the K-65 silos, the perched water, and the thorium storage barrels are considered, but are independent of the scenarios outlined above. These sources are evaluated for potential exposure in terms of their current, present day, conditions.

F. Arendt, G.J. Annokkée, R. Bosman and W.J. van den Brink (eds.), Contaminated Soil '93, 1647–1653.
© 1993 *Kluwer Academic Publishers. Printed in the Netherlands.*

ONSITE EXPOSURES

The Resident Farmer

Scenario Description. The resident farmer is considered to be the most the most conservative exposure setting; this is primarily due to the fact that this setting activates all possible pathways of exposure. In addition, the environment in which the resident farmer resides is contaminated with the highest radionuclide concentrations. Under the resident farmer scenario, a family is assumed to move onto the site after it has been released for use without any restrictions. The family is expected to construct a home on site, and to raise crops and livestock for family consumption. Water for all purposes is assumed to be supplied 100% from on-site wells. This resident farm scenario will activate all the exposure pathways contained in the RESRAD code. It should maximize the individual dose calculation and provide a conservative assumption for site use after release from regulatory control.

To add to the conservatism of this model, the family farm is assumed to be located on what is presently operable unit 3, the region of the site believed to have the most severely contaminated soils. The working assumption here is that all buildings have been decommissioned and removed, and the site has been returned to a "green" site. However, it is envisioned that the residual levels of radioactive material presently in the soil will remain and that no cover material will be employed on top of the contaminated soil. Even though this is not a viable alternative, and site release would not be permissible under these circumstances, this assumption allows for a complete radiological study with the current site conditions.

Analysis Description. Comparative exposures are best achieved using Dose to Source Ratios (DSR). These factors express the relative importance of each pathway and radionuclide regardless of initial soil concentrations. DSRs for the on-site resident farmer are evaluated using the RESRAD methodology (Gilbert, 1989); the code evaluates the transport, removal, and uptake of radionuclides via each pathway and evaluates a DSR in terms of CEDE in mRem/yr. per pCi of contaminant in one gram of soil [(mRem/yr)/(pCi/g)]. Migration from the contaminated zone to the aquifer is also modeled using a one-dimensional vertical leaching method. RESRAD evaluates the buildup of radionuclides in the groundwater and utilizes these concentrations for the water dependent pathways.

The initial analysis uses a 1 pCi/g source term for the calculation. This will yield a Dose to Source Ratio (DSR) in terms of mRem/year per pCi/g which allows the analysis to be independent of actual contaminant concentration and will aid in evaluating the relative importance of each radionuclide. The time frames for this analysis were assumed to be near term, as longer time frames are vulnerable to considerable uncertainties. Also, since the maximum DSR occurs in year 54 (Tmax), little is gained by extending the analysis over very long times. Time steps are in years after residence begins.

Results. DSR's at each time frame and for each radionuclide are tabulated below in Table 1. It is important to note the some radionuclides will not peak at Tmax; case in point is uranium 238 which is more dependent upon water transport and hence will not peak until year 660. Conversely, cesium 137 peaks at year 1 as it's exposure pathway is dominated by ground radiation.

Table 1: DSR (mRem/yr per pCi/g) At Year Shown By Nuclide, Soil Contamination

	1	10	25	50	100	Tmax
U-235	0.703	0.678	0.638	0.579	0.478	0.571
U-238	0.243	0.234	0.219	0.197	0.159	0.194
Th-228	7.021	0.269	0.001	0.000	0.000	0.000
Th-230	0.507	0.598	0.809	1.240	2.120	1.310
Th-232	3.330	11.70	16.40	17.30	17.30	17.30
Ra-226	18.700	27.60	36.40	41.50	38.60	41.70
Ra-228	7.420	5.510	0.919	0.041	0.000	0.026
Cs-137	2.994	2.392	1.644	0.881	0.253	0.800
Np-237	2.510	2.470	2.400	2.280	2.070	2.260
Pu-238	0.554	0.508	0.439	0.345	0.213	0.332
Pu-239	0.623	0.613	0.597	0.571	0.522	0.567

Radium and thorium are seen to dominate the DSR at all time frames. Exposure is not wholly due to the radionuclides themselves, but includes exposure from decay products. Ra-226, Ra-228, Th-228, and Th-232 will eventually decay to either Rn-222 or Rn-220 and are transported while in the gaseous phase. The external radiation dose from airborne radon decay products is negligible when compared with the internal inhalation dose to the bronchial epithelium region of the lung, or the external dose from the parent radium in the soil, or by the internal radiation dose received by the ingestion of plants grown in radium contaminated soils. The rate at which radionuclides are leached from the contaminated zone is significantly lower for thorium when compared to the other radionuclides considered. This has the effect of allowing thorium to remain resident in the top contaminated zone while other contributors are being leached to lower soil strata. The rate at which radionuclides are leached from the contaminated zone will have a direct effect upon the water independent pathways; these specifically include the ground and plant ingestion pathways which are shown to dominate the exposure routes.

The exposure by pathway was also analyzed for the on-site resident farmer. The percent contribution to the total dose at Tmax is shown for each pathway in Table 2.

Table 2: Percent Exposure by Pathway

Ground	Dust	Radon	Plants	Meat	Milk	Soil Ing.
33.9%	4.85%	11.72%	45.37%	3.00%	0.12%	1.03%

The dominance of ground and plant pathways can be directly attributed to the presence of radium, thorium, and their subsequent daughters. Thus, any effective remediation of the site must include alternatives to deal with radium and thorium in the top soil. Soil capping is considered to be a viable alternative for initial reduction of external ground exposure; for example, a 0.5 meter deep clay cover will reduce the DSR for Ra-226 by half, and a 1 meter thick layer will further reduce the DSR to 3.3 (mRem/yr.)/(pCi/g). However, for long term free release of the site soil contaminants must be cleaned up to concentrations which will keep the resident farmer exposure below any applicable dose standards.

Using the DSR at Tmax provided by this scenario, a level of soil cleanup can be evaluated. An example of such soil guidelines for a few selected isotopes are given in Table 3 assuming 25 mRem as the limiting allowed committed effective dose equivalent. This value of 25 mRem is consistent with the U.S. EPA regulations from 40 CFR Part 190.

Table 3: Typical Soil Guidelines

Radionuclide	Soil Guideline (pCi/g)
U-235	43.78
U-238	128.87
Th-228	3.56*
Th-230	19.08
Th-232	1.45**
Ra-226	0.56**
Cs-137	31.25
Np-237	11.06
Pu-238	75.30
Pu-239	44.09

Note:

* DSR for Th-228 is zero at Tmax; hence the DSR at year 1 is used.
** Soil guidelines for these radionuclides have been established at 5 pCi/g in the first 15 cm of soil, and 15 pCi/g for every subsequent 15 cm of soil; this follows DOE Order 5400.5

An exposure estimate from the existing soil contamination levels was also calculated for the resident farmer scenario. Four separate analyses were performed for the estimate, each with a different source term assumption. The four cases provide a range for risk from the site without soil cleanup, including upper and lower bounds. Case 1 assumes that the *maximum* production unit soil concentration for each isotope exists everywhere on the resident farm. Case 2 averages the case 1 maximum concentrations over the entire FEMP site by weighting them by the percent of the site area that is in the production unit. Case 3 assumes the *average* of the measured production unit concentrations exist everywhere on the resident farm. Case 4 assumes the case 3 concentrations, weighted for the percent of the site that is in the production area, exist on the resident farm. Case 1 is obviously a very conservative upper bound case, cases 2 and 3 are somewhat conservative, and case 4 is in this instance the most probable average contamination source for a site-wide resident farm. Table 4 presents the doses calculated for the four source cases considered.

Table 4: CEDE (millirem) By Nuclide For 4 Cases

	CASE 1	CASE 2	CASE 3	CASE 4
U-235	90.3	5.9	13.96	0.91
U-238	632.7	41.1	55.37	3.6
Th-228	689.5	44.9	55.47	3.58
Th-230	3588.9	233.3	32.88	2.14
Th-232	2238.3	145.3	131.48	8.55
Ra-226	59426.7	3861.4	258.54	16.68
Cs-137	11.5	0.75	NA	NA
Np-237	5.88	0.38	NA	NA
Pu-238	2.03	0.13	NA	NA
Pu-239	0.425	0.028	NA	NA

The variation in dose between each case reflects the nature and extent of the contamination present. The linearity of the Dose to Source Ratios with concentration permits simple evaluation of dose for any given soil 0contamination level. Thus any site release criteria, multiplied by the DSR will produce a dose (CEDE) for that radionuclide. The flexibility of these results lies in this linearity of dose with respect to source concentration, and the fact that in general risk is roughly linear with respect to dose. Therefore, to arrive at an estimate of risk, a simple multiplication of dose (in Rem) by the new risk number (risk/Rem) will yield lifetime risk.

OFFSITE EXPOSURES

Offsite Resident Farmer
Scenario Description. The offsite resident farmer scenario was created for this study in order to analyze the exposure potential to a neighboring resident farm family assuming no site cleanup. The methods employed to calculate DSRs for the offsite resident farm scenario were necessarily different from those used for the onsite resident farmer, as the offsite scenario involves initial radioactive contamination of air, water, and soil rather than just soil. The DSRs for the offsite farmer case are time independent since contamination is presumed to exist in the three environmental media (air, water, and soil) at all times. Thus, there is no need for transport between different environmental media, and the DSRs become time independent. Total doses, as opposed to DSRs, for an offsite resident farmer are time dependent and will vary with the total releases from onsite sources coupled to the environmental transport.   All climactic, agricultural, residential, and ingestion parameters used for analysis of the off-site farmer case are identical to those used in the on-site farmer case.

Analysis Method. Exposure and dose for the offsite farmer and the 50 mile population were calculated using the GENII code package written by Pacific Northwest Laboratory (Napier, 1988) . GENII was used because it allows direct input of measured radionuclide concentrations in air, soil, water, and foods. This data was available for the FEMP so the environmental transport steps of the analysis could be bypassed, thus reducing the uncertainty in the results. The calculations of both DSR (Rem per pCi/g) and total dose (Rem) are 50 year CEDE assuming a 1 year exposure. DSR results for the offsite resident farmer are presented as functions of the environmental media initially containing the radionuclides. Total dose is reported for both an average and a maximum case. The average case uses source concentrations which are the average of all the measurements around the site, while the maximum case uses source concentrations equal to the highest measured concentrations around the site.

Results. Table 5 shows the offsite farmer DSRs for radionuclides initially in air. Table 6 breaks these results down by major exposure pathway. The results show thorium isotopes, plutonium isotopes, and neptunium 237 to be significant hazards even in low concentrations when airborne, and that the majority of the dose is caused, as expected, by inhalation. While neptunium and plutonium are present on the site in only small quantities, thorium is present in large amounts in the K-65 silos and the thorium storage buildings. Any remediation work performed on these sources could thus present an exposure risk to nearby residents unless stringent controls are in place to control airborne emissions.

Table 5: DSR By Nuclide, Airborne Contaminants, CEDE (Rem per pCi/m³)

U 238	U 235	TH 230	PU 238	PU 239	TH 228	NP 237	TH 232
1.0E-6	1.1E-6	2.2E-6	2.4E-6	2.6E-6	2.9E-6	5.5E-6	9.9E-6

Table 6: DSR Percentage By Pathway, Airborne Contaminants

Inhale	Plants	Others
99.1	0.67	0.23

Table 7 details the offsite resident farmer DSRs by nuclide for isotopes initially in the soil at a concentration of 1 pCi per gram of soil. The DSRs in Table 7 are summed over all pathways. Thorium 232 and the plutonium isotopes are again seen to be very effective at delivering dose when initially present in soil. The pathway rankings by percent of total dose are shown in Table 8. The pathway rankings show that inhalation is still the single most important pathway for soil contamination, but the total of the ingestion pathways is approximately equal to the inhalation pathway. The high inhalation dose is the result of resuspended dust, which is present in fairly high concentrations in an agricultural environment.

Table 7: DSR By Nuclide, Soil Contamination, CEDE (Rem per pCi/g)

NP 237	U 238	U 235	TH 228	TH 230	PU 238	PU 239	RA 226	TH 232
1.1E-4	4.5E-4	6.6E-4	9.3E-4	1.2E-3	1.8E-3	2.0E-3	2.1E-3	6.1E-3

Table 8: DSR Percentage By Pathway, Soil Contamination

Inhale	Plants	Meat	Milk	Soil Ing.
49.13	39.12	1.44	0.89	10.06

Table 9 shows the nuclides most effective at delivering dose when initially present in the offsite farm's water source (assumed here as groundwater). Thorium isotopes, radium isotopes, and neptunium-237 are seen to be the radionuclides of greatest concern. Table 10 describes the relative effect each pathway has on exposure for a sum of all the considered isotopes. The most important pathway by far is the drinking water ingestion pathway. The other food pathways and inhalation are caused here by use of the groundwater for irrigation, which deposits the radionuclides in soil and on plants. Note that many years of irrigation will raise the contribution from these other pathways since the nuclides will tend to build up in the soil. The buildup is not strong enough to cause these other pathways to approach the effectiveness of the drinking water pathway, however, since downward leaching and harvest removal of the nuclides in the soil slow their buildup. These results are important not only for the offsite farmer scenario, but also for onsite workers since contaminated drinking water is a potential strong dose contributor to onsite workers. At present little documentation exists of radionuclides other than uranium being present in groundwater around or under the site. However, the presence of other nuclides, particularly thorium, in large quantities onsite would argue for testing groundwater both on and off the site for these other nuclides (if it is not currently being done), particularly since they are very effective in delivering dose through drinking water.

Table 9: DSR's By Nuclide, Water Contamination, CEDE (Rem per pCi/l)

U 238	U 235	PU 238	PU 240	PU 239	TH 228	TH 230	RA 226	TH 232	NP 237
2.5E-5	2.9E-5	5.0E-5	5.2E-5	5.2E-5	4.0E-4	5.5E-4	9.6E-4	2.8E-3	5.5E-3

Table 10: DSR Percentage By Pathway, Water Contamination

Inhale	Plants	Meat	Milk	Water	Soil Ing.
0.17	28.19	0.98	0.04	70.60	0.01

Total dose to the offsite resident farmer was calculated identically to the DSRs, except that actual measured source concentrations were used rather than unit concentrations. These total dose results are values from the site as it exists today with no remediation.

The average case is intended to provide a conservative estimate for the dose an offsite farmer could presently be receiving, while the maximum case is intended as a bounding estimate on the dose an offsite farmer could receive from the site under no remediation. The maximum case dose not attempt to include the potential doses from accidents. The 50 year CEDE from a 1 year exposure is calculated for the average case as 25 mRem, and for the maximum case as 140 mRem. The prime contributors to the average case CEDE are uranium 238 and thorium 232, with the primary pathway for both being inhalation. Uranium becomes important here not because of a high DSR but because of the large quantity available. The maximum case prime contributors are uranium 238 and uranium 234, both through the ingestion pathways. The most important change between the average and maximum cases is the large uranium concentration in the groundwater under the maximum case, which then dominates the maximum case total CEDE since the groundwater is used for drinking water.

Population Exposure

Scenario Description. The offsite population scenario calculates DSR by isotope and pathway and total dose for the total population in a 50 mile radius of the FEMP. All food consumed in the region is assumed grown in the 50 mile radius zone, thus maximizing the potential dose. Although this assumption seems highly conservative, it in fact makes little difference to the total dose since almost all the dose is received via the inhalation pathway. This assessment was intended to estimate dose from the site as it exists today without any cleanup or remediation work. The population of the 50 mile radius circle about the FEMP in 1990 was approximately 1.5 million.

Analysis Method. The population dose calculations were performed using the Gaussian diffusion air transport model contained in GENII (Napier, 1988). The site was modeled as a ground level release from a point source. The source for the DSR calculations was a unit release of 1 microcurie per year. The source for the total dose calculation was the estimated site release inventory for 1989 (Dugan, 1990), which is after production had stopped. The site model was chosen because almost all the airborne releases since the end of production have been resuspended particles from contaminated areas. The GENII air transport model was cross-checked for this study using the AIRDOS-EPA code written by Oak Ridge National Lab (ORNL, 1979) and by comparison to the off-site population dose results published by the FEMP (Dugan, 1990). Both these checks showed excellent agreement with differences less than 5%.

Results. The DSRs by isotope are shown in Table 11 for the isotopes having the highest DSRs. A chart by pathway is not included here since essentially all the dose was through inhalation. The nuclide rankings are, as expected, similar to those for air contaminants in the offsite farmer scenario. The total population dose calculated for the 50 mile population based on estimated 1989 airborne emissions from the FEMP was 22 person-Rem. The FEMP published result is 21 person-Rem (Dugan, 1990).

Table 11: DSR by Isotope For 50 mile Population, CEDE (man-Rem per microcurie/yr.)

U 238	U 235	TH 228	TH 230	PU 238	PU 240	NP 237	TH 232
6.0E-4	6.3E-4	1.3E-3	1.6E-3	2.0E-3	2.2E-3	3.3E-3	8.5E-3

SPECIAL EXPOSURE CONCERNS

K-65 Storage Silos 1 and 2

The major radiological concern regarding the K-65 silos is the release of radon present in the head of the silos, and resuspension of particulate matter into the atmosphere. These concerns were addressed in a Baseline Risk Assessment of the silos performed by the Radiological Environmental Assessment Group, University of Cincinnati; (Eckart, 1990). As well as calculating the probability of silo failure, several release scenarios were considered.

Acute Release A1. Complete removal of silo dome due to severe weather condition (tornado); subsequent release and redistribution of 8.5% of waste mass, and release of 50 Curies of radon from the free space of the silo. Major deposition in region surrounding silos with only 1% being available for atmospheric transport.

Acute Release A2. Failure of the dome structure in a manner which would permit only the release of radon (50 Curies).

Chronic Release. This case is characterized by the chronic release due to continual emanation of radon through cracks in the silo dome. Release term used is 650 Curies during a one year time frame.

For each release mode, three exposure points were considered; the doses in CEDE are listed below in Table 12 for each release event (all doses in Rem).

Table 12: Potential Exposures From K-65 Silo Releases (Rem)

Exposure Point	Acute Release A1	Acute Release A2	Chronic
Work Force	31.0	1.5	0.0013
Nearest Resident	2.8	0.13	2.3E-4
Population	2.57	0.21	0.0012

The current condition of the silos and the potential dose by accidental release, suggest that the silos require priority remedial action. The silos are in a advanced stage of deterioration suggesting that release scenario A2 is a highly probable situation. Case A2 could also be considered as a probable accident case during silo remediation.

Thorium Storage Barrels

The FEMP is the Department of Energy's interim storage faculty for thorium and thorium compounds. At present the equivalent of 15,000 fifty-five gallon drums of thorium are located at the FEMP. Some of the storage facilities are aging and such are susceptible to failure and release. In addition, it is known that some barrels may contain pyrophoric forms of thorium. Any possible fire could lead to significant on and off-site doses. The greatest threat concluded by the Removal Site Evaluation (FSO, 1991) is from a tornado. The off-site doses evaluated by the Removal Site Evaluation, using AIRDOS-EPA and assuming a 1 hour release, was 28.6 rem in the first year. The significance of this dose suggest that some form of remedial action or, most likely, stock removal is required.

REFERENCES

International Commission on Radiological Protection (1977), <u>ICRP Publication 26, Recommendations of the International Commission on Radiological Protection</u>, Pergamon Press, Oxford

Gilbert, T.L., et.al., (1989, June), <u>A Manual For Implementing Residual Radioactive Material Guidelines</u>, Report DOE/CH/8901, U.S. Department of Energy, Washington

Napier, B.A., et.al., (1988), GENII- <u>The Hanford Environmental Radiation Dosimetry Software System</u>, Report PNL-6584, Pacific Northwest Laboratory, Richland

Dugan, T.A., et.al., (1990), <u>Feed Materials Production Center Annual Environmental Report for Calendar Year 1989</u>, Westinghouse Materials Company of Ohio, Cincinnati

Oak Ridge National Lab, (1979), <u>AIRDOS-EPA: A Computerized Methodology For Estimating Environmental Concentrations and Doses to Man from Airborne Releases</u>, Report 5532, Oak Ridge National Lab, Oak Ridge

Eckart, L.E., et.al., (1990), <u>A Baseline Risk Assessment For The K-65 Silos Using EPA Methodology For Application To The EE/CA</u>, University of Cincinnati, Cincinnati

Fernald Site Office (FSO), (1991, July), <u>Removal Site Evaluation: Thorium Storage Buildings</u>, U.S. Department of Energy, Cincinnati

U.S. DEPARTMENT OF ENERGY, OFFICE OF TECHNOLOGY DEVELOPMENT, MIXED-WASTE TREATMENT RESEARCH, DEVELOPMENT, DEMONSTRATION, TESTING, AND EVALUATION

J. B. Berry
Mixed Waste Integrated Program
Oak Ridge National Laboratory*
Oak Ridge, Tennessee 37831

P. W. Lurk and G. J. Coyle, Jr.
U.S. Department of Energy
Washington, D.C. 20874

ABSTRACT

Both chemically hazardous and radioactive species contaminate mixed waste. Historically, technology has been developed to treat either hazardous or radioactive waste. Technology specifically designed to produce a low-risk final waste form for mixed low-level waste has not been developed, demonstrated, or tested. Site-specific solutions to management of mixed waste have been initiated; however, site-specific programs result in duplication of technology development effort between various sites. There is a clear need for technology designed to meet the unique requirements for mixed-waste processing and a system-wide integrated strategy for developing technology and managing mixed waste. This paper discusses the U.S. Department of Energy (DOE) approach to addressing these unique requirements through a national technology development effort.

INTRODUCTION

A comprehensive and consistent approach to the complex issue of mixed-waste management has been established by the DOE Office of Environmental Restoration and Waste Management. The Office of Waste Management established the Mixed Low-Level Waste Program with the primary objective of identifying and implementing the optimum treatment, storage, and disposal options for mixed waste (Bassi et al., 1992). The Office of Technology Development established the Mixed Waste Integrated Program to develop mixed-waste treatment technology in support of the Mixed Low-Level Waste Program.

The approach of the Mixed Low-Level Waste Program to the management of DOE mixed waste can be illustrated by using requirements set forth in U.S. Environmental Protection Agency (EPA) regulations. These requirements include a hierarchy of activities that constitute the format of this paper (U.S. EPA, 1985; U.S. EPA, 1988; Berry, 1992): identify and characterize the problem, establish system requirements, select alternatives, and develop technology to treat the waste. The Mixed Waste Integrated Program is clearly defining mixed-waste processing issues and developing technical resolution of these issues such as the lack of acceptance criteria for waste disposal, limits for recycle of decontaminated material, and public acceptance of incineration. These issues are being resolved by developing appropriate waste treatment technologies. The status of the following technical initiatives is described in this paper: destruction/stabilization technology, off-gas treatment, final waste form production/assessment.

---

*Research sponsored by the Office of Technology Development, U.S. Department of Energy with Martin Marietta Energy Systems, Inc., under contract DE-AC05-84OR21400.

1655

F. Arendt, G.J. Annokkée, R. Bosman and W.J. van den Brink (eds.), Contaminated Soil '93, 1655–1662.
© 1993 *Kluwer Academic Publishers. Printed in the Netherlands.*

IDENTIFY AND CHARACTERIZE THE PROBLEM

A total of approximately 70,000 m³ of mixed waste is being stored at DOE sites (Ross and Elmore, 1992), but such storage does not comply with specific requirements established by EPA. In addition, the generation of mixed waste continues at the rate of 7700 m³/year.

DOE wastes cover the gamut of possible waste management concerns. In fact, DOE has classified 700 waste streams into 7 broad categories. Waste streams that require similar processing steps have been combined to establish baseline treatment steps. The problem can be further characterized by a lack of treatment capacity and "a lack of existing proven technology to treat and dispose of . . . mixed waste to meet treatment standards" (U.S. DOE, 1991). Needs for technology development have been broadly categorized into the following technical areas: front-end waste handling, chemical/physical treatment, destruction/stabilization technology, off-gas treatment, and final waste form production/assessment.

ESTABLISH SYSTEM REQUIREMENTS

Key issues that have not been resolved and that affect the ability of DOE to assign definitive system requirements to the treatment of mixed waste include waste acceptance criteria for disposal (U.S. DOE, 1988; Duffy, 1991); limits for recycle of decontaminated material; public acceptance of incineration of organic material; and a lack of waste treatment capability and capacity.

SELECT ALTERNATIVES

One strategy being used by the Mixed Low-Level Waste Program is to "establish a standardized approach to mixed waste management activities throughout the U.S. DOE system that is cost effective, and sets a technically sound standard of excellence for environmental protection in the U.S. DOE waste management programs" (Bassi et al., 1992). To implement this strategy, a generic baseline mixed-waste treatment scheme has been established using currently available technologies. The logic for processing mixed waste is comprehensive and covers the range of wastes requiring treatment within the DOE complex. By using a comprehensive national approach to the mixed waste problem rather than focusing on a specific site, the major treatment needs have been identified (see Fig. 1). Functional and operating requirements have been published to describe the

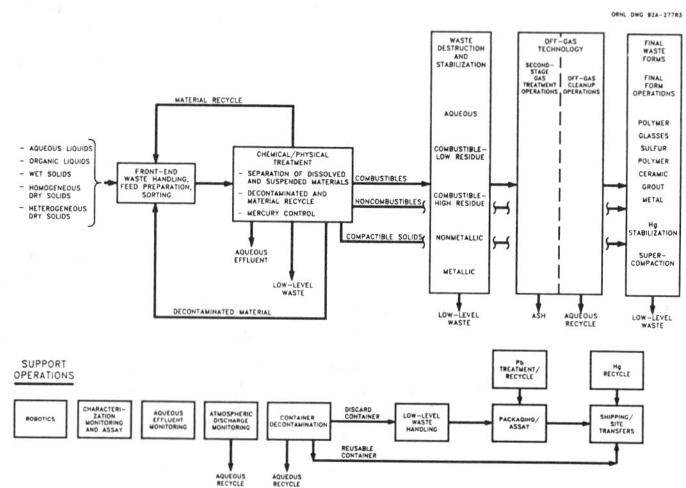

Fig. 1. Mixed Waste Integrated Program Technology Areas.

purpose of each unit operation used in this treatment scheme (T. K. Thompson, Incorporated, 1992). Rough order-of-magnitude costs and mass and energy balances have been calculated for this baseline flow sheet (Bechtel, 1992).

DOE plans to construct a production-scale, prototype treatment facility that will employ relevant aspects of the generic flow sheet at a specific site. This first treatment plant is scheduled for operation in 1997. Mixed-waste treatment facilities with common technologies will be built subsequently at additional DOE sites.

Commercially available equipment has been identified for the majority of processing steps; however, no currently available techniques were found for some steps in the processing logic. Further, this baseline treatment scheme includes disposal of radioactive material, recycle of decontaminated material, and incineration — activities for which system requirements have not been established. Technology development is required to investigate the application of existing technologies to the treatment of mixed waste and to the development of integrated treatment modules. Further development is required to obtain data to support the establishment of system requirements and to meet the identified needs of the treatment scheme. Additionally, there are opportunities to improve upon the baseline treatment process by applying innovative waste treatment operations. New technologies under development have the potential for major process impacts by simplifying and increasing the throughput rate of the baseline treatment process. Technologies will be developed to enhance or improve the baseline technologies such that the implemented treatment scheme is more cost-effective, has better performance, and has lower risk than the baseline.

A systematic approach is being employed to determine how the baseline treatment scheme should be modified to incorporate innovative technologies. An example of the application of this systems approach is seen by comparing Figs. 2 and 3. Figure 2 illustrates the baseline flow sheet for thermal treatment technologies. An innovative alternative to this baseline thermal treatment system includes a plasma-arc furnace that has the potential to treat numerous waste streams in one unit operation. The two systems are being compared for performance (i.e., material and energy balance, operability, maintainability, reliability); risk (i.e., environmental, health, and safety); and life-cycle cost. Those subsystems that are superior to commercially available subsystems will be included in mixed-waste treatment facility design.

DEVELOP TECHNOLOGY TO TREAT THE WASTE

Technology development needs have been identified by analyzing the baseline flow sheet as described above. The goal of the program is to develop a suite of technologies that will treat mixed waste to acceptable disposal criteria and provide design and reliability data on the schedule required to support implementation of mixed-waste treatment. In order to meet this goal, technical areas have been identified and technology development is proceeding in each area as described below.

DESTRUCTION/STABILIZATION TECHNOLOGY

DOE is investigating the use of thermal treatment technologies for waste destruction and stabilization (Dalton, Harris, DeWitt, 1992a). Innovative technologies, loosely defined as those technologies that are not currently being used on a large scale to treat wastes, are being demonstrated, and issues regarding full-scale operation are being resolved. Development of the following processes is currently in progress: metal-melting technologies, photochemical organic destruction processes, thermal reactors, and plasma-arc incineration. Metal-melting technologies are basically adapted from the metals industry (e.g., induction furnaces and plasma-arc melters) and the glass industry (e.g., fuel-fired and joule-heated melters). Although the operating principles for these processes are not new, there is only limited operational experience in the waste management area. In addition, new concepts in melting processes are being researched as waste management tools (e.g., the microwave melter) (White and Berry, 1990).

1658

Melter technologies hold the promise of being highly effective for waste treatment because the ash residue may not require additional treatment prior to disposal. The final waste form is physically and chemically stable and will likely pass regulatory performance standards. Because of the high temperature of melting operations, melters can be used to destroy residual

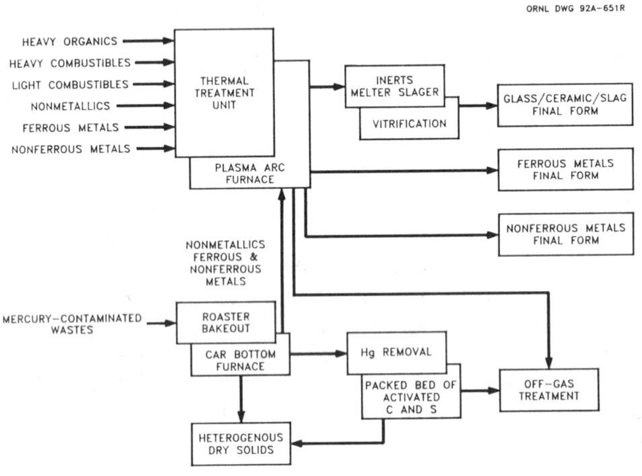

Fig. 2. MWIP Flow Sheet—Thermal Treatment Baseline.

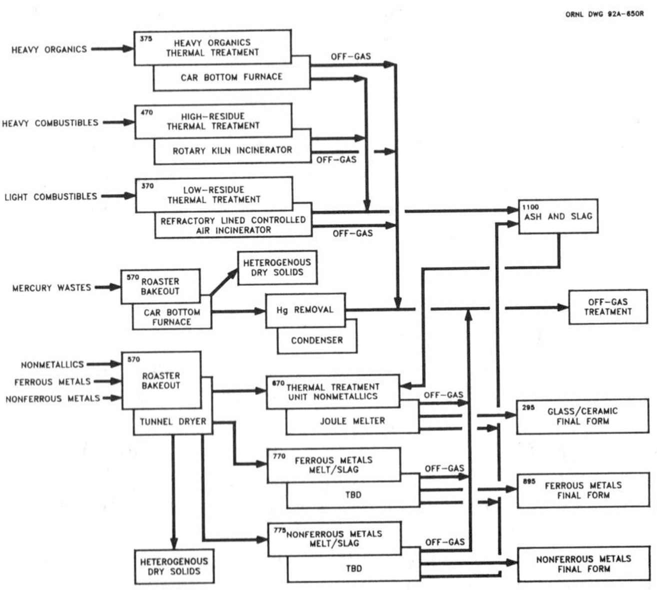

Fig. 3. MWTP Flow Sheet—Thermal Treatment.

organics. Process modifications will be required to ensure the destruction of organic material during the metal-melting process.

Melters are ideally suited for inorganic waste streams such as inorganic oxides and elemental metals. Furthermore, the chemistry in the melt can be reducing or oxidizing depending on the type of waste form desired. When processing oxides, the final waste form will be a glass or a ceramic, primarily depending on the rate of cooling. When processing metals, the melt will form a top layer of slag and a bottom layer of molten metal. The slag can then be separated from the molten metal, allowing for the recycle of the molten metal. Depending on operating parameters, it is possible to oxidize the majority of the radionuclides in the waste so the radionuclides become part of the slag, resulting in a decontaminated molten metal.

Research and development are needed in the waste destruction area. Generally, these needs can be classified as requiring more operational experience, better materials of construction, improved materials handling techniques, less waste pretreatment, control of the chemistry in the process, and detailed analysis of the resulting residue and off-gas to determine the constituents that are in the process effluent streams.

OFF-GAS TREATMENT (Dalton, et al., 1992b)

DOE is evaluating air pollution control equipment for use in treatment of gaseous effluent from mixed-waste stream processing based on criteria generally relevant to DOE facilities. These criteria include primary pollutant removal, secondary waste stream generation, safety, versatility, experience, simplicity, and cost. The preliminary evaluation resulted in ranking the spray dryer absorber best for acid-gas removal, high-efficiency particulate air filters best for particulate removal, activated carbon adsorption best for removal of both toxic metals and residual hydrocarbons, and selective catalytic reduction best for nitrogen oxide abatement. However, selection of off-gas components for a given application is highly dependent on waste streams, location, and thermal technology.

A systematic analysis is used to select the appropriate off-gas system for a given waste stream. For the purpose of illustration, the expected performance of two hypothetical DOE waste streams and thermal treatment technologies is presented here.

The first waste stream is defined to contain a large variety of waste types (i.e., solids, sludges, and liquids) and is, therefore, suitable for processing in a rotary kiln, which is a versatile thermal treatment device. The rotary kiln process is mated with wet and semidry off-gas systems in series. The wet off-gas system is designed with a rotary atomizer as the principal component and will accomplish excellent removal of acid gases and good removal of particulate. Other pollutants would either not be affected, or the removal efficiency for such pollutants is not known. The semidry off-gas system for the rotary kiln is designed with a spray dryer absorber/baghouse combination for good removal efficiency of acid gases and excellent removal efficiency of particulate and toxic metals.

The second waste stream is defined to contain a large percentage of contaminated soils and inorganic constituents, making it amenable to the plasma-arc system, which will not only destroy the organic constituents but also vitrify and stabilize the inorganic constituents. The plasma-arc process is mated with wet and dry off-gas systems in series. The wet off-gas system incorporates a venturi/packed-bed scrubber combination to accomplish excellent acid-gas and nitrogen oxides removal and to moderate removal of particulate and toxic metals. The dry off-gas system incorporates selective noncatalytic reduction followed by dry sorbent injection and a baghouse. This system accomplishes moderate acid-gas and NO_x removal and excellent removal of toxic metals.

Particular areas in which further research and development are necessary include treatment and disposal of secondary wastes, ability to remove multiple pollutants, mass-transfer rates, optimization of multiple air pollution control devices used in an integrated off-gas train, process control, catalyst activity and resistance to degradation, and materials of construction for air pollution control devices.

FINAL WASTE FORM PRODUCTION/ASSESSMENT

DOE is evaluating the use of the following groups of final waste forms for mixed waste disposal: hydraulic cement, sulfur polymer cement, glass, ceramic, and organic binders. The current status of the development of enhanced waste forms has been determined, gaps and deficiencies in what is known about technologies have been identified, and a course of action to alleviate these deficiencies has been recommended (Mayberry, Harris, DeWitt, 1992). Treatment of mixed waste is intended to protect human health and the environment from risks associated with the release of hazardous and radioactive components from the waste. DOE Order 5820.2A requires a performance assessment for each disposal site, which shows by analysis that the waste treatment process and other disposal controls adequately meet this objective. At present the assessment does not credit the waste form with any capacity to restrict contaminant mobility; all containment is attributed to physical barriers such as vaults. In order to rate the waste form as a physicochemical barrier to contaminant release, performance criteria must be developed. Criteria can be based solely on regulatory requirements, solely on technical parameters, or on a mixture of both. The latter approach has been selected, recognizing that current regulatory criteria do not address all of the properties of a waste form that determine its ability to reduce contaminant mobility and acknowledging that the properties measured may not, in fact, be indicative of the performance of the waste form in the specific disposal setting. The most important technical requirement affecting whether a waste form can be disposed of is the long-term stability of the waste form in the disposal setting. This information is critical in developing the performance assessment, which documents the ability of the entire waste control system to prevent unacceptable releases. Tests to measure and predict release rates of hazardous and radioactive constituents have not been developed and verified for either short- or long-term stability. There is a need to establish uniform testing requirements for waste form performance. Because of the volume and heterogeneity of mixed wastes, compositional flexibility, minimal volume increase, and low unit cost are all desirable properties of waste forms. Difficulties experienced with the long-term integrity and performance of traditional waste form technologies (e.g., cementation and grout) have led DOE to consider evaluation of technologies for production of enhanced waste forms (e.g., glass and ceramics) with the expectation that these waste forms will exhibit comparatively superior performance characteristics (e.g., leach resistance and durability), which will facilitate eventual disposal.

Evaluation of vitrification processes for mixed-waste streams is in progress in an attempt to assess alternative technologies for inclusion in the proposed prototype treatment plant for mixed waste. Vitrification-related treatability studies include the following waste streams: incinerator ash (including air pollution control equipment sludge), wastewater treatment sludge, soils, and off-specification cemented wastes. The treatability studies will involve characterization of the wastes, laboratory-scale or "crucible" studies to identify appropriate glass compositional formulations and additive requirements, engineering-scale process studies (e.g., steady-state operation) of small-scale melters to identify process concerns for scaleup, and pilot-scale demonstrations of the process in an integrated fashion including feed and air pollution control system evaluation.

CONCLUSION

The DOE Office of Technology Development has established a national approach to technology development for mixed waste treatment. This multi-functional approach reduces duplication of technology development effort as compared to a site-specific approach. A comprehensive approach to the solution of DOE mixed radioactive and toxic waste problems includes an analysis of alternative waste treatment systems. The implemented treatment schemes will be more cost effective, have better performance, and have lower risk than baseline technologies.

Technology development for mixed waste treatment has been categorized by evaluating the required treatment steps. A mixed waste treatment plant includes unit operations in the following technical categories: front-end waste handling, chemical/physical treatment, destruction/stabilization technology, off-gas treatment, and final waste form production/assessment. DOE has initiated activities in each of these technical areas. This paper describes significant progress to date and additional research needs in the areas of waste destruction/stabilization technology, off-gas treatment, and final forms production/assessment.

REFERENCES

Bassi, J., Chakraborti, S., DeBiase, T., Devarakonda, M., and Abashian, M. S., 1992, "A Decision Methodology for the Evaluation of Mixed Low-Level Waste Management Options for DOE Sites," to be published in Proceedings of 14th Annual DOE LLW Management Conference, Phoenix, AZ.

Bechtel, 1992, "Mixed Waste Treatment Project Process Systems and Facilities, Design Study and Cost Estimates," Lawrence Livermore National Laboratory, Livermore, CA.

Berry, J. B., 1992, "Methodology to Remediate a Mixed-Waste Site," M.S. Thesis, University of Tennessee, Knoxville, TN.

Dalton, J. D., Harris, T. L., and DeWitt, L. M., 1992a, "Technical Area Status Report for Waste Destruction and Stabilization," Draft DOE/MWIP-4, Oak Ridge National Laboratory, Oak Ridge, TN.

Dalton, J. D., Gillins, R. L., Harris, T. L., and Wollerman, A. L., 1992b, "An Assessment of Off-gas Treatment Technologies for Application to Thermal Treatment of Department of Energy Waste," DOE/MWIP-1, Oak Ridge National Laboratory, Oak Ridge, TN.

Duffy, L. P., 1991, Director, Office of Environmental Restoration and Waste Management, Department of Energy, "Commercial Disposal of Department of Energy Radioactive (By-product and Low-Level) and Mixed Wastes," letter to W. A. William, Environmental Restoration Division, Department of Energy.

Mayberry, J., Harris, T. L., and DeWitt, L. M., 1992, "Technical Area Status Report for Low-Level Mixed Waste Final Waste Forms," Draft, DOE/MWIP-3, Oak Ridge National Laboratory, Oak Ridge, TN.

Ross, W. A., and Elmore, M. R., 1992, "Locations, Volumes, and Characteristics of DOE's Mixed Low-Level Wastes," Proceedings of Waste Management 1992, Tucson, AZ.

Thompson, T. K., Incorporated, 1992, "Functional and Operational Requirements for an Integrated Facility, 1992," Mixed Waste Treatment Project, Los Alamos National Laboratory, Los Alamos, NM.

U.S. Department of Energy, 1991, "Land Disposal Restriction Case-by-Case Extension Application for Radioactive Mixed Wastes," U.S. Department of Energy, Washington, D.C.

U.S. Department of Energy, Nevada Operations Office, and Reynolds Electrical and Engineering Co., Inc., 1988, "Nevada Test Site Defense Waste Acceptance Criteria, Certification, and Transfer Requirements," NOV-325, U.S. Department of Energy.

U.S. Environmental Protection Agency, 1985, "Guidance on Feasibility Studies Under CERCLA," EPA/540/G-85/03, U.S. Environmental Protection Agency.

U.S. Environmental Protection Agency, 1988, "Guidance on Feasibility Studies Under CERCLA," EPA/540/G-89/004, U.S. Environmental Protection Agency.

White, T. L., and Berry, J. B., 1990, "Status of Microwave Process Development for RH-TRU Wastes at Oak Ridge National Laboratory," presented at the Waste Management 1990 Conference, Tucson, AZ.

NATIONAL AND INTERNATIONAL COOPERATION: THE KEY TO DEPLOYMENT OF
ADVANCED TECHNOLOGY FOR CLEANUP

Presented by
Dennis F. Miller
EG&G Idaho, Inc.
Environmental Restoration and Waste Management
U.S. Department of Energy
at the
Fourth International Kfk/TNO Conference on Contaminated Soil

Berlin, Germany
May 4, 1993

INTRODUCTION:

Private and public institutions in the United States are in the process of creating R&D programs and conducting environmental restoration and waste management activities in order to deal with extraordinary quantities of hazardous wastes. Public sector cleanup costs are dwarfed by the challenge of restoring contaminated industrial sites. Indeed, industrial environmental restoration produced $130 billion in business activity during 1990 and waste management produced $120 billion in business activity during 1991. In the public sector, the effort may not be as large, but it is massive. One U.S. government agency, the Department of Energy's (DOE) Office of Environmental Restoration and Waste Management, plans to spend over $6 billion in FY 1993 and may spend in excess of $300 billion over 30 years in the cleanup of over 4,000 sites and the enhancement of waste prevention and waste minimization capabilities at production sites. Among the Federal agencies, R&D and cleanup programs exist in the Environmental Protection Agency (EPA); Department of Interior (DOI), which also includes the Bureau of Mines, Geological Survey, Bureau of Reclamation and eight other Bureaus; National Aeronautics and Space Administration (NASA); and Department of Defense (DOD), which includes of the Departments of Army, Navy, and Air Force.

While the magnitude of the soil and water contamination on sites controlled by the other government agencies may not be as vast as the approximately 4,000 DOE sites, nevertheless, the total Federal agency level of effort is significant. The question facing all of these institutions is how to decrease costs, insure safety, and complete the work as quickly as possible while at the same time insuring quality and technical efficiency. The answer is to employ advanced and innovative technologies. This approach can be enhanced through collaboration on common problems and seeking common solutions by sharing in the development, use, and costs of advanced technologies. For many Federal agencies, such cooperation is beneficial. This is especially true for those concerned with duplication of effort, program redundancy, and who have a need to leverage expertise and funding. However, even with the emergence of advanced and innovative technologies, a major concern is how to convince risk adverse regulators and vendors to accept such technologies. I will provide an overview of collaboration and its importance among these agencies and discuss the role of the private sector in the cleanup. Even though my focus is within the U.S., non-United States private and public institutions should understand that collaboration can be conducted with them.

F. Arendt, G.J. Annokkée, R. Bosman and W.J. van den Brink (eds.), Contaminated Soil '93, 1663–1671.
© 1993 *Kluwer Academic Publishers. Printed in the Netherlands.*

REQUIREMENTS FORCING COOPERATION:

A major objective of U.S. government agency cleanup is to ensure a technically efficient and cost effective remediation of contaminated sites. At the agency level, this approach shapes stewardship programs. Furthermore, responsibilities for environmental restoration and waste management differ from agency to agency. This means that the information pertaining to the technologies required by each and their assessments of emerging technologies are shaped by individual agency missions, organizational structures, and functions. The following provides an overview of these differing perspectives and involvement for some Federal agencies.

U.S. Environmental Protection Agency (EPA)

The EPA was created by Congress in 1970 to control and abate pollution in the areas of air, water, solid waste, pesticides, radiation, and toxic substances. It endeavors to do this systematically by integration of various research, monitoring, standard setting, and enforcement activities. The EPA is equally dedicated to improving and preserving the global environment. EPA's challenges are national in scope and deal with two broad issues.

One challenge is to respond to environmental contamination caused by past mismanagement of hazardous waste. Over 34,000 potential sites have been evaluated. Of these, about 1200 are listed on the National Priorities List. In 1986, $8.6 billion was provided over five years to pay costs for overseeing work by those responsible for remediating waste sites to pay costs not assumed by responsible parties for cleanup. In 1990, another $5.1 billion was provided to extend the cleanup effort through September, 1994.

Another major challenge is to reduce current and future hazardous waste streams. The U.S. generates about 150 million metric tons of hazardous industrial wastes annually, while other sectors produce another 130 million tons of non-hazardous wastes. Over 4,400 treatment, storage, and disposal facilities, 20,000 transporters, and 240,000 generators are regulated by EPA and the States. There are 327 Federal treatment, storage, and disposal facilities and over 5,000 Federal facilities that generate hazardous wastes. Further, EPA and the States regulate more than 1.8 million underground storage tanks.

U.S. Department of Defense (DOD)

DOD is responsible for all U.S. military properties and installations located throughout the States, territories, and abroad. Almost all the facilities are a military service's responsibility, and they have their own independent missions, regulations, cultures, and organizational structures. Land disposal of munitions at some sites dates back to the mobilization for World War One. The environmental issues faced by DOD are dominated by the treatment, storage, and disposal of contaminants and hazardous wastes from industrial production and extensive training operations, most notably residues from fuels and explosives. Remediating toxic waste sites will be one of DOD's largest challenges and fastest growing areas of activity over the next ten years. Evidence of this is shown in the FY 1992 budget request for the Defense Environmental Restoration Program (DERP), which was $1.562 billion. This is about half again as large as the previous year.

The number of sites included in the DERP has increased steadily since the program began in 1984. By the end of 1991, 17,600 potential sites at 1,877 installations had been identified. However, only 90 sites are on EPA's National Priorities List. Current cleanup cost estimates (environmental restoration only) reach about $25 billion over the next ten years. Because of the tight budgets, DOD's approach is like that of DOE's, that is, in order to reduce costs and gain in efficiency there is a great emphasis on the development of innovative and advanced technologies. Under Project Reliance a Tri-Service Group developed plans to fully integrate the research capabilities of each service, which makes use of tri-service teams to perform research using each service's unique capabilities. In addition, a strategic plan integrating all services is being written at the DOD level.

To facilitate State participation, DOD reimburses States for up to one percent of the total cleanup costs. To expedite this program, DOD has agreements with 36 States. In FY 1991, DOD provided about $17 million to State environmental regulatory agencies to support their involvement in the DOD program.

U.S. Department of Interior (DOI)

DOI is the largest landowner in the United States. It manages about 440 million acres or 20 percent of the nation's surface area. In addition, DOI is trustee of an additional 50 million acres of Indian Trust Lands. Further, DOI holds about 32 percent of the nation's total subsurface and mineral estates in trust for the public. Organizationally DOI is made up of ten different Bureaus, each of which has its own mission. Among the Bureaus, there are about 422 sites listed on the Federal Facilities Hazardous Waste Compliance Docket. Among these, only three are on EPA's National Priorities List.

DOI's approach is based on five fundamental principles:

- o Prevention of hazardous waste generation
- o Reduction of wastes generated
- o Management of waste materials
- o Cleanup of contaminated areas
- o Restoration of injured natural resources

The total FY 1992 DOI budget for oversight, investigation, and cleanup of hazardous materials was $70.4 million. The FY 1993 request is $79.6 million, which is an eight-fold increase over that for FY 1988 ($8.7 million).

U.S. Department of Energy (DOE)

DOE faces the largest cleanup requirement in U.S. history. In addition, it must enhance waste prevention capabilities at its production plants.

These major tasks were adopted because of the almost five decades of defense-related production activities. Recognition of this cleanup responsibility was the creation in 1989 of the Office of Environmental Restoration and Waste Management (EM), which is headed by an Assistant Secretary. Three Deputy Assistant Secretaries are each responsible for one of the major programmatic areas: Environmental Restoration, Waste Management, and Technology Development.

The challenge is to restore nearly 4,000 contaminated sites covering tens of thousands of hectares with hazardous or radioactive waste, soil, groundwater, or structures. There are more than 1.4 million drums of buried or stored waste. And there are thousands of facilities awaiting decontamination, decommissioning, and dismantling. In addition and at the same time, DOE must manage wastes at those sites still operating and generating wastes. The goal is to solve the technical difficulties and to ensure that the nation receives maximal economic benefits for its remediation and waste management expenditures. To perform this massive cleanup, DOE is seeking to use the best cutting edge technologies, consistent with compliance schedules. The adoption of such technologies helps insure that remediation expenditures are no greater than actually necessary. Moreover, these technologies can assist the U.S. in gaining international leadership in a growing environmental industry. In this manner, DOE will be able to generate related public economic benefits. These can come through international sales of environmental goods and services, and through lower costs (and, hence, more competitive prices) for environmental compliance by other American industries competing in the world market.

The type of contamination found at DOE sites is common to metals and electronics production sites in the private sector, e.g., waste chemicals, organics, and heavy metals. However, remediation of DOE sites is more difficult because of mixed wastes, which are combinations of organics and inorganics and radioactive materials. In addition, there are hosts of problems, for example,

- 5,200 uranium mill tailing sites
- about 5000 peripheral properties that have soil contaminated with uranium mill tailings
- one building at a production site is about one mile long; several buildings at the same site contain about 4.3 million square feet and contain 700 tons of sheet metal, 650 tons of angle iron, 65 tons of sprinkler piping, 130 tons of exhaust fans, and thousands of tons of concrete.
- 2.5 million cubic meters of low-level radioactive waste
- more than 300,000 cubic meters of transuranic waste (generally contaminated with plutonium)
- the Hanford site holds enough contaminated soil to fill 33 football stadiums. The volume of waste there represents about 25 percent of DOE's environmental challenge today.

DOE planning for its EM activities has been published annually in a series of Five Year Plans. These documents, supported by programmatic and installation-level detail, report on progress-to-date toward meeting compliance and cleanup goals. The plans also describe near-term plans and over-all long-term strategy goals for mission accomplishment. The budget for EM is about $6 billion in FY 1992.

COMMON THEMES:

While each agency has its unique needs and responsibilities, there are common themes. Cost reduction, speed, and ease of operation (or other means of being compatible with work-force constraints) are important technology features for all the Federal agencies. Mixed wastes pose the most intractable challenges, though the volumes of such wastes are very much agency-dependent. Groundwater and other sub-surface pathways for contamination spread are generally of more concern than surface or airborne pathways, primarily because of the lack of access for remediation. All agencies recognize the importance of preventing pollution as preferable to removing contamination. What are some of the technology needs for selected agencies:

DOD

Cleanup Needs:

o   Improvements in site characterization and monitoring, including advancement of remote sensing, field sampling and analysis, and database development.

o   Development of remediation technologies that can be applied quickly and at less cost, which protect human health and the environment.

Pollution Prevention Needs:

o   Technologies to reduce hazardous waste and eliminate the need to purchase environmentally harmful materials.

o   Environmentally sound technological systems and platforms for carrying out ongoing DOD missions.

DOI

o   Development of technology to easily locate waste deposit sites and a fast inexpensive method for determining the wastes present.

o   Development of an inexpensive risk assessment methodology that allows a quick and economic-based assessment of a variety of risk scenarios.

o   Development of inexpensive, effective remediation technologies for control of contamination migration.

o   Development of inexpensive, efficient technologies for remediation of mixed waste.

DOE

Waste Management:

o   Development of treatment or disposal options for dealing
     with mixed wastes. This will require advanced material
     handling techniques and new technologies and process to
     separate mixed waste components, reduce their volume,
     and either destroy or immobilize them in new waste
     forms.

o   New technologies are needed to clean and decontaminate
     hazardous and radioactive materials from structures and
     equipment, so DOE may complete its modernization and
     decommissioning plans.

o   Waste avoidance/minimization requires the development of
     new technologies and methodologies to reduce or
     eliminate the generation of waste and new pretreatment
     techniques to minimize waste. Better separation and
     concentration processes must be developed to increase
     recycling. Non-toxic materials, particularly solvents,
     must be developed as substitutes for hazardous
     materials.

Environmental Restoration

o   Environmental remediation of contaminated soils and
     waters requires the development of innovative in-situ
     and ex-situ treatment technologies for the removal or
     destruction of contaminants to acceptable levels in both
     arid and non-arid soils. In-situ methods are preferred
     because they minimize public exposure and the volume of
     waste for disposal.

o   The need for faster characterization of contaminated
     areas requires the development of advanced technologies
     both to identify in situ and in real time the type and
     concentrations of contaminants and to monitor the site
     over time.

CURRENT PLANNING AND IMPLEMENTATION ACTIVITIES:

One forum for U.S. Government interagency cooperation is the Federal
Remediation Technologies Roundtable, which was formed in part to seek
common solutions to common problems experienced by government agencies.
Roundtable objectives include reducing costs through collaboration and
elimination of duplication in government R&D and cleanup programs.
Members currently include DOD, DOE, DOI, EPA, and NASA. The EPA divisions
participating include the Office of Federal Facilities Enforcement, Office
of R&D, and the Technology Innovation Office, which hosts the Roundtable.
The Roundtable provides a central forum to share the lessons that each
agency is learning in developing new technologies and to transfer this

information to other user communities. One significant private/public partnership initially started through the Roundtable is the Clean Sites/McClellan Air Force Base Project, which seeks to generate cost and performance data on innovative technologies. Joining forces in the project are the EPA, Air Force, California State EPA, and several private firms in order to demonstrate and evaluate hazardous waste site remediation technologies. The private firms stand to benefit by gaining cost and performance data on technologies targeted for problems similar to their own without the risks associated with testing on their own property. Among other Roundtable activities one panel is exploring the feasibility of conducting a joint demonstration of an innovative treatment technology for contaminated groundwater.

Another forum furthering government level interagency cooperation is the Strategic Environmental Research and Development Program (SERDP), which was established in FY 1991. The purposes of SERDP are to: (1) address environmental issues of concern to DOD and DOE through support for basic and applied research and development of technologies; (2) identify and share research, technologies, and other information developed for national defense purposes that would be useful to governmental and private organizations involved in the development of technologies addressing environmental research, waste minimization, hazardous material substitution, and other environmental concerns; (3) furnish other government and private organizations with data and enhanced data collection, and analysis capabilities for use in the conduct of environmental research; and (4) identify technologies developed by the private sector that are useful for DOD and DOE environmental restoration and waste management activities. DOD, DOE, and EPA are to pursue these objectives in cooperation with other Federal agencies, to conduct joint research, development, and demonstration projects relating to innovative technologies.

A major initiative is a cooperative agreement with the Western Governors' Association and the Federal Government Departments of Defense, Interior, Energy, and the U.S. Environmental Protection Agency. The purpose of the memorandum of understanding, which was signed in 1991, is to establish a more cooperative approach to development of technical solutions to environmental restoration and waste management problems shared by States, corporations, and the Federal government. The regional approach will serve as a demonstration of principles and practices, which may be adopted nationally. In this case, the parties at interest, want to achieve a regionally integrated cooperative approach to identifying solutions to problems and to share information on emerging technologies and describe needs for new advanced technologies, so scarce R&D funds are focused on the major problems on Federal lands and sites to the extent permitted by law. All the parties are committed to fostering the development of better, faster, safer, and more cost effective site restoration and waste management technologies and methods. A research and demonstration plan is being considered for implementation in the near future.

There are examples of current specific technical cooperation:

o   DOE, DOD and EPA have formed an Interagency Cost Estimating Group (ICEG) for Hazardous, Toxic and Radiological Waste Remediation

(HTRW). Estimating costs for HTRW is technically difficult, yet necessary in order to judge the appropriateness of cleanup proposals and to plan funding. The purpose of the ICEG is to establish a network among member organizations for collecting and sharing HTRW remediation cost information and related experience.

o   DOE, EPA, and industry support five Hazardous Substance Research Centers (HSRC), which cover all the U.S. EPA's regions. The HSRCs are university based consortia that conduct cutting edge research, which is targeted at some of the fundamental problems in environmental cleanup and waste prevention. With more than 20 major universities and 100 industries as members, these Centers allow government and industry to have access to innovative and advanced characterization, monitoring, prevention, and remediation R&D, which directly deal with the major technical issues of concern.

o   The U.S. Bureau of Mines has increased its outreach with other Federal agencies and has modified its research efforts to include more work directed at the broader issue of hazardous wastes than the Bureau's traditional minerals waste treatment activities. Agreements have been signed with the USBM and EPA, the U.S. Forestry Service, organizations in DOD, and a number of DOI agencies. Additional agreements are being discussed with both DOE and other agencies.

o   Requests by other Federal agencies for the U.S. Geological Survey (USGS) to provide its ground-water and surface-water geophysical, hydrogeologic, and water-quality capabilities to address their hazardous waste problems are increasing. Currently, the USGS is assisting the DOD, DOE, EPA, and other Federal and State agencies in their hazardous waste programs. For example, the USGS works on about 45 DOD installations as part of the Installation Restoration Program to characterize and cleanup past spills. With EPA, it has investigated bioremediation techniques at the Champion Mill site near Libby, Montana, to neutralize wood treating fluids and their constituents, including creosote and pentachlorophenol.

SUMMARY:

The current and future cleanup problems and issues facing the public and private sectors in the United States are among the largest and most complex challenges that the nation has ever faced. It is clear that hazardous wastes must be remediated and waste prevention must become a basic principle in the design of any manufacturing system. The long-term costs will be immense. However, it is important to realize that with the use of advanced and innovative technologies the cleanup period can be reduced, the remediation process can be made safer; and, through more technically efficient and cost effective technologies, the over-all cleanup efficiency can be enhanced and the over-all costs will be much less.

One emerging lesson is that the complexity and costs of the cleanup effort require avoidance of duplication of R&D and demonstration efforts through cooperation among the government agencies and with industry and universities. Being aware of the need for coordination and cooperation, interagency cooperation and collaboration has and is taking place in the U.S. government. This includes interactions at the Federal, State, and Regional levels. A great deal of time and effort is being made to involve private industry in cooperative cleanup programs with the government agencies. Furthermore, while time and money can be saved by having the public and private sectors work together, it must be realized that the private sector will carry out the cleanup work and it must be brought into the government R&D processes at an early stage as a partner, so it shares technologies and costs with the government. Collaboration will help all parties. DOE will benefit from existing advanced technologies, which may be under development in the private sector for more efficient and cost effective cleanup. In addition, while technologies are being developed and applied, the pool of available technologies will increase for both government and industry.

Through this cooperation, and ultimately technology transfer from the government to the private sector, success in the development and commercialization of innovative and advanced technologies will be achieved for the betterment of the nation. Furthermore, through increased international cooperation these benefits can be transferred from the United States to other nations and corporations and their advanced technology can be transferred to the U.S. With such cooperation, the Earth can become a safer environment for man and animals.

<u>The Need for a Sound Philosophical Basis for Environmental Restoration for Mixed Wastes</u>

A. John Ahlquist
Office of Southwestern Area Programs
Office of Environmental Restoration and Waste Management
U.S. Department of Energy
Washington, DC

In establishing a program to investigate and remediate environmental pollution from past practices, the philosophy, goals, and implementation much be understood by the stakeholders in the program and a balance must be established to ensure satisfactory restoration of the environment and good stewardship and conservation of financial and human resources.

The ideal goal of an environmental restoration program for mixed waste is for post-restoration activities to provide an environment free of contamination. Rarely is this goal practically achievable and to insist upon it ignores many natural phenomena and may lead to the false theology of "When the earth is from contamination free, we will live eternally."

In the United States environmental restoration programs have become very prescriptive and are strongly driven by environmental laws and regulations - particularly in the area of mixed waste - such that regulatory risk has become the driving force for these programs. Human health and ecological risk from contamination play a secondary role in evaluating the need for environmental assessment and restoration. The consequence of this approach is a very expensive program that often leads to little actual risk reduction but is popular with those who don't fully understand the consequences of this approach.

This paper provides recommendations for an approach that can take advantage of lessons learned from the United States prescriptive approach to environmental restoration.

F. Arendt, G.J. Annokkée, R. Bosman and W.J. van den Brink (eds.), Contaminated Soil '93, 1673.
© 1993 *Kluwer Academic Publishers. Printed in the Netherlands.*

"Contaminated Land Treatment- Derelict Land Reclamation in Europe"

The exchange of information on reclamation of derelict land and the development of joint initiatives at a European level is the task of an expert group from the Association of Technologie Regions (RETI), supported by the European Community.
Therefore to support the expert group, a specialist panel with members from Great-Britain, Belgium, France The Netherlands and Germany began work in 1991. teh aim is to provide practical support and expert knowledge for local bodys carrying out derelict land reclamation on the basis of the latest information from international research and development. In this way, the competence of existing scientific networks will be extended by including planning aspects of derelict land reclamation and by direct connection to the authorities in the regions affected.

Introduction: Prof. Dr.-Ing. Hans Reiner Böhm/TH-Darmstadt

Interregional Cooperation "Derelict Land Reclamation"
Jean-Marie Ernecq/Directeur de l'EPF Nord-Pas de Calais,
Secretaire Permanent de l'Association RETI

Integration of contaminated land treatment in derelict land
reclamation strategies
Dipl.-Ing. Uwe Ferber/TH-Darmstadt

Case Studys

Belgium: Jaqueline Miller; Christophe Rasumny/Université Libre de
Bruxelles

Great-Britain: John Palmer/Richards, Moorhead & Laing

France: Dominique Darmendrail/BRGM-Nord-Pas de Calais

Gemany: Dipl.-Ing. Wolfgang Selke/Stadtverband Saarbrücken
Dipl.-Ing. Christian Weingram/Standort und
Strukturberatung Chemnitz

Discussion: Priorities in international cooperation

F. Arendt, G.J. Annokkée, R. Bosman and W.J. van den Brink (eds.), Contaminated Soil '93, 1675.
© 1993 *Kluwer Academic Publishers. Printed in the Netherlands.*

DERELICT AND CONTAMINATED LAND - POLICY AND PRACTICE IN WALES

J P PALMER AND I G RICHARDS

RICHARDS, MOOREHEAD AND LAING LTD, 55 WELL STREET, RUTHIN, CLWYD, LL15 1AF,
UNITED KINGDOM

1. ABSTRACT

   Since 1966 Wales has embarked on what has been one of the biggest
derelict land reclamation programmes in Europe. Initially concentrating on
safety the emphasis has changed over the years so that dealing with
contamination has become more important. Research has been funded
alongside development and Wales is the only constituent country of the
United Kingdom to have both a national register of potentially contaminated
sites and to have assessed the effectiveness of its reclamation schemes for
contaminated land. Whereas covering systems have been the principle method
of reclaiming contaminated land in the past attention is now turning
towards on-site decontamination as an alternative. The example of national
coordination and funding of reclamation being carried out in harmony with
strategies for economic regeneration in Wales could be a model for other
European regions.

2. INTRODUCTION

   Wales is an integral part of the United Kingdom but many of the elements
of Wales' central government are administered through the Welsh Office
which is a separate government department answerable to a Secretary of
State. Aspects of UK government policy and control experience a change in
degree or emphasis when they are applied in Wales, and this is particularly
so with respect to the clearance and treatment of derelict or contaminated
land.

   Wales played an important part in the early industrialisation of the UK.
The hundred years between 1750 and 1850 saw the development of major
industries based principally on iron, coal and chemicals in the two
coalfields, one in the north and one in the south of the country. Slate
extraction and metal mining and refining were important in North and West
Wales. As in all industrial areas of the time, and indeed up until the
middle of this century in the UK, little thought was given to the quality
of life of the people as measured by the physical quality of their local
surroundings. The valleys of the South Wales Coalfield represented some of
the worst and densest urban sprawls in Europe with houses, churches, mines
and factories succeeding each other from valley head to mouth. The
villages and towns in the South Wales valleys have been described by one
writer as a "glacial flow of buildings, a masonry drift, mile after mile,
from the mountains to the sea".

F. Arendt, G.J. Annokkée, R. Bosman and W.J. van den Brink (eds.), Contaminated Soil '93, 1677–1685.
© 1993 *Kluwer Academic Publishers. Printed in the Netherlands.*

3. THE RECLAMATION PROGRAMME

The awakening of Wales to the problems of derelict land can be traced on one tragic event in October 1966 when the village schools at Aberfan were engulfed by a slide of colliery waste. One hundred and forty-four people died. The majority of them were children under the age of 11. From that moment, the Secretary of State for Wales and the whole population determined that their quality of life should be measured at least in part by the quality of their surroundings. The government embarked on what has been the biggest derelict land clearance programme in Europe. The total spent to date has been approximately £200 million with a supplementary £1000 million spent on development. The first priority in reclamation was safety, but it was immediately apparent that carrying out works of safety could create land which would be available for the development of new industries and housing in an area where land suitable for modern development was virtually non-existent.

The earliest reclamation schemes involved:

1. Controlling ground and surface water which were the main causes of instability in coal tips.

2. Reducing the angle of tip faces to increase stability and enable machines to move across the new landform.

3. Developing vegetative cover, usually grass, on the new landform.

4. A very nominal and generally inadequate management and maintenance programme in respect of the vegetation.

From the outset the Welsh Office looked to district authorities, that is local government, as the promoters of schemes. The Welsh Office also established a principle which is still applied today. This principle is that the whole programme should be administered by a very small team of specialists, who look to other organisations such as local government and private consultants to design schemes and private contractors to carry out the work.

In 1976, the government established the Welsh Development Agency and gave it the specific task of developing a new industrial base for the Welsh economy. Regenerating Welsh industry, clearing industrial dereliction and improving the environment were to be the Agency's main tasks. The establishment of an agency to assist the public and private sectors to both clear up and develop former industrial land has been a very successful model providing the following benefits:

* the development of regional strategies for clean-up and development over a long period - the so-called 'rolling programme'

* the accumulation of expertise and experience by a core team on a regional basis

* the targeting of clean-up and development to particular areas of need

* the funding of research projects of particular applicability to the
  region's problems.

The scale of the work has been such that the Agency confidently predicts
that all sites of major dereliction will have been removed by 1996. At
central government, Agency and local government levels in Wales, it is
recognised that it is in Wales' best business interests to improve the
quality of our environment. The implications of this target date of 1996
is that the pace of reclamation work is actually increasing. The programme
since 1976 has however already borne fruit: with only 5% of the UK
population Wales has attracted up to 20% of inward investment into the UK
in recent years.

4. CONTAMINATED LAND

Both the Welsh Office and Welsh Development Agency have promoted
research projects and surveys in relation to contaminated land. These have
ranged from regional surveys to assess the extent of contaminated land and
methods for its treatment to research on specific problem areas.

4.1. <u>Surveys</u>

Surveys to create a record of contaminated sites would do much to
avoid their blind development. Such a survey has been carried out in Wales
by the Welsh Office (Welsh Office 1988). Surveys have also been
recommended as a valuable approach by the House of Commons Environment
Committee (House of Commons 1990) and have been incorporated into the
Environmental Protection Act in 1991 as a duty which all Local Authorities
must perform. The Welsh Office survey of Contaminated Land in Wales was
originally published in March 1984 and revised in 1986 and 1988. The
survey was restricted by the following criteria:

* no site under 0.5 ha in site was included
* no sites in 'active' use were included.

The survey, carried out on the basis of archival work and consultation
only, identified over 700 potentially contaminated sites. The information
recorded for each site is shown in Table 1.

TABLE 1. Information on each potentially contaminated site recorded in the Welsh Office survey (1988).

1.	Code	11.	Scale of contamination
2.	Grid reference	12.	Toxicity of contaminants
3.	Grid square	13.	Proximity to housing
4.	Activity	14.	Status
5.	Name	15.	Hazard factor
6.	Location	16.	Development factor
7.	Topography	17.	Confidence
8.	Contaminants	18.	Comments
9.	Period of use	19.	Update
10.	Area	20.	Source

The types of sites recorded are shown in Table 2.

TABLE 2. Types of sites recorded in the Welsh Office survey (1988)

	Number	% of Total
gasworks/coke ovens	100	13.4
metal mines	120	16.1
waste tips	279	37.4
iron/steel/tinplate works	106	14.2
chemical works	30	4.0
transit areas	43	5.8
metal smelters	22	3.0
other	45	6.1

Although this survey was limited it was until recently the only survey of its kind in the United Kingdom. The information is published and is being used by local authorities in Wales.

All local authorities in the United Kingdom, including Wales, will be required to compile registers of land which may be contaminated under regulations to be enforced as part of the Environmental Protection Act 1990. The draft regulations contain a list of 'contaminative uses' (Table 3) and it is only sites where such uses have been demonstrated which will be recorded on the register. Unlike the Welsh Office Survey there is no proposed restriction on site size and active sites will be included. It is intended that the registers will be available to the public and will aid development control.

TABLE 3. Schedule of contaminative uses (Department of the Environment 1992)

Coal carbonisation
Steel or lead production
Asbestos manufacture
Manufacture, refining or recovery of petroleum
Manufacture, refining or recovery of other chemicals
(excluding minerals)
Final deposit of household, commercial or industrial waste on land
Treatment at a fixed installation of household, commercial or industrial
waste by chemical or thermal means
Storage of scrap metal

4.2.<u>Reclamation techniques</u>

The effectiveness and long term reliability of remedial systems can be assessed by a study of performance of past reclamation schemes, so that given a sufficient number, feedback resulting from this study can lead to improvement in design and practice.
In order that future reclamation schemes for contaminated land could learn from the past the Welsh Office commissioned Richards, Moorehead and Laing Ltd to carry out a study of the effectiveness of past reclamation schemes on contaminated land in Wales. The approach adopted was as follows:

(i)   A survey of reclaimed contaminated sites in Wales to provide
      information on the techniques used.
(ii) The development of criteria to assess the effectiveness of the
      techniques identified.
(iii) An assessment of the effectiveness of the techniques used at selected
      sites.
(iv) An appraisal of the lessons learned from this study to provide
      guidelines for future techniques.

The survey compiled information on site history, site investigation techniques, reclamation methods and post-reclamation monitoring practice. Domestic landfills were excluded. On the basis of their industrial association, 94 of the sites identified in the survey fell within the definition of contaminated land (Table 4). In only 29 of the sites had a reclamation scheme been carried out in the knowledge that the site was contaminated Table 5. Most of the others had either been assumed to be uncontaminated or found to be uncontaminated as a result of a site investigation. The 29 sites included chemical works and gas works, former heavy industrial sites such as iron, steel and tinplate producers; abandoned lead/zinc and metal mines and industrial complexes where many of the above activities took place in close proximity to each other.

1682

TABLE 4 Reclaimed sites by industrial association

Site type	No
Gas works	15
Industrial Waste tips	5
Iron/steel and tinplate	30
Metal mines	13
Chemical works	10
Transit areas	8
Smelters	4
Others	9

The survey showed that Wales has experience in the treatment of a wide range of contaminated land. Reclamation has taken place for two reasons.

- Solely to clear up contamination - the end use of a site was often a secondary concern. Metal mine reclamation frequently fell into this category.
- As part of a development project - reclamation was needed before development could be attracted onto a site or an intended development allowed to proceed. On occasions contamination was discovered whilst development was proceeding.

Some large areas of industrial land which may have been contaminated have been apparently satisfactorily reclaimed without the benefit of detailed site investigation. The following reasons may provide an explanation:

1. Any contaminated areas would have been small compared with the overall size of the site.
2. The most seriously contaminated land could be identified by industrial association, smell or colour and was buried or moved off site.
3. There was a large volume of uncontaminated material on site to use as covering material.
4. There was a large volume of uncontaminated material on site into which the contaminated material was dispersed and rendered harmless.

The lack of a detailed site investigation does however mean that it is not known what is buried or its location. These situations raise some misgivings since these sites could still create contamination problems at some time in the future.

Table 5 Classification of reclaimed sites in Wales on the basis of their
contamination status prior to reclamation

Class 1) **Assumed to be uncontaminated prior to reclamation Sites: 12**
On Class 1 sites reclamation principally involved the demolition of disused
buildings and levelling of the rubble or waste. The industrial use was not
thought to have caused contamination on-site. In addition this group can
include these sites which have been levelled, but on which no detailed site
investigation has yet been performed. Here, reclamation has involved site
clearance, and the site is in a transient state, and therefore is an "ongoing" scheme.

Class 2) **As Class 1 but with development/re-use Sites: 38**
Class 2 sites received the same treatment as those in Class 1 but
immediately, or some time after site clearance, the site was developed. No
site investigation to determine contamination was undertaken at any stage.
This class also includes sites where there was a change of ownership/usage,
without the requirement of major engineering. For example, buildings re-
occupied by new owner/users.

Class 3) **Deemed uncontaminated after site investigation and**
         reclaimed/redeveloped Sites: 4
On Class 3 sites, a detailed site investigation was performed on the
suspicion that the ground may be contaminated. Results showed it to be
non-contaminated and no measures to deal with contamination were undertaken
prior to development.

Class 4) **Identified as contaminated and reclaimed Sites: 16**
Class 4 sites are those on which remedial action was necessary to prevent
the spread of pollution to the surrounding environment eg. prevention of
dust blow from mine spoil. Survey was not always undertaken where
contaminated materials such as mine spoil were easily recognised.

Class 5) **As 4 but the emphasis was placed upon the final development**
         Sites: 13
Class 5 sites were identified as being contaminated and the site was needed
for development. The contamination had to be dealt with to allow
development to proceed. If the development had not provided the need,
reclamation may not have taken place.

Class 6) **On-going schemes Sites: 11**
Within this class grouped all those sites where reclamation or development
was in progress. The difference between Class 1 and Class 6 is that in
Class 6 the site is recognised as being potentially contaminated, or has
even been shown to be contaminated whereas in Class 1 this is not so.

Three principal techniques of reclamation have been used. These are:

- Excavation of contaminated material and removal off-site.
- Encapsulation of the contaminated material.
- Covering of the contaminated material undisturbed on site.
  Covering has been used in the majority of sites in Wales and has

apparently been successful on most sites. Where it has not been successful water borne contamination moving out through the covering material and migration of volatile organics have been the major reasons for failure.

In some cases not all potential pollutants and targets were considered at the time of reclamation because one appeared to be the overriding problem. Failure to consider all pollutants and targets may result in the reclamation scheme solving only part of the problem. For example at one site a scheme to reclaim a metalliferous mine with a dust problem has been successful but there has been an increase in water borne contamination since the scheme was completed.

Of the sites examined in this survey waste tips have proved the most difficult to assess and the greatest cause for concern. This is because of poor record keeping. Long term problems associated with waste tips are likely to manifest themselves and will be unpredictable because of these factors and because any site investigation which was carried out may not have identified all the contaminants. Unidentified encapsulated waste such as that in buried drums or plastic bags is potentially a very serious hazard as their contents will only leak out after the container degrades or is punctured. This could occur many years after the site has been reclaimed.

Since completion of the survey of reclamation techniques attention has turned towards on-site decontamination rather than covering as a method of reclamation. Vacuum extraction has been used to remove volatile organics from a site already reclaimed by covering and experimentation with other novel techniques is continuing. One such technique is the decontamination of metalliferous tailings using mineral processing techniques (Palmer 1992).

4. CONCLUSIONS

Wales has developed an approach to land reclamation and development which has concentrated on the promotion of environmental quality as the key to future prosperity. The strategy developed in Wales is applicable to other European regions and can be summarised as follows:

*   identify problems
*   establish principles at the highest level
*   set up a programme
*   give free rein to a motivated team
*   be flexible
*   be ready to research and incorporate technical innovations
*   expect difficulties

*   have patience and take the long-term view, environmental regeneration
       cannot be rushed, even if startling improvements are possible
*   involve the public.
The technical success of reclamation schemes and future technique
development depends on thorough site investigation, good record keeping
good reclamation scheme design and adequate monitoring.

6. REFERENCES

Welsh Office. (1988) <u>Survey of Contaminated Land in Wales</u>. Cardiff,
United Kingdom.

Environment Committee, House of Commons. <u>Contaminated Land, First
Report, Volume 1.</u>(1990) HMSO London, United Kingdom.

Welsh Office.(1988) <u>An assessment of the effectiveness of methods and
systems used to reclaim contaminated sites in Wales</u>, Cardiff, United
Kingdom.

Department of the Environment (1992 July). <u>Environmental Protection Act
1990 (Section 143 Registers) Regulations 1992 (draft)</u>. Department of the
Environment, London.

Palmer, J P (1992 October) <u>Decontamination of metalliferous mine spoil</u>.
Paper presented at Nato/CCMS Pilot study on 'Demonstration of Remedial
Action Technologies for Contaminated Land and Groundwater' Budapest.

<u>Meulen, G.R.B. ter</u>   523
RIVM, Postbus 1, 3720 BA Bilthoven. *The Netherlands.*

<u>Milanovich, F.P.</u>   1633
Lawrence Livermore National Laboratory, 7000 East Ave. P.O.
Box 808, L-590, 94550 Livermore, CA. *USA.*

<u>Miller, D.F.</u>   1663
U.S. Department of Energy, Environmental Restoration and Waste
Management, 2219 Lakeshire Drive, Alexandria V.A. *USA.*

<u>Misra, M.</u>   1623
Mackay School of Minise, Department of Chemical and
Metallurgical Engineering, Mail Stop 168, 98557 Reno, Nevada.
USA.

<u>Mohs, B.</u>   525
AHU - Büro für Hydrogeologie und Umwelt GmbH, Kirberichshofer
weg 6, 5100 Aachen. *Germany.*

<u>Müller, M.</u>   1419
Institut fur Siedlungswasserwirtschaft der RWTH Aachen,
Templograben 55, W-5100 Aachen. *Germany.*

<u>Mueller, H.</u>   957
Strasse der Einheit 22, 0-9200 Freiberg. *Germany.*

<u>Müller-Hurtig, R.</u>   529
Institut fur Biochemie und Biotechnologie Bereich A, TU
Braunschweig, Konstantin Uhde Str. 5, 3300 Brauschweig.
Germany.

<u>Muller-Kirchenbauer, H.</u>   1421
Universitat Hannover, Welfengarten 1, 3000 Hannover 1.
Germany.

<u>Munk, C.</u>   959
Harress Pickel Consult GmbH.,Niedervellmarsche Strasse 30,
3501 Fuldatal 1. *Germany.*

<u>Munz, Chr.</u>   1119
MBT Umwelttechnik AG, Vulkanstrasse 110, 8048 Zürich.
Switzerland.

<u>Neteler, Th.</u>   819
Lehrstuhl für Grundbau und Bodenmechanik, Ruhr-Universität
Bochum, Universitätsstrasse 150, 4630 Bochum 1. *Germany.*

<u>Nickel, U.</u>   1189
Umweltbehörde Hamburg, Altlastensanierung, Amelungstrasse 3,
2000 Hamburg 36. *Germany.*

<u>Nielsen, P. H.</u>   531
Department of Environmental Engineering/Groundwater research
Centre, Technical University of Denmark, building 115, 2800
Lyngby. *Denmark.*

<u>Nijhof, A.G.</u>   539
TAUW Infra Consult B.V., Postbus 479, 7400 AL Deventer.
The Netherlands.

<u>Nowak, K.</u>   1423
TU Hamburg-Harburg, AB Verfahrenstechnik II, Eissendorfer
Strasse 38, 2100 Hamburg 90. *Germany.*

<u>Oberbremer, A.</u>   1191
Biodetox Gesellschaft zur biologischen Schadstoffentsorgung,
Feldstrasse 2, 3061 Ahnsen. *Germany.*

<u>Okx, J.P.</u>   781
TAUW Infra Consult B.V., Postbus 479, 7400 AL Deventer.
The Netherlands.

<u>Ondruschka, J.</u>   1195, 1425
Center of Environmental Research Leipzig-Halle,
Permoserstrasse 15, O-7050 Leipzig. *Germany.*

<u>Osterkamp. G.</u>   541
Niedersächsisches Landesamt für Bodenforschung, Stilleweg 2,
3000 Hannover 51. *Germany.*

<u>Paech, W.</u>   543
Umwelt- und Wirtschaftsinstitut Berlin Gmbh, Rudower Chaussee
5, Berlin. *Germany.*

<u>Palmer, J.P.</u>   1677
Richards, Moorehead and Laing Ltd., 55 Well Street, Ruthin,
Clwyd, LL15 1AF. *United Kingdom.*

<u>Palsma, A.J.</u>   1609
Institute of Environmental Sciences TNO, P.O. Box 6011, 2600
JA Delft. *The Netherlands.*

<u>Pearl, M.</u>   1295
Warren Spring Laboratory, Gunnels Wood Road, Stevenage, Herts,
SG1 2BX. *United Kingdom.*

<u>Pedall, K. G.</u>   1197
HPC Harress Pickel Consult, Gustavstr. 65, 8510 Furth.
Germany.

<u>Pentinghaus, H.</u>   1537
Kernforschungszentrum Karlsruhe, Institut fur Nukleare
Entsorgungstechnik, P.O. Box 3640, 7500 Karlsruhe 1. *Germany.*

Siegrist, R.L.   1447
Oark Ridge National Laboratory, P.O. Box 2008, 37831-6038 Oak
Ridge, Tennessee. *USA*.

Sihler, A.   1449
Universitat Stuttgart, Inst. fur Sielungswasserbau,
Wassergute- und Abfallwirtschaft, Bandaele 1, W-7000 Stuttgart
80. *Germany*.

Sinz, S.H.   1451
Kohlenstrasse 32, 4300 Essen 17. *Germany*.

Sobisch, T.   1453
Adlershofer Umwetschutztechnik- und Forschungs- GmbH, Rudower
Chaussee 5, O-1199 Berlin. *Germany*.

Soczó, E.R.   1571
National institute of Public Health and Environmental
Protection, P.O. Box 1, 3720 BA Bilthoven. *The Netherlands*.

Sokollek, V.   991
Umweltbehörde Hamburg, Amt für Umweltschutz,
Altlastensanierung, Amelungsstrasse 3, 2000 Hamburg 36.
Germany.

Sonnen, H.D.   757
Ingenieurbüro für Altlastensanierung, Oraniendamm 64-72, 1000
Berlin 28. *Germany*.

Spence, L.   421
American Petroleum Institute, 1220 L. St. NW, Washington, D.C.
20005. *USA*.

Steffens, K.   835
Probiotec GmbH, Schillingsstrasse 333, 5160 Düren-Gürzenich.
Germany.

Steketee, J.J.   1559
Tauw Infra Consult, P.O. Box 479, 7400 AL Deventer.
The Netherlands.

Stief, K.   1479
Umweltbundesamt, Bismarckplatz 1, 1000 Berlin 33. *Germany*.

Stoffers, H.   1455
Contracon, Peter Henlein Str. 2-4, 2190 Cuxhaven. *Germany*.

Stottmeier, U.   1219
UFZ Umweltforschungszentrum Leipzig-Halle GmbH,
Permoserstrasse 15, O-7050 Leipzig. *Germany*.

Straalen, N.M. van   315
Vrije Universiteit, Dep. of Ecology and Ecotoxicology,
De Boelelaan 1087, 1081 HV Amsterdam. *The Netherlands*.

<u>Wollner, H.</u>   1469
Ludwig-Renn-Strasse 48, O-1142 Berlin. *Germany.*

<u>Wömmel, S.</u>   1287
Technische Universität Hamburg-Harburg, Umweltschutztechnik,
Eissendorfer Strasse 40, 2100 Hamburg 90. *Germany.*

<u>Zarth, M.</u>   737, 1473
Umweltbehörde, Amt für Umweltschutz, Altlastensanierung,
Amelungstrasse 3, 2000 Hamburg 36. *Germany.*

<u>Zeddel, A.</u>   1473
Forstbotanisches Institut der Universität Gottingen, Busgenweg
2, 3400 Gottingen. *Germany.*

<u>Zenneck, C.</u>   1475
TU Hamburg-Harburg, Arbeitsbereich Bioprozess- und
Bioverfahrenstechnik, Denickestr. 15, 2100 Hamburg 90.
Germany.

Note:
Postal codes in *Germany* will change in the course of 1993.

abandoned polluted areas 967
abandoned waste disposal sites 1479
abandoned waste sites 975
absorption 1204
absorption process 1451
acid-mine drainage 231, 235
action groups 99, 119
adsorption 163, 515, 516, 517, 245, 487, 1399, 1432
aerial photography evaluation 730
aftercare 1535
agriculture 955, 1511, 1561, 1593
amino acids 1157
ammonia 1511
amount of production 185
analysis 69, 89, 499, 509, 929
analytical methods 703, 911, 941
approximate values 58
aquifer 153, 163, 175, 257, 473, 483, 570, 937m 1229
areas of transgression 130
areas suspected of being contaminated 129
aromatic hydrocarbon contamination 959
aromatic hydrocarbons 531
aromatics 625, 657
arsenic 563, 1133, 1411
assessment of danger 459
atmospheric deposition 173
authorities 63, 133, 1609
authorities' role 69, 120, 123, 137
authority to issue orders 446
availability 997

background levels 173, 347, 499, 571, 781–786, 876, 1543
bacteria 966
barrier construction 69, 287, 925
barriers 434, 507, 577, 578, 1183, 1563
base sealings 1545
behaviour of contaminants 164, 185, 499, 509, 516, 571, 1204, 1413
behaviour 209, 955, 1559
bentonite-cement 507
benzene 1087
bilateral German-American agreement 799

bio-piles 799
bio-reactor test 1042
bioassays 315, 393, 571, 649
bioavailability 347, 393, 571, 649, 1050, 1151, 1220, 1440, 1453, 1457
bioavailability of PAHs 337
biochemical reactors 267
biodegradability 345, 1187, 1285
biodegradation 338, 485, 753, 1173, 1229, 1265, 1277, 1337, 1345, 1349, 1373, 1385, 1423, 1433, 1449, 1473, 1547, 1567
biogeochemical cycles 237
biological 1257
biological conversion 1235, 1511
biological decomposition 1042, 1565
biological degradation 1407
biological in-situ remediation 1047
biological pretreatment 1565
biological reclamation 1191
biological remediation 1439
biological techniques 1277
biological test procedure 951
biological treatment 123, 137, 649, 799, 1009, 1093, 1095, 1175, 1185, 1215, 1235, 1371, 1377, 1381, 1413, 1449, 1457, 1489, 1511, 1567
biomonitoring 1052
bioreactor test system 1406
bioreactors 1409, 1439
bioremediation 1151
biosensor 993
biotechnological treatment 1265
biotoxicity 345
bitumerrization 1643
BLK 1205
BMFT 799
borings 830
bottom sealing 925, 1533
bound residues 1253
break-off 135
BTEX 1205
BTEX-aromates 1135
buffering capacity 232, 238
building sites 891
buildings 865, 930, 936, 981